Stress Response in Microbiology

Edited by

Jose M. Requena

Centro de Biología Molecular 'Severo Ochoa' (CSIC-UAM)
Madrid
Spain

Caister Academic Press

Caister Academic Press
Norfolk, UK

www.caister.com

British Library Cataloguing-in-Publication Data
A catalogue record for this book is available from the British Library

ISBN: 978-1-908230-04-1

Cover design adapted from Figure 13.3

Printed and bound in Great Britain

Contents

Contributors

Jacqueline Abranches
Center for Oral Biology and Department of Microbiology and Immunology
University of Rochester Medical Center
Rochester, NY
USA

jacqueline_abranches@urmc.rochester.edu

Maria L. Alomar
Instituto Tecnológico de Chascomús (IIB-INTECH)
CONICET – Universidad Nacional de General San Martin
Chascomús
Argentina

lis@intech.gov.ar

Sergio O. Angel
Instituto Tecnológico de Chascomús (IIB-INTECH)
CONICET – Universidad Nacional de General San Martin
Chascomús
Argentina

sangel@intech.gov.ar

Juan A. Ayala
Centro de Biología Molecular 'Severo Ochoa' (CSIC-UAM)
Universidad Autónoma de Madrid
Madrid
Spain

jayala@cbm.uam.es

Jacek Bania
Department of Food Hygiene and Consumer Health Protection
Wrocław University of Environmental and Life Sciences
Wrocław
Poland

jacek.bania@up.wroc.pl

E. Fidelma Boyd
Department of Biological Sciences
University of Delaware
Newark, DE
USA

fboyd@udel.edu

Julien Brillard
UMR408 Sécurité et Qualité des Produits d'Origine Végétale
INRA – Université d'Avignon
Avignon
France

julien.brillard@avignon.inra.fr

Véronique Broussolle
UMR408 Sécurité et Qualité des Produits d'Origine Végétale
INRA – Université d'Avignon
Avignon
France

veronique.broussolle@avignon.inra.fr

Felipe Cava
Centro de Biología Molecular 'Severo Ochoa' (CSIC-UAM)
Universidad Autónoma de Madrid
Madrid
Spain

fcava@cbm.uam.es

Verónica M. Cóceres
Instituto Tecnológico de Chascomús (IIB-INTECH)
CONICET – Universidad Nacional de General San Martin
Chascomús
Argentina

coceres@intech.gov.ar

Marcelo A. Comini
Group Redox Biology of Trypanosomes
Institut Pasteur de Montevideo
Montevideo
Uruguay

mcomini@pasteur.edu.uy

Andrew J. Darwin
Department of Microbiology
New York University School of Medicine
New York, NY
USA

andrew.darwin@med.nyu.edu

Isabel Delany
Novartis Vaccines and Diagnostics
Siena
Italy

isabel.delany@novartis.com

Christian Doerig
Department of Microbiology
Monash University
Clayton, VIC
Australia

christian.doerig@monash.edu

Maria J. Figueras
Instituto Tecnológico de Chascomús (IIB-INTECH)
CONICET – Universidad Nacional de General San Martin
Chascomús
Argentina

mjfigueras@intech.gov.ar

N. Kaye Horstman
Department of Microbiology
New York University School of Medicine
New York, NY
USA

nkhorstman410@gmail.com

Suzanne Humphrey
National Centre for Zoonosis Research
University of Liverpool
Wirral
UK

s.humphrey@liverpool.ac.uk

Tom J. Humphrey
National Centre for Zoonosis Research
University of Liverpool
Wirral
UK

t.j.humphrey@liverpool.ac.uk

Mark A. Jepson
School of Biochemistry
University of Bristol
Bristol
UK

m.a.jepson@bristol.ac.uk

José A. Lemos
Center for Oral Biology and Department of Microbiology and Immunology
University of Rochester Medical Center
Rochester, NY
USA

jose_lemos@urmc.rochester.edu

Erika Rubí Luis-García
Departamento de Medicina Experimental
Facultad de Medicina
Universidad Nacional Autónoma de México
México

erikarubi_84@yahoo.com.mx

Melissa L. Madsen
College of Veterinary Medicine
Iowa State University
Ames, IA
USA

mlmadsen@mail.iastate.edu

Bruno Manta
Group Redox Biology of Trypanosomes
Institut Pasteur de Montevideo
Montevideo
Uruguay

bmanta@pasteur.edu.uy

Andrea Medeiros
Group Redox Biology of Trypanosomes
Institut Pasteur de Montevideo
Biochemistry Department
Faculty of Medicine
Universidad de La República
Montevideo
Uruguay

amedeiros@pasteur.edu.uy

F. Chris Minion
College of Veterinary Medicine
Iowa State University
Ames, IA
USA

fcminion@iastate.edu

Sylke Müller
Wellcome Trust Centre for Molecular Parasitology
Institute of Infection, Immunity & Inflammation
College of Medical, Veterinary and Life Sciences
University of Glasgow
Glasgow
UK

sylke.muller@glasgow.ac.uk

Eulàlia de Nadal
Departament de Ciències Experimentals i de la Salut
Cell Signaling Research Group
Universitat Pompeu Fabra
Barcelona
Spain

eulalia.nadal@upf.edu

Mario Nequiz
Departamento de Medicina Experimental
Facultad de Medicina
Universidad Nacional Autónoma de México
México

manequiz@yahoo.com.mx

Alfonso Olivos-García
Departamento de Medicina Experimental
Facultad de Medicina
Universidad Nacional Autónoma de México
México

olivosa@yahoo.com

Miguel A. de Pedro
Centro de Biología Molecular 'Severo Ochoa' (CSIC-UAM)
Universidad Autónoma de Madrid
Madrid
Spain

madepedro@cbm.uam.es

Ruy Pérez-Tamayo
Departamento de Medicina Experimental
Facultad de Medicina
Universidad Nacional Autónoma de México
México

ruypt@hotmail.com

Francesc Posas
Departament de Ciències Experimentals i de la Salut
Cell Signaling Research Group
Universitat Pompeu Fabra
Barcelona
Spain

francesc.posas@upf.edu

Jose M. Requena
Centro de Biología Molecular 'Severo Ochoa' (CSIC-UAM)
Universidad Autónoma de Madrid
Madrid
Spain

jmrequena@cbm.uam.es

Deivid C. Rodrigues
Instituto de Biofísica Carlos Chagas Filho
Universidade Federal do Rio de Janeiro
Rio de Janeiro
Brazil

deivid@biof.ufjr.br

Edson Rondinelli
Instituto de Biofísica Carlos Chagas Filho
Departamento de Clínica Médica
Faculdade de Medicina
Universidade Federal do Rio de Janeiro
Rio de Janeiro
Brazil

edrondin@biof.ufrj.br

Emma Saavedra
Departamento de Bioquímica
Instituto Nacional de Cardiología Ignacio Chávez
México

emma_saavedra2002@yahoo.com

Kate L. Seib
Novartis Vaccines and Diagnostics
Siena
Italy

kate.seib@novartis.com

Rosane Silva
Instituto de Biofísica Carlos Chagas Filho
Universidade Federal do Rio de Janeiro
Rio de Janeiro
Brazil

silvaros@biof.ufrj.br

Richard W. Stokes
Department of Microbiology & Immunology
Life Sciences Centre
University of British Columbia
Vancouver, BC
Canada

rstokes@mail.ubc.ca

Turán P. Ürményi
Instituto de Biofísica Carlos Chagas Filho
Universidade Federal do Rio de Janeiro
Rio de Janeiro
Brazil

turmenyi@me.com

Ewa Wałecka
Department of Food Hygiene and Consumer Health Protection
Wrocław University of Environmental and Life Sciences
Wrocław
Poland

ewa.walecka@up.wroc.pl

W. Brian Whitaker
Department of Biological Sciences
University of Delaware
Newark, DE
USA

wbwhit@udel.edu

Preface

Every living organism must cope with environmental changes that may represent stress situations, including elevated temperature, chemical stress or oxidative injury. Cells respond to stress stimuli through coordinated changes in gene expression, leading to the synthesis of specialized molecules that counteract the deleterious environmental insults. Bacteria and eukaryotic microorganisms are very useful for studying the stress response and its regulation as they have developed systems to constantly monitor the changing environment. One group of organisms that is subjected to dramatic environmental challenges throughout their life cycle, including large changes in temperature, availability of nutrients and exposure to host immune defenses, is the pathogenic microorganisms. For pathogenic bacteria and parasites that are transmitted from the environment (or by invertebrate vectors) to mammalian hosts, sudden changes in pH, osmotic pressure and temperature occur. Additionally, inside the body, invading microorganisms soon encounter the innate and adaptive defenses. The pathogens, as a group, have evolved a variety of mechanisms to circumvent the otherwise lethal effects of these defenses. On the other hand, pathogenic microorganisms have integrated the stress response into their life cycles, in which stress signalling pathways and the selfsame stress proteins play specific functions in the differentiation programme.

Stress Response in Microbiology comprises 17 excellent chapters, each one dedicated to a particular microorganism or group of microorganisms; most of the selected organisms represent important health threats for humans. With its coverage of a broad range of model organisms, the book gives a complete overview of the stress response in both prokaryotic and eukaryotic microorganisms, providing detailed information for researchers, as well as for teachers and students in the fields of microbiology and parasitology. The chapter authors, among the best in their respective fields, have done an excellent job of synthesizing data from numerous studies and making the book a well-referenced work. Thus, we hope that this work will serve as an informative resource for researchers and students at all levels.

The first chapter provides a complete description of the cell envelope stress responses and the stress-sensing regulatory systems, mainly in Gram-negative bacteria. Chapter 2 gives an overview on the stress responses in several pathogenic species of the genus *Streptococcus*; acid, oxidative and nutritional stresses are presented here in depth. Chapter 3 is devoted to oxidative and nitrosative defenses in pathogenic *Neisseria* species. In addition, the authors have included detailed information about biologically relevant oxidants and the chemical reactions involving oxidants in biological systems that are of considerable basic scientific interest. The relationship between stress response and virulence in the food-borne pathogen *Listeria monocytogenes* is the main focus of Chapter 4. Chapter 5 focuses on current knowledge and research activity about low-temperature adaptation of the spore former and human pathogen *Bacillus cereus*. Chapter 6 gives a complete overview of the main stress response mechanisms employed by *Salmonella* for survival in nutrient-limited conditions and during osmotic and acid stress exposure. In the next chapter, devoted to *Yersinia*, the authors review the responses of this pathogen to heat and cold shocks, encounter with macrophages and macrophage-like conditions. Also, the extracytoplasmic stress responses are covered in detail in Chapter 7. Chapter 8 describes how the stress response systems are vitally important for the vibrios to successfully establish in the host. Chapter 9 describes the function of the major stress proteins within mycobacteria, paying special attention to the interaction between the bacterial heat shock proteins and the host's cell-mediated immune response. Chapter 10 focuses on the types of stresses that mycoplasmas encounter *in vivo*, such as heat shock, oxidative stress, osmolarity shifts, hormone exposure, iron deprivation and biofilm formation. In Chapter 11, the authors describe the different mechanisms used by model yeasts *Saccharomyces cerevisiae* and *Schizosaccharomyces pombe*, as well as the pathogenic fungus *Candida albicans*, to sense and transduce stress signals to stress-activated protein kinases pathways in response to osmotic, heat and oxidative stresses. Among eukaryotic microorganisms, one group that is subjected to dramatic environmental changes throughout their complex life cycle are the parasitic protozoa, which are the focus of the remaining chapters. Chapter 12 summarizes the current knowledge about the responses of the malaria parasite *Plasmodium falciparum* to a variety of stresses: drug treatments, changes in temperature and elevation of oxidative stress. Chapter 13 summarizes the recent findings on the *Toxoplasma gondii* stress responses and the implication of these processes in the biology and pathogenesis of this parasite. The focus of Chapter 14 is the stress response in *Leishmania*, containing a comprehensive view on the implications of the stress response in parasite survival, in cytodifferentiation and in apoptotic processes. Chapter 15 reviews the components of the *Trypanosoma cruzi* stress response with emphasis on its relevance to the parasite biology and to Chagas' disease transmission, pathogenesis and treatment. In Chapter 16, the authors have compiled the most significant molecular and biological aspects related to the mechanisms and components of the stress response of *T. brucei* to adapt and survive in the bloodstream of mammals. The final chapter, devoted to *Entamoeba histolytica*, gives special emphasis to the oxidative and nitrosative stresses experienced by this protozoan parasite.

The acknowledgements list must start with our authors, who have generously expended time and effort in preparing and revising their chapters. Publication of a book requires the effort of many people besides the authors, and I wish to express special appreciation to the editorial and production staffs. In particular, I would like to thank Hugh Griffin for his guidance and support, and Emma Needs, who worked extremely hard coordinating production. Finally, without the comprehension and support of our families, this work could not be achieved. In particular, I would like to dedicate this book to my daughter Carmen.

Jose M. Requena

Cell Wall Stress-sensing Regulatory Systems in Gram-negative Bacteria

1

Juan A. Ayala, Felipe Cava and Miguel A. de Pedro

Abstract

The cell envelope is the major line of defence against environmental threats. It is an essential but vulnerable structure that shapes the cell and counteracts the turgor pressure. It provides a sensory interface, a molecular sieve and a structural support, mediating information flow, transport and assembly of supramolecular structures. Therefore, maintenance of cell envelope integrity in the presence of deleterious conditions is crucial for survival.

Several envelope stress responses, including two components regulatory systems (TCRS), of *Escherichia coli* are involved in the maintenance, adaptation and protection of the bacterial cell wall in response to a variety of stresses. Recent studies indicate that these stress responses exist in many Gram-negative pathogens. Particular emphasis has been made on the identified TCRS and their activating signals.

Another aspect of stress response is the generation of morphological modifications. Most bacteria alter shape when growth conditions change and upon symbiotic or parasitic processes. Any modification in cell shape is connected with cell wall metabolism and requires specific regulatory mechanisms. Recent advances support the existence of complex mechanisms mediating morphological responses to stress involving inter- and intra-specific signalling.

Introduction

Mechanisms for responding to environmental changes are universally present in living beings. Most bacteria can survive under diverse conditions, and particularly those associated with food-borne infections, as some *E. coli* serotypes, both inside and outside of the host. To overcome these adverse and fluctuating conditions, bacteria must perceive the environmental changes and respond accordingly with appropriate alterations in gene expression and protein activity. The stress response in bacteria involves a complex network of elements that acts against the external stimulus providing a change in some parameter of the cell envelope. Bacteria frequently react simultaneously to a wide variety of stresses which determine specific physiological changes. Therefore, the various stress response systems have to interact (cross-talk) with each other. A complex network of global regulatory systems displaying multiple molecular components ensures a coordinated and effective answer. These regulatory systems govern the expression of a plethora of further effectors that aim at maintaining stability of the cellular equilibrium under the various conditions.

In *E. coli*, five of the most important stress response systems are (1) the heat shock response, controlled by the sigma factor sigma32; (2) the envelope stress response, controlled mainly by the sigma factor sigma E and the Cpx two-component system; (3) the cold shock response, which governs expression of RNA chaperones and ribosomal factors; (4) the general stress response, which depends on the sigma factor sigma S; and (5) the (p)ppGpp-dependent stringent response which reduces the cellular protein synthesis capacity and controls further global responses upon nutritional downshift (Wick and Egli, 2004).

The Gram-negative envelope is composed of an inner membrane (IM), an outer membrane

(OM), and a peptidoglycan (PG) containing periplasmic space (P) between them (Luirink *et al.*, 2005; Ruiz *et al.*, 2006). A detailed description of the biosynthesis and chemical composition of the cell envelope is beyond the scope of this review, and readers are referred to updated reviews on this topic, and references therein (Vollmer and Seligman, 2010).

The envelope is a unique compartment of the bacterial cell and acts as the interface between the intracellular space and the extracellular environment. As such, the envelope is exposed to environmental changes and stresses, including changes in temperature, pH, and osmolarity; to toxic compounds; and to oxygen reactive species. The idea of stress response in bacteria is mostly understood as a response against short-lived, sudden changes in environmental conditions. However, from our perspective, progressive adaptation to prolonged periods under unfavourable conditions should also be considered as stress response. Indeed these adaptive responses are often associated to important alteration in shape, size or development of new cellular structures which in all cases require severe modifications at the cell wall level. To adapt to such stresses, bacteria have evolved a number of envelope stress responses that sense specific cues and prompt the cell to respond appropriately, usually by regulating gene expression. In this regard, at least two cell wall adaptive mechanisms have been described in the last few years: *bol*A and D-amino acid-mediated responses (Santos *et al.*, 1999; Lam *et al.*, 2009).

Amongst the 30 or so two-component systems known in *E. coli*, until now most attention has been focussed on the response produced by changes in the six envelope stress responses that have been identified in *E. coli*: the Cpx, sigma E, Bae, Psp (phage shock protein), Rcs (regulator of capsular synthesis), and vesicle release responses (Ruiz and Silhavy, 2005).

While some inducing signals and downstream regulatory components are common to two or more of these responses, each pathway also recognizes unique cues and appears to have distinct physiological roles in the cell.

Contributions of these systems to stress response have previously been examined individually; however, the possible interconnections between these pathways are largely unknown. Dynamics between five of the six stress response pathways have been analysed (Bury-Moné *et al.*, 2009) by determining the specificities of each system with respect to signal-inducing conditions, and monitoring global transcriptional changes in response to transient overexpression of each of the effectors.

Data show that different extracytoplasmic stress conditions elicit a combined response of these pathways, but transcriptional responses induced by the individual systems show little overlap (Fig. 1.1). So, the extracytoplasmic stress signalling pathways in *E. coli* regulate mainly complementary functions whose discrete contributions are integrated to mount the full adaptive response (Bury-Moné *et al.*, 2009).

Components and regulatory mechanisms of the stress response

Several stresses can be present in the environment that could trigger changes in the whole cell wall envelope, ranging from lipopolysaccharide (LPS) composition, membrane fluidity, misfolded proteins and peptidoglycan proper structure.

Most studies are focused on the effect of environmental agents on the composition or structure of the outer membrane, the periplasm content and changes on inner membrane proteins. Only a few papers describe the connection among the structural changes in peptidoglycan and the stress response. We will summarize first and very briefly the different cell envelope systems identified in Gram-negative bacteria, and next we will discuss extensively those involved in different aspect of peptidoglycan metabolism.

Cell envelope stress systems identified in Gram-negative bacteria

CpxAR regulon

The Cpx envelope stress response was initially identified by Silverman and colleagues (McEwen and Silverman, 1980), and workers in Silhavy's

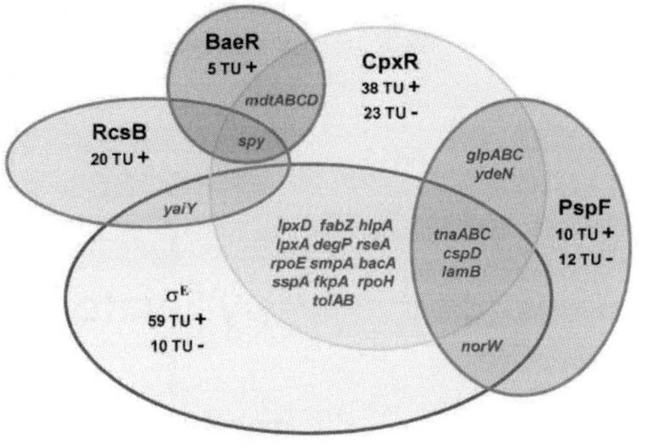

Figure 1.1 Venn diagram representation of all five extracytoplasmic response pathways. The total number of genes/operons (i.e. transcription units; TU) affected in each condition, positive (+) and negative (–) regulation is indicated for each regulon. The Cpx regulon comprises genes affected by CpxR overproduction or deletion (genes repressed or induced by the absence of *cpxR* are counted as + and –, respectively). Only genes that are common to at least two regulons and confirmed by qRT-PCR and/or the literature are listed (genes *glpABC*, *ydeN*, *tnaABC*, *cspD* and *lamB* are repressed, all others are induced). Genes described in the literature as dependent on CpxR and σ^E or BaeR are also included. (Adapted from Bury-Moné *et al.* 2009.)

laboratory, who identified a role for the Cpx two-component system in this process (Cosma *et al.*, 1995). The response is controlled by the two-component system CpxAR made up of the membrane-localized sensor histidine kinase (HK) CpxA and the response regulator CpxR. CpxA is induced by a variety of envelope stresses, all of which are predicted to result in membrane protein misfolding (Fig. 1.2). These include alkaline pH, the overexpression of the lipoprotein NlpE or misfolded pilus subunits, alterations in membrane composition, cell adhesion, P pili synthesis and others (Wolfe *et al.*, 2008).

Although the exact number is unknown, the regulon of CpxR consists of at least 50 genes, many still undefined or annotated to be unrelated to envelope protein maintenance, distributed within 34 operons (Price and Raivio, 2009). Up-regulated genes include proteins involved in, for example, protein secretion (*secA*) and folding (*dsbA*), lipid biosynthesis (*psd*) or virulence and adherence (*pap*), while genes involved in chemotaxis and motility (*aer, motAB, cheAW*) or cell division (*minCDE*) are down-regulated.

It has also been shown that CpxR_P, of the CpxAR system can bind to the *mdtABCD, baeSR, acrD* and *spy* promoters, but in the cases of *mdtABCD* and *acrD*, this binding is thought to be auxiliary to BaeR_P binding, and in the case of *spy*, these pathways are believed to act independently of one another (Hirakawa *et al.*, 2005).

It has been proposed that acetyl phosphate, the intermediate of the phosphotransacetylase (Pta)-acetate kinase (AckA) pathway, can activate the transcription of *cpxP* in a CpxA-independent manner by donating its phosphoryl group to CpxR. While CpxR could receive a phosphoryl group from acetyl phosphate, this global signal was not the primary trigger for CpxR activation associated with the Pta-AckA pathway. On the strength of these results, Wolfe and collaborators (Wolfe *et al.*, 2008) conclude that the interactions between central metabolism and signal transduction can be quite complex and that successful investigation of such interactions must include a complete epistatic analysis.

A new transcriptional regulation pathway in which the Cpx system modulates the integrity of the cell envelope, in part by controlling peptidoglycan amidase activity, has been identified in *Salmonella* Typhimurium and *E. coli* (Weatherspoon-Griffin *et al.*, 2011). A footprinting analysis revealed that the CpxR protein can interact with *amiA* and *amiC* promoters at the CpxR box, which is localized between the predicted –10 and –35 regions but present on different strands in these two genes.

Collectively, it has been suggested that the Cpx system plays a role in the early stages of infection and in the transition from a planktonic to sessile lifestyle within a biofilm, where the Psp response is induced.

Sigma E regulon

The extracytoplasmic function (ECF) Group IV sigma factors are conserved across both Gram-positive and Gram-negative species (Lonetto *et al.*, 1994), and comprise a large, phylogenetically distinct subfamily within the sigma^{70} family. Sigma^E, an ECF sigma factor that was initially recognized as a heat-shock sigma factor in *E. coli* (Erickson and Gross, 1989), responds to accumulation of specific unfolded proteins in the periplasm (Alba and Gross, 2004). Members of

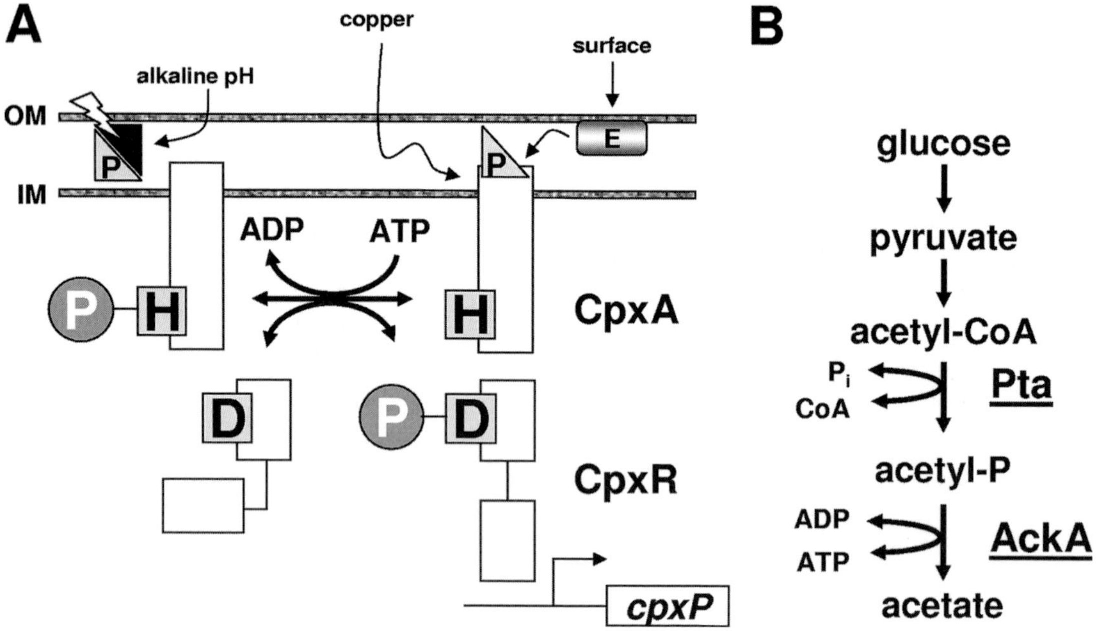

Figure 1.2 Cpx two-component signal transduction and Pta-AckA pathways. (A) Schematic diagram of the Cpx pathway. OM, outer membrane; IM, inner or cytoplasmic membrane; H, site of autophosphorylation of the histidine kinase CpxA; D, site of autophosphorylation of the cognate response regulator CpxR; bent arrow, transcription initiation site of the CpxR target gene, *cpxP*. CpxP (P), a periplasmic chaperone (grey triangle), inhibits the activity of CpxA. CpxP binds to a substrate (black triangle). The CpxP–substrate complex is degraded by the ATP-independent periplasmic protease DegP (thunderbolt). Released from CpxP, CpxA can autophosphorylate using ATP as its phosphoryl donor. Phospho-CpxA then acts as a phosphoryl donor to CpxR. Phosphorylation exposes the CpxR DNA-binding domain (Tapparel *et al.*, 2006), which binds its target genes, including *cpxP*. CpxA also possesses phosphatase activity, which removes phosphoryl groups from phospho-CpxR in the form of P_i. (B) Schematic diagram of the Pta-AckA pathway that converts acetyl-CoA to acetate via an acetyl-P intermediate. (Adapted from Wolfe *et al.* 2008).

the ECF subfamily are different from the rest of the σ^{70} family in that they regulate a wide range of functions involved in sensing and reacting to conditions in the membrane, periplasm, or extracellular environment (Raivio and Silhavy, 2001). Sensing of the extracellular environment is achieved via a signal transduction mechanism in which the ECF sigma factor is bound to a cognate inner membrane-bound anti-sigma factor. The σ^E response is induced by temperature fluctuations, exposure to ethanol or detergents, osmotic and oxidative stresses as well as during biofilm formation, stationary-phase growth and the overproduction of OM proteins, conditions all are thought to cause misfolding of OM proteins, resulting in the activation of the alternative factor σ^E via regulated intramembrane proteolysis.

The σ^E-directed envelope stress response maintains outer membrane homeostasis and is an important virulence determinant upon host infection in *E. coli* and related bacteria. The main regulated gene products are mostly involved in the folding and degradation of proteins in the cell envelope; examples include DsbC, SurA, FkpA and DegP (Fig. 1.3) (Hayden and Ades, 2008).

Sigma^E is activated by at least two distinct mechanisms: accumulation of OM porin precursors and an increase in the alarmone ppGpp upon transition to stationary phase.

The σ^E stress response was characterized as a single-tier activation network comprising ~100 genes, including the MicA and RybB non-coding sRNAs. These highly expressed sRNAs were thought to carry out the specialized function of halting *de novo* synthesis of several abundant porins when envelope homeostasis was perturbed.

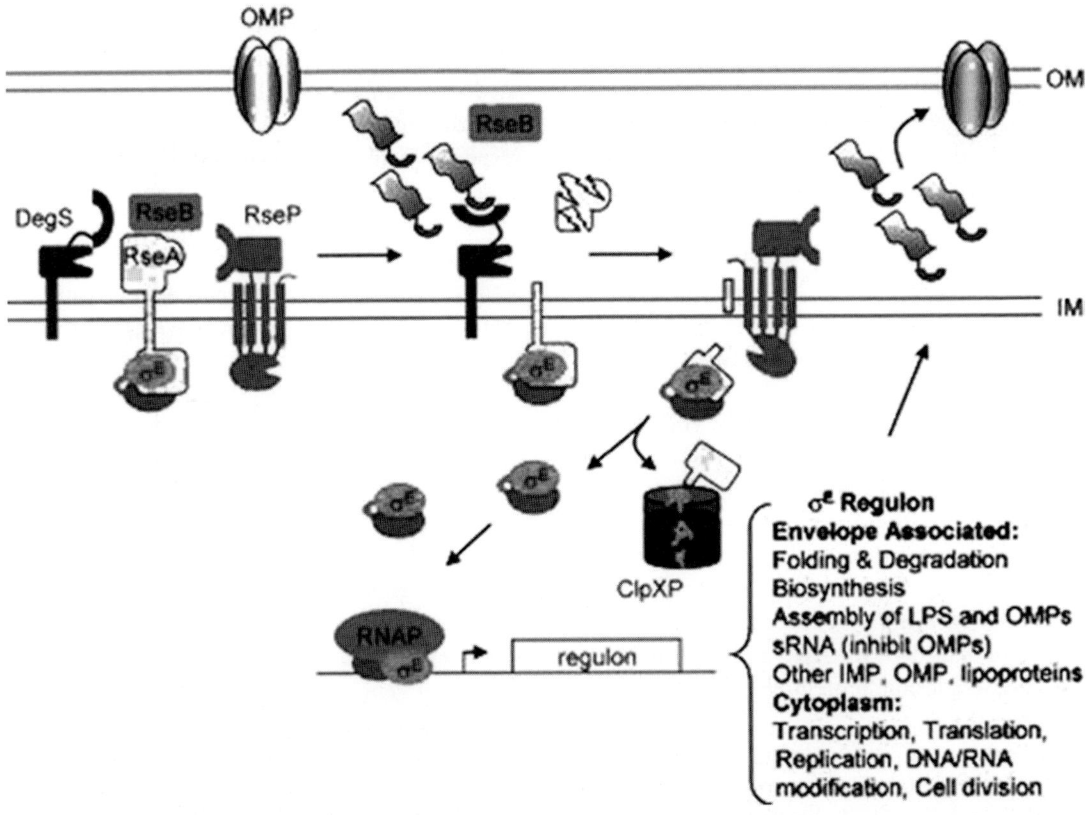

Figure 1.3 The σ^E-dependent extracytoplasmic stress response. σ^E is held at the membrane by the antisigma factor RseA. RseB binds to the periplasmic domain of RseA and protects RseA from proteolysis. Unfolded OMPs activate the protease DegS, which cleaves the periplasmic domain of RseA. The partially degraded RseA is now a substrate for RseP. RseP cleaves RseA further, releasing the cytoplasmic domain of RseA bound to σ^E. This remaining domain of RseA is degraded by ClpXP thereby freeing σ^E to interact with RNA polymerase and direct transcription of its regulon. The major classes of known genes in the σ^E regulon are indicated. The alarmone ppGpp and protein DksA can activate σ^E-dependent transcription once σ^E is released from RseA, but are not shown for clarity. IM, inner membrane; OM, outer membrane. (Adapted from Hayden and Ades, 2008.)

Recent data (Gogol *et al.*, 2011) have shown that MicA and RybB are each global mRNA repressors of both distinct and shared targets, and that the two sRNAs constitute a post-transcriptional repression arm whose regulatory scope rivals that of the protein-based σ^E activation arm. So, σ^E is a paradigm of a single-tier stress response with a clear division of labour in which highly expressed non-coding RNAs (MicA, RybB) endow a transcriptional factor intrinsically restricted to gene activation (σ^E) with the opposite repressor function.

Approximately 60 σ^E-dependent promoters are involved in driving the expression of the σ^E regulon. Mutalik *et al.* (2009) have shown that the σ^E promoters exhibit a large dynamic range, with a few strong and many weak promoters. The strongest promoters control either transcriptional regulators or functions related to porin homeostasis, the much conserved functions among *E. coli* and its close relatives. σ^E promoters are generally activated by ppGpp, and they are responsive to changing σ^E holoenzyme levels, reinforcing the idea that the σ^E regulon is extremely dynamic, enabling cellular adaptation to a constantly changing environment.

BaeSR regulon

The BaeSR pathway is also a classical two-component system and as such consists of a sensor histidine kinase (HK), BaeS, and a cytoplasmic response regulator (RR), BaeR. BaeS communicates with BaeR using phosphotransfer between the histidine of BaeS and the conserved aspartate residue in the BaeR receiver domain. Phosphorylation of BaeR to BaeR_P results in an active output (DNA binding) domain, and BaeR_P acts as a dimer to regulate transcription of the BaeR regulon.

The BaeSR-regulated response was first discovered in a screen for additional two-component systems in *E. coli* and was later classified as an envelope stress response (Raffa and Raivio, 2002). The confirmed regulon of BaeR consists of the *acrD* gene, the *spy* gene, and the *mdtABCD-baeSR* operon. These genes encode two RND (resistance, nodulation, and cell division) multidrug efflux pumps, MdtABC and AcrD; a predicted MFS (major facilitator superfamily) multidrug efflux pump, MdtD; the BaeS and BaeR proteins; and the periplasmic protein Spy, whose function is unknown.

Induction of *spy* gene expression during spheroplasting, exposure to indole, and overexpression of the pilin subunit PapG in the absence of its chaperone were partially attributable to BaeSR pathway activity. All three of these inducers have been shown to also induce other stress responses, such as the sigma^E response (PapG overexpression) and the Cpx pathway (spheroplasting, indole, and PapG overexpression) (Jones *et al.*, 1997). However, in the case of indole at least, the majority of *spy* gene induction occurs through BaeSR.

Flavonoids and sodium tungstate have been shown to produce a much stronger induction of the BaeSR response in an *mdtA* efflux pump mutant, while indole did not. These findings are consistent with the hypothesis that flavonoids and sodium tungstate are natural substrates of the MdtABC efflux pump. Indole has recently been implicated in cell–cell signalling and biofilm repression through a putative interaction with the LuxR homologue SdiA. Using genetic analyses, it has been found that induction of the BaeSR response by indole occurs via a pathway separate from the SdiA biofilm pathway (Leblanc *et al.*, 2011). Further, the authors demonstrated that the BaeSR response does not influence biofilm formation, nor is it involved in indole-mediated inhibition of biofilm formation. They hypothesized that the main function of the Bae response is to up-regulate efflux pump expression in response to specific envelope-damaging agents.

A genome-wide analysis of *E. coli* gene expression responsive to the BaeSR system reveals that BaeR overproduction activates 59 genes related to two-component signal transduction, chemotactic responses, flagellar biosynthesis, maltose transport, and multidrug transport, and BaeR overproduction also repressed the expression of the *ibpA and ibpB* genes (Nishino *et al.*, 2005). The expression levels of 15 of the 59 genes were decreased by deletion of *baeSR*, and 11 genes were identified as indole-regulated genes. The seven genes in the overlap of all three data sets were identified as components of the BaeSR regulon.

Several genes up-regulated by BaeR overproduction, including genes for maltose transport, chemotactic responses, and flagellar biosynthesis, required an intact PhoBR or CreBC two-component regulatory system for up-regulation. These data indicated that there is cross-regulation among the BaeSR, PhoBR, and CreBC two-component regulatory systems.

A widespread feature in the genomes of most bacteria and archaea is an array of clustered, regularly interspaced short palindromic repeats (CRISPRs) that, together with a group of CRISPR-associated (Cas) proteins, mediate immunity against invasive nucleic acids. Perez-Rodriguez *et al.* (2011) have recently shown that expression of a Tat substrate resulted in up-regulation of the Cas enzymes and subsequent silencing of the encoding plasmid in a manner that required the BaeSR two-component regulatory system.

They confirm that the CasCDE enzymes form a stable ternary complex and appear to function as the catalytic core of the Cas system to process CRISPR RNA into its mature form. Taken together, those results indicate that the CRISPR-Cas system targets DNA directly as part of a defence mechanism in bacteria, that involves the BaeSR regulon, and that is overlapping with but not limited to phage infection.

Rcs regulon

Detection of membrane integrity through the use of membrane-localized proteins is a logical but undefined mechanism for mediating signal transduction. The sensing of specific types of membrane disruption may be important for all organisms but is certainly logical for Gram-negative bacteria, which evolved a two-membrane system as a protection against environmental stresses and to promote environmental sensing and protection.

The Rcs regulon was initially described as a two-component system to control production of capsular cholanic acid in *E. coli*. Currently it is known as a more complex mechanism that it is involved in the regulation of several cellular process including cell division, flagella synthesis and Vi antigen expression, among others. The Rcs regulon in *E. coli* is composed of more than 150 genes (Ferrieres and Clarke, 2003).

One of the bacterial mechanisms used to sense and respond to envelope stress includes the RcsFCDB phosphorelay. One of such stresses includes amphipathic cationic antimicrobial peptides (CAMPs), and resistance to these peptides is an important property for microbial virulence for animals. The Rcs phosphorelay includes two inner membrane (IM) proteins, RcsC and RcsD; the response regulator RcsB; the accessory coregulator RcsA; and an outer membrane-bound lipoprotein, RcsF.

Rcs transcriptional activation by CAMPs required RcsF and preservation of its two internal disulfide linkages. The rerouting of RcsF to the inner membrane or its synthesis as an unanchored periplasmic protein resulted in constitutive activation of the Rcs regulon and RcsCD-dependent phosphorylation (Farris *et al.*, 2010). That finding suggests that RcsFCDB activation in response to CAMP-induced membrane disorder is a result of a change in structure or availability of RcsF to the IM signalling constituents of the Rcs phosphorelay. How does RcsF transfer the signal of outer membrane disorder by CAMPs? Once RcsF is available to the inner membrane signalling components, RcsC, RcsD, and the negative regulator IgaA are all candidates for a direct interaction with RcsF. Therefore, either direct interaction of RcsF with the periplasmic domains of RcsC or RcsD or binding to IgaA and release of inhibition could result in activation. It has not been excluded that an unknown additional periplasmic or inner membrane component of the regulon directly mediates the RcsF effect.

VR regulon

Production of outer membrane vesicles by Gram-negative bacteria has been described by microscopic observations for decades, but only recently have genetic and biochemical studies begun to reveal its physiological utility (McBroom and Kuehn, 2007). Vesicle release by both pathogenic and non-pathogenic Gram-negative bacteria is a ubiquitous process that occurs over the course of normal growth (McBroom and Kuehn, 2005). Biochemical and microscopic observations indicate that vesicles are formed by fission of outer membrane protrusions that enclose periplasmic components (Kulp and Kuehn, 2010). Vesicle mediated delivery of toxins to eukaryotic cells also suggests a role for vesicles in pathogenesis. While the composition of vesicles and their interactions with prokaryotic and eukaryotic cells have been investigated, the fundamental physiological role of vesicle production by bacterial cells has remained largely unknown.

Recently, McBroom and Kuehn (2007) have found a novel function for outer membrane vesicle release as a regulated stress response. The process of vesiculation is fully independent from the known stress-responsive systems, and is poised to complement them by offering an effective mechanism for removal of undesirable soluble and insoluble envelope components. Vesicle production is modulated in response to the state of the envelope, decreasing under low stress conditions and increasing with accumulation of overexpressed or misfolded envelope components. Increased vesiculation correlates with increased bacterial survival upon exposure to chemical stressors or accumulation of toxic protein species, and specific packaging of a stress protein into vesicles for removal from the cell demonstrates existence of a preferential cargo selection.

Cell wall stress-sensing regulatory systems in Gram-negative bacteria

Only two TCS regulons and one TA system have been identified as responsible for control of gene expression under changes in structure or metabolism of peptidoglycan, or under beta-lactam challenge.

CreBC regulon

CreBC is a two-component system that controls the expression of a number of genes in *E. coli* (called the *cre* regulon) that encode diverse functions, including intermediary metabolic enzymes. It has been shown that *cre* regulon gene expression is activated during growth in minimal media when glycolytic carbon sources are being fermented. It also is activated during aerobic growth when fermentation products are being used as carbon sources. CreB and CreC are essential for the activation of *cre* regulon gene expression, but CreA and CreD, encoded as part of the *creABCD* gene cluster, are not. CreB binds to a TTCACnnnnnnTTCAC direct repeat (the *cre* tag) *in vitro*, and this sequence, which is associated with *cre* regulon gene promoters, is required for the control of gene expression *in vivo*. So, CreBC is a functional two-component system involved, at least, in the metabolic control of transcription in *E. coli* and it has been confirmed that CreB is a DNA binding transcriptional regulator (Cariss *et al.*, 2008).

CreD is overproduced in a Cet2 (ColicinE2-tolerant) strain of *E. coli* as the result from hyperactivation of the CreBC system and it was believed to cause the Cet2 phenotype. Through microarray analysis and gene knockout and overexpression studies, it was shown that overexpression of another CreBC-regulated gene, *yieJ* (also known as *cbrC*), causes the Cet2 phenotype (Cariss *et al.*, 2010).

Orthologous TCS creBC is found in *Aeromonas* strains, called BlrAB, that responses in a similar manner to that of *E. coli*. Tayler and colleagues (2010) have studied the mechanism by which beta-lactam challenge leads to beta-lactamase induction in *Aeromonas hydrophila* through transposon-insertion mutagenesis. Disruption of the DD-carboxypeptidases/endopeptidases, penicillin-binding protein 4 or BlrY leads to elevated monomer disaccharide-pentapeptide levels in *A. hydrophila* peptidoglycan and concomitant overproduction of beta-lactamase through activation of the BlrAB two-component regulatory system. During beta-lactam challenge, monomer disaccharide-pentapeptide levels increase proportionately with beta-lactamase production and beta-lactamase induction is inhibited by vancomycin, which binds muro-pentapeptides. Taken together, these data strongly suggest that the *Aeromonas* spp. beta-lactamase regulatory sensor kinase, BlrB, responds to the concentration of monomer-disaccharide-pentapeptide in peptidoglycan (Fig. 1.4) (Tayler *et al.*, 2010). So, this mechanism of resistance is based on sensing the peptidoglycan fine structure directly, and transducing the signal from periplasm.

BlaR regulon

The BlaR regulon is a direct mechanism for bacteria to sense the antibiotic outside by a specific receptor and transduction to the cytoplasm.

The beta-lactam sensor-transducer (BlaR), an integral membrane protein, binds beta-lactam antibiotics on the cell surface and transduces the information to the cytoplasm, where gene expression is derepressed for both beta-lactamase and penicillin-binding protein. Expression of these proteins is controlled by similar regulation systems, including a repressor (BlaI/MecI) and a multidomain transmembrane receptor (BlaR/MecR). Resistance is triggered in response to a covalent binding event between a beta-lactam antibiotic and the extracellular sensor domain of BlaR/MecR by transduction of the binding signal to an intracellular protease domain capable of repressor inactivation. This mechanism has been described only in Gram-positive strain where a direct interaction of the antibiotic with inner membrane components is achievable.

It has been shown that beta-lactam antibiotics covalently modify the BlaR protein, and the protein contains the unusual carboxylated lysine that activates the active-site serine residue for acylation by the beta-lactam antibiotics. The protein undergoes acylation by beta-lactam antibiotics with microscopic rate constants, yet the deacylation process is essentially irreversible. The protein

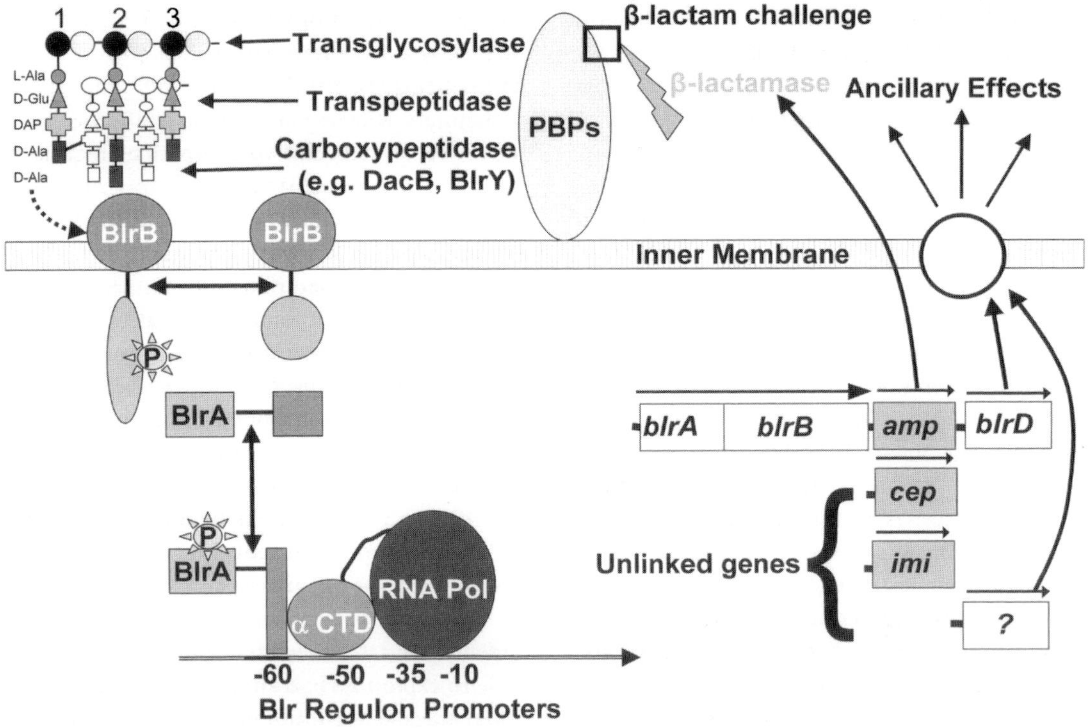

Figure 1.4 Model for β-lactamase induction in *Aeromonas* spp. and the role of the *blr* regulon. Penicillin-binding proteins (PBPs) are enzymes that have three major roles in peptidoglycan synthesis. High molecular mass essential PBPs are transglycosylases, which join the disaccharide pentapeptides together, and/or transpeptidases, which cross-link pentapeptides from different strands of peptidoglycan through formation of a bond between the fourth amino acid, a D-alanine (D-Ala), of a donor peptide and the third, diaminopimelic acid (DAP) moiety on a recipient peptide. The result is removal of the terminal D-Ala from the donor peptide (see the cross-linked peptide labelled '1', top left). Not all pentapeptides are cross-linked (see the peptide labelled '2'). Carboxypeptidases (non-essential low-molecular-mass PBPs) remove the terminal D-Ala from a peptide. The resulting tetrapeptides (see the peptide labelled '3') cannot be cross-linked by transpeptidases, so the role of these enzymes is to fine-tune the amount of cross-linking in peptidoglycan. We propose that inhibition of PBPs by β-lactams causes an increase in the concentration of the BlrB activatory ligand. This ligand then interacts with BlrB (dotted arrow), causing it to auto-phosphorylate ('P' in the diagram). This phosphate is transferred to BlrA, which binds to the *cre/blr* tag sequence found upstream of all *blr* regulon gene promoters. The effect of this is to recruit RNA polymerase and activate *blr* regulon transcription. Known *blr* regulon genes encode three β-lactamases, Amp, Cep and Imi, which directly hydrolyse the β-lactam, helping to prevent further peptidoglycan damage. The *blr* regulon is also known to include non-β-lactamase-encoding genes, e.g. *blrD*, and it is proposed that their products have a role in protecting the cell from β-lactam challenge through an ancillary mechanism. (Adapted from Tayler *et al.*, 2010.)

undergoes a significant conformational change on binding with beta-lactam antibiotics, a process that commences at the preacylation complex and reaches its full effect after protein acylation has been accomplished. These conformational changes are likely to be central to the signal transduction events when the organism is exposed to the beta-lactam antibiotic.

Comparison of the BlaR sensor domain structure with those of class D beta-lactamase complexes and penicillin-binding proteins provides interesting elements to hypothesize on possible signal transmission mechanisms.

PezAT system

Toxin–antitoxin (TA) systems consist of bicistronic operons that encode for a bacterial toxin and its cognate inhibitor, which neutralizes the toxin under dormant conditions. Once *de novo* synthesis from the operon is impaired, continuous proteolytic degradation of the antitoxin eventually releases the toxin and, depending on the functional mechanism, induces bacteria stasis or cell death. Thus, TA systems have been linked with numerous cellular functions, including programmed cell death, maintenance of mobile genetic elements, stress response, persistence, and biofilm formation (Mutschler *et al.*, 2011).

Members of the epsilon/zeta TA family have been shown to stabilize resistance plasmids in major human pathogens. The epsilon/zeta antitoxin toxin (PezAT) system is a chromosomally encoded, class II toxin-antitoxin system originally found in the human pathogen *S. pneumoniae*. However, the broad distribution of epsilon/zeta systems implies that zeta toxins must utilize a ubiquitous bacteriotoxic mechanism. Whereas all other TA families known to date poison macromolecules involved in translation or replication, the target of zeta toxin PezT impairs cell wall synthesis and triggers autolysis in *E. coli*. No induction signals have been already identified.

In vitro zeta toxins phosphorylate the ubiquitous peptidoglycan precursor uridine diphosphate-N-acetylglucosamine (UNAG) and this activity is counteracted by binding of antitoxin. Phosphorylated UNAG inhibits MurA, the enzyme catalysing the initial step in bacterial peptidoglycan biosynthesis. Thus, PezT and zeta toxin activity impair muramic acid synthesis in two different ways: first, it depletes the pool of UNAG precursors and, second, UNAG-3P is a competitive inhibitor of MurA. So, zeta toxins are novel kinases that poison bacteria through global inhibition of peptidoglycan synthesis (Mutschler *et al.*, 2010). Neutralization of the bacteriotoxic protein PezT is carried out by complex formation with its cognate antitoxin PezA. A kinetic analysis of PezAT assembly revealed that strong thermodynamic stabilization of both proteins upon binding is based on a strong, electrostatically guided binding mechanism leading to a stable toxin antitoxin complex with femtomolar affinity. These data show that PezAT complex formation is distinct to all other conventional toxin antitoxin modules and a controlled mode of toxin release is required for activation.

Stress response of cell envelope components mediated by BolA

In natural environments bacteria alternate periods of active growth and quiescent life because of nutrient deprivation or suboptimal environmental conditions. The ability of cells to quickly respond to such changes is important in terms of survival. Similarly, when conditions turn favourable, their ability to resume fast growth is also critical to outgrow and eventually displace competitor species. A number of general stress and starvation response mechanisms (stringent response, SOS, metabolic repression, *rpoS*-dependent stationary phase gene regulation) are involved in the coordination of cellular adaptation to unfavourable conditions as extensively discussed in other chapters of the present book. In general these mechanisms slow down growth and metabolic activity, shift to alternative metabolic pathways able to exploit second choice nutrients, and very often affect cell size and shape. How cells modify their dimensions in response to stress conditions is still far from understood.

An important element for the correct adaptation of cell size to stress conditions is the BolA protein. BolA was initially described in the 1980s as a morphogene in *E. coli* (Aldea *et al.*, 1988). However, later research showed that BolA is indeed widely conserved with homologues found both in prokaryotes as well as in eukaryotes giving rise to the BolA-like protein family (Kasai *et al.*, 2004). BolA-like proteins are thought to be generally involved in cell morphogenesis modulating the cell cycle, but is still a matter of debate in most cases.

The *E. coli* BolA is a 13.5-kDa protein whose gene locus maps in the −10 region of the chromosomal genetic map. It was first identified because when overexpressed it induced formation of osmotically stable spherical cells (Aldea *et al.*, 1988). The round phenotype was dependent on the cytoskeletal protein FtsZ, one of the earliest elements required for the correct assembly of the division machinery, but not on the late division protein FtsI (Aldea *et al.*, 1988). The structural gene is preceded by two promoters, P1 immediately adjacent to the structural sequence, and P2 about 70 bases upstream of P1. Whilst P2 has the properties of a week housekeeping promoter, P1 is a so-called 'gearshift' promoter, indicating that expression is inversely proportional to the rate of growth, and seems to be the predominant promoter. In fact, the *bolA* gene is growth-phase regulated (Aldea *et al.*, 1989), and expression from P1 is induced by the stress and stationary phase-specific sigma factor RpoS (Lange and Hengge-Aronis, 1991). Interestingly, expression from both promoters is, in turn, down-regulated by the direct binding of the histone like protein H-NS although the meaning of this regulation is still unknown (Moreira *et al.*, 2011). Expression of *bolA* results in a long lived, poly adenylated mRNA regulated by RNAseIII (Cao and Sarkar, 1997; Freire *et al.*, 2006a). Therefore, cellular levels of BolA rise strongly when cells enter stationary phase and remain high until growth resumes (Aldea *et al.*, 1989; Lange and Hengge-Aronis, 1991). Overexpression of *bolA* from its own promoters leads to a characteristic rounding of cells when growth rate goes down. This together with observations that *bolA* null mutants remain larger than the corresponding wild-type strains in stationary phase, led to the idea that BolA could be responsible for the adaptation of cell size and shape in response to growth rate variations (Santos *et al.*, 1999). This idea has found solid support in the confirmation of a direct effect of BolA on cell elongation modulating the levels of the key cytoskeletal protein MreB in *E. coli* (Freire *et al.*, 2009). MreB is a key protein which polymerizes on the inner surface of the cytoplasmic membrane in a helical fashion and is essential for proper elongation of rod-like bacteria (Mattei *et al.*, 2010).

The morphological effects triggered by variations in the expression levels of BolA are probably a consequence of the modulation of the expression of cell wall synthesis enzymes. BolA has a helix–turn–helix motif containing a putative DNA binding domain which would enable the protein to bind to DNA and regulate expression of a number of genes (Aldea *et al.*, 1989). Indeed, BolA has been shown to bind to the promoters of penicillin-binding proteins (PBPs) 5 and 6, as well as to the promoter of AmpC beta-lactamase in *E. coli* (Guinote *et al.*, 2011; Sochacki *et al.*, 2011). PBPs 5 and 6 are DD-carboxypeptidases directly involved in cell wall biosynthesis. Alterations in the levels of these proteins affect the morphogenetic process and often results in aberrant morphologies; either spherical cells upon overproduction (Markiewicz *et al.*, 1982) or branching and deformation when inactivated (Nelson and Young, 2000; de Pedro *et al.*, 2003). Therefore a direct connection between the expression levels of BolA and cell wall metabolic activity looks quite convincing. Furthermore, AmpC, a chromosomally encoded beta-lactamase, has also been linked with the bacterial response to cell-wall damaging agents and therefore could be considered an additional stress response element (Jacobs *et al.*, 1997). The coordinated modulation of cell wall biosynthesis via PBPs 5 and 6, and cell wall elongation via MreB, and the dependency of BolA activity on the cell division protein FtsZ and the growth dependent RpoS protein, makes the BolA-mediated regulatory network a credible mechanism to link growth rate and cell size.

Expression of BolA is not exclusive of stationary phase, or slow growth. Many environmental stress conditions do stimulate expression of the corresponding gene, and at least under specific conditions the responses of *rpoS* and *bolA* are not strictly parallel indicating that *bolA* and *rpoS* respond in differential ways to stimuli (Santos *et al.*, 2002; Adnan *et al.*, 2010, 2011).

A remarkable aspect of *bolA* physiology is its involvement in biofilm formation under stress conditions. Biofilm development is an adaptation of many bacteria to harsh environments, and confers extreme resistance to many otherwise deleterious agents and conditions (Costerton *et al.*, 1987). BolA overexpression favours biofilm development and impairment of its function reduces biofilm production by *E. coli*, therefore suggesting a direct implication in the formation of biofilms under stress conditions (Vieira *et al.*, 2004; Adnan *et al.*, 2011).

The effects of alteration in BolA levels not only affect the cell wall itself, but also affect the properties of other cell envelope components, in particular the OM (Freire *et al.*, 2006b). Quite unexpectedly, BolA has a protective effect on the endurance of the OM to damaging agents such as detergents in *E. coli*. Overexpression of the protein leads to a higher resistance to SDS. This phenotype is linked to a higher resistance to antibiotics such as vancomycin. The primary reason seems to be an alteration in the balance and abundance of the different kinds of porins in the OM, possibly through OmpR modulation. Interestingly the variation in BolA levels also results in drastic alteration of the accessibility of OM proteins to externally added impermeable reagents, suggesting alterations in the general ordering of the OM. Such reinforcing fits well with the general protective role that BolA seems to play for the cell, in particular with its biofilm promoting activity.

Non-canonical D-amino acids govern cell wall remodelling in the stationary phase

One of the most striking features of PG composition is the presence of D-amino acids in the stem peptides (Nagata *et al.*, 1998). D-Ala and D-Glu are by far the most common D-amino acids present in the bacterial cell wall; however, the PG of some bacteria include other D-amino acids, such as D-Asp in *Lactococcus* (Veiga *et al.*, 2006) and *Enterococcus* (Bellais *et al.*, 2006), and D-Ser in vancomycin-resistant *Staphylococcus aureus* (Sieradzki and Tomasz, 1996; De Jonge *et al.*, 2002; Reynolds and Courvalin, 2005). In some cases, such substitutions enable bacteria to tolerate bactericidal agents from the environment (de Lencastre *et al.*, 2007).

Conversion of L- to D- amino acid occurs by the action of racemases that change the stereochemistry of the chiral alpha-carbon atom in amino acids (Tanner, 2002).

As outlined above, the roles of D-amino acids in bacteria were thought to be fairly limited. However, recently it has become clear that D-amino acids are synthesized and released by bacteria from diverse phyla, at up to millimolar concentrations (Lam *et al.*, 2009). Furthermore, released D-amino acids function to regulate cell wall chemistry and architecture as well as biofilm development in bacteria (Kolodkin-Gal *et al.*, 2010; Cava *et al.*, 2011).

Although PG functions as an exoskeleton, bacterial growth and survival depends on PG plasticity. Under stress, PG also can be remodelled. For example, many rod-shaped bacteria reorganize their PG in stationary phase, becoming smaller and more coccoid in shape (Nystrom, 2004). Currently, there is little understanding of the factors that govern PG remodelling. However, D-amino acids were recently shown to control changes in PG composition and architecture in *Vibrio cholerae* as cells enter stationary phase (Lam *et al.*, 2009).

Many phyla of bacteria produce D-amino acids that were not previously known to be synthesized (named non-canonical D-amino acids; in short NCDAAs), and release them into the environment reaching concentration as high as a ~1 mM total concentration of four NCDAAs: D-Met, D-Leu, D-Ile and D-Val in *V. cholerae* (Lam *et al.*, 2009). Genomic analysis identified a novel *V. cholerae* racemase, BsrV (broad-spectrum racemase in *Vibrio*), responsible for the synthesis of the unusual D-amino acids which exhibits an unusually broad substrate specificity. BsrV is probably active only during stationary phase and was found exclusively in the periplasm, suggesting that NCDAAs produced by BsrV primarily act on periplasmic targets (Lam *et al.*, 2009). Release of NCDAAs into the media is not limited to *V. cholerae* and diverse phyla including *Bacillus subtilis, S. aureus, Pseudomonas aeruginosa, and Deinococcus radiodurans* do release different D-amino acids. Consistent with this observation, most bacterial genomes appear to encode one or more amino acid racemases in addition to the enzymes required to generate D-Ala and D-Glu for PG synthesis (Lam *et al.*, 2009). This suggests that production of broad spectrum racemases may be a conserved trait amongst diverse bacteria.

Impairment of the *bsrV* gene in *V. cholerae* leads to increased cell peptidoglycan content, which doubles the amount in the wild type. Addition of physiological concentrations of D-amino acids reduced the amount of PG to normal. Thus, NCDAAs down-regulate the amount of PG produced by *V. cholerae* in stationary phase. Incorporation of NCDAAs into the cell wall during stationary phase is instrumental for *V. cholerae* to resist osmotic stress and to control the amount of PG per cell (Cava *et al.*, 2011). Therefore,

NCDAAs may enable coordination of metabolic slowing in cell wall and cytoplasmic compartments when resources become scarce. Exactly how NCDAA-dependent cell wall remodelling occurs remains to be determined. However, at least two mechanisms are involved: incorporation of NCDAAs into the murein and direct modulation of PBPs.

HPLC-based analyses of muropeptides from wild-type cells demonstrated that D-Met and D-Leu were present in PG in stationary phase (Cava *et al.*, 2011). The capacity to incorporate such amino acids apparently is widespread among bacteria; hence NCDAAs can also act as paracrine-like effectors to influence PG physiology in species other than those that produce them (Cava *et al.*, 2011). However, multiple incorporation mechanisms seem to exist. Incorporation was detected at two sites in PG subunits, and seems to be mediated by at least three processes, including both periplasmic PG editing and cytoplasmic incorporation into PG precursors.

NCDAAs, such as D-Met, are incorporated at two distinct sites within *V. cholerae* PG subunits. The majority is present as the C-terminal amino acid within monomeric or dimeric tetramers (muro4^{M}), while the remainder comprises the C-terminal amino acid within pentapeptides (muro5^{M}). Formation of muro4^{M} peptides is dependent upon L,D-transpeptidases (Ldts), inner membrane anchored-periplasmic proteins that catalyse L,D crosslinking of muropeptides and linkage of lipoprotein Lpp to PG (Magnet *et al.*, 2007a,b, 2008). Both of *V. cholerae* two Ldts (LdtA and LdtB) contribute to incorporation of D-Met, although the two proteins play exclusive roles with respect to muropeptide cross-linking. Ldts also mediate incorporation of D-Met into PG tetrapeptides in *E. coli* and *P. aeruginosa*, even though *E. coli*, unlike the majority of bacterial species tested, does not produce NCDAAs. Ldts have a broad substrate specificity, a flexibility that may enable reaction to the range of D-amino acids produced during stationary phase by diverse organisms in the natural environment (Cava *et al.*, 2011).

Incorporation of D-Met into muro5^{M} probably occurs within *V. cholerae* cytoplasm, via PG precursors. Both *in vivo* and *in vitro* analyses indicate that D-Met is an alternate substrate for Ddl and that MurF can utilize dipeptides that contain D-Met, although at reduced rates (Cava *et al.*, 2011). Such inhibition might contribute to reduced PG accumulation in response to growth arrest. Muro5^{M} peptides also appear to alter periplasmic PG editing. As previously observed in *E. coli* and other species, canonical pentapeptides are scarce in *V. cholerae* PG, as they are presumably transformed into tetrapeptides by D,D-carboxypeptidases. In contrast, muro5^{M} is readily detected in stationary phase, suggesting that it is a poor substrate for these enzymes. It is possible that the alteration in the balance of pentapeptides and tetrapeptides also influences the accumulation and/or properties (e.g. strength) of PG.

Muro5^{M} was also detected in PG from other bacteria; however, such muropeptides were generated by at least two different pathways. Formation of muro5^{M} in *C. crescentus* is inhibited by D-cycloserine as in *V. cholerae*, suggesting that it is likewise mediated by Ddl and MurF.

D-amino acids also affect the activity of the periplasmic enzymes that synthesize and modify the PG polymer. Waldor and colleagues found that D- but not L-Met blocks the binding of a fluorescent derivative of penicillin G to several *V. cholerae* PBPs (Lam *et al.*, 2009). This result suggests that controlling the concentration of free D-amino acids in the periplasm could constitute a reversible mechanism to control PBP activity, reducing PG metabolism under challenging environmental conditions.

Production and incorporation of NCDAAs in *V. cholerae*, is governed by the sigma factor RpoS (Cava *et al.*, 2011), which becomes active in stationary phase and in response to numerous stresses (Hengge, 2008). Therefore, production and response to D-amino acids must be considered as one of the mechanisms involved in the general stress response of bacteria. Because D-amino acids down-regulate PG synthesis, they may well intervene in the coupling of cell wall synthesis and growth rate. Additionally, rapid diffusion of small molecule regulators like D-amino acids in aqueous environments can signal to the whole population to regulate PG amount, composition and strength. In consequence, D-amino acids may alter the fitness of bacteria within mixed microbial communities, including bacteria that do not produce NCDAAs, as the capacity for incorporation seems to be widely spread. Such alteration could lead to a variety of phenotypes. For example, incorporation of D-amino acids has been linked to dissolution of biofilms (see below) (Kolodkin-Gal *et al.*, 2010) and antibiotic resistance (Reynolds and Courvalin, 2005). Future studies are likely to reveal additional roles for D-amino acids in stress responses and bacterial communication.

Moreover, since bacteria grow in polymicrobial communities D-amino acids may mediate interspecies relations among bacteria or other organisms that occupy the same niche. It remains to be seen whether D-amino acids mediate both mutualistic and competitive behaviours among different bacteria that share a niche.

Resistance as a selection determinant

The driving force behind resistance formation is antibiotic stress, and under clinical conditions selection for resistance is continuously competing with selection for bacterial fitness. Both aspects of antibiotic stress and bacterial fitness are not described on this chapter, but surely they will be worthy of another chapter.

Future trends

The main challenge for the future is to achieve an understanding of the interactions of all these components, under different physiological conditions, and to identify the initial signal of each cell wall stress response.

Also, there are numerous examples demonstrating that the bacterial cell is highly organized in a spatial sense. Polar localized proteins provide classic examples, and the bacterial chromosome is known to have a particular orientation within the cell. Sequences of the replication origin can be found at the cell poles. The spatial organization of several complexes involved in the stress response may provide a clue for the mechanism of action. The dynamic formation of signalling complexes and the protein self-association may well generate pockets of locally concentrated signal processing complexes. Whether or not those signalling complexes are preassembled or whether their assembly itself is a part of the switching event remains to be determined. The curvature and lipid composition of the membrane at the pole and at the junction of the lateral and polar membrane may play a role in retaining such complexes. How the polar complexes initially arise may involve some distributed self-association events in the membrane; these in themselves may be favoured by well-separated (to reduce local competition) nucleating lower-order complexes capturing their partner proteins. Such distributions may 'drive' complexes to be formed far apart (at the poles) and subsequently stabilized by membrane curvature.

Acknowledgements

Because of the scope of this review, it was not possible to cite the multitude of recent studies that address this topic. The authors apologize to those whose work was not included.

Web resources

Two component system (TCS):
http://www.biochem.mpg.de/en/eg/oesterhelt/web_page_list/ShortDesc_two_component/index.html

A web page to understand the basic principles of two-component signal transduction: a short description and a more complex view.

P2CS (prokaryotic two-component systems): http://www.p2cs.org/

P2CS is an integrated and comprehensive database of two-component system (TCS) signal transduction proteins, which contains a compilation of the TCS genes within completely sequenced genomes and metagenomes. It provides detailed annotation of each TCS gene including sequence features, functional domains, as well as chromosomal localization.

Two-component system (KEGG):
http://www.genome.jp/kegg/pathway/ko/ko02020.html

This is the official web page of KEGG (Kyoto Encyclopaedia of Genes and Genomes) for the reference pathway of two-component systems.

Two-component system (SENTRA): http://www.mcs.anl.gov/research/project_detail.php?id=40

Sentra is a database of prokaryotic signal transduction proteins. Contains automated prediction of signal transduction proteins for all publicly available prokaryotic genomes as well as manually curated collection of signal transduction proteins from 202 completely sequenced genomes from Michael Galperin (NCBI)

Two-component system (MisT2): http://mistdb.com

The Microbial Signal Transduction database contains the signal transduction proteins for 1202 complete and 459 draft bacterial and archaeal genomes. These are identified using various domain profiles that directly or indirectly implicate a particular protein in participating in signal transduction.

Signal transduction pathways: http://biocyc.org/ECOLI/NEW-IMAGE?object=Signaling-Pathways

This is a database at BioCyc, showing data on *Escherichia coli* K12 signal transduction pathways. BioCyc is a collection of 1692 pathway/genome databases. Each database in the BioCyc collection describes the genome and metabolic pathways of a single organism.

CpxAR system: http://biocyc.org/ECOLI/NEW-IMAGE?type=PATHWAY&object=PWY0-1485

This is the particular web page displaying data on *Escherichia coli* K12 CpxAR two-component signal transduction system.

Sigma E regulon: http://biocyc.org/ECOLI/NEW-IMAGE?type=ENZYME&object=RPOE-MONOMER

This is the particular web page displaying data on *Escherichia coli* K12 RNA polymerase, sigma 24 (sigma E) factor.

Rcs regulon: http://ecocyc.org/ECOLI/NEW-IMAGE?type=PATHWAY&object=PWY0–1493

This is the particular web page at EcoCyc displaying data on *Escherichia coli* K12 RcsCDB two-component signal transduction system.

TA systems: http://bioinfo-mml.sjtu.edu.cn/TADB/

TADB is a web-based resource for Type 2 toxin–antitoxin loci in Bacteria and Archaea. The aim of this PostgreSQL-based TADB database is to create and maintain a resource of all currently available Type 2 TA loci-relevant data ranging from 'wet' experimental data to information garnered by bioinformatics analyses.

References

Adnan, M., Morton, G., Singh, J., and Hadi, S. (2010). Contribution of *rpoS* and *bolA* genes in biofilm formation in *Escherichia coli* K-12 MG1655. Mol. Cell Biochem. *342*, 207–213.

Adnan, M., Morton, G., and Hadi, S. (2011). Analysis of *rpoS* and *bolA* gene expression under various stress-induced environments in planktonic and biofilm phase using 2(-DeltaDeltaCT) method. Mol. Cell. Biochem. *357*, 275–282.

Alba, B.M., and Gross, C.A. (2004). Regulation of the *Escherichia coli* sigma-dependent envelope stress response. Mol. Microbiol. *52*, 613–619.

Aldea, M., Hernandez-Chico, C., de la Campa, A.G., Kushner, S.R., and Vicente, M. (1988). Identification, cloning, and expression of *bolA*, an *ftsZ*-dependent morphogene of *Escherichia coli*. J. Bacteriol. *170*, 5169–5176.

Aldea, M., Garrido, T., Hernandez-Chico, C., Vicente, M., and Kushner, S.R. (1989). Induction of a growth-phase-dependent promoter triggers transcription of *bolA*, an *Escherichia coli* morphogene. Embo J. *8*, 3923–3931.

Bellais, S., Arthur, M., Dubost, L., Hugonnet, J.E., Gutmann, L., van Heijenoort, J., Legrand, R., Brouard, J.P., Rice, L., and Mainardi, J.L. (2006). Aslfm, the D-aspartate ligase responsible for the addition of D-aspartic acid onto the peptidoglycan precursor of *Enterococcus faecium*. J. Biol. Chem. *281*, 11586–11594.

Bury-Moné, S., Nomane, Y., Reymond, N., Barbet, R., Jacquet, E., Imbeaud, S., Jacq, A., and Bouloc, P. (2009). Global analysis of extracytoplasmic stress signaling in *Escherichia coli*. PLoS Genet. *5*, e1000651.

Cao, G.J., and Sarkar, N. (1997). Stationary phase-specific mRNAs in *Escherichia coli* are polyadenylated. Biochem. Biophys. Res. Commun. *239*, 46–50.

Cariss, S.J., Tayler, A.E., and Avison, M.B. (2008) Defining the growth conditions and promoter-proximal DNA sequences required for activation of gene expression by CreBC in *Escherichia coli*. J. Bacteriol. *190*, 3930–3939.

Cariss, S.J., Constantinidou, C., Patel, M.D., Takebayashi, Y., Hobman, J.L., Penn, C.W., and Avison, M.B., (2010). YieJ (CbrC) mediates CreBC-dependent colicin E2 tolerance in *Escherichia coli*. J. Bacteriol. *192*, 3329–3336.

Cava, F., de Pedro, M.A., Lam, H., Davis, B.M., and Waldor, M.K. (2011). Distinct pathways for modification of the bacterial cell wall by non-canonical D-amino acids. EMBO J. *30*, 3442–3453.

Cosma, C.L., Danese, P.N., Carlson, J.H., Silhavy, T.J., and Snyder, W.B. (1995). Activation of the Cpx two-component signal transduction pathway in *Escherichia coli* suppresses envelope associated stresses. Mol. Microbiol. *18*, 491–505.

Costerton, J.W., Cheng, K.J., Geesey, G.G., Ladd, T.I., Nickel, J.C., Dasgupta, M., and Marrie, T.J. (1987). Bacterial biofilms in nature and disease. Annu. Rev. Microbiol. *41*, 435–464.

De Jonge, B.L., Gage, D., and Xu, N. (2002). The carboxyl terminus of peptidoglycan stem peptides is a determinant for methicillin resistance in *Staphylococcus aureus*. Antimicrob. Agents. Chemother. *46*, 3151–3155.

Erickson, J.W., and Gross, C.A. (1989) Identification of the sigma E subunit of *Escherichia coli* RNA polymerase: a second alternative sigma factor involved in high-temperature gene expression. Genes Dev. *3*, 1462–1471.

Farris, C., Sanowar, S., Bader, M.W., Pfuetzner, R., and Miller, S.I. (2010). Antimicrobial peptides activate the Rcs regulon through the outer membrane lipoprotein RcsF. J. Bacteriol. *192*, 4894–4903.

Ferrières, L., and Clarke, D.J. (2003). The RcsC sensor kinase is required for normal biofilm formation in *Escherichia coli* K-12 and controls the expression of a regulon in response to growth on a solid surface. Mol. Microbiol. *50*, 1665–1682.

Freire, P., Amaral, J.D., Santos, J.M., and Arraiano, C.M. (2006a). Adaptation to carbon starvation, RNase III ensures normal expression levels of *bolA*1p mRNA and sigma(S). Biochimie *88*, 341–346.

Freire, P., Vieira, H.L., Furtado, A.R., de Pedro, M.A., and Arraiano, C.M. (2006b) Effect of the morphogene *bolA* on the permeability of the *Escherichia coli* outer membrane. FEMS Microbiol. Lett. *260*, 106–111.

Freire, P., Moreira, R.N., and Arraiano, C.M. (2009). BolA inhibits cell elongation and regulates MreB expression levels. J. Mol. Biol. *385*, 1345–1351.

Gogol, E.B., Rhodius, V.A., Papenfort, K., Vogel, J., and Gross, C.A. (2011). Small RNAs endow a transcriptional activator with essential repressor functions for single-tier control of a global stress regulon. Proc.Natl. Acad. Sci. U.S.A. *108*, 12875–12880.

Guinote, I.B., Matos, R.G., Freire, P., and Arraiano, C.M. (2011). BolA affects cell growth, and binds to the promoters of penicillin-binding proteins 5 and 6 and regulates their expression. J. Microbiol. Biotechnol. *21*, 243–251.

Hayden, J.D., and Ades, S.E. (2008). The extracytoplasmic stress factor, sigmaE, is required to maintain cell envelope integrity in *Escherichia coli*. PLoS One *3*, e1573.

Hengge, R. (2008). The two-component network and the general stress sigma factor RpoS (sigma S) in *Escherichia coli*. Adv. Exp. Med. Biol. *631*, 40–53.

Hirakawa, H., Inazumi, Y., Masaki, T., Hirata, T., and Yamaguchi, A. (2005). Indole induces the expression of multidrug exporter genes in *Escherichia coli*. Mol. Microbiol. *55*, 1113–1126.

Jacobs, C., Frere, J.M., and Normark, S. (1997). Cytosolic intermediates for cell wall biosynthesis and degradation control inducible beta-lactam resistance in Gram-negative bacteria. Cell *88*, 823–832.

Jones, C.H., Danese, P.N., Pinkner, J.S., Silhavy, T.J., and Hultgren, S.J. (1997). The chaperone-assisted membrane release and folding pathway is sensed by two signal transduction systems. EMBO J. *16*, 6394–6406.

Kasai, T., Inoue, M., Koshiba, S., Yabuki, T., Aoki, M., Ninokawa, E., Seki, E., Matsuda, T., Matsuda, N., Tomo, Y., *et al.* (2004). Solution structure of a BolA-like protein from *Mus musculus*. Protein. Sci. *13*, 545–548.

Kolodkin-Gal, I., Romero, D., Cao, S., Clardy, J., Kolter, R., and Losick, R. (2010). D-amino acids trigger biofilm disassembly. Science *328*, 627–629.

Kulp, A., and Kuehn. M.J. (2010). Biological functions and biogenesis of secreted bacterial outer membrane vesicles. Annu. Rev. Microbiol. *64*, 163–184.

Lam, H., Oh, D.C., Cava, F., Takacs, C.N., Clardy, J., de Pedro, M.A., and Waldor, M.K. (2009). D-amino acids govern stationary phase cell wall remodeling in bacteria. Science *325*, 1552–1555.

Lange, R., and Hengge-Aronis, R. (1991). Growth phase-regulated expression of *bolA* and morphology of stationary-phase *Escherichia coli* cells are controlled by the novel sigma factor sigma S. J. Bacteriol. *173*, 4474–4481.

Leblanc, S.K.D., Oates, C.W., and Raivio, T.L. (2011). Characterization of the induction and cellular role of the baesr two-component envelope stress response of *Escherichia coli*. J. Bacteriol. *193*, 3367–3375.

de Lencastre, H., Oliveira, D., and Tomasz, A. (2007). Antibiotic resistant *Staphylococcus aureus*, a paradigm of adaptive power. Curr. Opin. Microbiol. *10*, 428–435.

Lonetto, M.A., Brown, K.L., Rudd, K.E., and Buttner, M.J. (1994) Analysis of the *Streptomyces coelicolor sigE* gene reveals the existence of a subfamily of eubacterial RNA polymerase sigma factors involved in the regulation of extracytoplasmic functions. Proc. Natl. Acad. Sci. U.S.A. *91*, 7573–7577.

Luirink, J., von Heijne, G., Houben, E., and de Gier, J.W. (2005). Biogenesis of inner membrane proteins in *Escherichia coli*. Annu. Rev. Microbiol. *59*, 329–355.

Magnet, S., Arbeloa, A., Mainardi, J.L., Hugonnet, J.E., Fourgeaud, M., Dubost, L., Marie, A., Delfosse, V., Mayer, C., Rice, L.B., *et al.* (2007a). Specificity of L,D-transpeptidases from Gram-positive bacteria producing different peptidoglycan chemotypes. J. Biol. Chem. *282*, 13151–13159.

Magnet, S., Bellais, S., Dubost, L., Fourgeaud, M., Mainardi, J.L., Petit-Frere, S., Marie, A., Mengin-Lecreulx, D.,

Arthur, M., and Gutmann, L. (2007b). Identification of the L,D-transpeptidases responsible for attachment of the Braun lipoprotein to *Escherichia coli* peptidoglycan. J. Bacteriol. *189*, 3927–3931.

Magnet, S., Dubost, L., Marie, A., Arthur, M., and Gutmann, L. (2008). Identification of the L,D-transpeptidases for peptidoglycan cross-linking in *Escherichia coli*. J. Bacteriol. *190*, 4782–4785.

Markiewicz, Z., Broome-Smith, J.K., Schwarz, U., and Spratt, B.G. (1982). Spherical *E. coli* due to elevated levels of D-alanine carboxypeptidase. Nature *297*, 702–704.

Mattei, P.J., Neves, D., and Dessen, A. (2010). Bridging cell wall biosynthesis and bacterial morphogenesis. Curr. Opin. Struct. Biol. *20*, 749–755.

McBroom, A.J., and Kuehn, M.J. (2005). Outer Membrane Vesicles. In EcoSal – *Escherichia coli* and *Salmonella*, Cellular and Molecular Biology, Curtiss, R., III ed. (ASM Press, Washington).

McBroom, A.J., and Kuehn, M.J. (2007). Release of outer membrane vesicles by Gram-negative bacteria is a novel envelope stress response. Mol. Microbiol. *63*, 545–558.

McEwen, J., Silverman, P.M. (1980). Chromosomal mutations of *Escherichia coli* that alter expression of conjugative plasmid functions. Proc. Natl. Acad. Sci. U.S.A. *77*, 513–517.

Moreira, R.N., Dressaire, C., Domingues, S., and Arraiano, C.M. (2011). A new target for an old regulator, H-NS represses transcription of *bolA* morphogene by direct binding to both promoters. Biochem. Biophys. Res. Commun. *411*, 50–55.

Mutalik, V.K., Nonaka, G., Ades, S.E., Rhodius, V.A., and Gross, C.A. (2009). Promoter strength properties of the complete sigma E regulon of *Escherichia coli and Salmonella enterica*. J. Bacteriol. *191*, 7279–7287.

Mutschler, H., Reinstein, J., and Meinhart, A. (2010). Assembly dynamics and stability of the pneumococcal epsilon zeta antitoxin toxin (PezAT) system from *Streptococcus pneumoniae*. J. Biol. Chem. *285*, 21797–21806.

Mutschler, M., Gebhardt, M., Shoeman, R.L., and Meinhart, A. (2011). A novel mechanism of programmed cell death in bacteria by toxin–antitoxin systems corrupts peptidoglycan synthesis. PLoS Biol. *9*, e1001033.

Nagata, Y., Fujiwara, T., Kawaguchi-Nagata, K., Fukumori, Y., and Yamanaka, T. (1998). Occurrence of peptidyl D-amino acids in soluble fractions of several eubacteria, archaea and eukaryotes. Biochim. Biophys. Acta. *1379*, 76–82.

Nelson, D.E., and Young, K.D. (2000). Penicillin binding protein 5 affects cell diameter, contour, and morphology of *Escherichia coli*. J. Bacteriol. *182*, 1714–1721.

Nishino, K., Honda, T., and Yamaguchi, A. (2005). Genome-wide analyses of *Escherichia coli* gene expression responsive to the BaeSR two-component regulatory system. J. Bacteriol. *187*, 1763–1772.

Nystrom, T. (2004). Stationary-phase physiology. Annu. Rev. Microbiol. *58*, 161–181.

de Pedro, M.A., Young, K.D., Holtje, J.V., and Schwarz, H. (2003). Branching of *Escherichia coli* cells arises from multiple sites of inert peptidoglycan. J. Bacteriol. *185*, 1147–1152.

Perez-Rodriguez, R., Haitjema, C., Huang, Q., Hyun Nam, K., Bernardis, S., Ke, A., and DeLisa, M.P. (2011). Envelope stress is a trigger of CRISPR RNA-mediated DNA silencing in *Escherichia coli*. Mol. Microbiol. *79*, 584–599.

Price, N.L., and Raivio, T.L. (2009). Characterization of the Cpx regulon in *Escherichia coli* strain MC4100. J. Bacteriol. *191*, 1798–1815.

Raffa, R.G., and Raivio, T.L. (2002). A third envelope stress signal transduction pathway in *Escherichia coli*. Mol. Microbiol. *45*, 1599–1611.

Raivio, T.L., and Silhavy, T.J. (2001). Periplasmic stress and ECF sigma factors. Annu. Rev. Microbiol. *55*, 591–624.

Reynolds, P.E., and Courvalin, P. (2005). Vancomycin resistance in enterococci due to synthesis of precursors terminating in D-alanyl-D-serine. Antimicrob. Agents. Chemother. *49*, 21–25.

Ruiz, N., and Silhavy, T.J. (2005). Sensing external stress, watchdogs of the *Escherichia coli* cell envelope. Curr. Opin. Microbiol. *8*, 122–126.

Ruiz, N., Kahne, D., and Silhavy, T.J. (2006). Advances in understanding bacterial outer-membrane biogenesis. Nat. Rev. Microbiol. *4*, 57–66.

Santos, J.M., Freire, P., Vicente, M., and Arraiano, C.M. (1999). The stationary-phase morphogene *bolA* from *Escherichia coli* is induced by stress during early stages of growth. Mol. Microbiol. *32*, 789–798.

Santos, J.M., Lobo, M., Matos, A.P., De Pedro, M.A., and Arraiano, C.M. (2002). The gene *bolA* regulates *dacA* (PBP5), *dacC* (PBP6) and *ampC* (AmpC), promoting normal morphology in *Escherichia coli*. Mol. Microbiol. *45*, 1729–1740.

Sieradzki, K., and Tomasz, A. (1996). A highly vancomycin-resistant laboratory mutant of *Staphylococcus aureus*. FEMS Microbiol. Lett. *142*, 161–166.

Sochacki, K.A., Shkel, I.A., Record, M.T., and Weisshaar, J.C. (2011). Protein diffusion in the periplasm of *E. coli* under osmotic stress. Biophys. J. *100*, 22–31.

Tanner, M.E. (2002). Understanding nature's strategies for enzyme-catalyzed racemization and epimerization. Acc. Chem. Res. *35*, 237–246.

Tapparel, C., Monod, A., and Kelley, W.L. (2006). The DNA-binding domain of the *Escherichia coli* CpxR two-component response regulator is constitutively active and cannot be fully attenuated by fused adjacent heterologous regulatory domains. Microbiology *152*, 431–441.

Tayler, A.E., Ayala, J.A., Niumsup, P., Westphal, K., Baker, J.A., Zhang, L., Walsh, T.R., Wiedemann, B., Bennett, P.M., and Avison, M.B. (2010). Induction of beta-lactamase production in *Aeromonas hydrophila* is responsive to beta-lactam-mediated changes in peptidoglycan composition. Microbiology *156*, 2327–2335.

Veiga, P., Piquet, S., Maisons, A., Furlan., S., Courtin, P., Chapot-Chartier, M.P., and Kulakauskas, S. (2006).

Identification of an essential gene responsible for D-Asp incorporation in the *Lactococcus lactis* peptidoglycan crossbridge. Mol. Microbiol. *62*, 1713–1724.
Vieira, H.L., Freire, P., and Arraiano, C.M. (2004). Effect of *Escherichia coli* morphogene *bolA* on biofilms. Appl. Environ. Microbiol. *70*, 5682–5684.
Vollmer, W., and Seligman, S.J. (2010). Architecture of peptidoglycan, more data and more models. Trends Microbiol. *18*, 59–66.
Weatherspoon-Griffin, N., Zhao, G., Kong, W., Kong, Y., Morigen, Andrews-Polymenis, H., McClelland, M., and Shi, Y. (2011). The CpxR/CpxA two-component system up-regulates two Tat-dependent peptidoglycan amidases to confer bacterial resistance to antimicrobial peptide. J. Biol. Chem. *286*, 5529–5539.
Wick, L.M., and Egli, T. (2004). Molecular components of physiological stress responses in *Escherichia coli*. Adv. Biochem. Eng. Biotechnol. *89*, 1–45.
Wolfe, A.J., Parikh, N., Lima, B.P., and Zemaitaiti, B. (2008). Signal integration by the two-component signal transduction response regulator CpxR. J. Bacteriol. *190*, 2314–2322.

Stress Responses in *Streptococcus*

2

Jacqueline Abranches and José A. Lemos

Abstract

The genus *Streptococcus* comprises a diverse group of organisms which includes food-associated, commensal and pathogenic species. The importance of this genus to the food industry and the capacity of certain species to infect animals and humans make streptococci one of the best-studied Gram-positive bacteria. In this chapter, we will describe the stress responses of the four major pathogenic streptococcal species: *Streptococcus mutans*, the aetiological agent of dental caries, *S. pyogenes* (commonly known as group A *Streptococcus* or GAS), responsible for a variety of suppurative diseases as well as life-threatening invasive infections and post-infection sequelae, *S. agalactiae* (group B *Streptococcus* or GBS), a major bacterial pathogen associated with neonatal infections, and *S. pneumoniae*, the leading causative agent of bacterial pneumonia. In the following pages, the description of the stress response mechanisms for each individual species is presented in the context of the environmental stress condition. In addition to highlighting the cross-species conservation of certain stress responses, this organization will allow the reader to follow the progresses obtained in each species, and, at the same time, identify areas that have been poorly explored.

Introduction

The genus *Streptococcus* comprises a diverse group of low GC (<50% GC content) Gram-positive bacteria that includes food-related species, several commensal species and a relatively small, but important, number of pathogenic species. Owing to their medical relevance and economical importance in the food industry, streptococcal species are among the best-studied Gram-positive bacteria, second only to the bacterial paradigm *Bacillus subtilis*. In recent years, the availability of complete genome sequences and the introduction of sophisticated genetic techniques for the manipulation of their genomes have accelerated our understanding of the biology of this important group of organisms. However, progress in the area has been uneven. The biology of a small number of species, such as the pathogenic *Streptococcus mutans, S. pneumoniae* and *S. pyogenes* (group A *Streptococcus,* herein designated GAS), or the food-associated *Streptococcus thermophilus* have been scrutinized in detail whereas commensal streptococci and even important pathogenic species, such as *Streptococcus agalactiae* (group B *Streptococcus,* herein designated GBS) have not received the necessary attention.

It is well established that the ability of an organism to sense adverse environmental changes and to mount an appropriate response to that stimuli, the so-called stress response, is a key trait allowing bacteria to colonize and thrive in specific niches. In some cases, such as in *S. mutans,* stress responses are intertwined with virulence. Likewise, the ability of food-related streptococci to withstand extreme conditions in temperature, salinity and pH is what separates them from other streptococcal species.

In the next few pages, we will discuss the stress physiology of *Streptococcus mutans*, a dental pathogen that serves as a model organism for studying the stress responses of low GC Gram-positive bacteria. As stated above, the rationale for studying the stress responses of *S. mutans* resides in the nature of the disease dental caries itself, which is directly linked to the organism's ability to thrive under adverse conditions, in particular low pH. In addition, we will also review the literature on the stress response mechanisms used by *Streptococcus pneumoniae*, GAS and GBS, the three most frequent streptococcal species associated with severe human infections. While not covered in this chapter owing to space constraints, a substantial body of work on the stress responses of food-related *S. thermophilus* and, to a less extent, *S. macedonicus* is also available. However, apart from a handful of studies that have characterized alkali generation pathways of oral streptococci, the stress response mechanisms of commensal species have been largely overlooked.

Relevance of stress responses for the pathogenesis of *S. mutans*

The natural habitat of *S. mutans* is the human oral cavity, more specifically in dental plaque, a multispecies biofilm formed on the hard surface of the tooth. While a normal inhabitant of dental plaque when in low numbers, *S. mutans* is mostly known for its importance in the aetiology of dental caries. Dental caries remains among the most common infectious diseases afflicting 60–90% of the population worldwide. In addition, it is not unusual for *S. mutans* to gain access to the bloodstream during dental procedures (Drangsholt, 1998; Hill *et al.*, 2006). If a sufficient number of cells enter the circulation, transient bacteraemia followed by adhesion to endothelial cells can lead to infective endocarditis (IE) (Moreillon and Que, 2004), particularly in persons with predisposing cardiac conditions.

The oral cavity is a dynamic environment that undergoes rapid and large fluctuations in pH, nutrient availability and source, oxygen temperature and osmolality (Lemos *et al.*, 2005). After intake of sweetened foodstuffs, sugar levels in plaque and saliva can increase up to 1000-fold followed by a dramatic decrease in pH to values as low as 4. In general, aciduric organisms like *S. mutans*, which can grow and carry out glycolysis at low pH values, gain a selective advantage over less aciduric species when oral biofilms become acidified. Sustained acidification of plaque leads to significant increases in the proportions of acid tolerant and strongly acidogenic species, which drive the development of caries. Therefore, the abilities to survive extreme and rapid changes in pH, as well as sustained acidification, are considered one of the most important virulence attributes of *S. mutans* and other cariogenic bacteria.

The investigation of the *S. mutans* stress responses have served as an excellent example to reveal critical differences in the way that host-associated bacteria cope with environmental stresses when compared with bacteria that have both a free-living and host-associated lifestyles. Although the intermittent ingestion of food by the human host has been suggested to expose bacteria in oral biofilms to a 'feast-and-famine' lifestyle (Carlsson, 1983), the oral cavity does not really expose microorganisms to severely oligotrophic environments such as those encountered by bacteria with a free-living lifestyle. The studies with *S. mutans* have demonstrated that the biology of microorganisms that evolved in obligate close association with a mammalian host may be significantly different than the biology of bacterial paradigms with a free-living lifestyle. An important outcome from those investigations is the realization that the stress regulon appears to control a broader set of biological functions including sugar metabolism, biofilm formation and genetic competence.

Relevance of stress responses for the pathogenesis of group A *Streptococcus*

It has been estimated that, each year, more than half million deaths worldwide are due to GAS infections (Carapetis *et al.*, 2005). GAS is a strict human pathogen that causes a broad spectrum of diseases, ranging from relatively mild throat and skin infections to severe invasive infections such as necrotizing fasciitis (flesh-eating disease), toxic shock syndrome and septicaemia. In addition, a large number of deaths caused by GAS are associated with the development of post-streptococcal infections sequelae, which includes rheumatic fever, acute glomerulonephritis and reactive arthritis (Cunningham, 2000).

The ability of GAS to survive and replicate in diverse anatomic sites such as the skin, throat, and blood implies that this organism can efficiently adapt to different microenvironments, including different temperatures, pH, oxygen tension (Po_2), as well as fluctuations in nutrient source and availability. For example, temperatures in the human body can range from the high 20s at the extremes of superficial skin to 40°C during fever episodes (Smoot *et al.*, 2001). During invasive infection, GAS must cope with the host innate immune response, which includes the oxidative burst mediated by macrophages and polymorphonuclear neutrophils. These professional phagocytic cells produce a number of highly bactericidal reactive oxygen species (ROS), including superoxide anions ($O_2^{\cdot-}$), H_2O_2, and hydroxyl radicals ($OH^{\cdot}$). While the pH of the skin is normally acidic, the release of perspiration results in increased salt concentrations that increase osmolarity and pH. During invasive infections, the pH may fall to 6 or less following formation of necrotic lesions or abscesses (Simmen and Blaser, 1993). Finally, glucose concentrations in the host can range from 3–6 mM in blood to as low as 0.02–0.4 mM in saliva (Shelburne *et al.*, 2008b). In light of its remarkable ability to adapt to a variety of human tissues, numerous studies have specifically demonstrated that GAS sense and respond to environmental changes by regulating expression of important virulence factors (Chaussee *et al.*, 1997; Graham *et al.*, 2005; Virtaneva *et al.*, 2005; Loughman and Caparon, 2006).

Relevance of stress responses for the pathogenesis of group B *Streptococcus*

GBS colonizes the recto-vaginal tract of women and can be transmitted to infants during birth, causing of neonatal sepsis and meningitis. It is also implicated in pregnancy-related morbidity and in severe invasive diseases in elderly and in immunocompromised patients. Like GAS, the pathophysiology of GBS infections indicates that this bacterium is capable to adapt to large environmental fluctuations within the host, including changes in pH, osmolarity, Po_2 and temperature. For example, GBS is well-suited to survive the acidic pH environment of the vagina or in intracellular endocytic compartments. The levels of Po_2 in tissues infected by GBS can also vary dramatically, ranging from atmospheric O_2 levels to near anaerobiosis (Johri *et al.*, 2003). During invasive infection, GBS must cope with the host innate immune response, which includes the oxidative burst mediated by macrophages and polymorphonuclear neutrophils.

Relevance of stress responses for the pathogenesis of *S. pneumoniae*

The human pathogen *S. pneumoniae*, also known as pneumococcus, is the causative agent of acute respiratory infections, otitis media, sepsis and meningitis. Pneumococci are the leading cause of bacterial pneumonia, meningitis and sepsis in children and are estimated to cause over 1 million child deaths per year worldwide (WHO, 2007). In addition, *S. pneumoniae* is also the leading agent of severe community-acquired pneumonia and meningitis in the elderly and in immunocompromised individuals (WHO, 2007). Although multivalent capsular vaccines and antibiotic therapies are useful to control *S. pneumoniae* infections, multiple antibiotic resistant strains and evasion of current vaccines by serotype selection are still on the rise. Normal disease progression begins with colonization of the nasopharyngeal environment followed by invasion of normally sterile sites such as the lung, meninges and bloodstream. During the course of infection, *S. pneumoniae* encounter numerous stress conditions, including changes in temperature (from 30–34°C at the upper respiratory tract to 37–40°C at deeper tissues), pH fluctuations, exposure to ROS generated by host immune cells and nutrient deprivation.

General stress survival pathways

Bacteria often utilize overlapping mechanisms to cope with environmental insults. The central pathways of stress responses in streptococci involve the up-regulation of molecular chaperones DnaK and GroEL, proteases such as HtrA and the Clp proteolytic system and DNA repair enzymes.

The ubiquitously distributed DnaK and GroEL molecular chaperones are central to stress tolerance and participate in a variety of cellular processes, including protein folding, protein translocation, presentation of proteins for degradation and assembly and disassembly of protein complexes (Hartl and Hayer-Hartl, 2002). The GroEL and DnaK complexes, which include GroES and DnaJ-GrpE, respectively, also regulate signal transduction pathways by controlling the stability and activities of transcriptional regulators and protein kinases (Hartl and Hayer-Hartl, 2002). In most Gram-positive bacteria, transcription of the *groE* (*groES-groEL*) and *dnaK* (*hrcA-grpE-dnaK-dnaJ*) operons is negatively controlled by HrcA, which binds to a highly conserved *cis*-acting element (CIRCE for controlling inverted repeat of chaperone expression) located in the regulatory regions of these operons (Zuber and Schumann, 1994). In some cases, the *groE* and *dnaK* operons can also be under negative control of CtsR, a repressor that binds to a conserved direct repeat sequence and was initially identified for its role in regulating *clp* gene expression (see below).

In *S. mutans*, the DnaK and GroEL chaperones appear to be indispensable (Lemos *et al.*, 2001), and the essential nature of these chaperones was confirmed by a study using down-regulation of *groEL* and *dnaK* expression (Lemos *et al.*, 2007b). Lowering of DnaK levels resulted in impaired capacity to form biofilms in the presence of glucose and rendered the strain more sensitive to stresses including low pH, high temperatures and oxidative stress (Lemos *et al.*, 2007b). The acid sensitivity of the strain with low levels of DnaK was attributed, at least in part, to DnaK participating in the biogenesis or stabilization of the proton-translocating F_1F_0-ATPase complex (Lemos *et al.*, 2007b). Down-regulation of GroEL also resulted in high temperature sensitivity and impaired capacity to form biofilms, but did not affect growth at low pH or under oxidative stress conditions (Lemos *et al.*, 2007b). Apart from microarray studies that have identified DnaK and GroEL as being differentially expressed during certain environmental stress conditions, the molecular chaperones have not been characterized in GAS, GBS and *S. pneumoniae*. However, an isolated study with an *hrcA* mutant confirmed that HrcA controls, to some extent, the levels of DnaK and GroEL in GAS. Interestingly, the *hrcA* mutant strain still accumulated DnaK and GroEL proteins when cells were subjected to heat stress suggesting that there is another, yet undisclosed, regulator contributing to DnaK and GroEL expression in GAS (Woodbury and Haldenwang, 2003).

One consequence of the exposure to environmental stresses is the accumulation of abnormal proteins due to increased errors in transcription and translation. Moreover, ageing cells present in mature biofilms are prone to mistranslation and aggregation. Thus, the activity of proteases by preventing the accumulation of misfolded proteins is central to physiological homeostasis. In the Gram-negative paradigm *E. coli*, there are four classes of energy-dependent cytoplasmic proteases: Lon, FtsH, ClpAP/XP and ClpYQ (also known as HslUV) (Gottesman, 2003). In Lon and FtsH, a single protein contains both the ATPase and proteolytic sites, whereas in the ClpAP/XP and ClpYQ classes the ATPase and proteolytic domains are located in different proteins. While the metalloprotease FtsH is ubiquitous in eubacteria, low GC Gram-positive bacteria lack the Lon protease (Butler *et al.*, 2006; Frees *et al.*, 2007). In addition, the available genome sequences of streptococci indicate that they also lack ClpYQ that is found in the genome of some other Gram-positive bacteria, including the pathogenic *Staphylococcus aureus* (Butler *et al.*, 2006; Frees *et al.*, 2007). Therefore, it has been proposed that ClpP complexes are more central for stress tolerance and global regulation in streptococci than in other bacterial groups (Frees *et al.*, 2007).

To function as a protease, the ClpP peptidase must associate with a Clp ATPase partner that possesses nucleotide binding domains characteristic of the AAA+ superfamily of ATPases to form a functional complex (Gottesman, 2003; Butler *et al.*, 2006; Frees *et al.*, 2007). In addition to forming protease complexes, Clp ATPases can also perform essential housekeeping functions, including protein reactivation activities typical of molecular chaperones. In most Gram-positive organisms, most *clp* genes are negatively controlled by CtsR, which is the product of the first gene of the *clpC* operon and was shown to bind to a directly repeated heptad sequence present in the regulatory region of *clp* genes (Derre *et al.*, 1999).

The genome of *S. mutans* encodes five orthologues of Clp ATPases (ClpB, ClpC, ClpE, ClpL and ClpX) and a single ClpP peptidase (Lemos and Burne, 2002; Kajfasz *et al.*, 2009). Among the five Clp ATPases, ClpC, ClpE, and ClpX have the recognition tripeptide that allows the proteins to interact with ClpP, while ClpB and ClpL are expected to function as protein chaperones independent of ClpP (Frees *et al.*, 2007). Systemic inactivation of all members of the Clp system in *S. mutans* indicated that inactivation of the genes encoding for the ClpP peptidase and the ClpL and ClpX ATPases affected the expression of virulence attributes in *S. mutans* (Kajfasz *et al.*, 2009). Compared with its parent strain the Δ*clpL* strain displayed an enhanced capacity to form biofilms in the presence of sucrose, had reduced long-term viability, and was more sensitive to acid killing (Kajfasz *et al.*, 2009). The Δ*clpP* and Δ*clpX* strains displayed several phenotypes in common, including enhanced capacity to form biofilms in sucrose, reduced autolysis, and reduced ability to grow under stress. Unexpectedly, the Δ*clpP* and Δ*clpX* mutants were more resistant to acid killing and demonstrated enhanced viability in long-term survival assays (Kajfasz *et al.*, 2009).

The Clp system has also been the subject of extensive studies in pneumococcus. The *S. pneumoniae* core genome encodes four Clp ATPases (ClpC, ClpE, ClpL and ClpX) and a single ClpP peptidase (Tettelin *et al.*, 2001). ClpL is mainly found in Gram-positive organisms and while it does not interact with ClpP, it is expected to function as a molecular chaperone performing essential housekeeping functions, including protein reactivation activities, folding and translocation. Although growth of the *clpL* mutant in *S. pneumoniae* was not affected at 30°C or 37°C, growth of Δ*clpL* was significantly impaired at 43°C (Kwon *et al.*, 2003). The Δ*clpL* strain was able to adhere and invade epithelial cells much more efficiently than the parent strain, was more susceptible to macrophage killing and showed a similar level of virulence when compared to the parent strain in three different murine models of infection (Kwon *et al.*, 2003; Tu le *et al.*, 2007). Among the remaining four *clp* genes of *S. pneumoniae* encoding for ClpP, and the ClpC, ClpE and ClpX ATPases, *clpX* was found to be essential whereas *clpC*, *clpE* and *clpP* were dispensable (Robertson *et al.*, 2003). The essentiality of *clpX* but not of *clpP* was unexpected given that ClpX functions as a regulatory subunit for the ClpXP protease. Transcriptional analysis of *clpC*, *clpE* and *clpP* revealed that expression of these genes was induced in response to high temperatures, a stress response that was mediated by the activity of the CtsR repressor (Charpentier *et al.*, 2000; Chastanet *et al.*, 2001). The ClpC ATPase was found to participate in several important aspects of pneumococcal physiology, including growth under heat stress, cell division, autolysis, adherence, and competence development (Charpentier *et al.*, 2000). However, independent studies suggested that these findings might be strain dependent (Chastanet *et al.*, 2001; Ibrahim *et al.*, 2005). In addition to being required for growth under heat stress, ClpE was shown to contribute to pneumococcal virulence as assessed in a mouse intraperitoneal infection model (Zhang *et al.*, 2009). Several studies have showed that ClpP-mediated proteolysis participates in many processes of pneumococcal physiology, including stress responses, competence development, and virulence (Kwon *et al.*, 2003). Subsequent work linked the attenuated virulence of *clpP* mutants to an increased sensitivity to oxidative stress caused by macrophages, in particular to reactive nitric oxide (NO) intermediates (Park *et al.*, 2010). The virulence of Δ*clpP* strains was found highly attenuated in four different murine infection models; colonization of the nasopharynx, lung survival, intraperitoneal infection, and sepsis (Robertson *et al.*, 2002; Kwon *et al.*, 2003, 2004). Notably, immunization of mice with the pneumococcal ClpP elicited a protective immune response against invasive infections suggesting that ClpP may be a useful antigen for the development of a broad-range pneumococcal vaccine (Cao *et al.*, 2007, 2009; Kwon *et al.*, 2004).

The GBS core genome encodes the ClpP peptidase and four Clp ATP-binding subunits (ClpC, ClpE, ClpL and ClpX) (Glaser *et al.*, 2002; Tettelin *et al.*, 2005). A mutant lacking the *clpP* gene was more sensitive to multiple stresses, a phenotype that was linked to increases in total protein oxidation (Nair *et al.*, 2003). The Clp system remains to be investigated in GAS.

HtrA (high-temperature requirement A) is a surface-associated protein that can function as both a protease and a chaperone. Many of these functions are central to physiological homeostasis, including growth, stress tolerance, transport and secretion and signal transduction (Gottesman, 2003). Although many bacteria have more than one copy of HtrA (Pallen and Wren, 1997), Gram-positive cocci have a single *htrA* gene. Inactivation of *htrA* generated multiple stress-sensitive phenotypes in *S. mutans*, and was also linked to altered biofilm formation and reduced genetic competence (Ahn *et al.*, 2005; Biswas and Biswas, 2005). HtrA was shown to play a role in the ability of pneumococcus to grow at high temperatures, resist oxidative stress, and to undergo genetic transformation (Ibrahim *et al.*, 2004b). Also, inactivation of *htrA* in *S. pneumoniae* resulted in major attenuation of virulence in mouse infection models (Ibrahim *et al.*, 2004b).

DNA repair enzymes are ubiquitous to all organisms. Pathogenic bacteria encounter many hostile environments that could lead to DNA damage. Many of the stresses encountered by bacteria induce DNA damage, in particular acid and oxidative stresses increase the formation of abasic sites in DNA. Hahn and co-workers identified an AP endonuclease in *S. mutans* that showed higher levels of activity in cells grown at low pH compared with cells grown at neutral pH (Hahn *et al.*, 1999), providing the first evidence linking DNA repair enzymes with acid adaptation in oral streptococci. In *E. coli*, UvrA participates in the nucleotide excision repair pathway, which is responsible for excising larger DNA lesions caused by acid and other DNA-damaging agents. An *S. mutans uvrA* mutant was extremely sensitive to UV irradiation and had enhanced sensitivity to acid killing when compared to the parent strain (Hanna *et al.*, 2001). Likewise, Faustoferri and colleagues characterized the Smx exonuclease of *S. mutans* and showed that the *smx* mutant was highly sensitive to DNA damage caused by the production of ROS via Fenton reaction (Faustoferri *et al.*, 2005). The RpoE subunit of the RNA polymerase is required for specificity of the whole enzyme and has been demonstrated to be a global modulator of acid and oxidative stress responses in *S. mutans* (Xue *et al.*, 2010). Deletion of *rpoE* rendered the cells more sensitive to acid and peroxide damage, suggesting that RpoE is critical for optimal growth and adaptation to changing environmental conditions (Xue *et al.*, 2010).

Genome wide analysis of the presence of DNA repair genes in GAS and in *S. pneumoniae* have been performed (Ambur *et al.*, 2009) although detailed studies characterizing the role of DNA repair in these organisms are still missing. Similarly, the relevance of DNA repair in the stress responses of GBS has not been explored.

Acid stress

In all lactic acid bacteria, including all streptococcal species, the production of lactic acid from fermentable sugars through glycolysis occurs at very rapid rates. In order to survive, these bacteria must be able to mount an appropriate response to the environmental acidification caused by their own metabolism. Pathogenic streptococci must also be able to withstand low pH values that are encountered following the formation of necrotic lesions or abscesses (Simmen and Blaser, 1993). In the case of acid stress responses, the overwhelming majority of information available in streptococci comes from studies with *S. mutans*. Yet, the ability to survive in low-pH environments is a relevant trait of all pathogenic streptococci. In the paragraphs below, we will highlight the most relevant pathways utilized by *S. mutans* to cope with acidification of the environment.

Following consumption of sugars by the human host, *S. mutans* can acidify the plaque pH to values below 4.0. Considering that the resting pH of a healthy dental plaque is close to neutrality, an abrupt change in pH requires a fast and effective response to survive acidification. In order to survive to the detrimental effects of acidifications, *S. mutans* has evolved sophisticated mechanisms to cope with acid challenges.

Genome-wide transcriptional and proteomic approaches have facilitated the identification of proteins and pathways that participate in the *S. mutans* acid tolerance and adaptation. These studies revealed that several proteins involved in protection, repair and degradation of macromolecules, including molecular chaperones, proteases and DNA repair enzymes discussed above, are an integral part of the organism general stress response (Wilkins *et al.*, 2002; Welin *et al.*, 2003; Len *et al.*, 2004a,b; Gong *et al.*, 2009). By coupling proteomic data with measurements of end products of carbon utilization, metabolic pathways expressed during growth at neutral and acidic conditions were mapped. Some of the major findings indicate that *S. mutans* tolerates growth at low pH by expending energy to extrude H^+, by modulating the production of acid end products to reduce H^+ production, and possibly by using branched chain amino acid (BCAA) biosynthesis to rearrange the membrane composition and reduce acid production and, at the same time, buffer the intracellular pH (Len *et al.*, 2004a). Collectively, these studies have showed that *S. mutans* uses a sophisticated network of regulatory strategies to properly respond and adapt against the sudden and frequent fluctuations in the plaque environment. As it will be discussed in the next pages, *S. mutans* utilizes a number of signal transductions systems, quorum-sensing (QS) signalling molecules, carbon catabolite repression and nutrient-sensing regulators to cope with acid stress.

In contrast to enteric bacteria, which usually maintain internal pH (pH_i) fairly constant slightly above 7.0 through the use of antiporters, pH_i in streptococci fluctuates in response to extracellular pH, with the organisms using a variety of pathways to maintain a ΔpH of 0.5 to 1 unit above the external environment. The membrane-bound F_1F_0-ATPase is the primary mechanism of proton extrusion to maintain pH homeostasis of all streptococci. Notably, acid tolerance in oral streptococci correlates well with the pH optimum for activity of the F-ATPase enzyme (Sturr and Marquis, 1992). For example, in the highly acid-tolerant *S. mutans*, the optimal pH for the F-ATPase activity is 6.0, whereas in the less acid-tolerant *Streptococcus salivarius* and *Streptococcus sanguinis*, the pH optima of the ATPase enzymes are 7.0 and 7.5, respectively. In addition to functioning as a proton pump, it was recently demonstrated that the F-ATPase of *S. mutans*, and other LAB, could also function as an ATP synthase in starved cells grown at low pH (Sheng and Marquis, 2006). Thus, the F-ATPase may play a dual role in acid tolerance, by extruding protons out of the cells and, under certain conditions, by generating ATP for growth and maintenance.

Although *S. mutans* is not capable of generating significant quantities of alkali as it lacks urease and the arginine deiminase system (ADS) pathways (Ajdic *et al.*, 2002), a functional agmatine deiminase system (AgDS), analogous to the ADS has been found in the genome of *S. mutans* (Griswold *et al.*, 2006). The AgDS converts agmatine, a decarboxylated derivative of arginine that is found in human dental plaque, to putrescine, ammonia and CO_2. Whereas the ADS and urease pathways catalyse substantial environmental alkalinization and appear to be associated with caries resistance, the AgDS of *S. mutans* is expressed at relatively low levels and does not produce enough ammonia to raise the extracellular environment. However, the production of ammonia from agmatine, is believed to contribute to the competitive fitness of *S. mutans* at low pH by increasing the pH_i and by providing ATP, which could be used for growth or to extrude protons via the F-ATPase enzyme (Griswold *et al.*, 2006).

Another contributor to acid tolerance of *S. mutans* is malolactic fermentation (MLF), which catalyses the conversion of dicarboxylic L-malate to the monocarboxylic lactic acid and CO_2. The net effect of the transport and enzymatic reaction of malate is the formation of a pH-gradient thereby increasing the proton motive force (PMF) available for the cell to take up other nutrients. Because malate utilization contributes to maintenance of ATP pools and the CO_2 molecule generated can be rapidly converted to bicarbonate via carbonic anhydrase, MLF is thought to enhance the buffering capacity of the bacterial cell. In fact, while malate did not serve as a catabolite for growth of *S. mutans*, it did serve to protect the organism against acid killing as well as other stress conditions (Sheng and Marquis, 2007; Sheng *et al.*, 2010).

The relevance of cell membrane integrity and composition in relation to changes that affect proton permeability and F-ATPase activity in *S. mutans* has been well documented. By using cells grown in continuous culture at pH 7.0 or 5.0, it was demonstrated that *S. mutans* grown at pH

5.0 had increased levels of monounsaturated fatty acids and longer chain fatty acids than cells grown at neutral pH (Quivey *et al.*, 2000). Inactivation of the gene responsible for biosynthesis of monounsaturated fatty acids, *fabM*, resulted in a strain that was extremely sensitive to low pH and unable to maintain ΔpH (Fozo and Quivey, 2004). In an animal model for dental caries, rats infected with the *fabM* mutant exhibited substantially reduced caries, as compared to the parent strain (Fozo *et al.*, 2007). The main mechanism by which changes in membrane composition are thought to affect acid tolerance is through decreased proton permeability, although an equally important potential mechanism is that the altered membrane fatty acid composition may affect the activity of the F-ATPase and protein transport in a manner that influence acid resistance. In *S. pneumoniae*, *fabM* was showed to be essential and mutation in the transcriptional regulator of the *fab* operon, *fabT*, resulted in increased acid sensitivity and higher levels of saturated fatty acids (Lu and Rock, 2006).

An isolate investigation of the acid stress responses of other (non-*S. mutans*) pathogenic streptococci was a microarray study that compared the transcriptome of GBS at neutral and acidic pH. An important observation from this study was that several genes involved in the GBS response to low pH were under the control of the CsrRS two-component system (TCS) (Santi *et al.*, 2009). More details on the importance of TCSs in stress tolerance are provided below.

Oxidative stress and metal homeostasis

Oxygen is one of the most abundant elements on Earth. Because of its small size and biradical structure ($^{\cdot}O{-}O^{\cdot}$), it can easily diffuse across biological membranes and interact with the catalytic sites of enzymes that could lead to enzyme inactivation. However, O_2 itself is not as detrimental to bacterial metabolism as other reduced forms of oxygen such as hydrogen peroxide (H_2O_2), hydroxyl radicals ($HO^{\cdot}$) and superoxide radicals ($O_2^{\cdot -}$).

Bacteria are constantly exposed to oxidative agents from either intracellular or extracellular origin. During respiration, reactive oxygen species (ROS) are generated inside the cell from single electron reductions of oxygen and the cells must cope with oxygen metabolites in order to survive. External sources of ROS include host defences, such as the lactoperoxidase system, as well as peroxide-containing hygiene products. In the oral cavity, the production of H_2O_2 by few commensal streptococci, such as *S. sanguinis*, *S. gordonii* and *S. oligofermentans*, was shown to provide a competitive advantage to the H_2O_2 producing bacteria by adding an oxidative stress burden to the oral biofilm bacteria (Ryan and Kleinberg, 1995; Tong *et al.*, 2007).

Streptococci do not possess cytochromes and therefore do not carry out oxidative phosphorylation. Instead, the bulk of the respiration activity in these organisms is due to NADH oxidases (Nox), a flavoprotein that catalyses the four-electron reduction of O_2 to H_2O (Marquis, 1995). Streptococci also lack catalase, a major protective enzyme against oxidative insults that converts H_2O_2 into H_2O and O_2. However, streptococcal species possess several detoxification enzymes that includes, among others, up to three peroxidases and at least one superoxide dismutase. Superoxide dismutases catalyse the conversion of superoxide molecules ($O_2^{\cdot -}$) to H_2O_2 and O_2 whereas peroxidases function by breaking down toxic H_2O_2 compounds into H_2O.

The genome of *S. mutans* encodes a NADH oxidase (Nox), an alkyl hydroperoxide reductase (AhpCF) and a thioperoxidase (Tpx), a manganese-dependent superoxide dismutase (SodA), a glutathione reductase (Gor) and a thioredoxin complex (TrxA/TrxB) (Ajdic *et al.*, 2002). Although the *S. mutans* H_2O_2-forming NADH oxidase and alkyl hydroperoxide reductase enzymes were shown to complement mutants for the corresponding enzymes in *E. coli*, inactivation of one or both enzymes did not affect peroxide tolerance or the ability of *S. mutans* to grow aerobically (Higuchi *et al.*, 1999). Similarly, a strain lacking the superoxide dismutase enzyme was able to grow aerobically, albeit more slowly than the parent strain (Nakayama, 1992). Interestingly, inactivation of the iron-binding Dpr significantly impaired oxygen tolerance in *S. mutans* (Yamamoto *et al.*, 2000). In particular, double Δ*dpr* Δ*sodA* and triple Δ*ahpC* Δ*nox* Δ*dpr* mutants were completely unable to grow under aerobic conditions (Yamamoto *et al.*, 2000). Based on its iron-binding ability, Dpr is thought to contribute to oxygen tolerance through the titration of free intracellular iron ions, which in turn prevent the formation of toxic hydroxyl radicals via the Fenton reaction. Few other gene products have been implicated in oxidative stress tolerance in *S. mutans*. Recently, random insertion mutagenesis was used to identify genes involved in superoxide tolerance (Zhang and Biswas, 2009). One of the identified genes was shown to encode a 3′-phosphoadenosine-5′-phosphate phosphatase (pAp) but the molecular mechanisms by which pAp confers superoxide tolerance remains unclear. In another study, the cell envelope-associated eukaryotic serine/threonine protein kinase (STPK) was shown to play a protective role against H_2O_2 (Zhu and Kreth, 2010).

The growth mode of bacteria, i.e. planktonic or as biofilms, largely influences the respiratory rates and the levels of activity of protective enzymes. In a study comparing the levels of Nox and SodA in planktonic and biofilm populations of several oral streptococci species, Nguyen and co-authors demonstrated that both enzymes were repressed in the biofilm state (Nguyen *et al.*, 2002). It was suggested that oxidative stress appears to be minimized in streptococcal biofilms owing to reduced oxygen metabolism by the bacteria. Reducing oxygen metabolism and consequently the production of ROS could be an important strategy to live in a slow growth mode and crowded environment such as a biofilm.

Given the nature of GAS infections, which are highly inflammatory resulting in the production of ROS, the oxidative stress responses of GAS have been investigated in some detail. In fact, metal regulation and oxidative stress responses have been directly implicated in GAS pathogenesis. Metal ions are essential micronutrients to bacterial homeostasis. For example, iron is an essential co-factor of bacterial metabolism that regulates virulence gene expression of many pathogenic bacteria. In the human body, very little free iron is present as most of the iron in the body is found intracellularly, or bound to glycoproteins, such as transferrins and lactoferrins. At least three ABC transporters (FtsABCD, HtsABC, and MtsABC) have been associated with iron uptake in GAS (Ferretti *et al.*, 2001). However, an excess of iron is toxic as a result of hydroxyl radicals that are produced via Fenton reaction. Thus, to prevent cellular damage, the processes associated with iron acquisition must be tightly regulated. The GAS genome encodes two metalloregulators PerR and MtsR that regulate expression of proteins involved in metal ions uptake. MtsR was shown to regulate the expression of *mtsABC* and *htsABC*, two iron uptake gene clusters, and it was suggested that this regulator may play an important role in GAS virulence (Hanks *et al.*, 2006). Inactivation of *mtsR* reduces GAS virulence in a zebrafish infection model (Bates *et al.*, 2005), but not in a mouse model of subcutaneous infection (Hanks *et al.*, 2006). PerR is a transcriptional repressor that has been linked to peroxide stress resistance and metalloregulation (Brenot *et al.*, 2005). Although a *perR* mutant was hyper-resistant to peroxide, animal studies have shown that inactivation of *perR* results in attenuated GAS virulence in skin, soft tissue and oropharynx infections (Ricci *et al.*, 2002; Brenot *et al.*, 2007; Gryllos *et al.*, 2008). In the case of oropharynx infections, the attenuated virulence was associated with reduced resistance to phagocytic killing (Gryllos *et al.*, 2008). Surprisingly, the PerR regulon appears to include genes involved in sugar uptake and utilization, suggesting a link among central metabolic processes, oxidative stress and virulence (Gryllos *et al.*, 2008).

As seen in *S. mutans*, the iron-binding protein Dpr (also known as MrgA in GAS) also participates in the oxidative stress response of GAS (Brenot *et al.*, 2005; Tsou *et al.*, 2008). Specifically, *in vitro* studies have shown that PerR regulates *dpr* expression, and that Dpr provides protection against stresses by preventing the Fenton reaction (Tsou *et al.*, 2008). Although Dpr functions as a general stress protein protecting the cells against multiple stresses, a *dpr* mutant of GAS was fully virulent in two different animal models (Brenot *et al.*, 2005). Few other genes involved in oxidative stress management have been studied in GAS. Inactivation of the H_2O-forming NADH oxidase gene (*nox*) resulted in reduced aerotolerance and increased sensitivity to the superoxide-generating agent paraquat (Gibson *et al.*, 2000). Functional analysis identified two peroxidases in the GAS genome, alkyl hydroperoxide reductase (AhpCF) and glutathione peroxidase (GpoA) (King *et al.*, 2000). The inactivation of one of both peroxidases resulted in increased sensitivity to superoxide-generating agents but neither enzyme was essential for growth under aerobic conditions (King *et al.*, 2000). Notably GpoA was essential for GAS pathogenesis in several murine models that mimic streptococcal suppurative diseases (Brenot *et al.*, 2004) but, on the other hand, was dispensable for virulence in the zebrafish model of streptococcal myositis, an infection characterized by the absence of the oxidative stress normally triggered by inflammatory responses (Brenot *et al.*, 2004).

Because the upper respiratory tract is an oxygen-rich environment, the mechanisms used by *S. pneumoniae* to cope with oxygen and its reactive derivatives have been explored in some detail. Transcriptome analysis of *S. pneumoniae* cells grown under aerobic or anaerobic conditions revealed that expression of approximately 5% of the genome is altered by the presence of oxygen (Bortoni *et al.*, 2009). Among the genes up-regulated under aerobiosis was the gene encoding for the transcriptional repressor Rgg. Inactivation of *rgg* resulted in increased sensitivity to oxygen and paraquat, but not to H_2O_2. In addition, virulence of the *rgg* mutant strain was attenuated in mice (Bortoni *et al.*, 2009). The *S. pneumoniae* genome encodes several enzymes that participate in oxidative stress responses, including, two types of superoxide dismutase enzymes (one Mn-dependent and one Fe-dependent), an alkylhydroperoxidase, and the H_2O-forming NADH oxidase. A pneumococcus strain lacking the Nox protein displayed normal growth under anaerobic conditions but was unable to grow under vigorous aeration, and showed attenuated virulence in an animal model (Yu *et al.*, 2001). Inactivation of MnSOD affected growth under aerobic conditions and resulted in attenuated virulence in mice (Yesilkaya *et al.*, 2000). A strain lacking the *S. pneumoniae ahpCF* genes was attenuated in competitive infections with the parent strain (Paterson *et al.*, 2006). Few other gene products have been linked to the ability of *S. pneumoniae* to survive oxidative stress conditions. Inactivation of *psaA* and *psaD*, members of an ABC Mn^{2+} transport complex and a thiol peroxidase, respectively, resulted in strains highly sensitive to oxidative stress, with significant reduction in the expression of the SOD and Nox enzymes, and attenuated virulence (Tseng *et al.*, 2002; McAllister *et al.*, 2004;).

The Spx (suppressor for *clpP* and *clpX*) global regulator is conserved among low-GC Gram-positive bacteria whose stability is controlled by ClpXP proteolysis (Zuber, 2004). Spx was first identified in *B. subtilis*, when it was demonstrated that accumulation of Spx was responsible for the pleiotropic phenotypes associated with *clpP* and *clpX* mutations (Zuber, 2004). Spx was shown to participate in the oxidative stress response by interacting with the C-terminal domain of the RNA polymerase α-subunit (α-CTD). Interestingly, *S. mutans* encodes two Spx orthologues (designated SpxA and SpxB). Inactivation of *spxA* and of *spxB* bypassed phenotypes of the *clpP* and *clpX* mutants, suggesting that the underlying mechanisms by which *S. mutans* ClpXP protease affects the expression of virulence traits are associated with accumulation of these two Spx orthologues (Kajfasz *et al.*, 2009). The subsequent characterization of strains lacking the *spx* gene revealed that Spx, as observed in *B. subtilis*, function as a positive regulator for the expression of genes involved in oxidative stress, including *ahpCF, dpr, gor, nox, sod, tpx* and *trxB* (Kajfasz *et al.*, 2010). In addition, *in vivo* studies strongly supported the association of Spx regulation with the virulence of *S. mutans* (Kajfasz *et al.*, 2010). In *S. pneumoniae*, two Spx-like proteins, designated SpxA1 and SpxA2 (homologous to the *S. mutans* SpxA and SpxB global regulators, respectively) were identified and the gene encoding SpxA1 was partially characterized (Turlan *et al.*, 2009). Single mutations in either of the *S. pneumoniae spx* genes could be stably isolated, but simultaneous inactivation of *spxA1* and *spxA2* was lethal. Although the association between the accumulation of Spx and the underlying mechanisms by which ClpP proteolysis affects virulence traits in *S. pneumoniae* was not explored, it was demonstrated that transcriptional repression by SpxA1 has a negative effect in the development of the X-state (competence), a proposed general stress response mechanism in *S. pneumoniae* (Claverys *et al.*, 2006), also linked to cannibalism or fratricide (Claverys *et al.*, 2007). However, the role of Spx in the regulation of the *S. pneumoniae* oxidative stress responses, and the possible involvement of SpxA1 and SpxA2 with the phenotypes observed in the *clpP* mutant strain remain to be explored. The recent identification and characterizations of Spx in *S. mutans* and *S. pneumoniae* clearly show that Spx is a major regulator in streptococci. Thus, we expect that the role of Spx regulation to the biology of other streptococcal species should be soon under investigation.

In comparison with other medically relevant *Streptococcus*, studies on the oxidative stress responses of GBS have been more limited. Still, few of the mechanisms used by GBS to cope with oxygen and ROS have been uncovered. Liu and co-workers demonstrated that a characteristic GBS pigment is used as a physical barrier at the bacterial surface to protect against ROS (Liu *et al.*, 2004). Another study revealed that ROS detoxification is highly dependent on the Mn-dependent superoxide dismutase as a *sodA* mutant strain was highly susceptible to oxidative stresses, showed increased susceptibility to macrophage killing, and survived poorly in the blood of infected mice (Poyart *et al.*, 2001). The H_2O-forming Nox was also shown to participate in GBS virulence (Yamamoto *et al.*, 2006). Interestingly, while inactivation of *nox* resulted in an aerobic growth arrest, this defect was not due to accumulation of toxic ROS but to an underlying defect in fatty acid biosynthesis. It was speculated that NAD^+ depletion in the *nox* mutant resulted in reduced acetyl-CoA production, which in turn affected fatty acid biosynthesis. Another interesting observation was that GBS, while primarily a fermentative organism, undergoes respiration metabolism given that exogenous sources of quinone and haem become available (Yamamoto *et al.*, 2005). In addition, AhpC was demonstrated to function as a haem-binding protein and was required for optimal respiration (Lechardeur *et al.*, 2010). This finding appears to be particularly important as respiration metabolism was shown to facilitate GBS dissemination and virulence by promoting survival in blood (Yamamoto *et al.*, 2005).

Nutrient limitation

Despite periods of nutrient limitation encountered during a bacterial life cycle, abrupt exposure to an excess amount of carbohydrate in the diet can result in the rapid accumulation of toxic glycolytic intermediates, acidification of the environment, and osmotic stress. To survive periods of starvation, to cope with the detrimental effects of glycolytic intermediates, and to maintain proper $NAD/NADH^+$ balances, most lactic acid bacteria developed sophisticated regulatory networks that combine transcriptional regulation with allosteric modulation of enzymatic activities to coordinate optimal flow of energy.

In some cases, carbohydrate source and availability is a key factor affecting the pathogenic potential of *Streptococcus*. The sugar phosphotransferase system (PTS) is the major carbohydrate transport system of streptococci, especially under carbohydrate-limiting conditions. The PTS consists of two proteins that are common to all PTS substrates, Enzyme I (EI) and HPr, the heat stable phosphocarrier protein, as well as sugar-specific permeases known as Enzyme II (EII) complexes. The EII complexes usually consist of three domains, A, B and C, but a fourth domain may also be present. The cytoplasmic A and B domains are directly responsible for the phosphorylation of the incoming sugar, whereas the C and D domains comprise the membrane permeases. The PTS internalizes a wide variety of sugars, including glucose, fructose and mannose, and disaccharides such as sucrose and lactose (Postma *et al.*, 1993). In addition to participating in sugar uptake, PTS components influence many other cellular processes in particular carbon catabolite repression (CCR). By tightly controlling expression and activity of non-preferred carbohydrate metabolism when in the presence of preferred substrates such as glucose and fructose, bacteria can optimize the pathways for energy generation (Abranches *et al.*, 2006; Gorke and Stulke, 2008; Zeng and Burne, 2008, 2010).

In Gram-positive bacteria, the metabolism of a non-preferred sugar source is repressed by CCR in the presence of a preferred sugar, usually glucose and fructose. In this case, CCR is exerted through HPr and the global regulatory protein catabolite control protein A (CcpA). The accumulation of glycolytic intermediates, reflecting the carbon flow through the glycolytic pathway, activates an HPr kinase, which phosphorylates HPr at Ser-46. The Ser-46 phosphate group can not be transferred to EII enzymes, allowing Hpr (Ser-P) to accumulate in the cell and interact with the global regulatory protein CcpA to form a complex that efficiently bind to catabolite response elements (CRE) located in the promoter region of genes under CCR control (Saier *et al.*, 1996).

In *S. mutans*, mutations in the EIIAB^{Man}, the major glucose/mannose- PTS permease, influenced CCR, biofilm development, regulation of acid tolerance and global control of gene expression, (Abranches *et al.*, 2003, 2006). CcpA was also shown to impact acid tolerance and the expression of other virulence traits of *S. mutans* (Abranches *et al.*, 2008; Lemos *et al.*, 2008). The enhanced acid tolerance of the CcpA mutant was associated with increases in the expression of PTS enzymes that resulted in higher rates of ATP generation through glycolysis.

Although the role of CCR in the stress survival of GAS and *S. pneumoniae* has not been directly tested, CCR and CcpA have been implicated in controlling the expression of virulence genes. In GAS, a *ccpA* mutant was significantly less virulent than the wild-type strain (Shelburne *et al.*, 2008a). In fact, transcriptome studies have revealed that carbohydrate metabolism has an important role in the pathogenesis of invasive GAS diseases where substrate source and availability is used as an environmental cue to distinguish between specific stages of infection (Shelburne *et al.*, 2008b). In *S. pneumoniae*, virulence of *ccpA* mutants was significantly attenuated in different mouse models of infection (Giammarinaro and Paton, 2002; Iyer *et al.*, 2005).

The stringent response is a global bacterial response to nutritional stress that is mediated by accumulation of the alarmones guanosine tetraphosphate and guanosine pentaphosphate, collectively known as (p)ppGpp (Potrykus and Cashel, 2008). These nucleotides are synthesized by enzymatic phosphorylation of GDP and GTP using ATP as a phosphate donor, and are produced at low basal levels during favourable growth conditions. Accumulation of (p)ppGpp results in strong down-regulation of genes for ribosomal RNAs and anabolic processes, and up-regulation of genes involved in amino acid biosynthesis and stress survival (Potrykus and Cashel, 2008). Thus, (p)ppGpp function as a chemical messenger that allow bacteria to switch their metabolism from a 'growth mode' to a 'survival mode'. For several decades, a single bifunctional enzyme known as RelA (or Rsh) was considered the only enzyme controlling (p)ppGpp production and degradation in Gram-positive bacteria. However, the initial finding that the *S. mutans* Δ*relA* mutant was still able to produce (p)ppGpp (Lemos *et al.*, 2004) led to the discovery of two small (p) ppGpp-synthetase enzymes in *S. mutans* (Lemos *et al.*, 2007a). BLAST search analysis indicated one or two small (p)ppGpp-synthetase could be commonly found in related Gram-positive bacteria (Lemos *et al.*, 2007a), and subsequent studies confirmed the functionality of these enzymes in *B. subtilis, S. pneumoniae* and *E. faecalis* (Battesti and Bouveret, 2008; Nanamiya *et al.*, 2008; Abranches *et al.*, 2009).

S. mutans, RelA enzyme was shown to participate in the regulation of properties that are required for establishment, persistence and survival (Lemos *et al.*, 2004; Nascimento *et al.*, 2008), supporting an overlap between circuits that govern nutrient starvation, general stress tolerance and biofilm formation. An (p)ppGpp^0 strain of *S. mutans* (Δ*relA*Δ*relP*Δ*relQ* triple mutant) was auxotrophic for branched chain amino acids (BCAA) leucine and valine, a phenotype that was directly related to CodY-dependent repression of genes involved in the synthesis of BCAA synthesis (see below) (Lemos *et al.*, 2008). Interestingly, RelP is co-transcribed with, and apparently regulated by, the RelRS two-component signal transduction system (TCS) (Lemos *et al.*, 2007a) suggesting that *S. mutans* may use environmental signals to optimize cell growth and survival in a manner that allows the organism to balance growth during dietary intake by the host with the capacity to rapidly mount an adaptive response during fasting periods. Consistent with the role of (p)ppGpp in bacteria, homologues of RelRS in GAS, designated SptRS (saliva persistence), were shown to be critical for this bacterium to survive in saliva (Shelburne *et al.*, 2005).

CodY is a global nutritional repressor that

is highly conserved in low GC Gram-positive bacteria (Sonenshein, 2005). In *B. subtilis,* CodY is strongly activated by the presence of GTP and BCAA and was shown to strongly repress the expression of BCAA operons (Sonenshein, 2005). However, studies with the CodY protein in *S. mutans* and *S. pneumoniae* revealed that CodY is not activated by GTP (Hendriksen *et al.*, 2008; Lemos *et al.*, 2008). Microarrays revealed that CodY is indeed a global regulator of gene expression in *S. mutans* and has a major role in the control of amino acid metabolism. In addition, phenotypic studies revealed that the *codY* mutant had reduced capacities to form biofilms and was more sensitive to growth at low pH (Lemos *et al.*, 2008).

The global nutrient sense regulators (p)ppGpp and CodY have also been linked to GAS survival and virulence gene expression under nutritional stress conditions. Studies using *relA* and *codY* mutants of GAS revealed that (p)ppGpp and CodY regulate the expression of key metabolic enzymes, transporters and virulence factors (Malke *et al.*, 2006; Malke and Ferretti, 2007; Kreth *et al.*, 2011).

The role of (p)ppGpp and CodY have also been explored in *S. pneumoniae.* Inactivation of the bifunctional (p)ppGpp synthetase/hydrolase RelA homologue did not affect growth in complex medium but the Δ*rel* strain did not grow in chemically defined medium unless supplemented with the metals cooper and manganese (Kazmierczak *et al.*, 2009). As a result, virulence of the *S. pneumoniae* Δ*rel* strain was significantly diminished in a murine pneumonia/sepsis model of infection (Kazmierczak *et al.*, 2009). CodY appears to be essential in *S. pneumoniae* (Caymaris *et al.*, 2010) and a CodY-deficient strain carrying compensatory mutations had significantly diminished colonization of the nasopharynx, a finding that was supported by the low ability of the *codY* mutant to adhere to nasopharyngeal cells *in vitro* (Hendriksen *et al.*, 2008).

In summary, streptococcal nutrient regulators, such as CcpA, CodY and (p)ppGpp, coordinate metabolism in a way that environmental cues such as nutrient source and availability will be used to optimize energy generation and to adjust growth and expression of virulence genes in response to changes in the host environment. However, the mechanisms by which these bacteria fine-tune the control of gene expression during fluctuation of nutrients and infection may vary within species.

Signalling pathways in stress tolerance

Quorum-sensing (QS) systems and TCS coordinate the regulation of gene expression in response to environmental changes playing a critical role in stress adaptation. QS systems coordinate gene expression in a cell density-dependent manner by releasing and responding to the accumulation of chemical signal molecules called autoinducers (AI). TCS consist of a sensor protein, which is a membrane-associated histidine kinase, and a response regulator, which is a DNA binding protein that modulates expression of target genes when phosphorylated by the kinase. Both QS and TCS have been shown to play important roles in bacterial adaptation, survival and virulence by sensing changes in the environment and altering expression of specific sets of genes involved in stress management and biofilm formation.

The universal LuxS QS system, which produces AI-2, a furanosyl borate diester has been shown to regulate a large panel of genes in a variety of microorganisms. A strain of *S. mutans* lacking the *luxS* gene was shown to have altered biofilm architecture and to have increased sensitivity to acid killing (Wen and Burne, 2004). By inactivating *luxS* in GAS, Siller and co-authors demonstrated that loss of AI-2 signalling leads to increased acid tolerance, increased cell invasion and intracellular survival (Marouni and Sela, 2003; Siller *et al.*, 2008). However, the exact role of the LuxS/AI-2 system in GAS pathogenesis remains to be elucidated. Although the effects of *luxS* inactivation in stress tolerance have not been directly evaluated in *S. pneumoniae,* LuxS was recently shown to mediate iron-dependent biofilm formation, competence and fratricide in this organism (Trappetti *et al.*, 2011). The LuxS QS system has not been investigated in GBS.

Sequence analysis identified 14 TCSs in *S. mutans* UA159 (Ajdic *et al.*, 2002; Biswas *et al.*, 2008). Global transcriptional analysis of acid-inducible genes revealed that genes encoding multiple TCSs, including CiaHR, LevSR, ScnKR, LiaSR and ComDE, were induced during acid adaptation (Gong *et al.*, 2009). Several studies have evaluated the role of TCS in *S. mutans* (Li *et al.*, 2001, 2002a,b; Qi *et al.*, 2004; Chen *et al.*, 2007; Deng *et al.*, 2007; Levesque *et al.*, 2007; Biswas *et al.*, 2008; Senadheera *et al.*, 2009) revealing that TCSs regulate virulence gene expression, induction of competence, biofilm development, bacteriocin production and stress tolerance. In particular, two independent groups systematically inactivated the genes encoding sensor kinases of the TCS and evaluated their role in stress tolerance (Levesque *et al.*, 2007; Biswas *et al.*, 2008). In the study by Lévesque and colleagues, *scnK* and *levS* mutants displayed significantly slower growth at pH 5.5, whereas the *ciaH* mutant grew better than the parental strain in the presence of NaCl or H_2O_2 (Levesque *et al.*, 2007). Biswas and co-workers found that inactivation of three sensor kinases, *liaS*, *ciaH* and *vicK*, affected stress tolerance of strain *S. mutans* UA159 strain (Biswas *et al.*, 2008). The *liaS* and *ciaH* mutants showed increased sensitivity to the aminonucleoside antibiotics puromycin (a translation inhibitor) and reduced growth when incubated in aerobic conditions or on agar medium supplemented with H_2O_2 (Biswas *et al.*, 2008). The *ciaH* mutant also showed significant reduction of growth at pH 5 and displayed increased sensitivity to DNA damage caused by mitomycin C (Biswas *et al.*, 2008). Notably, previous reports have also shown that inactivation of *ciaH* resulted in an acid-sensitive phenotype in strains UA159 and UA140 (Qi *et al.*, 2004; Ahn *et al.*, 2006). The *S. mutans* VicRK TCS system was shown to respond to, and protect against, oxidative stress in one particular study (Deng *et al.*, 2007), but not in the studies discussed above (Levesque *et al.*, 2007; Biswas *et al.*, 2008). A role in oxidative stress response was also assigned to ScnRK, as *scnRK* mutants were more sensitive to H_2O_2 and more susceptible to phagocytic killing in non-activated macrophages (Chen *et al.*, 2007). Finally, studies from the Cvitkovitch lab have shown that inactivation of *liaS* or *comDE* conferred an acid-sensitive phenotype to strains NG8 and BM71, respectively (Li *et al.*, 2001, 2002a), although *comD* or *comE* do not appear to affect acid tolerance in strain UA159 (Ahn *et al.*, 2006). Collectively, these data support the idea that there may be substantial heterogeneity among strains in the role of specific TCS, not only in the genes they regulate, but also in the external stimuli to which they respond. Nevertheless, the CiaHR TCS has been consistently found to play a role in the stress responses by *S. mutans*, and has also been implicated in competence development, bacteriocin production and biofilm formation (Qi *et al.*, 2004; Ahn *et al.*, 2006; Levesque *et al.*, 2007). More recently, it was demonstrated that the *ciaRH* operon of *S. mutans* consists of three genes with the first gene, named *ciaX*, encoding a small, double-glycine signalling peptide that allows CiaRH to modulate its own expression in response to calcium (He *et al.*, 2008). Human saliva is saturated in calcium, and calcium is the main mineral in tooth enamel, thus sensing calcium cations, through CiaRH, in the oral cavity may be an important mechanism for coping with different stresses (mainly acid and osmotic) in *S. mutans*.

Microarray analysis has been used to identify genes that contribute to GAS survival in human polymorphonuclear leucocytes, saliva and blood, to different temperatures, during mouse soft tissue infection and during experimental pharyngitis in macaques (Smoot *et al.*, 2001; Graham *et al.*, 2002, 2005, 2006; Virtaneva *et al.*, 2003, 2005; Shelburne *et al.*, 2005). These studies have highlighted the importance of several TCSs, in particular CovRS (control of virulence), in coordinating expression of genes encoding virulence factors, as well as catabolic and biosynthetic enzymes (Graham *et al.*, 2002, 2005; Virtaneva *et al.*, 2005). As mentioned in the 'Nutrient limitation' section, the SptRS TCS was shown to contribute to GAS survival in saliva by regulating metabolic pathways and virulence factor production (Shelburne *et al.*, 2005).

The *S. pneumoniae* genome encodes 13 TCSs and an additional orphan unpaired response regulator (Throup *et al.*, 2000). Attempts to individually inactivate all 14 response regulators revealed that VicRK was essential whereas seven of the other 12 TCS gene pairs and the orphan response regulator were important for growth and pathogenicity in the mouse respiratory tract infection model (Throup *et al.*, 2000). Like in *S.*

mutans, ComDE and CiaHR were also implicated in competence development (X-state) (Cheng *et al.*, 1997; Dagkessamanskaia *et al.*, 2004), and a third, LiaSR, was shown to coordinate protection against fratricide-induced self-lysis (Eldholm *et al.* 2010). The CiaHR TCS has been the subject of extensive research and, in addition to competence, it has been implicated in β-lactam resistance, autolysis, and virulence (Throup *et al.*, 2000; Dagkessamanskaia *et al.*, 2004; Mascher *et al.*, 2006). The CiaHR-dependent virulence was attributed, at least in part, to the regulation of the surface-associated HtrA protease by CiaHR (Ibrahim *et al.*, 2004a). As described above, HtrA has been implicated in virulence by influencing the ability of pneumococcus to colonize the nasopharynx of rats (Ibrahim *et al.*, 2004b). Notably, the reduced virulence of the *ciaR* mutant could be restored by increasing expression of HtrA (Ibrahim *et al.*, 2004a).

The GBS genome encodes a higher number of TCSs (17 to 21 depending on the sequenced strain) in comparison to related streptococcal species. Only a few TCSs in GBS have been studied to date, and only two have been assigned a role in stress adaptation. The CsrRS (also called CovRS), a major regulator of virulence in GAS (see GAS section for details), was shown to also regulate virulence expression in GBS. Inactivation of *csrRS* affected the expression of several virulence factors, such as reduced CAMP activity and increased haemolytic activity, and resulted in a significant reduction of virulence in rodents (Lamy *et al.*, 2004; Jiang *et al.*, 2005). As described in the 'Acid sress' section, a number of genes involved in pH adaptation are controlled by CsrRS, suggesting that CsrRS participates in the adaptation of GBS to different pH environments (Santi *et al.*, 2009). It has been hypothesized that the translocation of GBS from the acidic milieu of the vagina to the neutral pH of neonatal tissues signals the conversion from a colonizing to an invasive phenotype in a CsrRS-dependent manner (Santi *et al.*, 2009). The CiaRH TCS was also implicated in adaptation to the host environment. Specifically, inactivation of the *ciaR* response regulator resulted in a significant decrease in intracellular survival within neutrophils, macrophages and brain microvascular endothelial cells, which was linked to increased susceptibility to antimicrobial peptides, lysozyme, and ROS (Quach *et al.*, 2009).

Future trends

It has been well established that stress responses can be often found intertwined with virulence. Thus, there is a major need to continue to identify and characterize the stress response pathways of streptococci. Data generated from functional genomic, proteomic and transcriptomic studies have helped to unveil many aspects of the stress regulon in different streptococcal species. While some species have been thoroughly investigated, as in the case of the dental pathogen *S. mutans*, other species have been largely overlooked. In particular, there is a clear gap in the knowledge of the stress responses of GAS and GBS, two major human pathogens.

A major challenge for future research is to integrate novel research tools in the context of stress response studies. More specifically, with next-generation sequencing and the continuous advances in automated mass-spectrometry technologies, high-throughput cDNA sequencing (RNA-Seq) and an unprecedented ability to separate, identify and quantify proteins in complex mixtures can now be used to study the behaviour of microbial communities in their natural environment (metagenomics and metaproteomics). These types of technologies offer an exciting opportunity to study, in a meaningful way, the bacterial stress responses *in vivo* and may shed new light onto the mechanisms by which changes in the microbial community composition transform a healthy site into a diseased site.

Acknowledgements

This work was supported in part by grant DE019783 from the National Institute of Dental and Craniofacial Research.

Web resources

Oralgen: http://www.oralgen.lanl.gov

Molecular information pertaining the genome of oral pathogens, including *S. mutans* and other oral streptococci. Genome information of GAS, GBS and *S. pneumoniae* is also available.

NCBI: http://www.ncbi.nlm.nih.gov

Provides access to biomedical and genomic information. Access to several genomic, functional genomic and many other useful tools.

CDC: http://www.cdc.gov/ncidod/biotech/strep/strepindex.htm

Reference laboratory at the Center for Disease Control in the US the for identification of streptococcal isolates.

BROP: http://www.brop.org

Bioinformatic resources for oral pathogens at the Forsyth Institute.

References

Abranches, J., Chen, Y.Y., and Burne, R.A. (2003). Characterization of *Streptococcus mutans* strains deficient in EIIAB Man of the sugar phosphotransferase system. Appl. Environ. Microbiol. *69*, 4760–4769.

Abranches, J., Candella, M.M., Wen, Z.T., Baker, H.V., and Burne, R.A. (2006). Different roles of EIIABMan and EIIGlc in regulation of energy metabolism, biofilm development, and competence in *Streptococcus mutans*. J. Bacteriol. *188*, 3748–3756.

Abranches, J., Nascimento, M.M., Zeng, L., Browngardt, C.M., Wen, Z.T., Rivera, M.F., and Burne, R.A. (2008). CcpA regulates central metabolism and virulence gene expression in *Streptococcus mutans*. J. Bacteriol. *190*, 2340–2349.

Abranches, J., Martinez, A.R., Kajfasz, J.K., Chavez, V., Garsin, D.A., and Lemos, J.A. (2009). The molecular alarmone (p)ppGpp mediates stress responses, vancomycin tolerance and virulence in *Enterococcus faecalis*. J. Bacteriol. *191*, 2248–2256.

Ahn, S.J., Lemos, J.A., and Burne, R.A. (2005). Role of HtrA in growth and competence of *Streptococcus mutans* UA159. J. Bacteriol. *187*, 3028–3038.

Ahn, S.J., Wen, Z.T., and Burne, R.A. (2006). Multilevel control of competence development and stress tolerance in *Streptococcus mutans* UA159. Infect. Immun. *74*, 1631–1642.

Ajdic, D., McShan, W.M., McLaughlin, R.E., Savic, G., Chang, J., Carson, M.B., Primeaux, C., Tian, R., Kenton, S., Jia, H., *et al.* (2002). Genome sequence of *Streptococcus mutans* UA159, a cariogenic dental pathogen. Proc. Natl. Acad. Sci. U.S.A. *99*, 14434–14439.

Ambur, O.H., Davidsen, T., Frye, S.A., Balasingham, S.V., Lagesen, K., Rognes, T., and Tonjum, T. (2009). Genome dynamics in major bacterial pathogens. FEMS Microbiol. Rev. *33*, 453–470.

Bates, C.S., Toukoki, C., Neely, M.N., and Eichenbaum, Z. (2005). Characterization of MtsR, a new metal regulator in group A *Streptococcus*, involved in iron acquisition and virulence. Infect. Immun. *73*, 5743–5753.

Battesti, A., and Bouveret, E. (2009). Bacteria possessing two RelA/SpoT-like proteins have evolved a specific stringent response involving the ACP–SpoT interaction. J. Bacteriol. *191*, 616–624.

Biswas, S., and Biswas, I. (2005). Role of HtrA in surface protein expression and biofilm formation by *Streptococcus mutans*. Infect. Immun. *73*, 6923–6934.

Biswas, I., Drake, L., Erkina, D., and Biswas, S. (2008). Involvement of sensor kinases in the stress tolerance response of *Streptococcus mutans*. J. Bacteriol. *190*, 68–77.

Bortoni, M.E., Terra, V.S., Hinds, J., Andrew, P.W., and Yesilkaya, H. (2009). The pneumococcal response to oxidative stress includes a role for Rgg. Microbiology *155*, 4123–4134.

Brenot, A., King, K.Y., Janowiak, B., Griffith, O., and Caparon, M.G. (2004). Contribution of glutathione peroxidase to the virulence of *Streptococcus pyogenes*. Infect. Immun. *72*, 408–413.

Brenot, A., King, K.Y., and Caparon, M.G. (2005). The PerR regulon in peroxide resistance and virulence of *Streptococcus pyogenes*. Mol. Microbiol. *55*, 221–234.

Brenot, A., Weston, B.F., and Caparon, M.G. (2007). A PerR-regulated metal transporter (PmtA) is an interface between oxidative stress and metal homeostasis in *Streptococcus pyogenes*. Mol. Microbiol. *63*, 1185–1196.

Butler, S.M., Festa, R.A., Pearce, M.J., and Darwin, K.H. (2006). Self-compartmentalized bacterial proteases and pathogenesis. Mol. Microbiol. *60*, 553–562.

Cao, J., Chen, D., Xu, W., Chen, T., Xu, S., Luo, J., Zhao, Q., Liu, B., Wang, D., Zhang, X., *et al.* (2007). Enhanced protection against pneumococcal infection elicited by immunization with the combination of PspA, PspC, and ClpP. Vaccine *25*, 4996–5005.

Cao, J., Li, D., Gong, Y., Yin, N., Chen, T., Wong, C.K., Xu, W., Luo, J., Zhang, X., Lam, C.W., *et al.* (2009). Caseinolytic protease: a protein vaccine which could elicit serotype-independent protection against invasive pneumococcal infection. Clin. Exp. Immunol. *156*, 52–60.

Carapetis, J.R., Steer, A.C., Mulholland, E.K., and Weber, M. (2005). The global burden of group A streptococcal diseases. Lancet Infect. Dis. *5*, 685–694.

Carlsson, J. (1983). Regulation of sugar metabolism in relation to feast-and-famine existence of plaque. In Cariology Today, Guggenheim, B., ed. (Karger, Basel), pp. 205–211.

Caymaris, S., Bootsma, H.J., Martin, B., Hermans, P.W., Prudhomme, M., and Claverys, J.P. (2010). The global nutritional regulator CodY is an essential protein in the human pathogen *Streptococcus pneumoniae*. Mol. Microbiol. *78*, 344–360.

Charpentier, E., Novak, R., and Tuomanen, E. (2000). Regulation of growth inhibition at high temperature, autolysis, transformation and adherence in *Streptococcus pneumoniae* by *clpC*. Mol. Microbiol. *37*, 717–726.

Chastanet, A., Prudhomme, M., Claverys, J.P., and Msadek, T. (2001). Regulation of *Streptococcus pneumoniae clp* genes and their role in competence development and stress survival. J. Bacteriol. *183*, 7295–7307.

Chaussee, M.S., Phillips, E.R., and Ferretti, J.J. (1997). Temporal production of streptococcal erythrogenic toxin B (streptococcal cysteine proteinase) in response to nutrient depletion. Infect. Immun. *65*, 1956–1959.

Chen, P.M., Chen, H.C., Ho, C.T., Jung, C.J., Lien, H.T., Chen, J.Y., and Chia, J.S. (2007). The two-component system ScnRK of *Streptococcus mutans* affects hydrogen peroxide resistance and murine macrophage killing. Microbes Infect. *10*, 293–301.

Cheng, Q., Campbell, E.A., Naughton, A.M., Johnson, S., and Masure, H.R. (1997). The com locus controls genetic transformation in *Streptococcus pneumoniae*. Mol. Microbiol. *23*, 683–692.

Claverys, J.P., Prudhomme, M., and Martin, B. (2006). Induction of competence regulons as a general response to stress in Gram-positive bacteria. Annu. Rev. Microbiol. *60*, 451–475.

Claverys, J.P., Martin, B., and Havarstein, L.S. (2007). Competence-induced fratricide in streptococci. Mol. Microbiol. *64*, 1423–1433.

Cunningham, M.W. (2000). Pathogenesis of group A streptococcal infections. Clin. Microbiol. Rev. *13*, 470–511.

Dagkessamanskaia, A., Moscoso, M., Henard, V., Guiral, S., Overweg, K., Reuter, M., Martin, B., Wells, J., and Claverys, J.P. (2004). Interconnection of competence, stress and CiaR regulons in *Streptococcus pneumoniae*: competence triggers stationary phase autolysis of *ciaR* mutant cells. Mol. Microbiol. *51*, 1071–1086.

Deng, D.M., Liu, M.J., ten Cate, J.M., and Crielaard, W. (2007). The VicRK system of *Streptococcus mutans* responds to oxidative stress. J. Dent. Res. *86*, 606–610.

Derre, I., Rapoport, G., and Msadek, T. (1999). CtsR, a novel regulator of stress and heat shock response, controls clp and molecular chaperone gene expression in Gram-positive bacteria. Mol. Microbiol. *31*, 117–131.

Drangsholt, M.T. (1998). A new causal model of dental diseases associated with endocarditis. Ann. Periodontol. 3, 184–196.

Eldholm, V., Gutt, B., Johnsborg, O., Bruckner, R., Maurer, P., Hakenbeck, R., Mascher, T., and Havarstein, L.S. (2010). The pneumococcal cell envelope stress-sensing system LiaFSR is activated by murein hydrolases and lipid II-interacting antibiotics. J. Bacteriol. *192*, 1761–1773.

Faustoferri, R.C., Hahn, K., Weiss, K., and Quivey, R.G., Jr. (2005). Smx nuclease is the major, low-pH-inducible apurinic/apyrimidinic endonuclease in *Streptococcus mutans*. J. Bacteriol. *187*, 2705–2714.

Ferretti, J.J., McShan, W.M., Ajdic, D., Savic, D.J., Savic, G., Lyon, K., Primeaux, C., Sezate, S., Suvorov, A.N., Kenton, S., *et al.* (2001). Complete genome sequence of an M1 strain of *Streptococcus pyogenes*. Proc. Natl. Acad. Sci. U.S.A. *98*, 4658–4663.

Fozo, E.M., and Quivey, R.G., Jr. (2004). The *fabM* gene product of *Streptococcus mutans* is responsible for the synthesis of monounsaturated fatty acids and is necessary for survival at low pH. J. Bacteriol. *186*, 4152–4158.

Fozo, E.M., Scott-Anne, K., Koo, H., and Quivey, R.G., Jr. (2007). Role of unsaturated fatty acid biosynthesis in virulence of *Streptococcus mutans*. Infect. Immun. *75*, 1537–1539.

Frees, D., Savijoki, K., Varmanen, P., and Ingmer, H. (2007). Clp ATPases and ClpP proteolytic complexes regulate vital biological processes in low GC, Gram-positive bacteria. Mol. Microbiol. *63*, 1285–1295.

Giammarinaro, P., and Paton, J.C. (2002). Role of RegM, a homologue of the catabolite repressor protein CcpA, in the virulence of *Streptococcus pneumoniae*. Infect. Immun. *70*, 5454–5461.

Gibson, C.M., Mallett, T.C., Claiborne, A., and Caparon, M.G. (2000). Contribution of NADH oxidase to aerobic metabolism of *Streptococcus pyogenes*. J. Bacteriol. *182*, 448–455.

Glaser, P., Rusniok, C., Buchrieser, C., Chevalier, F., Frangeul, L., Msadek, T., Zouine, M., Couve, E., Lalioui, L., Poyart, C., *et al.* (2002). Genome sequence of *Streptococcus agalactiae*, a pathogen causing invasive neonatal disease. Mol. Microbiol. *45*, 1499–1513.

Gong, Y., Tian, X.L., Sutherland, T., Sisson, G., Mai, J., Ling, J., and Li, Y.H. (2009). Global transcriptional analysis of acid-inducible genes in *Streptococcus mutans:* multiple two-component systems involved in acid adaptation. Microbiology *155*, 3322–3332.

Gorke, B., and Stulke, J. (2008). Carbon catabolite repression in bacteria: many ways to make the most out of nutrients. Nat. Rev. Microbiol. *6*, 613–624.

Gottesman, S. (2003). Proteolysis in bacterial regulatory circuits. Annu. Rev. Cell. Dev. Biol. *19*, 565–587.

Graham, M.R., Smoot, L.M., Migliaccio, C.A., Virtaneva, K., Sturdevant, D.E., Porcella, S.F., Federle, M.J., Adams, G.J., Scott, J.R., and Musser, J.M. (2002). Virulence control in group A *Streptococcus* by a two-component gene regulatory system: global expression profiling and *in vivo* infection modeling. Proc. Natl. Acad. Sci. U.S.A. *99*, 13855–13860.

Graham, M.R., Virtaneva, K., Porcella, S.F., Barry, W.T., Gowen, B.B., Johnson, C.R., Wright, F.A., and Musser, J.M. (2005). Group A *Streptococcus* transcriptome dynamics during growth in human blood reveals bacterial adaptive and survival strategies. Am. J. Pathol. *166*, 455–465.

Graham, M.R., Virtaneva, K., Porcella, S.F., Gardner, D.J., Long, R.D., Welty, D.M., Barry, W.T., Johnson, C.A., Parkins, L.D., Wright, F.A., *et al.* (2006). Analysis of the transcriptome of group A *Streptococcus* in mouse soft tissue infection. Am. J. Pathol. *169*, 927–942.

Griswold, A.R., Jameson-Lee, M., and Burne, R.A. (2006). Regulation and physiologic significance of the agmatine deiminase system of *Streptococcus mutans* UA159. J. Bacteriol. *188*, 834–841.

Gryllos, I., Grifantini, R., Colaprico, A., Cary, M.E., Hakansson, A., Carey, D.W., Suarez-Chavez, M., Kalish, L.A., Mitchell, P.D., White, G.L., *et al.* (2008). PerR confers phagocytic killing resistance and allows pharyngeal colonization by group A *Streptococcus*. PLoS Pathog. *4*, e1000145.

Hahn, K., Faustoferri, R.C., and Quivey, R.G., Jr. (1999). Induction of an AP endonuclease activity in *Streptococcus mutans* during growth at low pH. Mol. Microbiol. *31*, 1489–1498.

Hanks, T.S., Liu, M., McClure, M.J., Fukumura, M., Duffy, A., and Lei, B. (2006). Differential regulation of iron- and manganese-specific MtsABC and heme-specific HtsABC transporters by the metalloregulator MtsR of group A *Streptococcus*. Infect. Immun. *74*, 5132–5139.

Hanna, M.N., Ferguson, R.J., Li, Y.H., and Cvitkovitch, D.G. (2001). *uvrA* is an acid-inducible gene involved in the adaptive response to low pH in *Streptococcus mutans*. J. Bacteriol. *183*, 5964–5973.

Hartl, F.U., and Hayer-Hartl, M. (2002). Molecular chaperones in the cytosol: from nascent chain to folded protein. Science *295*, 1852–1858.

He, X., Wu, C., Yarbrough, D., Sim, L., Niu, G., Merritt, J., Shi, W., and Qi, F. (2008). The *cia* operon of *Streptococcus mutans* encodes a unique component required for calcium-mediated autoregulation. Mol. Microbiol. *70*, 112–126.

Hendriksen, W.T., Bootsma, H.J., Estevao, S., Hoogenboezem, T., de Jong, A., de Groot, R., Kuipers, O.P., and Hermans, P.W. (2008). CodY of *Streptococcus pneumoniae*: link between nutritional gene regulation and colonization. J. Bacteriol. *190*, 590–601.

Higuchi, M., Yamamoto, Y., Poole, L.B., Shimada, M., Sato, Y., Takahashi, N., and Kamio, Y. (1999). Functions of two types of NADH oxidases in energy metabolism and oxidative stress of *Streptococcus mutans*. J. Bacteriol. *181*, 5940–5947.

Hill, E.E., Herijgers, P., Herregods, M.C., and Peetermans, W.E. (2006). Evolving trends in infective endocarditis. Clin. Microbiol. Infect. *12*, 5–12.

Ibrahim, Y.M., Kerr, A.R., McCluskey, J., and Mitchell, T.J. (2004a). Control of virulence by the two-component system CiaR/H is mediated via HtrA, a major virulence factor of *Streptococcus pneumoniae*. J. Bacteriol. *186*, 5258–5266.

Ibrahim, Y.M., Kerr, A.R., McCluskey, J., and Mitchell, T.J. (2004b). Role of HtrA in the virulence and competence of *Streptococcus pneumoniae*. Infect. Immun. *72*, 3584–3591.

Ibrahim, Y.M., Kerr, A.R., Silva, N.A., and Mitchell, T.J. (2005). Contribution of the ATP-dependent protease ClpCP to the autolysis and virulence of *Streptococcus pneumoniae*. Infect. Immun. *73*, 730–740.

Iyer, R., Baliga, N.S., and Camilli, A. (2005). Catabolite control protein A (CcpA) contributes to virulence and regulation of sugar metabolism in *Streptococcus pneumoniae*. J. Bacteriol. *187*, 8340–8349.

Jiang, S.M., Cieslewicz, M.J., Kasper, D.L., and Wessels, M.R. (2005). Regulation of virulence by a two-component system in group B *streptococcus*. J. Bacteriol. *187*, 1105–1113.

Johri, A.K., Padilla, J., Malin, G., and Paoletti, L.C. (2003). Oxygen regulates invasiveness and virulence of group B *streptococcus*. Infect. Immun. *71*, 6707–6711.

Kajfasz, J.K., Martinez, A.R., Rivera-Ramos, I., Abranches, J., Koo, H., Quivey, R.G., Jr., and Lemos, J.A. (2009). Role of Clp proteins in expression of virulence properties of *Streptococcus mutans*. J. Bacteriol. *191*, 2060–2068.

Kajfasz, J.K., Rivera-Ramos, I., Abranches, J., Martinez, A.R., Rosalen, P.L., Derr, A.M., Quivey, R.G., and Lemos, J.A. (2010). Two Spx proteins modulate stress tolerance, survival, and virulence in *Streptococcus mutans*. J. Bacteriol. *192*, 2546–2556.

Kazmierczak, K.M., Wayne, K.J., Rechtsteiner, A., and Winkler, M.E. (2009). Roles of rel in stringent response, global regulation and virulence of serotype 2 *Streptococcus pneumoniae* D39. Mol. Microbiol. *72*, 590–611.

King, K.Y., Horenstein, J.A., and Caparon, M.G. (2000). Aerotolerance and peroxide resistance in peroxidase and PerR mutants of *Streptococcus pyogenes*. J. Bacteriol. *182*, 5290–5299.

Kreth, J., Chen, Z., Ferretti, J., and Malke, H. (2011). Counteractive balancing of transcriptome expression involving CodY and CovRS in *Streptococcus pyogenes*. J. Bacteriol. *193*, 4153–4165.

Kwon, H.Y., Kim, S.W., Choi, M.H., Ogunniyi, A.D., Paton, J.C., Park, S.H., Pyo, S.N., and Rhee, D.K. (2003). Effect of heat shock and mutations in ClpL and ClpP on virulence gene expression in *Streptococcus pneumoniae*. Infect. Immun. *71*, 3757–3765.

Kwon, H.Y., Ogunniyi, A.D., Choi, M.H., Pyo, S.N., Rhee, D.K., and Paton, J.C. (2004). The ClpP protease of *Streptococcus pneumoniae* modulates virulence gene expression and protects against fatal pneumococcal challenge. Infect. Immun. *72*, 5646–5653.

Lamy, M.C., Zouine, M., Fert, J., Vergassola, M., Couve, E., Pellegrini, E., Glaser, P., Kunst, F., Msadek, T., Trieu-Cuot, P., *et al.* (2004). CovS/CovR of group B *streptococcus*: a two-component global regulatory system involved in virulence. Mol. Microbiol. *54*, 1250–1268.

Lemos, J.A., and Burne, R.A. (2002). Regulation and physiological significance of ClpC and ClpP in *Streptococcus mutans*. J. Bacteriol. *184*, 6357–6366.

Lemos, J.A., Chen, Y.Y., and Burne, R.A. (2001). Genetic and physiologic analysis of the *groE* operon and role of the HrcA repressor in stress gene regulation and acid tolerance in *Streptococcus mutans*. J. Bacteriol. *183*, 6074–6084.

Lemos, J.A., Brown, T.A., Jr., and Burne, R.A. (2004). Effects of RelA on key virulence properties of planktonic and biofilm populations of *Streptococcus mutans*. Infect. Immun. *72*, 1431–1440.

Lemos, J.A., Abranches, J., and Burne, R.A. (2005). Responses of cariogenic streptococci to environmental stresses. Curr. Issues Mol. Biol. *7*, 95–107.

Lemos, J.A., Lin, V.K., Nascimento, M.M., Abranches, J., and Burne, R.A. (2007a). Three gene products govern (*p*)ppGpp production by *Streptococcus mutans*. Mol. Microbiol. *65*, 1568–1581.

Lemos, J.A., Luzardo, Y., and Burne, R.A. (2007b). Physiologic effects of forced down-regulation of *dnaK* and *groEL* expression in *Streptococcus mutans*. J. Bacteriol. *189*, 1582–1588.

Lemos, J.A., Nascimento, M.M., Lin, V.K., Abranches, J., and Burne, R.A. (2008). Global regulation by (*p*)ppGpp and CodY in *Streptococcus mutans*. J. Bacteriol. *190*, 5291–5299.

Len, A.C., Harty, D.W., and Jacques, N.A. (2004a). Proteome analysis of *Streptococcus mutans* metabolic phenotype during acid tolerance. Microbiology *150*, 1353–1366.

Len, A.C., Harty, D.W., and Jacques, N.A. (2004b). Stress-responsive proteins are up-regulated in *Streptococcus mutans* during acid tolerance. Microbiology *150*, 1339–1351.

Levesque, C.M., Mair, R.W., Perry, J.A., Lau, P.C., Li, Y.H., and Cvitkovitch, D.G. (2007). Systemic inactivation and phenotypic characterization of two-component systems in expression of *Streptococcus mutans* virulence properties. Lett. Appl. Microbiol. *45*, 398–404.

Li, Y.H., Hanna, M.N., Svensater, G., Ellen, R.P., and Cvitkovitch, D.G. (2001). Cell density modulates acid adaptation in *Streptococcus mutans*: implications for survival in biofilms. J. Bacteriol. *183*, 6875–6884.

Li, Y.H., Lau, P.C., Tang, N., Svensater, G., Ellen, R.P., and Cvitkovitch, D.G. (2002a). Novel two-component regulatory system involved in biofilm formation and acid resistance in *Streptococcus mutans*. J. Bacteriol. *184*, 6333–6342.

Li, Y.H., Tang, N., Aspiras, M.B., Lau, P.C., Lee, J.H., Ellen, R.P., and Cvitkovitch, D.G. (2002b). A quorum-sensing signaling system essential for genetic competence in *Streptococcus mutans* is involved in biofilm formation. J. Bacteriol. *184*, 2699–2708.

Liu, G.Y., Doran, K.S., Lawrence, T., Turkson, N., Puliti, M., Tissi, L., and Nizet, V. (2004). Sword and shield: linked group B streptococcal beta-hemolysin/cytolysin and carotenoid pigment function to subvert host phagocyte defense. Proc. Natl. Acad. Sci. U.S.A. *101*, 14491–14496.

Loughman, J.A., and Caparon, M. (2006). Regulation of SpeB in *Streptococcus pyogenes* by pH and NaCl: a model for *in vivo* gene expression. J. Bacteriol. *188*, 399–408.

Lu, Y.J., and Rock, C.O. (2006). Transcriptional regulation of fatty acid biosynthesis in *Streptococcus pneumoniae*. Mol. Microbiol. *59*, 551–566.

McAllister, L.J., Tseng, H.J., Ogunniyi, A.D., Jennings, M.P., McEwan, A.G., and Paton, J.C. (2004). Molecular analysis of the psa permease complex of *Streptococcus pneumoniae*. Mol. Microbiol. *53*, 889–901.

Malke, H., and Ferretti, J.J. (2007). CodY-affected transcriptional gene expression of *Streptococcus pyogenes* during growth in human blood. J. Med. Microbiol. *56*, 707–714.

Malke, H., Steiner, K., McShan, W.M., and Ferretti, J.J. (2006). Linking the nutritional status of *Streptococcus pyogenes* to alteration of transcriptional gene expression: the action of CodY and RelA. Int. J. Med. Microbiol. *296*, 259–275.

Marouni, M.J., and Sela, S. (2003). The *luxS* gene of *Streptococcus pyogenes* regulates expression of genes that affect internalization by epithelial cells. Infect. Immun. *71*, 5633–5639.

Marquis, R.E. (1995). Oxygen metabolism, oxidative stress and acid-base physiology of dental plaque biofilms. J. Ind. Microbiol. *15*, 198–207.

Mascher, T., Heintz, M., Zahner, D., Merai, M., and Hakenbeck, R. (2006). The CiaRH system of *Streptococcus pneumoniae* prevents lysis during stress induced by treatment with cell wall inhibitors and by mutations in *pbp2x* involved in beta-lactam resistance. J. Bacteriol. *188*, 1959–1968.

Moreillon, P., and Que, Y.A. (2004). Infective endocarditis. Lancet *363*, 139–149.

Nair, S., Poyart, C., Beretti, J.L., Veiga-Fernandes, H., Berche, P., and Trieu-Cuot, P. (2003). Role of the *Streptococcus agalactiae* ClpP serine protease in heat-induced stress defence and growth arrest. Microbiology *149*, 407–417.

Nakayama, K. (1992). Nucleotide sequence of *Streptococcus mutans* superoxide dismutase gene and isolation of insertion mutants. J. Bacteriol. *174*, 4928–4934.

Nanamiya, H., Kasai, K., Nozawa, A., Yun, C.S., Narisawa, T., Murakami, K., Natori, Y., Kawamura, F., and Tozawa, Y. (2008). Identification and functional analysis of novel (*p*)ppGpp synthetase genes in *Bacillus subtilis*. Mol. Microbiol. *67*, 291–304.

Nascimento, M.M., Lemos, J.A., Abranches, J., Lin, V.K., and Burne, R.A. (2008). Role of RelA of *Streptococcus mutans* in global control of gene expression. J. Bacteriol. *190*, 28–36.

Nguyen, P.T., Abranches, J., Phan, T.N., and Marquis, R.E. (2002). Repressed respiration of oral streptococci grown in biofilms. Curr. Microbiol. *44*, 262–266.

Pallen, M.J., and Wren, B.W. (1997). The HtrA family of serine proteases. Mol. Microbiol. *26*, 209–221.

Park, C.Y., Kim, E.H., Choi, S.Y., Tran, T.D., Kim, I.H., Kim, S.N., Pyo, S., and Rhee, D.K. (2010). Virulence attenuation of *Streptococcus pneumoniae clpP* mutant by sensitivity to oxidative stress in macrophages via an NO-mediated pathway. J. Microbiol. *48*, 229–235.

Paterson, G.K., Blue, C.E., and Mitchell, T.J. (2006). An operon in *Streptococcus pneumoniae* containing a putative alkylhydroperoxidase D homologue contributes to virulence and the response to oxidative stress. Microb. Pathog. *40*, 152–160.

Postma, P.W., Lengeler, J.W., and Jacobson, G.R. (1993). Phosphoenolpyruvate:carbohydrate phosphotransferase systems of bacteria. Microbiol. Rev. *57*, 543–594.

Potrykus, K., and Cashel, M. (2008). (p)ppGpp: still magical? Annu. Rev. Microbiol. *62*, 35–51.

Poyart, C., Pellegrini, E., Gaillot, O., Boumaila, C., Baptista, M., and Trieu-Cuot, P. (2001). Contribution of Mn-cofactored superoxide dismutase (SodA) to the virulence of Streptococcus agalactiae. Infect. Immun. *69*, 5098–5106.

Qi, F., Merritt, J., Lux, R., and Shi, W. (2004). Inactivation of the *ciaH* Gene in *Streptococcus mutans* diminishes mutacin production and competence development, alters sucrose-dependent biofilm formation, and reduces stress tolerance. Infect. Immun. *72*, 4895–4899.

Quach, D., van Sorge, N.M., Kristian, S.A., Bryan, J.D., Shelver, D.W., and Doran, K.S. (2009). The CiaR response regulator in group B *Streptococcus* promotes intracellular survival and resistance to innate immune defenses. J. Bacteriol. *191*, 2023–2032.

Quivey, R.G., Jr., Faustoferri, R., Monahan, K., and Marquis, R. (2000). Shifts in membrane fatty acid profiles associated with acid adaptation of *Streptococcus mutans*. FEMS Microbiol. Lett. *189*, 89–92.

Ricci, S., Janulczyk, R., and Bjorck, L. (2002). The regulator PerR is involved in oxidative stress response and iron homeostasis and is necessary for full virulence of *Streptococcus pyogenes*. Infect. Immun. *70*, 4968–4976.

Robertson, G.T., Ng, W.L., Foley, J., Gilmour, R., and Winkler, M.E. (2002). Global transcriptional analysis of *clpP* mutations of type 2 *Streptococcus pneumoniae* and their effects on physiology and virulence. J. Bacteriol. *184*, 3508–3520.

Robertson, G.T., Ng, W.L., Gilmour, R., and Winkler, M.E. (2003). Essentiality of *clpX*, but not *clpP*, *clpL*, *clpC*, or *clpE*, in *Streptococcus pneumoniae* R6. J. Bacteriol. *185*, 2961–2966.

Ryan, C.S., and Kleinberg, I. (1995). Bacteria in human mouths involved in the production and utilization of hydrogen peroxide. Arch. Oral Biol. *40*, 753–763.

Saier, M.H., Jr., Chauvaux, S., Cook, G.M., Deutscher, J., Paulsen, I.T., Reizer, J., and Ye, J.J. (1996). Catabolite repression and inducer control in Gram-positive bacteria. Microbiology *142*, 217–230.

Santi, I., Grifantini, R., Jiang, S.M., Brettoni, C., Grandi, G., Wessels, M.R., and Soriani, M. (2009). CsrRS regulates group B *Streptococcus* virulence gene expression in response to environmental pH: a new perspective on vaccine development. J. Bacteriol. *191*, 5387–5397.

Senadheera, D., Krastel, K., Mair, R., Persadmehr, A., Abranches, J., Burne, R.A., and Cvitkovitch, D.G. (2009). Inactivation of VicK affects acid production and acid survival of *Streptococcus mutans*. J. Bacteriol. *191*, 6415–6424.

Shelburne, S.A., 3rd, Sumby, P., Sitkiewicz, I., Granville, C., DeLeo, F.R., and Musser, J.M. (2005). Central role of a bacterial two-component gene regulatory system of previously unknown function in pathogen persistence in human saliva. Proc. Natl. Acad. Sci. U.S.A. *102*, 16037–16042.

Shelburne, S.A., 3rd, Keith, D., Horstmann, N., Sumby, P., Davenport, M.T., Graviss, E.A., Brennan, R.G., and Musser, J.M. (2008a). A direct link between carbohydrate utilization and virulence in the major human pathogen group A *Streptococcus*. Proc. Natl. Acad. Sci. U.S.A. *105*, 1698–1703.

Shelburne, S.A., Davenport, M.T., Keith, D.B., and Musser, J.M. (2008b). The role of complex carbohydrate catabolism in the pathogenesis of invasive streptococci. Trends Microbiol. *16*, 318–325.

Sheng, J., and Marquis, R.E. (2006). Enhanced acid resistance of oral streptococci at lethal pH values associated with acid-tolerant catabolism and with ATP synthase activity. FEMS Microbiol. Lett. *262*, 93–98.

Sheng, J., and Marquis, R.E. (2007). Malolactic fermentation by *Streptococcus mutans*. FEMS Microbiol. Lett. *272*, 196–201.

Sheng, J., Baldeck, J.D., Nguyen, P.T., Quivey, R.G., Jr., and Marquis, R.E. (2010). Alkali production associated with malolactic fermentation by oral streptococci and protection against acid, oxidative, or starvation damage. Can. J. Microbiol. *56*, 539–547.

Siller, M., Janapatla, R.P., Pirzada, Z.A., Hassler, C., Zinkl, D., and Charpentier, E. (2008). Functional analysis of the group A *streptococcal luxS*/AI-2 system in metabolism, adaptation to stress and interaction with host cells. BMC Microbiol. *8*, 188.

Simmen, H.P., and Blaser, J. (1993). Analysis of pH and pO2 in abscesses, peritoneal fluid, and drainage fluid in the presence or absence of bacterial infection during and after abdominal surgery. Am. J. Surg. *166*, 24–27.

Smoot, L.M., Smoot, J.C., Graham, M.R., Somerville, G.A., Sturdevant, D.E., Migliaccio, C.A., Sylva, G.L., and Musser, J.M. (2001). Global differential gene expression in response to growth temperature alteration in group A *Streptococcus*. Proc. Natl. Acad. Sci. U.S.A. *98*, 10416–10421.

Sonenshein, A.L. (2005). CodY, a global regulator of stationary phase and virulence in Gram-positive bacteria. Curr. Opin. Microbiol. *8*, 203–207.

Sturr, M.G., and Marquis, R.E. (1992). Comparative acid tolerances and inhibitor sensitivities of isolated F-ATPases of oral lactic acid bacteria. Appl. Environ. Microbiol. *58*, 2287–2291.

Tettelin, H., Nelson, K.E., Paulsen, I.T., Eisen, J.A., Read, T.D., Peterson, S., Heidelberg, J., DeBoy, R.T., Haft, D.H., Dodson, R.J., *et al.* (2001). Complete genome sequence of a virulent isolate of *Streptococcus pneumoniae*. Science *293*, 498–506.

Tettelin, H., Masignani, V., Cieslewicz, M.J., Donati, C., Medini, D., Ward, N.L., Angiuoli, S.V., Crabtree, J., Jones, A.L., Durkin, A.S., *et al.* (2005). Genome analysis of multiple pathogenic isolates of *Streptococcus agalactiae*: implications for the microbial 'pan-genome'. Proc. Natl. Acad. Sci. U.S.A. *102*, 13950–13955.

Throup, J.P., Koretke, K.K., Bryant, A.P., Ingraham, K.A., Chalker, A.F., Ge, Y., Marra, A., Wallis, N.G., Brown, J.R., Holmes, D.J., *et al.* (2000). A genomic analysis of two-component signal transduction in *Streptococcus pneumoniae*. Mol. Microbiol. *35*, 566–576.

Tong, H., Chen, W., Merritt, J., Qi, F., Shi, W., and Dong, X. (2007). *Streptococcus oligofermentans* inhibits *Streptococcus mutans* through conversion of lactic acid into inhibitory H_2O_2: a possible counteroffensive strategy for interspecies competition. Mol. Microbiol. *63*, 872–880.

Trappetti, C., Potter, A.J., Paton, A.W., Oggioni, M.R., and Paton, J.C. (2011). LuxS Mediates Iron-Dependent Biofilm Formation, Competence, and Fratricide in *Streptococcus pneumoniae*. Infect. Immun. *79*, 4550–4558.

Tseng, H.J., McEwan, A.G., Paton, J.C., and Jennings, M.P. (2002). Virulence of *Streptococcus pneumoniae*: PsaA mutants are hypersensitive to oxidative stress. Infect. Immun. *70*, 1635–1639.

Tsou, C.C., Chiang-Ni, C., Lin, Y.S., Chuang, W.J., Lin, M.T., Liu, C.C., and Wu, J.J. (2008). An iron-binding protein, Dpr, decreases hydrogen peroxide stress and protects *Streptococcus pyogenes* against multiple stresses. Infect. Immun. *76*, 4038–4045.

Tu le, N., Jeong, H.Y., Kwon, H.Y., Ogunniyi, A.D., Paton, J.C., Pyo, S.N., and Rhee, D.K. (2007). Modulation of adherence, invasion, and tumor necrosis factor alpha secretion during the early stages of infection by *Streptococcus pneumoniae* ClpL. Infect. Immun. *75*, 2996–3005.

Turlan, C., Prudhomme, M., Fichant, G., Martin, B., and Gutierrez, C. (2009). SpxA1, a novel transcriptional regulator involved in X-state (competence) development in *Streptococcus pneumoniae*. Mol. Microbiol. *73*, 492–506.

Virtaneva, K., Graham, M.R., Porcella, S.F., Hoe, N.P., Su, H., Graviss, E.A., Gardner, T.J., Allison, J.E., Lemon, W.J., Bailey, J.R., *et al.* (2003). Group A *Streptococcus* gene expression in humans and cynomolgus macaques with acute pharyngitis. Infect. Immun. *71*, 2199–2207.

Virtaneva, K., Porcella, S.F., Graham, M.R., Ireland, R.M., Johnson, C.A., Ricklefs, S.M., Babar, I., Parkins, L.D., Romero, R.A., Corn, G.J., *et al.* (2005). Longitudinal analysis of the group A Streptococcus transcriptome in experimental pharyngitis in cynomolgus macaques. Proc. Natl. Acad. Sci. U.S.A. *102*, 9014–9019.

Welin, J., Wilkins, J.C., Beighton, D., Wrzesinski, K., Fey, S.J., Mose-Larsen, P., Hamilton, I.R., and Svensater, G. (2003). Effect of acid shock on protein expression by biofilm cells of *Streptococcus mutans*. FEMS Microbiol. Lett. *227*, 287–293.

Wen, Z.T., and Burne, R.A. (2004). LuxS-mediated signaling in *Streptococcus mutans* is involved in regulation of acid and oxidative stress tolerance and biofilm formation. J. Bacteriol. *186*, 2682–2691.

WHO (2007). Pneumococcal conjugate vaccine for childhood immunization – World Health Organization position paper. Wkly. Epidemiol. Rec. *82*, 93–104.

Wilkins, J.C., Homer, K.A., and Beighton, D. (2002). Analysis of *Streptococcus mutans* proteins modulated by culture under acidic conditions. Appl. Environ. Microbiol. *68*, 2382–2390.

Woodbury, R., and Haldenwang, W.G. (2003). HrcA is a negative regulator of the *dnaK* and *groESL* operons of *Streptococcus pyogenes*. Biochem. Biophys. Res. Commun. *302*, 722–727.

Xue, X., Tomasch, J., Sztajer, H., and Wagner-Dobler, I. (2010). The delta subunit of RNA polymerase, RpoE, is a global modulator of *Streptococcus mutans* environmental adaptation. J. Bacteriol. *192*, 5081–5092.

Yamamoto, Y., Higuchi, M., Poole, L.B., and Kamio, Y. (2000). Role of the dpr product in oxygen tolerance in *Streptococcus mutans*. J. Bacteriol. *182*, 3740–3747.

Yamamoto, Y., Poyart, C., Trieu-Cuot, P., Lamberet, G., Gruss, A., and Gaudu, P. (2005). Respiration metabolism of Group B *Streptococcus* is activated by environmental haem and quinone and contributes to virulence. Mol. Microbiol. *56*, 525–534.

Yamamoto, Y., Pargade, V., Lamberet, G., Gaudu, P., Thomas, F., Texereau, J., Gruss, A., Trieu-Cuot, P., and Poyart, C. (2006). The Group B *Streptococcus* NADH oxidase Nox-2 is involved in fatty acid biosynthesis during aerobic growth and contributes to virulence. Mol. Microbiol. *62*, 772–785.

Yesilkaya, H., Kadioglu, A., Gingles, N., Alexander, J.E., Mitchell, T.J., and Andrew, P.W. (2000). Role of manganese-containing superoxide dismutase in oxidative stress and virulence of *Streptococcus pneumoniae*. Infect. Immun. *68*, 2819–2826.

Yu, J., Bryant, A.P., Marra, A., Lonetto, M.A., Ingraham, K.A., Chalker, A.F., Holmes, D.J., Holden, D., Rosenberg, M., and McDevitt, D. (2001). Characterization of the *Streptococcus pneumoniae* NADH oxidase that is required for infection. Microbiology *147*, 431–438.

Zeng, L., and Burne, R.A. (2008). Multiple sugar: phosphotransferase system permeases participate in catabolite modification of gene expression in *Streptococcus mutans*. Mol. Microbiol. *70*, 197–208.

Zeng, L., and Burne, R.A. (2010). Seryl-phosphorylated HPr Regulates CcpA-Independent Carbon Catabolite Repression in Conjunction with PTS Permeases in *Streptococcus mutans*. Mol. Microbiol. 75,1145–1158.

Zhang, J., and Biswas, I. (2009). 3′-Phosphoadenosine-5′-phosphate phosphatase activity is required for superoxide stress tolerance in *Streptococcus mutans*. J. Bacteriol. *191*, 4330–4340.

Zhang, Q., Xu, S.X., Wang, H., Xu, W.C., Zhang, X.M., Wu, K.F., Liu, L., and Yin, Y.B. (2009). Contribution of ClpE to virulence of *Streptococcus pneumoniae*. Can. J. Microbiol. *55*, 1187–1194.

Zhu, L., and Kreth, J. (2010). Role of *Streptococcus mutans* eukaryotic-type serine/threonine protein kinase in interspecies interactions with *Streptococcus sanguinis*. Arch. Oral Biol. *55*, 385–390.

Zuber, P. (2004). Spx-RNA polymerase interaction and global transcriptional control during oxidative stress. J. Bacteriol. *186*, 1911–1918.

Zuber, U., and Schumann, W. (1994). CIRCE, a novel heat shock element involved in regulation of heat shock operon *dnaK* of *Bacillus subtilis*. J. Bacteriol. *176*, 1359–1363.

Oxidative and Nitrosative Stress Responses in Pathogenic *Neisseria*

3

Isabel Delany and Kate L. Seib

Abstract

Mechanisms to sense, avoid and scavenge oxidants as well as repair damaged biomolecules are important survival and virulence factors of the obligate human pathogens *Neisseria meningitidis* and *Neisseria gonorrhoeae*. These bacteria are routinely exposed to several forms of oxidative and nitrosative stress during colonization and interaction with the host, of which superoxide, hydrogen peroxide and nitric oxide are some of the key oxidants that result in damage to the bacteria. However, the pathogenic *Neisseria* express an array of defence mechanisms to combat oxidative and nitrosative stress, such as catalase, superoxide dismutase, nitric oxide reductase, as well as thiol-based defences and proteins involved in metal homeostasis and repair of damage to DNA and proteins. The expression of these defences is tightly regulated by a series of transcription factors containing redox-sensitive active sites, including OxyR, Fur, PerR/Zur, FNR, MseR, LexA NsrR, NmlR, which sense and maintain the redox homeostasis of the cell.

Introduction

The pathogenic *Neisseria*, *N. meningitidis* and *N. gonorrhoeae*, are the causative agents of meningococcal disease and gonorrhoea, respectively. These bacteria encounter oxidative stress during respiration and interactions with host phagocytic cells and competing microflora, resulting in damage to DNA, lipids and proteins. Furthermore, they are exposed to nitrosative stress from nitric oxide (NO) produced from host immune cells as an early line of defence against pathogenic bacteria. The pathogenic *Neisseria* have evolved to be highly adapted to their host environment, and contain oxidative stress defences that are tightly regulated to enable rapid and coordinated responses to stress signals. The mechanisms used to avoid and remove oxidants play an important role in the survival and virulence of human pathogens (Wilks *et al.*, 1998; Yesilkaya *et al.*, 2000; Dhandayuthapani *et al.*, 2001; Sansone *et al.*, 2002; Hwang *et al.*, 2002; Harris *et al.*, 2003; Brenot *et al.*, 2004; Halsey *et al.*, 2004; Piacenza *et al.*, 2009; Roop *et al.*, 2009) and are increasingly been noted as important survival and virulence factors of obligate human pathogens, such as *N. meningitidis* and *N. gonorrhoeae*.

N. meningitidis and *N. gonorrhoeae* are highly adapted to their sole host, the human, and are typically able to persist without causing severe disease. Only rarely do the pathogenic *Neisseria* penetrate into deeper tissues or the blood stream and cause invasive disease. There are many similarities in the oxidative stress responses of *N. meningitidis* and *N. gonorrhoeae*, which are closely related organisms with high similarity with respect to their DNA sequence, with 80–100% DNA identity (Guibourdenche *et al.*, 1986) and conservation of the majority of genes and synteny (Klee *et al.*, 2000; Dillard and Seifert, 2001). However, in spite of this similarity, *N. meningitidis* and *N. gonorrhoeae* inhabit different ecological niches and cause distinct diseases.

N. meningitidis (also known as the meningococcus) is carried asymptomatically in the mucosa of the nasopharynx in 5–30% of healthy individuals (Cartwright, 1995; Maiden and Stuart, 2002). Carriage of *N. meningitidis* typically does not elicit an overt inflammatory response (DeVoe, 1982). However, in a small number of cases and for reasons not entirely understood, infection may lead to life threatening meningococcal meningitis and sepsis and there is an estimated 1.2 million cases of meningococcal disease worldwide each year and 135,000 deaths (WHO, 1998; CDNA, 2001). The majority of disease is seen in the developing world, most notably in a region south of the Saharan desert known as the 'meningitis belt', where large epidemics of meningococcal disease occur every 5–12 years with disease rates as high as 10–1000 per 100,000. Sporadic/endemic disease is the typical pattern of disease in the developed world with a rate of < 1–20 cases per 100,000 (WHO, 1998).

Disease caused by *N. gonorrhoeae* is a significant health problem worldwide, with an estimated global incidence of 62 million cases a year, and an average of 22 million cases at any given time, most occurring in South and South-East Asia, sub-Saharan Africa and South and Central America (Gerbase *et al.*, 1998). *N. gonorrhoeae* is typically characterized by a symptomatic localized inflammatory response of the urethra in men (urethritis) and the endocervix in women (endocervicitis). However, ascending reproductive tract infections in women can lead to complications such as pelvic inflammatory disease (PID; inflammation of the uterus, fallopian tubes, ovaries or other pelvic structures) in 10–40% of infected women. Sequelae of PID include infertility, ectopic pregnancy, foetal wastage, neonatal blindness and disseminated gonococcal infections (DGI)(Hook and Holmes, 1985). Infection with *N. gonorrhoeae* also significantly increases the risk of infection with HIV, with increased susceptibility to infection in HIV negative patients and increased transmission by HIV-positive patients (Cohen *et al.*, 1997; Levine *et al.*, 1998).

Within the host, pathogenic bacteria such as *N. meningitidis* and *N. gonorrhoeae* are exposed to oxidants that are generated in three main ways: (1) as an unavoidable by-product of their own metabolic processes, (2) as an active antimicrobial defence mechanism and (3) as a result of exposure to other factors within the host environment, such as metals, or commensal organisms that generate oxidants, or reduction-oxidation (redox) active drugs. Oxidants (or oxidizing agents) are molecules that are capable of removing an electron from other molecules. The most commonly found oxidants in biological systems are the reactive oxygen species (ROS), which include the superoxide anion ($O_2^{\cdot-}$), hydrogen peroxide (H_2O_2) and the hydroxyl radical ($HO^{\cdot}$) (see Table 3.1 for properties). In addition, several reactive nitrogen species (RNS), including nitric oxide ($NO^{\cdot}$) and peroxynitrite ($ONOO^-$), as well as reactive sulfur and chloride agents are found in biological systems and are described in Table 3.1. The major reactions by which these molecules are formed and removed within biological systems are shown in Table 3.2. Oxidants exhibit a broad spectrum of toxic effects in biological systems, with particular species having different reactivity with DNA, proteins and membranes (reviewed by Storz and Imlay, 1999). This chapter details the main forms of oxidative stress encountered by the pathogenic *Neisseria*, the defence mechanisms used to avoid and repair oxidative damage, as well as the ways in which they sense and respond to oxidative stress.

Sources of oxidative stress encountered by *N. meningitidis* and *N. gonorrhoeae*

Endogenous sources of oxidative stress generated from bacterial respiration

Organisms oxidize a variety of substrates to generate the energy used for metabolism and the majority of bacteria contain several electron transport pathways that enable them to grow in a range of environments under aerobic and anaerobic conditions. For example, the respiratory chain of *Escherichia coli* consists of 15 primary dehydrogenases, 10 terminal reductases and three different quinones (Unden and Bongaerts, 1997). However, the pathogenic *Neisseria* are restricted to the human host and have less respiratory pathways. *N. gonorrhoeae* and *N meningitidis* have highly related respiratory chains that allow them to co-metabolize oxygen and nitrite as electron acceptors under microaerobic conditions and as such are adapted to similar growth environments at distinct ecological sites (i.e.

Table 3.1 Properties of biologically relevant oxidants

Oxidant		Properties
Superoxide	$O_2^{\cdot -}$	Moderately reactive oxidant and reductant; inactivates iron/sulfur proteins and releases iron enabling hydroxyl radical production by the Fenton reaction; pH-dependent membrane permeability
Hydrogen peroxide	H_2O_2	Moderately reactive oxidizing agent; reacts slowly with reducing agents such as thiols; reacts with reduced iron and copper to generate hydroxyl radicals via the Fenton reaction; membrane permeable
Hydroxyl radical	$HO^{\cdot}$	Extremely reactive oxidizing agent with resulting low diffusibility and short-range action; causes DNA modification and strand breaks, enzyme inactivation, lipid peroxidation; formed by the Fenton reaction
Singlet oxygen	$^1\Delta gO_2/^1\Sigma^+gO_2$	Electronically excited state of oxygen with unpaired electrons in opposing spins, eliminating the spin restriction of the ground-state O_2; reacts with a number of biological molecules
Nitric oxide	$NO^{\cdot}$	Gaseous free radical; reacts very rapidly with superoxide to give peroxynitrite; forms complexes with haem proteins; inactivates iron/sulfur centres; forms nitrosothiols
Peroxynitrite	$ONOO^-$	Unstable short-lived strong oxidant with properties similar to hydroxyl radical; hydroxylates and nitrates compounds; reacts rapidly with thiols: breaks down to nitrate; produced by reaction of nitric oxide and superoxide
Hypochlorous acid	$HOCl/^-OCl$	Strong oxidant; reacts with thiols and thioethers; converts amines to chloramines; chlorinates phenols and unsaturated bonds; oxidizes iron centres; cross-links proteins; in the presence of metal ions and superoxide can produce hydroxyl radicals; membrane permeable
Chloramines	ClN	Milder and longer-lived oxidants than HOCl; react with thiols, thioethers, iron centres; membrane permeability depends on polarity
Thiyl radical	$RS^{\cdot}$	Results from the oxidation of glutathione (GSH)
Nitrosothiol	RSNO	Results from reaction of GSH with $NO^{\cdot}$
Carbon-centred radical	$R^{\cdot}$	Lipid, nucleic acid, protein or carbohydrate formed from the attack of an oxidizing radical; can react with oxygen to form peroxyl radicals ($ROO^{\cdot}$), which in turn participate in reactions that form alkoxyl radicals ($RO^{\cdot}$)

Table adapted from Hampton (1998).

urogenital vs. oropharyngeal mucosal sites). They are oxidase-positive bacteria that contain a mitochondrial-like respiratory chain (Bovre, 1984), with a single cytochrome oxidase of the cbb_3-type (CcoNOQP) (Seib *et al.*, 2006; Aspholm *et al.*, 2010). The cbb_3-type oxidases are characterized by their high oxygen affinity and their ability to pump protons (Pitcher *et al.*, 2002). The pathogenic *Neisseria* spp. encode multiple small *c*-type cytochromes which are thought to be involved in the passage of electrons from the cytochrome bc_1 complex to the cbb_3-type oxidase as part of a multiply branched respiratory chain (Deeudom *et al.*, 2008; Li *et al.*, 2010). Deletion of either the *c*4 or *c*5 small cytochrome in gonococcus, and *cx* or *c*4 in meningococcus, results in reduced rates of respiration and increased sensitivity to high levels of aeration (Deeudom *et al.*, 2008; Li *et al.*, 2010) indicating that the cytochromes are enabling the *Neisseria* spp. to utilize oxygen and grow aerobically. *Neisseria* have a highly active cytochrome oxidase (Jurtshuk and Milligan, 1974) and high levels of respiration (Archibald and Duong, 1986) indicating the potential for generation of significant endogenous oxidative stress. It has also been hypothesized that the high respiratory capacity and excess capacity for oxygen reduction of *Neisseria* with microaerobic lifestyles, prevents the accumulation of oxygen and acts as a defence against ROS generated as by-products of their own metabolism (Li *et al.*, 2010). Expression of the *ccoNOQP* in *N. meningitidis,* as well as the *c*4 and *c*5 cytochrome-encoding genes, has been shown to be regulated by the oxygen-responsive FNR (fumarate and nitrate reductase regulator) (Bartolini *et al.*, 2006; Fantappie *et al.*, 2011)

Table 3.2 Chemical reactions involving the production and removal of oxidants in biological systems

	Reaction	Comment
Generation of oxidants		
1	$O_2 + e^- \rightarrow O_2^{\cdot -}$	Superoxide generation
2	$Fe^{2+} + O_2 \rightarrow Fe^{3+} + O_2^{\cdot -}$ $Cu^{1+} + O_2 \rightarrow Cu^{2+} + O_2^{\cdot -}$	Superoxide generation (autoxidation of transition metals)
3	$O_2 + 2e^- + 2H^+ \rightarrow H_2O_2$	Hydrogen peroxide generation
4	$2O_2^{\cdot -} + 2H^+ \rightarrow H_2O_2 + O_2$	Superoxide dismutation
5	$H_2O_2 + e^- + H^+ \rightarrow H_2O + {}^{\cdot}OH$	Hydroxyl radical generation
6	$O_2^{\cdot -} + HOCl \rightarrow {}^{\cdot}OH + Cl^- + O_2$	–
7	$H_2O_2 + O_2^{\cdot -} \rightarrow O_2 + {}^{\cdot}OH + OH^-$	Haber–Weiss reaction
8	$H_2O_2 + Fe^{2+} \rightarrow {}^{\cdot}OH + OH^- + Fe^{3+}$	Fenton reaction
9	$H_2O_2 + Cl^- \rightarrow HOCl + {}^{\cdot}OH$	Hypochlorous acid generation
10	$O_2^{\cdot -} + H^+ \rightarrow HO_2^{\cdot}$	Hydroperoxyl radical–superoxide interconversion
11	L-Arginine → L-citrulline + $NO^{\cdot}$	Nitric oxide generation
12	$O_2^{\cdot}- + NO^{\cdot} \rightarrow ONOO-$	Peroxynitrite generation
13	LH PUFA + $R^{\cdot}$ radical → $L^{\cdot}$ FA radical + RH $L^{\cdot} + O_2 \rightarrow LOO^{\cdot}$; $LOO^{\cdot} + L'H \rightarrow LOOH + L'^{\cdot}$ $LOOH \rightarrow LO^{\cdot}$, $LOO^{\cdot}$, aldehydes	Lipid peroxidation chain reaction (FA, fatty acid; PUFA, polyunsaturated FA; L, lipid, protein or DNA)
14	$[4Fe–4S]^{2+} \rightarrow [4Fe–4S]^{3+} \rightarrow [3Fe–4S]^{+} + Fe^{2+}$	Oxidation of iron–sulfur centre
Removal of oxidants		
15	$2O_2^{\cdot -} + 2H^+ \rightarrow H_2O_2 + O_2$	Superoxide dismutase (SOD)
16	$e- + O_2^{\cdot -} + 2H^+ \rightarrow H_2O_2$	Superoxide reductase (SOR)
17	$2H_2O_2 \rightarrow 2H_2O + O_2$	Catalase
18	$H_2O_2 + XH_2 \rightarrow 2H_2O + X$ $2 R'\text{-}SH + ROOH \rightarrow R'\text{-}S\text{-}S\text{-}R' + H_2O + ROH$	Peroxidase Peroxiredoxin
19	$2GSH + H_2O_2 \rightarrow G\text{-}S\text{-}S\text{-}G + 2H_2O$	Glutathione peroxidase (Gpx)
20	G-S-S-G + 2NADPH → 2GSH + 2NADP	Glutathione reductase (Gor)
21	$O_2^{\cdot -} + 2H^+ + Mn(II) \rightarrow H_2O_2 + O_2$ $O_2^{\cdot -} + Mn(III) \rightarrow O_2 + Mn(II)$	Mn quenching of superoxide
	$2O_2^{\cdot -} + 2H^+ \rightarrow H_2O_2 + O_2$	–
22	$2O_2^{\cdot -} + 4H^+ + 2Mn(II) \rightarrow 2H_2O_2 + 2Mn(III)$ $2H_2O_2 + 2Mn(III) \rightarrow O_2 + 2H^+ + 2Mn(II)$	Mn quenching of superoxide
	$2O_2 + 2H^+ \rightarrow H_2O_2 + O_2$	–
23	$2H_2O_2 + 2Mn(III) \rightarrow O_2 + 2H^+ + 2Mn(II)$ $H_2O_2 + Mn(II) \rightarrow O_2 + {}^{\cdot}OH + OH^- + Mn(III)$ ${}^{\cdot}OH + Mn(II) \rightarrow OH– + Mn(III)$	Mn quenching of hydrogen peroxide
	$2H_2O_2 \rightarrow 2H_2O + O_2$	–
24	$H_2O_2 + Mn(II) \rightarrow O_2 + {}^{\cdot}OH + OH^- + Mn(III)$ ${}^{\cdot}OH + H_2O_2 \rightarrow O_2^{\cdot -} + H^+ + H_2O$ $O_2^{\cdot -} + Mn(III) \rightarrow O_2 + Mn(II)$	Mn quenching of hydrogen peroxide
	$2H_2O_2 \rightarrow 2H_2O + O_2$	–

regulator, as seen in other organisms (Cosseau and Batut, 2004).

Several microenvironments inhabited by *N. gonorrhoeae* and *N. meningitidis* are microaerobic or anaerobic; mucosal surfaces, underlying epithelial cells, inflammatory cells, exudates and impaired tissues are all low oxygen environments. Human mucous membranes have a low Po_2, e.g. 0.4 kPa for the buccal folds, presumably less for the genital mucosae (described in Archibald and Duong, 1986), and 0.13 kPa for the vagina (Rashad *et al.*, 1992). *N. meningitidis* would be exposed to highly divergent partial pressures of oxygen as they move form the surface of the upper respiratory tract (21 kPa), to mucus membranes (0.4 kPa for buccal folds), blood (4 kPa central venous) and cerebrospinal fluid (CSF; 5 kPa) (cited in Archibald and Duong, 1986). *N. gonorrhoeae* is commonly isolated in the presence of strict anaerobes (cited in Short *et al.*, 1982). While the ability to grow anaerobically confers adaptive advantages, reactive intermediates are produced during anaerobic respiration.

N. gonorrhoeae and *N. meningitidis* use nitrite to support anaerobic growth (Knapp and Clark, 1984; Clark *et al.*, 1987). The pathogenic *Neisseria* are minimal denitrifiers that can use only the two central reactions of the denitrification pathway (Berks *et al.*, 1995; Zumft, 1997), converting nitrite to nitrous oxide catalysed by nitrite reductase (AniA) and nitric oxide reductase (NorB). Nitrite is present in biological fluids such as urine, blood and saliva (Knapp and Clark, 1984) and patients with meningococcal meningitis have elevated levels of nitrite and nitrate in the CSF (presumably from NO synthesized by the host in response to bacterial invasion) (Visser *et al.*, 1994). In addition, bacteria that can reduce nitrate to nitrite, including *E. coli* and *Lactobacillus* species (Knapp and Clark, 1984) are present in sites infected by pathogenic *Neisseria*.

Exogenous host-mediated sources of oxidative stress

Professional phagocytes including neutrophils, eosinophils, monocytes and macrophages use oxygen-dependent killing mechanisms as well as NO to control pathogenic organisms (Klebanoff, 1999; Bogdan, 2001). Bacterial infections involving the mucosal surfaces of the urogenital, respiratory and gastrointestinal tracts are characterized by the transepithelial migration of polymorphonuclear neutrophils (PMNs) (reviewed in Burg and Pillinger, 2001; Liu *et al.*, 2004). The purulent exudate characteristic of gonorrhoea predominantly contains PMNs (Archibald and Duong, 1986). Similarly, accumulation of PMNs in the subarachnoidal space (Leib and Tauber, 1999) and in purpuric lesions (Sotto *et al.*, 1976) is characteristic of meningococcal meningitis and acute meningococcemia, respectively.

N. gonorrhoeae is able to survive and replicate within PMNs (reviewed by Rest and Shafer, 1989). In light of this resistance to PMN killing, the relative significance of the oxygen-dependent and oxygen-independent antimicrobial mechanisms of PMNs during their interactions with the pathogenic *Neisseria* has been questioned. *N. gonorrhoeae* (Virji and Heckels, 1986; Naids and Rest, 1991; Belland *et al.*, 1992) and *N. meningitidis* (DeLeo *et al.*, 1999) stimulate the PMN respiratory burst. However several studies have indicated that oxygen-dependent mechanisms are not involved in the killing of *N. gonorrhoeae* (Rest *et al.*, 1982; Britigan *et al.*, 1985; Britigan and Cohen, 1986; Casey *et al.*, 1986; Britigan *et al.*, 1988; Fu *et al.*, 1989; Seib *et al.*, 2005). It has recently been shown that PMNs kill *N. gonorrhoeae* independently of ROS production and live gonococci suppress the PMN oxidative burst (Criss and Seifert, 2008; Criss *et al.*, 2009). *N. gonorrhoeae* are also able to replicate within the phagosomes of PMNs and modulate apoptosis, which may be a virulence strategy used by *N. gonorrhoeae* to preserve an intracellular niche (Simons *et al.*, 2005).

Other host cells also generate oxidants, including NO generated by epithelial and endothelial cells via the activity of constitutive endothelial and neuronal nitric oxide synthases (eNOS and nNOS, respectively) and the inducible nitric oxide synthase (iNOS) (Bogdan, 2001). NO can react with superoxide to form peroxynitrite ($ONOO^-$; Table 3.2, reaction 12), which is much more reactive than its parent molecules (Beckman and Koppenol, 1996). The human nasopharynx contains high levels of NO, and the major NO sources are believed to be epithelial cells (iNOS/eNOS) and macrophages (iNOS) (Gerlach *et al.*, 1994; Lundberg and Weitzberg, 1999; Busch *et al.*, 2000). NO synthesis is induced in cells within the central nervous system compartment in response to *N. meningitidis* infection (Visser *et al.*, 1994;

Constantin *et al.*, 2002). Cervical endothelial and epithelial cells also produce NO (Tschugguel *et al.*, 1999; Edwards, 2010). However, it has been shown that NO promotes the intracellular survival of *N. gonorrhoeae* during *ex vivo* infection of human primary cervical epithelial (Pex) cells (Edwards, 2010).

Microfloral environment inhabited by *N. meningitidis*

The upper respiratory tract, with approximately 10,000–15,000 l of air inhaled each day, is exposed to a variety of microorganisms and antigenic substances and a high oxygen concentration. Bacterial flora of the upper respiratory tract frequently consists of aerobic and anaerobic organisms including streptococci (*S. pneumoniae, S. pyogenes, S. mitis, S. oralis, S. salivalius*), staphylococci (*S. epidermidis, S. aureus*), diphtheroids, *Corynebacterium* and *Haemophilus influenzae* (Proctor and Andersen, 1982; Tanaka *et al.*, 1996). *Mycoplasma pneumoniae* produces superoxide (Lynch and Cole, 1980; Almagor *et al.*, 1984) and *S. pneumoniae* produces H_2O_2 (Duane *et al.*, 1993; Spellerberg *et al.*, 1996). Nitric oxide synthase, is also found in several prokaryotes including *Bacillus subtilis* (Adak *et al.*, 2002a), *Deinococcus radiodurans* (Adak *et al.*, 2002b) and a *Nocardia* species (Chen and Rosazza, 1995). Therefore, *Neisseria* may also be exposed to bacterially produced exogenous NO.

Microfloral environment inhabited by *N. gonorrhoeae*

The female genital tract contains between 10^8 and 10^9 bacteria per gram of fluid, composed of aerobic (e.g. lactobacilli, streptococci, *Staphylococcus epidermidis, E. coli*) and anaerobic (e.g. lactobacilli, peptococci and *Bacteroides* spp.) organisms (reviewed in Holmes *et al.*, 1999; Larsen and Monif, 2001). Of key interest is the presence of lactobacilli, which greatly affect the female genital tract environment and are believed to control the vaginal microflora, and have been proposed to inhibit *N. gonorrhoeae* growth by production of H_2O_2, bacteriocins and organic acids (e.g. L-lactate), which lower the vaginal pH (Hughes and Saigh *et al.*, 1978; Hillier, 1990, 1992; Zheng *et al.*, 1994; Aroutcheva *et al.*, 2001; St Amant *et al.*, 2002). *L. fermentum* also produces NO (Morita *et al.*, 1997; Xu and Verstraete, 2001). Metabolism of L-lactate (produced by lactobacilli) by *N. gonorrhoeae* also greatly enhances oxygen consumption (Britigan *et al.*, 1988). The low pH environment of the vagina (pH 4.5) influences the forms of ROS present; in slightly acidic environments superoxide is protonated to form perhydroxyl radicals (reaction 10, Table 3.2), which spontaneously dismutate to form H_2O_2. These ROS species have greater membrane permeability than superoxide and can penetrate and attack cytoplasmic targets (Korshunov and Imlay, 2002).

Streptococci (group A, *S. faecalis, S. mitis, S. sanguis, S. viridans*), staphylococci (*S. epidermidis, S. aureus*), Corynebacterium species, *Neisseria* subflava, strains of *Neisseria, Candida* and *Bifidobacterium* as well as anaerobic bacteria belonging to the genera *Bacteroides, Peptostreptococcus* and *Eubacterium* isolated from the urogenital tract all inhibit growth of *N. gonorrhoeae* (Shtibel, 1976; Kaye and Levison, 1977; Morin *et al.*, 1980; Dubreuil *et al.*, 1984, 1985; Saiki *et al.*, 1988) possibly due to ROS or redox-cycling antibiotics.

Molecular mechanisms used to counteract oxidative stress

In this part of the chapter we will review the mechanisms used by the pathogenic *Neisseria* to avoid and scavenge oxidants as well as to repair damaged biomolecules (Table 3.3 and Fig. 3.1). Brief comments regarding the regulation of these oxidative stress defences are included, but are dealt with in detail in the following section and in Table 3.4.

Superoxide dismutase

SOD catalyses the disproportionation of $O_2^{\cdot-}$ to H_2O_2 and H_2O (Table 3.2, reaction 15) (McCord and Fridovich, 1969; Fridovich, 1995), and three main classes of SOD exist, SodA (cytoplasmic, Mn-cofactored), SodB (cytoplasmic, Fe-cofactored) and SodC (periplasmic, Cu/Zn-cofactored) (Storz and Imlay, 1999). At least one class of SOD is present in most aerotolerant organisms but SOD is often absent in obligate anaerobes (McCord *et al.*, 1971). *E. coli* contains all three of the main types of SOD, and the importance of these enzymes is indicated by the fact that *E. coli* grown in rich medium have

Table 3.3 Oxidative and nitrosative stress defence mechanisms of *Neisseria gonorrhoeae* and *Neisseria meningitidis*

Name	Function	Gene ID[a]	Sensitivity of mutant strain[b]	Regulation[c]	Reference
Oxidative stress defences					
SodB	Superoxide dismutase B	NMB0884 NGO0405	*x* PQ, X/XO (*Ng*) *S* PQ (*Nm*)	Fur/high Fe, H_2O_2 Mn Hfq	Norrod and Morse (1979), Archibald and Duong (1986), Tseng *et al.* (2001), Sebastian *et al.* (2002), Seib *et al.* (2004), Stohl *et al.* (2005), Fantappie *et al.* (2009); Metruccio *et al.* (2009), Pannekoek *et al.* (2009), Mellin *et al.* (2010), Wu *et al.* (2010)
SodC	Superoxide dismutase C	NMB1398	*S* X/XO, Macrophage	Hfq	Kroll *et al.* (1995), Wilks *et al.* (1998), Seib *et al.*, (2004), Pannekoek *et al.* (2009), Fantappie *et al.* (2011)
KatA	Catalase	NMB0216 NGO1767	*S* PQ, X/XO, H_2O_2	OxyR, H_2O_2 Fur/high Fe	Norrod and Morse (1979), Archibald and Duong (1986), Hassett *et al.* (1990), Zheng *et al.* (1992), Johnson *et al.* (1993, 1996), Grifantini *et al.* (2003), Tseng *et al.* (2003), Grifantini *et al.* (2004), Seib *et al.* (2004, 2007), Soler-Garcia and Jerse (2004), Stohl *et al.* (2005), Ieva *et al.* (2008)
MntABC/Mn	ABC type Mn transporter	NMB0586–88 NGO0168–70	*S* PQ, H_2O_2, Pex, Biofilm	PerR/Zur, Mn	Tseng *et al.* (2001), Seib *et al.* (2004), Wu *et al.* (2006), Lim *et al.* (2008)
Ccp	Cytochrome c peroxidase	NGO1769	*S* H_2O_2	FNR/low O_2, OxyR, H_2O_2	Lissenden *et al.* (2000), Turner *et al.* (2003), Seib *et al.* (2004, 2007); Stohl *et al.* (2005), Falsetta *et al.* (2009)
MsrA/B	Methionine sulfoxide reductase	NMB0044 NGO2059	*S* H_2O_2, X/XO	Ecf σ factor H_2O_2 (Ng) Mn	Taha *et al.* (1991), Wizemann *et al.* (1996), Kauffmann *et al.* (2002), Lowther *et al.* (2002), Olry *et al.* (2002); Skaar *et al.* (2002), Stohl *et al.* (2005), Gunesekere *et al.* (2006), Hopman *et al.* (2010), Wu *et al.* (2010)
BfrAB	Bacterioferritin	NMB1207/6 NGO0794/5	*S* H_2O_2, PQ	MisR, MtrR Mn, O_2	Chen and Morse (1999), Tzeng *et al.* (2008), Falsetta *et al.* (2009), Isabella and Clark (2011)
Sco	Putative thiol:disulfide oxidoreductase/ peroxiredoxin	NMB1578 NGO1237	*S* PQ	ND	Seib *et al.* (2003)
GSH Glutathione	GSH synthetase (GshB) Glutamate-cysteine ligase (GshA)	NMB1559/1037 NGO1217/0680	ND	ND	Archibald and Duong (1986)
	ABC-type glutamate transporter (GltT)	NMB1965 NGO2117	*S* H_2O_2, Infant rat, PMNs	ND	Tala *et al.* (2011)
Gor	Glutathione reductase	NMB0947 NGO0925	*R* H_2O_2 *S* Pex, Biofilm *x* PQ, X/XO, CHP	OxyR, H_2O_2	Stohl *et al.* (2005), Seib *et al.* (2007)

Table 3.3 Continued

Name	Function	Gene ID[a]	Sensitivity of mutant strain[b]	Regulation[c]	Reference
Prx/Grx	Peroxiredoxin	NMB0946 NGO0926	*R* H_2O_2 *S* Biofilm *x* PQ, X/XO, CHP	OxyR, Hfq	Rouhier and Jacquot (2003), Seib *et al*. (2007), Fantappie *et al*. (2009)
Laz	Azurin	NMB1533 NGO0994	*S* H_2O_2, Cu, Pex *x* PQ, X/XO	MisR Mn	H.J. Wu *et al*. (2005), Tzeng *et al*. (2008), Wu *et al*. (2010)
PriA	DNA replication restart helicase	NMB0551 NGO1437	*S* H_2O_2, CHP, NO	ND	Kline and Seifert (2005), Tala *et al*. (2008)
RecN	DNA repair	NMB0740 NGO0318	*S* H_2O_2, PMNs *x* PQ, CHP, Diamide	H_2O_2, LexA, Fur	Sebastian *et al*. (2002), Stohl *et al*. (2005), Schook *et al*. (2011)
RecABCD RecJOQ RuvACG UvrABD	DNA repair	NA	*S* H_2O_2		Davidsen and Tonjum (2006), Stohl and Seifert (2006), LeCuyer *et al*. (2010)
NGO1686	Putative zinc metalloprotease	NMB0315 NGO1686	*S* H_2O_2, CH, PMNs	H_2O_2	Stohl *et al*. (2005)
NGO0554	*N. gonorrhoeae* specific predicted protein	NGO0554	*S* H_2O_2	H_2O_2	Stohl *et al*. (2005)
NExo	3′-phosphodiesterase exonuclease	NMB0399 NGO1561	*S* H_2O_2, PQ, Infant rat	ND	Carpenter *et al*. (2007)
Nape	apurinic or apyrimidinic site (AP) endonuclease	NMB2082 NGO1994	*S* H_2O_2, PQ, Infant rat	ND	Carpenter *et al*. (2007)
Ppa	Pyrophosphatase	NMB0641 NGO0223	*R* PQ, X/XO, H_2O_2 *S* Pex	Mn	Wu *et al*. (2009)
Ppk	Polyphosphate kinase	NMB1900 NGO0003	*R* PQ, H_2O_2 *S* Pex	Mn	Wu *et al*. (2009)
NMB1436–8	Oxidoreductase homologues	NMB1436–8 NGO0904–6	*S* H_2O_2, Mouse, PMNs	Fur, Fe	Grifantini *et al*. (2004)

Nitrosative stress defences					
CycP	Cytochrome c′ (binds NO)	NGO1080 NMB0923	*S* SNP, SNAP, Macrophage, Nasopharyngeal epithelia		Anjum *et al.* (2002), Stevanin *et al.* (2005)
DnrN	Di-iron protein (repair of iron centre (RIC) protein)	NMB1365 NGO0653	*S* H_2O_2, NO	NsrR, NO O_2	Isabella *et al.* (2008), Overton *et al.* (2006, 2008)
NorB	Nitric oxide reductase	NMB1622 NGO1275	*S* NO, GSNO, SNP, SNAP, Macrophage, Biofilm, Nasopharyngeal explants	Fur, ArsR, NsrR NO, Nitrite	Householder *et al.* (2000), Anjum *et al.* (2002), Sebastian *et al.* (2002), Grifantini *et al.* (2003), Delany *et al.* (2004), Stevanin *et al.* (2005, 2007), Rock *et al.* (2007), Falsetta *et al.* (2009)
AhpC	Alkly hydroperoxide reductase	NMB0750 NGO0328	*x* PN		Barth *et al.* (2009b)
AdhC	Alcohol dehydrogenase/GSNO reductase	NMB1304 NGO0601*	ND	NmlR	Kidd *et al.* (2005), Potter *et al.* (2007)
EstD	Carboxylic ester hydrolase	NMB1305 NGO0600	*S* GSNO, Nitrite, Pex, Biofilm *x* NO	NmlR	Kidd *et al.* (2005), Potter *et al.* (2009a)
TlpA	Thioredoxin-like protein	NMB0006 NGO1923	*S* PQ, H_2O_2, Pex	ND	Achard *et al.* (2009)
DsbA1, 2, 3/ DsbD	Thiol disulphide interchange proteins	NMB0278,0294, 0407/NMB1519 NGO1717,1548/ NGO0978	*S* DTT (DsbA1/A2) *S* PQ, H_2O_2, Cu (DsbA1/ A2/D)	MisR/S DTT	Tinsley and Nassif (1996), Sinha *et al.* (2004), Tinsley *et al.* (2004), Newcombe *et al.* (2005), Tala *et al.* (2008), Tzeng *et al.* (2008), Kumar *et al.* (2011)
TrxB	Thioredoxin reductase	NMB1324 NGO0580	*S* PQ, CHP, Diamide, Pex, Biofilm *R* H_2O_2	NmlR, diamide	Kidd *et al.* (2005), Potter *et al.* (2009b)

[a]Gene identification according to the annotation of the *N. meningitidis* MC58 (NMBxxxx) and *N. gonorrhoeae* FA1090 (NGOxxxx) genome strains (http://cmr.jcvi.org/cgi-bin/CMR/shared/Genomes.cgi). NA, not applicable. *Pseudogene.
[b]Sensitivity of mutant strains to *in vitro* oxidative stress killing assays relative to the wild-type strain: ***x***, same phenotype as wild type; ***S***, sensitive to killing; ***R***, resistant to killing (increased survival relative to wild type). These assays typically involve exposure of a suspension of 10^4–10^7 cells over 1–2 hours to either oxidative stress [paraquat (PQ), xanthine/xanthine oxidase (X/XO) or hydrogen peroxide (H_2O_2), cumene hydroperoxide (CHP)], thiol-specific oxidants [diamide, dithiothreitol (DTT)], nitrosative stress [nitric oxide (NO), peroxynitrite (PN), sodium nitroprusside (SNP), S-nitrosoglutathione (GSNO) or S-nitrosopenicillamine (SNAP)], or cells [human primary cervical epithelial cells (Pex), polymorphonuclear neutrophils (PHNs)]. For further details see the references cited in the table.
[c]Regulation of defences, either activated/up-regulated or repressed. Genes of interest under the control of regulators are also shown. For details see cited references and Table 3.4.

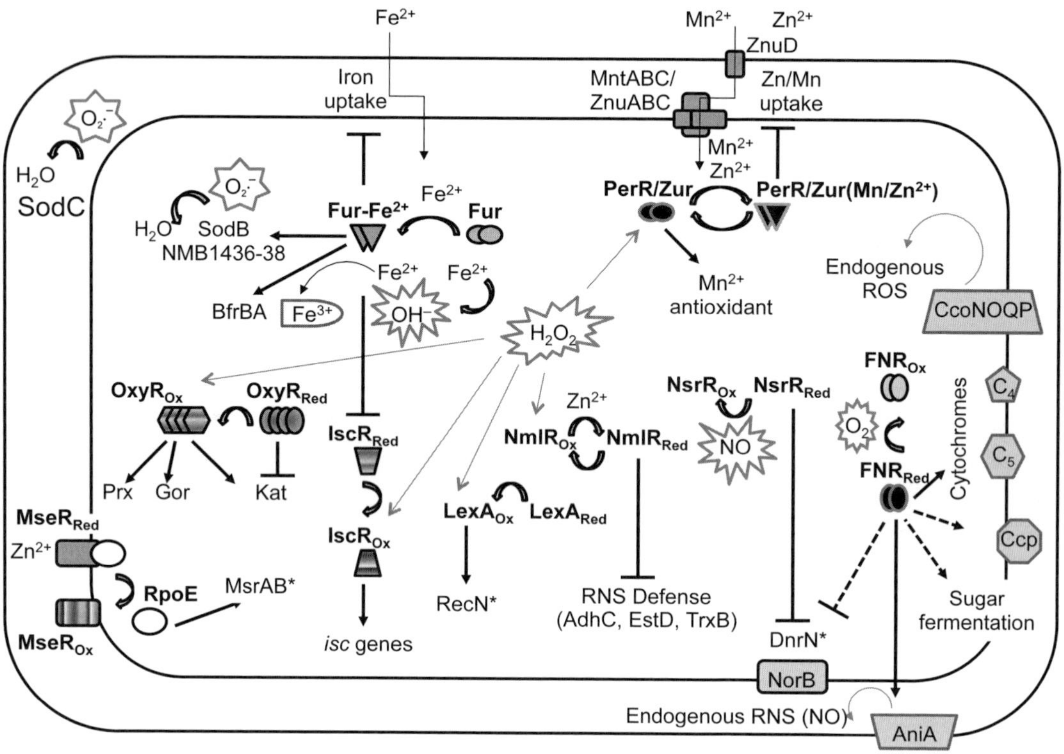

Figure 3.1 Coordinated sensing, regulation and defence against ROS and RNS in the pathogenic *Neisseria*. ROS and RNS are shown in grey stars and include superoxide ($O_2^{\cdot-}$), hydrogen peroxide (H_2O_2), hydroxyl radical (OH^-) and nitric oxide (NO) and oxygen itself (O_2). The respiratory chain cytochromes involved in the endogenous generation of ROS (CcoNOQP, C_4, C_5) and RNS [the nitrite reductase (AniA)] are also shown. Redox-sensing regulators of ROS and RNS defences are shown in bold and include FNR, PerR/Zur, OxyR, Fur, LexA, NmlR, IscR, RpoE/MseR and NsrR. These regulators are shown in their active and inactive forms (i.e. curled arrows show either the transition of redox-active cysteines from reduced (red) to oxidized (ox) forms, or the binding of a metal ion), and the stress signals that are sensed by these regulators are indicated. The activation (↓) or repression (⊥) of ROS and RNS defences by these regulators is also shown, dashed lines indicate distinct results specific to either *N. gonorrhoeae* or *N. meningitidis*. Proteins involved in ROS and RNS defences are shown and include manganese (Mn) and/or zinc (Zn) transporters (MntABC, ZnuABC, ZnuD), superoxide dismutase B (SodB), superoxide dismutase C (SodC), bacterioferritin (Bfr), catalase (Kat), cytochrome c peroxidase (Ccp), peroxiredoxin (Prx), glutathione reductase (Gor), methionine sulfoxide reductase (MsrA/B), DNA repair enzyme (RecN), repair of iron centre protein (DnrN), nitric oxide reductase (NorB), alcohol dehydrogenase/GSNO reductase (AdhC), thioredoxin reductase (TrxB), and esterase D (EstD) and the proteins of unknown function encoded by the meningococcal operon NMB1436–38. An asterisk indicates proteins involved in repairing oxidative damage to biomolecules. Other defence mechanisms (Table 3.3), which are not currently known to be regulated by the redox-sensing regulators, are discussed in the text but may be not be included in this figure. For more details, see the text and Tables 3.1 and 3.2 for ROS and RNS, and Tables 3.3 and 3.4 for the stress defences and regulators.

intracellular SOD activities of ~3600 units/ml, while endogenous superoxide is present in levels as low as one molecule per cell (Imlay and Fridovich, 1991). Furthermore, SOD plays a role in virulence in several pathogenic bacteria including *N. meningitidis* (SodC, discussed below) (Dunn *et al.*, 2003), *S. pneumoniae* (SodA) (Yesilkaya *et al.*, 2000), *Shigella flexneri* (SodB) (Franzon *et al.*, 1990), *Salmonella choleraesuis* (SodC1 and SodC2) (Sansone *et al.*, 2002), *Salmonella enterica* serovar Typhimurium (SodC) (Farrant *et al.*, 1997), *Vibrio vulnificus* (SodA, SodB and SodC)

Table 3.4 Regulators of oxidative and nitrosative stress responses of *Neisseria gonorrhoeae* and *Neisseria meningitidis*

Regulator	(Name) Function	Gene ID[a]	Regulatory site[b]	Sensitivity of mutant strain[c]	Regulation[d]	Reference
Oxidative stress regulators						
OxyR	H_2O_2-dependent regulator	NMB0173 NGO1813	Cysteine disulphide	*R* X/XO, H_2O_2 *S* Pex, Biofilm	*katA*, *prx/grx*, *gor*	Tseng *et al*. (2003), Seib *et al*. (2007), Ieva *et al*. (2008)
PerR/Zur	Mn/Zn-dependent repressor	NMB1266 NGO0542	Mn^{2+}/Zn^{2+}-binding site	*R* X\XO, H_2O_2 *S* Pex	*mntABC/znABC*, *znuD*, *rpmEJ*, *adhA*	Wu *et al*. (2006), Stork *et al*. (2010)
Fur	(Ferric uptake regulation protein) Fe-responsive regulator	NMB0205 NGO1779	Fe^{2+}-binding site	*S* H_2O_2, Blood	*fur*, *sodB*, *bfrAB* *katA*, *recN*, *aniA*, *norB*	Berish *et al*. (1993), Sebastian *et al*. (2002), Delany *et al*. (2003, 2004, 2006), Grifantini *et al*. (2003, 2004), Seib and Delany (unpublished data)
FNR	(Fumarate/nitrate reductase regulator), oxygen-sensitive regulator	NMB0318 NGO1579	Fe–S cluster	*S* Anaerobiosis, Mouse, Infant rat	*ccp*, *aniA*, *aniS*, *norB*, *dnrN*, *ccoNOPQ*, *cycA*, *cycB*	Lissenden *et al*. (2000, Turner *et al*. (2003), Bartolini *et al*. (2006), Whitehead *et al*. (2007), Fantappie *et al*. (2011), Isabella and Clark (2011)
NmlR	(*Neisseria* MerR-like regulator) Zn- and redox-responsive repressor	NMB1303 NGO0602	Cysteine Zn^{2+}-binding site	*S* CHP, Diamide	*adhC*, *trxB*, *copA*, *estD*, *nmlR*	Kidd *et al*. (2005)
IscR	Iron–sulfur cluster assembly repressor	NMB1378 NGO0637	Fe–S cluster	ND	*iscRSUA*, *norB*	Delany (personal communication), Isabella *et al*. (2009)
MseR	(Meningococcal sigma E regulator), redox-responsive anti-sigma factor	NMB2145 NGO1943	Zn-binding cysteine	ND	*msrAB*	Gunesekere *et al*. (2006), Hopman *et al*. (2010)
LexA	*Neisseria* LexA-orthologue, redox-responsive repressor	NMB0556 NGO1427	Cysteine	*x* H_2O_2	*recN*	Schook *et al*. (2011)
Nitrosative stress regulators						
NsrR	(Nitrite-sensitive repressor) NO-responsive repressor	NMB0437 NGO1519	Fe–S cluster	*S* Aerobiosis *R* Anaerobiosis	*aniA*, *norB*, *dnrN*, *nsrR*	Rock *et al*. (2007), Heurlier *et al*. (2008), Isabella *et al*. (2009), Tucker *et al*. (2010)

[a]Gene identification according to the annotation of the *N. meningitidis* MC58 (NMBxxxx) and *N. gonorrhoeae* FA1090 (NGOxxxx) genome strains (http://cmr.jcvi.org/cgi-bin/CMR/shared/Genomes.cgi).
[b]Regulatory site responsible for sensing ROS or RNS. ND, not determined.
[c]Sensitivity of mutant strains to *in vitro* oxidative stress killing assays relative to the wild-type strain (see Table 3.3 for details).
[d]Genes of interest under the control of regulators. For details see cited references and Table 3.3.

(Kang *et al.*, 2007) and *Burkholderia pseudomallei* (SodC)(Vanaporn *et al.*, 2011).

N. gonorrhoeae: SodB

N. gonorrhoeae has very low levels of SOD activity, attributed to the activity of a SodB enzyme (Norrod and Morse, 1979; Archibald and Duong, 1986; Hassett *et al.*, 1990; Tseng *et al.*, 2001). While all *N. gonorrhoeae* strains investigated possess a gene predicted to encode a complete Fe-dependent SodB (*sodB*) (Broad Institute, 2011; Tseng *et al.*, 2001), 80–100% of strains have no SOD activity and the remainder have very low activity (Norrod and Morse, 1979; Archibald and Duong, 1986). Genes encoding SodA or SodC are not present in *N. gonorrhoeae* (Tseng *et al.*, 2001). Interestingly, *N. gonorrhoeae* does not appear to rely on SodB for defence against oxidative stress even in the few strains that do have SodB activity. The *N. gonorrhoeae* strain 1291 *sodB* mutant and the wild-type strain have the same level of sensitivity to killing by chemical inducers of oxidative stress, in paraquat (PQ; generates $O_2^{\cdot-}$ in the bacterial cytoplasm) and xanthine/xanthine oxidase (X/XO; generates extracellular $O_2^{\cdot-}$ and H_2O_2) assays (Tseng *et al.*, 2001). Furthermore, the *sodB* mutant strain has the same phenotype as the wild-type strain in an *ex vivo* survival assay using human polymorphonuclear leucocytes (Seib *et al.*, 2005). In *E. coli* SodB has a limited role in protection against oxidative stress, which is restricted to the transition from anaerobic to aerobic conditions (Kargalioglu and Imlay, 1994). The role of SodB in *N. gonorrhoeae* may also be more relevant under specific conditions that have not yet been identified. In fact, *sodB* expression is increased under several conditions: during growth in high iron concentrations (via positive regulation by Fur) (Sebastian *et al.*, 2002); during growth in high manganese conditions (2.5-fold increase of SodB via a post-transcriptional mechanism) (Wu *et al.*, 2010); and after exposure to H_2O_2 (2.8-fold increase after exposure to 5 mM H_2O_2 for 15 minutes) (Stohl *et al.*, 2005).

N. meningitidis: SodB and SodC

N. meningitidis displays moderate levels of SOD activity relative to *N. gonorrhoeae* and other prokaryotes (Norrod and Morse, 1979; Archibald and Duong, 1986). *N. meningitidis* strain MC58 has been shown to have approximately 3.5-fold higher levels of SOD activity than *N. gonorrhoeae* strain 1291 (Seib *et al.*, 2004). The SOD activity of *N. meningitidis* has been attributed to a cytoplasmic SodB and a periplasmic SodC (Norrod and Morse, 1979; Archibald and Duong, 1986; Kroll *et al.*, 1995; Wilks *et al.*, 1998). The meningococcal SodB and SodC are involved in protection against ROS, with *sodB* and *sodC* mutant strains having increased sensitivity to ROS generated by PQ (Seib *et al.*, 2004) and X/XO (Wilks *et al.*, 1998), respectively, in accordance with their cytoplasmic and periplasmic locations. SodC is also reported to be associated with enhanced early virulence in an intraperitoneal mouse model, with animals infected with a *sodC* mutant strain having increased survival with respect to animals infected with the wild-type strain (Wilks *et al.*, 1998). Studies in human monocytes/macrophages revealed that SodC also contributes to the protection of opsonized *N. meningitidis* against phagocytosis, with significantly higher numbers of *sodC* mutant organisms being endocytosed compared to wild-type organisms. ROS have been shown to enhance Fc receptor-mediated phagocytosis (Gresham *et al.*, 1988; Pricop *et al.*, 1999), and it has been suggested that the meningococcal SodC may modulate cellular interactions at the onset of phagocytosis by altering the concentration of superoxide and H_2O_2 at the phagocytic cell surface (Dunn *et al.*, 2003). Furthermore, the *N. meningitidis* SodC contains a histidine rich N-terminal extension that has been proposed to be involved in a novel strategy for Cu chaperoning and enhanced protection from phagocytic killing (Battistoni *et al.*, 2001). SodC is absent in *N. gonorrhoeae* and commensal *Neisseria* and appears to have been acquired by *N. meningitidis* via horizontal transfer, most likely from commensal *Haemophilus* species (Kroll *et al.*, 1995, 1998).

The meningococcal *sodB* and the *sodC* mutant strains displayed near wild-type levels of SOD activity, indicating that both SodB and SodC may be up-regulated to compensate for the absence of the other Sod (Seib *et al.*, 2004). However, it has not been reported that these genes are up-regulated in response to ROS. The meningococcal *sodB* gene is induced by iron, which is dependent on both the iron regulator Fur (Delany *et al.*, 2006) and the RNA chaperone Hfq (Fantappie *et al.*, 2009; Metruccio *et al.*, 2009; Pannekoek *et al.*, 2009; Mellin *et al.*, 2010). Similarly, *sodC* is up-regulated in a *hfq* mutant strain (Pannekoek *et al.*, 2009; Fantappie *et al.*, 2011).

Catalase and peroxidase

Catalase and peroxidase are haem-cofactored enzymes that are responsible for removal of H_2O_2 (Seaver and Imlay, 2001a,b). Catalases are widespread in aerobic bacteria (McCord *et al.*, 1971) and convert H_2O_2 to O_2 and H_2O (Table 3.2, reaction 17). On the other hand, peroxidases convert H_2O_2 to H_2O by oxidation of an organic compound (Table 3.2, reaction 18).

N. gonorrhoeae: catalase (KatA) and cytochrome c peroxidase

N. gonorrhoeae possess very high constitutive levels of catalase, near 100 times greater than *N. meningitidis* (Archibald and Duong, 1986) and *E. coli* (Hassett *et al.*, 1990). *N. gonorrhoeae* contains a single catalase, encoded by the *katA* gene, which is primarily located in the cytoplasm with a small concentration potentially located in the cytoplasmic membrane (Hassett *et al.*, 1990; Zheng *et al.*, 1992). Bacterial catalases are classified into three groups based on their sequence and enzymatic properties (Zamocky *et al.*, 2008), and the gonococcal catalase belongs to the typical monofunctional group of bacterial catalases. *N. gonorrhoeae* also have a high peroxidase activity (Table 3.2, reaction 18), and it has been suggested that some of this activity may be associated with catalase (Archibald and Duong, 1986; Johnson *et al.*, 1993). However, the dual catalase and peroxidase activity of KatA has been questioned (Soler-Garcia and Jerse, 2004). Despite similar catalase concentrations in several *N. gonorrhoeae* strains, different sensitivities to H_2O_2 have been reported, indicating additional mechanisms of H_2O_2 removal (Alcorn *et al.*, 1994). However, an aerobically expressed gonococcal peroxidase has not been reported.

Gonococcal *katA* mutants have no detectable levels of catalase activity and have increased sensitivity to DNA damage and killing by H_2O_2, PQ and X/XO (Johnson *et al.*, 1993, 1996; Turner *et al.*, 2003; Seib *et al.*, 2004; Soler-Garcia and Jerse, 2004). Catalase also significantly increases the ability of *N. gonorrhoeae* to resist killing by H_2O_2-producing lactobacilli *in vitro* (Zheng *et al.*, 1994). Furthermore, expression of the gonococcal catalase in *E. coli* conferred increased resistance to H_2O_2 with respect to the parent strain (Soler-Garcia and Jerse, 2004).

Catalase activity of *N. gonorrhoeae* is inducible, increasing 2- to 5-fold when exposed to PMNs or 1 mM H_2O_2 (Seib *et al.*, 2004, 2007; Zheng *et al.*, 1992) and 1.5-fold when exposed to H_2O_2 producing *L. acidophilus* (Zheng *et al.*, 1994). A more profound increase in *katA* expression was seen by microarray analysis of strain FA1090 after a 15 minute exposure to 5 mM H_2O_2, which showed a ~52-fold up-regulation of the *katA* gene (Stohl *et al.*, 2005). The H_2O_2-inducible expression of catalase in *Neisseria* spp. is regulated by OxyR (Tseng *et al.*, 2003; Seib *et al.*, 2007; Ieva *et al.*, 2008).

N. gonorrhoeae expresses a cytochrome c peroxidase (Ccp) under anaerobic conditions, which is regulated by the transcription factor FNR, and is associated with the periplasmic side of the inner membrane (Lissenden *et al.*, 2000). Ccp is a di-haem *c*-type cytochrome that catalyses the reduction of H_2O_2 to H_2O, using a periplasmic c-type cytochrome as the electron donor. The gonococcal Ccp protects against H_2O_2 killing, with a *ccp/katA* double mutant having increased sensitivity to H_2O_2 when grown anaerobically, with respect to the *katA* mutant strain (Turner *et al.*, 2003; Seib *et al.*, 2004). In addition, *katA* and *ccp/katA* strains have increased sensitivity to H_2O_2-producing *L. crispatus in vitro,* but not to the *L. murinus* strain that does not produce H_2O_2 (Muench *et al.*, 2009). Gonococcal *ccp* is also induced approximately 7-fold during growth in a biofilm, probably owing to FNR activation in the biofilm (Falsetta *et al.*, 2009).

Analysis of these strains in model systems has failed to reveal a phenotype for either catalase or Ccp. The gonococcal *katA* and *ccp/katA* mutants were no more susceptible that the wild-type strain to killing in an *ex vivo* survival assay using human PMNs (Seib *et al.*, 2005) and the *katA* strain did not have decreased survival in a human Pex cell model (Seib *et al.*, 2007). Also, in a mouse coinfection model, no difference was seen in duration of gonococcal infection or bacterial numbers recovered when *N. gonorrhoeae* (wild-type, *katA* or *ccp/katA* strains) were introduced to the vagina of mice colonized with *L. crispatus* (Muench *et al.*, 2009).

N. meningitidis: catalase (KatA)

N. meningitidis displays moderate/low levels of catalase activity compared to *N. gonorrhoeae* (consumption of ~9 nmol H_2O_2/min compared to 615 nmol H_2O_2/min for *N. gonorrhoeae*) (Norrod and Morse, 1979; Archibald and Duong, 1986) and the meningococcus has been reported to have 10- to 100-fold lower H_2O_2 tolerance than *N. gonorrhoeae*. Like *N. gonorrhoeae, N. meningitidis* contains a single *katA* gene that is regulated by OxyR in a H_2O_2-dependent manner (Ieva *et al.*, 2008). The meningococcal *katA* is also up-regulated under high iron conditions and by PQ (Grifantini *et al.*, 2003, 2004).

N. meningitidis displays lower levels of peroxidase activity than *N. gonorrhoeae* (~11 nmol oxidation of KCN-sensitive 3,3′-diaminobenzidine (DAB) compared to 198 nmol for *N. gonorrhoeae*)(Archibald and Duong, 1986). Ccp is absent in *N. meningitidis* (Seib *et al.*, 2004).

Thiol-based defences against oxidative stress

Thiols, owing to the ability of their sulfhydryl groups (-SH) to be reversibly oxidized, play an important role in maintaining the cell's redox balance and protecting against oxidative stress (Winyard *et al.*, 2005). The thiol-dependent oxidative stress defences include glutathione and several proteins with thiol-containing amino acids, cysteine (Cys) or methionine (Met), at their active sites.

Glutathione and glutathione oxidoreductase

The low molecular weight compound glutathione (γ-L-glutamyl-L-cysteinylglycine; GSH) is considered one of the first lines of oxidative stress defence (Pomposiello and Demple, 2002). GSH, typically present in cells in millimolar concentrations (5 mM in *E. coli*) (Prinz *et al.*, 1997), is a chemical scavenger of radicals and acts as a hydrogen donor to restore oxidized macromolecules (Carmel-Harel and Storz, 2000). *N. gonorrhoeae* contains very high GSH concentrations (17.3 mM), which may have a significant antioxidant potential (Archibald and Duong, 1986). Although *N. meningitidis* also has high GSH concentrations (7.4 mM), it is less than half that seen in *N. gonorrhoeae* (Archibald and Duong, 1986).

Recent work has highlighted the importance of GSH and glutamate uptake in *N. meningitidis* (Tala *et al.*, 2011). The meningococcus has no glutathione uptake system and must synthesize its own glutathione from glutamate, using the enzymes GshA and GshB. Furthermore, L-glutamate is an essential amino acid for the meningococcus, which lacks a glutamine synthase and so must rely on uptake of exogenous L-glutamate via the GltT ABC-type transporter and the GltS Na^+-dependent symporter (Monaco *et al.*, 2006; Colicchio *et al.*, 2009). It was shown that acquisition of L-glutamate by *N. meningitidis* increases its virulence by enhancing resistance to ROS generated by human PMNs. The intracellular glutathione levels in a *gshB* mutant are less than 5% of levels in the wild-type strain, and 10–25% of wild-type levels in a *gltT* mutant strain (Tala *et al.*, 2011). The *gltT* mutant had decreased survival in an infant rat model of bacteraemia, as well as in H_2O_2 and PMN-killing assays (Tala *et al.*, 2011).

The GSH oxidoreductase (Gor; also called NAD(P)H:oxidized-glutathione oxidoreductase) typically controls redox balance by maintaining the reduced pool of GSH using NADPH as a reductant (Table 3.2, reaction 20) (Carmel-Harel and Storz, 2000). The neisserial Gor was identified as part of the gonococcal OxyR regulon (Seib *et al.*, 2007) and is annotated in the *N. gonorrhoeae* genome as a dihydrolipoamide dehydrogenase (DldH), owing to its similarity to this family of proteins. No significant GSH oxidoreductase activity was present in the *gor* mutant strain relative to the wild-type strain, and this mutant was slightly more resistant to H_2O_2, but behaved like the wild-type strain in the X/XO, PQ and cumene hydroperoxide assays (Seib *et al.*, 2007). *gor* is cotranscribed with the upstream gene *prx* (described below) and both genes are up-regulated by OxyR (Seib *et al.*, 2007) and H_2O_2

(Stohl *et al.*, 2005; Seib *et al.*, 2007). Investigation of gonococcal association, invasion and survival with human Pex cells revealed a small but significant difference in the ability of the *gor* and *oxyR* mutant strains to associate with cervical epithelia, and a more significant decrease in invasion and survival over 2 hours, when compared with the wild-type gonococci (Seib *et al.*, 2007). The *N. gonorrhoeae gor, prx* and *oxyR* mutant strains also displayed a significant decrease in biofilm formation relative to the wild-type strain after two days of growth in continuous flow chambers; *gor, prx* and *oxyR* formed approximately 9%, 3% and 7% of the wild-type biofilm biomass, respectively (Seib *et al.*, 2007).

Peroxiredoxin

Peroxiredoxins (Prx) are non-haem enzymes that catalyse alkyl hydroperoxide reduction via conserved reactive cysteines (Table 3.2, reaction 18), which are typically regenerated by a thioredoxin (Trx) or glutaredoxin (Grx) thiol reductant (Hofmann *et al.*, 2002). A hybrid peroxiredoxin/glutaredoxin has been isolated in *N. meningitidis* that is active in the reduction of various peroxides, including H_2O_2 and dehydroascorbate, in the presence of GSH (Rouhier and Jacquot, 2003). The *N. meningitidis* Prx is an atypical 2-Cys class of Prx (Wood *et al.*, 2003) and the hybrid protein contains a Prx module in the N-terminal and a Grx module in the C-terminal joined by an eight amino acid long linker peptide (Rouhier and Jacquot, 2003). The Grx domain uses the reducing power of GSH to regenerate the catalytic cysteines of Prx (Rouhier and Jacquot, 2003).

In *N. gonorrhoeae*, the *prx* mutant strain was significantly more resistant to killing with H_2O_2 than the wild-type strain (but not to X/XO, PQ or cumene hydroperoxide), most probably as a result of the >2-fold increased levels of catalase seen in the *prx* mutant strain (Seib *et al.*, 2007). No difference in association, invasion or survival was seen in human Pex cells, although the *prx* mutant had a significant decrease in biofilm formation relative to the wild-type strain (Seib *et al.*, 2007). As described above, the Prx homologue in *N. gonorrhoeae* is cotranscribed with the adjacent gene encoding Gor, and these genes are both activated by H_2O_2 in an OxyR-dependent manner (Seib *et al.*, 2007). Prx expression is also increased in a *hfq* mutant strain (Fantappie *et al.*, 2009).

Thioredoxin and glutaredoxin

The thiol donors thioredoxin (Trx) and glutaredoxin (Grx) are small proteins with conserved cysteine pairs that are oxidized to cysteine disulfides upon reduction of cellular proteins, maintaining a reduced cytosol. These thiol proteins are members of the cytoplasmic thioredoxin superfamily, which includes thioredoxins 1 and 2 (TrxA and TrxC, respectively) and glutaredoxins 1, 2, and 3 (GrxA, GrxB and GrxC, respectively) (Aslund and Beckwith, 1999). Reduced Trx and Grx are regenerated by an NADPH-dependent thioredoxin reductase and a GSH-dependent Gor, respectively (Aslund and Beckwith, 1999). Several genes in the genome of *N. gonorrhoeae* have been annotated as thioredoxins and glutaredoxins (Seib *et al.*, 2006), although this group of proteins has not been extensively studied in *N. gonorrhoeae* or *N. meningitidis*. However, a periplasmic thioredoxin-like protein (NG_TlpA), which is a reducing thioredoxin with disulfide reductase activity, has been characterized as having an antioxidant role in *N. gonorrhoeae* (Achard *et al.*, 2009). A *tlpA* mutant of *N. gonorrhoeae* was found to be highly sensitive to oxidative killing by PQ and H_2O_2. The *tlpA* mutant also had reduced intracellular survival in human Pex cells (Achard *et al.*, 2009). It has been suggested that TlpA may mediate electron flow between MsrAB and DsbD (described in detail below) (Achard *et al.*, 2009). It is interesting to note that *trxA, grxB,* and *grxC* are up-regulated in *N. gonorrhoeae* after exposure to H_2O_2 (~14-, 3- and 5-fold, respectively) (Stohl *et al.*, 2005), suggesting they may also play a role in defence against H_2O_2.

Sco

A homologue of the Sco (synthesis of cytochrome c oxidase) proteins is present in *N. meningitidis* and *N. gonorrhoeae* and is involved in defence against oxidative stress via an unknown mechanism. The gonococcal and meningococcal *sco* mutant strains have increased sensitivity to PQ (Seib *et al.*, 2003). Sco homologues in prokaryotes and eukaryotes contain a highly conserved CxxxC motif and are associated with a wide range of activities, including redox signalling, regulation of copper homeostasis, copper delivery to enzymes and the well-established role in formation of the Cu_A centre of cytochrome oxidase aa_3 (Banci *et al.*, 2011). However, *Neisseria* contain only the cbb_3 oxidase which does not have a Cu_A centre, and the *Neisseria sco* mutants are unaffected in cytochrome oxidase activity (Seib *et al.*, 2003). The gonococcal *sco* strain also has the same phenotype as the wild-type strain in an *ex vivo* survival assay using human PMNs (Seib *et al.*, 2005).

In terms of its possible mechanism for defence against oxidative stress, it has been shown that Sco has structural homology to peroxiredoxins/thiol:disulfide oxidoreductases (Chinenov, 2000; Seib *et al.*, 2003). In addition, the human Sco homologue (hSCO1) has high similarity to family members of the peroxiredoxin (Prx) and glutathione peroxidase (Gpx) families, however purified hSCO1 did not display disulfide reductase activity or peroxidase activity (Williams *et al.*, 2005).

Methionine sulfoxide reductase

Methionine residues are highly susceptible to oxidation by superoxide, H_2O_2 and NO, resulting in formation of methionine sulfoxide residues, which have been implicated in conformational changes and inactivation of proteins (Vogt, 1995). Methionine sulfoxide reductase (Msr) is capable of catalysing the reduction of methionine sulfoxide residues back to methionine in damaged proteins (Brot *et al.*, 1981; Brot and Weissbach, 1983; Levine *et al.*, 1996; St John *et al.*, 2001). As such, methionine residues are considered endogenous antioxidants in proteins owing to their ability to be reduced and regenerated by Msr (Levine *et al.*, 1996). Msr is in turn reduced by the thioredoxin system (St John *et al.*, 2001). Msr is also believed to be involved in maintenance of adhesins in bacterial pathogens, which are vital for colonization and virulence (Wizemann *et al.*, 1996; Hassouni *et al.*, 1999).

Methionine sulfoxide reductase (MsrA/B), responsible for reducing oxidized methionine residues in proteins, is present in *N. gonorrhoeae* (Taha *et al.*, 1991; Wizemann *et al.*, 1996; Skaar *et al.*, 2002) and *N. meningitidis* (Kauffmann *et al.*, 2002; Lowther *et al.*, 2002; Olry *et al.*, 2002). MsrA and MsrB, specific for different methionine sulfoxide epimers, are often separate proteins, however they are encoded as a fused protein from a single open reading frame in *N. gonorrhoeae* and *N. meningitidis*, with an N-subdomain disulfide oxidoreductase, a central MsrA and a C-subdomain MsrB (Olry *et al.*, 2002; Skaar *et al.*, 2002). *N. gonorrhoeae* produces two forms of MsrA/B, a cytoplasmic form and a form with a signal sequence that is secreted to the periplasmic side of the outer membrane (Skaar *et al.*, 2002). However, the outer membrane localization of MsrA/B has been questioned, since the majority of similar proteins are located on the periplasmic side of the inner membrane (Brot *et al.*, 2006). The gonococcal MsrA/B is involved in protection from oxidative stress, with a null mutant strain as well as a strain producing only the cytoplasmic form of MsrA/B having increased sensitivity to H_2O_2 and X/XO, with respect to the wild-type strain (Skaar *et al.*, 2002).

The sulfoxide reductase reactions require the reduction of the disulfide bond within the Msr molecule to regenerate the active site and enable additional catalysis. Typically, for cytoplasmic Msr proteins, this reduction is performed by thioredoxin (Trx), which is then recycled by thioredoxin reductase (TrxR) in an NADPH dependent reaction. However, for the periplasmic neisserial MsrA/B, the reduction of the MsrA and MsrB domains is performed by the N-terminal thioredoxin-like domain, which is in turn believed to be regenerated by electrons provided by DsbD (J. Wu *et al.*, 2005; Brot *et al.*, 2006; Ranaivoson *et al.*, 2006).

MsrA/B is up-regulated by sigma E (σ^E) in both *N. gonorrhoeae* (Gunesekere *et al.*, 2006) and *N. meningitidis* (Hopman *et al.*, 2010). However, it has been suggested that there are different modes of σ^E regulation since the *N. gonorrhoeae* MsrA/B is also regulated by H_2O_2, up-regulated ~4-fold (Stohl *et al.*, 2005), while MsrA/B of *N. meningitidis* is not regulated by H_2O_2, nor diamide or superoxide (Hopman *et al.*, 2010). MsrA/B was initially designated as PilB, a regulator of pilin expression (Taha *et al.*, 1988), however this has since been disproven (Skaar *et al.*, 2002).

A third protein, a free methionine-R-sulfoxide reductase (fRMsr), has been identified in *N. meningitidis*, that like MsrA and MsrB catalyses the reduction of methionine sulfoxide, however fRMsr selectively reduces the *R* isomer at the sulfoxide function of free methionine sulfoxide (Gruez *et al.*, 2010). The structure and mechanisms of these three methionine sulfoxide reductases have been studied in detail (Ranaivoson *et al.*, 2006, 2008, 2009; Neiers *et al.*, 2007; Quinternet *et al.*, 2008; Gruez *et al.*, 2010).

DsbABCD

In many Gram-negative bacteria, the enzymes DsbABCD are involved in the formation and maintenance of intramolecular disulfide bonds in many membrane and exported proteins, preventing their misfolding, loss of function and proteolytic degradation (Aslund and Beckwith, 1999). The protein oxidation pathway comprises the periplasmic thiol-disulphide oxidoreductase DsbA that forms intramolecular disulphide bonds between two cysteine residues in the protein. DsbA is then reoxidized by the inner membrane protein DsbB. The disulfide isomerase DsbC catalyses the shuffling of multiple disulfide bonds within a misfolded protein until the correct conformation is obtained, and the inner membrane protein DsbD reduces DsbC, using the cytoplasmic thioredoxin for reducing power (Collet and Bardwell, 2002; Kadokura *et al.*, 2003). DsbD also provides reducing power to MrsAB to enable repair of oxidatively damaged proteins (see above).

N. meningitidis contains three *dsbA* genes, which encode two membrane-anchored lipoproteins, DsbA1 and DsbA2, and the soluble periplasmic DsbA3 (Sinha *et al.*, 2004; Tinsley *et al.*, 2004). *dsbA1* is meningococcal specific, while *dsbA2* and *dsbA3* are also present in *N. gonorrhoeae* (Tinsley and Nassif, 1996; Sinha *et al.*, 2004; Tinsley *et al.*, 2004). Genes encoding DsbB, DsbC and DsbD have also been identified in the meningococcal genomes (Sinha *et al.*, 2004; Tinsley *et al.*, 2004). Analysis of the sensitivity of various mutant strains to the reducing agent dithiothreitol (DTT) indicated that these enzymes are not redundant in their function and that the expression of at least one of the membrane-associated proteins, DsbA1 or DsbA2, but not DsbA3, is essential for maintaining the periplasmic redox equilibrium (Tinsley *et al.*, 2004; Sinha *et al.*, 2008). Mutational analysis demonstrated that the deletion of any single enzyme was compensated by the action of the remaining two on bacterial growth, whereas the triple mutant was unable to grow (Tinsley *et al.*, 2004). Also, *dsbD* could only be inactivated in the presence of minimal DsbA oxidative activity (i.e. in a *dsbA1A2* double mutant or a *dsbA1A2A3* triple mutant strain), indicating that DsbD is required for DsbA catalysed oxidative protein folding in *N. meningitidis* (Kumar *et al.*, 2011). A meningococcal *dsbA1A2* mutant was more sensitive to PQ and Cu than the wild-type strain, but did not display increased sensitivity to H_2O_2. Further deletion of *dsbD* (i.e. a *dsbA1A2D* triple mutant) led to an even higher sensitivity to PQ and Cu than the *dsbA1A2* mutant, and also resulted in increased sensitivity to H_2O_2 (Kumar *et al.*, 2011). This sensitivity to ROS is proposed to be a result of the slower oxidation process in the meningococcal *dsbA1A2* mutant owing to the increase of free thiol groups in periplasmic proteins that are sensitive to oxidative stresses (Kumar *et al.*, 2011). Furthermore, the additional deletion of *dsbD* may further increase sensitivity owing to the role DsbD plays in providing reducing power to MrsAB, which is also important for oxidative stress defence (Skaar *et al.*, 2002; Brot *et al.*, 2006).

The expression of *dsbD* is induced by DTT, through the MisR/S regulatory system (Kumar *et al.*, 2011). *dsbD* is significantly down-regulated (8- to 13-fold) in *misR*, *misS* and *misRS* mutants (Tzeng *et al.*, 2008; Kumar *et al.*, 2011) and MisR has been shown to interact directly with the promoter of *dsbD* (Kumar *et al.*, 2011). A separate microarray analysis of a *misR* mutant in a serogroup C strain reported that expression of *dsbA2*, *dsbA3* and *dsbC* were also deregulated (Newcombe *et al.*, 2005), although these genes were not seen to be regulated in the serogroup B strain NMB used in the study by Kumar *et al.* (2011). In *N. meningitidis*, *dsbC* was shown to be up-regulated during *in vitro* infection of HeLa cells (Tala *et al.*, 2008).

Metal homeostasis and oxidative stress defences

Manganese and MntABC

Oxidative stress and bacterial metal homeostasis are closely linked, and in many organisms a high Mn:Fe ratio has emerged as an important correlate of protection against ROS, and induction or repression of ion transporters by oxidative stress leads to decreased uptake of iron and elevated levels of Mn(II) and or Zn(II) (Faulkner and Helmann, 2011). Many lactic acid bacteria lack SOD but use Mn, accumulated to high intracellular concentrations (±30 mM) chelated to polyphosphate, to scavenge superoxide (Archibald and Fridovich, 1981a,b, 1982a,b; Archibald and Duong, 1984). Mn(II) and Mn(III) can scavenge the superoxide radical non-enzymatically (Table 3.2, reactions 21–22) (Archibald and Fridovich, 1982a) and the rate constant of the interaction of Mn(II) with peroxyl radicals has been shown to be high enough to provide an antioxidant mechanism (Coassin *et al.*, 1992). Direct dismutation of H_2O_2 by Mn(II) or Mn(III) complexed with bicarbonate has also been described (Table 3.2, reactions 23–24) (Archibald and Fridovich, 1982b; Stadtman *et al.*, 1990). In addition, some Mn complexes, including biologically relevant Mn(II)-pyrophosphate and Mn(II)-polyphosphate, can be effective antioxidants by indirectly decreasing or blocking hydroxyl radical production via Fenton, Haber–Weiss, xanthine oxidase-Fe-EDTA or Fe(III)-H_2O_2-type reactions (Cheton and Archibald, 1988). Furthermore, Mn concentrations vary up to 1000-fold between different sites in the human body; for example, the Mn concentration in saliva is 36 mM (Chicharro *et al.*, 1999) compared with 0.02 mM in serum (Krachler *et al.*, 1999), providing a potential signal for microorganisms to adapt to different host microenvironments.

N. gonorrhoeae possesses an ABC cassette-type Mn uptake system (MntABC) that is involved in Mn accumulation and protection against $O_2^{\cdot-}$ and H_2O_2 killing by a mechanism that is independent of SodB (Tseng *et al.*, 2001) and catalase (Seib *et al.*, 2004), respectively. Supplementation of media with Mn(II), but not Co(II), Mg(II) or Zn(II), confers resistance to *in vitro* oxidative killing assays (Tseng *et al.*, 2001). The MntABC Mn uptake system, like typical ABC transporters, consists of three components: the periplasmic substrate-binding protein (MntC), two integral membrane permeases (MntB), and two peripheral membrane proteins that bind and hydrolyse ATP (MntA) (Higgins, 1992). The gonococcal *mntC* mutant has decreased Mn accumulation (approximately 30% of wild-type levels) and was found to be more sensitive to superoxide (Tseng *et al.*, 2001) and H_2O_2 (Wu *et al.*, 2006) killing than the wild type. As such, *N. gonorrhoeae* appears to use Mn as a chemical quenching agent in a similar way to the already established process in *Lactobacillus plantarum* (Archibald and Fridovich, 1981a).

The *mntC* mutant has a reduced growth rate and enters stationary phase later than the wild-type strain, while the *mntAB* mutant strain has a severely defective growth rate (Wu *et al.*, 2006). Further analysis revealed that growth of the gonococcal *mntC* mutant could be stimulated by addition of either Mn(II) or Zn(II) ions, and that the recombinant MntC protein binds Zn and Mn at the same binding site with almost equal affinity (dissociation constant of 0.1 μM), indicating that the MntABC system can transport both ions (Lim *et al.*, 2008). On the other hand, growth of the *mntAB* mutant in liquid culture could only be rescued when the medium was supplemented with an antioxidant, such as mannitol (Lim *et al.*, 2008).

Both the *mntC* and *mntAB* mutants had reduced intracellular survival in a human Pex cell model (Wu *et al.*, 2006; Lim *et al.*, 2008) and showed reduced ability to form a biofilm (Lim *et al.*, 2008) indicating that the MntABC Mn transporter is important for gonococcal survival in the host. The *mntC* and *mntAB* mutants had a similar level of association to Pex cells as the wild type; however, there was a significant reduction in the levels of invasion and survival over 3 hours, in particular for the *mntC* mutant. The *mntC* and *mntAB* mutants had lower average biofilm thickness than the wild type, and had 7.0% and 19.8% of the biomass of the wild-type strain, respectively (Lim *et al.*, 2008). MntABC expression in *N. gonorrhoeae* is repressed by Mn and PerR, a transcriptional repressor from the Fur family (Wu *et al.*, 2006).

Transcriptomic and proteomic analyses

showed that the level of several oxidative stress defence proteins of *N. gonorrhoeae* was altered, via a post-transcriptional mechanism, under high Mn conditions. Increased levels of bacterioferritin (Bfr), azurin (Laz), and SodB were seen, while there were decreased levels of methionine sulfoxide reductase (MsrAB), peroxiredoxin (Prx) and pyrophosphatase (Ppa) (Wu *et al.*, 2010). Analysis of *ppa* [hydrolyses pyrophosphate (PPi) to orthophosphate (Pi)] and *ppk* (polyphosphate kinase, syntheses PolyP from Pi) mutant strains revealed they had increased resistance to PQ and H_2O_2, but decreased survival in Pex cells (Wu *et al.*, 2010). Mn-mediated protection against ROS was reduced in the *ppa* mutant strain, and a role for Mn(II)–PPi complexes in ROS defences was proposed, rather than the Mn(II)–polyP complexes seen in lactic acid bacteria (Wu *et al.*, 2010).

The meningococcal MntC protein is also involved in protection against ROS, with a *mntC* mutant displaying increased sensitivity to both PQ and X/XO. However, supplementation of media with Mn(II) does not confer increased resistance to ROS (Seib *et al.*, 2004). *N. meningitidis* can grow on $MnSO_4$ concentrations 50–100 times higher than *N. gonorrhoeae* (5 mM vs. 0.1 mM), suggesting that the mechanism of Mn transport varies between these species (Seib *et al.*, 2004).

Iron sequestration and BfrAB

Free iron, although required by aerobic cells, is dangerous owing to its ability to catalyse Fenton reactions that generate ROS (Table 3.2, reaction 8). Overloading of cells with iron leads to accelerated DNA damage (Touati *et al.*, 1995; Keyer and Imlay, 1996). Consequently, iron transport is tightly regulated and free iron (Fe^{2+}) is low in cells (Emerit *et al.*, 2001). In human cells, intracellular iron is mainly bound as ferritin while extracellular iron is bound to lactoferrin and transferrin. Pathogenic *Neisseria* are able to use multiple host iron-binding proteins as a direct source of iron, and contain several high affinity transport as well as iron storage proteins that maintain low levels of free intracellular iron (reviewed in Genco and Desai, 1996; Rohde and Dyer, 2003). Intracellular bacterial iron is primarily complexed to ferritin or the haem-containing bacterioferritin (Neilands, 1981) and many bacteria regulate iron uptake and metabolism via the ferric uptake regulator, Fur (Escolar *et al.*, 1999).

Iron deprivation has been shown to protect *N. gonorrhoeae* from ROS (Cohen *et al.*, 1987). The expression of the majority of iron binding proteins, as well as several proteins involved in oxidative stress defences are regulated by Fur, including *sodB* and *recN* in *N. gonorrhoeae* (Berish *et al.*, 1993; Sebastian *et al.*, 2002; Jackson *et al.*, 2010) and *sodB, recN aniA* and *norB* in *N. meningitidis* (Grifantini *et al.*, 2003; Perkins-Balding *et al.*, 2004; Delany *et al.*, 2006).

Bacterioferritin (Bfr) of *N. gonorrhoeae*, composed of two subunits BfrA (ferroxidase centre) and BfrB (haem-binding ligand), is involved in iron storage, growth under iron-limited conditions and defence against iron-mediated oxidative stress generated by H_2O_2 and PQ (Chen and Morse, 1999). Most bacteria possess either a ferritin or Bfr homologue (Chen and Morse, 1999), although *E. coli* possess both (Yariv *et al.*, 1981; Izuhara *et al.*, 1991). Investigation of genome databases has not revealed a ferritin homologue in *N. gonorrhoeae* or *N. meningitidis* (Chen and Morse, 1999). The meningococcal *bfrB* gene is induced by iron in a Fur dependent manner (Chen and Morse, 1999; Delany *et al.*, 2006). It has also recently been shown that *bfrAB* in *N. gonorrhoeae* is induced under anaerobic conditions (Isabella and Clark, 2011) and that *bfrB* is repressed by MtrR (Folster *et al.*, 2009). BfrA/B is regulated by the MisR/MisS two-component regulatory system in *N. meningitidis* (Tzeng *et al.*, 2008; Folster *et al.*, 2009).

Repair of oxidative DNA damage

Oxidative stress is a main cause of DNA damage (Lindahl, 1993). Oxidative DNA lesions include single- and double-strand breaks, abasic (apurinic/apyrimidinic) sites, and base damages, of which one of the most common is the oxidation product of guanine, 7,8-dihydro-8-oxo-2′-deoxyguanosine (8oxoG) that can mispair with adenine during replication and cause G:C→T:A transversions (Cheng *et al.*, 1992). Bacterial stress responses include mechanisms to repair cell components that have been damaged by oxidative stress. For example, *E. coli* possesses several DNA repair systems, many of which are up-regulated by oxidative stress (e.g. *recA* of the SOS response, exonuclease III and endonuclease IV) (Demple and Halbrook, 1983; Demple *et al.*, 1986; Imlay and Linn, 1986, 1987). Comparison of potential *Neisseria* DNA repair genes to the well-characterized DNA repair mechanisms of *E. coli* has indicated that *Neisseria* contains a variety of genes involved in, or predicted to be involved in, DNA repair, including base excision repair (BER), nucleotide excision repair (NER), recombination repair, mismatch repair and replication restart, and there are no gross differences in the repair capabilities of these organisms, with the exception of the reduced complexity of genes participating in each pathway in *Neisseria* (reviewed by Kline *et al.*, 2003; Davidsen and Tonjum, 2006; Davidsen *et al.*, 2007a).

In *N. gonorrhoeae,* mutant strains lacking the recombinase encoding gene *recA*, or genes in the RecBCD pathway (*recB*, *recC*, and *recD*), the RecF-like pathway (*recJ*, *recO*, and *recQ*), the Holliday junction processing pathway (*ruvA*, *ruvC*, and *recG*) (Stohl and Seifert, 2006), the NER pathway (*uvrA*, *uvrB* and *uvrD*) (LeCuyer *et al.*, 2010) or *recN* (Stohl *et al.*, 2005) all showed decreased resistance to H_2O_2 resistance. However, a *recX* mutant did not show decreased resistance in the oxidative stress assays used (Stohl and Seifert, 2006). The mutants in the RecBCD pathway had increased sensitivity to H_2O_2 compared to the RecF-like pathway mutants, indicating that RecBCD, which is responsible for repairing double-strand DNA breaks, is the predominant repair pathway for oxidative DNA damage in *N. gonorrhoeae.* Mutants of the NER pathway showed increased sensitivity to H_2O_2 killing at relatively low H_2O_2 (2.5 and 5 mM) but not higher H_2O_2 concentrations (10 mM), suggesting that the NER system plays a role in repairing low levels of oxidative DNA damage (LeCuyer *et al.*, 2010). On the other hand, a *recN* mutant showed increased sensitivity to increasing H_2O_2 concentrations, and also had decreased survival in PMNs (Stohl *et al.*, 2005). Interestingly, *recN* is the only gene in these pathways identified as being up-regulated under oxidative stress conditions, with 3- to 4-fold increased expression after exposure to H_2O_2 in *N. gonorrhoeae* (Stohl *et al.*, 2005) and *N. meningitidis* (Grifantini *et al.*, 2004).

The PriA helicase of *N. gonorrhoeae* is involved in protection against oxidative damage of DNA, based on its function in restarting replication from stalled replication forks resulting from DNA damage (Kline and Seifert, 2005). A gonococcal *priA* mutant was ~50-fold more sensitive to H_2O_2 and ~10-fold more sensitive to cumene hydroperoxide than the parental strain. The *priA* mutant also displayed a growth and transformation deficiency, and a more than a 10,000-fold reduction in ability to repair double-strand DNA breaks (Kline and Seifert, 2005). A meningococcal *priA* mutant strain had increased sensitivity to H_2O_2 and NO, reduced growth and viability during late logarithmic growth phase (especially when incubated under oxygen limitation in the presence of nitrite), and impaired intracellular replication in the HeLa cell infection model (Tala *et al.*, 2008).

Interestingly, unlike *N. gonorrhoeae,* a panel of meningococcal mutants strains lacking genes involved in BER, MMR, NER, translesion synthesis and recombinational repair pathways all displayed increased spontaneous mutation, UV and alkylation-induced DNA damage; however, none of these strains had increased sensitivity to ROS compared to the wild-type strain (Davidsen *et al.*, 2007b). This finding was proposed to be due to overlapping functions of these enzymes, and to the strong antioxidant mechanisms of *N. meningitidis* (Davidsen *et al.*, 2007b). However, two meningococcal enzymes have been identified, a typical neisserial AP endonuclease (NApe) and a specialized 3′-phosphodiesterase neisserial exonuclease (NExo), which have distinct DNA repair activities and play an important role in the defence of *N. meningitidis* against oxidative stress (Carpenter *et al.*, 2007). Mutant strains lacking NExo or NApe were significantly more sensitive to H_2O_2 and PQ, and a double mutant lacking both enzymes was significantly more impaired than the single mutants, indicating an additive response of these enzymes in defence against ROS. NApe and NExo are believed to be involved in meningococcal virulence, since the mutant strains were recovered from the bloodstream of infected infant rats at significantly lower levels than the wild-type strain (Carpenter *et al.*, 2007).

Oxidative stress defences with uncharacterized mechanisms of action

Azurin

The lipid-modified azurin (Laz) of pathogenic *Neisseria* is an outer membrane lipoprotein that possesses an N-terminal domain found in the H.8 protein and a C-terminal domain similar to bacterial azurins (Kawula *et al.*, 1987; Cannon, 1989; Woods *et al.*, 1989). Azurin is a small, blue, copper-containing protein that functions in electron transport during respiration in several microorganisms (Ryden, 1984), but is not required for growth of *Neisseria* (Cannon, 1989). The periplasmic azurin of *Pseudomonas aeruginosa* is not essential for denitrification but is involved in protection from oxidative stress (Vijgenboom *et al.*, 1997). Similarly, in the pathogenic *Neisseria* Laz plays a role in protection against oxidative stress, and *laz* mutant strains of *N. meningitidis* and *N. gonorrhoeae* are sensitive to H_2O_2 killing (but not PQ or X/XO) and Cu (H.J. Wu *et al.*, 2005). Cu, like Fe, can react with H_2O_2 to form extremely reactive hydroxyl radicals ($HO^{\cdot}$) via the Fenton reaction (Koppenol, 1993). The copper sensitivity of the *laz* mutant suggests that the protein may play an important role in Cu sequestration. The *N. gonorrhoeae laz* knockout also has decreased survival in human Pex cells (H.J. Wu *et al.*, 2005). In MC58 *laz* is up-regulated when incubated in blood (cited in Ray *et al.*, 2011). Interestingly, the H.8 moiety of Laz is also involved in entry into brain glioblastoma cells (Hong *et al.*, 2006) and is the target of blocking IgG antibodies in the sera of individuals with decreased serum bactericidal antibody function (Ray *et al.*, 2011). Laz is ~2-fold down regulated in the *N. meningitidis misR* mutant, but a direct interaction was not shown (Tzeng *et al.*, 2008).

NGO1686 and NGO0554

The transcriptional response to H_2O_2 in *N. gonorrhoeae* involves up-regulation of 75 genes and down-regulation of a further 80 genes (Stohl *et al.*, 2005). Several of the genes that were up-regulated encode proteins with a known role in oxidative stress defences, including catalase, Ccp, SodB, MsrA/B, Gor, thioredoxin and glutaredoxin (discussed above). This study also led to the identification of several new oxidative stress defences including RecN (described above), and two uncharacterized proteins encoded by the most highly up-regulated genes in the H_2O_2 regulon, NGO1686 (a putative zinc metalloprotease that was up-regulated 19-fold by microarray analysis, 17 by qRT-PCR) and NG0554 (a *N. gonorrhoeae* specific protein of unknown function; up-regulated 71-fold by microarray analysis, 37 by qRT-PCR). A *NGO1686* mutant strain was 30–300-fold more sensitive to H_2O_2 at concentrations of 10–50 mM, and 4-fold more sensitive to 0.01% cumene hydroperoxide than the wild-type strain. A *NGO0554* knockout was greater than 100-fold more sensitive to high levels (50 mM) of H_2O_2. The *NGO1686* mutant strain also had increased susceptibility to PMN killing relative to the wild type (Stohl *et al.*, 2005), as a result of extracellularly directed non-oxidative PMN killing mechanisms (Criss *et al.*, 2009).

NMB1436–8

Analysis of Fur and iron regulation in *N. meningitidis* revealed the presence of an operon, *NMB1436–8*, that is positively regulated by Fur during high-Fe conditions and that is involved in oxidative stress defence (Grifantini *et al.*, 2004). Deletion of this operon resulted in increased sensitivity to H_2O_2 and PMN killing, and led to decreased survival in the blood of adult mice, with respect to the wild-type strain. The cytoplasmic proteins NMB1436 and NMB1438 have homology to oxidoreductases and NMB1438 contains putative Fe–S clusters that are present in many oxidoreductases (Grifantini *et al.*, 2004).

Defences against reactive nitrogen species

Bacterial responses to RNS have been increasingly characterized in the past few years, and in many cases RNS resistance mechanisms overlap defences against ROS. The pathogenic *Neisseria* lack several of the genes that are typically involved in defence against RNS in other bacteria, including the flavohaemoglobin *hmp, noxR1, noxR3, cysH* (Barth *et al.*, 2009b). However, several defence mechanisms have been defined in the pathogenic *Neisseria* and in particular for *N. gonorrhoeae*. The importance of several of these enzymes during natural infection is demonstrated by the fact that mutant strains lacking these defences have decreased survival within human cervical epithelial cells, as well as reduced biofilm formation on these cells (described in detail below).

N. gonorrhoeae and *N. meningitidis* use nitrite to support anaerobic growth, converting nitrite to NO and then to nitrous oxide via a truncated denitrification pathway, catalysed by nitrite reductase (AniA) and nitric oxide reductase (NorB), respectively (Knapp and Clark, 1984; Clark *et al.*, 1987; Berks *et al.*, 1995; Zumft, 1997). The toxic nature of NO requires that the genes involved in its production and removal are tightly regulated. It is interesting to note that the genes involved in NO reduction (NorB, NsrR), are highly conserved across *Neisseria* species, while other parts of the denitrification pathway including nitrite reduction (AniA, NarQ) have a higher degree of divergence. These evolutionary differences most probably reflect the different ecological niches that these bacteria inhabit (Barth *et al.*, 2009a) but also suggest the importance of NO reduction in metabolism and or RNS defence for *Neisseria* species.

AniA

AniA (formerly Pan1), the major anaerobically induced outer membrane protein of *N. gonorrhoeae* and *N. meningitidis*, is a class II Cu-containing nitrite reductase that terminates an anaerobic respiratory chain reducing nitrite to NO (Clark *et al.*, 1987; Hoehn and Clark, 1992b; Mellies *et al.*, 1997; Lissenden *et al.*, 2000; Boulanger and Murphy, 2002).

There are several differences between *N. gonorrhoeae* and *N. meningitidis* in terms of their growth under anaerobic conditions using nitrite reductase. Gonococcal nitrite reductase activity has been reported as ~3-fold higher than meningococcal activity (Barth *et al.*, 2009a). Furthermore, all gonococcal strains investigated have a fully intact *aniA* gene (Barth *et al.*, 2009a), and a gonococcal *aniA* mutant is severely impaired in its ability to grow microaerophilically in the presence of nitrite (Mellies *et al.*, 1997). On the other hand, a functional AniA is found only in some strains of *N. meningitidis* and it is not required for virulence (Stefanelli *et al.*, 2008). In one study, 32% (13/41) of strains investigated contained a frameshift mutation in *aniA*, which resulted in premature termination of translation (Ku *et al.*, 2009).

NorB

NorB, a nitric oxide reductase, converts NO to the less toxic compound nitrous oxide (Householder *et al.*, 2000; Anjum *et al.*, 2002). Nitric oxide reductases play a major role in protecting organisms from NO (reviewed in Zumft, 1997). Typically, nitric oxide reductases are two subunits enzymes encoded by *norBC*; however, *norC* is not present in the pathogenic *Neisseria* (Householder *et al.*, 2000; Anjum *et al.*, 2002). NorC typically donates electrons to the NorB subunit containing the active site; however, the N-terminal extension of NorB in *N. meningitidis* may act as a quinol oxidase providing electrons to the active site (Anjum *et al.*, 2002).

Gonococcal *norB* mutants can no longer grow but can survive anaerobically, suggesting that NO is not as toxic to *N. gonorrhoeae* as it is to other organisms and indicating that there may be additional detoxification pathways present in the bacterium (Householder *et al.*, 2000). Indeed, examination of nine gonococcal strains revealed a high level of resistance to peroxynitrite, which is generated from $O_2^{\cdot-}$ and NO, at concentrations that led to a four- to seven-log decrease in viability for *N. meningitidis* (Barth *et al.*, 2009b). Gonococcal *ahpC, dnrN* and *msrA* mutants did not have increased sensitivity, and peroxynitrite treatment did not lead to accumulation of nitrated proteins in *N. gonorrhoeae* suggesting that an active peroxynitrite reductase is responsible for the peroxynitrite resistance (Barth *et al.*, 2009b).

N. gonorrhoeae can rapidly establish a steady-state NO level during anaerobic growth, and is able to reduce proinflammatory NO concentrations of >1 μM to anti-inflammatory levels of ~100 nM (Cardinale and Clark, 2000), and inactivation of the gonococcal denitrification pathway prevents establishment of the anti-inflammatory NO steady-state levels (Cardinale and Clark, 2005; Barth and Clark, 2008). As such it is believed that *N. gonorrhoeae* is able to rapidly remove both endogenously produced NO, the product of AniA, as well as NO produced by phagocytic cells, and that the ability to establish anti-inflammatory NO steady-state levels may play a role in the high rates of asymptomatic gonococcal infection observed in women (Cardinale and Clark, 2005; Barth and Clark, 2008). This is further supported by the finding that genital secretions from women concurrently infected with *N. gonorrhoeae* and *Trichomonas vaginalis* or *C. trachomatis* do not have the high cytokine levels that are normally associated with these infections (Hedges *et al.*, 1998), indicating that *N. gonorrhoeae* may be creating an immunosuppressive environment that protects itself and other bacteria from clearance by the host (Hedges *et al.*, 1998).

Growth of *N. gonorrhoeae* in biofilms during natural cervical infection has been suggested as a mechanism used by the bacteria to enable persistent infection (Steichen *et al.*, 2008). During growth of *N. gonorrhoeae* in biofilms on glass and transformed cervical epithelial cells, *aniA, norB* and *ccp* are 7- to 18-fold up-regulated, with respect to expression in planktonic growth from biofilm run off (Falsetta *et al.*, 2009). Gonococcal *aniA, norB* and *ccp* mutants were attenuated for biofilm growth on both glass slides and human cervical epithelial cells (Falsetta *et al.*, 2009). Addition of an NO quencher enhanced *norB* mutant biofilm formation, indicating that accumulation of NO is responsible for the biofilm-deficient phenotype (Falsetta *et al.*, 2009). On the other hand, NO treatment partially restores *aniA* mutant biofilm formation, and while addition of high NO concentrations to the biofilm early in its development inhibits biofilm formation, addition of NO once anaerobic respiration is initiated leads to enhanced biofilm formation (Falsetta *et al.*, 2010). Based on these findings it was proposed that *N. gonorrhoeae* in biofilms uses NO for anaerobic growth but prefers nitrite, and that this anaerobic respiration can reduce NO levels and contributes to nitrosative stress tolerance (Falsetta *et al.*, 2010).

Meningococcal strains have ~3-fold higher nitric oxidase activity than *N. gonorrhoeae* (Barth *et al.*, 2009a). *N. meningitidis* is able to deplete NO during growth in culture broth and during infection of macrophages, and meningococcal NorB-mediated removal of NO modifies the release of cytokines and chemokines from human macrophages (Stevanin *et al.*, 2007). Interestingly, NO decreases the toxicity of H_2O_2 in *N. meningitidis* via an unknown mechanism (Dyet and Moir, 2006). NorB mutants are sensitive to NO donors such as sodium nitroprusside (SNP), S-nitrosopenicillamine (SNAP), and S-nitrosoglutathione (GSNO) (Anjum *et al.*, 2002), and infection of macrophages with the *norB* mutant leads to NO accumulation, decreased ability for intracellular survival and more rapid killing during early stages of infection, as well as inhibition of macrophage apoptosis (Anjum *et al.*, 2002; Stevanin *et al.*, 2005; Tunbridge *et al.*, 2006). Furthermore, *norB* mutants display decreased survival over 18 hours in human nasopharyngeal explants (Stevanin *et al.*, 2005).

Unlike the situation seen for *N. gonorrhoeae,* meningococcal infection seems to lead to the development of a pro-inflammatory response (Stevanin *et al.*, 2007). Furthermore, meningococcal NorB increases the decomposition rate in murine macrophages of the low-molecular-weight compound *S*-nitrosothiol (SNO), which is formed by the interaction of NO with thiol residues in cellular proteins and serves as a reversible post-translational modification that plays a role in host cell signalling (Laver *et al.*, 2010). This action has been proposed as a novel mechanism of host cell injury by the meningococcus (Laver *et al.*, 2010).

The *aniA* and *norB* genes are divergently transcribed from two promoters in the same locus and regulation of their expression has been well studied in *Neisseria* and provides an example of a highly complex promoter area where multiple regulators and signals converge to control and fine-tune their expression. AniA synthesis is activated by FNR in response to anaerobiosis and is up-regulated by the two-component regulatory system NarP/NarQ and in response to nitrite (Hoehn and Clark, 1992a; Householder *et al.*, 1999; Lissenden *et al.*, 2000). The NarP/Q system is thought not to directly sense nitrite, rather its activation is blocked by the repressor NsrR, which is induced by NO and binds downstream of NarP (Rock *et al.*, 2007). NsrR senses NO directly and controls *aniA* and *norB* regulation, (Overton *et al.*, 2006; Rock *et al.*, 2007; Heurlier *et al.*, 2008; Isabella *et al.*, 2008, 2009). However, in *N. meningitidis* NsrR-dependent repression of *aniA* (but not *norB*) remains in the presence of NO and it is proposed that FNR retains partial activity aerobically, and that NsrR prevents FNR-dependent expression of *aniA* during aerobic growth (Rock *et al.*, 2007). The *norB* gene is induced by nitrite and NO but is not regulated by NarP, or FNR and anaerobiosis (Hoehn and Clark, 1992a; Householder *et al.*, 2000). Instead, *norB* expression is derepressed by NO in an NsrR-dependent manner (Rock *et al.*, 2007). Expression of NorB and AniA is up-regulated during growth with iron (Sebastian *et al.*, 2002; Grifantini *et al.*, 2003), and it has been proposed that this iron dependent regulation is directly mediated by Fur (Grifantini *et al.*, 2003; Delany *et al.*, 2004). However, another study suggested that Fur may regulate *norB* by a novel indirect activation method, by preventing the binding of a neisserial ArsR homologue, a second repressor whose putative binding site overlaps the Fur binding site (Isabella *et al.*, 2008). So at least five transcriptional factors have been implicated in the direct binding of this regulatory promoter region which controls the truncated denitrification pathway and NO-detoxification of *Neisseria* in response to multiple signals (nitrite, NO, anaerobiosis, and iron levels).

Cytochrome c

Cytochrome *c'* (CycP), a haem-containing outer membrane lipoprotein that binds NO *in vitro*, is believed to protect *N. gonorrhoeae* (Turner *et al.*, 2005) and *N. meningitidis* (Anjum *et al.*, 2002) from endogenously and/or exogenously generated nitrosative stress. The exact mechanisms of CycP is unknown, however it has been suggested that its constitutive synthesis provides an instant defence by binding NO until NorB has accumulated to sufficient levels, preventing the generation of peroxynitrite from free NO and superoxide (Turner *et al.*, 2005). A *cycP* mutant strain of *N. gonorrhoeae* has an extended lag phase during microaerobic growth in the presence of nitrite, indicating a role for CycP in defence against NO generated by nitrite reduction (Turner *et al.*, 2005). Similar growth was seen between the meningococcal wild type and the *cycP* mutant strain (Anjum *et al.*, 2002). A meningococcal *cycP* mutant strain is sensitive to the NO donors SNP and SNAP, however, unlike NorB, CycP does not contribute to depletion of NO during microaerophilic respiration (Anjum *et al.*, 2002). The *N. meningitidis cycP* mutant strains had decreased survival in human nasopharyngeal mucosa and in primary human monocyte-derived macrophages *in vitro*, which could be reversed by the addition of a nitric oxide synthase inhibitor (Stevanin *et al.*, 2005). In addition, infection of macrophages with a *cycP* mutant resulted in an increased level of macrophage NO as well as increased macrophage apoptosis relative to the wild type, indicating that NO detoxification plays a role in preventing macrophage apoptosis in wild-type meningococcus (Tunbridge *et al.*, 2006).

DnrN

DnrN is a di-iron protein that protects bacteria from damage caused by nitrosative and oxidative stress (Overton *et al.*, 2008). The growth of a gonococcal *dnrN* mutant strain was inhibited by addition of nitrite to oxygen-limited cultures (as a result of the consequent pulse of NO generation), while the wild-type strain recovered after an initial inhibition of growth. In addition, deletion of *dnrN* led to increased sensitivity to H_2O_2, both in the wild-type strain as well as in a *kat* mutant strain (Overton *et al.*, 2008).

DnrN is believed to be required for the repair of NO damage to the gonococcal transcription factors, FNR and NsrR, and it has been proposed that members of this protein family should be named *r*epair of *i*ron *c*entres, or RIC, proteins. Exposure to NO resulted in loss of FNR-dependent activation of *aniA* expression as well as NsrR-dependent repression of *norB* and *dnrN* expression, which was overcome in the wild-type strain but not the *dnrN* mutant strain (Overton *et al.*, 2008). *dnrN* is induced when the NsrR repressor is inactivated by NO (Overton *et al.*, 2006; Isabella *et al.*, 2008). However, anaerobic growth of the *nsrR* mutant led to a further 4-fold higher expression than aerobically, suggesting the presence of additional regulation of this gene (Isabella *et al.*, 2008).

TrxB

TrxB is a thioredoxin reductase that catalyses the NADPH-dependent reduction of thioredoxin. The gonococcal *trxB* mutant displays decreased thioredoxin reductase activity as well as increased sensitivity to disulfide stress, organic hydroperoxides, superoxide and NO with respect to the wild type. However, the *trxB* strain was significantly more resistant to H_2O_2 than the wild type, which has also been seen for an *E. coli trxB* mutant strain (Takemoto *et al.*, 1998). The *trxB* mutant also had a longer doubling time under aerobic conditions than wild-type cells (3 hours compared with 1 hour) (Potter *et al.*, 2009b). Diffuse biofilms were formed by the *trxB* mutant over transformed cervical cells or glass, with large gaps present between biofilm clusters, compared to the compact biofilms formed by the wild type. In addition, a small decrease in the ability of the *trxB* mutant to associate with and invade Pex cells was observed, whereas the *trxB* mutant had a severely attenuated ability to survive and replicate within Pex cells compared with the wild-type strain (Potter *et al.*, 2009b).

Interestingly, the expression of several genes was altered in the *trxB* mutant. Firstly, the OxyR regulon was seen to be constitutively active in the gonococcal *trxB* mutant, which may explain the increased resistance of this strain to H_2O_2 (Potter *et al.*, 2009b). Secondly, the microaerobic induction of *aniA* and *norB* was lost in the *trxB* mutant, and it was proposed this may be a result of inactivation of Fur activity (Potter *et al.*, 2009b). Since regions within biofilms are microaerobic (Falsetta *et al.*, 2010), the biofilm deficiency of the *trxB* mutant may be due to the lack of induction of AniA and NorB (Potter *et al.*, 2009b). *trxB* is a monocistronic operon under the control of NmlR (for Neisseria MerR-like regulator), with NmlR acting as a repressor of *trxB, estD, adhC* and *copA* but with activation occurring in the presence diamide, a thiol-specific oxidant (Kidd *et al.*, 2005).

EstD

EstD is a carboxylic ester hydrolase that is part of the NmlR regulon in *N. gonorrhoeae* (Kidd *et al.*, 2005). The substrate for EstD has not yet been identified, however it has been suggested that EstD may catalyse the hydrolysis of a toxic intermediate produced from glutathione *N*-hydroxysulfenamide (GSNHOH) (Potter *et al.*, 2009a). A gonococcal *estD* mutant has increased sensitivity to killing by GSNO and nitrite relative to the wild-type strain, but does not have increased sensitivity to NO. Microaerobic growth of wild-type gonococci, but not the *estD* mutant strain, is stimulated by nitrite. Furthermore, addition of nitrite or GSNO to a growing culture led to instant cessation of growth of the *estD* mutant strain but not the wild-type strain (Potter *et al.*, 2009a). The *estD* mutant strain had a small but significant decreased ability to associate with Pex cells, and a severely attenuated ability to survive and replicate within Pex cells, compared to the wild-type strain. In addition, the *estD* mutant was markedly deficient in biofilm formation on transformed human cervical epithelial cells (THCECs), yet formed a biofilm on glass similar to the wild-type strain with a minor decrease in maximum biofilm thickness (Potter *et al.*, 2009a).

AdhC

AdhC, a class III alcohol dehydrogenase, is associated with NADH-dependent GSNO oxidoreductase activity in *N. meningitidis* but not in *N. gonorrhoeae.* A meningococcal *adhC* mutant displayed decreased GSNO reductase activity with respect to the wild type (Potter *et al.*, 2007). However, the gonococcal *adhC* mutant had wild-type levels of GSNO reductase activity, as a result of the activity of an as yet unidentified enzyme(s) (Potter *et al.*, 2007). The gonococci *adhC* is a pseudogene due to a single base insertion that causes a premature stop codon, however it is transcriptionally active and cotranscribed with *estD* as part of the NmlR regulon (Kidd *et al.*, 2005; Potter *et al.*, 2007, 2009a).

It is interesting to note that the *ggt* gene, which encodes a γ-glutamyl transpeptidase that catalyses the hydrolysis of γ-glutamyl compounds and can accelerate the decomposition of GSNO (Hogg *et al.*, 1997), is also a pseudogene in *N. gonorrhoeae* but not in *N. meningitidis* (Takahashi and Watanabe, 2005). The loss of functional *ggt* and *adhC* gene in *N. gonorrhoeae* indicates the presence of distinct mechanisms for dealing with NO stress between the gonococcus and the meningococcus (Potter *et al.*, 2007).

Regulation of oxidative and nitrosative stress responses

The ability to sense environmental O_2 is essential for many bacteria, in particular pathogens, both for protection from oxidative stress and switching between anaerobic and aerobic respiration. Basal defences are sufficient to cope with the amount of stress generated during routine aerobiosis. Instead, oxidative response regulons are activated whenever cells encounter environments containing exogenous ROS or RNS (Zheng and Storz, 2000; Mongkolsuk and Helmann, 2002). Furthermore, during infection bacterial pathogens encounter host anatomical sites with low oxygen levels, and the fact that most pathogens have anaerobic pathways capable of providing energy and eventually using nitrate and fumarate as final electron acceptors strongly suggests that anaerobic metabolism plays an important role in the infection process. Therefore, pathogenic bacteria must also be able to respond to anaerobiosis or oxygen limitation and induce the alternatives to aerobic respiration to survive and grow in the absence of oxygen. The ability to adapt to changes in O_2 availability and oxidative stress by expressing different groups of genes is controlled at the level of transcription by redox-sensing regulatory proteins. In this part of the chapter we will review the regulatory proteins that sense and respond to the two extremes of oxygen stress, oxidative stress (when ROS levels are high) and hypoxia or oxygen limitation (Table 3.4 and Fig. 3.1). We will also include regulators that respond to the closely related signals of iron and to nitrosative stress, such as NO levels, as these regulate the ability of the pathogenic *Neisseria* to tolerate oxidative stress and to perform anaerobic respiration.

Regulatory proteins sense redox signals, which activate or deactivate their function as transcriptional regulators through reversible covalent modification of redox active amino acids or their associated metalloprotein centres. These redox sensors measure the intracellular concentrations of ROS and, through induction of the stress response, maintain them below a critical toxic level, as well as repairing any damage caused. In this way, reactive species that are causing the stress also function as a direct signal to the inducible stress response controlled by these dedicated regulators. This includes induction of the expression of detoxification systems and antioxidants, as well as proteins and molecules to assist in restoring redox homeostasis, and repression of genes encoding deleterious functions under the sensed stress condition. As such, redox proteins usually function directly as transcriptional activators or repressors, or both.

Reactive cysteines: thiol-based switches

Thiol-based switches are commonly involved in sensing and responding to redox signals in prokaryotic and eukaryotic systems. Owing to their facile redox chemistry cysteines play a central role in the active regulatory sites of the redox sensing regulators. Intra- or intermolecular disulfide bond formation serves as a direct regulatory switch for several bacterial transcription factors (e.g. OxyR, OhrR, Spx, YodB, CrtJ, and CprK) and indirectly regulates others (e.g. the RsrA anti-sigma factor and RegB sensory histidine kinase) (Antelmann and Helmann, 2011).

OxyR

The OxyR transcription regulator, a member of the LysR family of bacterial transcription factors, is well characterized in *E. coli* and is a prototype for proteins whose function is activated by ROS modification (Storz *et al.*, 1990). Upon exposure to elevated levels of H_2O_2, OxyR is oxidized, leading to disulfide bond formation between two cysteine residues (C199 and C208), and activates the expression of a regulon of genes encoding defence activities including antioxidant genes (Zheng and Storz, 2000). Both the oxidized and reduced forms of OxyR bind DNA; however, the different redox forms of the tetrameric transcription factor make different contacts along the promoter, enabling it to act as both an activator and a repressor (Toledano *et al.*, 1994). Crystallization of both the oxidized and reduced forms of the protein have shown that disulphide bond formation between the two conserved cysteine residues in the oxidized form results in a significant structural change in the regulatory domain (Choi *et al.*, 2001) resulting in a repositioning of the protein on the DNA, which triggers the transcriptional activation thought to occur via direct contact with RNA polymerase (Tao *et al.*, 1993; Wang *et al.*, 2006).

Studies of the pathogenic *Neisseria* suggested that their response to oxidative damage is fundamentally different from that of *E. coli.* While in the pathogenic *Neisseria* spp. transcription of the catalase gene is induced in response to H_2O_2 (Grifantini *et al.*, 2004; Stohl *et al.*, 2005), it was found that in *N. gonorrhoeae* and *N. meningococcus* an *oxyR* mutant strain has considerably higher levels of catalase expression and is significantly more resistant to H_2O_2 killing than is the wild type (Tseng *et al.*, 2003; Ieva *et al.*, 2008), suggesting that OxyR functions as a repressor of antioxidant genes. This was seemingly in contrast to the situation in *E. coli* and *Salmonella enterica* serovar Typhimurium, in which increased sensitivity to H_2O_2 is seen in *oxyR* mutants and in which OxyR is a positive regulator of H_2O_2-inducible genes (Christman *et al.*, 1985, 1989). Consequently it was shown that in *N. meningitidis* OxyR both represses basal catalase activity in the absence of ROS and activates expression of the enzyme in response to H_2O_2 by binding directly to the promoter upstream of the catalase gene (Ieva *et al.*, 2008). In fact, pre-treatment of the wild type with sublethal doses of H_2O_2 resulted in induction of catalase and resistance to oxidative killing to levels close to that of the OxyR mutant (Ieva *et al.*, 2008). In the *E. coli* model system OxyR tends to either activate or repress at a single promoter, however OxyR from other systems has been shown to exhibit similar dual regulation at one promoter. While the gonococcal OxyR can complement an *E. coli oxyR* mutant and restore H_2O_2 resistance (Tseng *et al.*, 2003), the respective meningococcal and *E. coli* proteins cannot complement for catalase regulation in the opposing *oxyR* mutant backgrounds suggesting that there are distinct differences between the analogues of the two systems (Ieva *et al.*, 2008). Certainly the OxyR binding site, revealed by DNA footprinting, upstream of the catalase gene in meningococcus, bares very little similarity to the consensus OxyR binding sequence of *E. coli* (Ieva *et al.*, 2008).

The crystal structure of the regulatory domain of the OxyR protein of *N. meningitidis* has been solved (Sainsbury *et al.*, 2010) and molecular modelling with the reduced and oxidized OxyR structures indicated that formation of an intramolecular disulphide bond would affect the dimer–dimer interface of the neisserial OxyR tetramer which could affect DNA binding.

There is a controversy in the literature about the sensing specificity of the two cysteines in the *E. coli* protein and two different mechanisms have been proposed (reviewed in Paget and Buttner, 2003). One consists of an on/off redox regulation between oxidized (S-S) and reduced (S-H) forms of the OxyR protein (Zheng *et al.*, 1998) while the other proposes that there may be a number of potential reaction intermediates to disulphide formation, including S-OH, S-NO and S-SG, which may have different activities (Kim *et al.*, 2002). The phenotypic differences between the neisserial strains expressing site-directed mutants of the OxyR protein (single C199S and C208A mutant) suggests that the single cysteines play distinct roles in activity of the protein (Ieva *et al.*, 2008). Both conserved cysteines, C199 and C208, are essential for the activation of *kat* expression in response to H_2O_2, presumably by the formation of the C199–C208 disulphide bond. However, repression of the *kat* promoter only requires C199 since the C208A mutant retains repressive activity but cannot activate. This seemingly 'intermediate' phenotype of the C208A mutant has been proposed to indicate a functional role for a partially reduced form of the neisserial protein (Ieva *et al.*, 2008), similar to the second proposed mechanism describing potential reaction intermediates for the *E. coli* OxyR (Kim *et al.*, 2002). In this hypothesis, varying oxidation states of the protein could result in a graded response to the oxidative signal. Instead of an on/off redox regulation between oxidized and reduced forms of the OxyR protein, there may be potential reaction intermediates such as a sulfenic acid, OxyR-SOH, with oxidation of a single cysteine, most likely C199, and this form may have different activities on different promoters.

A microarray study revealed a rather limited

regulon for OxyR in *N. gonorrhoeae*, with catalase and only two other genes being deregulated in the *oxyR* mutant under normal growth conditions, and these genes, *prx* and *gor*, were seen to be activated by the neisserial OxyR protein (Seib *et al.*, 2007). Prx is a glutathione-based system that reduces H_2O_2, as well as alkylhydroperoxides, and is likely to play a role as the major peroxidase for low concentrations of H_2O_2 in *Neisseria* as it does in *H. influenzae* (Pauwels *et al.*, 2004), while the role of catalase is in the detoxification of elevated levels of H_2O_2 (Archibald and Duong, 1986). In this way, it has been proposed that neisserial OxyR could process different redox-related signals into distinct functional responses and activate a tiered hierarchical response, repressing catalase and activating Prx under low H_2O_2 before also activating catalase expression under high levels of H_2O_2 (Ieva *et al.*, 2008). The other member of the neisserial OxyR regulon is the Gor protein, which may play a role in the reversible reduction of the OxyR protein (Carmel-Harel and Storz, 2000; Seib *et al.*, 2007). Intriguingly, the OxyR regulon appears to have an important role in pathogenesis since *N. gonorrhoeae* strains lacking either *oxyR*, *prx* or *gor* are deficient in both biofilm formation and show reduced survival in cervical epithelial models (Seib *et al.*, 2007), and in another report the *oxyR* mutant was attenuated in its ability to bind to and invade urethral epithelial cells without being altered in its potential to induce IL6 and IL8 release (Parti *et al.*, 2011). A common theme in the characterized regulons of transcriptional regulators of the *Neisseria* spp. is the small number of genes apparently regulated in *Neisseria* when compared to similar regulons in *E. coli* or other model organisms. This may reflect the adaptation of *Neisseria* to their human host, which acts as the only known reservoir for these bacteria. Nonetheless, the OxyR regulon may be more extensive in *Neisseria* spp. than what was previously reported as the protein may not have been activated fully without elevated levels of H_2O_2.

LexA

One of the major consequences of ROS generated either from endogenous cellular metabolism or exogenously from competing microflora or immune cells, is the oxidative damage of DNA. When DNA is damaged and DNA synthesis is inhibited, a diverse set of metabolic functions including increased DNA repair capacity are induced in many bacteria, known collectively as the SOS response. This stress regulon is controlled by the LexA repressor protein and the RecA protein, which acts as a co-protease aiding the autolytic self-cleavage of LexA (Kelley, 2006; Butala *et al.*, 2009). The cleavage of LexA disrupts its dimerization and DNA binding activity and ultimately results in derepression of the LexA regulon.

The pathogenic *Neisseria* were long thought to lack an SOS system, owing to the failure to identify LexA homologues or classic SOS boxes in the neisserial genomes, as well as the fact that exposure to UV light or chemical mutagens did not lead to an inducible SOS response (Black *et al.*, 1998), or up-regulation of *recA* (reviewed by Kline *et al.*, 2003). However, recently a LexA homologue has been identified in *N. gonorrhoeae* that has been proposed to play a role in a non-classical SOS response important in regulating genes involved in defence against ROS (Schook *et al.*, 2011). NG1427 of gonococcus is predicted to encode a S24 peptidase with similarity to LexA of *E. coli* (22% amino acid identity), and its autoproteolytic cleavage at the conserved LexA cleavage site was shown to be facilitated *in vitro* by RecA. It controls a small regulon in gonococci that includes itself, a DNA repair gene *recN* known to be required for oxidative damage survival, and a gene of unknown function, NGO1428. In this study treatment with DNA-damaging agents, such as mitomycin C and methyl methanesulfonate, activated RecA-dependent cleavage of LexA and the derepression of its regulon. Furthermore, LexA was demonstrated to sense H_2O_2 directly through thiol modification of a single cysteine residue, in a RecA-independent manner, which results in induction of its regulon (Schook *et al.*, 2011). However, the *ΔlexA* mutant does not have increased sensitivity to H_2O_2 (Schook *et al.*, 2011). Therefore, the LexA homologue NG1427 encodes a self-repressing transcriptional repressor, and regulates a much smaller SOS response in gonococcus, of which *recN* is the only gene that has previously been found in an SOS regulon.

Furthermore in a manner distinct from other homologues, neisserial LexA acts as a redox sensing protein that senses and responds to oxidative signals directly through thiolation of a non-conserved cysteine.

NmlR (*Neisseria* MerR-like regulator)

NmlR is the only MerR family regulator in the *Neisseria* spp. and has a distinct pattern of cysteine residues within its sequence which distinguishes it from other metal-binding MerR proteins. NmlR was shown to regulate a small regulon, consisting of itself and three loci in *N. gonorrhoeae* including an operon encoding a glutathione-dependent dehydrogenase and esterase D protein (AdhC, EstD), and genes encoding a CPx-type ATPase (CopA) and a thioredoxin reductase (TrxB), all involved in defence against oxidative stress (Kidd *et al.*, 2005). Zinc (Zn^{2+}) is required for the action of NmlR as a repressor of these genes (Kidd *et al.*, 2005). A mutant lacking NmlR had decreased growth under microaerophilic conditions and was more susceptible to killing by cumene hydroperoxide, an organic peroxide, and diamide, a reagent that causes oxidative stress by depleting the reduced thiol pool in cells. This suggested that NmlR responds to oxidative stress related to a thiol-disulphide redox imbalance. Indeed, NmlR is thought to act as a thiol-based regulatory switch and four cysteine residues may function as thiolate ligands to coordinate a Zn^{2+} ion. While mutagenesis of the cysteine residues indicated that these amino acids were important for correct regulatory activity of the protein, the precise mechanism through which NmlR is activated is unknown. Subsequently, it was shown that a NO defence system was present in *Streptococcus pneumoniae* based on a NmlR-like regulator and including an AdhC target gene (Stroeher *et al.*, 2007). Therefore, NmlR is the prototype for a new subgroup of MerR proteins which regulate the expression of genes in response to disulphide stress caused by NO and organic hydroperoxides.

Anti-sigma factor, MseR (meningococcal sigma E regulator)

Another protein controlling gene transcription which may be regulated through thiol-sensitive cysteine residues is a novel meningococcal anti-sigma factor recently described encoded in a polycistronic operon with the gene encoding the sigma factor which it regulates (Hopman *et al.*, 2010). While the *Neisseria* spp. express an ECF sigma factor similar to σ^E, there are no homologues of the periplasmic protease DegS or the anti-sigma factor RseA, which are responsible for activation of RpoE in response to periplasmic or envelope stress (De Las Penas *et al.*, 1997) Furthermore, *Neisseria* does not have an envelope stress response that senses and responds to the accumulation of misfolded proteins in the periplasm. RpoE is an alternative sigma factor that controls, in *E. coli*, the periplasmic heat shock regulon that is activated by unfolded protein in the periplasm. The *rpoE* gene of *Neisseria* is located in an autoregulated operon that also contains, directly downstream, the gene encoding its cognate anti-sigma factor. To date the RpoE regulon has been identified though overexpression of RpoE, and consists of only one other target, the *msrAB* gene, which encodes methionine sulfoxide reductase involved in reducing oxidized proteins after oxidative stress (Gunesekere *et al.*, 2006; Hopman *et al.*, 2010).

The antisigma factor protein MseR, encoded by NMB2145, is structurally related to anti-sigma domain (ASD) proteins and contains the ZAS motif ($His^{30}X_3Cys^{34}X_2Cys^{37}$). Mutagenesis of the Cys residues in the ZAS motif, as well as a Cys at position 4, are important (Cys4 and Cys37) or essential (Cys34) for anti-σE activity of NMB2145 (Hopman *et al.*, 2010). The *Streptomyces coelicolor* anti-sigma factor RsrA is an example of a thiol-dependent regulatory switch using a Zn^{2+} metal centre coordinated by redox-active cysteine residues (Paget *et al.*, 1998). Owing to similarity in structure and predicted function it has been proposed that the NMB2145 encoded sigma factor is a Zn-containing metalloprotein which responds, similarly to RsrA, to the depletion of cellular thiol pools and oxidation of cysteine groups leading to the release of the coordinated Zn^{2+} which in turn triggers release of the ECF sigma factor (Li *et al.*, 2003; Hopman *et al.*, 2010). While the *msrAB* gene is induced by H_2O_2 in *N. gonorrhoeae* (Stohl *et al.*, 2005), there is as yet no evidence that the meningococcal orthologues respond to oxidative or envelope stress (Hopman *et al.*, 2010).

Iron–sulfur [Fe–S] cluster regulatory switches

The most common Fe–S clusters found in nature are [2Fe–2S] and [4Fe–4S] clusters and these are usually coordinated to their cognate proteins by cysteine ligands. Fe–S clusters are inherently sensitive to oxidation and therefore are frequently employed as sensors of oxidants. It is generally accepted that the protein environment surrounding these Fe–S clusters greatly influences the redox properties of these ligands, which explains why these proteins operate over a wide range of oxidation–reduction potentials (Kiley and Beinert, 2003). Therefore these regulatory proteins may respond to different oxidants or levels of specific oxidant. Fe–S clusters also respond to iron and RNS.

There are three transcriptional regulatory proteins in *Neisseria* that utilize an Fe–S cluster as their regulatory sensor, FNR, IscR and NsrR, and these proteins mediate cellular responses to specific oxidants, generally O_2, ROS and NO, respectively. A fourth well-known Fe–S regulated transcriptional factor, SoxR, which senses and responds to superoxide stress, is absent from *Neisseria*.

FNR

To persist and cause disease in the human host, pathogenic *Neisseria* spp. encounter and must survive in numerous complex extracellular and intracellular environments. Both *N. meningitidis* and *N. gonorrhoeae* are able to live in anaerobic niches as part of their normal transmission, colonization and infectious cycles within the host. It has been shown that although *N. meningitidis* fails to grow under strictly anaerobic conditions, under oxygen limitation the bacterium expresses a denitrification pathway that supplements growth (Anjum *et al.*, 2002; Rock and Moir, 2005) while *N. gonorrhoeae* uses the same pathway to grow anaerobically in the presence of nitrite (Knapp and Clark, 1984). The distinct niches colonized by the pathogenic *Neisseria* spp., the nasopharynx and urogenital tract, can generally be considered to be environments that have fluctuating oxygen concentrations, and both strict aerobes and strict anaerobes co-colonize these niches (Newkirk, 1996; Brook, 2003). In bacteria, aerobic respiration occurs at oxygen concentrations above 5 mbar (equivalent to approximately 0.5% v/v, compared with atmospheric concentrations of around 200 mbar or 21% v/v), while anaerobic respiration occurs between 1 and 5 mbar (0.1–0.5% v/v), and fermentation is initiated at concentrations lower than 1 mbar (0.1% v/v) (Unden and Bongaerts, 1997).

Pathogenic *Neisseria* spp., in a similar manner to that of *E. coli*, use the FNR (Fumarate nitrate regulator) global transcription factor to control the responses involved in switching from aerobic respiration to the denitrification and sugar fermentation pathways. FNR is the only member of the CRP/FNR family of global regulators expressed by *Neisseria* spp. FNR functions as a redox-responsive transcriptional activator, and FNR is activated by oxygen concentrations below 0.5 kPa (Sawers, 1999). FNR is activated under anaerobic conditions by the acquisition of a [4Fe–4S] cluster. The presence of the [4Fe–4S] cluster promotes FNR dimerization and activates DNA-binding activity for specific sites in promoters called FNR-boxes. When these are positioned immediately upstream of the −35 element of the promoter, FNR acts as an activator protein. However, FNR can also directly repress gene expression, usually when the FNR-box is overlapping or downstream of the promoter. In the presence of O_2, the [4Fe–4S] cluster is converted to a [2Fe–2S] cluster with consequent inactivation of FNR as a transcriptional regulator.

In transcriptome studies of the pathogenic *Neisseria* spp., only a limited number of genes have been identified as being responsive to the FNR regulator. A total of nine transcriptional units in meningococci (Bartolini *et al.*, 2006) and 14 genes in the gonococci (Whitehead *et al.*, 2007) are induced under anaerobiosis in a FNR-dependent manner, in contrast to *E. coli* where FNR acts as a real global regulator coordinating the switch from aerobic to anaerobic metabolism and altering the expression of greater than 200 operons (Salmon *et al.*, 2003; Kang *et al.*, 2005; Constantinidou *et al.*, 2006). In more recent studies, 44 genes were deregulated in a meningococcal strain expressing a constitutively active FNR mutant (Fantappie *et al.*, 2011) and using next generation Deep sequencing technology, about 10% of the gonococcal genome was found to be differentially regulated after growth and adaptation to anaerobiosis (Isabella and Clark, 2011) suggesting that the response and adaptation to anaerobiosis does not involve merely the limited number of genes regulated by FNR, and that the anaerobiosis stimulon is actually much larger than predicted. The finding that the meningococcal FNR mutant is attenuated in the mouse and infant rat animal models (Bartolini *et al.*, 2006) indicates the importance of these responses to the virulence and pathogenesis of the organism and highlights the need for sensing and responding to oxygen limitation for survival *in vivo*.

From microscopic analyses of biopsies from tonsils and the urogenital tract, it has been reported that both *N. gonorrhoeae* and *N. meningitidis* form biofilms in their respective human niches *in vivo* (Sim *et al.*, 2000; Steichen *et al.*, 2008). Biofilm formation by *N. gonorrhoeae* and *N. meningitidis* is thought to play an important role in asymptomatic carriage and in the increasing problem of antibiotic resistance (Greiner *et al.*, 2005; Lappann *et al.*, 2006; Steichen *et al.*, 2008, 2011). Interestingly, recent transcriptomic studies have shown that many FNR-regulated genes are up-regulated in the neisserial biofilm (Falsetta *et al.*, 2009, 2010; O'Dwyer *et al.*, 2009) and furthermore the function of these genes contributed to biofilm formation, suggesting that the biofilm is an anaerobic, or at least microaerobic, environment. On the other hand, the ability of the *Neisseria* to tolerate oxidative stress is critical for normal biofilm formation, with several stress tolerance genes being required for optimal biofilm formation, including the OxyR regulator and genes of its operon (Seib *et al.*, 2007), as well as gene of the PerR and NmlR regulons including MntC (Lim *et al.*, 2008), EstD and TrxB (Potter *et al.*, 2009a). Therefore, the ability to sense and respond to changes in oxygen potential and oxidative stress appears to be paramount in biofilm formation for the pathogenic *Neisseria* spp.

Recently, a small regulatory RNA (sRNA) was identified within the FNR stress regulon in meningococcus and gonococcus (Fantappie *et al.*, 2011; Isabella and Clark, 2011). AniS is an anaerobically induced sRNA whose synthesis is strictly dependent on the FNR anaerobic transcriptional regulator. FNR binds the *aniS* promoter directly and activates synthesis of the sRNA in response to oxygen limitation. The identification of this novel sRNA underpins the need to rapidly coordinate gene expression in response to changing oxygen environments in the host. In *E. coli* an analogous sRNA, FnrS, plays a major role in the immediate metabolic need of adapting to anaerobic growth by shutting down aerobic respiration in response to oxygen limitation (Boysen *et al.*, 2010; Durand and Storz, 2010). Although at least two genes have been identified as direct targets of AniS (Fantappie *et al.*, 2011), neither of these are implicated in aerobic respiration and so the role of AniS within the *Neisseria* may differ from that of FnrS in *E. coli*. Regulation by AniS may have more to do with a coordinated response to niche-relevant anaerobic environments within the host for the pathogenic *Neisseria* spp., which are exclusively human-adapted pathogens.

IscR

Neisseria species express two members of the Rrf2 family of proteins which contain Fe–S clusters, IscR and NsrR, which are likely to respond to different redox signal species. The *Neisseria* homologue of the IscR protein is expressed in an operon with homologues of the iron sulphur-cluster assembly (*isc*) genes. The assembly and insertion of Fe–S clusters into apo-proteins *in vivo* requires assembly or scaffold proteins (reviewed in Frazzon and Dean, 2003) and these are generally encoded by three operons in bacteria, the *iscRSUA* (housekeeping Fe–S assembly) operon, and often one of two other specialized Fe–S assembly operons, the *nif* operon (specialized for Fe–S assembly of nitrogenase enzyme) or the *suf* operon (originally involved in Fe–S assembly in the FhuF ferric reductase protein; Patzer and Hantke, 1999). Only the *iscRSUA* system is encoded in the pathogenic *Neisseria* spp. The ORFs NMB1378 and NGO0637 encode a homologue with 46% identity to the IscR protein of *E. coli* and are present as the first gene of an *iscRSUA*-like operon, in meningococcus and gonococcus, respectively, and are followed by the downstream *hscBAfdx* encoding molecular chaperone proteins as well as the ferridoxin that has also been shown to be involved in Fe–S assembly.

The IscR protein of *E. coli* was characterized as the repressor of itself and the other *isc* genes of the operon (Schwartz *et al.*, 2001) and is involved in a negative feedback loop allowing Fe–S biogenesis to be sensitive to the Fe–S cluster status of cells. IscR repression is relieved when a [2Fe–2S] cluster present in IscR is damaged or fails to form, causing induction of the *isc* promoter and hence expression of the assembly proteins (Schwartz *et al.*, 2001). When the Isc and Hsc proteins increase cellular Fe–S biogenesis adequately, IscR will itself acquire an Fe–S cluster, enabling it to repress expression of the assembly proteins. It is now recognized that IscR regulates a much larger set of genes in *E. coli* through both repressive and activation mechanisms (Giel *et al.*, 2006; Yeo *et al.*, 2006) and is involved in stress response and pathogenicity in many bacteria (Choi *et al.*, 2007; Rincon-Enriquez *et al.*, 2008; Kim *et al.*, 2009). IscR in its apo-form (i.e. cluster-free) recognizes different DNA binding sites (Nesbit *et al.*, 2009) and can activate genes (Yeo *et al.*, 2006).

Although no detailed study of the function of IscR in *Neisseria* has been reported to date, genes of the *isc* operon in both *N. meningitidis* and *N. gonococcus* are induced during exposure to H_2O_2 (Grifantini *et al.*, 2004; Stohl *et al.*, 2005) suggesting that the repressive activity of IscR responds to oxidation by H_2O_2 as in *E. coli* (Zheng *et al.*, 2001). Furthermore, in a microarray study of iron-regulated genes, the *isc* operon was amongst a small number of genes whose expression responded to iron-limitation in a Fur-independent manner (Delany *et al.*, 2006) suggesting that IscR may also respond to intrinsic iron levels under certain conditions. The presence of a Fur binding site upstream of the *iscR* gene implicates Fur in classical iron-regulation of this operon (Delany *et al.*, 2006). In addition, the *iscR* mutant of gonococcus was reported to express higher levels of the *norB* gene, implicating IscR in nitrosative stress responses in *Neisseria*, possibly indirectly through Fe–S homeostasis as has been reported in other systems (Hyduke *et al.*, 2007; Pullan *et al.*, 2007). It will be interesting to understand the genes affected by this regulator in the *Neisseria* spp. and the role it plays in stress responses and homeostasis that is important for the infection process.

NsrR

NsrR, the second member of the Rrf2 family, is found in a wide variety of bacteria, and functions as a Fe–S-containing dedicated NO sensor. While other regulatory proteins with primary functions in sensing and responding to iron availability or oxidative stress may contain active sites that can be nitrosylated by NO, including Fur (D'Autreaux *et al.*, 2002), FNR (Cruz-Ramos *et al.*, 2002), OxyR (Hausladen *et al.*, 1996), IscR and SoxR (Pullan *et al.*, 2007), NsrR is a dedicated NO sensor and regulates genes involved in NO detoxification. The mechanism by which the Fe–S cluster of NsrR senses NO is unresolved, since, as reviewed recently (Tucker *et al.*, 2010) it is uncertain whether the active sensor is a [4Fe–4S] cluster or a [2Fe–2S] cluster. The *Neisseria* NsrR protein is thought to respond to NO through an oxygen stable Fe–S cluster that is sensitive to NO and is required for DNA binding activity. MALDI-TOF mass spectroscopy analysis of the gonococcal protein revealed a mass consistent with the presence of a [2Fe–2S] cluster (Isabella *et al.*, 2008, 2009). The mutagenesis of three conserved cysteines inactivates the protein *in vitro* and *in vivo*, implicating these amino acids in the coordination of the Fe–S cluster (Isabella *et al.*, 2008, 2009) and as such the overall structure/function of NsrR reflects that of the IscR protein of *E. coli*.

The involvement of NsrR in regulation of nitrosative stress and denitrifying genes in *Neisseria* spp. was first predicted through comparative genomic analyses (Rodionov *et al.*, 2005). However many studies in meningococcus and gonococcus have now demonstrated this role. NsrR has been shown to regulate the *aniA* and *norB* genes encoding proteins involved in the denitrification pathway and NO detoxification; *dnrN*, a gene involved in the repair of RNS or ROS damaged Fe–S centres; *mobA*, a gene involved in molybdopterin guanine dinucleotide biosynthesis; and also it is involved in an auto-regulatory feedback loop of itself (Justino *et al.*, 2007; Heurlier *et al.*, 2008; Isabella *et al.*, 2008; Overton *et al.*, 2008). While NsrR of *E. coli* appears to regulate a global stress response to NO, the *Neisseria* NsrR has a more limited role in regulating NO damage repair and key steps in the denitrification pathway to avoid NO build-up (Tucker *et al.*, 2010).

Redox and RNS sensitive metalloprotein centres

Another class of redox sensing regulators responds through redox active metal centres or changes in the stability of a metal centre. Many bacteria express more than one Fur-like protein and in the past decade it has become clear that these represent a family of proteins with different metal specificity and which includes sensors of zinc (Zur), manganese (Mur), and nickel (Nur) and can utilize metal-catalysed oxidation reactions to sense peroxide stress (PerR) or the availability of haem (Irr) (reviewed in Lee and Helmann, 2007). There are two homologues of the Fur family of metalloprotein regulators in the pathogenic *Neisseria* spp. that have been implicated in the regulation of the ROS and RNS stress responses.

Fur

Fur (the ferric uptake regulator) is a global regulator common to Gram-negative and Gram-positive bacteria which responds to intracellular environmental iron concentrations. Initially, its regulatory role was defined in the expression of proteins for acquisition and metabolism of iron, which, being both essential for growth and highly toxic in bacteria, requires a delicate homeostasis. In aerobic environments oxygen chemically oxidizes ferrous iron in the environment to its ferric form, which rapidly precipitates or forms insoluble complexes with anionic salts and therefore is unavailable in nutrient limiting concentrations. Bacteria have evolved high affinity iron uptake systems and *Neisseria* are exquisitely adapted to their human host expressing systems that uptake human iron-proteins such as transferrin, lactoferrin and haemoglobin and also scavenge xenosiderophores produced by other microflora (Schryvers and Stojiljkovic, 1999; Perkins-Balding *et al.*, 2004).

Iron overload results in toxicity, mainly through the ability of ferrous iron (Fe^{2+}) to catalyse Fenton reactions (Table 3.2, reaction 8), thereby converting H_2O_2 to the more reactive oxygen species, hydroxyl radical and ferryl iron, and potentiating oxygen stress. Therefore iron uptake is strictly regulated in bacteria, often by the Fur regulator, to avoid accumulation of intracellular iron and therefore oxidative stress. The role of Fur is now known to extend outside iron metabolism to oxidative stress resistance (reviewed in Touati, 2000; Cornelis *et al.*, 2011) and other functions in which iron is intimately involved, including nitrosative stress resistance (D'Autreaux *et al.*, 2002; Mukhopadhyay *et al.*, 2004), acid resistance (Hall and Foster, 1996; Bijlsma *et al.*, 2002; Zhu *et al.*, 2002) and virulence (Litwin and Calderwood, 1993; Vasil and Ochsner, 1999; Horsburgh *et al.*, 2001).

Fur proteins typically contain a regulatory iron-binding site, characterized by histidine and aspartate residues and a Zn-binding site, typically a CXXC motif, which is thought to be involved in structural integrity of the protein. The binding of ferrous iron at the regulatory site, which in general acts as a co-repressor, causes a conformational change in the dimeric protein and activates the DNA-binding activity of the protein for Fur binding sites (Fur box) in iron-regulated promoters (reviewed in Escolar *et al.*, 1999). While in the structural site located in the dimerization domain, the coordination of the zinc ion and the redox state of the cysteines are essential to maintain the quaternary structure of Fur dimers (Jacquamet *et al.*, 1998; D'Autreaux *et al.*, 2002; Pecqueur *et al.*, 2006). Therefore, while its regulatory activity responds to ferrous iron concentrations, oxidation of the zinc structural site by ROS, and nitrosylation of its regulatory site by NO (D'Autreaux *et al.*, 2002) can also inhibit Fur function.

Classical Fur-regulated promoters, such as those of the iron uptake genes, are repressed under high-iron conditions, helping to prevent iron overload. However, these promoters are derepressed or induced when Fur is inactivated by reversion to the apo-form as a result of iron limitation (or another inactivating stimulus). Microarray studies have shown that over 10% of the genes respond to iron in meningococcus and gonococcus (Grifantini *et al.*, 2003; Ducey *et al.*, 2005). Furthermore it has been reported that many Fur-regulated genes are induced under oxidative stress, in particular H_2O_2 treatment, probably through inactivation of the repressive activity of Fur through metal-catalysed oxidation (Grifantini *et al.*, 2004; Stohl *et al.*, 2005).

Fur activates the expression of many genes by either indirect or direct mechanisms. In general, induction of expression of genes involved in iron storage and oxidative stress defence in response to increasing iron concentrations has been shown in many bacteria (Touati, 2000; Hantke, 2001). For a long time, the unavailability of a *fur* deletion mutant hindered the investigation of the direct role of Fur in regulation in the pathogenic *Neisseria* spp. and *fur* was hypothesized to be an essential gene in *Neisseria* (Thomas and Sparling, 1994, 1996). However, this was subsequently overcome, firstly in meningococcus (Delany *et al.*, 2003) where microarray experiments with the *fur*-null mutant indicated that Fur positively regulates at least 39 genes in response to iron (Delany *et al.*, 2006). The *fur* mutant is highly susceptible to oxidative stress assays *in vitro* in H_2O_2 assays (Seib and Delany, unpublished data), and is also sensitive to serum-mediated killing in an *ex vivo* whole human blood model of bacteraemia (Echenique-Rivera *et al.*, 2011) demonstrating the close relationship between iron homeostasis and oxidative stress resistance and virulence.

A small subset of positively Fur-regulated genes, involved in aerobic and anaerobic respiration(Delany *et al.*, 2004) exhibit high affinity Fur-binding sites in their promoters implicating a direct role of Fur in activation (Delany *et al.*, 2004; Isabella *et al.*, 2008). Another member of this set of Fur-positively regulated genes, the NMB1436–1438 operon, was subsequently shown to express genes with a role in survival during peroxide-induced oxidative stress (Grifantini *et al.*, 2004). A much larger set of Fur- and iron-positively regulated genes show no evidence for direct binding of Fur in their promoter regions and are therefore candidates for indirect regulation, possibly through regulatory sRNA-mediated mechanisms such as those described in *E. coli* and *P. aeruginosa* (Masse and Gottesman, 2002; Wilderman *et al.*, 2004). These include iron storage and oxidative stress resistance genes bacterioferritin and *sodB*, as well as iron-containing proteins such as aconitase, fumarase and succinate dehydrogenase and the cytochrome C oxidase and cytochrome bc_1 complexes (Delany *et al.*, 2006). A Fur-regulated sRNA, NrrF, was the first regulatory sRNA to be identified in meningococcus and was shown to be responsible for the Fur-mediated regulation of the succinate dehydrogenase operon (Mellin *et al.*, 2007; Metruccio *et al.*, 2009). NrrF is synthesized under iron-limiting conditions in meningococcus and also gonococcus (Ducey *et al.*, 2009) and post-transcriptionally down-regulates succinate dehydrogenase (Sdh) and possibly other as yet unidentified targets. Nrrf is directly repressed by Fur under high-iron conditions thereby mediating Fur and iron activation of Sdh through a double negative loop. Sdh is a key enzyme in the tricarboxylic acid cycle (TCA). The coupling of the expression of iron-containing proteins such as Sdh to the presence of internal iron in the cell has been hypothesized as an iron-saving mechanism (Masse *et al.*, 2003) and may ensure that these protein complexes are not translated unless there is sufficient iron in the cell to support Fe–S cluster assembly and incorporation into the newly synthesized metalloprotein active sites. Bacteria need near millimolar concentrations of intracellular iron primarily for Fe–S assembly (Outten and O'Halloran, 2001). Furthermore, *sodB* and another subset of iron-activated genes are activated by Fur and iron and this is mediated by the Hfq RNA chaperone, but is independent of NrrF in *N. meningitidis* (Fantappie *et al.*, 2009; Metruccio *et al.*, 2009; Pannekoek *et al.*, 2009; Mellin *et al.*, 2010). This finding suggests that additional sRNA regulators may be involved in iron homeostasis and oxidative stress in this bacterium. The activation of the oxidative stress resistance genes in response to increased intracellular iron ensures the detoxification of ROS that can result from iron overload.

PerR

B. subtilis PerR is the prototype for the H_2O_2-sensing metalloproteins of the Fur family, which regulates inducible peroxide-defence genes and as such is functionally analogous to OxyR. PerR represses transcription when associated with either Mn(II) or Fe(II), in its regulatory site, but only the Fe(II)-containing form of the repressor can sense H_2O_2 (Herbig and Helmann, 2001; Mongkolsuk and Helmann, 2002) through metal-catalysed oxidation of the histidines (Lee and Helmann, 2006). The pathogenic *Neisseria* spp. encode a second Fur-like protein which has been nominated PerR in gonococcus as it was found to play a key role in regulation of the peroxide stress response of *N. gonorrhoeae* via a mechanism that is linked to Mn(II) metabolism (Wu *et al.*, 2006). The gonococcal PerR mutant was found to be more resistant to X/XO and H_2O_2 oxidative stress and PerR was found to be responsible for Mn^{2+}-dependent repression of the MntABC manganese transporter, which plays a role in superoxide and H_2O_2 stress resistance (Tseng *et al.*, 2001; Wu *et al.*, 2006). However, the *perR* mutant has reduced invasion and intracellular survival in a human Pex cell model (Wu *et al.*, 2006).

Several roads of investigation suggest that this second Fur homologue of *Neisseria* instead functions as a Zur orthologue, possibly with specificity for both Mn^{2+} and Zn^{2+} cations. Firstly, the protein shows 20% identity with the neisserial Fur protein, and 27% identity to the Zur protein from *E. coli* while only showing 12% identity to PerR from *B. subtilis* or *Campylobacter jejuni*. Secondly, the identified regulon, contains an arsenal of genes involved in Zn metabolism (Wu *et al.*, 2006), in particular the newly identified ZnuD outer membrane receptor involved in high affinity Zn-uptake (Stork *et al.*, 2010). Indeed in meningococcus, the NMB1266 PerR/Zur homologue was shown to be responsible for the Zn-dependent regulation of ZnuD (Stork *et al.*, 2010) and the ABC transporter orthologue to MntABC, which has been called ZnuABC in *N. meningitidis*. Furthermore, the ZnuA gene was shown to be required to sustain intracellular Zn concentrations (Stork *et al.*, 2010), suggesting that this functions as a Zn-transporter, and so this operon may encode a transporter with specificity for both Mn and Zn ions. It is still unknown whether this Fur-like protein functions in a manner similar to PerR or to Zur, with either Zn or Mn specificity; however, it appears to play a role in regulating metal ion uptake that may be fundamental for the oxidative stress responses of the pathogenic *Neisseria* spp.

Non-coding small RNAs involved in stress response regulons

Non-coding sRNAs are crucial regulatory elements in bacterial stress responses and virulence (reviewed in Gottesman, 2004; Waters and Storz, 2009; Papenfort and Vogel, 2010). Unlike regulatory proteins, non-coding regulatory sRNAs have not been readily identified and annotated in the bacterial genome sequences that are available in the databases, and their existence is still largely unexplored in many bacteria, including *Neisseria* spp. As reported above, to date there are two regulatory sRNA that have been identified in the pathogenic *Neisseria* spp., which are members of the Fur and FNR stress regulons (Mellin *et al.*, 2007; Ducey *et al.*, 2009; Metruccio *et al.*, 2009; Fantappie *et al.*, 2011; Isabella and Clark, 2011). However we predict that many more sRNAs may be responsible for regulation of gene expression and stress responses, which will be elucidated with time.

The *trans*-acting antisense class of sRNAs often require the RNA chaperone Hfq as a cofactor, which facilitates the interaction between sRNAs and target mRNAs (Valentin-Hansen *et al.*, 2004; Aiba, 2007; Chao and Vogel, 2010). Characterization of a *hfq* null mutant, which gives rise to pleiotropic phenotypes of stress sensitivity and attenuation (Dietrich *et al.*, 2009; Fantappie *et al.*, 2009; Pannekoek *et al.*, 2009) suggests that there is a large network of sRNAs in *Neisseria* that is undiscovered to date. Proteomic and transcriptomic studies of the *hfq* mutant in both meningococcus and gonococcus have highlighted a large number of proteins and genes that are differentially regulated in the absence of the Hfq chaperone, and therefore candidate targets of the Hfq-mediated RNA network (Dietrich *et al.*, 2009; Fantappie *et al.*, 2009, 2011; Pannekoek *et al.*, 2009; Mellin *et al.*, 2010). In particular, in meningococcus proteomic analyses SodB and SodC were found to be up-regulated in the *hfq* mutant (Fantappie *et al.*, 2009; Pannekoek *et al.*, 2009) along with Prx and the H_2O_2-inducible putative peroxidase encoded by NMB1590. The sensitivity of the *hfq* mutant to oxidative stress killing assays implicates a major role of this chaperone and unidentified sRNAs in a coordinated response to ROS and possibly RNS stresses.

Conclusions and future trends

The stress defences of the pathogenic *Neisseria* are of major importance to their survival and virulence within the human host. Reactive oxygen and nitrogen species (Tables 3.1 and 3.2) arise during normal aerobic metabolism and the pathogenic *Neisseria* are also continuously faced with exogenous oxidant stress during their interactions with host cells and other microflora. Bacteria cope with ROS and RNS by expressing an inducible stress response including a range of detoxifying enzymes and proteins involved in maintaining the redox homeostasis of the cell. We have highlighted how the gonococcal and meningococcal defence mechanisms play defined roles in the removal of ROS/RNS or the repair of damaged cellular molecules (Table 3.3). In many cases these defences are also involved in processes that are important for infection, such as biofilm production, colonization potential and intracellular survival.

Pathogenic *Neisseria* species have a relatively small genome (~2.2 Mb) encoding just over 2000 genes and yet many of these genes have been characterized as playing a role in the defence of and maintenance of the oxidative status of the cell (Table 3.3). In particular, of the approximately 35 transcriptional regulators annotated in *Neisseria*, nine of these (Table 3.4) to date have been reported to sense redox signals within the cell and respond by regulating the appropriate stress response regulons. Therefore, a major part of the signal transduction potential of *Neisseria* is primed to respond to these signals, highlighting the importance of adaptation to changing oxygen and oxidative environments for the infectious process in the human host. Fig. 3.1 summarizes the stress responses controlled by these transcription factors. We propose that the transcriptional response of the pathogenic *Neisseria* to reactive oxygen and nitrogen species is a composite response mediated by the modification of multiple transcription factors containing redox-active cysteines or metal centres, some specifically designed to sense ROS or NO and their derivatives and others that are collaterally activated by these reactive species. Furthermore, as research in the area of non-coding regulatory sRNAs continues to develop, we expect that additional levels of regulation will be identified for several of the neisserial stress defences described in this chapter. The interconnection of the complex regulatory networks involved in the stress responses of the pathogenic *Neisseria* is not yet fully understood, and should be a key focus of future work in this field.

Web resources

Meningococcal and gonococcal information

http://Neisseria.org

Neisseria.org provides resources about research and clinical aspects of *N. meningitidis* and *N. gonorrhoeae*, including typing, diagnosis, management and surveillance. Information is also posted about past and future conferences, including the biannual International Pathogenic Neisseria Conference. This site also has links to several useful sites under the 'Public information' section, including the CDC, the Meningitis Trust, and the International Collaboration on Gonococci (ICG).

Comprehensive Microbial Resource (CMR; J. Craig Venter Institute)

http://cmr.jcvi.org/tigr-scripts/CMR/GenomePage.cgi?database=gnm

http://cmr.jcvi.org/tigr-scripts/CMR/GenomePage.cgi?database=ntng03

The CMR provides the complete genome sequence of *N. meningitidis* and *N. gonorrhoeae* strains, as well as detailed information on functional classes of proteins.

Oxidative stress

http://en.wikipedia.org/wiki/Oxidative_stress

Provides a general overview of oxidative stress in several biological settings, including links to human diseases.

References

Achard, M.E., Hamilton, A.J., Dankowski, T., Heras, B., Schembri, M.S., Edwards, J.L., Jennings, M.P., and McEwan, A.G. (2009). A periplasmic thioredoxin-like protein plays a role in defense against oxidative stress in *Neisseria gonorrhoeae*. Infect. Immun. 77, 4934–4939.

Adak, S., Aulak, K.S., and Stuehr, D.J. (2002a). Direct evidence for nitric oxide production by a nitric-oxide synthase-like protein from *Bacillus subtilis*. J. Biol. Chem. 277, 16167–16171.

Adak, S., Bilwes, A.M., Panda, K., Hosfield, D., Aulak, K.S., McDonald, J.F., Tainer, J.A., Getzoff, E.D., Crane, B.R., and Stuehr, D.J. (2002b). Cloning, expression, and characterization of a nitric oxide synthase protein from Deinococcus radiodurans. Proc. Natl. Acad. Sci. U.S.A. 99, 107–112.

Aiba, H. (2007). Mechanism of RNA silencing by Hfq-binding small RNAs. Curr. Opin. Microbiol. *10*, 134–139.

Alcorn, T.M., Zheng, H.Y., Gunther, M.R., Hassett, D.J., and Cohen, M.S. (1994). Variation in hydrogen peroxide sensitivity between different strains of *Neisseria gonorrhoeae* is dependent on factors in addition to catalase activity. Infect. Immun. *62*, 2138–2140.

Almagor, M., Kahane, I., and Yatziv, S. (1984). Role of superoxide anion in host cell injury induced by *Mycoplasma pneumoniae* infection. A study in normal and trisomy 21 cells. J. Clin. Invest. *73*, 842–847.

Anjum, M.F., Stevanin, T.M., Read, R.C., and Moir, J.W. (2002). Nitric oxide metabolism in *Neisseria meningitidis*. J. Bacteriol. *184*, 2987–2993.

Antelmann, H., and Helmann, J.D. (2011). Thiol-based redox switches and gene regulation. Antioxid Redox Signal *14*, 1049–1063.

Archibald, F.S., and Duong, M.N. (1984). Manganese acquisition by *Lactobacillus plantarum*. J. Bacteriol. *158*, 1–8.

Archibald, F.S., and Duong, M.N. (1986). Superoxide dismutase and oxygen toxicity defenses in the genus *Neisseria*. Infect. Immun. *51*, 631–641.

Archibald, F.S., and Fridovich, I. (1981a). Manganese and defenses against oxygen toxicity in *Lactobacillus plantarum*. J. Bacteriol. *145*, 442–451.

Archibald, F.S., and Fridovich, I. (1981b). Manganese, superoxide dismutase, and oxygen tolerance in some lactic acid bacteria. J. Bacteriol. *146*, 928–936.

Archibald, F.S., and Fridovich, I. (1982a). Investigations of the state of the manganese in *Lactobacillus plantarum*. Arch. Biochem. Biophys. *215*, 589–596.

Archibald, F.S., and Fridovich, I. (1982b). The scavenging of superoxide radical by manganous complexes: *in vitro*. Arch. Biochem. Biophys. *214*, 452–463.

Aroutcheva, A., Gariti, D., Simon, M., Shott, S., Faro, J., Simoes, J.A., Gurguis, A., and Faro, S. (2001). Defense factors of vaginal lactobacilli. Am. J. Obstet. Gynecol. *185*, 375–379.

Aslund, F., and Beckwith, J. (1999). The thioredoxin superfamily: redundancy, specificity, and gray-area genomics. J. Bacteriol. *181*, 1375–1379.

Aspholm, M., Aas, F.E., Harrison, O.B., Quinn, D., Vik, A., Viburiene, R., Tonjum, T., Moir, J., Maiden, M.C., and Koomey, M. (2010). Structural alterations in a component of cytochrome c oxidase and molecular evolution of pathogenic *Neisseria* in humans. PLoS Pathog. *6*, e1001055

Banci, L., Bertini, I., Cavallaro, G., and Ciofi-Baffoni, S. (2011). Seeking the determinants of the elusive functions of Sco proteins. FEBS J. *278*, 2244–2262.

Barth, K., and Clark, V.L. (2008). Differences in nitric oxide steady states between arginine, hypoxanthine, uracil auxotrophs (AHU) and non-AHU strains of *Neisseria gonorrhoeae* during anaerobic respiration in the presence of nitrite. Can. J. Microbiol. *54*, 639–646.

Barth, K.R., Isabella, V.M., and Clark, V.L. (2009a). Biochemical and genomic analysis of the denitrification pathway within the genus *Neisseria*. Microbiology *155*, 4093–4103.

Barth, K.R., Isabella, V.M., Wright, L.F., and Clark, V.L. (2009b). Resistance to peroxynitrite in *Neisseria gonorrhoeae*. Microbiology *155*, 2532–2545.

Bartolini, E., Frigimelica, E., Giovinazzi, S., Galli, G., Shaik, Y., Genco, C., Welsch, J.A., Granoff, D.M., Grandi, G., and Grifantini, R. (2006). Role of FNR and FNR-regulated, sugar fermentation genes in *Neisseria meningitidis* infection. Mol. Microbiol. *60*, 963–972.

Battistoni, A., Pacello, F., Mazzetti, A.P., Capo, C., Kroll, J.S., Langford, P.R., Sansone, A., Donnarumma, G., Valenti, P., and Rotilio, G. (2001). A histidine-rich metal binding domain at the N terminus of Cu,Zn-superoxide dismutases from pathogenic bacteria: a novel strategy for metal chaperoning. J. Biol. Chem. *276*, 30315–30325.

Belland, R.J., Chen, T., Swanson, J., and Fischer, S.H. (1992). Human neutrophil response to recombinant *Neisserial* Opa proteins. Mol. Microbiol. *6*, 1729–1737.

Berish, S.A., Subbarao, S., Chen, C.Y., Trees, D.L., and Morse, S.A. (1993). Identification and cloning of a fur homolog from *Neisseria gonorrhoeae*. Infect. Immun. *61*, 4599–4606.

Berks, B.C., Ferguson, S.J., Moir, J.W., and Richardson, D.J. (1995). Enzymes and associated electron transport systems that catalyse the respiratory reduction of nitrogen oxides and oxyanions. Biochim. Biophys. Acta *1232*, 97–173.

Bijlsma, J.J., Waidner, B., Vliet, A.H., Hughes, N.J., Hag, S., Bereswill, S., Kelly, D.J., Vandenbroucke-Grauls, C.M., Kist, M., and Kusters, J.G. (2002). The *Helicobacter pylori* homologue of the ferric uptake regulator is involved in acid resistance. Infect. Immun. *70*, 606–611.

Black, C.G., Fyfe, J.A., and Davies, J.K. (1998). Absence of an SOS-like system in *Neisseria gonorrhoeae*. Gene *208*, 61–66.

Bogdan, C. (2001). Nitric oxide and the immune response. Nat. Immunol. *2*, 907–916.

Boulanger, M.J., and Murphy, M.E. (2002). Crystal structure of the soluble domain of the major anaerobically induced outer membrane protein (ania) from pathogenic *Neisseria*: a new class of copper-containing nitrite reductases. J. Mol. Biol. *315*, 1111–1127.

Bovre, K. (1984). Family VIII. Neisseriaceae Prevot *1933*, 199. In Bergey's Manual of Systemic Bacteriology, Krieg, N.R., and Hold, J.G., eds (Williams and Wilkins, Baltimore, MD), pp. 288–297.

Boysen, A., Moller-Jensen, J., Kallipolitis, B., Valentin-Hansen, P., and Overgaard, M. (2010). Translational regulation of gene expression by an anaerobically induced small non-coding RNA in *Escherichia coli*. J. Biol. Chem. *285*, 10690–10702.

Brenot, A., King, K.Y., Janowiak, B., Griffith, O., and Caparon, M.G. (2004). Contribution of glutathione peroxidase to the virulence of *Streptococcus pyogenes*. Infect. Immun. *72*, 408–413.

Britigan, B.E., and Cohen, M.S. (1986). Effects of human serum on bacterial competition with neutrophils for molecular oxygen. Infect. Immun. *52*, 657–663.

Britigan, B.E., Chai, Y., and Cohen, M.S. (1985). Effects of human serum on the growth and metabolism of *Neisseria gonorrhoeae*: an alternative view of serum. Infect. Immun. *50*, 738–744.

Britigan, B.E., Klapper, D., Svendsen, T., and Cohen, M.S. (1988). Phagocyte-derived lactate stimulates oxygen consumption by *Neisseria gonorrhoeae*. An unrecognized aspect of the oxygen metabolism of phagocytosis. J. Clin. Invest. *81*, 318–324.

Broad Institute (2011). *N. gonorrhoeae* group Sequencing Project. Broad Institute of Harvard and MIT. Available at: http://www.broadinstitute.org/.

Brook, I. (2003). Effects of antimicrobial therapy on the microbial flora of the adenoids. J. Antimicrob. Chemother. *51*, 1331–1337.

Brot, N., and Weissbach, H. (1983). Biochemistry and physiological role of methionine sulfoxide residues in proteins. Arch. Biochem. Biophys. *223*, 271–281.

Brot, N., Weissbach, L., Werth, J., and Weissbach, H. (1981). Enzymatic reduction of protein-bound methionine sulfoxide. Proc. Natl. Acad. Sci. U.S.A. *78*, 2155–2158.

Brot, N., Collet, J.F., Johnson, L.C., Jonsson, T.J., Weissbach, H., and Lowther, W.T. (2006). The thioredoxin domain of *Neisseria gonorrhoeae* PilB can use electrons from DsbD to reduce downstream methionine sulfoxide reductases. J. Biol. Chem. *281*, 32668–32675.

Burg, N.D., and Pillinger, M.H. (2001). The neutrophil: function and regulation in innate and humoral immunity. Clin. Immunol. *99*, 7–17.

Busch, T., Kuhlen, R., Knorr, M., Kelly, K., Lewandowski, K., Rossaint, R., Falke, K.J., and Gerlach, H. (2000). Nasal, pulmonary and autoinhaled nitric oxide at rest and during moderate exercise. Intensive Care Med. *26*, 391–399.

Butala, M., Zgur-Bertok, D., and Busby, S.J. (2009). The bacterial LexA transcriptional repressor. Cell. Mol. Life Sci. *66*, 82–93.

Cannon, J.G. (1989). Conserved lipoproteins of pathogenic *Neisseria* species bearing the H.8 epitope: lipid-modified azurin and H.8 outer membrane protein. Clin. Microbiol. Rev. *2*, S1–4.

Cardinale, J.A., and Clark, V.L. (2000). Expression of AniA, the major anaerobically induced outer membrane protein of *Neisseria gonorrhoeae*, provides protection against killing by normal human sera. Infect. Immun. *68*, 4368–4369.

Cardinale, J.A., and Clark, V.L. (2005). Determinants of nitric oxide steady-state levels during anaerobic respiration by *Neisseria gonorrhoeae*. Mol. Microbiol. *58*, 177–188.

Carmel-Harel, O., and Storz, G. (2000). Roles of the glutathione- and thioredoxin-dependent reduction systems in the *Escherichia coli* and *Saccharomyces cerevisiae* responses to oxidative stress. Annu. Rev. Microbiol. *54*, 439–461.

Carpenter, E.P., Corbett, A., Thomson, H., Adacha, J., Jensen, K., Bergeron, J., Kasampalidis, I., Exley, R., Winterbotham, M., Tang, C., *et al.* (2007). AP endonuclease paralogues with distinct activities in DNA repair and bacterial pathogenesis. EMBO J. *26*, 1363–1372.

Cartwright, K. (1995). Meningococcal Disease (Wiley, Chichester, UK).

Casey, S.G., Shafer, W.M., and Spitznagel, J.K. (1986). *Neisseria gonorrhoeae* survive intraleukocytic oxygen-independent antimicrobial capacities of anaerobic and aerobic granulocytes in the presence of pyocin lethal for extracellular gonococci. Infect. Immun. *52*, 384–389.

CDNA (2001). Communicabe Diseases Network Australia: Guidelines for the early clinical and public health management of meningococcal disease in Australia. (Commonwealth Dept. of Health and Aged Care, Canberra, Australia).

Chao, Y., and Vogel, J. (2010). The role of Hfq in bacterial pathogens. Curr. Opin. Microbiol. *13*, 24–33.

Chen, C.Y., and Morse, S.A. (1999). *Neisseria gonorrhoeae* bacterioferritin: structural heterogeneity, involvement in iron storage and protection against oxidative stress. Microbiology *145*, 2967–2975.

Chen, Y., and Rosazza, J.P. (1995). Purification and characterization of nitric oxide synthase (NOSNoc) from a Nocardia species. J. Bacteriol. *177*, 5122–5128.

Cheng, K.C., Cahill, D.S., Kasai, H., Nishimura, S., and Loeb, L.A. (1992). 8-Hydroxyguanine, an abundant form of oxidative DNA damage, causes G→T and A→C substitutions. J. Biol. Chem. *267*, 166–172.

Cheton, P.L., and Archibald, F.S. (1988). Manganese complexes and the generation and scavenging of hydroxyl free radicals. Free Radic. Biol. Med. *5*, 325–333.

Chicharro, J.L., Serrano, V., Urena, R., Gutierrez, A.M., Carvajal, A., Fernandez-Hernando, P., and Lucia, A. (1999). Trace elements and electrolytes in human resting mixed saliva after exercise. Br. J. Sports Med. *33*, 204–207.

Chinenov, Y.V. (2000). Cytochrome c oxidase assembly factors with a thioredoxin fold are conserved among prokaryotes and eukaryotes. J. Mol. Med. *78*, 239–242.

Choi, H., Kim, S., Mukhopadhyay, P., Cho, S., Woo, J., Storz, G., and Ryu, S.E. (2001). Structural basis of the redox switch in the OxyR transcription factor. Cell *105*, 103–113.

Choi, Y.S., Shin, D.H., Chung, I.Y., Kim, S.H., Heo, Y.J., and Cho, Y.H. (2007). Identification of *Pseudomonas aeruginosa* genes crucial for hydrogen peroxide resistance. J. Microbiol. Biotechnol. *17*, 1344–1352.

Christman, M.F., Morgan, R.W., Jacobson, F.S., and Ames, B.N. (1985). Positive control of a regulon for defenses against oxidative stress and some heat-shock proteins in *Salmonella typhimurium*. Cell *41*, 753–762.

Christman, M.F., Storz, G., and Ames, B.N. (1989). OxyR, a positive regulator of hydrogen peroxide-inducible genes in *Escherichia coli* and *Salmonella typhimurium*, is homologous to a family of bacterial regulatory proteins. Proc. Natl. Acad. Sci. U.S.A. *86*, 3484–3488.

Clark, V.L., Campbell, L.A., Palermo, D.A., Evans, T.M., and Klimpel, K.W. (1987). Induction and repression of outer membrane proteins by anaerobic growth of *Neisseria gonorrhoeae*. Infect. Immun. *55*, 1359–1364.

Coassin, M., Ursini, F., and Bindoli, A. (1992). Antioxidant effect of manganese. Arch. Biochem. Biophys. *299*, 330–333.

Cohen, M.S., Chai, Y., Britigan, B.E., McKenna, W., Adams, J., Svendsen, T., Bean, K., Hassett, D.J., and Sparling, P.F. (1987). Role of extracellular iron in the action of the quinone antibiotic streptonigrin: mechanisms of killing and resistance of *Neisseria gonorrhoeae*. Antimicrob. Agents Chemother. *31*, 1507–1513.

Cohen, M.S., Hoffman, I.F., Royce, R.A., Kazembe, P., Dyer, J.R., Daly, C.C., Zimba, D., Vernazza, P.L., Maida, M., Fiscus, S.A., *et al.* (1997). Reduction of concentration of HIV-1 in semen after treatment of urethritis: implications for prevention of sexual transmission of HIV-1. AIDSCAP Malawi Research Group. Lancet *349*, 1868–1873.

Colicchio, R., Ricci, S., Lamberti, F., Pagliarulo, C., Pagliuca, C., Braione, V., Braccini, T., Tala, A., Montanaro, D., Tripodi, S., *et al.* (2009). The meningococcal ABC-Type L-glutamate transporter GltT is necessary for the development of experimental meningitis in mice. Infect. Immun. *77*, 3578–3587.

Collet, J.F., and Bardwell, J.C. (2002). Oxidative protein folding in bacteria. Mol. Microbiol. *44*, 1–8.

Constantin, D., Ala'Aldeent, D., and Murphy, S. (2002). Transcriptional activation of nitric oxide synthase-2, and NO-induced cell death, in mouse cerebrovascular endothelium exposed to *Neisseria meningitidis*. J. Neurochem. *81*, 270–276.

Constantinidou, C., Hobman, J.L., Griffiths, L., Patel, M.D., Penn, C.W., Cole, J.A., and Overton, T.W. (2006). A reassessment of the FNR regulon and transcriptomic analysis of the effects of nitrate, nitrite, NarXL, and NarQP as *Escherichia coli* K12 adapts from aerobic to anaerobic growth. J. Biol. Chem. *281*, 4802–4815.

Cornelis, P., Wei, Q., Andrews, S.C., and Vinckx, T. (2011). Iron homeostasis and management of oxidative stress response in bacteria. Metallomics *3*, 540–549.

Cosseau, C., and Batut, J. (2004). Genomics of the ccoNOQP-encoded cbb(3) oxidase complex in bacteria. Arch. Microbiol. *181*, 89–96.

Criss, A.K., and Seifert, H.S. (2008). *Neisseria gonorrhoeae* suppresses the oxidative burst of human polymorphonuclear leukocytes. Cell. Microbiol. *10*, 2257–2270.

Criss, A.K., Katz, B.Z., and Seifert, H.S. (2009). Resistance of *Neisseria gonorrhoeae* to non-oxidative killing by adherent human polymorphonuclear leucocytes. Cell. Microbiol. *11*, 1074–1087.

Cruz-Ramos, H., Crack, J., Wu, G., Hughes, M.N., Scott, C., Thomson, A.J., Green, J., and Poole, R.K. (2002). NO sensing by FNR: regulation of the *Escherichia coli* NO-detoxifying flavohaemoglobin, Hmp. EMBO J. *21*, 3235–3244.

D'Autreaux, B., Touati, D., Bersch, B., Latour, J.M., and Michaud-Soret, I. (2002). Direct inhibition by nitric oxide of the transcriptional ferric uptake regulation protein via nitrosylation of the iron. Proc. Natl. Acad. Sci. U.S.A. *99*, 16619–16624.

Davidsen, T., and Tonjum, T. (2006). Meningococcal genome dynamics. Nat. Rev. Microbiol. *4*, 11–22.

Davidsen, T., Koomey, M., and Tonjum, T. (2007a). Microbial genome dynamics in CNS pathogenesis. Neuroscience *145*, 1375–1387.

Davidsen, T., Tuven, H.K., Bjoras, M., Rodland, E.A., and Tonjum, T. (2007b). Genetic interactions of DNA repair pathways in the pathogen *Neisseria meningitidis*. J. Bacteriol. *189*, 5728–5737.

De Las Penas, A., Connolly, L., and Gross, C.A. (1997). The sigmaE-mediated response to extracytoplasmic stress in *Escherichia coli* is transduced by RseA and RseB, two negative regulators of sigmaE. Mol. Microbiol. *24*, 373–385.

Deeudom, M., Koomey, M., and Moir, J.W. (2008). Roles of c-type cytochromes in respiration in *Neisseria meningitidis*. Microbiology *154*, 2857–2864.

Delany, I., Ieva, R., Alaimo, C., Rappuoli, R., and Scarlato, V. (2003). The iron-responsive regulator fur is transcriptionally autoregulated and not essential in *Neisseria meningitidis*. J. Bacteriol. *185*, 6032–6041.

Delany, I., Rappuoli, R., and Scarlato, V. (2004). Fur functions as an activator and as a repressor of putative virulence genes in *Neisseria meningitidis*. Mol. Microbiol. *52*, 1081–1090.

Delany, I., Grifantini, R., Bartolini, E., Rappuoli, R., and Scarlato, V. (2006). Effect of *Neisseria meningitidis* fur mutations on global control of gene transcription. J. Bacteriol. *188*, 2483–2492.

DeLeo, F.R., Allen, L.A., Apicella, M., and Nauseef, W.M. (1999). NADPH oxidase activation and assembly during phagocytosis. J. Immunol. *163*, 6732–6740.

Demple, B., and Halbrook, J. (1983). Inducible repair of oxidative DNA damage in *Escherichia coli*. Nature *304*, 466–468.

Demple, B., Daikh, Y., Greenberg, J., and Johnson, A. (1986). Alkylation and oxidative damages to DNA: constitutive and inducible repair systems. Basic Life Sci. *39*, 205–217.

DeVoe, I.W. (1982). The meningococcus and mechanisms of pathogenicity. Microbiol. Rev. *46*, 162–190.

Dhandayuthapani, S., Blaylock, M.W., Bebear, C.M., Rasmussen, W.G., and Baseman, J.B. (2001). Peptide methionine sulfoxide reductase (MsrA) is a virulence determinant in *Mycoplasma genitalium*. J. Bacteriol. *183*, 5645–5650.

Dietrich, M., Munke, R., Gottschald, M., Ziska, E., Boettcher, J.P., Mollenkopf, H., and Friedrich, A. (2009). The effect of hfq on global gene expression and virulence in *Neisseria gonorrhoeae*. FEBS J. *276*, 5507–5520.

Dillard, J.P., and Seifert, H.S. (2001). A variable genetic island specific for *Neisseria gonorrhoeae* is involved in providing DNA for natural transformation and is found more often in disseminated infection isolates. Mol. Microbiol. *41*, 263–277.

Duane, P.G., Rubins, J.B., Weisel, H.R., and Janoff, E.N. (1993). Identification of hydrogen peroxide as a *Streptococcus pneumoniae* toxin for rat alveolar epithelial cells. Infect. Immun. *61*, 4392–4397.

Dubreuil, D., Bisaillon, J.G., and Beaudet, R. (1984). Inhibition of *Neisseria gonorrhoeae* growth due to hydrogen peroxide production by urogenital streptococci. Microbios *39*, 159–167.

Dubreuil, D., Bisaillon, J.G., Beaudet, R., and Portelance, V. (1985). *In vitro* inhibition of *Neisseria gonorrhoeae* growth by a urogenital strain of *Streptococcus faecalis*. Exp. Biol. *43*, 243–250.

Ducey, T.F., Carson, M.B., Orvis, J., Stintzi, A.P., and Dyer, D.W. (2005). Identification of the iron-responsive genes of *Neisseria gonorrhoeae* by microarray analysis in defined medium. J. Bacteriol. *187*, 4865–4874.

Ducey, T.F., Jackson, L., Orvis, J., and Dyer, D.W. (2009). Transcript analysis of *nrrF*, a Fur repressed sRNA of *Neisseria gonorrhoeae*. Microb. Pathog. *46*, 166–170.

Dunn, K.L., Farrant, J.L., Langford, P.R., and Kroll, J.S. (2003). Bacterial [Cu,Zn]-cofactored superoxide dismutase protects opsonized, encapsulated *Neisseria meningitidis* from phagocytosis by human monocytes/macrophages. Infect. Immun. *71*, 1604–1607.

Durand, S., and Storz, G. (2010). Reprogramming of anaerobic metabolism by the FnrS small RNA. Mol. Microbiol. *75*, 1215–1231.

Dyet, K., and Moir, J. (2006). Effect of combined oxidative and nitrosative stress on *Neisseria meningitidis*. Biochem. Soc. Trans. *34*, 197–199.

Echenique-Rivera, H., Muzzi, A., Del Tordello, E., Seib, K.L., Francois, P., Rappuoli, R., Pizza, M., and Serruto, D. (2011). Transcriptome analysis of *Neisseria meningitidis* in human whole blood and mutagenesis studies identify virulence factors involved in blood survival. PLoS Pathog. *7*, e1002027.

Edwards, J.L. (2010). *Neisseria gonorrhoeae* survival during primary human cervical epithelial cell infection requires nitric oxide and is augmented by progesterone. Infect. Immun. *78*, 1202–1213.

Emerit, J., Beaumont, C., and Trivin, F. (2001). Iron metabolism, free radicals, and oxidative injury. Biomed. Pharmacother. *55*, 333–339.

Escolar, L., Perez-Martin, J., and de Lorenzo, V. (1999). Opening the iron box: transcriptional metalloregulation by the Fur protein. J. Bacteriol. *181*, 6223–6229.

Falsetta, M.L., Bair, T.B., Ku, S.C., Vanden Hoven, R.N., Steichen, C.T., McEwan, A.G., Jennings, M.P., and Apicella, M.A. (2009). Transcriptional profiling identifies the metabolic phenotype of gonococcal biofilms. Infect. Immun. *77*, 3522–3532.

Falsetta, M.L., McEwan, A.G., Jennings, M.P., and Apicella, M.A. (2010). Anaerobic metabolism occurs in the substratum of gonococcal biofilms and may be sustained in part by nitric oxide. Infect. Immun. *78*, 2320–2328.

Fantappie, L., Metruccio, M.M., Seib, K.L., Oriente, F., Cartocci, E., Ferlicca, F., Giuliani, M.M., Scarlato, V., and Delany, I. (2009). The RNA chaperone Hfq is involved in stress response and virulence in *Neisseria meningitidis* and is a pleiotropic regulator of protein expression. Infect. Immun. *77*, 1842–1853.

Fantappie, L., Oriente, F., Muzzi, A., Serruto, D., Scarlato, V., and Delany, I. (2011). A novel Hfq-dependent sRNA that is under FNR control and is synthesized in oxygen limitation in *Neisseria meningitidis*. Mol. Microbiol. *80*, 507–523.

Farrant, J.L., Sansone, A., Canvin, J.R., Pallen, M.J., Langford, P.R., Wallis, T.S., Dougan, G., and Kroll, J.S. (1997). Bacterial copper- and zinc-cofactored superoxide dismutase contributes to the pathogenesis of systemic salmonellosis. Mol. Microbiol. *25*, 785–796.

Faulkner, M.J., and Helmann, J.D. (2011). Peroxide stress elicits adaptive changes in bacterial metal ion homeostasis. Antioxid. Redox Signal *15*, 175–189.

Folster, J.P., Johnson, P.J., Jackson, L., Dhulipali, V., Dyer, D.W., and Shafer, W.M. (2009). MtrR modulates *rpoH* expression and levels of antimicrobial resistance in *Neisseria gonorrhoeae*. J. Bacteriol. *191*, 287–297.

Franzon, V.L., Arondel, J., and Sansonetti, P.J. (1990). Contribution of superoxide dismutase and catalase activities to *Shigella flexneri* pathogenesis. Infect. Immun. *58*, 529–535.

Frazzon, J., and Dean, D.R. (2003). Formation of iron–sulfur clusters in bacteria: an emerging field in bioinorganic chemistry. Curr. Opin. Chem. Biol. *7*, 166–173.

Fridovich, I. (1995). Superoxide radical and superoxide dismutases. Annu. Rev. Biochem. *64*, 97–112.

Fu, H.S., Hassett, D.J., and Cohen, M.S. (1989). Oxidant stress in *Neisseria gonorrhoeae*: adaptation and effects on L-(+)-lactate dehydrogenase activity. Infect. Immun. *57*, 2173–2178.

Genco, C.A., and Desai, P.J. (1996). Iron acquisition in the pathogenic *Neisseria*. Trends Microbiol. *4*, 179–184.

Gerbase, A.C., Rowley, J.T., and Mertens, T.E. (1998). Global epidemiology of sexually transmitted diseases. Lancet *351 (Suppl 3)*, 2–4.

Gerlach, H., Rossaint, R., Pappert, D., Knorr, M., and Falke, K.J. (1994). Autoinhalation of nitric oxide after endogenous synthesis in nasopharynx. Lancet *343*, 518–519.

Giel, J.L., Rodionov, D., Liu, M., Blattner, F.R., and Kiley, P.J. (2006). IscR-dependent gene expression links iron–sulphur cluster assembly to the control of O2-regulated genes in *Escherichia coli*. Mol. Microbiol. *60*, 1058–1075.

Gottesman, S. (2004). The small RNA regulators of *Escherichia coli*: roles and mechanisms*. Annu. Rev. Microbiol. *58*, 303–328.

Greiner, L.L., Edwards, J.L., Shao, J., Rabinak, C., Entz, D., and Apicella, M.A. (2005). Biofilm formation by *Neisseria gonorrhoeae*. Infect. Immun. *73*, 1964–1970.

Gresham, H.D., McGarr, J.A., Shackelford, P.G., and Brown, E.J. (1988). Studies on the molecular mechanisms of human Fc receptor-mediated phagocytosis. Amplification of ingestion is dependent on the generation of reactive oxygen metabolites and is deficient in polymorphonuclear leukocytes from patients with chronic granulomatous disease. J. Clin. Invest. *82*, 1192–1201.

Grifantini, R., Frigimelica, E., Delany, I., Bartolini, E., Giovinazzi, S., Balloni, S., Agarwal, S., Galli, G., Genco, C., and Grandi, G. (2004). Characterization of a novel *Neisseria meningitidis* Fur and iron-regulated operon required for protection from oxidative stress: utility of DNA microarray in the assignment of the biological role of hypothetical genes. Mol. Microbiol. *54*, 962–979.

Grifantini, R., Sebastian, S., Frigimelica, E., Draghi, M., Bartolini, E., Muzzi, A., Rappuoli, R., Grandi, G., and Genco, C.A. (2003). Identification of iron-activated and -repressed Fur-dependent genes by transcriptome analysis of *Neisseria meningitidis* group B. Proc. Natl. Acad. Sci. U.S.A. *100*, 9542–9547.

Gruez, A., Libiad, M., Boschi-Muller, S., and Branlant, G. (2010). Structural and biochemical characterization of free methionine-R-sulfoxide reductase from *Neisseria meningitidis*. J. Biol. Chem. *285*, 25033–25043.

Guibourdenche, M., Popoff, M.Y., and Riou, J.Y. (1986). Deoxyribonucleic acid relatedness among *Neisseria gonorrhoeae, N. meningitidis, N. lactamica, N. cinerea* and *'Neisseria polysaccharea'*. Ann. Inst. Pasteur Microbiol. *137B*, 177–185.

Gunesekere, I.C., Kahler, C.M., Ryan, C.S., Snyder, L.A., Saunders, N.J., Rood, J.I., and Davies, J.K. (2006). Ecf, an alternative sigma factor from *Neisseria gonorrhoeae*, controls expression of msrAB, which encodes methionine sulfoxide reductase. J. Bacteriol. *188*, 3463–3469.

Hall, H.K., and Foster, J.W. (1996). The role of fur in the acid tolerance response of *Salmonella typhimurium* is physiologically and genetically separable from its role in iron acquisition. J. Bacteriol. *178*, 5683–5691.

Halsey, T.A., Vazquez-Torres, A., Gravdahl, D.J., Fang, F.C., and Libby, S.J. (2004). The ferritin-like Dps protein is required for *Salmonella enterica* serovar Typhimurium oxidative stress resistance and virulence. Infect. Immun. 72, 1155–1158.

Hampton, M.B., Kettle, A.J., and Winterbourn, C.C. (1998). Inside the neutrophil phagosome: oxidants, myeloperoxidase, and bacterial killing. Blood *92*, 3007–3017.

Hantke, K. (2001). Iron and metal regulation in bacteria. Curr. Opin. Microbiol. *4*, 172–177.

Harris, A.G., Wilson, J.E., Danon, S.J., Dixon, M.F., Donegan, K., and Hazell, S.L. (2003). Catalase (KatA) and KatA-associated protein (KapA) are essential to persistent colonization in the *Helicobacter pylori* SS1 mouse model. Microbiology *149*, 665–672.

Hassett, D.J., Charniga, L., and Cohen, M.S. (1990). *recA* and catalase in H2O2-mediated toxicity in *Neisseria gonorrhoeae*. J. Bacteriol. *172*, 7293–7296.

Hassouni, M.E., Chambost, J.P., Expert, D., Van Gijsegem, F., and Barras, F. (1999). The minimal gene set member *msrA*, encoding peptide methionine sulfoxide reductase, is a virulence determinant of the plant pathogen *Erwinia chrysanthemi*. Proc. Natl. Acad. Sci. U.S.A. *96*, 887–892.

Hausladen, A., Privalle, C.T., Keng, T., DeAngelo, J., and Stamler, J.S. (1996). Nitrosative stress: activation of the transcription factor OxyR. Cell *86*, 719–729.

Hedges, S.R., Sibley, D.A., Mayo, M.S., Hook, E.W., 3rd, and Russell, M.W. (1998). Cytokine and antibody responses in women infected with *Neisseria gonorrhoeae*: effects of concomitant infections. J. Infect. Dis. *178*, 742–751.

Herbig, A.F., and Helmann, J.D. (2001). Roles of metal ions and hydrogen peroxide in modulating the interaction of the *Bacillus subtilis* PerR peroxide regulon repressor with operator DNA. Mol. Microbiol. *41*, 849–859.

Heurlier, K., Thomson, M.J., Aziz, N., and Moir, J.W. (2008). The nitric oxide (NO)-sensing repressor NsrR of *Neisseria meningitidis* has a compact regulon of genes involved in NO synthesis and detoxification. J. Bacteriol. *190*, 2488–2495.

Higgins, C.F. (1992). ABC transporters: from microorganisms to man. Annu. Rev. Cell Biol. *8*, 67–113.

Hillier, S.L., Krohn, M.A., Klebanoff, S.J., and Eschenbach, D.A. (1992). The relationship of hydrogen peroxide-producing lactobacilli to bacterial vaginosis and genital microflora in pregnant women. Obstet. Gynecol. *79*, 369–373.

Hoehn, G.T., and Clark, V.L. (1992a). Isolation and nucleotide sequence of the gene (*aniA*) encoding the major anaerobically induced outer membrane protein of *Neisseria gonorrhoeae*. Infect. Immun. *60*, 4695–4703.

Hoehn, G.T., and Clark, V.L. (1992b). The major anaerobically induced outer membrane protein of *Neisseria gonorrhoeae*, Pan 1, is a lipoprotein. Infect. Immun. *60*, 4704–4708.

Hofmann, B., Hecht, H.J., and Flohe, L. (2002). Peroxiredoxins Biol. Chem. *383*, 347–364.

Hogg, N., Singh, R.J., Konorev, E., Joseph, J., and Kalyanaraman, B. (1997). S-Nitrosoglutathione as a substrate for gamma-glutamyl transpeptidase. Biochem. J. *323*, 477–481.

Holmes, K.K., Sparling, P.F., Mardh, P., Lemon, S.M., Stamm, W.E., Piot, P., and Wasserheit, J.N. (1999). Sexually Transmitted Diseases, 3rd edn (McGraw-Hill, New York).

Hong, C.S., Yamada, T., Hashimoto, W., Fialho, A.M., Das Gupta, T.K., and Chakrabarty, A.M. (2006). Disrupting the entry barrier and attacking brain tumors: the role of the *Neisseria* H.8 epitope and the Laz protein. Cell Cycle *5*, 1633–1641.

Hook, E.W., 3rd, and Holmes, K.K. (1985). Gonococcal infections. Ann. Intern. Med. *102*, 229–243.

Hopman, C.T., Speijer, D., van der Ende, A., and Pannekoek, Y. (2010). Identification of a novel anti-sigmaE factor in *Neisseria meningitidis*. BMC Microbiol. *10*, 164.

Horsburgh, M.J., Clements, M.O., Crossley, H., Ingham, E., and Foster, S.J. (2001). PerR controls oxidative stress resistance and iron storage proteins and is required for virulence in *Staphylococcus aureus*. Infect. Immun. *69*, 3744–3754.

Householder, T.C., Belli, W.A., Lissenden, S., Cole, J.A., and Clark, V.L. (1999). cis- and trans-acting elements involved in regulation of *aniA*, the gene encoding the major anaerobically induced outer membrane protein in *Neisseria gonorrhoeae*. J. Bacteriol. *181*, 541–551.

Householder, T.C., Fozo, E.M., Cardinale, J.A., and Clark, V.L. (2000). Gonococcal nitric oxide reductase is encoded by a single gene, *norB*, which is required for anaerobic growth and is induced by nitric oxide. Infect. Immun. *68*, 5241–5246.

Hughes, V.L., and Hillier, S.L. (1990). Microbiologic characteristics of Lactobacillus products used for colonization of the vagina. Obstet. Gynecol. *75*, 244–248.

Hwang, C.S., Rhie, G.E., Oh, J.H., Huh, W.K., Yim, H.S., and Kang, S.O. (2002). Copper- and zinc-containing superoxide dismutase (Cu/ZnSOD) is required for the protection of *Candida albicans* against oxidative stresses and the expression of its full virulence. Microbiology *148*, 3705–3713.

Hyduke, D.R., Jarboe, L.R., Tran, L.M., Chou, K.J., and Liao, J.C. (2007). Integrated network analysis identifies nitric oxide response networks and dihydroxyacid dehydratase as a crucial target in *Escherichia coli*. Proc. Natl. Acad. Sci. U.S.A. *104*, 8484–8489.

Ieva, R., Roncarati, D., Metruccio, M.M., Seib, K.L., Scarlato, V., and Delany, I. (2008). OxyR tightly regulates catalase expression in *Neisseria meningitidis* through both repression and activation mechanisms. Mol. Microbiol. *70*, 1152–1165.

Imlay, J.A., and Fridovich, I. (1991). Assay of metabolic superoxide production in *Escherichia coli*. J. Biol. Chem. *266*, 6957–6965.

Imlay, J.A., and Linn, S. (1986). Bimodal pattern of killing of DNA-repair-defective or anoxically grown *Escherichia coli* by hydrogen peroxide. J. Bacteriol. *166*, 519–527.

Imlay, J.A., and Linn, S. (1987). Mutagenesis and stress responses induced in *Escherichia coli* by hydrogen peroxide. J. Bacteriol. *169*, 2967–2976.

Isabella, V.M., and Clark, V.L. (2011). Deep sequencing-based analysis of the anaerobic stimulon in *Neisseria gonorrhoeae*. BMC Genomics *12*, 51.

Isabella, V., Wright, L.F., Barth, K., Spence, J.M., Grogan, S., Genco, C.A., and Clark, V.L. (2008). cis- and trans-acting elements involved in regulation of *norB* (*norZ*), the gene encoding nitric oxide reductase in *Neisseria gonorrhoeae*. Microbiology *154*, 226–239.

Isabella, V.M., Lapek, J.D., Jr., Kennedy, E.M., and Clark, V.L. (2009). Functional analysis of NsrR, a nitric oxide-sensing Rrf2 repressor in *Neisseria gonorrhoeae*. Mol. Microbiol. *71*, 227–239.

Izuhara, M., Takamune, K., and Takata, R. (1991). Cloning and sequencing of an *Escherichia coli* K12 gene which encodes a polypeptide having similarity to the human ferritin H subunit. Mol. Gen. Genet. *225*, 510–513.

Jackson, L.A., Ducey, T.F., Day, M.W., Zaitshik, J.B., Orvis, J., and Dyer, D.W. (2010). Transcriptional and functional analysis of the *Neisseria gonorrhoeae* Fur regulon. J. Bacteriol. *192*, 77–85.

Jacquamet, L., Aberdam, D., Adrait, A., Hazemann, J.L., Latour, J.M., and Michaud-Soret, I. (1998). X-ray absorption spectroscopy of a new zinc site in the fur protein from *Escherichia coli*. Biochemistry *37*, 2564–2571.

Johnson, S.R., Steiner, B.M., Cruce, D.D., Perkins, G.H., and Arko, R.J. (1993). Characterization of a catalase-deficient strain of *Neisseria gonorrhoeae*: evidence for the significance of catalase in the biology of *N. gonorrhoeae*. Infect. Immun. *61*, 1232–1238.

Johnson, S.R., Steiner, B.M., and Perkins, G.H. (1996). Cloning and characterization of the catalase gene of *Neisseria gonorrhoeae*: use of the gonococcus as a host organism for recombinant DNA. Infect. Immun. *64*, 2627–2634.

Jurtshuk, P., and Milligan, T.W. (1974). Quantitation of the tetramethyl-*p*-phenylenediamine oxidase reaction in *Neisseria* species. Appl. Microbiol. *28*, 1079–1081.

Justino, M.C., Almeida, C.C., Teixeira, M., and Saraiva, L.M. (2007). *Escherichia coli* di-iron YtfE protein is necessary for the repair of stress-damaged iron–sulfur clusters. J. Biol. Chem. *282*, 10352–10359.

Kadokura, H., Katzen, F., and Beckwith, J. (2003). Protein disulfide bond formation in prokaryotes. Annu. Rev. Biochem. *72*, 111–135.

Kang, I.H., Kim, J.S., and Lee, J.K. (2007). The virulence of *Vibrio vulnificus* is affected by the cellular level of superoxide dismutase activity. J. Microbiol. Biotechnol. *17*, 1399–1402.

Kang, Y., Weber, K.D., Qiu, Y., Kiley, P.J., and Blattner, F.R. (2005). Genome-wide expression analysis indicates that FNR of *Escherichia coli* K-12 regulates a large number of genes of unknown function. J. Bacteriol. *187*, 1135–1160.

Kargalioglu, Y., and Imlay, J.A. (1994). Importance of anaerobic superoxide dismutase synthesis in facilitating outgrowth of *Escherichia coli* upon entry into an aerobic habitat. J. Bacteriol. *176*, 7653–7658.

Kauffmann, B., Favier, F., Olry, A., Boschi-Muller, S., Carpentier, P., Branlant, G., and Aubry, A. (2002). Crystallization and preliminary X-ray diffraction studies of the peptide methionine sulfoxide reductase B domain of *Neisseria meningitidis* PILB. Acta Crystallogr. D Biol. Crystallogr. *58*, 1467–1469.

Kawula, T.H., Spinola, S.M., Klapper, D.G., and Cannon, J.G. (1987). Localization of a conserved epitope and an azurin-like domain in the H.8 protein of pathogenic *Neisseria*. Mol. Microbiol. *1*, 179–185.

Kaye, D., and Levison, M.E. (1977). In vitro inhibition of growth of *Neisseria gonorrhoeae* by genital microorganisms. Sex. Transm. Dis. *4*, 1–3.

Kelley, W.L. (2006). Lex marks the spot: the virulent side of SOS and a closer look at the LexA regulon. Mol. Microbiol. *62*, 1228–1238.

Keyer, K., and Imlay, J.A. (1996). Superoxide accelerates DNA damage by elevating free-iron levels. Proc. Natl. Acad. Sci. U.S.A. *93*, 13635–13640.

Kidd, S.P., Potter, A.J., Apicella, M.A., Jennings, M.P., and McEwan, A.G. (2005). NmlR of *Neisseria gonorrhoeae*: a novel redox responsive transcription factor from the MerR family. Mol. Microbiol. *57*, 1676–1689.

Kiley, P.J., and Beinert, H. (2003). The role of Fe–S proteins in sensing and regulation in bacteria. Curr. Opin. Microbiol. *6*, 181–185.

Kim, S.H., Lee, B.Y., Lau, G.W., and Cho, Y.H. (2009). IscR modulates catalase A (KatA) activity, peroxide resistance and full virulence of *Pseudomonas aeruginosa* PA14. J. Microbiol. Biotechnol. *19*, 1520–1526.

Kim, S.O., Merchant, K., Nudelman, R., Beyer, W.F., Jr., Keng, T., DeAngelo, J., Hausladen, A., and Stamler,

J.S. (2002). OxyR: a molecular code for redox-related signaling. Cell *109*, 383–396.

Klebanoff, S.J. (1999). Oxygen metabolites from phagocytes. In Inflammation: Basic Principles and Clinical Correlates, Gallin, J.I., and Snyderman, R., eds (Lippincott Williams & Wilkins, Philadelphia), pp. 721–768.

Klee, S.R., Nassif, X., Kusecek, B., Merker, P., Beretti, J.L., Achtman, M., and Tinsley, C.R. (2000). Molecular and biological analysis of eight genetic islands that distinguish *Neisseria meningitidis* from the closely related pathogen *Neisseria gonorrhoeae*. Infect. Immun. *68*, 2082–2095.

Kline, K.A., and Seifert, H.S. (2005). Mutation of the *priA* gene of *Neisseria gonorrhoeae* affects DNA transformation and DNA repair. J. Bacteriol. *187*, 5347–5355.

Kline, K.A., Sechman, E.V., Skaar, E.P., and Seifert, H.S. (2003). Recombination, repair and replication in the pathogenic *Neisseriae*: the 3 R's of molecular genetics of two human-specific bacterial pathogens. Mol. Microbiol. *50*, 3–13.

Knapp, J.S., and Clark, V.L. (1984). Anaerobic growth of *Neisseria gonorrhoeae* coupled to nitrite reduction. Infect. Immun. *46*, 176–181.

Koppenol, W.H. (1993). The centennial of the Fenton reaction. Free Radic. Biol. Med. *15*, 645–651.

Korshunov, S.S., and Imlay, J.A. (2002). A potential role for periplasmic superoxide dismutase in blocking the penetration of external superoxide into the cytosol of Gram-negative bacteria. Mol. Microbiol. *43*, 95–106.

Krachler, M., Rossipal, E., and Micetic-Turk, D. (1999). Concentrations of trace elements in sera of newborns, young infants, and adults. Biol. Trace Elem. Res. *68*, 121–135.

Kroll, J.S., Langford, P.R., Wilks, K.E., and Keil, A.D. (1995). Bacterial [Cu,Zn]-superoxide dismutase: phylogenetically distinct from the eukaryotic enzyme, and not so rare after all! Microbiology *141*, 2271–2279.

Kroll, J.S., Wilks, K.E., Farrant, J.L., and Langford, P.R. (1998). Natural genetic exchange between *Haemophilus* and *Neisseria*: intergeneric transfer of chromosomal genes between major human pathogens. Proc. Natl. Acad. Sci. U.S.A. *95*, 12381–12385.

Ku, S.C., Schulz, B.L., Power, P.M., and Jennings, M.P. (2009). The pilin O-glycosylation pathway of pathogenic *Neisseria* is a general system that glycosylates AniA, an outer membrane nitrite reductase. Biochem. Biophys. Res. Commun. *378*, 84–89.

Kumar, P., Sannigrahi, S., Scoullar, J., Kahler, C.M., and Tzeng, Y.L. (2011). Characterization of DsbD in *Neisseria meningitidis*. Mol. Microbiol. *79*, 1557–1573.

Lappann, M., Haagensen, J.A., Claus, H., Vogel, U., and Molin, S. (2006). Meningococcal biofilm formation: structure, development and phenotypes in a standardized continuous flow system. Mol. Microbiol. *62*, 1292–1309.

Larsen, B., and Monif, G.R. (2001). Understanding the bacterial flora of the female genital tract. Clin. Infect. Dis. *32*, e69–77.

Laver, J.R., Stevanin, T.M., Messenger, S.L., Lunn, A.D., Lee, M.E., Moir, J.W., Poole, R.K., and Read, R.C. (2010). Bacterial nitric oxide detoxification prevents host cell S-nitrosothiol formation: a novel mechanism of bacterial pathogenesis. FASEB J. *24*, 286–295.

LeCuyer, B.E., Criss, A.K., and Seifert, H.S. (2010). Genetic characterization of the nucleotide excision repair system of *Neisseria gonorrhoeae*. J. Bacteriol. *192*, 665–673.

Lee, J.W., and Helmann, J.D. (2006). The PerR transcription factor senses H2O2 by metal-catalysed histidine oxidation. Nature *440*, 363–367.

Lee, J.W., and Helmann, J.D. (2007). Functional specialization within the Fur family of metalloregulators. Biometals *20*, 485–499.

Leib, S.L., and Tauber, M.G. (1999). Pathogenesis of bacterial meningitis. Infect. Dis. Clin. North. Am. *13*, 527–548, v–vi.

Levine, R.L., Mosoni, L., Berlett, B.S., and Stadtman, E.R. (1996). Methionine residues as endogenous antioxidants in proteins. Proc. Natl. Acad. Sci. U.S.A. *93*, 15036–15040.

Levine, W.C., Pope, V., Bhoomkar, A., Tambe, P., Lewis, J.S., Zaidi, A.A., Farshy, C.E., Mitchell, S., and Talkington, D.F. (1998). Increase in endocervical CD4 lymphocytes among women with nonulcerative sexually transmitted diseases. J. Infect. Dis. *177*, 167–174.

Li, W., Bottrill, A.R., Bibb, M.J., Buttner, M.J., Paget, M.S., and Kleanthous, C. (2003). The Role of zinc in the disulphide stress-regulated anti-sigma factor RsrA from *Streptomyces coelicolor*. J. Mol. Biol. *333*, 461–472.

Li, Y., Hopper, A., Overton, T., Squire, D.J., Cole, J., and Tovell, N. (2010). Organization of the electron transfer chain to oxygen in the obligate human pathogen *Neisseria gonorrhoeae*: roles for cytochromes c4 and c5, but not cytochrome c2, in oxygen reduction. J. Bacteriol. *192*, 2395–2406.

Lim, K.H., Jones, C.E., vanden Hoven, R.N., Edwards, J.L., Falsetta, M.L., Apicella, M.A., Jennings, M.P., and McEwan, A.G. (2008). Metal binding specificity of the MntABC permease of *Neisseria gonorrhoeae* and its influence on bacterial growth and interaction with cervical epithelial cells. Infect. Immun. *76*, 3569–3576.

Lindahl, T. (1993). Instability and decay of the primary structure of DNA. Nature *362*, 709–715.

Lissenden, S., Mohan, S., Overton, T., Regan, T., Crooke, H., Cardinale, J.A., Householder, T.C., Adams, P., O'Conner, C.D., Clark, V.L., *et al.* (2000). Identification of transcription activators that regulate gonococcal adaptation from aerobic to anaerobic or oxygen-limited growth. Mol. Microbiol. *37*, 839–855.

Litwin, C.M., and Calderwood, S.B. (1993). Role of iron in regulation of virulence genes. Clin. Microbiol. Rev. *6*, 137–149.

Liu, Y., Shaw, S.K., Ma, S., Yang, L., Luscinskas, F.W., and Parkos, C.A. (2004). Regulation of leukocyte transmigration: cell surface interactions and signaling events. J. Immunol. *172*, 7–13.

Lowther, W.T., Weissbach, H., Etienne, F., Brot, N., and Matthews, B.W. (2002). The mirrored methionine sulfoxide reductases of *Neisseria gonorrhoeae pilB*. Nat. Struct. Biol. *9*, 348–352.

Lundberg, J.O., and Weitzberg, E. (1999). Nasal nitric oxide in man. Thorax *54*, 947–952.

Lynch, R.E., and Cole, B.C. (1980). *Mycoplasma pneumoniae*: a prokaryote which consumes oxygen and generates superoxide but which lacks superoxide dismutase. Biochem. Biophys. Res. Commun. *96*, 98–105.

McCord, J.M., and Fridovich, I. (1969). Superoxide dismutase. An enzymic function for erythrocuprein (hemocuprein). J. Biol. Chem. *244*, 6049–6055.

McCord, J.M., Keele, B.B., Jr., and Fridovich, I. (1971). An enzyme-based theory of obligate anaerobiosis: the physiological function of superoxide dismutase. Proc. Natl. Acad. Sci. U.S.A. *68*, 1024–1027.

Maiden, M.C., and Stuart, J.M. (2002). Carriage of serogroup C meningococci 1 year after meningococcal C conjugate polysaccharide vaccination. Lancet *359*, 1829–1831.

Masse, E., and Gottesman, S. (2002). A small RNA regulates the expression of genes involved in iron metabolism in *Escherichia coli*. Proc. Natl. Acad. Sci. U.S.A. *99*, 4620–4625.

Masse, E., Escorcia, F.E., and Gottesman, S. (2003). Coupled degradation of a small regulatory RNA and its mRNA targets in *Escherichia coli*. Genes Dev. *17*, 2374–2383.

Mellies, J., Jose, J., and Meyer, T.F. (1997). The *Neisseria gonorrhoeae* gene *aniA* encodes an inducible nitrite reductase. Mol. Gen. Genet. *256*, 525–532.

Mellin, J.R., Goswami, S., Grogan, S., Tjaden, B., and Genco, C.A. (2007). A novel fur- and iron-regulated small RNA, NrrF, is required for indirect fur-mediated regulation of the *sdhA* and *sdhC* genes in *Neisseria meningitidis*. J. Bacteriol. *189*, 3686–3694.

Mellin, J.R., McClure, R., Lopez, D., Green, O., Reinhard, B., and Genco, C. (2010). Role of Hfq in iron-dependent and -independent gene regulation in *Neisseria meningitidis*. Microbiology *156*, 2316–2326.

Metruccio, M.M., Fantappie, L., Serruto, D., Muzzi, A., Roncarati, D., Donati, C., Scarlato, V., and Delany, I. (2009). The Hfq-dependent small noncoding RNA NrrF directly mediates Fur-dependent positive regulation of succinate dehydrogenase in *Neisseria meningitidis*. J. Bacteriol. *191*, 1330–1342.

Monaco, C., Tala, A., Spinosa, M.R., Progida, C., De Nitto, E., Gaballo, A., Bruni, C.B., Bucci, C., and Alifano, P. (2006). Identification of a meningococcal L-glutamate ABC transporter operon essential for growth in low-sodium environments. Infect. Immun. *74*, 1725–1740.

Mongkolsuk, S., and Helmann, J.D. (2002). Regulation of inducible peroxide stress responses. Mol. Microbiol. *45*, 9–15.

Morin, A., Saheb, S.A., Bisaillon, J.G., Beaudet, R., and Sylvestre, M. (1980). *In vitro* inhibition of *Neisseria gonorrhoeae* growth by strict anaerobes. Infect. Immun. *28*, 766–770.

Morita, H., Yoshikawa, H., Sakata, R., Nagata, Y., and Tanaka, H. (1997). Synthesis of nitric oxide from the two equivalent guanidino nitrogens of L-arginine by Lactobacillus fermentum. J. Bacteriol. *179*, 7812–7815.

Muench, D.F., Kuch, D.J., Wu, H., Begum, A.A., Veit, S.J., Pelletier, M.E., Soler-Garcia, A.A., and Jerse, A.E. (2009). Hydrogen peroxide-producing lactobacilli inhibit gonococci *in vitro* but not during experimental genital tract infection. J. Infect. Dis. *199*, 1369–1378.

Mukhopadhyay, P., Zheng, M., Bedzyk, L.A., LaRossa, R.A., and Storz, G. (2004). Prominent roles of the NorR and Fur regulators in the *Escherichia coli* transcriptional response to reactive nitrogen species. Proc. Natl. Acad. Sci. U.S.A. *101*, 745–750.

Naids, F.L., and Rest, R.F. (1991). Stimulation of human neutrophil oxidative metabolism by nonopsonized *Neisseria gonorrhoeae*. Infect. Immun. *59*, 4383–4390.

Neiers, F., Sonkaria, S., Olry, A., Boschi-Muller, S., and Branlant, G. (2007). Characterization of the amino acids from *Neisseria meningitidis* methionine sulfoxide reductase B involved in the chemical catalysis and substrate specificity of the reductase step. J. Biol. Chem. *282*, 32397–32405.

Neilands, J.B. (1981). Microbial iron compounds. Annu. Rev. Biochem. *50*, 715–731.

Nesbit, A.D., Giel, J.L., Rose, J.C., and Kiley, P.J. (2009). Sequence-specific binding to a subset of IscR-regulated promoters does not require IscR Fe–S cluster ligation. J. Mol. Biol. *387*, 28–41.

Newcombe, J., Jeynes, J.C., Mendoza, E., Hinds, J., Marsden, G.L., Stabler, R.A., Marti, M., and McFadden, J.J. (2005). Phenotypic and transcriptional characterization of the meningococcal PhoPQ system, a magnesium-sensing two-component regulatory system that controls genes involved in remodeling the meningococcal cell surface. J. Bacteriol. *187*, 4967–4975.

Newkirk, G.R. (1996). Pelvic inflammatory disease: a contemporary approach. Am. Fam. Physician *53*, 1127–1135.

Norrod, P., and Morse, S.A. (1979). Absence of superoxide dismutase in some strains of *Neisseria gonorrhoeae*. Biochem. Biophys. Res. Commun. *90*, 1287–1294.

O'Dwyer, C.A., Li, M.S., Langford, P.R., and Kroll, J.S. (2009). Meningococcal biofilm growth on an abiotic surface – a model for epithelial colonization? Microbiology *155*, 1940–1952.

Olry, A., Boschi-Muller, S., Marraud, M., Sanglier-Cianferani, S., Van Dorsselear, A., and Branlant, G. (2002). Characterization of the methionine sulfoxide reductase activities of PILB, a probable virulence factor from *Neisseria meningitidis*. J. Biol. Chem. *277*, 12016–12022.

Outten, C.E., and O'Halloran, T.V. (2001). Femtomolar sensitivity of metalloregulatory proteins controlling zinc homeostasis. Science *292*, 2488–2492.

Overton, T.W., Whitehead, R., Li, Y., Snyder, L.A., Saunders, N.J., Smith, H., and Cole, J.A. (2006). Coordinated regulation of the *Neisseria gonorrhoeae*-truncated denitrification pathway by the nitric oxide-sensitive repressor, NsrR, and nitrite-insensitive NarQ-NarP. J. Biol. Chem. *281*, 33115–33126.

Overton, T.W., Justino, M.C., Li, Y., Baptista, J.M., Melo, A.M., Cole, J.A., and Saraiva, L.M. (2008). Widespread distribution in pathogenic bacteria of di-iron proteins that repair oxidative and nitrosative damage to iron–sulfur centers. J. Bacteriol. *190*, 2004–2013.

Paget, M.S., and Buttner, M.J. (2003). Thiol-based regulatory switches. Annu. Rev. Genet. *37*, 91–121.

Paget, M.S., Kang, J.G., Roe, J.H., and Buttner, M.J. (1998). sigmaR, an RNA polymerase sigma factor that modulates expression of the thioredoxin system in response to oxidative stress in *Streptomyces coelicolor* A3(2). EMBO J. *17*, 5776–5782.

Pannekoek, Y., Huis in 't Veld, R., Hopman, C.T., Langerak, A.A., Speijer, D., and van der Ende, A. (2009). Molecular characterization and identification of proteins regulated by Hfq in *Neisseria meningitidis*. FEMS Microbiol. Lett. *294*, 216–224.

Papenfort, K., and Vogel, J. (2010). Regulatory RNA in bacterial pathogens. Cell Host Microbe *8*, 116–127.

Parti, R.P., Biswas, D., Helgeson, S., Michael, F.S., Cox, A., and Dillon, J.A. (2011). Attenuated virulence of min operon mutants of *Neisseria gonorrhoeae* and their interactions with human urethral epithelial cells. Microbes Infect. *13*, 545–554.

Patzer, S.I., and Hantke, K. (1999). SufS is a NifS-like protein, and SufD is necessary for stability of the [2Fe–2S] FhuF protein in *Escherichia coli*. J. Bacteriol. *181*, 3307–3309.

Pauwels, F., Vergauwen, B., and Van Beeumen, J.J. (2004). Physiological characterization of *Haemophilus influenzae* Rd deficient in its glutathione-dependent peroxidase PGdx. J. Biol. Chem. *279*, 12163–12170.

Pecqueur, L., D'Autreaux, B., Dupuy, J., Nicolet, Y., Jacquamet, L., Brutscher, B., Michaud-Soret, I., and Bersch, B. (2006). Structural changes of *Escherichia coli* ferric uptake regulator during metal-dependent dimerization and activation explored by NMR and X-ray crystallography. J. Biol. Chem. *281*, 21286–21295.

Perkins-Balding, D., Ratliff-Griffin, M., and Stojiljkovic, I. (2004). Iron transport systems in *Neisseria meningitidis*. Microbiol. Mol. Biol. Rev. *68*, 154–171, table of contents.

Piacenza, L., Zago, M.P., Peluffo, G., Alvarez, M.N., Basombrio, M.A., and Radi, R. (2009). Enzymes of the antioxidant network as novel determiners of *Trypanosoma cruzi* virulence. Int. J. Parasitol. *39*, 1455–1464.

Pitcher, R.S., Brittain, T., and Watmough, N.J. (2002). Cytochrome cbb(3) oxidase and bacterial microaerobic metabolism. Biochem. Soc. Trans. *30*, 653–658.

Pomposiello, P.J., and Demple, B. (2002). Global adjustment of microbial physiology during free radical stress. Adv. Microb. Physiol. *46*, 319–341.

Potter, A.J., Kidd, S.P., Jennings, M.P., and McEwan, A.G. (2007). Evidence for distinctive mechanisms of S-nitrosoglutathione metabolism by AdhC in two closely related species, *Neisseria gonorrhoeae* and *Neisseria meningitidis*. Infect. Immun. *75*, 1534–1536.

Potter, A.J., Kidd, S.P., Edwards, J.L., Falsetta, M.L., Apicella, M.A., Jennings, M.P., and McEwan, A.G. (2009a). Esterase D is essential for protection of *Neisseria gonorrhoeae* against nitrosative stress and for bacterial growth during interaction with cervical epithelial cells. J. Infect. Dis. *200*, 273–278.

Potter, A.J., Kidd, S.P., Edwards, J.L., Falsetta, M.L., Apicella, M.A., Jennings, M.P., and McEwan, A.G. (2009b). Thioredoxin reductase is essential for protection of *Neisseria gonorrhoeae* against killing by nitric oxide and for bacterial growth during interaction with cervical epithelial cells. J. Infect. Dis. *199*, 227–235.

Pricop, L., Gokhale, J., Redecha, P., Ng, S.C., and Salmon, J.E. (1999). Reactive oxygen intermediates enhance Fc gamma receptor signaling and amplify phagocytic capacity. J. Immunol. *162*, 7041–7048.

Prinz, W.A., Aslund, F., Holmgren, A., and Beckwith, J. (1997). The role of the thioredoxin and glutaredoxin pathways in reducing protein disulfide bonds in the *Escherichia coli* cytoplasm. J. Biol. Chem. *272*, 15661–15667.

Proctor, D.F., and Andersen, I.B., eds (1982). The Nose: Upper Airway Physiology and the Atmospheric Environment (Elsevier Biomedical Press B.V., Amsterdam).

Pullan, S.T., Gidley, M.D., Jones, R.A., Barrett, J., Stevanin, T.M., Read, R.C., Green, J., and Poole, R.K. (2007). Nitric oxide in chemostat-cultured *Escherichia coli* is sensed by Fnr and other global regulators: unaltered methionine biosynthesis indicates lack of S nitrosation. J. Bacteriol. *189*, 1845–1855.

Quinternet, M., Tsan, P., Neiers, F., Beaufils, C., Boschi-Muller, S., Averlant-Petit, M.C., Branlant, G., and Cung, M.T. (2008). Solution structure and dynamics of the reduced and oxidized forms of the N-terminal domain of PilB from *Neisseria meningitidis*. Biochemistry *47*, 8577–8589.

Ranaivoson, F.M., Kauffmann, B., Neiers, F., Wu, J., Boschi-Muller, S., Panjikar, S., Aubry, A., Branlant, G., and Favier, F. (2006). The X-ray structure of the N-terminal domain of PILB from *Neisseria meningitidis* reveals a thioredoxin-fold. J. Mol. Biol. *358*, 443–454.

Ranaivoson, F.M., Antoine, M., Kauffmann, B., Boschi-Muller, S., Aubry, A., Branlant, G., and Favier, F. (2008). A structural analysis of the catalytic mechanism of methionine sulfoxide reductase A from *Neisseria meningitidis*. J. Mol. Biol. 377, 268–280.

Ranaivoson, F.M., Neiers, F., Kauffmann, B., Boschi-Muller, S., Branlant, G., and Favier, F. (2009). Methionine sulfoxide reductase B displays a high level of flexibility. J. Mol. Biol. *394*, 83–93.

Rashad, A.L., Toffler, W.L., Wolf, N., Thornburg, K., Kirk, E.P., Ellis, G., and Whitehead, W.E. (1992). Vaginal PO2 in healthy women and in women infected with *Trichomonas vaginalis*: potential implications for metronidazole therapy. Am. J. Obstet. Gynecol. *166*, 620–624.

Ray, T.D., Lewis, L.A., Gulati, S., Rice, P.A., and Ram, S. (2011). Novel blocking human IgG directed against the pentapeptide repeat motifs of *Neisseria meningitidis* Lip/H.8 and Laz lipoproteins. J. Immunol. *186*, 4881–4894.

Rest, R.F., and Shafer, W.M. (1989). Interactions of *Neisseria gonorrhoeae* with human neutrophils. Clin. Microbiol. Rev. *2*, S83–91.

Rest, R.F., Fischer, S.H., Ingham, Z.Z., and Jones, J.F. (1982). Interactions of *Neisseria gonorrhoeae* with human neutrophils: effects of serum and gonococcal opacity on phagocyte killing and chemiluminescence. Infect. Immun. *36*, 737–744.

Rincon-Enriquez, G., Crete, P., Barras, F., and Py, B. (2008). Biogenesis of Fe/S proteins and pathogenicity: IscR plays a key role in allowing *Erwinia chrysanthemi* to adapt to hostile conditions. Mol. Microbiol. *67*, 1257–1273.

Rock, J.D., and Moir, J.W. (2005). Microaerobic denitrification in *Neisseria meningitidis*. Biochem. Soc. Trans. *33*, 134–136.

Rock, J.D., Thomson, M.J., Read, R.C., and Moir, J.W. (2007). Regulation of denitrification genes in *Neisseria meningitidis* by nitric oxide and the repressor NsrR. J. Bacteriol. *189*, 1138–1144.

Rodionov, D.A., Dubchak, I.L., Arkin, A.P., Alm, E.J., and Gelfand, M.S. (2005). Dissimilatory metabolism of nitrogen oxides in bacteria: comparative reconstruction of transcriptional networks. PLoS Comput. Biol. *1*, e55.

Rohde, K.H., and Dyer, D.W. (2003). Mechanisms of iron acquisition by the human pathogens *Neisseria meningitidis* and *Neisseria gonorrhoeae*. Front Biosci. *8*, d1186–1218.

Roop, R.M., 2nd, Gaines, J.M., Anderson, E.S., Caswell, C.C., and Martin, D.W. (2009). Survival of the fittest: how Brucella strains adapt to their intracellular niche in the host. Med. Microbiol. Immunol. *198*, 221–238.

Rouhier, N., and Jacquot, J.P. (2003). Molecular and catalytic properties of a peroxiredoxin–glutaredoxin hybrid from *Neisseria meningitidis*. FEBS Lett. *554*, 149–153.

Ryden, L. (1984). Structure and Evolution of the Small Blue Proteins (CRC Press Inc., Boca Raton, FL).

Saigh, J.H., Sanders, C.C., and Sanders, W.E., Jr. (1978). Inhibition of *Neisseria gonorrhoeae* by aerobic and facultatively anaerobic components of the endocervical flora: evidence for a protective effect against infection. Infect. Immun. *19*, 704–710.

Saiki, R.K., Gelfand, D.H., Stoffel, S., Scharf, S.J., Higuchi, R., Horn, G.T., Mullis, K.B., and Erlich, H.A. (1988). Primer-directed enzymatic amplification of DNA with a thermostable DNA polymerase. Science *239*, 487–491.

Sainsbury, S., Ren, J., Nettleship, J.E., Saunders, N.J., Stuart, D.I., and Owens, R.J. (2010). The structure of a reduced form of OxyR from *Neisseria meningitidis*. BMC Struct. Biol. *10*, 10.

Salmon, K., Hung, S.P., Mekjian, K., Baldi, P., Hatfield, G.W., and Gunsalus, R.P. (2003). Global gene expression profiling in *Escherichia coli* K12. The effects of oxygen availability and FNR. J. Biol. Chem. *278*, 29837–29855.

Sansone, A., Watson, P.R., Wallis, T.S., Langford, P.R., and Kroll, J.S. (2002). The role of two periplasmic copper- and zinc-cofactored superoxide dismutases in the virulence of *Salmonella choleraesuis*. Microbiology *148*, 719–726.

Sawers, G. (1999). The aerobic/anaerobic interface. Curr. Opin. Microbiol. *2*, 181–187.

Schook, P.O., Stohl, E.A., Criss, A.K., and Seifert, H.S. (2011). The DNA-binding activity of the *Neisseria gonorrhoeae* LexA orthologue NG1427 is modulated by oxidation. Mol. Microbiol. *79*, 846–860.

Schryvers, A.B., and Stojiljkovic, I. (1999). Iron acquisition systems in the pathogenic *Neisseria*. Mol. Microbiol. *32*, 1117–1123.

Schwartz, C.J., Giel, J.L., Patschkowski, T., Luther, C., Ruzicka, F.J., Beinert, H., and Kiley, P.J. (2001). IscR, an Fe–S cluster-containing transcription factor, represses expression of *Escherichia coli* genes encoding Fe–S cluster assembly proteins. Proc. Natl. Acad. Sci. U.S.A. *98*, 14895–14900.

Seaver, L.C., and Imlay, J.A. (2001a). Alkyl hydroperoxide reductase is the primary scavenger of endogenous hydrogen peroxide in *Escherichia coli*. J. Bacteriol. *183*, 7173–7181.

Seaver, L.C., and Imlay, J.A. (2001b). Hydrogen peroxide fluxes and compartmentalization inside growing *Escherichia coli*. J. Bacteriol. *183*, 7182–7189.

Sebastian, S., Agarwal, S., Murphy, J.R., and Genco, C.A. (2002). The gonococcal fur regulon: identification of additional genes involved in major catabolic, recombination, and secretory pathways. J. Bacteriol. *184*, 3965–3974.

Seib, K.L., Jennings, M.P., and McEwan, A.G. (2003). A Sco homologue plays a role in defence against oxidative stress in pathogenic *Neisseria*. FEBS Lett. *546*, 411–415.

Seib, K.L., Tseng, H.J., McEwan, A.G., Apicella, M.A., and Jennings, M.P. (2004). Defenses against oxidative stress in *Neisseria gonorrhoeae* and *Neisseria meningitidis*: distinctive systems for different lifestyles. J. Infect. Dis. *190*, 136–147.

Seib, K.L., Simons, M.P., Wu, H.J., McEwan, A.G., Nauseef, W.M., Apicella, M.A., and Jennings, M.P. (2005). Investigation of oxidative stress defenses of *Neisseria gonorrhoeae* by using a human polymorphonuclear leukocyte survival assay. Infect. Immun. *73*, 5269–5272.

Seib, K.L., Wu, H.J., Kidd, S.P., Apicella, M.A., Jennings, M.P., and McEwan, A.G. (2006). Defenses against oxidative stress in *Neisseria gonorrhoeae*: a system tailored for a challenging environment. Microbiol. Mol. Biol. Rev. *70*, 344–361.

Seib, K.L., Wu, H.J., Srikhanta, Y.N., Edwards, J.L., Falsetta, M.L., Hamilton, A.J., Maguire, T.L., Grimmond, S.M., Apicella, M.A., McEwan, A.G., *et al.* (2007). Characterization of the OxyR regulon of *Neisseria gonorrhoeae*. Mol. Microbiol. *63*, 54–68.

Short, H.B., Clark, V.L., Kellogg, D.S., Jr., and Young, F.E. (1982). Anaerobic survival of clinical isolates and laboratory strains of *Neisseria gonorrhoea*: use in transfer and storage. J. Clin. Microbiol. *15*, 915–919.

Shtibel, R. (1976). Inhibition of growth of *N. gonorrhoeae* by bacterial interference. Can. J. Microbiol. *22*, 1430–1436.

Sim, R.J., Harrison, M.M., Moxon, E.R., and Tang, C.M. (2000). Underestimation of meningococci in tonsillar tissue by nasopharyngeal swabbing. Lancet *356*, 1653–1654.

Simons, M.P., Nauseef, W.M., and Apicella, M.A. (2005). Interactions of *Neisseria gonorrhoeae* with adherent polymorphonuclear leukocytes. Infect. Immun. *73*, 1971–1977.

Sinha, S., Ambur, O.H., Langford, P.R., Tonjum, T., and Kroll, J.S. (2008). Reduced DNA binding and uptake in the absence of DsbA1 and DsbA2 of *Neisseria meningitidis* due to inefficient folding of the outer-membrane secretin PilQ. Microbiology *154*, 217–225.

Sinha, S., Langford, P.R., and Kroll, J.S. (2004). Functional diversity of three different DsbA proteins from *Neisseria meningitidis*. Microbiology *150*, 2993–3000.

Skaar, E.P., Tobiason, D.M., Quick, J., Judd, R.C., Weissbach, H., Etienne, F., Brot, N., and Seifert, H.S. (2002). The outer membrane localization of the *Neisseria gonorrhoeae* MsrA/B is involved in survival against reactive oxygen species. Proc. Natl. Acad. Sci. U.S.A. *99*, 10108–10113.

Soler-Garcia, A.A., and Jerse, A.E. (2004). A *Neisseria gonorrhoeae* catalase mutant is more sensitive to hydrogen peroxide and paraquat, an inducer of toxic oxygen radicals. Microb. Pathog. *37*, 55–63.

Sotto, M.N., Langer, B., Hoshino-Shimizu, S., and de Brito, T. (1976). Pathogenesis of cutaneous lesions in acute meningococcemia in humans: light, immunofluorescent, and electron microscopic studies of skin biopsy specimens. J. Infect. Dis. *133*, 506–514.

Spellerberg, B., Cundell, D.R., Sandros, J., Pearce, B.J., Idanpaan-Heikkila, I., Rosenow, C., and Masure, H.R. (1996). Pyruvate oxidase, as a determinant of virulence in *Streptococcus pneumoniae*. Mol. Microbiol. *19*, 803–813.

St Amant, D.C., Valentin-Bon, I.E. and Jerse, A.E. (2002). Inhibition of *Neisseria gonorrhoeae* by Lactobacillus species that are commonly isolated from the female genital tract. Infect. Immun. *70*, 7169–7171.

St John, G., Brot, N., Ruan, J., Erdjument-Bromage, H., Tempst, P., Weissbach, H., and Nathan, C. (2001). Peptide methionine sulfoxide reductase from *Escherichia coli* and *Mycobacterium tuberculosis* protects bacteria against oxidative damage from reactive nitrogen intermediates. Proc. Natl. Acad. Sci. U.S.A. *98*, 9901–9906.

Stadtman, E.R., Berlett, B.S., and Chock, P.B. (1990). Manganese-dependent disproportionation of hydrogen peroxide in bicarbonate buffer. Proc. Natl. Acad. Sci. U.S.A. *87*, 384–388.

Stefanelli, P., Colotti, G., Neri, A., Salucci, M.L., Miccoli, R., Di Leandro, L., and Ippoliti, R. (2008). Molecular characterization of nitrite reductase gene (*aniA*) and gene product in *Neisseria meningitidis* isolates: is *aniA* essential for meningococcal survival? IUBMB Life *60*, 629–636.

Steichen, C.T., Shao, J.Q., Ketterer, M.R., and Apicella, M.A. (2008). Gonococcal cervicitis: a role for biofilm in pathogenesis. J. Infect. Dis. *198*, 1856–1861.

Steichen, C.T., Cho, C., Shao, J.Q., and Apicella, M.A. (2011). The *Neisseria gonorrhoeae* biofilm matrix contains DNA, and an endogenous nuclease controls its incorporation. Infect. Immun. *79*, 1504–1511.

Stevanin, T.M., Moir, J.W., and Read, R.C. (2005). Nitric oxide detoxification systems enhance survival of *Neisseria meningitidis* in human macrophages and in nasopharyngeal mucosa. Infect. Immun. *73*, 3322–3329.

Stevanin, T.M., Laver, J.R., Poole, R.K., Moir, J.W., and Read, R.C. (2007). Metabolism of nitric oxide by *Neisseria meningitidis* modifies release of NO-regulated cytokines and chemokines by human macrophages. Microbes Infect. *9*, 981–987.

Stohl, E.A., and Seifert, H.S. (2006). *Neisseria gonorrhoeae* DNA recombination and repair enzymes protect against oxidative damage caused by hydrogen peroxide. J. Bacteriol. *188*, 7645–7651.

Stohl, E.A., Criss, A.K., and Seifert, H.S. (2005). The transcriptome response of *Neisseria gonorrhoeae* to hydrogen peroxide reveals genes with previously uncharacterized roles in oxidative damage protection. Mol. Microbiol. *58*, 520–532.

Stork, M., Bos, M.P., Jongerius, I., de Kok, N., Schilders, I., Weynants, V.E., Poolman, J.T., and Tommassen, J. (2010). An outer membrane receptor of *Neisseria meningitidis* involved in zinc acquisition with vaccine potential. PLoS Pathog. *6*, e1000969.

Storz, G., and Imlay, J.A. (1999). Oxidative stress. Curr. Opin. Microbiol. *2*, 188–194.

Storz, G., Tartaglia, L.A., Farr, S.B., and Ames, B.N. (1990). Bacterial defenses against oxidative stress. Trends Genet. *6*, 363–368.

Stroeher, U.H., Kidd, S.P., Stafford, S.L., Jennings, M.P., Paton, J.C., and McEwan, A.G. (2007). A pneumococcal MerR-like regulator and S-nitrosoglutathione reductase are required for systemic virulence. J. Infect. Dis. *196*, 1820–1826.

Taha, M.K., Dupuy, B., Saurin, W., So, M., and Marchal, C. (1991). Control of pilus expression in *Neisseria gonorrhoeae* as an original system in the family of two-component regulators. Mol. Microbiol. *5*, 137–148.

Taha, M.K., So, M., Seifert, H.S., Billyard, E., and Marchal, C. (1988). Pilin expression in *Neisseria gonorrhoeae* is under both positive and negative transcriptional control. EMBO J. *7*, 4367–4378.

Takahashi, H., and Watanabe, H. (2005). A gonococcal homologue of meningococcal gamma-glutamyl transpeptidase gene is a new type of bacterial pseudogene that is transcriptionally active but phenotypically silent. BMC Microbiol. *5*, 56.

Takemoto, T., Zhang, Q.M., and Yonei, S. (1998). Different mechanisms of thioredoxin in its reduced and oxidized forms in defense against hydrogen peroxide in *Escherichia coli*. Free Radic. Biol. Med. *24*, 556–562.

Tala, A., De Stefano, M., Bucci, C., and Alifano, P. (2008). Reverse transcriptase-PCR differential display analysis of meningococcal transcripts during infection of human cells: up-regulation of *priA* and its role in intracellular replication. BMC Microbiol. *8*, 131.

Tala, A., Monaco, C., Nagorska, K., Exley, R.M., Corbett, A., Zychlinsky, A., Alifano, P., and Tang, C.M. (2011). Glutamate utilization promotes meningococcal survival *in vivo* through avoidance of the neutrophil oxidative burst. Mol. Microbiol. *81*, 1330–1342.

Tanaka, I., Suzuki, K., Tanaka, E., and Baba, S. (1996). Investigation of normal bacterial flora in the upper respiratory tract. Acta Otolaryngol. Suppl. *525*, 44–50.

Tao, K., Fujita, N., and Ishihama, A. (1993). Involvement of the RNA polymerase alpha subunit C-terminal region in co-operative interaction and transcriptional activation with OxyR protein. Mol. Microbiol. *7*, 859–864.

Thomas, C.E., and Sparling, P.F. (1994). Identification and cloning of a fur homologue from *Neisseria meningitidis*. Mol. Microbiol. *11*, 725–737.

Thomas, C.E., and Sparling, P.F. (1996). Isolation and analysis of a fur mutant of *Neisseria gonorrhoeae*. J. Bacteriol. *178*, 4224–4232.

Tinsley, C.R., and Nassif, X. (1996). Analysis of the genetic differences between *Neisseria meningitidis* and *Neisseria gonorrhoeae*: two closely related bacteria expressing two different pathogenicities. Proc. Natl. Acad. Sci. U.S.A. *93*, 11109–11114.

Tinsley, C.R., Voulhoux, R., Beretti, J.L., Tommassen, J., and Nassif, X. (2004). Three homologues, including two membrane-bound proteins, of the disulfide oxidoreductase DsbA in *Neisseria meningitidis*: effects on bacterial growth and biogenesis of functional type IV pili. J. Biol. Chem. *279*, 27078–27087.

Toledano, M.B., Kullik, I., Trinh, F., Baird, P.T., Schneider, T.D., and Storz, G. (1994). Redox-dependent shift of OxyR-DNA contacts along an extended DNA-binding site: a mechanism for differential promoter selection. Cell *78*, 897–909.

Touati, D. (2000). Iron and oxidative stress in bacteria. Arch. Biochem. Biophys. *373*, 1–6.

Touati, D., Jacques, M., Tardat, B., Bouchard, L., and Despied, S. (1995). Lethal oxidative damage and mutagenesis are generated by iron in delta *fur* mutants of *Escherichia coli*: protective role of superoxide dismutase. J. Bacteriol. *177*, 2305–2314.

Tschugguel, W., Schneeberger, C., Lass, H., Stonek, F., Zaghlula, M.B., Czerwenka, K., Schatten, C., Kaider, A., Husslein, P., and Huber, J.C. (1999). Human cervical ripening is associated with an increase in cervical inducible nitric oxide synthase expression. Biol. Reprod. *60*, 1367–1372.

Tseng, H.J., Srikhanta, Y., McEwan, A.G., and Jennings, M.P. (2001). Accumulation of manganese in *Neisseria gonorrhoeae* correlates with resistance to oxidative killing by superoxide anion and is independent of superoxide dismutase activity. Mol. Microbiol. *40*, 1175–1186.

Tseng, H.J., McEwan, A.G., Apicella, M.A., and Jennings, M.P. (2003). OxyR acts as a repressor of catalase expression in *Neisseria gonorrhoeae*. Infect. Immun. *71*, 550–556.

Tucker, N.P., Le Brun, N.E., Dixon, R., and Hutchings, M.I. (2010). There's NO stopping NsrR, a global regulator of the bacterial NO stress response. Trends Microbiol. *18*, 149–156.

Tunbridge, A.J., Stevanin, T.M., Lee, M., Marriott, H.M., Moir, J.W., Read, R.C., and Dockrell, D.H. (2006). Inhibition of macrophage apoptosis by *Neisseria meningitidis* requires nitric oxide detoxification mechanisms. Infect. Immun. *74*, 729–733.

Turner, S.M., Reid, E.G., Smith, H., and Cole, J.A. (2003). A novel cytochrome c peroxidase from *Neisseria gonorrhoeae*, a lipoprotein from a Gram-negative bacterium. Biochem. J. *373*, 865–873.

Turner, S.M., Moir, J.W., Griffiths, L., Overton, T.W., Smith, H., and Cole, J.A. (2005). Mutational and biochemical analysis of cytochrome c', a nitric oxide-binding lipoprotein important for adaptation of *Neisseria gonorrhoeae* to oxygen-limited growth. Biochem. J. *388*, 545–553.

Tzeng, Y.L., Kahler, C.M., Zhang, X., and Stephens, D.S. (2008). MisR/MisS two-component regulon in *Neisseria meningitidis*. Infect. Immun. *76*, 704–716.

Unden, G., and Bongaerts, J. (1997). Alternative respiratory pathways of *Escherichia coli*: energetics and transcriptional regulation in response to electron acceptors. Biochim. Biophys. Acta *1320*, 217–234.

Valentin-Hansen, P., Eriksen, M., and Udesen, C. (2004). The bacterial Sm-like protein Hfq: a key player in RNA transactions. Mol. Microbiol. *51*, 1525–1533.

Vanaporn, M., Wand, M., Michell, S.L., Tyson, M.S., Ireland, P.M., Goldman, S., Kewcharoenwong, C., Rinchai, D., Lertmemongkolchai, G., and Titball, R.W. (2011). Superoxide dismutase C is required for intracellular survival and virulence of Burkholderia pseudomallei. Microbiology *157*, 2392–2400.

Vasil, M.L., and Ochsner, U.A. (1999). The response of *Pseudomonas aeruginosa* to iron: genetics, biochemistry and virulence. Mol. Microbiol. *34*, 399–413.

Vijgenboom, E., Busch, J.E., and Canters, G.W. (1997). *In vivo* studies disprove an obligatory role of azurin in denitrification in *Pseudomonas aeruginosa* and show that *azu* expression is under control of RpoS and ANR. Microbiology *143*, 2853–2863.

Virji, M., and Heckels, J.E. (1986). The effect of protein II and pili on the interaction of *Neisseria gonorrhoeae* with human polymorphonuclear leucocytes. J. Gen. Microbiol. *132*, 503–512.

Visser, J.J., Scholten, R.J., and Hoekman, K. (1994). Nitric oxide synthesis in meningococcal meningitis. Ann. Intern. Med. *120*, 345–346.

Vogt, W. (1995). Oxidation of methionyl residues in proteins: tools, targets, and reversal. Free Radic. Biol. Med. *18*, 93–105.

Wang, X., Mukhopadhyay, P., Wood, M.J., Outten, F.W., Opdyke, J.A., and Storz, G. (2006). Mutational analysis to define an activating region on the redox-sensitive transcriptional regulator OxyR. J. Bacteriol. *188*, 8335–8342.

Waters, L.S., and Storz, G. (2009). Regulatory RNAs in bacteria. Cell *136*, 615–628.

Whitehead, R.N., Overton, T.W., Snyder, L.A., McGowan, S.J., Smith, H., Cole, J.A., and Saunders, N.J. (2007). The small FNR regulon of *Neisseria gonorrhoeae*: comparison with the larger *Escherichia coli* FNR regulon and interaction with the NarQ-NarP regulon. BMC Genomics *8*, 35.

WHO (1998). Control of Epidemic Meningococcal Disease, WHO Practical Guidelines, 2nd Edition (WHO/EMC/BAC/98.3).

Wilderman, P.J., Sowa, N.A., FitzGerald, D.J., FitzGerald, P.C., Gottesman, S., Ochsner, U.A., and Vasil, M.L. (2004). Identification of tandem duplicate regulatory small RNAs in *Pseudomonas aeruginosa* involved in iron homeostasis. Proc. Natl. Acad. Sci. U.S.A. *101*, 9792–9797.

Wilks, K.E., Dunn, K.L., Farrant, J.L., Reddin, K.M., Gorringe, A.R., Langford, P.R., and Kroll, J.S. (1998). Periplasmic superoxide dismutase in meningococcal pathogenicity. Infect. Immun. *66*, 213–217.

Williams, J.C., Sue, C., Banting, G.S., Yang, H., Glerum, D.M., Hendrickson, W.A., and Schon, E.A. (2005). Crystal structure of human SCO1: implications for redox signaling by a mitochondrial cytochrome c oxidase 'assembly' protein. J. Biol. Chem. *280*, 15202–15211.

Winyard, P.G., Moody, C.J., and Jacob, C. (2005). Oxidative activation of antioxidant defence. Trends Biochem. Sci. *30*, 453–461.

Wizemann, T.M., Moskovitz, J., Pearce, B.J., Cundell, D., Arvidson, C.G., So, M., Weissbach, H., Brot, N., and Masure, H.R. (1996). Peptide methionine sulfoxide reductase contributes to the maintenance of adhesins in three major pathogens. Proc. Natl. Acad. Sci. U.S.A. *93*, 7985–7990.

Wood, Z.A., Schroder, E., Robin Harris, J., and Poole, L.B. (2003). Structure, mechanism and regulation of peroxiredoxins. Trends Biochem. Sci. *28*, 32–40.

Woods, J.P., Dempsey, J.F., Kawula, T.H., Barritt, D.S., and Cannon, J.G. (1989). Characterization of the Neisserial lipid-modified azurin bearing the H.8 epitope. Mol. Microbiol. *3*, 583–591.

Wu, H., Soler-Garcia, A.A., and Jerse, A.E. (2009). A strain-specific catalase mutation and mutation of the metal-binding transporter gene *mntC* attenuate *Neisseria gonorrhoeae in vivo* but not by increasing susceptibility to oxidative killing by phagocytes. Infect. Immun. *77*, 1091–1102.

Wu, H.J., Seib, K.L., Edwards, J.L., Apicella, M.A., McEwan, A.G., and Jennings, M.P. (2005). Azurin of pathogenic *Neisseria* spp. is involved in defense against hydrogen peroxide and survival within cervical epithelial cells. Infect. Immun. *73*, 8444–8448.

Wu, H.J., Seib, K.L., Srikhanta, Y.N., Kidd, S.P., Edwards, J.L., Maguire, T.L., Grimmond, S.M., Apicella, M.A., McEwan, A.G., and Jennings, M.P. (2006). PerR controls Mn-dependent resistance to oxidative stress in *Neisseria gonorrhoeae*. Mol. Microbiol. *60*, 401–416.

Wu, H.J., Seib, K.L., Srikhanta, Y.N., Edwards, J., Kidd, S.P., Maguire, T.L., Hamilton, A., Pan, K.T., Hsiao, H.H., Yao, C.W., *et al.* (2010). Manganese regulation of virulence factors and oxidative stress resistance in *Neisseria gonorrhoeae*. J. Proteomics *73*, 899–916.

Wu, J., Neiers, F., Boschi-Muller, S., and Branlant, G. (2005). The N-terminal domain of PILB from *Neisseria meningitidis* is a disulfide reductase that can recycle methionine sulfoxide reductases. J. Biol. Chem. *280*, 12344–12350.

Xu, J., and Verstraete, W. (2001). Evaluation of nitric oxide production by lactobacilli. Appl. Microbiol. Biotechnol. *56*, 504–507.

Yariv, J., Kalb, A.J., Sperling, R., Bauminger, E.R., Cohen, S.G., and Ofer, S. (1981). The composition and the structure of bacterioferritin of *Escherichia coli*. Biochem. J. *197*, 171–175.

Yeo, W.S., Lee, J.H., Lee, K.C., and Roe, J.H. (2006). IscR acts as an activator in response to oxidative stress for the suf operon encoding Fe–S assembly proteins. Mol. Microbiol. *61*, 206–218.

Yesilkaya, H., Kadioglu, A., Gingles, N., Alexander, J.E., Mitchell, T.J., and Andrew, P.W. (2000). Role of manganese-containing superoxide dismutase in oxidative stress and virulence of *Streptococcus pneumoniae*. Infect. Immun. *68*, 2819–2826.

Zamocky, M., Furtmuller, P.G., and Obinger, C. (2008). Evolution of catalases from bacteria to humans. Antioxid. Redox Signal. *10*, 1527–1548.

Zheng, H.Y., Hassett, D.J., Bean, K., and Cohen, M.S. (1992). Regulation of catalase in *Neisseria gonorrhoeae*. Effects of oxidant stress and exposure to human neutrophils. J. Clin. Invest. *90*, 1000–1006.

Zheng, H.Y., Alcorn, T.M., and Cohen, M.S. (1994). Effects of H2O2-producing lactobacilli on *Neisseria gonorrhoeae* growth and catalase activity. J. Infect. Dis. *170*, 1209–1215.

Zheng, M., and Storz, G. (2000). Redox sensing by prokaryotic transcription factors. Biochem. Pharmacol. *59*, 1–6.

Zheng, M., Aslund, F., and Storz, G. (1998). Activation of the OxyR transcription factor by reversible disulfide bond formation. Science *279*, 1718–1721.

Zheng, M., Wang, X., Templeton, L.J., Smulski, D.R., LaRossa, R.A., and Storz, G. (2001). DNA microarray-mediated transcriptional profiling of the *Escherichia coli* response to hydrogen peroxide. J. Bacteriol. *183*, 4562–4570.

Zhu, C., Ngeleka, M., Potter, A.A., and Allan, B.J. (2002). Effect of fur mutation on acid-tolerance response and *in vivo* virulence of avian septicemic *Escherichia coli*. Can. J. Microbiol. *48*, 458–462.

Zumft, W.G. (1997). Cell biology and molecular basis of denitrification. Microbiol. Mol. Biol. Rev. *61*, 533–616.

Stress Response in *Listeria monocytogenes*

4

Ewa Wałecka and Jacek Bania

Abstract

Listeria monocytogenes is a food-borne pathogen widespread in the environment. The majority of human listeriosis is associated with consumption of contaminated food. It has the ability to invade many types of non-phagocytic cells and spread from cell to cell, crossing important barriers in host organism. Despite intensified surveillance in food manufacturing serious cases of listeriosis are still reported. Before *L. monocytogenes* causes disease it has to endure adverse conditions encountered in food during its processing and storage, such as supraoptimal temperatures, low pH, high osmolarity, presence of oxidants. In the human intestinal tract *L. monocytogenes* must overcome another set of challenges as the low pH of the stomach, volatile fatty acids, low oxygen levels, osmotic stress, nutrient variability, bile stress and natural flora in the intestine. To survive in hostile environment bacteria adjust their metabolism which involves expression of stress response genes. Consequently, bacteria synthesize proteins that repair damages, maintain the cell stability, eliminate the stress factor, and restore homeostasis. The stress response not only affects *L. monocytogenes* resistance to subsequent doses of stress factors, but can also alter the pathogen's virulence.

Introduction

Characteristic of the pathogen

Listeria monocytogenes was first time described, as a cause of lethal disease in rodents, in England by E.G.D. Murray in 1926, who called it *Bacterium monocytogenes* (Swaminathan and Gerner-Smidt, 2007, Starvu *et al.*, 2011). One year later Pirie isolated the same microorganism from the liver of gerbil in South Africa. He proposed the name *Listerella hepatolytica*, honouring British surgeon Lister, a pioneer of antisepsis in surgery. Finally, the name *Listeria monocytogenes* was adopted (Farber and Peterkin, 1991). The bacterium was officially recognized as a human pathogen in the late 1970s, and as a food-borne pathogen in the 1980s (Low and Donachie 1997; Stavru *et al.*, 2011).

L. monocytogenes is a small, non-sporulating, facultatively anaerobic, Gram-positive bacterium (Farber and Peterkin, 1991). *L. monocytogenes* is relatively nutritionally undemanding and may grow over wide range of temperatures (3–45°C) and pH (5–9.6), but also tolerates up to 10% of salt (Bajard *et al.*, 1996). The bacterium is widespread in the environment, including soil, water, sewage, farming effluents, the faeces of humans and animals. Its natural habitat is decaying vegetation, where it lives as a saprophyte (Vázquez-Boland *et al.*, 2001). The species comprises 13 serotypes, which are classified into three genetic lineages (Ragon *et al.*, 2008). Lineage I contains serotypes 1/2b, 3b, 4b, 4d, 4e, whereas serotypes 1/2a, 1/2c/3a/and 3c are found in lineage II. Lineage III comprises serotypes rarely isolated from humans, i.e. 4a and 4c (Zhang *et al.*, 2003a). Over 90% of human listeriosis is caused by serotypes 1/2a, 1/2b, and 4b, from which 1/2a and 1/2b are the most frequently isolated from food and the food production environment, and 4b is responsible for the majority of human listeriosis outbreaks (Swaminathan and Gerner-Smidt, 2007).

Epidemiology

Because of its wide distribution, the pathogen frequently contaminates food. The majority of human listeriosis is associated with consumption of food containing this pathogen, especially dairy products, soft cheeses, smoked fish, seafood and ready to eat food (RTE food) (Vázquez-Boland *et al.*, 2001). Infectious dose of *L. monocytogenes* has not been established, as it may vary depending on the strain virulence and individual susceptibility of the host. However, according to estimates, even dose of 100 CFU/ml may be sufficient to trigger the disease (Fretz *et al.*, 2010). Clinical symptoms of listeriosis range from mild infections, such as gastroenteritis in immunocompetent individuals, to severe disseminated form of infection, such as septicaemia, abortion, meningitis and encephalitis in immunocompromised individuals, i.e. pregnant women, neonates, and elderly people (McGann *et al.*, 2007). In 47% of infected patients *L. monocytogenes* causes central nervous system infections. The pathogen is one of the most frequent causes of meningitis in North America and Western Europe (Drevets and Bronze, 2008). Therefore, the US Department of Agriculture (USDA) and the Food and Drug Administration, Centre for Food Safety established zero tolerance for *L. monocytogenes* in RTE food (Voetsch *et al.*, 2007). Despite increased surveillance in the food industry *L. monocytogenes* becomes still a serious threat. The USDA registered 333 food recalls from 2002 to 2006, 32.4% of which were due to *L. monocytogenes*. The incidence of listeriosis ranges from 0.1 to 11.3 per million people in different countries, with the highest incidence among patients with AIDS (Swaminathan and Gerner-Smidt, 2007; Drevets and Bronze, 2008). In the USA the pathogen is responsible for 28% of all deaths from food-borne diseases (Mead *et al.*, 1999; Lynch *et al.*, 2007). The 2008 Canadian listeriosis outbreak has accounted for 57 cases of listeriosis, with 22 of lethal outcome (http://www.fda.gov/oc/po/firmrecalls/archive.html). In Europe outbreaks of listeriosis have been also reported. Recent outbreak in Austria and Germany, comprised 33 patients hospitalized from June 2009 to February 2010, including eight fatalities. The outbreak was caused by two different clones of serotype1/2a, and was due to consumption of a sour milk curd cheese (Fretz *et al.*, 2010).

Pathogenesis

L. monocytogenes is an intracellular pathogen that is capable of crossing three important host barriers: the intestinal barrier, placental barrier and blood–brain barrier (Vázquez-Boland *et al.*, 2001). Moreover, the bacterium is able to survive within macrophages and to resist intracellular killing (Cossart and Toledo-Arana, 2008). The pathogen invades the intestinal mucosa, using the M cells of Peyer's patches and intestinal epithelial cells as an entry portal, and subsequently disseminates via the lymph and blood vessels to deeper tissues. The primary target organs for *L. monocytogenes* are the liver and the spleen, leading to a microabcess formation. Finally, conquering two further barriers the bacterium may trigger septicaemia, meningoencephalitis, and abortion (Vázquez-Boland *et al.*, 2001).

L. monocytogenes has the ability to invade many types of non-phagocytic cells and spread from cell to cell. The cycle begins from the adhesion to the surface of eukaryotic cell (Vázquez-Boland *et al.*, 2001). Following adhesion the pathogen induces its own internalization via zipper-like phagocytosis, involving interaction between bacterial surface ligands and their cellular receptors (Lecuit *et al.*, 1997). The most important bacterial ligands are internalin inlA and inlB. Internalins belong to a superfamily of proteins containing an N-terminal domain with leucine-rich repeats (LRRs) that interact with their respective host-cell receptors. InlA promotes entry into epithelial cells, endothelial cells, and some hepatocytes by interacting with E-cadherin (Bierne *et al.*, 2007). The carboxyl-terminal domain of E-cadherin binds β-catenin, which subsequently interacts with α-catenin, leading to actin cytoskeleton rearrangement and thus allowing the pathogen to enter the host cell (Lecuit *et al.*, 2000). InlB is indispensable for internalization of epithelial cells, endothelial cells, hepatocytes, and fibroblasts. InlB has three receptors, i.e. gC1qR/p32, Met/HGFR (hepatocyte growth factor receptor) and GAGs (glycosaminoglycans), from which the most important is Met protein. InlB binding to Met activates tyrosine kinase and, finally, PI-3 kinase, leading to the activation of the Arp2/3 complex, actin rearrangement, and phagocytosis (Bierne *et al.*, 2007).

Upon entry the bacterium is encapsulated in a vacuole from which it escapes after about 30 minutes (Tilney *et al.*, 1989). The most important virulence determinant, which enables the pathogen to disrupt phagosomal membranes, is listeriolysin O (LLO). LLO is a haemolysin that belongs to family of cholesterol-dependent, pore-forming toxins. The enzyme is active in a narrow range of pH, with the optimum at pH 5.5 (Vázquez-Boland *et al.*, 2001). The N-terminal region of LLO contains PEST-like sequence (region enriched in proline, glutamic acid, serine, and threonine), which is required for targeting eukaryotic proteins to phosphorylation and degradation. It was suggested that PEST-like sequence targets LLO for phosphorylation and degradation in the host cytosol, restricting its pore-forming activity to the phagosome (Decatur and Portnoy, 2000). However, the recent study has showed that the presence of the PEST-like region does not influence the ubiquitination and intracellular half-life of the protein, but regulates the amount of LLO in the cytosol (Schnupf *et al.*, 2006). The study by Singh *et al.* (2008) has reported the role of γ-interferon-inducible lysosomal thioloxidoreductase (GILT) in listeriolysin activation. GILT, present in phagosomes, reduces the single cysteine residue of LLO, at position 485, thereby promoting its pore-forming activity (Singh *et al.*, 2008). Phagosomal escape is additionally determined by activity of two phospholipases. PI-PLC, displays high specificity for phosphatidylinositol and is responsible for the efficient escape from phagosomes of primary murine macrophages (Camilli *et al.*, 1993). The second phospholipase, PC-PLC (lecithinase), is highly specific for phosphatidylcholine, and is indispensable for the lysis of the double-membrane phagosome (Smith *et al.*, 1995). The activation of the lecithinase requires zinc metalloprotease (Poyart *et al.*, 1996).

After escape to the cytosol, *L. monocytogenes* needs to adjust its metabolism to replicate efficiently. This adaptation requires expression of several proteins that are crucial for the intracellular growth, e.g. a sugar uptake system, the hexose phosphate transporter (Hpt) (Starvu *et al.*, 2011). The movement towards adjacent cells, hence spread within the host, is mostly due to ActA protein. ActA is a surface protein with three distinct domains: N-terminal domain rich in cationic residues, central region of proline-rich repeats, and C-terminal domain. The N-terminal region is responsible for F-actin assembly and movement (Lasa *et al.*, 1995). ActA mimics the host Wiskott–Aldrich syndrome protein (WASP), thereby being able to recruit actin-related proteins 2 and 3 (Arp2/3 complex), indispensable for the nucleation of actin polymerization. In turn, the central region of ActA binds vasodilator-stimulated phosphoprotein (VASP), which interacts with the G-actin binding protein, called profilin, which supports filament elongation. Additionally, α-actinin cross-links actin filaments, stabilizing the tail, and actin depolymerizing factor (ADF or cofilin) disassembles old filaments, preventing their elongation (Cameron *et al.*, 2000; Starvu *et al.*, 2011) (Fig. 4.1).

Stress and virulence

Before *L. monocytogenes* causes disease, it has to endure stress encountered outside and inside the host. Stress in this context means any adverse conditions or factors that negatively affect bacterial growth or survival. During food preservation, preparation, and storage the pathogen is exposed to a wide range of stresses of different magnitude and outcome, including physical treatments, e.g. heat, radiation, high pressure processing, pulsed electric field, and ultrasonic waves, as well as the presence of chemicals such as salts, acids, and oxidants (Yousef and Polly, 2002). Following ingestion, the bacterium encounters another challenge, the low pH of the stomach. Bacteria that survive these adverse conditions, enter the small intestine, where they need to deal with volatile fatty acids, low oxygen levels, osmotic stress, nutrient variability and intestinal flora. In the human intestine *L. monocytogenes* must be able to overcome bile stress, as the gall bladder secrets up to 1 l of bile into the intestinal tract each day (Gahan *et al.*, 1999; Begley *et al.*, 2002; Lungu *et al.*, 2009). To survive in such a deleterious environment bacteria must adjust their metabolism, which involves expression of a new set of genes. Consequently, bacteria synthesize proteins

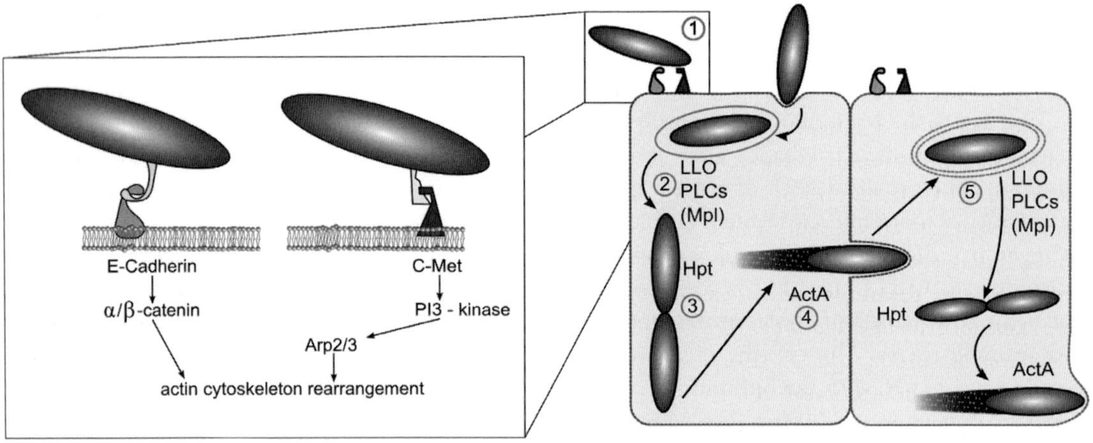

Figure 4.1 Intracellular life cycle of *L. monocytogenes*. (1) *L. monocytogenes* adheres to the host cell surface and induces its own internalization. The entry is promoted by internalins: inlA and inlB, which interact with their specific ligands, E-cadherin and c-Met, respectively. (2) Internalized bacteria are encapsulated in a phagosome, from which they escape owing to expression of listeriolysin O (LLO), and phospholipases PlcA and PlcB (activated by metalloprotease Mpl). (3) In the cytosol the pathogen adapts its metabolism and replicates, expressing several genes including hexose phosphate transporter (Hpt). (4) The bacterium spreads to the adjacent cells producing ActA protein, where it ends up in a double-membrane vacuole, from which it escapes via LLO and PLCs expression (5).

that repair damage, maintain the cell stability, eliminate the stress factor and restore homeostasis (Yousef and Polly, 2002).

Taking into account the fact that *L. monocytogenes* is highly resistant to environmental factors such as low temperature, acidic pH, and high salt concentration, some preservative procedures applied during food processing may not reduce effectively the pathogen's number, allowing it to survive. Bacteria under stress conditions initiate the stress response. The stress response may not only increase the pathogen's resistance to subsequent doses of stress factors, but can also contribute to enhancement of bacterial virulence. Therefore, bacteria with more effective mechanism of stress response are thought to be more virulent (Roche *et al.*, 2005). Virulence of *L. monocytogenes* strains seems to be heterogeneous. Around 8–21% of *L. monocytogenes* population is weakly virulent or non-virulent (Roche *et al.*, 2005). Lineage I strains were found to more efficiently invade epithelial cells, and spread from cell to cell. They had also significantly shorter intracellular doubling time than lineage II strains (Wiedmann *et al.*, 1997; Zhou *et al.*, 2005).

Virulence heterogeneity of *L. monocytogenes* strains may be partially explained by defects in virulence genes. The invasion efficiency can be influenced by the presence of premature stop codons in *inlA* gene, which lead to the production of truncated InlA (Nightingale *et al.*, 2005, 2008). In addition to internalins few other genes are involved in adhesion and invasion, e.g. Ami (amidase), p60 (invasion-associated protein), ActA, FbpA, and FlaA (flagellin), and mutations in genes encoding these proteins may negatively affect adhesion ability (Chatterjee *et al.*, 2005). Loss of *hly* gene, encoding LLO, or its overexpression may also contribute to impaired virulence. Increased level of LLO may be due to mutations in the PEST-like sequence or mutations that change the LLO pH optimum, increasing its activity in cytoplasm leading to damage of the host cell membrane and destruction of the pathogen by the host immune system (Roberts *et al.*, 2005). Virulence attenuation may also be caused by nucleotide insertions, introducing an early termination codon or the nucleotide substitution in *prfA* gene, encoding positive regulatory factor of virulence genes (Roche *et al.*, 2005).

Heat stress

Heating is one of the most influential hurdles that bacteria have to deal with during food preservation. Heating may be mediated by a range of means such as water, steam, hot air, as well as electrical, light, ultrasound or microwave energy. Thermal energy affects not only the cell as a whole, but also its particular constituents, the structures and the molecules. Hence, the basal heat resistance may be due to the intrinsic stability of macromolecules, i.e. RNA, ribosomes, nucleic acids, enzymes and proteins inside the cell and the membrane (Earnshaw *et al.*, 1995).

The mechanism of heat resistance is complex and not fully understood. Nonetheless, it involves production of heat shock proteins (Hsps) (Lundén *et al.*, 2008). An increase in temperature above the optimum of growth results in accumulation of unfolded and non-functional proteins that can either regain its native structure, or undergo proteolysis (Skowyra *et al.*, 1990). Hsps and their cohorts are present at normal temperatures, performing essential functions in the cell. In response to elevated temperatures bacteria synthesize increased amounts of these proteins (Lindquist and Craig, 1988). Hsps play a role in the repair of heat denatured proteins and stabilization of polypeptides already existing in the cell, preventing their misfolding or aggregation (Yura and Nakahigashi, 1999). Molecular chaperones being expressed on the cell surface can also act as adhesins or mitogens. Therefore they are thought to function as signalling virulence factors (Henderson *et al.*, 2006). There are three classes of heat-inducible genes in *L. monocytogenes*, class I and III which are negatively regulated by HrcA, and CtsR, respectively, and class II genes positively regulated by σ^B (Hu *et al.*, 2007a,b).

Molecular chaperones and their role in virulence

The best known molecular chaperones, involved in heat stress response, are GroEL, GroES, DnaK and DnaJ. According to the molecular weight chaperones are classified into different families (Muga and Moro, 2008). GroEL belongs to chaperonins, i.e. Hsp60 family. In conjunction with its co-chaperonin GroES (Hsp10) GroEL is required for bacterial growth, playing role in fundamental cellular processes both at elevated and suboptimal temperatures (Fayet *et al.*, 1989). Another Hsps important for viability at elevated temperatures are DnaK (Hsp70) and its co-chaperone DnaJ (Hsp40). These four chaperones cooperate in maintaining appropriate structure of proteins at high temperatures (Langer *et al.*, 1992). The mechanism of chaperones action includes a sequence of events. Accumulation of unfolded proteins generated in supraoptimal temperatures is thought to be primary signal to trigger heat shock response. DnaK and DnaJ interact with unfolded proteins, stabilizing their conformation, then GrpE mediates the transfer of proteins to GroE which is responsible for proper folding (Craig *et al.*, 1993). An increased synthesis of GroE and DnaK has also been observed in response to ethanol, salt and cold stresses, indicating their role in general stress response (Salotra *et al.*, 1995; Kilstrup *et al.*, 1997).

Some of chaperones have been found to play a role in *L. monocytogenes* virulence. Genes encoding class I stress proteins, i.e. *gro*EL, *gro*ES, *dna*J, *dna*K and *grp*E were shown to be up-regulated during vacuolar survival of *L. monocytogenes* but only *dna*J, *dna*K and *grp*E were still induced in cytosol (Chatterjee *et al.*, 2006). Increased transcription of *gro*ESL was found during infection of murine J774 macrophage cell line (Gahan *et al.*, 2001). DnaK was shown to influence macrophage phagocytosis, but not intracellular survival of *L. monocytogenes*. *dna*K deletion affects adhesion of *L. monocytogenes* mutants. However, all already characterized DnaK proteins were found only in the cytoplasm and do not possess sequences targeting them to the surface, suggesting that bacterial adhesion is influenced indirectly by this protein (Hanawa *et al.*, 1999). It was demonstrated that DnaK is involved in the transcription of virulence factors, i.e. flagellin shown to facilitate the contact with epithelia contributing to invasion, and LmaB involved in induction of delayed hypersensitivity in *L. monocytogenes*-infected mice (Hanawa *et al.*, 2002). DnaK was shown to be involved in biofilm formation and disinfectant resistance in *L. monocytogenes* (van der Veen and Abee, 2010).

Regulation of molecular chaperones

Genes encoding molecular chaperones of *L. monocytogenes* are organized in hexacistronic *dnaK* operon (*hrcA-grpE-dnaK-dnaJ-orf35-orf29*) (Hanawa *et al.*, 2000) and bicistronic *groESL* operon (*groES-groEL*) (Gahan *et al.*, 2001). Upstream *hrcA* and *groES* the presence of controlling inverted repeat of chaperone expression (CIRCE) has been reported (Segal and Ron, 1996; Haley *et al.*, 2007). Both operons are negatively regulated by the HrcA repressor, interacting with an operator CIRCE (Zuber and Schumann, 1994). The activity of HrcA is modulated by GroE proteins. Only active HrcA is able to bind to its operators and repress transcription (Mogk *et al.*, 1997). At optimal growth temperatures HrcA is kept in active form via interaction with GroE (Schumann *et al.*, 2003). Unfolded proteins, appearing immediately after temperature upshift binds GroE leading to its titration. Consequently HrcA repressor is inactivated and dissociates from its operator transiently allowing elevated expression of the *dnaK* and *groESL* operons (Hecker *et al.*, 1996). Removal of unfolded proteins from the cytoplasm leads to increase of pool of free GroE which is observed around 8 minutes after heat shock. Hence, HrcA repressor returns to its active conformation and interacts with the CIRCE operator (Fig. 4.2) (Schulz and Schumann, 1996).

Clp proteins and their role in virulence

Clp (caseinolytic protein) is a large ATP-dependent protein complex that regulates proteolysis and acts as a molecular chaperone participating in protein folding and assembly (Gottesman *et al.*, 1997). In *L. monocytogenes* four different Clp proteins, i.e. ClpB, ClpC, ClpE, and ClpP have been reported (Rouquette *et al.*, 1996; Nair *et al.*, 1999; Gaillot *et al.*, 2000; Chastanet *et al.*, 2004). ClpB, ClpC, and ClpE belong to Hsp100/Clp ATPases

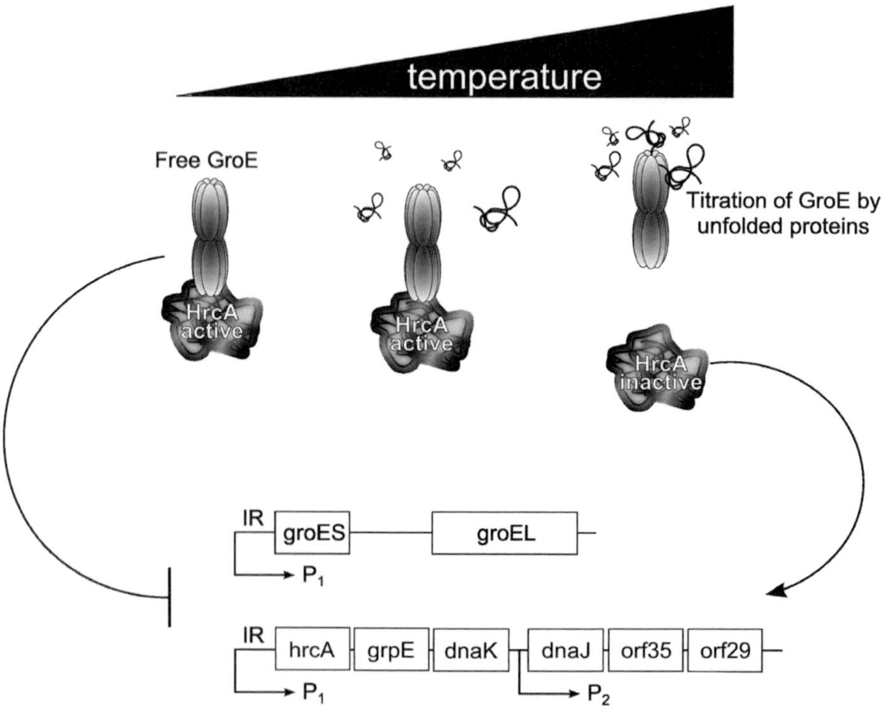

Figure 4.2 The HrcA–GroE regulation system. Molecular chaperones are negatively regulated by the HrcA repressor, which interacts with the CIRCE operator (IR), located upstream *hrcA* and *groES* operons. The HrcA activity is modulated by GroE. At optimal temperatures GroE converts HrcA into active state, contributing to repression of *hrcA* and *groES* operons. After heat shock GroE is titrated by unfolded proteins, and the repressor is inactivated, leading to expression of the operons.

which play various housekeeping functions in bacterial cell, and target proteins to degradation by ClpP proteinase. Although designated as Clp protein ClpP is unrelated to Clp ATPase family. The functional ClpP proteinase is composed of a proteolytic subunit ClpP associated with one of Hsp-100/Clp ATPases (Maurizi *et al.*, 1990). When ATPase component is absent, ClpP itself displays only peptidase activity and is not able to degrade peptide longer than seven amino acids (Woo *et al.*, 1989). The association of ClpP with the regulatory ATPase component in ATP-dependent manner confers proteinase activity to ClpP (Wang *et al.*, 1997). Active Clp participates in degradation of misfolded proteins potentially toxic to bacteria. ClpP has been found to be required for growth at elevated temperatures, cell motility, synthesis of degradative enzymes, or cell competence (Porankiewicz *et al.*, 1999). ClpP has been also shown to be important for intracellular survival and replication, as well as modulation of LLO expression, contributing to the *L. monocytognes* virulence (Gaillot *et al.*, 2000, 2001). ClpB has been shown to play a role in the pathogen's thermotolerance and virulence (Chastanet *et al.*, 2004). ClpC has been found to be required for adhesion and invasion of *L. monocytogenes*, and to modulate expression of virulence factors such as InlA, InlB, and ActA (Nair *et al.*, 2000). ClpC probably protects live bacteria internalized in macrophages by dissolving protein aggregates and participating in refolding of virulence factors (e.g. LLO, PlcA, PlcB), PrfA or other proteins indispensable for intracellular survival. Promoting early escape from the phagosomal compartments of macrophages ClpC contributes to intracellular survival in the host cells (Rouquette *et al.*, 1996, 1998). ClpE was shown to interact synergistically with ClpC during cell division at elevated temperature. Unlike ClpC, ClpE expression is not stimulated by temperature and salt (Nair *et al.*, 1999). Experiments on mice demonstrated involvement of ClpE and ClpC in *L. monocytogenes* virulence, as mutants growth was limited in the spleen and liver. Since expression of InlA, InlB, and ActA remained unchanged in the absence of ClpE, only ClpC seems to be involved in invasion of hepatocytes (Nair *et al.*, 2000).

Regulation of Clp proteins

Genes encoding Clp are located in four transcriptional units. Their expression is negatively regulated by CtsR (class III stress gene repressor) (Nair *et al.*, 2000; Chastanet *et al.*, 2004). The repressor is encoded by first gene of the *clpC* operon. It specifically binds to tandemly arranged heptanucleotide direct repeat of operator sequence. The repressor overlaps start sequence of transcription or −35 and −10 regions, thereby suggesting its action by competing or interfering with RNA polymerase $E\sigma^A$ holoenzyme (Derré, 1999). CtsR acts in dimeric form and comprises at least three functional domains: a dimerization domain (first 25 amino acids), a helix–turn–helix DNA-binding motif (amino acids 29–48) and the central glycine-rich region (amino acids 59–71) probably involved in heat sensing (Derré *et al.*, 2000). It has been suggested that the last domain is essential for proper conformation and activity of CtsR, since the deletion of a single glycine residue results in the loss of DNA binding activity (Derré *et al.*, 2000; Karatzas *et al.*, 2003). CtsR regulon is expressed at low level at 37°C being up-regulated after heat stress. The regulatory mechanism seems to rely on maintaining a certain steady-state level of CtsR at 37°C, followed by its transient inactivation after exposure to heat (Schumann *et al.*, 2003). Zinc finger protein (McsA) and a tyrosine kinase (McsB) play important role in the regulation of Clp (Kirstein *et al.*, 2005). At ambient temperatures, under non-stress conditions, McsA positively influences CtsR, stabilizing its activity. This results in repression of CtsR-dependent genes. After temperature upshift DNA binding capacity of CtsR is modified by McsB, the repressor is directed for ClpCP-mediated degradation, which results in up-regulation of *clp* genes (Krüger *et al.*, 2001) (Fig. 4.3). CtsR has been shown to regulate expression of genes involved in resistance to heat, H_2O_2, and high pressure (Karatzas and Bennik, 2002; Karatzas *et al.*, 2003; Joerger *et al.*, 2006).

HtrA – high-temperature requirement A

HtrA is a heat shock-induced serine proteinase (Poquet *et al.*, 2000). The protein contains the catalytic residues and PDZ domains, involved

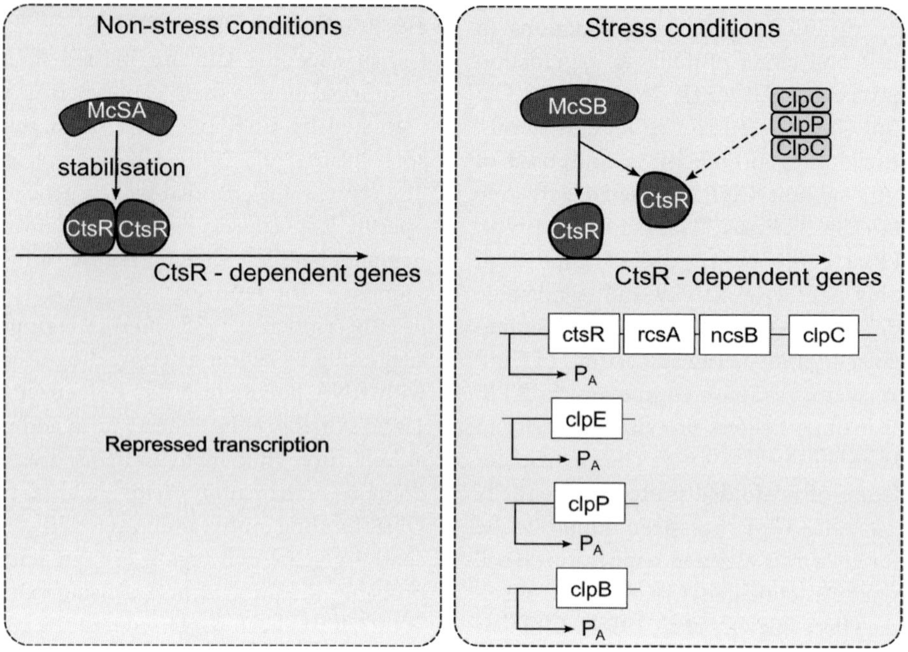

Figure 4.3 Regulation of *clp* genes. *clp* genes are located in four transcriptional units, and are negatively regulated by CtsR. The regulatory mechanism seems to rely on maintenance of steady-state level of CtsR at 37°C, followed by transient inactivation after heat exposure. Under non-stress conditions CtsR is stabilized by McsA, thereby binding promoter sequences and repressing *clp* genes transcription. After heat stress, binding capacity of CtsR is modified by McsB. Consequently McsB is directed for ClpCP-mediated degradation, and *clp* genes are expressed.

in substrate recognition, and located in its C-terminal region (Pallen and Wren, 2000). HtrA displays both chaperone and proteolytic activity, which switches from one to another in temperature-dependent manner (Spiess *et al.*, 1999). At optimal growth temperatures chaperone conformation masks the proteolytic residues abolishing substrate binding and catalysis. At supraoptimal temperatures, the catalytic residues become accessible and proteolytic conformation dominates (Krojer *et al.*, 2002). The role of HtrA is degradation of misfolded proteins that accumulate during heat stress, in order to maintain the bacterial growth. It is also possible that fragments produced by HtrA proteolytic activity serve as signals inducing stress adaptation (Wonderling *et al.*, 2004). The protease has been also shown to play pivotal role for survival of *L. monocytogenes* under adverse conditions including acid, heat, osmotic and oxidative stress (Sleator *et al.*, 2005; Stack *et al.*, 2005; Wilson *et al.*, 2006). It has been shown that *htrA* mutants exhibit about 1-log reduction of pathogenicity, indicating that HtrA is required for full virulence of the pathogen (Stack *et al.*, 2005; Wilson *et al.*, 2006).

The expression of *htrA* gene is regulated by the two-component regulatory system LisRK. The HK component (encoded by *lisK*) is a membrane-associated histidine kinase, which monitors environmental parameters, and RR (encoded by *lisR*) is a cytoplasmic regulator that mediates adaptive response to stress conditions (Parkinson, 1993). The LisRK has been shown to play a role in acid, oxidative and ethanol tolerance and is essential for *L. monocytogenes* virulence and growth in the presence of antimicrobial agents (Cotter *et al.*, 1999; Kallipolitis and Ingmer, 2001; Cotter *et al.*, 2002).

Heat resistance of *L. monocytogenes*

Heat resistance of *L. monocytogenes* has been shown to be impacted by many factors, including the strain tested, its growth phase, and composition of the medium (Doyle *et al.*, 2001). The decimal reduction time (*D*) for *L. monocytogenes* was reported to range from 53 minutes at 52°C to 0.017 minutes at 75°C (Mackey and Bratchell, 1989). It has been found that previous exposure to sodium chloride increases heat resistance (Jørgensen *et al.*, 1995). Low a_w and high concentration of solutes, for example sucrose, glucose, galactose, mannose, lactose, were also found to increase heat resistance of the pathogen (Smith and Hunters, 1988).

Effect of heat stress on *L. monocytogenes* virulence

It has been found that heat stress alters *L. monocytogenes* invasiveness in a time- and growth phase-dependent manner, and is heterogeneous in bacterial population. Heating at 54°C for 20 minutes is accompanied by a decrease of the invasiveness of *L. monocytogenes*, which is more pronounced in exponential-phase bacteria. Prolongation of heating time to 60 minutes results in significant induction of invasiveness especially in the stationary phase bacteria (Wałecka *et al.*, 2011a).

Osmotic stress

Osmotic strength of the environment is one of the main factors affecting bacterial ability to proliferate in particular habitat. Osmotic shock is the change in the osmotic strength of the bacterial environment (Csonka, 1989). Exposure to this type of stress results in rapid water fluxes across the cytoplasmic membrane in direction defined by the external osmolarity of the surrounding (Wood, 1999). Osmotic downshift triggers a water influx, and thus an increase in the turgor, what in extreme cases may lead to the cell burst. To avoid it, and compensate the water entry solutes and solvents are rapidly released to the environment by mechanosensitive channels. In turn, osmotic upshift causes water efflux, contributing to dehydration of the cytoplasm, decrease in the turgor and reduction in the intracellular water activity (Ziegler *et al.*, 2010). *L. monocytogenes* encounters osmotic stress in the food environment, where it is able to withstand 10% NaCl (McClure *et al.*, 1989), as well as in the upper small intestine, in which the NaCl concentration may reach 0.3 M (Chowduhry *et al.*, 1996). To forestall these conditions the pathogen actively regulates distribution of specific solutes across the membrane (Fig. 4.4). Since these molecules accumulate in the cytoplasm only in response to stress, and do not impinge on vital functions of the cell, they are called compatible solutes (Wood, 2007). The ability to accumulate these compounds defines the pathogen's capability to deal with the increased osmotic strength (Angelidis *et al.*, 2003b).

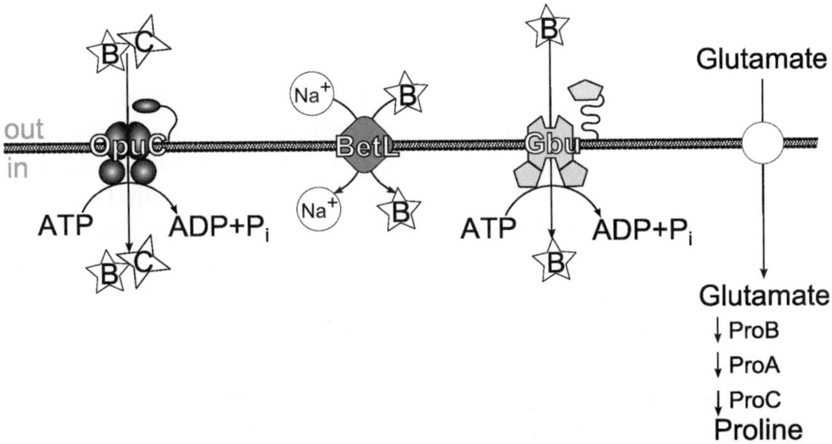

Figure 4.4 Compatible solutes uptake systems in *L. monocytogenes* (details are discussed in the text). B, betaine; C, carnitine; P, proline.

Compatible solutes

Compatible solutes are molecules indispensable for cell turgor and volume restoration that are able to pass across the membrane only with aid of active transport. They can be divided into electrolytes, e.g. K^+ ions or glutamate, and organic solutes such as glycine-betaine or carnitine (Csonka, 1989). It seems that *L. monocytogenes* has evolved the biphasic response to elevated osmolarity, where the accumulation of K^+ ions and glutamate represent the primary stress response, and organic solutes constitute the secondary response (Sleator *et al.*, 2003a). Potassium ions are the most frequent cations in the cytoplasm and play a key role in turgor maintenance. Furthermore, their accumulation under osmotic shock may act as the regulatory signal for glutamate synthesis. However, K^+ seems to be not effective under hyperosmotic conditions (Csonka, 1989). Being highly soluble, either neutral or zwitterionic molecules, organic solutes stabilize the native conformation of macromolecules and provide protection against heating, freezing and drying (Lippert and Galinski, 1992; Wood *et al.*, 2001). The intracellular level of compatible solutes depends on degree of osmotic stress scented by the bacteria. In bacteria they can be accumulated in the cytoplasm by the novo synthesis or by transport from the environment. *L. monocytogenes* does not seem to be able to synthesize compatible solutes *de novo,* thus their uptake occurs only via active transport from the environment (Angelidis *et al.*, 2003a). Transport systems serve to capture compatible solutes from scarce environmental sources, as well as to their re-import when they leak from bacterial cells under high osmolarity (Ziegler *et al.*, 2010). Survival of *L. monocytogenes* under elevated osmotic strength is attributed mainly to accumulation of three solutes: glycine-betaine, carnitine and proline (Beumer *et al.*, 1994). Glycine-betaine is the most prevalent compatible solute present in high concentrations in sugar beets and other plant food products (Beumer *et al.*, 1994). Glycine-betaine acts as stabilizer preventing aggregation and maintaining solubility of cellular proteins. It can alter physical properties of the cell membrane, contributing to both osmoprotection and cryoprotection (Ko *et al.*, 1994). Glycine-betaine cannot be transported directly via the membrane, thus its uptake involves two transport systems. The betaine porter I (BetL) relies on the presence of a hyperosmotic gradient provided by NaCl (Gerhardt *et al.*, 1996). Coupling of betaine transport to the influx of Na^+ ions induces the electric gradient, enabling betaine transport. Na^+ cations are exchanged for protons and thereby removed from the cell (Smith *et al.*, 1998). BetL is highly specific to glycine-betaine and is not involved in transport of other osmolytes such as carnitine or choline (Sleator *et al.*, 1999). This porter is responsible for immediate response to osmotic upshift and long-term adaptation to this kind of stress. In turn, porter II (Gbu) belongs to the family of ATP-binding cassette (ABC) transporters and is Na^+ independent (Ko and Smith, 1999). It couples ATP hydrolysis to substrate translocation across biological membranes (Sleator *et al.*, 1999). Gbu is energized by ascorbate-phenazine methosulfate and its activation depends on the presence of solute gradient imposed by either KCl or sucrose (Gerhardt *et al.*, 2000). This system is required for long-term adaptation and survival under extreme salt stress conditions. Porter II, unlike porter I, contributes to cryotolerance and may be involved in regulation of osmoprotectants level (Gerhardt *et al.*, 2000; Mendum *et al.*, 2002a,b). Gbu is also involved in the alternate carnitine pathway (Angelidis *et al.*, 2003b). Carnitine is biosynthesized by mammals and eukaryotic microorganisms from lysine, where is required for the transport of fatty acids. This compound is widespread in materials of animal origin, and because of its vitamin-like activity serves as food additive (Ko *et al.*, 1994; Verheul *et al.*, 1995). Its uptake requires driving force provided by ATP or equivalent energy-rich phosphorylated intermediate. The major carnitine transporter, i.e. OpuC, is another member of ABC transporters family (Verheul *et al.*, 1995). It was shown that OpuC participates in response to osmotic and cold stresses in *L. monocytogenes* (Anegilidis *et al.*, 2002). Moreover, together with betaine it contributes to listerial barotolerance and may influence the maintenance of membrane fluidity (Smiddy *et al.*, 2004). Proline is another compatible solute linked to salt stress in *L. monocytogenes.* The transport mechanism of this solute has not been characterized. Nonetheless, this amino acid can be synthesized from glutamate via three enzymatic reactions (Sleator *et al.*, 2001c). It has been found that inhibition of the proline synthesis by inactivation of *proAB* locus influences listerial growth and survival under elevated osmolarity, but is not involved in osmoprotection (Patchett *et al.*, 1992; Sleator *et al.*, 2001b).

Only OpuC has been found to affect the *L. monocytogenes* virulence, since *opuC* mutants were impaired in colonization of upper small intestine and in triggering systemic infection after oral inoculation (Sleator *et al.*, 2001a; Wemekamp-Kamphuis *et al.*, 2002).

Regulation of compatible solutes transport

The *betL* encodes a single-component membrane-bound protein (BetL) (Sleator *et al.*, 1999), whereas *gbu* consists of three genes encoding the ATPase subunit (*gbuA*), a six-helix transmembrane channel (*gbuB*), and the substrate-binding protein (*gbuC*) (Ko and Smith, 1999). *opuC* operon is composed of four genes encoding ATP binding protein (*opuCA*), an extracellular substrate binding protein (*opuCC*), and two membrane-associated proteins (*opuCB*, *opuCD*) (Fraser *et al.*, 2000). The analysis of sequences upstream *betL* and *opuCA* genes revealed the presence of promoter binding sites at positions −33 and −58, respectively, exhibiting similarity to the σ^B-dependent promoters (Sleator *et al.*, 1999; Fraser *et al.*, 2000). Decreased level of *opuCA* transcripts has been found in *sigB* mutant in cells stressed with 0.5 M NaCl, indicating direct role of σ^B in regulation of *opuC* transcription (Fraser *et al.*, 2003). This has been also supported by Cetin *et al.* (2004), who indicated that *opuC* is transcribed exclusively from σ^B-dependent promoter. *sigB* mutant has also compromised the betaine uptake in response to osmotic stress (Becker *et al.*, 1998). However, study by Fraser *et al.* (2003) has excluded the direct role of σ^B in regulation of *betL* and *gbu* genes. It appears that the *gbuA* can be expressed from two promoters, one of which *gbuAP2* is under σ^B control (Cetin *et al.*, 2004). Thus, the betaine and carnitine uptake is modulated at the transcriptional level partially in σ^B-dependent manner. The BetL transporter can be additionally activated by NaCl at biochemical level, independently of transcriptional control (Sleator *et al.*, 2003b). Transport of betaine and carnitine can be also controlled at the level of enzyme activity. It has been demonstrated that the uptake of betaine and carnitine is inhibited by the presence of both osmolytes on the inner surface of the membrane. The fact that the carnitine transport is more strongly impeded may explain the preferred uptake of glycine-betaine over carnitine. The inhibition effect is reduced when the pathogen is subjected to osmotic shock. It is probably due to alterations in membrane structure affecting the transporters allosteric binding sites for betaine and carnitine (Verheul *et al.*, 1997). Regulation of proline synthesis has been proposed to occur by feedback inhibition of the first enzyme of the pathway encoded by the *proB* gene (Sleator *et al.*, 2001c).

Other proteins and their regulation

Mass spectrometry and microsequencing revealed 12 proteins induced in response to osmotic shock including general stress response proteins DnaK, ClpC ATPase and ClpP serine proteinases and Ctc. Ctc has been shown to be involved in the osmotolerance of *L. monocytogenes* in the absence of osmoprotectants such as carnitine or glycine-betaine (Duché *et al.*, 1994). The insertional mutant of *ctc* not only displays 2-fold lower growth rate but also shows abnormal morphology. The N-terminal part of this protein shows similarity with the 50s ribosomal protein L25. Being associated at least to some extent with the ribosome, Ctc probably binds 5S rRNA. It can be anticipated that ribosomes acts as a sensor of salt stress via Ctc activity. In turn, presence of a putative σ^B-dependent promoter upstream the *ctc* gene suggests a role of σ^B in its regulation (Gardan *et al.*, 2003). The (p)ppGpp synthetase, encoded by the *relA* gene, has been also demonstrated to play a role in listerial survival in high osmolarity environment (Okada *et al.*, 2002). Guanosine polyphosphates (p)ppGpp are accumulated in response to nutritional starvation, and elevated osmolality. Therefore, it is thought to be a stress response-related factor. Furthermore, studies on *relA* mutants indicate that this protein is required for biofilm formation, as well as virulence in a murine model of infection (Taylor *et al.*, 2002). Regulation of *relA* in *L. monocytogenes* has not been elucidated so far. Nevertheless, it seems that RelA is itself modulator of the alternative sigma factor B under nutritional stress (Gentry *et al.*, 1993; Zhang and Haldenwang, 2003).

Effect of osmotic stress on *L. monocytogenes* survival and virulence

The survival under osmotic stress has been reported to be affected by temperature. It appears that *L. monocytogenes* survives better in the environment of high salinity at lower temperatures since viable cells can be recovered after four weeks from the bacteria cultured in 4% NaCl at 30°C, 10% at 10°C, and 18% at 5°C (Cole *et al.*, 1990). In turn, low NaCl concentration, i.e. 2.2% did not influence the number of viable cells, regardless of temperature (Garner *et al.*, 2006).

It has been found that 10 minutes' exposure to 1.7% NaCl is sufficient to alter the *L. monocytogenes* invasiveness. Prolongation of stress exposure to 1 hour and increase of salt concentration up to 10% do not significantly affect the invasiveness (Wałecka *et al.*, 2011b). Studies by Garner and Olesen demonstrated that both short-term stress and long-term adaptation increase the pathogen ability to invade human cell line. In contrast, Jensen *et al.* (2008) working on the same cells did not find any correlation between osmotic stress and invasion. Our study has found that alteration of invasiveness in response to salt stress strongly depends on strain studied. It appears that the change of invasion efficiency after osmotic shock is impacted by growth phase, as stationary-phase bacteria exhibit greater capacity to enhance their pathogenicity in response to NaCl (Wałecka *et al.*, 2011b).

Acid stress

L. monocytogenes is a neutrophile that is not able to grow at pH below 4.5 (Koutsoumanis *et al.*, 2003). During its life cycle the pathogen is exposed to various conditions, including low pH. Similarly to many other enteric pathogens *L. monocytogenes* has evolved mechanisms of response to acidic environments that aid the pathogen to survive inside the host stomach. The level of acid resistance may greatly influence further progression of infection. Low pH is thought to act as a signal to the microorganism that it invaded host intestinal tract, leading to induction of virulence factors (Audia *et al.*, 2001). Weak acids are the main source of acid-related stress encountered by the bacteria in food. At low pH weak acids exist in uncharged form what enables them to freely diffuse across the membrane. Owing to higher value pH of cytoplasm weak acids dissociate, releasing a proton and decreasing the cytoplasm pH. Hence, the lower external pH is, the more weak acid crosses the membrane and acidifies the cytoplasm (Cotter and Hill *et al.*, 2003). Another acid challenges encountered by *L. monocytogenes* are high level of inorganic acids of gastric fluid and volatile fatty acids in the small intestine (Audia *et al.*, 2001). To survive at low pH, *L. monocytogenes* developed a range of strategies including the exponential-phase-dependent acid stress tolerance (ATR), F_0F_1-ATPase system and synthesis of amino acid decarboxylases (Fig. 4.5) (Foster *et al.*, 1994).

Acid tolerance response

The ATR is an adaptive response that helps bacteria to withstand harsh acidic conditions. The response involves two stages. The first stage, i.e. pre-shock, is activated after exposure to mild acid conditions (pH 5.5–5.8) and induces homeostasis that sustains the internal pH. The second stage, i.e. post-shock, is induced after exposure to acid stress of greater magnitude (pH 4.5 or below), and results in production of acid-shock proteins that take part in the prevention or repair of acid damage (Foster, 1993). Thus ATR serves as a protection mechanism that extends beyond constitutive acid tolerance (Bonnet and Montville, 2005). The effectiveness of this response depends on the initial pH and the length of the adaptive period (Davis *et al.*, 1996). The ATR involves multiple systems for maintenance pH homeostasis which defines the ability to maintain the cytoplasmic pH close to neutrality in spite of fluctuations in external pH. Homeostasis can be achieved by passive mechanisms such as increase of cytoplasm buffering capacity and low proton permeability of the cell membrane or

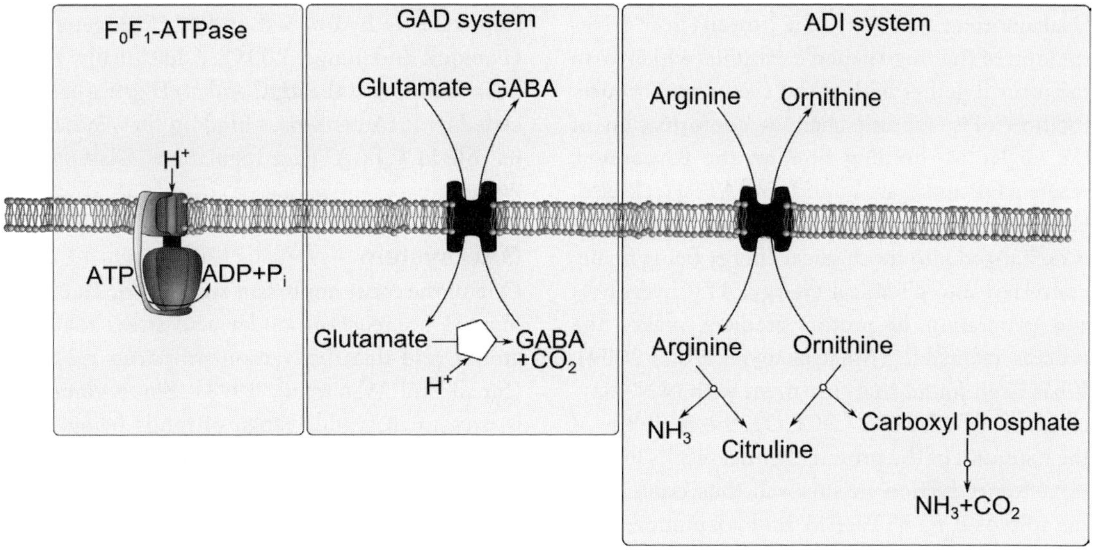

Figure 4.5 Selected acid resistance systems in *L. monocytogenes* (details are discussed in the text).

actively by extrusion of protons by a membrane-bound proton pump (Hill *et al.*, 1995). The active transport of H^+ can be prompted by the energy deriving either from the respiratory chain or ATP hydrolysis (Phan-Thanh and Mahouin, 1999).

The ATR can provide cross-protection against other stresses such as high temperature, ethanol, oxidative and osmotic stress, and nisin (O'Driscoll *et al.*, 1996; Lou *et al.*, 1997; van Schaik *et al.*, 1999; Phan-Thanh *et al.*, 2000). Under moderate acidic conditions ATR was shown to improve listerial attachment to surfaces used during food production, e.g. stainless steel. In extreme acidic environment ATR contributes to increase of *L. monocytogenes* persistence on surfaces (Chorianopoulos *et al.*, 2011). The ability to adapt to low pH influences the virulence of the pathogen (Marron *et al.*, 1997). The ATR affects expression of genes involved in host invasion and intracellular parasitism of *L. monocytogenes* (Conte *et al.*, 2002). It has been shown that natural strains, being constitutively acid tolerant, more efficiently invade enterocyte-like cells and better survive in a murine macrophage-like cells, as well as exhibit increased virulence in a mouse model (O'Driscoll *et al.*, 1996; Conte *et al.*, 2000).

Mechanisms of acid resistance

F_0F_1-ATPase

The F_0F_1-ATPase is a key enzyme in cellular energy conversion and maintenance of pH homeostasis, induced in response to mild acid treatment (Capaldi and Aggeler, 2002). This membrane-bound enzyme consists of cytoplasmic domain (F_1) and membrane domain (F_0), structurally and functionally distinct. F_1 is a water-soluble complex responsible for ATP synthesis/hydrolysis and is bound to F_0 membrane domain which acts as a channel for proton translocation (Fillingame, 1997). The role of F_0F_1-ATPase is to synthesize ATP aerobically when protons move into the cytoplasm, and to generate a proton motive force (PMF) anaerobically, via the expulsion of protons. Therefore, the F_0F_1-ATPase activity leads to increase of intracellular pH in response to pH drop. The F_0F_1-ATPase operon encompasses nine genes, located into a 7-kb DNA region. The F_0 sector is composed of four subunits a, c, b, and δ, encoded by *atpB, atpE, atpF, atpH* genes, respectively. The F_1 sector consists of four subunits α, γ, β and ε encoded by the *atpA, atpG, atpD, atpC* genes (Cotter *et al.*, 2000). The unique feature of this enzyme is rotary mechanism

(Nakamoto *et al.*, 1999). The protons flow drives rotation of the ring-shaped c subunits which form the rotor together with γ and ε subunits. In turn, rotation of γ subunit changes conformation of the nucleotide-binding sites on the β subunits, where ADP and P_i are bound, and ATP is released. Hence, the potential derived from proton gradient is exchanged into mechanical energy being finally converted into chemical energy. ATP hydrolysis and generation of proton gradient makes the process reversible (Mnatsakanyan *et al.*, 2009). It has been found that treatment with N,N'-dicyclohexylcarbodiimide (DCCD), the inhibitor of the c subunit of the proton channel results in over three-log reduction in survival, thus confirming the role of F_0F_1-ATPase in acid adaptation of *L. monocytogenes* (Cotter *et al.*, 2000).

The F_0F_1-ATPase role in listerial pathogenesis has not been established. The analysis of non-pathogenic *L. welshimeri* genome revealed the absence of virulence genes and genes required for intracellular survival, including F_0F_1-ATPase (Hain *et al.*, 2006). Therefore, it can be assumed that this enzyme may be important in virulence of *L. monocytogenes*.

Regulation of F_0F_1-ATPase

The regulation of F_0F_1-ATPase activity is complex and involves various mechanisms, one of which relies on ADP inhibition. When Mg^{2+} ADP binds the F_0F_1-ATPase catalytic site, an inactive complex is formed, resulting in ATP hydrolysis (Murataliev and Boyer, 1992). Energization of the membrane via acid-based transition causes the release of the bound ADP and increases the affinity of ATP synthase for phosphate. As a result ATP is synthesized. Hence, in energized state, elevated affinity for phosphate alleviates ADP inhibition (Feldman and Boyer, 1985). Another mechanism is associated with the ε subunit. The regulation is achieved by conformational transitions of the domain of ε subunit in response to membrane energization, change in ATP/ADP ratio or presence of inhibitors (Feniouk *et al.*, 2006). It has been suggested that decrease of PMF below the thermodynamic threshold for ATP synthesis reverses rotation of subunit in ATP hydrolysis direction, inducing the conformational transition of subunit from the extended to contracted form. In this state ATP is hydrolysed and PMF is generated (Feniouk and Junge, 2005). Additionally, it has been shown that the *atpC* and *atpD* genes are preceded by PrfA consensus binding sites, indicating its role in F_0F_1-ATPase regulation (Glaser *et al.*, 2001).

GAD system

One of the most important strategy for maintaining pH homeostasis under acid stress relies on amino acid decarboxylation-antiporter reactions (Small and Waterman, 1998). Since glutamate is present in a wide range of foods being commonly used to adjust the acidity and enhance the food flavour, glutamate decarboxylase (GAD) system play a key role in acid resistance of *L. monocytogenes*. The system consists of a cytoplasmic glutamate decarboxylase and a putative glutamate:GABA antiporter (Cotter *et al.*, 2001b). At low pH, cytoplasmic GAD enzyme irreversibly converts glutamate of extracellular origin to neutral compound γ-aminobutyrate (GABA). During this reaction an intracellular proton is consumed, resulting in increase of cytoplasmic pH and restoration of homeostasis. Subsequently intracellular GABA is transported out of the cell membrane by a glutamate:GABA antiporter in exchange for a molecule of glutamate (Cotter *et al.*, 2005). The GABA, being less acidic than glutamate, contributes additionally to alkalization of the environment (Olier *et al.*, 2004). In the presence of glutamate the GAD activity affects *L. monocytogenes* acid tolerance, being indispensable for survival in acidified food and the stomach, as well as for colonization of the gut (Cotter *et al.*, 2001a,b). Moreover it probably plays a role in combating the acid stress during intracellular dissociation of bile salts (Begley *et al.*, 2002).

The GAD system consists of three isoforms of glutamic acid decarboxylases (*gadD1*, *gadD2*, *gadD3*) and two antiporters (*gadT1*, *gadT2*). The first two decarboxylases are organized in two operons together with the proper antiporter (*gadD1T1*, *gadT2D2*), whereas *gadD3* is located separately and is transcribed monocistronically (Ryan and Gahan, 2008). GadD2 plays significant role in stationary phase and is responsible for the overall survival at low pH, while GadD1 is more important during exponential phase and facilitates growth at milder acidic conditions (Cotter *et al.*, 2001a). In turn, GadD3 apart from acid conditions, has been shown to be expressed also in response to osmotic shock and entry to stationary phase (Kazmierczak *et al.*, 2003). It has been found that serogroup 4 strains, including the main causative agent of listeriosis serotype 4b, do not posses *gadD1T1* operon (Call *et al.*, 2003). This fact can explain their rare frequency in the food and suggests that the presence of the operon is not essential for *L. monocytogenes* virulence (Cotter *et al.*, 2005).

Little is known about the impact of glutamate decarboxylase system on *L. monocytogenes* virulence. The direct role of GAD system genes in virulence has not been established. However, when taking into account the fact that this system facilitates survival during gastrointestinal passage, it could be stated that GAD influences bacterial pathogenesis indirectly.

Regulation of GAD system

Regulation of the GAD system has not been clearly elucidated. Analysis of the *gad* loci revealed the presence of multiple promoters and putative transcription termination sites, suggesting that multiple regulatory factors may be involved in regulation of the GAD system (Conte *et al.*, 2002). It has been shown that σ^B controls expression of *gad* genes (Wemekamp-Kamphuis *et al.*, 2004). In contrast, a study by Ferreira *et al.* (2003) demonstrated that regulation of the GAD system, as well as net proton movement across the cell membrane seems to be σ^B-independent. However, the recent study by Hu *et al.* (2007a,b) manifests that transcription of *gadT2D2* operon, is affected by σ^B, but also by HrcA and CtsR.

ADI system

Arginine deimination is another strategy for maintaining pH homeostasis, contributing to bacterial survival in acidic conditions. The ADI (arginine deiminase) system supplies ATP, as a result of conversion of arginine to ornithine, CO_2, and NH_3 (Marquis *et al.*, 1987). The ADI system includes three main enzymes: arginine deiminase, ornithine carbamoyltransferase, and carbamate kinase (Casiano-Colón *et al.*, 1988). Extracellular arginine is transported into the cell, where is deaminated via arginine deiminase to citrulline. Citrulline is converted by ornithine carbamoyltransferase to ornithine and carbamoyl-phosphate. Decarboxylation of carbamoyl-phosphate, by carbamate kinase leads to production of CO_2, and NH_3. Ornithine is transported out of the cell in exchange for another molecule of arginine in an energy-independent manner by a membrane-bound antiporter. Two moles of ammonia are produced as a by-product of catabolism of one mole of arginine. Ammonia binds an intracellular proton, yielding ammonium ions (NH_4^+). Thus, the intracellular pH increases, facilitating bacteria to withstand lethal effect of acid stress. Additionally, 1 mole of ATP is supplied by the system, which may be used for extrusion of cytoplasmic protons via F_0F_1-ATPase (Ryan *et al.*, 2009). The system can also work in a non-catabolic direction, such as providing carbamyl phosphate for synthesis of citrulline or pyrimidines, or synthesis of glucose-1-phosphate in the pathway to glycogen (Casiano-Colón *et al.*, 1988).

Three of the ADI genes, i.e. *arcB, arcD, arcC,* encoding ornithine carbamoyltransferase, arginine:ornithine antiporter, and carbamate kinase respectively, are organized in one operon, whereas *arcA* encoding arginine deiminase is separated from the operon by uncharacterized genes. Two of these flanking genes encode putative agmatine deiminases, probably involved in agmatine degradation, supplying carbamoyl phosphate, putrescine, and ATP.

The ADI system has been found to be crucial for acid tolerance at sublethal and lethal pH. The ADI genes have been shown to be transcribed in response to low pH, as well as anaerobiosis and arginine supplementation. The role of ADI genes in listerial pathogenesis has been also reported. *arc* genes and their regulator ArgR are essential for intracellular growth within the macrophage. Null mutants of *arcA* and *argR* genes display 10-fold decrease in survival in the murine model, indicating their role in systemic infection (Ryan *et al.*, 2009). However, neither of these genes contributes to listerial invasion. Joseph *et al.* (2006) have established that *argC* and *argD* genes, involved in arginine synthesis, are up-regulated in human cell line.

Regulation of ADI system

The arginine deiminase system being induced by arginine is regulated by catabolite repression (Casiano-Colón *et al.*, 1988). Expression of *arc* genes is regulated by ArgR, which plays a dual role of positive regulator of ADI system and negative regulator of the arginine biosynthesis pathway. In turn, arginine is essential for maximal efficiency of the system. Supplementation with this amino acid induces transcription of *arc* genes of the catabolic pathway, but also represses genes of anabolic pathway. Anabolic/catabolic switch may be mediated by ArgR. Furthermore, the presence of potential SigB binding sites and the PrfA consensus site has been found upstream ADI genes. The loss of σ^B and PrfA resulted in decreased transcription of *arcA* and *argR* genes, indicating their role in regulation of expression of ADI genes (Ryan *et al.*, 2009).

Effect of acid stress on survival and virulence

The influence of acidic conditions on listerial survival seems to be strain dependent. Studies on five *L. monocytogenes* strains by Werbrouck *et al.* (2009) have shown that time required for inactivation of 8 log CFU at pH 3.5 ranges from 20 to over 50 minutes. Moreover, both, short-term and long-term adaptation to sublethal pH (5.5) results in enhanced survival at lethal conditions (pH 3.5) (Olesen *et al.*, 2009; Werbrouck *et al.*, 2009). Prior exposure to mild acid increases listerial survival in acidified food and gastric fluid (Gahan *et al.*, 1996; Ferreira *et al.*, 2003).

Acid shock was found to impact *L. monocytogenes* virulence. Exposure to organic acids, as well as adaptation to mild acid environment has been demonstrated to increase the *L. monocytogenes* ability to invade CaCo-2 cells (Conte *et al.*, 2000; Garner *et al.*, 2006). Werbrouck *et al.* (2009) have found that exposure of three strains to sublethal acid conditions affects the invasiveness, resulting in inverse effects on invasion, including increase, decrease or no effect. Observed effects correlated with *inlA* expression. In turn, Olesen *et al.* (2009) working on two *L. monocytogenes* strains reported no changes in invasion efficiency upon long-term exposure to hydrochloric acid (pH 5.5).

Oxidative stress

Food-borne pathogens such as *L. monocytogenes* have a limited tolerance for oxygen. They could be exposed to oxidative stress during both intracellular and extracellular growth (Storz and Imlay, 1999). Molecular oxygen is small enough to freely penetrate the membrane. In its triplet state oxygen is a poor oxidant (Messner and Imlay, 1999). Since most biomolecules are poor univalent electron donors, molecular oxygen, as a relatively weak univalent acceptor, cannot efficiently oxidize nucleic acids or amino acids. Stronger univalent oxidants are reactive oxygen species (ROS) such as superoxide anions (O_2^-), hydrogen peroxide (H_2O_2) and hydroxyl radicals ($OH^·$) (Imlay, 2003). ROS constitute an inevitable aspect of aerobic growth being generated during sequential univalent reductions of molecular oxygen catalysed by membrane-associated respiratory chain enzymes. They are also produced in macrophages in response to invasion of pathogenic bacteria (Cabiscol *et al.*, 2000), as well as during industrial processing as a result of radiation, presence of heavy metals, treatments with sanitizers or repeated freezing (Rodriguez-Romo and Yousef, 2002). Moreover, formation of ROS, e.g. O_2^- and $OH^·$ can also be due to acid and heat stress exposure (Mols and Abee, 2011). ROS can be produced both by specific systems under fine cellular control and as side-products of cellular metabolism, simultaneously being degraded (Lushchak, 2011). Imbalance between ROS production and elimination leads to oxidative stress that may result in DNA, proteins and lipids damage (Imlay, 2002). To combat this stress the pathogen mobilizes antioxidant systems. Analysis of the *L. monocytogenes* genome has revealed presence of few genes encoding antioxidants such as superoxide dismutase, catalase, DNA-binding proteins, peroxiredoxins, thioredoxin and thioredoxin reductase (Glaser *et al.*, 2001).

ROS

Molecular oxygen crosses membranes and is converted into mixture of O_2^- and H_2O_2 by direct oxidation of flavoenzymes (Storz and Imlay, 1999) (Fig. 4.6). The initial electron transfer to oxygen occurs in one-electron step

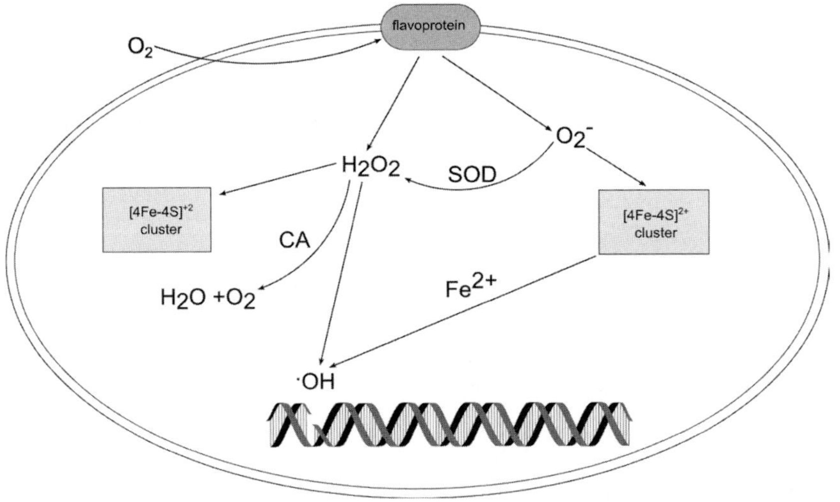

Figure 4.6 Mechanism of oxidative cell damage. Molecular oxygen crosses membranes and is converted into O_2^- and H_2O_2 by direct oxidation of flavoenzymes. O_2^- and H_2O_2 penetrate into the active sites of dehydratases containing $[4Fe–4S]^{2+}$ cluster. O_2^- oxidizes the cluster univalently, releasing the catalytic iron atom inactivating the enzyme. Subsequently iron is reduced to Fe^{2+}. Fe^{2+} reacts with H_2O_2 generating highly reactive hydroxyl radicals, which attacks DNA, blocking replication and contributing to the cell death. Degradation of O_2^- and H_2O_2 is due to activity of superoxide dismutase (SOD) and catalase (CA). SOD coverts O_2^- to H_2O_2 that is next cleaved into H_2O and O_2 by catalase.

generating O_2^- and flavosemiquinone. Nascent O_2^- or the flavosemiquinone species undergo a spin inversion, before release of O_2^-, contributing to orbital overlap and formation of a peroxy adduct to the isoalloxazine ring. As a consequence of heterolytic cleavage H_2O_2 rather than O_2^- is released (Imlay, 2003). The amount of oxidative stress that *L. monocytogenes* experiences depends on the rates at which its enzymes autoxidize and whether H_2O_2 or O_2^- predominates (Messner and Imlay, 2002). Dehydrogenases react very slowly with oxygen, forming mainly H_2O_2. Transferases act rapidly generating mainly O_2^-. In turn, the rapid activity of oxidases results in H_2O_2 predominance (Massey, 1994). The balance between these oxidants is critical for cell physiology and fitness. It seems that hydrogen peroxide is more tolerated by bacterial cell than superoxide anion (Messner and Imlay, 1999). Hydrogen peroxide serves as a first line of defence against microbes invading mammal macrophages. It is also used as a bleaching agent and disinfectant in households and industry (Mols and Abee, 2011). Exposure to H_2O_2 and O_2^- leads to protein and DNA damage. Hydrogen peroxide and superoxide penetrate into the active sites of dehydratases containing $[4Fe–4S]^{2+}$ cluster, binding the iron atom and oxidizing the cluster to unstable form (Imlay, 2008). The O_2^- oxidizes the cluster univalently, releasing the catalytic iron atom and thus inactivating dehydratases. Subsequently, by reaction with cysteine or reduced flavin, iron is reduced to Fe^{2+} that reacts with H_2O_2 (Fenton reaction) generating highly reactive hydroxyl radicals (Zuber, 2009). Since some of Fe^{2+} can be associated with phosphodiester backbone of nucleic acids, DNA is the main target of hydroxyl radicals (Storz and Imlay, 1999). Attack on the bases and sugar moieties causes single- and double-strand breaks in the backbone, blocking replication and contributing to the cell death (Cabiscol *et al.*, 2000).

Lipids are also the major target of ROS, which can attack directly polyunsaturated fatty acids in membrane leading to lipid peroxidation. Lipid peroxidation contributes to decrease in membrane fluidity, thus altering its properties and degrading membrane-bound proteins (Cabiscol *et al.*, 2000).

Antioxidants and their role in *L. monocytogenes* virulence

SOD and catalase

SOD and catalase work synergistically in ROS degradation (Fig. 4.6) (Dallmier and Martin, 1988). Superoxide dismutase (SOD) is a metalloprotein, first time described in 1969 by McCord and Fridovich that isolated the enzyme from bovine erythrocytes (McCord and Fridovich, 1969). All cells have one or more cytosolic SOD. They act by converting superoxide radical anions during aerobic metabolism into H_2O_2 (Kroll *et al.*, 1995). Superoxide radicals are generated during reduction of molecular oxygen, where four electrons are accepted univalently or in pairs. Despite their short half-lives, they are extremely reactive and may contribute to cell damage (Tally *et al.*, 1977). Activity of the enzyme decreases intracellular O_2^- concentration to the level that the cell is able to deal with (Keyer *et al.*, 1995). Three isozymes of SOD have been distinguished on the basis of their central metal cation: iron (Fe), manganese (Mn) and copper/zinc (Cu/Zn). Bacteria contain either Fe–SOD or Mn-SOD or both (Grace, 1990). It was previously suggested that *L. monocytogenes* bears only one form of SOD with iron as a cofactor (Welch *et al.*, 1979). However, results of subsequent research demonstrate that *L. monocytogenes* expresses exclusively manganese form of SOD (Schiavone and Hassan, 1987; Brehm *et al.*, 1992; Vasconcelos and Deneer, 1994). The expression of listerial SOD has been found to be induced by oxidation and the presence of superoxide radical anions, and its activity appears to be primarily extracytoplasmic (Welch *et al.*, 1979). The activity of listerial SOD is down-regulated in stationary phase by serine/threonine phosphorylation. The most active form of the enzyme is non-phosphorylated and in this form SOD is secreted by SecA2 pathway (Archambaud *et al.*, 2006).

H_2O_2 as a product of SOD activity is cleaved into water and oxygen by catalase (CA) (Mongkolsuk and Helmann, 2002). Catalase is an enzyme isolated for the first time from bovine liver in 1937 (Sumner and Dounce, 1937). Bacterial catalases can be divided into three groups: typical (monofunctional) catalases, catalases–peroxidases and non-haem manganese catalases. The second CA group in addition to the conventional catalytic function exhibits a peroxidase activity (Zamocky *et al.*, 2008). Typical catalases represent the largest group, with dual activity. They not only act as a scavenger of hydrogen peroxide, but at low concentration of H_2O_2 they also catalyse the oxidation of electron donors such as ethanol or phenols (Percy, 1984). Bacteria possess one or more isoenzymes of catalase, but usually only one is induced in response to oxidative stress. *Bacillus subtilis* contains two genes encoding typical catalases: oxidative stress-induced KatA and a general stress protein KatE (Engelmann and Hecker, 1996). *Escherichia coli* also have two enzymes, a catalase–peroxidase KatG (HPI), induced during oxidative stress and a typical catalase KatE (HPII) expressed during transition into stationary phase (Loewen, 1996). There is no data about catalase gene in *L. monocytogenes.* It has been demonstrated that CA may influence the ability of *L. monocytogenes* to grow aerobically at low temperatures, as the growth of catalase-deficient mutants is inhibited at 4°C. However, CA is not essential for listerial cryotolerance, since no difference in survival after repeated freezing and thawing between the wild-type and mutant strain was observed (Azizoglu and Kathariou, 2010). Moreover, loss of catalase activity seems to be compensated by elevated level of SOD, as CA-negative mutants produce twice as much of SOD as the wild type (Fisher *et al.*, 2000). The reason for such phenomenon is not clear. Nonetheless, it may involve yet uncharacterized compensatory mechanism (Leblond-Francillard *et al.*, 1989).

SOD and to a less extent catalase have been found to play a protective role during exposure to ozone (Fisher *et al.*, 2000). It has been suggested that the listerial virulence may be affected by the ability to achieve the proper balance between SOD and CA (Welch *et al.*, 1979). It was shown that virulent *L. monocytogenes* strains expressed high level of SOD and catalase, whereas low virulent strains exhibited either low activity of both enzymes or lacked catalase, simultaneously exhibiting high SOD activity. The role of SOD in bacterial virulence has been already postulated. The enzyme has been shown to help the anaerobic pathogens to survive within oxygenated tissues (Tally *et al.*, 1977). The recent study by Archambaud *et al.* (2006) has found that *L. monocytogenes* Δ*sod* mutant is defective in survival in macrophages, as well as exhibits attenuated ability to polymerize cytosolic actin and to escape from the vacuole. Moreover, its pathogenicity is impaired *in vivo,* since the number of mutants in spleen was 10-fold lower than the wild-type strain. Thus, it can be assumed that SOD is required for full listerial virulence (Archambaud *et al.*, 2006).

The role of CA in listerial pathogenesis is not clear. The early study by Leblond-Francillard *et al.* (1989) found no difference in survival in murine spleen and liver between CA mutant and isogenic strain, questioning its role in *L. monocytogenes* virulence. The involvement of CA-negative strains in human listeriosis cases supports previous findings (Swartz *et al.*, 1991; Elsner *et al.*, 1996; Bubert *et al.*, 1997; Cepeda *et al.*, 2006). In contrast, van Dissel *et al.* (1993) have shown that low concentration of mouse macrophages is sufficient to kill catalase-negative mutants, postulating the role of catalase in *L. monocytogenes* resistance to macrophages. Furthermore, induction of the catalase gene in cultured cell line infected by *L. monocytogenes* and in the spleen of infected mice indicates that the enzyme may play a role in the pathogen's virulence (Camejo *et al.*, 2009).

Peroxiredoxins

Peroxiredoxins (Prxs) are a large family of cysteine-based peroxidases (Poole, 2005). They exhibit moderate catalytic efficiency, which is compensated by the high abundance in microbial cell (Pascual *et al.*, 2010). They are divided in two classes, the 1-Cys and 2-Cys Prxs, based on the number of cysteine residues involved in catalysis (Chae *et al.*, 1994). They use reactive cysteine thiols to reduce peroxides with electrons ultimately derived from reduced pyridine dinucleotides. Its activity constitutes defence against hydrogen peroxide (Dubbs and Mongkolsuk, 2007). During catalytic attack on the peroxide, N-terminal peroxidatic cysteine is transiently oxidized to sulfenic acid, which subsequently forms a disulfide bridge with the second cysteine residue. The peroxidatic cysteine may undergo overoxidation from sulfenic to sulfinic acid, leading to the enzyme inactivation (Pascual *et al.*, 2010). 2-Cys Prxs have been found to contribute to bacterial survival in the presence of oxidants (Chen *et al.*, 1998; Seaver and Imlay, 2001). It has been demonstrated that they also play essential role during growth and survival of *L. monocytogenes* in the environment with high concentration of H_2O_2, as reduction in viable cell counts of 2-Cys Prxs deletion mutant was 2-log higher than in a wild strain. However, no difference in survival in macrophages, as well as in murine model has been reported. Thus, it can be assumed that Prxs do not play a crucial role in the course of listerial infection (Kim *et al.*, 2007). Prxs activity has been shown to be regulated in addition to overoxidation, by phosphorylation and proteolysis (Wood *et al.*, 2003).

The PerR regulator

Oxidative stress is modulated by metal ion homeostasis (Mongkolsuk and Helmann, 2002). Metals such as iron and manganese serve as cofactors for antioxidant enzymes of *L. monocytogenes,* for example catalase or superoxide dismutase (Agranoff and Krishna, 1998). Metal ion uptake has been found to be regulated by four main regulatory proteins: Fur, PerR, Zur and MntR (Rea *et al.*, 2005). PerR is a major regulator of peroxide-inducible stress response. It has been shown to mediate repression of catalase, thioredoxin, Fur and alkyl hydroperoxide reductase in *B. subtilis,* as well as to down-regulate transcription of catalase in *L. monocytogenes* (Rea *et al.*, 2005). Nonetheless, PerR has not been studied extensively in *L. monocytogenes.* The regulator is dimeric and contains Zn (II) atom binding site, probably playing a structural role, and site that binds regulatory metal ion, responsible for binding to the target operator sites (Mongkolsuk and Helmann, 2002). To repress transcription PerR requires divalent metal ion such as Mn(II) or Fe(II) (Herbig and Helmann, 2001). Moreover, these two metal ions seem to affect in a different way the transcription of PerR regulon genes in response to hydrogen peroxide. Manganese form of PerR is relatively insensitive to H_2O_2, whereas ferrous ion converts PerR into hydrogen peroxide-sensitive form (Herbig and Helmann, 2001). Presence of manganese in the environment decreases H_2O_2-dependent induction of PerR regulon, whereas iron induces PerR (Fuangthong *et al.*, 2002). The mechanism by which PerR senses hydrogen peroxide is not clear. One model postulates that H_2O_2 directly reacts with the regulatory metal ion, contributing to metal dissociation and derepression. The another model postulate that H_2O_2 reacts with reactive cysteine residues, leading to disulphide bond formation and loss of metal ion (Mongkolsuk and Helmann, 2002).

It has been demonstrated that PerR is essential for survival of *L. monocytogenes* under oxidative stress conditions. Deletion of *perR* gene results in increased expression of its regulon. As a consequence the mutant grows slowly, forms smaller colonies and exhibits greater sensitivity to H_2O_2. PerR also plays a pivotal role in full virulence of *L. monocytogenes*, since the number of PerR mutants recovered from the mouse spleen was significantly reduced compared to isogenic strain (Rea *et al.*, 2004, 2005).

Sigma B

Transcription redirection, mediated by association of alternative sigma factors present in a bacterial cell with the core RNA polymerase, leads to expression of stress response genes (Boor, 2006). Sigma factors are dissociable subunits of RNA polymerase which bind to its core, forming complex that recognizes promoters and melts DNA, thereby initiating transcription (Kazmierczak *et al.*, 2005). Alternative sigma factors recognize different promoters than standard sigma factors, contributing to the initiation of transcription of new set of genes, not expressed under normal conditions (Boor *et al.*, 2006; Wałecka *et al.*, 2009). Sigma factors are classified into two unrelated families, i.e. σ^{70} and σ^{54}. σ^{54} called also σ^{N} is involved in a range of processes including nitrogen metabolism, but has also been implicated in regulation of some virulence-associated bacterial genes. Factors belonging to σ^{70} family were divided into four groups, according to their structure and function. Group I contains primary sigma factors required for bacterial growth and metabolism under normal conditions. Group II and IV are alternative sigma factors, regulating expression of genes responsible for maintaining cell viability in the stationary phase and responding to accumulation of misfolded proteins in the periplasm, respectively. Group III sigma factors (σ^{B}) are involved in stress response, sporulation and flagella biosynthesis (Paget and Helmann, 2003; Kazmierczak *et al.*, 2005; Boor *et al.*, 2006).

One of the best-studied alternative sigma factors in bacteria, including *L. monocytogenes*, is σ^{B} (Kazmierczak *et al.*, 2005; van Schaik and Abee, 2005). The *sigB* gene encoding this factor is the seventh open reading frame in an operon consisted of eight genes regulating σ^{B} (*rsbR, rsbS, rsbT, rsbU, rsbV, rsbW, sigB, rsbX*) (Ferreira *et al.*, 2004). Activation of σ^{B} by Rsb proteins involves a complex phosphorylation/dephosphorylation cascade in response to energy or environmental stress (Chaturongakul and Boor, 2004). Under optimal growth conditions σ^{B} is present in an inactive form due to interaction with anti-sigma factor RsbW. In stressed cells the unphosphorylated form of anti-anti-sigma factor RsbV binds to RsbW, thereby releasing σ^{B}. In *Bacillus subtilis* the phosphorylation of RsbV is regulated by two different phosphatases. Under energy stress RsbQ activates RsbP, which activates RsbV. Environmental stress activates RsbU through involvement of stressosome, which consists of RsbS, RsbR, and four RsbR paralogues, i.e. YkoB, YojH, YqhA, and YtvA. STAS (sulfate transporter and anti-anti-sigma factor) domains of stressosome are phosphorylated during stress by RsbT kinase, releasing RsbT from the complex, which subsequently activates RsbU (Kazmierczak *et al.*, 2005; Marles-Wright and Lewis, 2007). However, a two-gene operon *rsbQ-rsbP* has not been identified in *L. monocytogenes*. Hence, in this microorganism activation of σ^{B} under energy and environmental stress probably occurs via a single pathway that includes RsbT, RsbU, RsbV, and RsbW (Fig. 4.7) (Chaturongakul and Boor, 2004, 2006).

Role of σ^{B} in stress adaptation

Microarray analyses identified 55 σ^{B}-dependent genes in *L. monocytogenes*, from which 54 were preceded by a putative σ^{B} promoter (Kazmierczak *et al.*, 2003). σ^{B} has been shown to control expression of genes involved in survival under stress conditions, such as high osmolarity, low and high temperature, low pH, high hydrostatic pressure, high ethanol concentrations, oxidative stress and

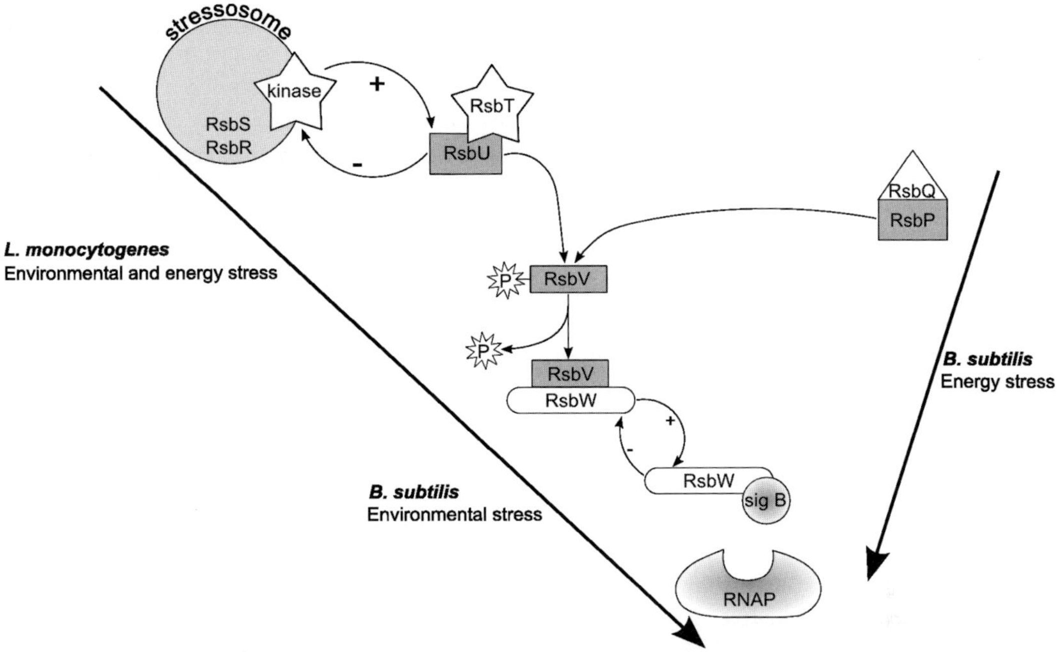

Figure 4.7 Activation of σ^B under energy and environmental stress in *L. monocytogenes* and *B. subtilis*. Under normal conditions σ^B is kept in inactive state via interaction with anti-sigma factor RsbW. Under stress conditions the unphosphorylated form of anti-anti-sigma factor RsbV binds RsbW, thereby releasing σ^B. In *B. subtilis* RsbV is dephosphorylated by RsbP or RsbU, depending on stress type. Under energy stress RsbQ activates RsbP, leading to RsbV activation. Under environmental stress RsbT kinase phosphorylates STAS domains of stressosome, releasing itself from the complex. Subsequently RsbT binds RsbU. Activated RsbU dephosphorylates RsbV. In *L. monocytogenes* activation of σ^B under both stress conditions occurs via a single pathway that includes RsbT, RsbU, RsbV, and RsbW.

carbon starvation (Ferreira *et al.*, 2001). Potential σ^B-dependent *opuC* and *gbuAP2* promoters have been identified in *L. monocytogenes*. It has been demonstrated that σ^B plays an essential role in adaptation to osmotolerance, since the level of *opuCA* transcripts was significantly reduced in *sigB* deletion mutants exposed to osmotic upshift (Fraser *et al.*, 2003; Sue *et al.*, 2004). The *sigB* mutants had also impaired ability to use carnitine and betaine as osmoprotectants (Becker *et al.*, 1998). However, the study of Fraser *et al.* (2003) found that only carnitine utilization was abolished in the absence of *sigB*. The impact of σ^B on *L. monocytogenes* acid resistance has also been reported. Putative σ^B-dependent promoter was found upstream from the *gadT2D2* and *gadD3* operons (Wemekamp-Kamphuis *et al.*, 2004), however, only σ^B-dependent *gadD3* promoter has been confirmed (Kazmierczak *et al.*, 2003). It has been shown that the acid-induced transcription in a *sigB* deletion mutant was significantly reduced for *gadD2, gadT2,* and slightly increased for *gadD3,* perhaps to compensate the reduction of expression of other GAD system genes. The stationary phase *sigB* null mutant exposed to pH 2.5 displayed essentially lower survival than its isogenic parent (Wiedmann *et al.*, 1998; Ferreira *et al.*, 2001; Wemekamp-Kamphuis *et al.*, 2004). Nonetheless, the loss of *sigB* gene did not abolish the ATR in *L. monocytogenes,* suggesting the involvement of two mechanisms of acid tolerance. One of them being pH-dependent should be at least partially σ^B-independent, and the second σ^B-dependent (Ferreira *et al.*, 2001, 2003). The influence of σ^B on heat tolerance has been investigated. The study by Ferreira *et al.* (2001) has found no correlation between heat resistance and σ^B, as the *sigB* null mutant and its isogenic parent displayed the same viability after exposure to 50°C. Thus, it was assumed that heat resistance is at least partially σ^B-independent. Results of a recent study demonstrate that σ^B may be, however, involved in heat tolerance, since over 50-fold decrease in survival of σ^B mutant exposed to 60°C for 30 seconds was observed. The protective role of σ^B was found to be more pronounced at neutral pH than acid pH (Somolis *et al.*, 2010). Moreover, identification of putative σ^B-dependent promoter sequence in the *clpP* gene confirmed the σ^B involvement in heat stress response (Wemekamp-Kamphuis *et al.*, 2004). It was found that 15 minutes' exposure of *sigB* mutant to 13.8 mM cumene hydroperoxide resulted in100-fold decrease in its viability when compared to the wild-type strain, suggesting the contribution of σ^B to survival in response to oxidative stress. The level of *sigB* transcript was significantly higher under these conditions (Ferreira *et al.*, 2001). It seems that the impact of σ^B on survival under acid and oxidative stress depends on genetic background, as its role is more pronounced in *L. monocytogenes* lineage I, II, and IIIb, but not in lineage IIIA (Oliver *et al.*, 2010). The involvement of alternative sigma B factor in starvation stress has not been clearly elucidated. Ferreira *et al.* (2001) showed that viability of *sigB* mutant during glucose limitation was over 3-fold lower than its parental strain, but in study by Lungu *et al.* (2008) viability of *sigB* mutant was comparable with its isogenic *L. monocytogenes* strain. In the absence of oxygen σ^B may be necessary for the long-term survival of nutrient-deprived bacteria subjected to chemical stresses (Lungu *et al.*, 2008). σ^B seems to play a role also in cryotolerance (Moorhead and Dykes, 2003). It has been found that temperature downshift applied to log-phase cultures of *sigB* mutant and the parental strain resulted in comparable lag periods, whereas the lag phase of stationary-phase cultures of the mutant was twice as long as its isogenic parent. Therefore, it was assumed that σ^B is crucial for low-temperature adaptation in stationary phase only. It is also essential for utilization of betaine and carnitine in log-phase cells, and betaine in stationary phase cells (Becker *et al.*, 2000).

The role of σ^B in *L. monocytogenes* virulence

σ^B has been found to control expression of genes associated with *L. monocytogenes* virulence, e.g. *bsh* or *inlAB*, as the level of these transcripts was essentially decreased in *sigB* mutant subjected to osmotic stress (Kazmierczak *et al.*, 2003). Bile salt hydrolase (Bsh), encoded by *bsh*, is a virulence factor that facilitates bacteria to colonize the gallbladder. InlA and InlB are leucine-rich repeat proteins responsible for invasion into various non-phagocytic mammalian cells (Bierne *et al.*, 2007). To date six promoters have been identified in the *inlAB* locus, including a PrfA-regulated promoter (P3inlA) and a σ^B-dependent promoter (P4inlA) upstream of *inlA*, and a putative σ^B-dependent promoter (P2inlB) upstream of *inlB* (Lingnau *et al.*, 1995; Kazmierczak *et al.*, 2003). σ^B may also indirectly regulate transcription of *inlAB* through the control of *prfA* expression (Kim *et al.*, 2005). PrfA is a key regulator of virulence gene expression, activated upon bacterial entry into the host cells (Garner *et al.*, 2006). It controls every step of the intracellular life cycle of *L. monocytogenes*, which includes genes required for phagosomal escape, such as *hly*, encoding listeriolysin O (LLO), *plcA* and *plcB*, encoding phospholipases C, *mpl*, encoding metalloproteinase, responsible for the activation of one of phospholipases, and *actA*, encoding the actin assembly protein, indispensable for cell-to-cell spread (Engelbrecht *et al.*, 1996; Portnoy *et al.*, 2002). One of *prfA* promoters has been shown to be under σ^B control (Freitag *et al.*, 1993). In the course of infection, *L. monocytogenes* sequentially activates a range of factors that allow it to survive in the host cell and spread within the tissue. It has been suggested that the pathogen entering the host cell, switches the predominantly σ^B-dependent expression of genes required for gastrointestinal survival to the PrfA-dependent expression required for phagosomal escape, multiplication, and cell-to-cell spread (Kazmierczak *et al.*, 2006).

σ^B plays a pivotal role in gastrointestinal stage of listeriosis and host cell invasion. The loss of *sigB* gene significantly decreases expression of *inlA* and *inlB* genes, thereby affecting the pathogen's ability to invade into CaCo-2 and HepG2 cells and the number of bacteria recovered from intragastrically inoculated guinea pigs (Kim *et al.*, 2004, 2005; Garner *et al.*, 2006). Nonetheless, it seems that σ^B does not have an impact on survival and replication of *L. monocytogenes* in intravenously infected animals, when bacteria are inside the cells and PrfA is fully activated (Garner *et al.*, 2006; de las Heras *et al.*, 2011). Thus, involvement of σ^B in listerial pathogenesis appears to be restricted to gastrointestinal stage before entry, through regulation of genes that help bacteria to withstand adverse conditions in the intestinal lumen (de las Heras *et al.*, 2011).

Stress response and cross protection

L. monocytogenes is able to survive under a range of adverse environmental conditions, including low temperatures, low pH, and elevated osmolarity (Bajard *et al.*, 1996). Such stresses elicit a strong response in the pathogen that increases its tolerance to the same or other type of stress. It has been found that, adaptation to sublethal dose of salt results in the higher resistance to hydrogen peroxide (Lou and Yousef, 1997), as well as help bacteria to deal with bile stress in the duodenum (Sleator *et al.*, 2007). Exposure to sodium chloride affects also heat resistance, since cells grown in the presence of low NaCl concentration, subjected to a short osmotic upshift, displayed elevated thermotolerance (Jørgensen *et al.*, 1995). Low salt concentrations provide also a cross-protection for bacteria grown at low pH, as the time required for the recovery of pH-injured cells was shorter in the presence of salt than in its absence (Cole *et al.*, 1990). On the contrary, Garner *et al.* (2006) has revealed that bacteria grown in the presence of salt or organic acids have decreased ability to survive in low pH of gastric fluid. Adaptation to sublethal pH (4.5–5.5) has been shown to enhance the survival at lethal acid conditions (pH 3.5) (Lou and Yousef, 1997; Werbrouck *et al.*, 2009), lethal doses of ethanol (17.5%) and H_2O_2 (0.1%). Moreover, exposure to mild acid increases *L. monocytogenes* survival in acidified food and gastric fluid (Gahan *et al.*, 1996; Ferreira *et al.*, 2003). Adaptation of the bacteria to elevated temperatures (45°C) provides protection against lethal doses of ethanol, H_2O_2, and salt (25%), whereas treatment with ethanol (5%) additionally protects them against acid shock (pH 3.5) (Lou and Yousef, 1997). It seems that adaptation to stresses encountered during food processing may help bacteria to overcome stresses associated with gastrointestinal passage, and thus increase their ability to cause a disease. Nonetheless, there are little data about the alteration of virulence of stress-adapted bacteria in response to subsequent dose of stress. Garner *et al.* (2006) demonstrated that bacteria grown in the broth containing salt or lactic acid subsequently subjected to gastric fluid exhibit reduced invasiveness, in comparison to bacteria exposed to lactic acid or NaCl. The invasiveness of double-stressed bacteria, however, does not change when comparing with non-stressed bacteria. Influence of stress on invasiveness appears to be heterogenic in *L. monocytogenes* population. Depending on strain, the same stress leads to increase or decrease of invasiveness. In some strains invasiveness is not affected by stress (Werbrouck *et al.*, 2009; Wałecka *et al.*, 2011a,b). Therefore, it is difficult to predict how stress will affect virulence of particular strain.

Conclusions and future trends

In 1980s *L. monocytogenes* was recognized as foodborne pathogen. Since then extensive research on the pathogen's cell biology, virulence mechanisms, and epidemiology has been initiated. At present, *L. monocytogenes* appears as a model in biology of intracellular pathogens. Despite intensified surveillance in food manufacturing, which resulted in improvement of microbial quality of food, serious cases of listeriosis are still reported. Before *L. monocytogenes* causes illness, it has to overcome various stresses encountered both in the environment and in the host. Over past decades the molecular mechanism of *L. monocytogenes* tolerance to heat, osmotic, acid and oxidative stress has been extensively explored. It is widely recognized that exposure to sublethal treatment can enhance *L. monocytogenes* resistance to subsequent doses of stress. As a consequence bacteria can survive better as if the same factors act independently. Stress response elicited during multistep inactivation procedures can result in increase of bacterial resistance and thereby reduce effectiveness of bacterial inactivation. Some recent data highlight the link between stress and the pathogen's virulence. It seems that the ability to invade epithelial cells may be significantly influenced by environmental factors. In turn, invasiveness is one of the key virulence determinants of *L. monocytogenes*. Food was recognized to be major source of human contamination by *L. monocytogenes*. Various conditions encountered by the pathogen both in train of food processing and during its storage were shown to impinge *L. monocytogenes* virulence. Therefore, conditions occurring in food could be thought the major factors modulating pathogenic potential of a given strain. It was proven that the stress type or its intensity may differentially impact the pathogen's invasiveness. Some of the factors investigated to date seem to enhance whereas the other decrease the invasiveness of *L. monocytogenes*. The effect of these factors has been studied separately. Nonetheless, in the course of infection *L. monocytogenes* is challenged by a sequence of stresses, first during food processing or preparation, then during the passage through the human gastrointestinal tract. There is practically no data on how sequential stresses affect *L. monocytogenes* invasiveness. Knowledge of microbial stress resistance and response to sequences of stresses may be of great value for food safety. It is possible that insight into role of factors affecting stress response could influence the design of modern food preservation methods. Design of appropriate sequence of technological approaches in which stress response is not activated may lead to development of preservation methods that do not induce the stress response, making reduction of the pathogen more effective.

Acknowledgements

The authors thank Maciej Zacharski for preparation of figures.

Web resources

NLM: http://www.nlm.nih.gov/medlineplu

US National Library of Medicine, National Institutes of Health, MEDLINE Plus – this service contains news, articles, and factsheets on *L. monocytogenes* and listeriosis.

FDA: http://www.fda.gov

US Food and Drug Administration – provides information about listeriosis, including symptoms, cause, treatment, risk factors, associated foods, and prevention.

USDA: http://fsrio.nal.usda.gov

United States Department of Agriculture National Agricultural Library – contains resources, articles, and news releases about listeriosis.

CDC: http://www.cdc.gov

Centers for Disease Control and Prevention – provides information on infection caused by *L. monocytogenes*.

References

Agranoff, D.D., and Krishna, S. (1998). Metal ion homeostasis and intracellular parasitism. Mol. Microbiol. *28*, 403–412.

Angelidis, A.S., and Smith, G.M. (2003a). Role of the glycine betaine and carnitine transporters in adaptation of *Listeria monocytogenes* to chill stress in defined medium. Appl. Environ. Microbiol. *69*, 7492–7498.

Angelidis, A.S., and Smith, G.M. (2003b). Three transporters mediate uptake of glycine betaine and carnitine by *Listeria monocytogenes* in response to hyperosmotic stress. Appl. Environ. Microbiol. *69*, 1013–1022.

Angelidis, A.S., Smith, L.T., Hoffman, L.M., and Smith, G.M. (2002). Identification of opuC as a chill-activated and osmotically activated carnitine transporter in *Listeria monocytogenes*. Appl. Environ. Microbiol. *68*, 2644–2650.

Archambaud, C., Nahori, M.A., Pizarro-Cerda, J., Cossart, P., and Dussurget, O. (2006). Control of *Listeria* superoxide dismutase by phosphorylation. J. Biol. Chem. *281*, 31812–31822.

Audia, J.P., Webb, C.C., and Foster, J.W. (2001). Breaking through the acid barrier: an orchestrated response to proton stress by enteric bacteria. Int. J. Med. Microbiol. *291*, 97–106.

Azizoglu, R.O., and Kathariou, S. (2010). Temperature-dependent requirement for catalase in aerobic growth of *Listeria monocytogenes* F2365. Appl. Environ. Microbiol. *76*, 6998–7003.

Bajard, S., Rosso, L., Fardel, G., and Flandrois, J.P. (1996). The particular behaviour of *Listeria monocytogenes* under sub-optimal conditions. Int. J. Food Microbiol. *29*, 201–211.

Becker, L.A., Cetin, M.S., Hutkins, R.W., and Benson, A.K. (1998). Identification of the gene encoding the alternative sigma factor sigmaB from *Listeria monocytogenes* and its role in osmotolerance. J. Bacteriol. *180*, 4547–4554.

Becker, L.A., Evans, S.N., Hutkins, R.W., and Benson, A.K. (2000). Role of sigma(B) in adaptation of *Listeria monocytogenes* to growth at low temperature. J. Bacteriol. *182*, 7083–7087.

Begley, M., Gahan, C.G., and Hill, C. (2002). Bile stress response in *Listeria monocytogenes* LO28: adaptation, cross-protection, and identification of genetic loci involved in bile resistance. Appl. Environ. Microbiol. *68*, 6005–6012.

Beumer, R.R., Te Giffel, M.C., Cox, L.J., Rombouts, F.M., and Abee, T. (1994). Effect of exogenous proline, betaine, and carnitine on growth of *Listeria monocytogenes* in a minimal medium. Appl. Environ. Microbiol. *60*, 1359–1363.

Bierne, H., Sabet, C., Personnic, N., and Cossart, P. (2007). Internalins: a complex family of leucine-rich repeat-containing proteins in *Listeria monocytogenes*. Microbes Infect. *9*, 1156–1166.

Bonnet, M., and Montville, T.J. (2005). Acid-tolerant *Listeria monocytogenes* persist in a model food system fermented with nisin-producing bacteria. Lett. Appl. Microbiol. *40*, 237–242.

Boor, K.J. (2006). Bacterial stress responses: what doesn't kill them can make them stronger. PLoS Biol *4*, e23.

Brehm, K., Haas, A., Goebel, W., and Kreft, J. (1992). A gene encoding a superoxide dismutase of the facultative intracellular bacterium *Listeria monocytogenes*. Gene. *118*, 121–125.

Bubert, A., Riebe, J., Schnitzler, N., Schönberg, A., Goebel, W., and Schubert, P. (1997). Isolation of catalase-negative *Listeria monocytogenes* strains from listeriosis patients and their rapid identification by anti-p60 antibodies and/or PCR. J. Clin. Microbiol. *35*, 179–183.

Cabiscol, E., Tamarit, J., Ros, J. (2000). Oxidative stress in bacteria and protein damage by reactive oxygen species. Int. Microbiol. *3*, 3–8.

Call, D.R., Borucki, M.K., and Besser, T.E. (2003). Mixed-genome microarrays reveal multiple serotype and lineage-specific differences among strains of *Listeria monocytogenes*. J. Clin. Microbiol. *41*, 632–639.

Camejo, A., Buchrieser, C., Couvé, E., Carvalho, F., Reis, O., Ferreira, P., Sousa, S., Cossart, P., and Cabanes, D. (2009). *In vivo* transcriptional profiling of *Listeria monocytogenes* and mutagenesis identify new virulence factors involved in infection. PLoS Pathog. *5*, e1000449.

Cameron, L.A., Giardini, P.A., Soo, F.S., and Theriot, J.A. (2000). Secrets of actin-based motility revealed by a bacterial pathogen. Nat. Rev. Mol. Cell. Biol. *1*, 110–119.

Camilli A., Tilney L.G., and Portnoy D.A. (1993). Dual roles of plcA in *Listeria monocytogenes* pathogenesis. Mol. Microbiol. *8*, 143–157.

Capaldi, R.A., and Aggeler, R. (2002). Mechanism of the F(1)F(0)-type ATP synthase, a biological rotary motor. Trends Biochem. Sci. *27*, 154–160.

Casiano-Colón, A., and Marquis, R.E. (1988). Role of the arginine deiminase system in protecting oral bacteria and an enzymatic basis for acid tolerance. Appl. Environ. Microbiol. *54*, 1318–1324.

Cepeda, J.A., Millar, M., Sheridan, E.A., Warwick, S., Raftery, M., Bean, D.C., and Wareham, D.W. (2006). Listeriosis due to infection with a catalase-negative strain of *Listeria monocytogenes*. J. Clin. Microbiol. *44*, 1917–1918.

Cetin, M.S., Zhang, C., Hutkins, R.W., and Benson, A.K. (2004). Regulation of transcription of compatible solute transporters by the general stress sigma factor, sigmaB, in *Listeria monocytogenes*. J. Bacteriol. *186*, 794–802.

Chae, H.Z., Robison, K., Poole, L.B., Church, G., Storz, G., and Rhee, S.G. (1994). Cloning and sequencing of thiol-specific antioxidant from mammalian brain: alkyl hydroperoxide reductase and thiol-specific antioxidant define a large family of antioxidant enzymes. Proc. Natl. Acad. Sci. U.S.A. *91*, 7017–7021.

Chastanet, A., Derre, I., Nair, S., and Msadek, T. (2004). *clpB*, a novel member of the *Listeria monocytogenes* CtsR regulon, is involved in virulence but not in general stress tolerance. J. Bacteriol. *186*, 1165–1174.

Chatterjee, S.S., Otten, S., Hain, T., Lingnau, A., Carl, U.D., Wehland, J., Domann, E., and Chakraborty, T. (2005). Invasiveness is a variable and heterogeneous phenotype in *Listeria monocytogenes* serotype strains. Int. J. Med. Microbiol. *296*, 277–286.

Chatterjee, S.S., Hossain, H., Otten, S., Kuenne, C., Kuchmina, K., Machata, S., Domann, E., Chakraborty, T., and Hain, T. (2006). Intracellular gene expression profile of *Listeria monocytogenes*. Infect. Immun. *74*, 1323–1338.

Chaturongakul, S., and Boor, K.J. (2004). RsbT and RsbV contribute to sigmaB-dependent survival under environmental, energy, and intracellular stress conditions in *Listeria monocytogenes*. Appl. Environ. Microbiol. *70*, 5349–5356.

Chaturongakul, S., and Boor, K.J. (2006). SigmaB activation under environmental and energy stress conditions in *Listeria monocytogenes*. Appl. Environ. Microbiol. *72*, 5197–5203.

Chen, L., Xie, Q.W., Nathan, C. (1998). Alkyl hydroperoxide reductase subunit C (AhpC) protects bacterial and human cells against reactive nitrogen intermediates. Mol. Cell *1*, 795–805.

Chorianopoulos, N., Giaouris, E., Grigoraki, I., Skandamis, P., and Nychas G.J. (2011). Effect of acid tolerance response (ATR) on attachment of *Listeria monocytogenes* Scott A to stainless steel under extended exposure to acid or/and salt stress and resistance of sessile cells to subsequent strong acid challenge. Int. J. Food Microbiol. *145*, 400–406.

Chowduhry, R., Sahu, G.K., and Das, J. (1996). Stress response in pathogenic bacteria. J. Biosci. *21*, 149–160.

Cole, M.B., Jones, M.V., and Holyoak, C. (1990). The effect of pH, salt concentration and temperature on the survival and growth of *Listeria monocytogenes*. J. Appl. Bacteriol. *69*, 63–72.

Conte, M.P., Petrone, G., Di Biase, A.M., Ammendolia, M.G., Superti, F., and Seganti, L. (2000). Acid tolerance in *Listeria monocytogenes* influences invasiveness of enterocyte-like cells and macrophage-like cells. Microb. Pathog. *29*, 137–144.

Conte, M.P., Petrone, G., Di Biase, A.M., Longhi, C., Penta, M., Tinari, A., Superti, F., Fabozzi, G., Visca, P.,and Seganti, L. (2002). Effect of acid adaptation on the fate of *Listeria monocytogenes* in THP-1 human macrophages activated by gamma interferon. Infect. Immun. *70*, 4369–4378.

Cossart, P., and Toledo-Arana, A. (2008). *Listeria monocytogenes*, a unique model in infection biology: an overview. Microbes Infect. *10*, 1041–1050.

Cotter, P.D., and Hill, C. (2003). Surviving the acid test: responses of Gram-positive bacteria to low pH. Microbiol. Mol. Biol. Rev. *67*, 429–453.

Cotter, P.D., Emerson, N., Gahan, C.G., and Hill, C. (1999). Identification and disruption of lisRK, a genetic locus encoding a two-component signal transduction system involved in stress tolerance and virulence in *Listeria monocytogenes*. J. Bacteriol. *181*, 6840–6843.

Cotter, P.D., Gahan, C.G., and Hill, C. (2000). Analysis of the role of the *Listeria monocytogenes* F0F1 -AtPase operon in the acid tolerance response. Int. J. Food Microbiol. *60*, 137–146.

Cotter, P.D., Gahan, C.G., and Hill, C. (2001a). A glutamate decarboxylase system protects *Listeria monocytogenes* in gastric fluid. Mol. Microbiol. *40*, 465–475.

Cotter, P.D., O'Reilly, K., and Hill, C. (2001b). Role of the glutamate decarboxylase acid resistance system in the survival of *Listeria monocytogenes* LO28 in low pH foods. J. Food. Prot. *64*, 1362–1368.

Cotter, P.D., Guinane, C.M., and Hill, C. (2002). The LisRK signal transduction system determines the sensitivity of *Listeria monocytogenes* to nisin and cephalosporins. Antimicrob. Agents Chemother. *46*, 2784–2790.

Cotter, P.D., Ryan, S., Gahan, C.G., and Hill, C. (2005). Presence of GadD1 glutamate decarboxylase in selected *Listeria monocytogenes* strains is associated with an ability to grow at low pH. Appl. Environ. Microbiol. *71*, 2832–2839.

Craig E.A., Gambill B.D., and Nelson R.J. (1993). Heat shock proteins: molecular chaperones of protein biogenesis. Microbiol. Rev. *57*, 402–414.

Csonka, L.N. (1989). Physiological and genetic responses of bacteria to osmotic stress. Microbiol. Rev. *53*, 121–147.

Dallmier, A.W., and Martin, S.E. (1988). Catalase and superoxide dismutase activities after heat injury of *Listeria monocytogenes*. Appl. Environ. Microbiol. *54*, 581–582.

Davis, M.J., Coote, P.J., and O'Byrne, C.P. (1996). Acid tolerance in *Listeria monocytogenes*: the adaptive acid tolerance response (ATR) and growth-phase-dependent acid resistance. Microbiology *142*, 2975–2982.

Decatur, A.L., and Portnoy, D.A. (2000). A PEST-like sequence in listeriolysin O essential for *Listeria monocytogenes* pathogenicity. Science *290*, 992–995.

Derré, I., Rapoport, G., and Msadek, T. (1999). CtsR, a novel regulator of stress and heat shock response, controls clp and molecular chaperone gene expression in Gram-positive bacteria. Mol. Microbiol. *31*, 117–131.

Derré, I., Rapoport, G., and Msadek, T. (2000). The CtsR regulator of stress response is active as a dimer and specifically degraded *in vivo* at 37 degrees C. Mol. Microbiol. 38, 335–347.

van Dissel, J.T., Stikkelbroeck, J.J., and van Furth, R. (1993). Differences in the rate of intracellular killing of catalase-negative and catalase-positive *Listeria monocytogenes* by normal and interferon-gamma-activated macrophages. Scand. J. Immunol. *37*, 443–446.

Doyle, M.E., Mazzotta, A.S., Wang, T., Wiseman, D.W., and Scott, V.N. (2001). Heat resistance of *Listeria monocytogenes*. J. Food Prot. *64*, 410–429.

Drevets, D.A., and Bronze, M.S. (2008). *Listeria monocytogenes*: epidemiology, human disease, and mechanisms of brain invasion. FEMS Immunol. Med. Microbiol. *53*, 151–165.

Dubbs, J.M., Mongkolsuk, S. (2007). Peroxiredoxins in bacterial antioxidant defense. Subcell. Biochem. *44*, 143–193.

Duché, O., Trémoulet, F., Glaser, P., and Labadie, J. (2002). Salt stress proteins induced in *Listeria monocytogenes*. Appl. Environ. Microbiol. *68*, 1491–1498.

Earnshaw, R.G., Appleyard, J., and Hurst, R.M. (1995). Understanding physical inactivation processes: combined preservation opportunities using heat, ultrasound and pressure. Int. J. Food Microbiol. *28*, 197–219.

Elsner, H.A., Sobottka, I., Bubert, A., Albrecht, H., Laufs, R., and Mack, D. (1996). Catalase-negative *Listeria monocytogenes* causing lethal sepsis and meningitis in an adult hematologic patient. Eur. J. Clin. Microbiol. Infect. Dis. *15*, 965–967.

Engelbrecht, F., Chun, S.K., Ochs, C., Hess, J., Lottspeich, F., Goebel, W., and Sokolovic, Z. (1996). A new PrfA-regulated gene of *Listeria monocytogenes* encoding a small, secreted protein which belongs to the family of internalins. Mol. Microbiol. *21*, 823–837.

Engelmann, S., and Hecker, M. (1996). Impaired oxidative stress resistance of *Bacillus subtilis sigB* mutants and the role of *katA* and *katE*. FEMS Microbiol. Lett. *145*, 63–69.

Farber, J.M., and Peterkin, P.I. (1991). *Listeria monocytogenes*, a food-borne pathogen. Microbiol. Rev. *55*, 476–511.

Fayet, O., Ziegelhoffer, T., and Georgopoulos, C. (1989). The groES and groEL heat shock gene products of *Escherichia coli* are essential for bacterial growth at all temperatures. J. Bacteriol. *171*, 1379–1385.

Feldman, R.I., and Boyer, P.D. (1985). The role of tightly bound ADP on chloroplast ATPase. J. Biol. Chem. *260*, 13088–13094.

Feniouk, B.A., and Junge, W. (2005). Regulation of the F0F1-ATP synthase: the conformation of subunit epsilon might be determined by directionality of subunit gamma rotation. FEBS Lett. *579*, 5114–5118.

Feniouk, B.A., Suzuki, T., and Yoshida, M. (2006). The role of subunit epsilon in the catalysis and regulation of FOF1-ATP synthase. Biochim. Biophys. Acta. *1757*, 326–338.

Ferreira, A., O'Byrne, C.P., and Boor, K.J. (2001). Role of sigma(B) in heat, ethanol, acid, and oxidative stress resistance and during carbon starvation in *Listeria monocytogenes*. Appl. Environ. Microbiol. *67*, 4454–4457.

Ferreira A., Sue D., O'Byrne C.P., Boor K.J. (2003). Role of *Listeria monocytogenes* sigma(B) in survival of lethal acidic conditions and in the acquired acid tolerance response. Appl. Environ. Microbiol. *69*, 2692–2698.

Ferreira, A., Gray, M., Wiedmann, M., and Boor, K.J. (2004). Comparative genomic analysis of the *sigB* operon in *Listeria monocytogenes* and in other Gram-positive bacteria. Curr. Microbiol. *48*, 39–46.

Fillingame, R.H. (1997). Coupling H+ transport and ATP synthesis in F1F0-ATP synthases: glimpses of interacting parts in a dynamic molecular machine. J. Exp. Biol. *200*, 217–224.

Fisher, C.W., Lee, D., Dodge, B.A., Hamman, K.M., Robbins, J.B., and Martin, S.E. (2000). Influence of catalase and superoxide dismutase on ozone inactivation of *Listeria monocytogenes*. Appl. Environ. Microbiol. *66*, 1405–1409.

Foster, J.W. (1993). The acid tolerance response of *Salmonella typhimurium* involves transient synthesis of key acid shock proteins. J. Bacteriol. *175*, 1981–1987.

Foster, J.W., Park, Y.K., Bang, I.S., Karem, K., Betts, H., Hall, H.K., and Shaw, E. (1994). Regulatory circuits involved with pH-regulated gene expression in *Salmonella typhimurium*. Microbiology *140*, 341–352.

Fraser, K.R., Harvie, D., Coote, P.J., O'Byrne, C.P. (2000). Identification and characterization of an ATP binding cassette L-carnitine transporter in *Listeria monocytogenes*. Appl. Environ. Microbiol. *66*, 4696–4704.

Fraser, K.R., Sue, D., Wiedmann, M., Boor, K., O'Byrne, C.P. (2003). Role of sigmaB in regulating the compatible solute uptake systems of *Listeria monocytogenes*: osmotic induction of *opuC* is sigmaB dependent. Appl. Environ. Microbiol. *69*, 2015–2022.

Freitag, N.E., Rong, L., and Portnoy, D.A. (1993). Regulation of the *prfA* transcriptional activator of *Listeria monocytogenes*: multiple promoter elements contribute to intracellular growth and cell-to-cell spread. Infect. Immun. *61*, 2537–2544.

Fretz, R., Pichler, J., Sagel, U., Much, P., Ruppitsch, W., Pietzka, A.T., Stöger, A., Huhulescu, S., Heuberger, S., Appl, G., *et al.* (2010). Update: Multinational listeriosis outbreak due to 'Quargel', a sour milk curd cheese, caused by two different *L. monocytogenes* serotype 1/2a strains, 2009–2010. Eur. Surveill. *15*, 19543.

Fuangthong M., Herbig A.F., Bsat N., Helmann J.D. (2002). Regulation of the *Bacillus subtilis* fur and *perR* genes by PerR: not all members of the PerR regulon are peroxide inducible. J. Bacteriol. *84*, 3276–3286.

Gahan, C.G., and Hill, C. (1999). The relationship between acid stress responses and virulence in *Salmonella typhimurium* and *Listeria monocytogenes*. Int. J. Food Microbiol. *50*, 93–100.

Gahan, C.G., O'Driscoll, B., and Hill, C. (1996). Acid adaptation of *Listeria monocytogenes* can enhance survival in acidic foods and during milk fermentation. Appl. Environ. Microbiol. *62*, 3128–3132.

Gahan, C.G., O'Mahony, J., and Hill, C. (2001). Characterization of the groESL operon in *Listeria monocytogenes*: utilization of two reporter systems (gfp and hly) for evaluating *in vivo* expression. Infect. Immun. *69*, 3924–3932.

Gaillot, O., Pellegrini, E., Bregenholt, S., Nair, S., and Berche, P. (2000). The ClpP serine protease is essential for the intracellular parasitism and virulence of *Listeria monocytogenes*. Mol. Microbiol. *35*, 1286–1294.

Gaillot, O., Bregenholt. S., Jaubert, F., Di Santo, J.P., and Berche, P. (2001). Stress-induced ClpP serine protease of *Listeria monocytogenes* is essential for induction of listeriolysin O-dependent protective immunity. Infect. Immun. *69*, 4938–4943.

Gardan, R., Duché, O., Leroy-Sétrin, S., and Labadie, J. (2003). Role of ctc from *Listeria monocytogenes* in osmotolerance. Appl. Environ. Microbiol. *69*, 154–161.

Garner, M.R., James, K.E., Callahan, M.C., Wiedmann, M., and Boor, K.J. (2006). Exposure to salt and organic acids increases the ability of *Listeria monocytogenes* to invade Caco-2 cells but decreases its ability to survive gastric stress. Appl. Environ. Microbiol. *72*, 5384–5395.

Gentry, D.R., Hernandez, V.J., Nguyen, L.H., Jensen, D.B., and Cashel, M. (1993). Synthesis of the stationary-phase sigma factor sigma s is positively regulated by ppGpp. J. Bacteriol. *175*, 7982–7989.

Gerhardt, P.N., Smith, L.T., and Smith, G.M. (1996). Sodium-driven, osmotically activated glycine betaine transport in *Listeria monocytogenes* membrane vesicles. J. Bacteriol. *178*, 6105–6109.

Gerhardt, P.N., Tombras Smith, L., and Smith, G.M. (2000). Osmotic and chill activation of glycine betaine porter II in *Listeria monocytogenes* membrane vesicles. J. Bacteriol. *182*, 2544–2550.

Glaser P., C., Rusniok C., Amend A., Baquero F., Berche P., Bloecker H., Brandt P., Chakraborty T., Charbit A., Chetouani F., *et al.* (2001). Comparative genomics of *Listeria* species. Science *294*, 849–852.

Gottesman, S., Wickner, S., Maurizi, M.R. (1997). Protein quality control: triage by chaperones and proteases. Genes Dev. *11*, 815–823.

Grace, S.C. (1990). Phylogenetic distribution of superoxide dismutase supports an endosymbiotic origin for chloroplasts and mitochondria. Life Sci. *47*, 1875–1886.

Hain, T., Steinweg, C., Kuenne, C.T., Billion, A., Ghai, R., Chatterjee, S.S., Domann, E., Kärst, U., Goesmann, A., Bekel, T., *et al.* (2006). Whole-genome sequence of Listeria welshimeri reveals common steps in genome reduction with *Listeria innocua* as compared to *Listeria monocytogenes*. J. Bacteriol. *188*, 7405–7415.

Hanawa, T., Fukuda, M., Kawakami, H., Hirano, H., Kamiya, S., and Yamamoto, T. (1999). The *Listeria monocytogenes* DnaK chaperone is required for stress tolerance and efficient phagocytosis with macrophages. Cell Stress Chaperones *4*, 118–128.

Hanawa, T., Kai, M., Kamiya, S., and Yamamoto, T. (2000). Cloning, sequencing, and transcriptional analysis of the *dnaK* heat shock operon of *Listeria monocytogenes*. Cell Stress Chaperones *5*, 21–29.

Hanawa, T., Yamanishi, S., Murayama, S., Yamamoto, T., and Kamiya, S. (2002). Participation of DnaK in expression of genes involved in virulence of *Listeria monocytogenes*. FEMS Microbiol. Lett. *214*, 69–75.

Hecker, M., Schumann, W., and Völker, U. (1996). Heat-shock and general stress response in *Bacillus subtilis*. Mol. Microbiol. *19*, 417–428.

Henderson, B., Allan, E., and Coates, A.R. (2006). Stress wars: the direct role of host and bacterial molecular chaperones in bacterial infection. Infect. Immun. *74*, 3693–3706.

de las Heras, A., Cain, R.J., Bielecka, M.K., and Vázquez-Boland, J.A. (2011). Regulation of *Listeria* virulence: PrfA master and commander. Curr. Opin. Microbiol. *14*, 118–127.

Herbig, A.F., and Helmann, J.D. (2001). Roles of metal ions and hydrogen peroxide in modulating the interaction of the *Bacillus subtilis* PerR peroxide regulon repressor with operator DNA. Mol. Microbiol. *41*, 849–859.

Hill, C., O'Driscoll, B., and Booth, I. (1995). Acid adaptation and food poisoning microorganisms. Int. J. Food Microbiol. *28*, 245–254.

Hu, Y., Raengpradub, S., Schwab, U., Loss, C., Orsi, R.H., Wiedmann, M., and Boor, K.J. (2007a). Phenotypic and transcriptomic analyses demonstrate interactions between the transcriptional regulators CtsR and Sigma B in *Listeria monocytogenes*. Appl. Environ. Microbiol. *73*, 7967–7980.

Hu, Y., Oliver, H.F., Raengpradub, S., Palmer, M.E., Orsi, R.H., Wiedmann, M., and Boor, K.J. (2007b). Transcriptomic and phenotypic analyses suggest a network between the transcriptional regulators HrcA and sigmaB in *Listeria monocytogenes*. Appl. Environ. Microbiol. *73*, 7981–7991.

Imlay J.A. (2002). How oxygen damages microbes: oxygen tolerance and obligate anaerobiosis. Adv. Microb. Physiol. *46*, 111–153.

Imlay J.A. (2003). Pathways of oxidative damage. Annu. Rev. Microbiol. *57*, 395–418.

Imlay J.A. (2008). Cellular defenses against superoxide and hydrogen peroxide. Annu. Rev. Biochem. *77*, 755–776.

Jensen, A., Thomsen, L.E., Jørgensen, R.L., Larsen, M.H., Roldgaard, B.B., Christensen, B.B., Vogel, B.F., Gram, L., and Ingmer, H. (2008). Processing plant persistent strains of *Listeria monocytogenes* appear to have a lower virulence potential than clinical strains in selected virulence models. In. J. Food Microbiol. *123*, 254–261.

Joerger, R.D., Chen, H., and Kniel, K.E. (2006). Characterization of a spontaneous, pressure-tolerant *Listeria monocytogenes* Scott A *ctsR* deletion mutant. Food-borne Pathog. Dis. *3*, 196–202.

Jørgensen, F., Stephens, P.J., Knøchel, S. (1995). The effect of osmotic shock and subsequent adaptation on the thermotolerance and cell morphology of *Listeria monocytogenes*. J. Appl. Bacteriol. 79,274–281.

Joseph, B., Przybilla, K., Stühler, C., Schauer, K., Slaghuis, J., Fuchs. T.M., and Goebel, W. (2006). Identification of *Listeria monocytogenes* genes contributing to intracellular replication by expression profiling and mutant screening. J. Bacteriol. *188*, 556–568.

Kallipolitis, B.H., and Ingmer, H. (2001). *Listeria monocytogenes* response regulators important for stress tolerance and pathogenesis. FEMS Microbiol. Lett. *204*, 111–115.

Karatzas, K.A., and Bennik, M.H. (2002). Characterization of a *Listeria monocytogenes* Scott A isolate with high tolerance towards high hydrostatic pressure. Appl. Environ. Microbiol. *68*, 3183–3189.

Karatzas, K.A., Wouters, J.A., Gahan, C.G., Hill, C., Abee, T., and Bennik. M.H. (2003). The CtsR regulator of *Listeria monocytogenes* contains a variant glycine repeat region that affects piezotolerance, stress resistance, motility and virulence. Mol. Microbiol. *49*, 1227–1238.

Kazmierczak, M.J., Mithoe, S.C., Boor, K.J., and Wiedmann M. (2003). *Listeria monocytogenes* sigma B regulates stress response and virulence functions. J. Bacteriol. *185*, 5722–5734.

Kazmierczak, M.J., Wiedmann, M., and Boor, K.J. (2005). Alternative sigma factors and their roles in bacterial virulence J. Bacteriol. *185*, 5722–5734.

Kazmierczak, M.J., Wiedmann, M., and Boor, K.J. (2006). Contributions of *Listeria monocytogenes* sigmaB and PrfA to expression of virulence and stress response genes during extra- and intracellular growth. Microbiology *152*, 1827–1838.

Keyer, K., Gort, A.S., and Imlay, J.A. (1995). Superoxide and the production of oxidative DNA damage. J. Bacteriol. *177*, 6782–6790.

Kilstrup, M., Jacobsen, S., Hammer, K., and Vogensen, F.K. (1997). Induction of heat shock proteins DnaK, GroEL, and GroES by salt stress in *Lactococcus* lactis. Appl. Environ. Microbiol. *63*, 1826–1837.

Kim, H., Boor, K.J., and Marquis, H. (2004). *Listeria monocytogenes* sigmaB contributes to invasion of human intestinal epithelial cells. Infect. Immun. *72*, 7374–7378.

Kim, H., Marquis, H., and Boor, K.J. (2005). SigmaB contributes to *Listeria monocytogenes* invasion by controlling expression of *inlA* and *inlB*. Microbiology *151*, 3215–3222.

Kim, K.P., Hahm, B.K., Bhunia, A.K. (2007). The 2-cys peroxiredoxin-deficient *Listeria monocytogenes* displays impaired growth and survival in the presence of hydrogen peroxide *in vitro* but not in mouse organs. Curr. Microbiol. *54*, 382–387.

Kirstein, J., Zühlke, D., Gerth, U., Turgay, K., and Hecker, M. (2005). A tyrosine kinase and its activator control the activity of the CtsR heat shock repressor in *B. subtilis*. EMBO J. *24*, 3435–3445.

Ko, R., and Smith, L.T. (1999). Identification of an ATP-driven, osmoregulated glycine betaine transport system in *Listeria monocytogenes*. Appl. Environ. Microbiol. *65*, 4040–4048.

Ko, R., Smith, L.T., and Smith, G.M. (1994). Glycine betaine confers enhanced osmotolerance and cryotolerance on *Listeria monocytogenes*. J. Bacteriol. *176*, 426–431.

Koutsoumanis, K.P., Kendall, P.A., and Sofos, J.N. (2003). Effect of food processing-related stresses on acid tolerance of *Listeria monocytogenes*. Appl. Environ. Microbiol. *69*, 7514–7526.

Krojer, T., Garrido-Franco, M., Huber. R., and Ehrmann, M., Clausen T. (2002). Crystal structure of DegP

(HtrA) reveals a new protease-chaperone machine. Nature *416*, 455–459.

Kroll, J.S., Langford, P.R., Wilks, K.E., and Keil, A.D. (1995). Bacterial [Cu,Zn]-superoxide dismutase: phylogenetically distinct from the eukaryotic enzyme, and not so rare after all! Microbiology *141*, 2271–2279.

Krüger, E., Zühlke, D., Witt, E., Ludwig, H., and Hecker, M. (2001). Clp-mediated proteolysis in Gram-positive bacteria is autoregulated by the stability of a repressor. EMBO J. *20*, 852–863.

Langer, T., Lu, C., Echols, H., Flanagan, J., Hayer, M.K., and Hartl, F.U. (1992). Successive action of DnaK, DnaJ and GroEL along the pathway of chaperone-mediated protein folding. Nature 356, 683–689.

Lasa, I., David, V., Gouin, E., Marchand, J.B., and Cossart, P. (1995). The amino-terminal part of ActA is critical for the actin-based motility of *Listeria monocytogenes*; the central proline-rich region acts as a stimulator. Mol. Microbiol. *18*, 425–436.

Leblond-Francillard, M., Gaillard, J.L., and Berche, P. (1989). Loss of catalase activity in Tn1545-induced mutants does not reduce growth of *Listeria monocytogenes in vivo.* Infect. Immun. 57, 2569–2573.

Lecuit, M., Ohayon, H., Braun, L., Mengaud, J., and Cossart, P. (1997). Internalin of *Listeria monocytogenes* with an intact leucine-rich repeat region is sufficient to promote internalization. Infect. Immun. *65*, 5309–5319.

Lecuit, M., Hurme, R., Pizarro-Cerda, J., Ohayon, H., Geiger, B., Cossart, P. (2000). A role for alpha-and beta-catenins in bacterial uptake. Proc. Natl. Acad. Sci. U.S.A. *18*, 10008–10013.

Lindquist, S., and Craig, E.A. (1998). The heat-shock proteins. Annu. Rev. Genet. *22*, 631–677.

Lingnau, A., Domann, E., Hudel, M., Bock, M., Nichterlein, T., Wehland, J., and Chakraborty, T. (1995). Expression of the *Listeria monocytogenes* EGD *inlA* and *inlB* genes, whose products mediate bacterial entry into tissue culture cell lines, by PrfA-dependent and -independent mechanisms. Infect. Immun. *63*, 3896–3903.

Lippert, E., and Galinski, A.E. (1992). Enzyme stabilization by ectoine-type compatible solutes: protection against heating, freezing and drying. Appl. Microbiol. Biotechnol. *37*, 61–65.

Loewen, P. (1996). Probing the structure of catalase HPII of *Escherichia coli* – a review. Gene *179*, 39–44.

Lou Y., and Yousef A.E. (1997). Adaptation to sublethal environmental stresses protects *Listeria monocytogenes* against lethal preservation factors. Appl. Environ. Microbiol. *63*, 1252–1255.

Low, J.C., and Donachie W. (1997). A review of *Listeria monocytogenes* and listeriosis. Vet. J. *153*, 9–29.

Lundén, J., Tolvanen, R., and Korkeala, H. (2008). Acid and heat tolerance of persistent and nonpersistent *Listeria monocytogenes* food plant strains. Lett. Appl. Microbiol. *46*, 276–280.

Lungu, B., Ricke, S.C., Johnson, M.G. (2008). Resistance of nutrient-deprived *Listeria monocytogenes* 10403S and a DeltasigB mutant to chemical stresses in the presence or absence of oxygen. J. Food Sci. 73, M339–345.

Lungu, B., Ricke, S.C., and Johnson, M.G. (2009). Growth, survival, proliferation and pathogenesis of *Listeria monocytogenes* under low oxygen or anaerobic conditions: a review. Anaerobe *1*, 7–17.

Lushchak V.I. (2011). Adaptive response to oxidative stress: Bacteria, fungi, plants and animals. Comp. Biochem. Physiol. C. Toxicol. Pharmacol. *153*, 175–190.

Lynch, M., Painter, J., Woodruff, R., Braden, C. (2006). Centers for Disease Control and Prevention. Surveillance for foodborne-disease outbreaks – United States, 1998–2002. MMWR Surveill. Summ. *55*, 1–42.

McClure, P.J., Roberts, T.A., and Oguru, P.O. (1989). Comparison of the effects of sodium chloride, pH and temperature on the growth of *Listeria monocytogenes* on gradient plates and in liquid medium. Lett. Appl. Microbiol. *9*, 95–99.

McCord J.M., and Fridovich I. (1969). Superoxide dismutase. An enzymic function for erythrocuprein (hemocuprein). J. Biol. Chem. *244*, 6049–6055.

McGann, P., Wiedmann, M., and Boor, K.J. (2007). The alternative sigma factor sigma B and the virulence gene regulator PrfA both regulate transcription of *Listeria monocytogenes* internalins. Appl. Environ. Microbiol. *73*, 2919–2930.

Mackey, B.M., and Bratchell, N. (1989) The heat resistance of *Listeria monocytogenes*. Lett. Appl. Microbiol. *9*, 89–94.

Marles-Wright, J., and Lewis, R.J. (2007). Stress responses of bacteria Curr. Opin. Struct. Biol. *17*, 755–760.

Marquis, R.E., Bender, G.R., Murray, D.R., and Wong, A. (1987). Arginine deiminase system and bacterial adaptation to acid environments. Appl. Environ. Microbiol. *53*, 198–200.

Marron, L., Emerson, N., Gahan, C.G., and Hill, C. (1997). A mutant of *Listeria monocytogenes* LO28 unable to induce an acid tolerance response displays diminished virulence in a murine model. Appl. Environ. Microbiol. *63*, 4945–4957.

Massey, V. (1994). Activation of molecular oxygen by flavins and flavoproteins. J. Biol. Chem. *269*, 22459–22462.

Maurizi, M.R., Clark, W.P., Kim S.H., andGottesman S. (1990). Clp P represents a unique family of serine proteases. J. Biol. Chem. *265*, 12546–12552.

Mead, P.S., Slutsker, L., Dietz, V., McCaig, L.F., Bresee, J.S., Shapiro, C., Griffin, P.M., and Tauxe, R.V. (1999). Food-related illness and death in the United States. Emerg. Infect. Dis. *5*, 607–625.

Mendum, M.L., Smith, L.T. (2002a). Characterization of glycine betaine porter I from *Listeria monocytogenes* and its roles in salt and chill tolerance. Appl. Environ. Microbiol. *68*, 813–819.

Mendum, M.L., Smith, L.T. (2002b). Gbu glycine betaine porter and carnitine uptake in osmotically stressed *Listeria monocytogenes* cells. Appl. Environ. Microbiol. *68*, 5647–5655.

Messner, K.R., and Imlay, J.A. (1999). The identification of primary sites of superoxide and hydrogen peroxide formation in the aerobic respiratory chain and sulfite reductase complex of *Escherichia coli*. J. Biol. Chem. *274*, 10119–10128.

Messner, K.R., and Imlay J.A. (2002). Mechanism of superoxide and hydrogen peroxide formation by fumarate reductase, succinate dehydrogenase, and aspartate oxidase. J. Biol. Chem. *277*, 42563–42571.

Mnatsakanyan, N., Hook, J.A., Quisenberry, L., and Weber, J. (2009). ATP synthase with its gamma subunit reduced to the N-terminal helix can still catalyze ATP synthesis. J. Biol. Chem. *284*, 26519–26525.

Mogk A., Homuth G., Scholz C., Kim L., Schmid F.X., and Schumann W. (1997). The GroE chaperonin machine is a major modulator of the CIRCE heat shock regulon of *Bacillus subtilis*. EMBO J. *16*, 4579–4590.

Mols, M., and Abee, T. (2011). Primary and secondary oxidative stress in *Bacillus*. Environ. Microbiol. *13*,1387–1394.

Mongkolsuk, S., and Helmann, J.D. (2002). Regulation of inducible peroxide stress responses. Mol. Microbiol. *45*, 9–15.

Moorhead, S.M., and Dykes, G.A. (2003). The role of the *sigB* gene in the general stress response of *Listeria monocytogenes* varies between a strain of serotype 1/2a and a strain of serotype 4c. Curr. Microbiol. *46*, 461–466.

Muga, A., and Moro, F. (2008). Thermal adaptation of heat shock proteins. Curr. Protein. Pept. Sci. *9*, 552–566.

Murataliev, M.B., and Boyer, P.D. (1992). The mechanism of stimulation of MgATPase activity of chloroplast F1-ATPase by non-catalytic adenine-nucleotide binding. Acceleration of the ATP-dependent release of inhibitory ADP from a catalytic site. Eur. J. Biochem. *209*, 681–687.

Nair, S., Frehe, L.C., Nguyen, L., Escuyer, V., and Berche, P. (1999). ClpE, a novel member of the HSP100 family, is involved in cell division and virulence of *Listeria monocytogenes*. Mol. Microbiol. *31*, 185–196.

Nair, S., Milohanic, E., and Berche, P. (2000). ClpC ATPase is required for cell adhesion and invasion of *Listeria monocytogenes*. Infect. Immun. *68*, 7061–7068.

Nakamoto, R.K., Ketchum, C.J., and al-Shawi, M.K. (1999). Rotational coupling in the F0F1 ATP synthase. Annu. Rev. Biophys. Biomol. Struct. *28*, 205–234.

Nightingale, K.K., Windham, K., Martin, K.E., Yeung, M., and Wiedmann, M. (2005). Select *Listeria monocytogenes* subtypes commonly found in foods carry distinct nonsense mutations in *inlA*, leading to expression of truncated and secreted internalin A, and are associated with a reduced invasion phenotype for human intestinal epithelial cells. Appl. Environ. Microbiol. *71*, 8764–8772.

Nightingale, K.K., Ivy, R.A., Ho, A.J., Fortes, E.D., Njaa, B.L., Peters, R.M., and Wiedmann, M. (2008). *inlA* premature stop codons are common among *Listeria monocytogenes* isolates from foods and yield virulence-attenuated strains that confer protection against fully virulent strains. Appl. Environ. Microbiol. *74*, 6570–6583.

O'Driscoll, B., Gahan, C.G., and Hill, C. (1996). Adaptive acid tolerance response in *Listeria monocytogenes*: isolation of an acid-tolerant mutant which demonstrates increased virulence. Appl. Environ. Microbiol. *62*, 1693–1698.

Okada, Y., Makino, S., Tobe, T., Okada, N., and Yamazaki, S. (2002). Cloning of rel from *Listeria monocytogenes* as an osmotolerance involvement gene. Appl. Environ. Microbiol. *68*, 1541–1547.

Olesen, I., Vogensen, F.K., and Jespersen, L. (2009). Gene transcription and virulence potential of *Listeria monocytogenes* strains after exposure to acidic and NaCl stress. Food-borne Pathog. Dis. *6*, 669–680.

Olier, M., Rousseaux, S., Piveteau, P., Lemaître, J.P., Rousset, A., and Guzzo, J. (2004). Screening of glutamate decarboxylase activity and bile salt resistance of human asymptomatic carriage, clinical, food, and environmental isolates of *Listeria monocytogenes*. Int. J. Food. Microbiol. *93*, 87–99.

Oliver, H.F., Orsi, R.H., Wiedmann, M., and Boor, K.J. (2010). *Listeria monocytogenes* {sigma}B has a small core regulon and a conserved role in virulence but makes differential contributions to stress tolerance across a diverse collection of strains. Appl. Environ. Microbiol. *76*, 4216–4232.

Paget, M.S., and Helmann, J.D. (2003). The sigma70 family of sigma factors. Genome Biol. *4*, 203.

Pallen, M.J., Wren, B.W. (1997). The HtrA family of serine proteases. Mol. Microbiol. *26*, 209–221.

Parkinson, J.S. (1993). Signal transduction schemes of bacteria. Cell 73, 857–871.

Pascual, M.B., Mata-Cabana, A., Florencio, F.J., Lindahl, M., Cejudo, F.J. (2010). Overoxidation of 2-Cys peroxiredoxin in prokaryotes: cyanobacterial 2-Cys peroxiredoxins sensitive to oxidative stress. J. Biol. Chem. *285*, 34485–34492.

Patchett, R.A., Kelly, A.F., Kroll, R.G. (1992). Effect of sodium chloride on the intracellular solute pools of *Listeria monocytogenes*. Appl. Environ. Microbiol. *58*, 3959–3963.

Percy, M.E. (1984). Catalase: an old enzyme with a new role? Can. J. Biochem. Cell Biol. *62*, 1006–1014.

Phan-Thanh, L., and Mahouin, F. (1999). A proteomic approach to study the acid response in *Listeria monocytogenes*. Electrophoresis. *20*, 2214–2224.

Phan-Thanh, L., Mahouin, F., and Aligé, S. (2000). Acid responses of *Listeria monocytogenes*. Int. J. Food Microbiol. *55*, 121–126.

Poole, L.B. (2005). Bacterial defenses against oxidants: mechanistic features of cysteine-based peroxidases and their flavoprotein reductases. Arch. Biochem. Biophys. *433*, 240–254.

Poquet, I., Saint, V., Seznec, E., Simoes, N., Bolotin, A., and Gruss, A. (2000). HtrA is the unique surface housekeeping protease in *Lactococcus lactis* and is required for natural protein processing. Mol. Microbiol. 35, 1042–1051.

Porankiewicz, J., Wang, J., and Clarke, A.K. (1999). New insights into the ATP-dependent Clp protease: *Escherichia coli* and beyond. Mol. Microbiol. *32*, 449–458.

Portnoy, D.A., Auerbuch, V., Glomski, I.J. (2002). The cell biology of *Listeria monocytogenes* infection:

the intersection of bacterial pathogenesis and cell-mediated immunity. J. Cell Biol. *158*, 409–414.

Poyart, C., Abachin, E., Razafimanantsoa, I., and Berche, P. (1993). The zinc metalloprotease of *Listeria monocytogenes* is required for maturation of phosphatidylcholine phospholipase C: direct evidence obtained by gene complementation. Infect. Immun. *61*, 1576–1580.

Ragon, M., Wirth, T., Hollandt, F., Lavenir, R., Lecuit, M., Le Monnier, A., and Brisse, S. (2008). A new perspective on *Listeria monocytogenes* evolution. PLoS Pathog. *4*, e1000146.

Rea, R., Hill, C., Gahan, C.G. (2005). *Listeria monocytogenes* PerR mutants display a small-colony phenotype, increased sensitivity to hydrogen peroxide, and significantly reduced murine virulence. Appl. Environ. Microbiol. *71*, 8314–8322.

Rea, R.B., Gahan, C.G., and Hill, C. (2004). Disruption of putative regulatory loci in *Listeria monocytogenes* demonstrates a significant role for Fur and PerR in virulence. Infect. Immun. *72*, 717–727.

Roberts, A., Chan, Y., and Wiedmann, M. (2005). Definition of genetically distinct attenuation mechanisms in naturally virulence-attenuated *Listeria monocytogenes* by comparative cell culture and molecular characterization. Appl. Environ. Microbiol. *71*, 3900–3910.

Roche, S.M., Gracieux, P., Milohanic, E., Albert, I., Virlogeux-Payant, I., Témoin, S., Grépinet, O., Kerouanton, A., Jacquet, C., Cossart, P., and Velge, P. (2005). Investigation of specific substitutions in virulence genes characterizing phenotypic groups of low-virulence field strains of *Listeria monocytogenes*. Appl. Environ. Microbiol. *71*, 6039–6048.

Rodriguez-Romo, L.A., and Yousef, A.E. (2002). Microbial stress adaptation and safety of produce. In Microbiology of Fruits and Vegetables, Gorny, J.R., Yousef, A.E., and Sapers, G.M., eds (CRC Press, Boca Raton, FL), pp. 95–114.

Rouquette, C., Ripio, M.T., Pellegrini, E., Bolla, J.M., Tascon, R.I., Vázquez-Boland, J.A., and Berche P. (1996). Identification of a ClpC ATPase required for stress tolerance and *in vivo* survival of *Listeria monocytogenes*. Mol. Microbiol. *21*, 977–987.

Rouquette, C., de Chastellier, C., Nair, S., and Berche, P. (1998). The ClpC ATPase of *Listeria monocytogenes* is a general stress protein required for virulence and promoting early bacterial escape from the phagosome of macrophages. Mol. Microbiol. *27*, 1235–1245.

Ryan, S., Hill, C., and Gahan, C.G. (2008). Acid stress responses in *Listeria monocytogenes*. Adv. Appl. Microbiol. *65*, 67–91.

Ryan, S., Begley, M., Gahan, C.G., and Hill, C. (2009). Molecular characterization of the arginine deiminase system in *Listeria monocytogenes*: regulation and role in acid tolerance. Environ. Microbiol. *11*, 432–445.

Salotra, P., Singh, D.K., Seal, K.P., Krishna, N., Jaffe, H., and Bhatnagar, R. (1995). Expression of DnaK and GroEL homologs in *Leuconostoc esenteroides* in response to heat shock, cold shock or chemical stress. FEMS Microbiol. Lett. *131*, 57–62.

van Schaik, W., and Abee, T. (2005). The role of sigmaB in the stress response of Gram-positive bacteria—targets for food preservation and safety. Curr. Opin. Biotechnol. *16*, 218–224.

van Schaik, W., Gahan, C.G., and Hill, C. (1999). Acid-adapted *Listeria monocytogenes* displays enhanced tolerance against the lantibiotics nisin and lacticin 3147. J. Food Prot. *62*, 536–539.

Schiavone, J.R., and Hassan, H.M. (1987). Biosynthesis of superoxide dismutase in eight prokaryotes: Effects of oxygen, paraquat and an iron chelator. FEMS Microbiol. Lett. *42*, 33–38.

Schnupf, P., Portnoy, D.A., and Decatur, A.L. (2006). Phosphorylation, ubiquitination and degradation of listeriolysin O in mammalian cells: role of the PEST-like sequence. Cell. Microbiol. *8*, 353–364.

Schulz, A., and Schumann, W. (1996). *hrcA*, the first gene of the *Bacillus subtilis dnaK* operon encodes a negative regulator of class I heat shock genes. J. Bacteriol. *178*, 1088–1093.

Schumann, W. (2003). The *Bacillus subtilis* heat shock stimulon. Cell Stress Chaperones *8*, 207–217.

Seaver, L.C., and Imlay, J.A. (2001). Alkyl hydroperoxide reductase is the primary scavenger of endogenous hydrogen peroxide in *Escherichia coli*. J. Bacteriol. *183*, 7173–7181.

Segal, R., and Ron, E.Z. (1996). Regulation and organization of the *groE* and *dnaK* operons in Eubacteria. FEMS Microbiol. Lett. *138*, 1–10.

Singh, R., Jamieson, A., and Cresswell, P. (2008). GILT is a critical host factor for *Listeria monocytogenes* infection. Nature *455*, 1244–1247.

Skowyra, D., Georgopoulos, C., and Zylicz, M. (1990). The E. coli *dnaK* gene product, the hsp70 homolog, can reactivate heat-inactivated RNA polymerase in an ATP hydrolysis-dependent manner. Cell *62*, 939–944.

Sleator, R.D., and Hill, C. (2005). A novel role for the LisRK two-component regulatory system in listerial osmotolerance. Clin. Microbiol. Infect. *11*, 599–601.

Sleator, R.D., Gahan, C.G., Abee, T., and Hill, C. (1999). Identification and disruption of BetL, a secondary glycine betaine transport system linked to the salt tolerance of *Listeria monocytogenes* LO28. Appl. Environ. Microbiol. *65*, 2078–2083.

Sleator, R.D., Wouters, J., Gahan, C.G., and Abee, T., Hill, C. (2001a). Analysis of the role of OpuC, an osmolyte transport system, in salt tolerance and virulence potential of *Listeria monocytogenes*. Appl. Environ. Microbiol. *67*, 2692–2698.

Sleator, R.D., Gahan, C.G., and Hill, C. (2001b). Identification and disruption of the proBA locus in *Listeria monocytogenes*: role of proline biosynthesis in salt tolerance and murine infection. Appl. Environ. Microbiol. *67*, 2571–2577.

Sleator, R.D., Gahan, C.G., and Hill, C. (2001c). Mutations in the listerial *proB* gene leading to proline overproduction: effects on salt tolerance and murine infection. Appl. Environ. Microbiol. *67*, 4560–4565.

Sleator, R.D., Gahanm, C.G., and Hill, C. (2003a). A postgenomic appraisal of osmotolerance in *Listeria monocytogenes*. Appl. Environ. Microbiol. *69*, 1–9.

Sleator, R.D., Wood, J.M., Hill, C. (2003b). Transcriptional regulation and posttranslational activity of the betaine transporter BetL in *Listeria monocytogenes* are controlled by environmental salinity. J. Bacteriol. *185*, 7140–7144.

Sleator, R.D., Clifford, T., and Hill, C. (2007). Gut osmolarity: a key environmental cue initiating the gastrointestinal phase of *Listeria monocytogenes* infection? Med. Hypotheses *69*, 1090–1092.

Small, P.L., and Waterman, S.R. (1998). Acid stress, anaerobiosis and gadCB: lessons from *Lactococcus* lactis and *Escherichia coli*. Trends Microbiol. *6*, 214–216.

Smiddy, M., Sleator, R.D., Patterson, M.F., Hill, C., and Kelly, A.L. (2004). Role for compatible solutes glycine betaine and L-carnitine in listerial barotolerance. Appl. Environ. Microbiol. *70*, 7555–7557.

Smith, G.A., Marquis, H., Jones, S., Johnston, N.C., Portnoy, D.A., and Goldfine, H. (1995). The two distinct phospholipases C of *Listeria monocytogenes* have overlapping roles in escape from a vacuole and cell-to-cell spread. Infect. Immun. *63*, 4231–4237.

Smith, G.M., Smith L.T., Gerhardt, P.N.M., Ko, R. (1998). Solute transport enzymes related to stress tolerance in *Listeria monocytogenes*. J. Food Biochem. *22*, 269–285.

Smith, J.T., and Hunters, E. (1988) Heat injury in *Listeria monocytogenes*: prevention by solutes. Lebens-Wiss. Technol. *21*, 307–311.

Somolinos, M., Espina, L., Pagán, R., and Garcia, D. (2010). *sigB* absence decreased *Listeria monocytogenes* EGD-e heat resistance but not its Pulsed Electric Fields resistance. Int. J. Food Microbiol. *141*, 32–38.

Spiess, C., Beil, A., and Ehrmann, M. (1999). A temperature-dependent switch from chaperone to protease in a widely conserved heat shock protein. Cell *97*, 339–347.

Stack, H.M., Sleator, R.D., Bowers, M., Hill C., and Gahan, C.G. (2005). Role for HtrA in stress induction and virulence potential in *Listeria monocytogenes*. Appl. Environ. Microbiol. *71*, 4241–4247.

Stavru F., Archambaud C., and Cossart P. (2011). Cell biology and immunology of *Listeria monocytogenes* infections: novel insights. Immunol. Rev. *240*, 160–184.

Storz, G., and Imlay, J.A. (1999). Oxidative stress. Curr. Opin. Microbiol. *2*, 188–194.

Sue, D., Fink, D., Wiedmann, M., and Boor, K.J. (2004). SigmaB-dependent gene induction and expression in *Listeria monocytogenes* during osmotic and acid stress conditions simulating the intestinal environment. Microbiology *150*, 3843–3855.

Sumner J.B., and Dounce A.L. (1937). Crystalline catalase. J. Biol. Chem. *121*, 417–424.

Swaminathan, B., and Gerner-Smidt, P. (2007). The epidemiology of human listeriosis. Microbes Infect. *9*, 1236–1243.

Swartz, M.A., Welch, D.F., and Narayanan, R.P., Greenfield R.A. (1991). Catalase-negative *Listeria monocytogenes* causing meningitis in an adult. Clinical and laboratory features. Am. J. Clin. Pathol. *96*, 130–133.

Tally F.P., Goldin, B.R., Jacobus, N.V., and Gorbach, S.L. (1977). Superoxide dismutase in anaerobic bacteria of clinical significance. Infect. Immun. *16*, 20–25.

Taylor, C.M., Beresford, M., Epton, H.A., Sigee, D.C., Shama, G., Andrew. P.W., and Roberts, I.S. (2002). *Listeria monocytogenes relA* and hpt mutants are impaired in surface-attached growth and virulence. J. Bacteriol. *184*, 621–628.

Tilney, L.G., and Portnoy, D.A. (1989). Actin filaments and the growth, movement, and spread of the intracellular bacterial parasite, *Listeria monocytogenes*. J. Cell. Biol. *109*, 1597–1608.

Vasconcelos, J.A., and Deneer, H.G. (1994). Expression of superoxide dismutase in *Listeria monocytogenes*. Appl. Environ. Microbiol. *60*, 2360–2366.

Vázquez-Boland, J.A., Kuhn, M., Berche, P., Chakraborty, T., Domínguez-Bernal, G., Goebel, W., González-Zorn, B., Wehland, J., and Kreft, J. (2001). Listeria pathogenesis and molecular virulence determinants. Clin. Microbiol. Rev. *14*, 584–640.

van der Veen, S., Abee, T. (2010). HrcA and DnaK are important for static and continuous-flow biofilm formation and disinfectant resistance in *Listeria monocytogenes*. Microbiology *156*, 3782–3790.

Verheul, A., Rombouts, F.M., Beumer, R.R., and Abee, T. (1995). An ATP-dependent L-carnitine transporter in *Listeria monocytogenes* Scott A is involved in osmoprotection. J. Bacteriol. *177*, 3205–3212.

Verheul, A., Glaasker, E., Poolman, B., and Abee, T. (1997). Betaine and L-carnitine transport by *Listeria monocytogenes* Scott A in response to osmotic signals. J. Bacteriol. *179*, 6979–6985.

Voetsch, A.C., Angulo, F.J., Jones, T.F., Moore, M.R., Nadon, C., McCarthy, P., Shiferaw, B., Megginson, M.B., Hurd S., Anderson, B.J., *et al.* (2007). Reduction in the incidence of invasive listeriosis in food-borne diseases active surveillance network sites, 1996–2003. Clin. Infect. Dis. *44*, 513–520.

Wałecka, E., Molenda, J., and Bania, J. (2009). The impact of environmental stress on *Listeria monocytogenes* virulence. Pol. J. Vet. Sci. *12*, 575–579.

Wałecka, E., Molenda, J., Karpíšková, R., and Bania, J. (2011a). Effect of heat exposure on invasiveness of *Listeria monocytogenes* strains. Food-borne Pathog. Dis. *8*, 839–841.

Wałecka, E., Molenda, J., Karpíšková, R., and Bania, J. (2011b). Effect of osmotic stress and culture density on invasiveness of *Listeria monocytogenes* strains. Int. J. Food Microbiol. *144*, 440–445.

Wang, J., Hartling, J.A., and Flanagan, J.M. (1997). The structure of ClpP at 2.3 A resolution suggests a model for ATP-dependent proteolysis. Cell *91*, 447–456.

Welch, D.F., Sword, C.P., Brehm, S., and Dusanic, D. (1979). Relationship between superoxide dismutase and pathogenic mechanisms of *Listeria monocytogenes*. Infect. Immun. *23*, 863–872.

Wemekamp-Kamphuis, H.H., Wouters, J.A., Sleator, R.D., Gahan, C.G., Hill, C., and Abee, T. (2002). Multiple deletions of the osmolyte transporters BetL, Gbu, and OpuC of *Listeria monocytogenes* affect virulence and growth at high osmolarity. Appl. Environ. Microbiol. *68*, 4710–4716.

Wemekamp-Kamphuis, H.H., Wouters, J.A., de Leeuw, P.P., Hain, T., Chakraborty, T., and Abee, T. (2004). Identification of sigma factor sigma B-controlled genes and their impact on acid stress, high hydrostatic pressure, and freeze survival in *Listeria monocytogenes* EGD-e. Appl. Environ. Microbiol. *70*, 3457–3466.

Werbrouck, H., Vermeulen, A., Van Coillie, E., Messens, W., Herman, L., Devlieghere, F., and Uyttendaele, M. (2009). Influence of acid stress on survival, expression of virulence genes and invasion capacity into Caco-2 cells of *Listeria monocytogenes* strains of different origins. Int. J. Food Microbiol. *31*, 140–146.

Wiedmann, M., Bruce, J.L., Keating, C., Johnson, A.E., McDonough, P.L., and Batt, C.A. (1997). Ribotypes and virulence gene polymorphisms suggest three distinct *Listeria monocytogenes* lineages with differences in pathogenic potential. Infect. Immun. *65*, 2707–2716.

Wiedmann, M., Arvik, T.J., Hurley, R.J., and Boor, K.J. (1998). General stress transcription factor sigmaB and its role in acid tolerance and virulence of *Listeria monocytogenes*. J. Bacteriol. *180*, 3650–3656.

Wilson, R.L., Brown, L.L., Kirkwood-Watts, D., Warren. T.K., Lund, S.A., King, D.S., Jones, K.F., and Hruby, D.E. (2006). *Listeria monocytogenes* 10403S HtrA is necessary for resistance to cellular stress and virulence. Infect. Immun. *74*, 765–768.

Wonderling, L.D., Wilkinson, B.J., and Bayles, D.O. (2004). The *htrA* (*degP*) gene of *Listeria monocytogenes* 10403S is essential for optimal growth under stress conditions. Appl. Environ. Microbiol. *70*, 1935–1943.

Woo, K.M., Chung, W.J., Ha, D.B., Goldberg, A.L., and Chung, C.H. (1989). Protease Ti from *Escherichia coli* requires ATP hydrolysis for protein breakdown but not for hydrolysis of small peptides. J. Biol. Chem. *264*, 2088–2091.

Wood, J.M. (1999). Osmosensing by bacteria: signals and membrane-based sensors. Microbiol. Mol. Biol Rev. *63*, 230–262.

Wood, J.M. (2007). Bacterial osmosensing transporters. Methods Enzymol. *428*, 77–107.

Wood, J.M., Bremer, E., Csonka, L.N., Kraemer, R., Poolman, B., van der Heide. T., Smith, L.T. (2001). Osmosensing and osmoregulatory compatible solute accumulation by bacteria. Comp. Biochem. Physiol. A Mol. Integr. Physiol. *130*, 437–460.

Wood, Z.A., Schröder, E., Robin Harris, J., and Poole, L.B. (2003). Structure, mechanism and regulation of peroxiredoxins. Trends Biochem Sci. *28*, 32–40.

Yousef, A.E., Polly D.C. (2002). Basics of stress adaptation and implications in new-generation foods. In Microbial Stress Adaptation and Food Safety, Yousef, A.E., and Juneja, V.K., eds (CRC Press LLC, Boca Raton, FL), pp. 1–30.

Yura, T., and Nakahigashi, K. (1999). Regulation of the heat-shock response. Curr. Opin. Microbiol. *2*, 153–158.

Zamocky, M., Furtmüller, P.G., and Obinger, C. (2008). Evolution of catalases from bacteria to humans. Antioxid. Redox Signal. *10*, 1527–1548.

Zhang, C., Zhang, M., Ju J., Nietfeldt, J., Wise, J., Terry, P.M., Olson, M., Kachman, S.D., Wiedmann, M., Samadpour, M., *et al.* (2003a). Genome diversification in phylogenetic lineages I and II of *Listeria monocytogenes*: identification of segments unique to lineage II populations. J. Bacteriol. *185*, 5573–5584.

Zhang, S., Haldenwang, W.G. (2003). RelA is a component of the nutritional stress activation pathway of the *Bacillus subtilis* transcription factor sigma B. J. Bacteriol. *185*, 5714–5721.

Zhou, X., Jiao, X., and Wiedmann, M. (2005). *Listeria monocytogenes* in the Chinese food system: strain characterization through partial *actA* sequencing and tissue-culture pathogenicity assays. J. Med. Microbiol. *54*, 217–224.

Ziegler, C., Bremer, E., and Krämer, R. (2010). The BCCT family of carriers: from physiology to crystal structure. Mol. Microbiol. *78*, 13–34.

Zuber, P. (2009). Management of oxidative stress in *Bacillus*. Annu. Rev. Microbiol. *63*, 575–597.

Zuber, U., and Schumann, W. (1994). CIRCE, a novel heat shock element involved in regulation of heat shock operon *dnaK* of *Bacillus subtilis*. J. Bacteriol. *176*, 1359–1363.

Mechanisms Involved in Low-temperature Adaptation in *Bacillus cereus*

5

Julien Brillard and Véronique Broussolle

Abstract

Among the soil bacteria of the spore former genus *Bacillus,* the human pathogens mostly belong to the *B. cereus* group. This species is divided in seven phylogenetic groups, with particular traits in virulence, and particular growth temperature ranges, where each of these seven phylogenetic groups corresponds to a specific 'thermotype', showing clear differences in ability to grow at low or high temperatures.

After a temperature downshift, changes that occur in the bacterial cell include a decrease of the membrane fluidity, a stabilization of secondary structures of nucleic acids which consequently causes a decreased efficiency in transcription and translation, a misfolding of some proteins, etc. The bacterial cell response involves various mechanisms which, among the *Bacillus* genus, have been mostly studied in *Bacillus subtilis*. This chapter focuses on current research about *B. cereus* low-temperature adaptation, compared to what is well described in *B. subtilis*.

Introduction

Among the soil bacteria of the genus *Bacillus,* the human pathogens mostly belong to the *B. cereus* group. This group displays a wide diversity of strains belonging to six closely related species: *B. cereus sensu stricto, B. anthracis, B. thuringiensis, B. mycoides, B. pseudomycoides* and *B. weihenstephanensis* (Helgason *et al.,* 2000; Guinebretiere *et al.,* 2008; Tourasse *et al.,* 2011). *B. cereus sensu stricto* is an opportunistic human pathogen causing various infections (Bottone, 2010), but is more studied for its role in food poisoning (Stenfors Arnesen *et al.,* 2008). *B. anthracis* is the causative agent of anthrax in grazing animals and other various mammalians including humans. *B. thuringiensis* is an insect pathogen widely used as biopesticide. A cluster of strains rare and phylogenetically remote from the rest of the *B. cereus* group has been identified and described as *Bacillus cereus* subsp. *cytotoxicus* (Lund *et al.,* 2000; Brillard and Lereclus, 2004; Fagerlund *et al.,* 2007; Auger *et al.,* 2008; Lapidus *et al.,* 2008). It should officially constitute soon a seventh species named *Bacillus cytotoxicus* (M.H. Guinebretière, personal communication). In this chapter, data available on various species of the *B. cereus* group will be given, but we will focus mainly on *B. cereus sensu stricto.*

B. cereus, like the other members of the *Bacillus* genus, is an endospore former bacterium. Spores and vegetative cells are ubiquitously distributed in the environment. They can be isolated from soil, water and from the gastrointestinal tract of dead or live eukaryotic organisms, either arthropods or mammalian (Jensen *et al.,* 2003; Swiecicka, 2008). Given all the changing and sometimes hostile environments this bacterium encounters during its life cycle, *B. cereus* must develop an adaptive strategy, elaborating different physiological responses depending on the stress encountered.

The ability of *B. cereus* to survive as spores and to colonize as vegetative cells highly diverse ecological niches explain the important role played by this bacterium in food safety issues (Stenfors Arnesen *et al.,* 2008; Carlin *et al.,* 2010). Because of their wide occurrence, of the abilities of their spores to adhere strongly to various surfaces, to resist to heat and other physical or chemical stresses, *B. cereus* strains are frequently associated with raw food materials and therefore sometimes found in final meals.

A review in a risk analysis perspective and including data on *B. cereus* ecology, prevalence and growth capacity in foods and resistance to process was published by the European Food Safety Authority (EFSA, 2005). This review indicates that *B. cereus* is a common source of food-borne poisoning, representing 1–33% of food-borne poisoning cases depending on the countries. It is difficult to trace precisely the food involved in food-borne poisoning because *B. cereus* is ubiquitous and might be present in several foods consumed by patients. Foods implicated in *B. cereus* food-poisoning events are frequently heat-treated foods, but other food categories (salad sprouts, juice, dressing) have also been implicated (EFSA, 2005).

Two types of syndromes of food-borne poisonings are caused by *B. cereus*. An emetic syndrome is caused by the production in the food of a highly thermoresistant toxin, a cyclic peptide called cereulide. The diarrhoeic syndrome is caused by a multiplication of *B. cereus* in the intestine where it secretes several enterotoxins such as Nhe, CytK or Hbl. For a review on this topic, see Stenfors Arnesen *et al.* (Stenfors Arnesen *et al.*, 2008). It also produces various haemolysins such as cholesterol-dependent cytolysins (Brillard and Lereclus, 2007; Bishop *et al.*, 2010) and other degradative enzymes (Stenfors Arnesen *et al.*, 2008). *B. cereus* food-poisoning events are usually mild. However, they also regularly lead to fatal cases (Mahler *et al.*, 1997; Lund *et al.*, 2000; Dierick *et al.*, 2005). In addition, the opportunistic virulence properties of *B. cereus* have also been reported (Bottone, 2010). *B. cereus* can cause various types of systemic and local infections in humans (i.e. endophthalmitis, wound infection, septicaemia etc.; Schoeni and Wong, 2005), which illustrates the ability of *B. cereus* to adapt to various environments.

In order to cause a food poisoning event, *B. cereus* must be ingested at a high level. Studies of outbreak related incriminated foods revealed *B. cereus* counts ranging from 200 to 10^9 per gram (or ml), with calculated infective doses ranging from 5×10^4 to 10^{11} cells (EFSA, 2005). The dose response for cereulide intoxination was mentioned to be 0.01 to 1.3 µg/g. Thus, spoiled food containing more than 10^4 cells or spores per gram may represent a risk (Granum and Lund, 1997; Nguyen-the and Broussolle, 2005).

In order to reach such a high level of bacteria in the food, either an extremely high contamination of the raw ingredient is present, or more probably, a multiplication of the bacteria occurs during food storage, prior to consumption. Storage conditions, such as refrigeration, are therefore critical steps to favour or repress *B. cereus* growth leading to food-poisoning events.

Among all the stresses encountered by *B. cereus* during its life cycle, the low temperature is a stress that can occur quickly and lasts for a long period (in soil during winter, in refrigerated food during conservation, etc.). Cold is a physical stress that drastically changes all physical and chemical parameters of the bacteria which subsequently needs to elaborate an adaptive response. Among the *Bacillus* genus, adaptation to low temperature has mostly been studied in *B. subtilis*. This chapter will illustrate what is known in *B. cereus* and compare to what is similar/different in *B. subtilis*.

Low-temperature adaptation in *Bacillus cereus*

Taxonomy of the *B. cereus* group and thermotolerance

Among the *B. cereus* group, strains display various phenotypical traits depending on the growth temperature. Previous findings showed that *B. cereus* group strains originating from cold environments belong mostly to psychrotrophic strains whereas strains originating from tropical environments are mostly mesophilic (von Stetten *et al.*, 1999). Interestingly, psychrotrophic strains display a particular sequence in their *cspA* gene that was proposed to be a psychrotolerance-associated signature (Francis *et al.*, 1998). Another psychrotolerant-associated signature was also proposed based on the *rrs* sequence (Pruss *et al.*, 1999). A new species, *B. weihenstephanensis* was therefore proposed corresponding to psychrotolerant strains (Lechner *et al.*, 1998). However, not all psychrotolerant *B. cereus* group strains belong to this species (Stenfors and Granum, 2001).

The phylogeny of the *B. cereus* group was recently redefined. About 400 *B. cereus* group strains, including all the seven species and originating from different geographic origins, from various environments (soil, food industry, food-poisoning etc.) were phylogenetically and phenotypically analysed. This study led to a proposed subdivision of the *B. cereus* group in seven phylogenetic groups (named group I to group VII) (Guinebretiere *et al.*, 2008). Several phenotypical traits of each of these 400 strains were determined, including their cytotoxic activities and temperature growth range (Table 5.1) (Guinebretiere *et al.*, 2008, 2010).

Interestingly, the taxonomical position among the *B. cereus* group (i.e. the belonging of a given strain to one of the seven proposed phylogenetic groups) was not correlated to the seven defined species, in agreement with previous taxonomic studies (Helgason *et al.*, 2000), but was correlated to the growth ability at a given temperature range (Guinebretiere *et al.*, 2008). In a context where multiplication of bacteria in refrigerated food is a key step to lead to a food poisoning event, this original property of the *B. cereus* group led us to focus on the importance of the mechanisms that could be responsible for growth at low temperature.

Only a limited number of studies deciphering the mechanisms with which *B. cereus* is coping with the cold were performed, and most of them dealt with mesophilic strains (particularly with the ATCC 14579 strain). However, the presence of psychotrophic strains among the *B. cereus* group may lead to new fields of investigation. The genome of a psychrotrophic strain (KBAB4) was completely sequenced a few years ago (Lapidus *et al.*, 2008). When comparing the genome of this strains to that of other *B. cereus* group strains (all mesophilic strains at the time of this genome sequencing project), no major clue of what would cause a better psychrotophy of this strain could arise. Although alignment of the genome of this strain with that of other *B. cereus* group genomes revealed that it was collinear, indicating a similar genome structure, additional sequence of 650 kb was identified. Most of this additional genetic data consisted of large plasmids and integrated or non-integrated temperate phages. A higher ability of genetic exchange was also suggested among the psychrotrophic population (Sorokin *et al.*, 2006). Whether these extrachromosomal DNA regions are involved in psychrotrophy will have to be investigated in the future.

The major mechanisms involved in bacterial adaptation to low temperature have been mostly

Table 5.1 Characteristics of the seven *B. cereus* phylogenetic groups, according to (Guinebretiere *et al.*, 2008, 2010)

Phylogenetic group	Growth range	Characteristic[a]	*cspA* Signature[b]	Species	Cytotoxicity
Group I	10–43°C	Mesophilic strains	Absence	*B. pseudomycoides*	Generally not or weakly cytotoxic
Group II	7–40°C	Mesophilic to psychrotolerant strains, with 50% of intermediate strains	Absence	*B. thuringiensis II* and *B. cereus II*	Contain cytotoxic strains
Group III	15–45°C	Mesophilic strains	Absence	*B. thuringiensis III*, *B. anthracis*, *B. cereus III* (emetic strains included)	Generally cytotoxic
Group IV	10–45°C	Mesophilic strains	Absence	*B. thuringiensis IV* and *B. cereus IV*	Generally cytotoxic
Group V	8–40°C	Intermediate	Absence	*B. thuringiensis V* and *B. cereus V*	Contain cytotoxic strains
Group VI	5–37°C	Psychrotolerant	Presence	*B. weihenstephanensis*, *B. mycoides B. thuringiensis VI*	Generally not or weakly cytotoxic
Group VII	20–50°C	Thermotolerant	Absence	*B. cytotoxicus* (new species)	Highly cytotoxic

[a]Based on phenotypic tests and on the psychrotolerant *rrs* sequence signature, as defined by Pruss *et al.* (1999).
[b]Psychrotolerance-associated-signature on *cspA* gene, as defined by Francis *et al.* (1998).

studied in model bacteria such as *B. subtilis* or in extremophile bacteria. The following parts of this chapter will focus on what is known in *B. cereus* compared to what has already been described in *B. subtilis*.

First changes in the bacterial cell after a temperature downshift: role of cold shock proteins

Cold is a physical stress that drastically changes physical and chemical parameters of the cell but an extreme variability of temperature range at which microorganisms can grow is observed. Thus, the word 'cold' by itself is not very informative. For a given microorganism, the difference between optimal temperature of growth and a lower temperature of the environment could be defined as ΔT. When ΔT increases, the bacteria elaborate an adaptive response called a 'cold shock response' (Thieringer *et al.*, 1998). At first, the growth strongly slows down or stops, representing a lag phase. It corresponds to a 'cold acclimation phase' when the mechanisms necessary for low temperature adaptation are settled (Graumann and Marahiel, 1996; Hebraud and Potier, 1999).

During such a 'cold-shock' response, a neo-synthesis of proteins involved in cell protection against the cold is observed, as well as a reduced synthesis of most cellular proteins, causing a growth inhibition (Fig. 5.1) (Thieringer *et al.*, 1998; Weber and Marahiel, 2002; Phadtare, 2004).

Two phases occur after a temperature downshift. First, a quick response is characterized by a massive and transient synthesis of cold shock proteins (CSP), concomitant to a dramatic decrease of non-CSP proteins synthesis, including housekeeping proteins. In addition, the amount of CAP proteins, already present in the cell even in optimal temperature condition, increases moderately and transiently during this acclimation phase. Later on, the second phase begins when the CSP and CAP synthesis decreases concomitantly to a non-CSP proteins neo-synthesis. CSPs and CAPs proteins are involved in various functions, such as DNA and RNA metabolism, cell metabolism, protein folding and degradation, and regulation of gene expression, maintain of an efficient translation etc. Previously described in *E. coli*, a transcriptomic and proteomic analysis of *B. subtilis* downshift from 37°C to 15°C revealed the involvement of similar functions in a Gram-positive organism (Budde *et al.*, 2006). At that step, the bacterial cells are adapted to low-temperature and growth can resume (Jones *et al.*, 1987; Thieringer *et al.*, 1998; Beckering *et al.*, 2002; Phadtare, 2004).

In *B. cereus*, five CSPs have been described, and two of them were shown to be highly synthesized

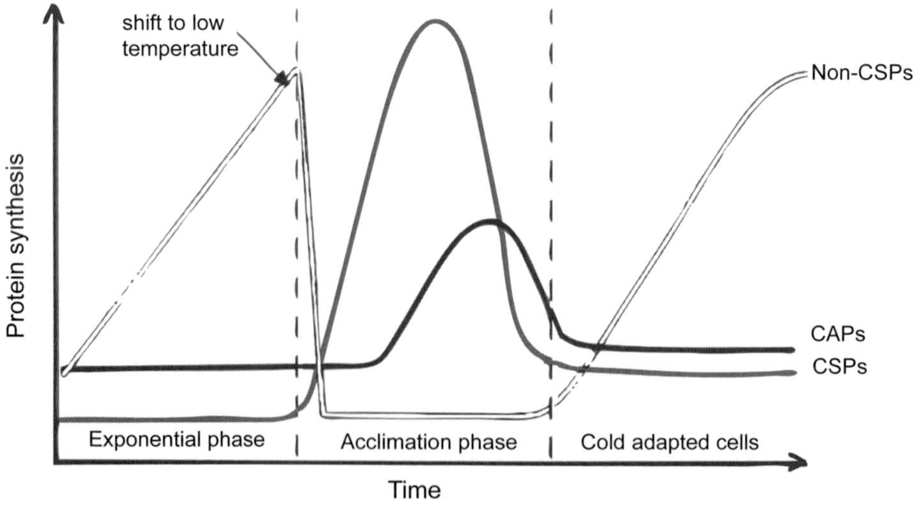

Figure 5.1 Protein synthesis evolution after a temperature downshift (adapted from Thieringer *et al.*, 1998; Horn *et al.*, 2007). CAP, cold acclimation protein; CSP, cold shock protein.

upon cold-shock (Mayr *et al.*, 1996). *B. cereus* CspA and CspE most probably act as RNA chaperones interacting with folded mRNA contributing to an efficient translation into proteins at low temperature, as described for other species (Hunger *et al.*, 2006; Iost and Dreyfus, 2006). The role of other *B. cereus* Csp at low temperature is unknown.

Interestingly, some *B. cereus* psychrotolerant strains (i.e. strains belonging to the *B. weihenstephanensis* species) display clear differences in their *cspA* (and also16S rDNA) sequences compared to the sequences of the orthologous genes in mesophilic strains (Lechner *et al.*, 1998). Such differences may therefore constitute signatures of a psychrotolerant character, although the exact function of such signatures in the growth potential of psychrotolerant *B. cereus* at low temperature is not clearly stated.

Compatible solute acquisition

In *B. subtilis*, glycine betaine was shown to be an effective cold stress protectant (Hoffmann and Bremer, 2011) where the addition of this compatible solute to growth medium allowed growth at low temperature. Surprisingly, the expression of some of the genes encoding the solute transporters (Opu) was repressed when *B. subtilis* was grown at low temperature (Budde *et al.*, 2006). However, *de novo* synthesis of glycine betaine from its precursor choline was also demonstrated to play a role in bacterial growth at such low-temperature conditions (Hoffmann and Bremer, 2011).

In the genome of *B. cereus* ATCC 14579, several glycine betaine transporter-encoding genes are present (Ivanova *et al.*, 2003). In this species, the involvement of such compounds in low-temperature adaptation has not been studied, but we could hypothesize similar roles as what has been described in *B. subtilis*.

It has been shown that swarming differentiation depends on the intracellular levels of the glycine betaine osmoprotectant in a strain belonging to the *B. cereus* group (Salvetti *et al.*, 2009). This study showed that a mutant obtained in a *B. thuringiensis* strain after the insertion of a transposon in a gene and causing an abnormal accumulation of intracellular glycine betaine displayed an impaired swarming ability. Because a reduced expression of genes involved in chemotaxis and motility at low temperature has been described in *B. subtilis* (Budde *et al.*, 2006), it would be interesting to study the swarming behaviour of *B. subtilis* and some *B. cereus* group strains at low temperature, and also to check for the ability of the *B. thuringiensis* mutant described by Salvetti *et al.* (2009) to grow at low temperature.

Membrane fluidity

A direct consequence of a drop in temperature is the modification of the fluidity of the cell membrane (Zhang and Rock, 2008, 2009). The bacterial cell membrane, mostly composed of a phospholipid bilayer, switches from a liquid fluid crystalline state to a rigid state (gel phase) when the temperature decreases. However, bacterial survival depends on membrane lipid homeostasis, as optimal membrane fluidity is required for physiological exchanges between the cytoplasm and the extracellular environment. In a given environmental condition, the fatty acids composition of the membrane determines its viscosity. The membrane viscosity, in turn, influences many membrane-associated functions such as the permeability of hydrophobic molecules, the active solute transport or some protein–protein interactions (Zhang and Rock, 2009). To illustrate the importance of low temperature on *B. cereus* membrane integrity, we can mention the study performed by den Besten *et al.*, where the loss of membrane integrity could partly explain a cold-induced heterogeneity among the bacterial population during low-temperature incubation (den Besten *et al.*, 2010).

It has been described in several bacteria that the composition in membrane fatty acids changes depending on the growth temperature. Adaptation of the membrane fluidity in *B. subtilis* has been described several times (Aguilar *et al.*, 1998; Weber *et al.*, 2001; Cybulski *et al.*, 2002; Mansilla and de Mendoza, 2005). Modification of expression in genes implicated in the adaptation of the membrane fluidity in *B. subtilis* was confirmed in two distinct studies using transcriptional profiling after a cold shock response (Beckering *et al.*, 2002; Kaan *et al.*, 2002).

In *B. cereus*, similar observations have been made (Kaneda, 1972). In the *B. cereus* ATCC

14579 strain, several changes in the fatty acids profiles of cells grown at 12°C instead of 37°C were observed (Brillard *et al.*, 2010). The fatty acids composition of cells grown at low temperature changed particularly by a decreased amount of two branched-chain compounds and an increased amount of five others (among which were found three branched-chain and one unsaturated fatty acids), as well as three long-chain (C >16) unsaturated compounds that were specifically found in cold grown cells grown. The proportion of unsaturated/saturated fatty acids also increased slightly for cells grown at low temperature. In another study, the fatty acids profiles of seven strains of *B. cereus* isolated from food also showed clear differences when grown at 15°C compared with 30°C (Haque and Russell, 2004). When grown at low temperature, the proportion of unsaturated fatty acids was increased, as well as the anteiso-/iso-branched ratio, while a decreased proportion of branched-chain fatty acids and their equivalent chain length were observed.

The production of membrane lipid components is energetically expensive for a bacterial cell. Therefore, bacteria must have evolved various mechanisms to tightly control the fatty acids synthesis or the modification of already present fatty acids in order to adjust the biophysical properties of the membrane (Zhang and Rock, 2009). The mechanism by which *B. subtilis* senses the decrease in temperature and consequently increases the proportion of unsaturated fatty acids in its membrane has been well characterized. It involves a signal transduction mechanism, where the two proteins of a so-called 'two-component system' play a key role to sense a decrease in membrane fluidity. Consequently, a regulation occurs at the transcriptional level of a gene encoding a desaturase which allows modification (unsaturation) of fatty acids already present in the membrane (Aguilar *et al.*, 2001). This mechanism of perception of a temperature downshift is detailed below, in the 'regulators' section of this chapter.

Nucleases and DNA gyrases

In *B. subtilis*, a low-temperature induction of genes involved in the excision of pyrimidine dimers in DNA was observed by a combined transcriptomic and proteomic study, suggesting a greater need for DNA excision repair during growth at low temperature (Budde *et al.*, 2006).

Also, a *B. subtilis* DNAse involved in chromosomal degradation after a cold (and also a heat) shock was identified (it was called 'yocF' by Sakamoto *et al.*, 2001). A mutant of this nuclease encoding gene displayed a higher survival rate after a cold shock (Sakamoto *et al.*, 2001). Whether a nuclease could play a similar role in *B. cereus* survival after a cold shock remains to be determined, but a search by a Blast analysis for an orthologous protein as the one described in *B. subtilis* by Sakamoto *et al.* (2001) matched no sequence among the genomes of *B. cereus* group strains.

A cold adaptive response is the DNA negative supercoiling that has been reported to occur in *B. subtilis* upon cold shock (Krispin and Allmansberger, 1995). DNA negative supertwist, that may promote transcription of bacterial cold-regulated genes, is due to the activity of the DNA gyrase. The encoding *gyrA* and *gyrB* genes were two-fold induced after a cold shock from 37°C to 15°C, while the *topA* gene encoding the DNA topoisomerase A was repressed, these antagonist activities resulting in the DNA negative supercoil observed after cold stress (Beckering *et al.*, 2002). However, the induction of the *gyr* genes was not observed when *B. subtilis* cells were continuously grown at low temperature (Budde *et al.*, 2006). DNA supercoiling may regulate unsaturated fatty acid synthesis in *B. subtilis* as desaturation of fatty acids occurring after exposition at low temperature is inhibited in presence of an antibiotic that alters the DNA gyrase activity (Grau *et al.*, 1994).

Translation machinery and RNA helicases

As already mentioned, a huge decrease in the translational process is observed in bacterial cells immediately after exposure to low temperature. The ribosome has been proposed as one of the bacterial physiological sensor for cold stress (Graumann and Marahiel, 1996). VanBogelen and Neidhart observed that C-antibiotics, such as chloramphenicol or erythromycin, which block the A-site of the ribosome with charged tRNA also induce a cold shock response (VanBogelen and Neidhardt, 1990). This so-called C-state of the ribosome lead to a decrease of the stringent response regulator (*p*)ppGpp in the cell which induces the expression of the CIP (Jones *et al*, 1992). Many genes encoding ribosomal or ribosomal-associated proteins, such as *rbfA*, *infA* and *infB*, are also induced in *B. subtilis* cold-shocked cells (Beckering *et al.*, 2002; Kaan *et al.*, 2002). Thus, the translational machinery is able to adapt to the change in the environmental temperature. Moreover, it has been shown that the SsrA-SmpB rescue system is also recruited in cold stressed *B. subtilis* to maintain the pool of actively translating ribosomes and to provide the needed amino acids (Shin and Price, 2007). This cluster of genes is also present in *B. cereus* genomes but its functionality remains to be demonstrated.

The highly reduced translational efficiency observed when bacteria are exposed to low temperature is also due to the presence of secondary structures in mRNA that are stabilized in cold conditions. To overcome this problem, cells synthesize RNA chaperones, such as DEAD box RNA helicases or CSP which resolve and prevent misfolding of RNA molecules by respectively destabilizing RNA duplexes or by binding to single-stranded RNA.

RNA helicases are enzymes found in most of prokaryote genomes (Iost and Dreyfus, 2006). They participate in a very specific way in a variety of processes involving RNA from transcription to RNA decay (Linder, 2010). The DEAD box family of RNA helicases is considered to group enzymes that catalyse *in vitro* the separation of double-stranded RNAs in an energy-dependent manner, usually using ATP (Cordin *et al.*, 2006). *In vivo*, their role in cold adaptation has been evidenced in many Archaea and bacteria (Owttrim, 2006), and more extensively studied in *E. coli* (Charollais *et al.*, 2003, 2004; Trubetskoy *et al.*, 2009) and in *B. subtilis*. Four RNA helicase-encoding genes are present in *B. subtilis* genome, *yfmL*, *dead (yxiN)*, *ydbR (cshA)* and *yqfR (cshB)*. The two latest ones were induced upon cold shock from 37°C to 15°C (Beckering *et al.*, 2002). One study reported that deletion of *B. subtilis cshA* gene leads to a cold-sensitive phenotype (Ando and Nakamura, 2006) while another paper showed that the absence of *cshA* or *cshB* gene alone had no impact on *B. subtilis* growth upon cold shock and that only a double deletion of *cshA* or *cshB* and *cspB* resulted in a cold-specific growth defect (Hunger *et al.*, 2006). *B. subtilis* CshA is an ATP-independent RNA-binding protein with RNA-dependent ATPase activity, associated to the 50S ribosomal subunit (Ando and Nakamura, 2006). As demonstrated for the *E. coli* RNA helicase CsdA (Charollais *et al.*, 2004), *B. subtilis* CshA may be involved in ribosome biogenesis. Using GFP-fusions, it was demonstrated that *B. subtilis* CshA and CshB co-localized together with CspB and the L1 ribosomal protein around the nucleoid at the pole of the cell, suggesting an active role of the RNA helicases and CSP in the transcription and translation processes (Hunger *et al.*, 2006). Moreover, a strong interaction and specific *in vivo* interaction was evidenced between CspB and CshB, in an active process occurring at the mRNA level (Hunger *et al.*, 2006). In this cooperative process, RNA helicases could rescue misfolding of mRNA by unwinding the unfavourable secondary structures leading to single-stranded RNA where CSPs could bind to prevent unfolding until translation occurs in the ribosome.

The *B. cereus* ATCC 14579 strain possesses five RNA-helicases named CshA to CshE and among them CshA, CshB and CshC were involved in low-temperature adaptation, as corresponding mutant strains showed no growth or altered growth compared to the wild-type strain when grown at 10°C (Pandiani *et al.*, 2010). The morphology of the cold-sensitive mutants cells were affected when grown at a permissive low temperature, however the viability of these cells was not significantly different from those of the wild-type strain. Complementation of the *cshA* and *cshB* mutant strains was obtained with an *in trans* plasmidic copy of *cshA*, in contrast to *cshC* mutant cells (Pandiani *et al.*, 2010) and each RNA helicase showed a role in a specific range of temperature (Pandiani *et al.*, 2011). These results suggested that these RNA helicases could have different role in the cold adaptive response of *B. cereus* that remains to be fully elucidated.

Recent studies described that *B. subtilis* RNA helicase CshA is present in a multiprotein complex of mRNA degradation, functionally equivalent to the *E. coli* degradosome (Carpousis, 2007), although the mechanisms of RNA decay in *E. coli* and *B. subtilis* are different (Condon, 2007). Thus, CshA is associated to the RNAse Y, the polynucleotide phosphorylase PNPase, the RNAses J1 and J2, the enolase and phosphofructokinase PfkA (Commichau *et al.*, 2009; Lehnik-Habrink *et al.*, 2010, 2011). The RNAse Y interacts with the components of the complex, with some doubts on the direct interaction with the RNAse J1 (Mathy *et al.*, 2010; Lehnik-Habrink *et al.*, 2011). The membrane-associated RNAse Y seems to be the central component of the *B. subtilis* degradosome, equivalent to the *E. coli* RNAse E as the scaffold of the complex (Lehnik-Habrink *et al.*, 2011). This could be also the case in the species, such as *B. cereus*, harbouring no RNAse E/G or truncated forms (Kaberdin *et al.*, 2011).

It has been described in *E. coli* that the composition of the degradosome complex changes according to the temperature, as the RNA helicase CsdA can replace RhlB, the resident RNA helicase of the degradosome, when cells are exposed to low temperature (Prud'homme-Genereux *et al.*, 2004). The exact composition of the *B. subtilis* degradosome at low temperature has not been studied so far but low temperature has an impact on CshB production while not on CshA (Lehnik-Habrink *et al.*, 2010). This interesting point needs further studies both in *B. subtilis* and in *B. cereus*, which also possesses all the components of *B. subtilis* degradosome.

Protein folding

The proper folding is required for a protein to assure its function. Well described after heat shock, the misfolding of proteins and the refolding of damaged proteins is also important after cold shock (Phadtare, 2004). Various proteins may be involved in the bacterial cell to counteract such misfoldings, by getting rid of them or by refolding damaged proteins.

For instance, peptidyl–prolyl bonds in proteins cannot rotate freely. They therefore have to exist either in a cis or a trans configuration and the isomerization of these two states is important step in the folding of many proteins (Graumann and Marahiel, 1999). In *B. subtilis*, synthesis of some enzymes (PPiB and trigger factor) catalysing this reaction is increased after cold shock (Graumann *et al.*, 1996) suggesting their importance in protein folding at low temperatures.

ATP-dependent proteases of the Clp family are complex of proteins involved in the degradation of several substrates, but can also function as molecular chaperones, preventing polypeptides from misfolding and aggregation. In *B. subtilis*, the ClpP subunit which belongs to the Clp protease complex is necessary for growth at high temperature and tolerance to various stresses (Msadek *et al.*, 1994; Gerth *et al.*, 1998).

Among the *B. cereus* group, a gene encoding a ClpP protein (*clpP1*) has been shown to play a major role during growth at low temperature. Indeed, a *clpP* mutant constructed in a *B. thuringiensis* strain showed a reduced ability to grow at 25°C compared with 37°C (Fedhila *et al.*, 2002). A possible explanation for this phenotype would be that an efficient Clp protease complex would degrade denatured proteins and solubilize aggregated proteins at a suboptimal temperature.

Shift in metabolism

It is well known that bacteria utilize different metabolic pathways depending on the environmental conditions, such as anaerobiosis (Nakano and Hulett, 1997). Rerouting of metabolic pathways at low temperature has also been observed. In a study combining transcriptomic and proteomic global approaches, a general reduction in *B. subtilis* metabolism was observed in cells adapted to growth at low temperature (Budde *et al.*, 2006). It included a repression of various genes illustrating a reduction of catabolic routes (glycolysis, oxidative phosphorylation, ATP synthesis) and anabolic routes (biosynthesis of purines, pyrimidines, fatty acids and haem). However, transcriptomic analysis revealed an increased expression of genes belonging to a few biosynthetic pathways, such as arginine biosynthetic pathway, genes encoding proteins belonging to the urea cycle, that uses arginine as a substrate for ammonia generation, or tryptophan biosynthetic pathway. Such a shift in metabolism can be observed very early after a temperature downshift, as illustrated by the repression of genes encoding enzymes involved in the biosynthesis of amino acids, nucleotides and coenzymes (Beckering *et al.*, 2002; Kaan *et al.*, 2002).

It has been shown in a *B. cereus* psychrotrophic strain, that the metabolism of glucose at low growth temperature involves the pentose phosphate pathway (PPP) preferentially to the classical glycolysis pathway (Chung *et al.*, 1976; Choma *et al.*, 2000). In the mesophilic strain *B. cereus* ATCC 14579, a Tn-mutant library was constructed and mutants were screened for their ability to grow at low temperature. Mutants encoding enzymes of the central metabolism, including a mutant for one of the PPP-enzyme encoding gene displayed a growth defect at low temperature (Broussolle *et al.*, 2010).

We have mentioned above the importance for a bacterial cell of the adaptation of the membrane fluidity to ensure an adequate membrane's function even at low temperature. Among the mechanisms involved, the synthesis of branched fatty acids is important and necessitates the activation of particular metabolic pathways (Klein *et al.*, 1999). The importance of isoleucine for *B. subtilis* has been demonstrated in order to survive cold shock from 37°C to 15°C. A cold-protective function for all intermediates of anteiso-branched fatty acid biosynthesis was shown, and an isoleucine-deficient strain displayed cold-sensitive phenotype when grown in the absence of isoleucine (Klein *et al.*, 1999). Whether similar mechanisms are employed by *B. cereus* remains to be determined, but a change in the fatty acid profile, including an increased proportion of branched fatty acids when grown at low temperature has already been observed (Haque and Russell, 2004; Brillard *et al.*, 2010).

In a recent study, several promoters have been shown to be specifically activated during *B. cereus* growth at low temperature (Brillard *et al.*, 2010). Among the 17 promoters identified, 11 were found upstream from genes encoding proteins related to metabolism, according to their broad functional COG category assigned (Tatusov *et al.*, 2000). This result illustrates the importance of metabolic pathways that can be rerouted during a change in temperature growth. Such rerouting was illustrated in this study by a differential expression of several genes, but we can also hypothesize a rerouting of metabolic pathways depending on the temperature by a better/lesser activity of some metabolic enzymes.

In the same study, a gene encoding for a putative lipase was shown to be necessary for optimal growth at low temperature (Brillard *et al.*, 2010). The lipase encoding gene was disrupted and the phenotype of the corresponding mutant strain was studied. Whether a change in metabolism occurred was not investigated, but the fatty acids profile of this mutant revealed a few differences (including a lower proportion of methyl-11-methyldodecanoate) when compared to that of the parental strain during growth at low temperature.

Existence of specific regulators

Perception of decreased membrane fluidity by a sensor/regulator system

Two-component systems (TCS) are frequently involved in signal transduction mechanisms encountered during response to changing environments (Hoch, 2000).

TCS are characterized by a sensor histidine kinase (HK), often located in the membrane and coupled with a cognate response regulator. Perception of a particular stimulus by the HK leads to its autophosphorylation, after hydrolysis of an ATP molecule. The phosphoryl group is then transferred to the receiver domain of the response regulator. This activation of the response regulator causes a change in conformation which usually leads to transcriptional regulation (activation or repression) of genes owing to its direct interaction with specific DNA targets (Fig. 5.2A and B).

TCS are widespread among bacteria, and such systems are able to sense changes of conditions as diverse as osmolarity, pH, temperature, lack of essential nutrients, or host interaction signals. Thus, some TCS are vital for bacterial virulence, biofilm formation, or antibiotic resistance.

In *B. subtilis*, the TCS DesKR is a thermosensor and regulator able to adapt the bacteria to a rapid decrease in temperature (Aguilar *et al.*, 2001). The lowered membrane fluidity caused by the drop in temperature (i.e. after a shift from 37°C to 15°C) is sensed by the membrane domain of the DesK histidine kinase.

The exact stimulus detected by DesK has been identified only recently and is related to the thickness of the cellular membrane (Fig. 5.2). Indeed, when the temperature decreases, the

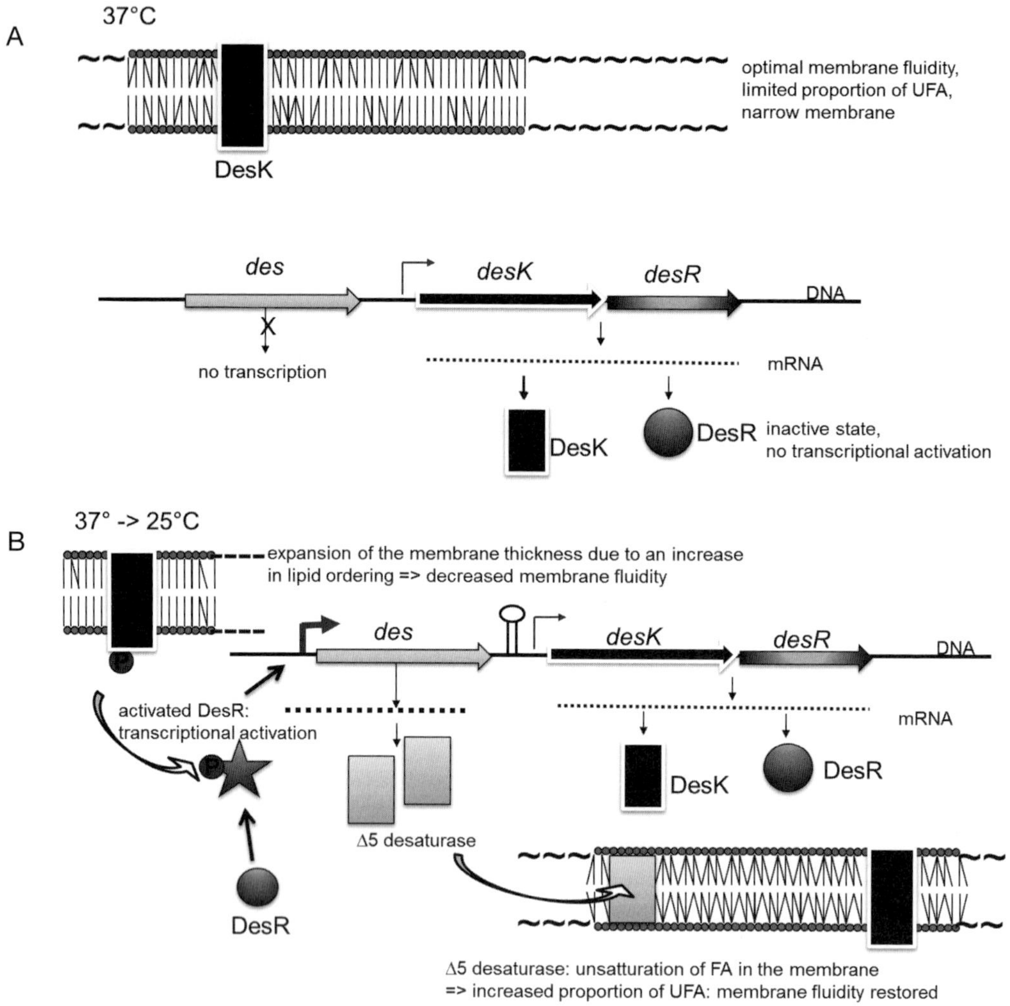

Figure 5.2 The *B. subtilis* DesKR two-component system, at optimal temperature (A) and after a temperature downshift (B). Adapted from Aguilar *et al.* (2001) and Cybulski *et al.* (2010).

membrane becomes thicker owing to an increase in lipid ordering (Nagle and Tristram-Nagle, 2000; Pan *et al.*, 2008). It was shown that the multimembrane spanning domain of DesK contains three hydrophilic amino acids near the lipid–water interface (Cybulski *et al.*, 2010). These amino acids display a hydration state which directly depends on the membrane thickness, and therefore create an instability hot spot. They are thus implicated in temperature sensing, with an unstable state caused by deprivation of hydration consequently to membrane expansion (at low temperature) turning on the system by activating the kinase activity of DesK, whereas stabilization of this instability hotspot through hydration of the three hydrophilic amino acids when the membrane is narrowing (at high temperature) promotes phosphatase activity of DesK, shutting the system off.

The autophosphorylation of DesK is then followed by the transfer of the phosphoryl group to an aspartate residue of the DesR response regulator, leading to its activation. DesR is a transcriptional activator able to recognize the promoter region of the gene encoding a Δ5-desaturase. This enzyme introduces double bounds in fatty acid moieties already present in the membrane (Fig. 5.2B). Thus, activation of expression of the desaturase gene causes an increase in the level of unsaturated fatty acid in the membrane, which leads to a regained fluidity (Aguilar *et al.*, 2001; Schumann, 2009).

Interestingly, this enzyme was shown to be active only in aerobiosis conditions (Beranova *et al.*, 2010). *B. subtilis* adapts its membrane fluidity in anaerobiosis by an increased anteiso/iso ratio of branched-chain fatty acids and not by the temporarily increased level of unsaturated fatty acids that is typical under aerobic conditions (Beranova *et al.*, 2010).

The size of the genome of a given bacterium usually correlates with the number of TCS encoding genes present, the bigger the genome, the most numerous TCS identified (Galperin, 2005; Alm *et al.*, 2006). In the *B. cereus* strain ATCC 14579, the genome sequence has revealed the presence of 55 sensor kinases (Ivanova *et al.*, 2003; Anderson *et al.*, 2005). Orthologues of most of them are found in other members of the *B. cereus* group, including *B. anthracis* and *B. thuringiensis* (de Been *et al.*, 2006). In the bacteria of the *B. cereus* group, only very few of them have been characterized. An *in silico* approach has proposed to predict the response regulator target genes and therefore to provide insight into the role of some TCS (de Been *et al.*, 2008). Up until now, some of the *B. cereus* TCS have been shown to play a role in bacterial adaptation to its environment. The ResDE TCS is conserved in many bacteria, including *B. cereus*, and is involved in adaptation to low oxidation–reduction potential (Duport *et al.*, 2006). The YvfTU TCS was shown to play a role in *B. cereus* virulence using an insect model (Brillard *et al.*, 2008). Although the promoter of the operon encoding the LytSR TCS was shown to be specifically activated during growth at 12°C (Brillard *et al.*, 2010), a mutant strain deleted of these genes did not display an impaired growth at low temperature (Brillard and Sanchis, unpublished). So, up to now, none of the numerous *B. cereus* TCS was shown to be necessary for low-temperature adaptation.

In the *B. cereus* ATCC 14579 genome, two genes orthologous to the *B. subtilis* desaturase encoding gene (*desA*) are present, and are annotated 'fatty acid desaturase' and 'Δ5-desaturase'. Whether the membrane adaptation observed in *B. cereus* after a temperature downshift (Haque and Russell, 2004; Brillard *et al.*, 2010) is controlled by a two-component system, as described for the DesKR system of *B. subtilis* (Aguilar *et al.*, 2001), remains to be determined. Nothing is known about the way the two desaturases genes are regulated (i.e. if a TCS might be responsible for their transcription), but preliminary experiments suggest that they are not overexpressed after a temperature downshift (Brillard, unpublished), suggesting a different regulation of such genes between *B. cereus* and *B. subtilis*.

The alternative sigma factors

Other specific regulators have been described in *B. subtilis* low-temperature adaptation and may play similar roles in *B. cereus*.

The alternative sigma B factor (σB) is involved in transcription of several stress-responsive genes in many Gram-positive bacteria. After activation of σB upon a stress exposure, σB replaces the housekeeping sigma factor in the RNA polymerase which therefore recognizes specific promoters and thus activates specific regulons. In *B. subtilis*, the σB is found in an inactive state when it is trapped by the anti-sigma factor RsbW to form a protein complex (de Been *et al.*, 2011). Under stressful conditions, the anti-anti-sigma factor RsbV is activated by dephosphorylation (by various specific phosphatase, which depend on the stress encountered) resulting in the sequestration of RsbW, therefore activating σB by releasing it from its complex with RsbW. In contrast, in *B. cereus* the σB activation is governed by a single phosphatase (RsbY) which is itself activated by an hybrid-kinase protein (RsbK) (de Been *et al.*, 2010). Such difference in σB activation between *B. subtilis* and *B. cereus* illustrates the need to perform mechanistic studies not only in model organisms such as *B. subtilis* but also in various pathogenic bacteria.

The alternative σB factor was shown to play a role in virulence in *Staphylococcus aureus* (Novick, 2003), in *Listeria monocytogenes* (Chaturongakul *et al.*, 2008) and in *B. anthracis* (Fouet *et al.*, 2000). But various stress conditions are responsible for σB -regulon activation, including the cold stress. For instance, the role of σB has been demonstrated in growth and survival at refrigeration temperature of the food-borne pathogen *Listeria monocytogenes* (Becker *et al.*, 2000) and in *B. subtilis* (Brigulla *et al.*, 2003), but not in *B. cereus*. In *B. cereus*, although *sigB* null mutant cells were less resistant than WT cells to high temperature exposure, upon osmotic up-shock or ethanol exposure (van Schaik *et al.*, 2004), a cold-sensitive phenotype of *B. cereus* σB mutants has not been reported yet.

The σL factor was also shown to play a role in *B. subtilis* cold-shock adaptation by regulation of the isoleucine metabolism implicated in the biosynthesis of the branched-chain fatty acid in the membrane (Wiegeshoff *et al.*, 2006). Although several sigma-factors are present in *B. cereus*, their exact function remains to be studied.

The FapR transcriptional regulator

FapR, is a transcription factor involved in global regulation of membrane lipid biosynthesis (Schujman *et al.*, 2003). It is a transcriptional regulator, able to recognize DNA binding sites at specific promoters in order to repress several genes involved in fatty acids biosynthesis and phospholipid metabolism (such genes were called the *fap* regulon). The authors have shown that the cellular pools of malonyl-CoA, an essential molecule for fatty acids elongation, are directly responsible for the FapR induction, therefore controlling the expression of the *fap* regulon. The binding of malonyl-CoA to the FapR protein induces a change in protein conformation which impairs its DNA binding ability, therefore releasing FapR from (or preventing its binding to) its DNA operator (Schujman *et al.*, 2006). Various mutations in the FapR protein disrupting the binding pocket where malonyl-CoA usually binds are deleterious for cell growth suggesting a major mechanism in *B. subtilis* physiology.

FapR is conserved in many Gram-positive bacteria, including *B. cereus* (Schujman *et al.*, 2003). In the *B. cereus* ATCC 14579 strain, the FapR homologue is encoded by the BC3852 gene. Orthologues of most of the genes that are controlled by FapR in *B. subtilis* are also present in *B. cereus* (Ivanova *et al.*, 2003). In addition, syntheny between the two bacterial species is conserved: in *B. cereus* ATCC 14579, downstream of the *fapR* orthologue are found genes encoding a fatty acid/phospholipid synthesis protein, a malonyl-CoA transacylase, a 3-oxoacyl-reductase and an acyl carrier protein.

TblastN alignments also indicate a high sequence homology between the *B. subtilis* FapR and a protein found in the genome of a psychrotrophic *B. cereus* strain (KBAB4) (Lapidus *et al.*, 2008), suggesting a general role of FapR in the *Bacillus* genus.

Transcriptional regulation of cold-induced genes

While many functional properties of bacterial cold-shock response have been analysed down to the protein structure level, much less is known about how the regulatory network controlling cold-dependent gene expression is organized.

Some studies gave insights on the regulation of cold-induced genes, in particular the genes coding for the *E. coli* CspA protein. The transcript possesses a long and structured 5′ UTR, containing regulatory elements, such as cold-boxes, involved in transcriptional and post-transcriptional regulation (Gualerzi *et al.*, 2003). The regulatory elements are involved in the stabilization of the *cspA* transcripts observed after a cold shock when compared to 37°C (Jiang *et al.*, 1996; Xia *et al.*, 2002). Recently, it has been demonstrated that the *cspA* mRNA is able to sense temperature downshift and to undergo a structural rearrangement that is temperature dependent, leading to a cold-shock mRNA structure more efficiently translated and less susceptible to degradation than the 37°C structure (Giuliodori *et al.*, 2010).

Such a new cold-induced RNA switch has not been described in *Bacillus* species up to now, but cold-box elements have been observed upstream of cold-expressed genes, such as the *B. subtilis cspB*, or the genes encoding the RNA-helicases CshA and CshB of *B. subtilis* (Hunger *et al.*, 2006) and CshA of *B. cereus* (Broussolle *et al.*, 2010).

As mentioned above, a recent study identified several promoters that were specifically activated during *B. cereus* growth at low temperature (Brillard *et al.*, 2010). However, no highly conserved motif was identified over the 17 promoter regions, strongly suggesting that not one but several transcriptional regulators may be implicated in the expression of those promoters at low temperature.

Sporulation and low temperature

Sporulation is a complex process of differentiation, involving numerous genes to transform a vegetative cell into a highly resistant spore, usually occurring when *Bacillus* cells are submitted to nutrient starvation in their environment (Stephenson and Lewis, 2005). Spore formation is a temporally and spatially regulated process, well conserved among the genomes of the bacilli and also in *Clostridia* species (de Hoon *et al.*, 2010). Sporulation may occur in very diverse environments and many factors, such as temperature, can greatly influence spore properties (like resistance and germination, for example) (Carlin, 2010).

Impact of low temperature on sporulation

Several recent studies reported that low temperature greatly influences the spore production of the psychrotrophic *B. weihenstephanensis* species, as it increases the elapsed time from the detection of the first phase-bright spores to the detection of the maximum of sporulated cells and decrease the spore yield (Garcia *et al.*, 2010; Baril *et al.*, 2011; Planchon *et al.*, 2011). The same observation has also recently been reported for *B. subtilis* spores produced in regulated conditions at 19°C versus 37°C (Minh *et al.*, 2011). Baril *et al.* investigated the influence of temperature on sporulation, independently of the growth temperature, using continuous and non-continuous processes: whatever the process, they observed that the sporulation temperature impacted significantly on the final spore concentration, whereas the temperature of cell growth before entry in the sporulation process has only a minor impact (Baril *et al.*, 2011).

It has been shown that the gene encoding the master transcription factor of sporulation Spo0A was induced after a temperature downshift of *B. subtilis* cells from 37°C to 20°C, revealing a new role for this regulator that could also be involved in cell adaptation to low temperature before sporulation (Mendez *et al.*, 2004).

Spore resistance

Bacterial spores possess a structure that enables them to survive to unfavourable conditions for extended periods (Setlow *et al.*, 2006). Spores are usually characterized by a high resistance to wet or dry heat, chemicals, UV radiations, thanks to multiple mechanisms that protect them from stresses, even if dormant spores are metabolically inactive. Studies mainly done on *B. subtilis* spores determined that several factors are involved in resistances, including a very low water content of the spore's core influenced by the presence of dipicolinic acid (DPA) which also impact on the level of minerals, the saturation of spore DNA by the spore-specific SASP proteins (small acid-soluble proteins) to protect it from UV radiations in addition to DNA repair systems, the impermeability of the inner spore membrane to toxic agents, the presence of coat layers that protect spores from lytic enzymes and chemicals (Setlow *et al.*, 2006).

The sporulation temperature has a significant impact on spore heat resistance, the spores formed at low temperature usually being less resistant to heat treatments than spores formed at higher temperatures, as shown in different studies both on *B. subtilis* (Melly *et al.*, 2002; Cortezzo and Setlow, 2005) and on spores of strains belonging to the *B. cereus* group (Gounina-Allouane *et al.*, 2008; Garcia *et al.*, 2010; Baril *et al.*, 2011; Planchon *et al.*, 2011). This could be due to a higher level of water content in the core of spores produced at low temperature. Although *B. subtilis* spores lacking DPA or with altered cations level showed a strong decrease in wet heat resistance (Paidhungat *et al.*, 2000; Melly *et al.*, 2002), no significant difference in the level of these compounds was observed in spores produced at low or high temperature (Atrih and Foster, 2001; Melly *et al.*, 2002). However, a decrease in the DPA content of spores produced at 10°C compared with spores produced at 30°C was reported for *B. cereus* (Planchon *et al.*, 2011). The sporulation temperature impacted significantly on the heat resistance of *B. weihenstephanensis* KBAB4 spores, but the low temperature of cell growth before entry in the sporulation process has only a minor impact on this property (Baril *et al.*, 2011).

For chemicals that kill spores by DNA damage, such as formaldehyde or nitrous acid, the *B. subtilis* spore resistance increased with the sporulation temperature (Melly *et al.*, 2002; Cortezzo and

Setlow, 2005), whereas no difference was observed for *B. cereus* spores (Planchon *et al.*, 2011). Both *B. subtilis* and *B. cereus* spores produced at a low sporulation temperature were more sensitive to hydrogen peroxide and NaOH treatments, maybe because of variations in protein composition of coat layers, modification of the cortex and the membranes layers (Melly *et al.*, 2002; Planchon *et al.*, 2011).

B. cereus spores produced at low temperature were also less resistant than spores produced at 30°C to UV treatment delivered by pulsed light and to a lesser extend to continuous UV-C and this could be correlated to a lower DPA content in those spores (Planchon *et al.*, 2011).

Spore germination

Dormant spores retain in alert a sensing system that allows them to return to vegetative growth if environmental conditions are more favourable. This process results in loss of the spore properties, such as heat resistance and core dehydration and dormancy. *Bacillus* spore germination can be triggered by the presence of some specific amino acids which recognize germinant receptors called Ger proteins, located in the spore inner membrane (Broussolle *et al.*, 2008; Paredes-Sabja *et al.*, 2010). Several studies reported the impact of sporulation temperature on spore germination with contrasted results: with 10 mM alanine, no clear difference in germination was observed for *B. subtilis* spores produced at different temperatures from 22°C to 48°C (Melly *et al.*, 2002). Psychrotrophic spores of *B. cereus* produced at 12°C germinated faster than spores produced at 20°C, only if non-heat activated (Garcia *et al.*, 2010) or also when heat-activated before germination triggering (Gounina-Allouane *et al.*, 2008; Planchon *et al.*, 2011). This could be correlated to the low level of DPA observed in the spores formed at 10°C.

Spore germination could also be induced by some mechanical processes, such as low (LHP) or high hydrostatic pressure (HHP) treatments. *B. subtilis* spores produced at 27°C germinated less than spores produced at 40°C in response to LHP (Black *et al.*, 2005). No effect of the sporulation temperature was observed on *B. weihenstephanensis* spore germination induced by HHP, while spores produced at 20°C showed a lower germination capacities than spores produced at 30°C when submitted to LHP (Garcia *et al.*, 2010), confirming that the mechanisms of spore germination in response to low or high pressures could be different (Paredes-Sabja *et al.*, 2010).

Spore outgrowth

Once germinated, the bacterial spore undergoes a process called outgrowth, which corresponds to the initiation of spore metabolism by using SASP protein degradation for the synthesis of new macromolecules and escape of the new cell from the spore coat layers (Setlow *et al.*, 2006). On a safety point of view, when considering food-borne pathogenic Bacilli, it seems very important to know if the history of the spore, and for example the sporulation temperature, could have any impact on the final steps of spore germination and on subsequent cell growth. Only little information is available but it was reported that growth of *B. weihenstephanensis* spores produced at 12°C, 20°C and 30°C was similar whatever the incubation temperature (Garcia *et al.*, 2010) but more data on several species are undoubtedly needed.

Structure of spores formed at low temperature

The structure and the composition of bacilli spores formed at low temperature are modified when compared to those of spores formed at higher temperatures. The size of *B. weihenstephanensis* spores produced at 12°C was smaller than that of spores formed at 20°C but the high hydrophobicity of these spores was clearly independent from the sporulation temperature (Garcia *et al.*, 2010). As previously mentioned, the DPA content in the *B. cereus* spores formed at low temperature was lower, which has a strong impact on their properties (Planchon *et al.*, 2011). The composition of certain fatty acids differed significantly between spores produced at low and high temperature, both for *B. subtilis* (Cortezzo and Setlow, 2005) and *B. cereus* species (Planchon *et al.*, 2011). These variations in fatty acids may trigger conformational changes in membrane layer and may lead to a different access to membrane proteins in the spore.

Modifications of the spore coat layers were also observed in *B. subtilis* spores according to the sporulation temperature: the level of CotA and CotS proteins was significantly higher in spores produced at low temperature but not related to a higher *cotA* transcription during the sporulation process (Melly *et al.*, 2002). This modification could be due to the CotE protein involved in the assembly of the coat layers. The expression of the sigma factor σ_G, which governs the expression of the *ger* genes in *B. subtilis* and probably in *B. cereus*, was increased in spores produced at low temperature (Garcia *et al.*, 2010), suggesting that sporulation temperature could impact on the expression of several genes involved in the sporulation and in the germination process.

Conclusion and future trends

Major changes that occur in the bacterial cell after a decrease in temperature are represented on Fig. 5.3A. *Bacillus* counteracts these stresses by various mechanisms which are illustrated on Fig. 5.3B.

Although some information exists about the *B. cereus* mechanisms implicated in low-temperature adaptation, many more needs to be learned in regard to what is known in other model bacteria such as *B. subtilis*. What is common among the *Bacillus* genus, or in contrast, what are the mechanisms that are specific to the *B. cereus* group, but also to some particular *B. cereus* strains that are well adapted to low-temperature environments.

Several studies revealed the pathogenic potential of strains belonging to the *B. weihenstephanensis* species, but also the existence of pathogenic strains that are psychrotrophic but do not belong to this cluster of strains (Stenfors and Granum, 2001; Stenfors *et al.*, 2002; Guinebretiere *et al.*, 2008). A recent study which led to the redefinition of the taxonomy of the *B. cereus* group by defining seven new genetic groups also suggested an evolution of the *B. cereus* group towards psychrotrophy (Guinebretiere *et al.*, 2008). Despite a probable counterbalanced selection of *B. cereus* strains to warmer temperature that may occur because of the climate change, such evolution might lead in the future to an increased proportion of pathogenic strains able to multiply in cold environments, representing therefore an additional issue in food safety, offering a justification in studying more thoroughly the mechanisms involved in *B. cereus* adaptation to low temperature.

Acknowledgements

This work was supported by INRA (Institut National de la Recherche Agronomique) and by a grant from *the Agence Nationale de la Recherche* as part of an ANR-05-PNRA-013 *B. cereus* contract.

Web resources

SymPrevius: https://www.tools.sympreviusorg/Bcereus/english.php

SymPrevius is a collection of tools for food safety inspections designed for food sector businesses. This page allows the affiliation to phylogenetic groups into the *Bacillus cereus* Group, by uploading the DNA sequence of a single PCR-amplified gene.

HyperCAT: http://mlstoslo.uio.no/

HyperCAT is a database for global multischeme and multidatatype phylogenetic analysis of the *Bacillus cereus* group population, which includes *B. cereus*, *B. anthracis*, *B. thuringiensis*, *B. mycoides*, *B. pseudomycoides*, and *B. weihenstephanensis*. It allows rapid supertree reconstructions.

JGI: http://img.jgi.doe.gov/cgi-bin/w/main.cgi?section=TaxonDetail&page=taxonDetail&taxon_oid=641228477

The Joint Genome Institute provides integrated high throughput sequencing and computational analysis. This page is dedicated to the first sequenced *B. cereus* group psychrotrophic strain (KBAB4).

NCBI: http://www.ncbi.nlm.nih.gov/bioproject/384

The National Center for Biotechnology Information provides several genomic biology tools and resources. This page is dedicated to the *B. cereus* type strain (ATCC 14579).

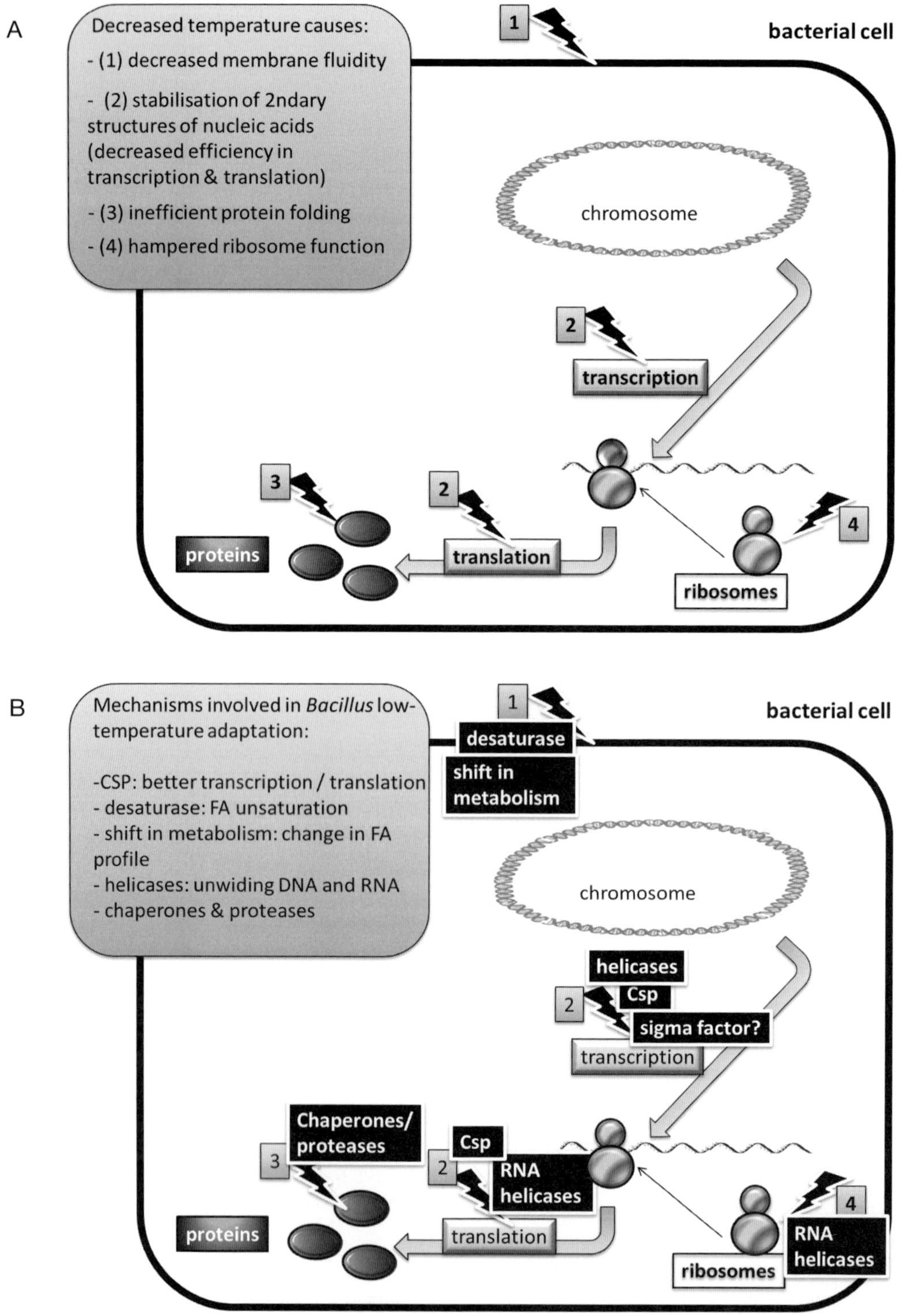

Figure 5.3 Major changes occurring in *Bacillus* after a temperature downshift (A) and mechanisms involved in its adaptation (B). CSp, cold shock protein; FA, fatty acid.

References

Aguilar, P.S., Cronan, J.E., Jr., and de Mendoza, D. (1998). A *Bacillus subtilis* gene induced by cold shock encodes a membrane phospholipid desaturase. J. Bacteriol. *180*, 2194–2200.

Aguilar, P.S., Hernandez-Arriaga, A.M., Cybulski, L.E., Erazo, A.C., and de Mendoza, D. (2001). Molecular basis of thermosensing: a two-component signal transduction thermometer in *Bacillus subtilis*. EMBO J. *20*, 1681–1691.

Alm, E., Huang, K., and Arkin, A. (2006). The evolution of two-component systems in bacteria reveals different strategies for niche adaptation. PLoS Comput. Biol. *2*, e143.

Anderson, I., Sorokin, A., Kapatral, V., Reznik, G., Bhattacharya, A., Mikhailova, N., Burd, H., Joukov, V., Kaznadzey, D., Walunas, T., *et al.* (2005). Comparative genome analysis of *Bacillus cereus* group genomes with *Bacillus subtilis*. FEMS Microbiol. Lett. *250*, 175–184.

Ando, Y., and Nakamura, K. (2006). *Bacillus subtilis* DEAD protein YdbR possesses ATPase, RNA binding, and RNA unwinding activities. Biosci. Biotechnol. Biochem. *70*, 1606–1615.

Atrih, A., and Foster, S.J. (2001). Analysis of the role of bacterial endospore cortex structure in resistance properties and demonstration of its conservation amongst species. J. Appl. Microbiol. *91*, 364–372.

Auger, S., Galleron, N., Bidnenko, E., Ehrlich, S.D., Lapidus, A., and Sorokin, A. (2008). The genetically remote pathogenic strain NVH391–98 of the *Bacillus cereus* group is representative of a cluster of thermophilic strains. Appl. Environ. Microbiol. *74*, 1276–1280.

Baril, E., Coroller, L., Postollec, F., Leguerinel, I., Boulais, C., Carlin, F., and Mafart, P. (2011). The wet-heat resistance of *Bacillus weihenstephanensis* KBAB4 spores produced in a two-step sporulation process depends on sporulation temperature but not on previous cell history. Int. J. Food Microbiol. *146*, 57–62.

Becker, L.A., Evans, S.N., Hutkins, R.W., and Benson, A.K. (2000). Role of sigma(B) in adaptation of *Listeria monocytogenes* to growth at low temperature. J. Bacteriol. *182*, 7083–7087.

Beckering, C.L., Steil, L., Weber, M.H., Volker, U., and Marahiel, M.A. (2002). Genomewide transcriptional analysis of the cold shock response in *Bacillus subtilis*. J. Bacteriol. *184*, 6395–6402.

de Been, M., Francke, C., Moezelaar, R., Abee, T., and Siezen, R.J. (2006). Comparative analysis of two-component signal transduction systems of *Bacillus cereus, Bacillus thuringiensis* and *Bacillus anthracis*. Microbiology *152*, 3035–3048.

de Been, M., Bart, M.J., Abee, T., Siezen, R.J., and Francke, C. (2008). The identification of response regulator-specific binding sites reveals new roles of two-component systems in *Bacillus cereus* and closely related low-GC Gram-positives. Environ. Microbiol. *10*, 2796–2809.

de Been, M., Tempelaars, M.H., van Schaik, W., Moezelaar, R., Siezen, R.J., and Abee, T. (2010). A novel hybrid kinase is essential for regulating the sigma(B)-mediated stress response of *Bacillus cereus*. Environ. Microbiol. *12*, 730–745.

de Been, M., Francke, C., Siezen, R.J., and Abee, T. (2011). Novel sigmaB regulation modules of Gram-positive bacteria involve the use of complex hybrid histidine kinases. Microbiology *157*, 3–12.

Beranova, J., Mansilla, M.C., de Mendoza, D., Elhottova, D., and Konopasek, I. (2010). Differences in cold adaptation of *Bacillus subtilis* under anaerobic and aerobic conditions. J. Bacteriol. *192*, 4164–4171.

den Besten, H.M., Garcia, D., Moezelaar, R., Zwietering, M.H., and Abee, T. (2010). Direct-imaging-based quantification of *Bacillus cereus* ATCC 14579 population heterogeneity at a low incubation temperature. Appl. Environ. Microbiol. *76*, 927–930.

Bishop, B.L., Lodolce, J.P., Kolodziej, L.E., Boone, D.L., and Tang, W.J. (2010). The role of anthrolysin O in gut epithelial barrier disruption during *Bacillus anthracis* infection. Biochem. Biophys. Res. Commun. *394*, 254–259.

Black, E.P., Koziol-Dube, K., Guan, D.S., Wei, H., Setlow, B., Cortezzo, D.E., Hoover, D.G., and Setlow, P. (2005). Factors influencing germination of *Bacillus subtilis* spores via activation of nutrient receptors by high pressure. Appl. Environ. Microbiol. *71*, 5879–5887.

Bottone, E.J. (2010). *Bacillus cereus*, a volatile human pathogen. Clin. Microbiol. Rev. *23*, 382–398.

Brigulla, M., Hoffmann, T., Krisp, A., Volker, A., Bremer, E., and Volker, U. (2003). Chill induction of the SigB-dependent general stress response in *Bacillus subtilis* and its contribution to low-temperature adaptation. J. Bacteriol. *185*, 4305–4314.

Brillard, J., and Lereclus, D. (2004). Comparison of cytotoxin *cytK* promoters from *Bacillus cereus* strain ATCC 14579 and from a *B. cereus* food-poisoning strain. Microbiology *150*, 2699–2705.

Brillard, J., and Lereclus, D. (2007). Characterization of a small PlcR-regulated gene co-expressed with cereolysin O. BMC Microbiol. *7*, 52.

Brillard, J., Susanna, K., Michaud, C., Dargaignaratz, C., Gohar, M., Nielsen-Leroux, C., Ramarao, N., Kolsto, A.B., Nguyen-the, C., Lereclus, D., *et al.* (2008). The YvfTU two-component system is involved in *plcR* expression in *Bacillus cereus*. BMC Microbiol. *8*, 183.

Brillard, J., Jehanno, I., Dargaignaratz, C., Barbosa, I., Ginies, C., Carlin, F., Fedhila, S., Nguyen-the, C., Broussolle, V., and Sanchis, V. (2010). Identification of *Bacillus cereus* genes specifically expressed during growth at low temperatures. Appl. Environ. Microbiol. *76*, 2562–2573.

Broussolle, V., Gauillard, F., Nguyen-The, C., and Carlin, F. (2008). Diversity of spore germination in response to inosine and L-alanine and its interaction with NaCl and pH in the *Bacillus cereus* group. J. Appl. Microbiol. *105*, 1081–1090.

Broussolle, V., Pandiani, F., Haddad, N., Michaud, C., Carlin, F., Nguyen-the, C., and Brillard, J. (2010). Insertional mutagenesis reveals genes involved in *Bacillus cereus* ATCC 14579 growth at low temperature. FEMS Microbiol. Lett. *306*, 177–183.

Budde, I., Steil, L., Scharf, C., Volker, U., and Bremer, E. (2006). Adaptation of *Bacillus subtilis* to growth at low temperature: a combined transcriptomic and proteomic appraisal. Microbiology *152*, 831–853.

Carlin, F. (2010). Origin of bacterial spores contaminating foods. Food Microbiol. *28*, 177–182.

Carlin, F., Brillard, J., Broussolle, V., Clavel, T., Duport, C., Jobin, M., Guinebretiere, M.H., Auger, S., Sorokine, A., and Nguyen-The, C. (2010). Adaptation of *Bacillus cereus*, an ubiquitous worldwide-distributed food-borne pathogen, to a changing environment. Food Res. Int. *43*, 1885–1894.

Carpousis, A.J. (2007). The RNA degradosome of *Escherichia coli*: An mRNA-degrading machine assembled on RNase E. Annu. Rev. Microbiol. *61*, 71–87.

Charollais, J., Pflieger, D., Vinh, J., Dreyfus, M., and Iost, I. (2003). The DEAD-box RNA helicase SrmB is involved in the assembly of 50S ribosomal subunits in *Escherichia coli*. Mol. Microbiol. *48*, 1253–1265.

Charollais, J., Dreyfus, M., and Iost, I. (2004). CsdA, a cold-shock RNA helicase from *Escherichia coli*, is involved in the biogenesis of 50S ribosomal subunit. Nucleic Acids Res. *32*, 2751–2759.

Chaturongakul, S., Raengpradub, S., Wiedmann, M., and Boor, K.J. (2008). Modulation of stress and virulence in *Listeria monocytogenes*. Trends Microbiol. *16*, 388–396.

Choma, C., Guinebretiere, M.H., Carlin, F., Schmitt, P., Velge, P., Granum, P.E., and Nguyen-The, C. (2000). Prevalence, characterization and growth of *Bacillus cereus* in commercial cooked chilled foods containing vegetables. J. Appl. Microbiol. *88*, 617–625.

Chung, B.H., Cannon, R.Y., and Smith, R.C. (1976). Influence of growth temperature on glucose metabolism of a psychotrophic strain of *Bacillus cereus*. Appl. Environ. Microbiol. *31*, 39–45.

Commichau, F.M., Rothe, F.M., Herzberg, C., Wagner, E., Hellwig, D., Lehnik-Habrink, M., Hammer, E., Volker, U., and Stulke, J. (2009). Novel activities of glycolytic enzymes in *Bacillus subtilis*. Mol. Cell. Proteomics *8*, 1350–1360.

Condon, C. (2007). Maturation and degradation of RNA in bacteria. Curr. Opin. Microbiol. *10*, 271–278.

Cordin, O., Banroques, J., Tanner, N.K., and Linder, P. (2006). The DEAD-box protein family of RNA helicases. Gene *367*, 17–37.

Cortezzo, D.E., and Setlow, P. (2005). Analysis of factors that influence the sensitivity of spores of *Bacillus subtilis* to DNA damaging chemicals. J. Appl. Microbiol. *98*, 606–617.

Cybulski, L.E., Albanesi, D., Mansilla, M.C., Altabe, S., Aguilar, P.S., and de Mendoza, D. (2002). Mechanism of membrane fluidity optimization: isothermal control of the *Bacillus subtilis* acyl-lipid desaturase. Mol. Microbiol. *45*, 1379–1388.

Cybulski, L.E., Martin, M., Mansilla, M.C., Fernandez, A., and de Mendoza, D. (2010). Membrane thickness cue for cold sensing in a bacterium. Curr. Biol. *20*, 1539–1544.

Dierick, K., Van Coillie, E., Swiecicka, I., Meyfroidt, G., Devlieger, H., Meulemans, A., Hoedemaekers, G., Fourie, L., Heyndrickx, M., and Mahillon, J. (2005). Fatal family outbreak of *Bacillus cereus*-associated food poisoning. J. Clin. Microbiol. *43*, 4277–4279.

Duport, C., Zigha, A., Rosenfeld, E., and Schmitt, P. (2006). Control of enterotoxin gene expression in *Bacillus cereus* F4430/73 involves the redox-sensitive ResDE signal transduction system. J. Bacteriol. *188*, 6640–6651.

EFSA (2005). Opinion of the scientific panel on biological hazards on *Bacillus cereus* and other *Bacillus* spp. in foodstuffs. EFSA J. *175*, 1–48.

Fagerlund, A., Brillard, J., Furst, R., Guinebretiere, M.H., and Granum, P.E. (2007). Toxin production in a rare and genetically remote cluster of strains of the *Bacillus cereus* group. BMC Microbiol. *7*, 43.

Fedhila, S., Msadek, T., Nel, P., and Lereclus, D. (2002). Distinct *clpP* genes control specific adaptive responses in *Bacillus thuringiensis*. J. Bacteriol. *184*, 5554–5562.

Fouet, A., Namy, O., and Lambert, G. (2000). Characterization of the operon encoding the alternative sigma(B) factor from *Bacillus anthracis* and its role in virulence. J. Bacteriol. *182*, 5036–5045.

Francis, K.P., Mayr, R., von Stetten, F., Stewart, G.S., and Scherer, S. (1998). Discrimination of psychrotrophic and mesophilic strains of the *Bacillus cereus* group by PCR targeting of major cold shock protein genes. Appl. Environ. Microbiol. *64*, 3525–3529.

Galperin, M.Y. (2005). A census of membrane-bound and intracellular signal transduction proteins in bacteria: bacterial IQ, extroverts and introverts. BMC Microbiol. *5*, 35.

Garcia, D., van der Voort, M., and Abee, T. (2010). Comparative analysis of *Bacillus weihenstephanensis* KBAB4 spores obtained at different temperatures. Int. J. Food Microbiol. *140*, 146–153.

Gerth, U., Kruger, E., Derre, I., Msadek, T., and Hecker, M. (1998). Stress induction of the *Bacillus subtilis clpP* gene encoding a homologue of the proteolytic component of the Clp protease and the involvement of ClpP and ClpX in stress tolerance. Mol. Microbiol. *28*, 787–802.

Giuliodori, A.M., Di Pietro, F., Marzi, S., Masquida, B., Wagner, R., Romby, P., Gualerzi, C.O., and Pon, C.L. (2010). The *cspA* mRNA is a thermosensor that modulates translation of the cold-shock protein CspA. Mol. Cell *37*, 21–33.

Gounina-Allouane, R., Broussolle, V., and Carlin, F. (2008). Influence of the sporulation temperature on the impact of the nutrients inosine and L-alanine on *Bacillus cereus* spore germination. Food Microbiol. *25*, 202–206.

Granum, P.E., and Lund, T. (1997). *Bacillus cereus* and its food poisoning toxins. FEMS Microbiol. Lett. *157*, 223–228.

Grau, R., Gardiol, D., Glikin, G.C., and Demendoza, D. (1994). DNA supercoiling and thermal regulation of unsaturated fatty-acid synthesis in *Bacillus subtilis*. Mol. Microbiol. *11*, 933–941.

Graumann, P., and Marahiel, M.A. (1996). Some like it cold: response of microorganisms to cold shock. Arch. Microbiol. *166*, 293–300.

Graumann, P., Schroder, K., Schmid, R., and Marahiel, M.A. (1996). Cold shock stress-induced proteins in *Bacillus subtilis*. J. Bacteriol. *178*, 4611–4619.

Graumann, P.L., and Marahiel, M.A. (1999). Cold shock response in *Bacillus subtilis*. J Mol. Microbiol. Biotechnol *1*, 203–209.

Gualerzi, C.O., Giuliodori, A.M., and Pon, C.L. (2003). Transcriptional and post-transcriptional control of cold-shock genes. J. Mol. Biol. *331*, 527–539.

Guinebretiere, M.H., Thompson, F.L., Sorokin, A., Normand, P., Dawyndt, P., Ehling-Schulz, M., Svensson, B., Sanchis, V., Nguyen-The, C., Heyndrickx, M., *et al.* (2008). Ecological diversification in the *Bacillus cereus* Group. Environ. Microbiol. *10*, 851–865.

Guinebretiere, M.H., Velge, P., Couvert, O., Carlin, F., Debuyser, M.L., and Nguyen-The, C. (2010). Ability of *Bacillus cereus* group strains to cause food poisoning varies according to phylogenetic affiliation (groups I to VII) rather than species affiliation. J. Clin. Microbiol. *48*, 3388–3391.

Haque, M.A., and Russell, N.J. (2004). Strains of *Bacillus cereus* vary in the phenotypic adaptation of their membrane lipid composition in response to low water activity, reduced temperature and growth in rice starch. Microbiology *150*, 1397–1404.

Hebraud, M., and Potier, P. (1999). Cold shock response and low temperature adaptation in psychrotrophic bacteria. J. Mol. Microbiol. Biotechnol. *1*, 211–219.

Helgason, E., Okstad, O.A., Caugant, D.A., Johansen, H.A., Fouet, A., Mock, M., Hegna, I., and Kolsto (2000). *Bacillus anthracis, Bacillus cereus,* and *Bacillus thuringiensis*--one species on the basis of genetic evidence. Appl. Environ. Microbiol. *66*, 2627–2630.

Hoch, J.A. (2000). Two-component and phosphorelay signal transduction. Curr. Opin. Microbiol. *3*, 165–170.

Hoffmann, T., and Bremer, E. (2011). Protection of *Bacillus subtilis* against cold stress via compatible-solute acquisition. J. Bacteriol. *193*, 1552–1562.

de Hoon, M.J.L., Eichenberger, P., and Vitkup, D. (2010). Hierarchical evolution of the bacterial sporulation network. Curr. Biol. *20*, R735–R745.

Horn, G., Hofweber, R., Kremer, W., and Kalbitzer, H.R. (2007). Structure and function of bacterial cold shock proteins. Cell. Mol. Life Sci. *64*, 1457–1470.

Hunger, K., Beckering, C.L., Wiegeshoff, F., Graumann, P.L., and Marahiel, M.A. (2006). Cold-induced putative DEAD box RNA helicases CshA and CshB are essential for cold adaptation and interact with cold shock protein B in *Bacillus subtilis*. J. Bacteriol. *188*, 240–248.

Iost, I., and Dreyfus, M. (2006). DEAD-box RNA helicases in *Escherichia coli*. Nucleic Acids Res. *34*, 4189–4197.

Ivanova, N., Sorokin, A., Anderson, I., Galleron, N., Candelon, B., Kapatral, V., Bhattacharyya, A., Reznik, G., Mikhailova, N., Lapidus, A., *et al.* (2003). Genome sequence of *Bacillus cereus* and comparative analysis with *Bacillus anthracis*. Nature *423*, 87–91.

Jensen, G.B., Hansen, B.M., Eilenberg, J., and Mahillon, J. (2003). The hidden lifestyles of *Bacillus cereus* and relatives. Environ. Microbiol. *5*, 631–640.

Jiang, W.N., Fang, L., and Inouye, M. (1996). The role of the 5′-end untranslated region of the mRNA for CspA, the major cold-shock protein of *Escherichia coli,* in cold-shock adaptation. J. Bacteriol. *178*, 4919–4925.

Jones, P.G., VanBogelen, R.A., and Neidhardt, F.C. (1987). Induction of proteins in response to low temperature in *Escherichia coli*. J. Bacteriol. *169*, 2092–2095.

Kaan, T., Homuth, G., Mader, U., Bandow, J., and Schweder, T. (2002). Genome-wide transcriptional profiling of the *Bacillus subtilis* cold-shock response. Microbiology *148*, 3441–3455.

Kaberdin, V.R., Singh, D., and Sue, L.C. (2011). Composition and conservation of the mRNA-degrading machinery in bacteria. J. Biomed. Sci. *18*, 23.

Kaneda, T. (1972). Positional preference of fatty acids in phospholipids of *Bacillus cereus* and its relation to growth temperature. Biochim. Biophys. Acta *280*, 297–305.

Klein, W., Weber, M.H., and Marahiel, M.A. (1999). Cold shock response of *Bacillus subtilis*: isoleucine-dependent switch in the fatty acid branching pattern for membrane adaptation to low temperatures. J. Bacteriol. *181*, 5341–5349.

Krispin, O., and Allmansberger, R. (1995). Changes in DNA supertwist as a response of *Bacillus subtilis* towards different kinds of stress. FEMS Microbiol. Lett. *134*, 129–135.

Lapidus, A., Goltsman, E., Auger, S., Galleron, N., Segurens, B., Dossat, C., Land, M.L., Broussolle, V., Brillard, J., Guinebretiere, M.H., *et al.* (2008). Extending the *Bacillus cereus* group genomics to putative food-borne pathogens of different toxicity. Chem. Biol. Interact. *171*, 236–249.

Lechner, S., Mayr, R., Francis, K.P., Pruss, B.M., Kaplan, T., Wiessner-Gunkel, E., Stewart, G.S., and Scherer, S. (1998). *Bacillus weihenstephanensis* sp. *nov.* is a new psychrotolerant species of the *Bacillus cereus* group. Int. J. Syst. Bacteriol. *48*, 1373–1382.

Lehnik-Habrink, M., Pfortner, H., Rempeters, L., Pietack, N., Herzberg, C., and Stulke, J. (2010). The RNA degradosome in *Bacillus subtilis*: identification of CshA as the major RNA helicase in the multiprotein complex. Mol. Microbiol. *77*, 958–971.

Lehnik-Habrink, M., Newman, J., Rothe, F.M., Solovyova, A.S., Rodrigues, C., Herzberg, C., Commichau, F.M., Lewis, R.J., and Stülke, J. (2011). RNase Y in *Bacillus subtilis*: a natively disordered protein that is the functional equivalent to RNase E from *Escherichia coli*. J. Bacteriol. *193*, 5431–5441.

Linder, P. (2010). The dynamic Life with DEAD-box RNA helicases. In RNA Helicases, Jankowsky, E., ed. (Royal Society of Chemistry, Cambridge), pp. 32–60.

Lund, T., De Buyser, M.L., and Granum, P.E. (2000). A new cytotoxin from *Bacillus cereus* that may cause necrotic enteritis. Mol. Microbiol. *38*, 254–261.

Mahler, H., Pasi, A., Kramer, J.M., Schulte, P., Scoging, A.C., Bar, W., and Krahenbuhl, S. (1997). Fulminant liver failure in association with the emetic toxin of *Bacillus cereus*. New Engl. J. Med. *336*, 1142–1148.

Mansilla, M.C., and de Mendoza, D. (2005). The *Bacillus subtilis* desaturase: a model to understand phospholipid modification and temperature sensing. Arch. Microbiol. *183*, 229–235.

Mathy, N., Hebert, A., Mervelet, P., Benard, L., Dorleans, A., de la Sierra-Gallay, I.L., Noirot, P., Putzer, H., and Condon, C. (2010). *Bacillus subtilis* ribonucleases J1 and J2 form a complex with altered enzyme behaviour. Mol. Microbiol. *75*, 489–498.

Mayr, B., Kaplan, T., Lechner, S., and Scherer, S. (1996). Identification and purification of a family of dimeric major cold shock protein homologs from the psychrotrophic *Bacillus cereus* WSBC 10201. J. Bacteriol. *178*, 2916–2925.

Melly, E., Genest, P.C., Gilmore, M.E., Little, S., Popham, D.L., Driks, A., and Setlow, P. (2002). Analysis of the properties of spores of *Bacillus subtilis* prepared at different temperatures. J. Appl. Microbiol. *92*, 1105–1115.

Mendez, M.B., Orsaria, L.M., Philippe, V., Pedrido, M.E., and Grau, R.R. (2004). Novel roles of the master transcription factors Spo0A and sigma(B) for survival and sporulation of *Bacillus subtilis* at low growth temperature. J. Bacteriol. *186*, 989–1000.

Minh, H.N.T., Durand, A., Loison, P., Perrier-Cornet, J.M., and Gervais, P. (2011). Effect of sporulation conditions on the resistance of *Bacillus subtilis* spores to heat and high pressure. Appl. Microbiol. Biotechnol. *90*, 1409–1417.

Msadek, T., Kunst, F., and Rapoport, G. (1994). MecB of *Bacillus subtilis*, a member of the ClpC ATPase family, is a pleiotropic regulator controlling competence gene expression and growth at high temperature. Proc. Natl. Acad. Sci. U.S.A. *91*, 5788–5792.

Nagle, J.F., and Tristram-Nagle, S. (2000). Structure of lipid bilayers. Biochim. Biophys. Acta *1469*, 159–195.

Nakano, M.M., and Hulett, F.M. (1997). Adaptation of *Bacillus subtilis* to oxygen limitation. FEMS Microbiol. Lett. *157*, 1–7.

Nguyen-the, C., and Broussolle, V. (2005). *Bacillus cereus*: factors affecting virulence. In Understanding Pathogen Behaviour: Virulence, Stress Response And Resistance, Griffiths, M., ed. (Woodhead Publishing Limited, Cambridge, UK), pp. 309–3330.

Novick, R.P. (2003). Autoinduction and signal transduction in the regulation of staphylococcal virulence. Mol. Microbiol. *48*, 1429–1449.

Owttrim, G.W. (2006). RNA helicases and abiotic stress. Nucleic Acids Res. *34*, 3220–3230.

Paidhungat, M., Setlow, B., Driks, A., and Setlow, P. (2000). Characterization of spores of *Bacillus subtilis* which lack dipicolinic acid. J. Bacteriol. *182*, 5505–5512.

Pan, J., Tristram-Nagle, S., Kucerka, N., and Nagle, J.F. (2008). Temperature dependence of structure, bending rigidity, and bilayer interactions of dioleoylphosphatidylcholine bilayers. Biophys. J. *94*, 117–124.

Pandiani, F., Brillard, J., Bornard, I., Michaud, C., Chamot, S., Nguyen-the, C., and Broussolle, V. (2010). Differential involvement of the five RNA helicases in adaptation of *Bacillus cereus* ATCC 14579 to low growth temperatures. Appl. Environ. Microbiol. *76*, 6692–6697.

Pandiani, F., Chamot, S., Brillard, J., Carlin, F., Nguyen-The, C., and Broussolle, V. (2011). Role of the five RNA helicases in adaptive response of *Bacillus cereus* ATCC 14579 cells to temperature, pH and oxidative stress. Appl. Environ. Microbiol. *77*, 5604–5609.

Paredes-Sabja, D., Setlow, P., and Sarker, M.R. (2010). Germination of spores of Bacillales and Clostridiales species: mechanisms and proteins involved. Trends Microbiol. *19*, 85–94.

Phadtare, S. (2004). Recent developments in bacterial cold-shock response. Curr. Issues Mol. Biol. *6*, 125–136.

Planchon, S., Dargaignaratz, C., Levy, C., Ginies, C., Broussolle, V., and Carlin, F. (2011). Spores of *Bacillus cereus* strain KBAB4 produced at 10 degrees C and 30 degrees C display variations in their properties. Food Microbiol. *28*, 291–297.

Prud'homme-Genereux, A., Beran, R.K., Iost, I., Ramey, C.S., Mackie, G.A., and Simons, R.W. (2004). Physical and functional interactions among RNase E, polynucleotide phosphorylase and the cold-shock protein, CsdA: evidence for a 'cold shock degradosome'. Mol. Microbiol. *54*, 1409–1421.

Pruss, B.M., Francis, K.P., von Stetten, F., and Scherer, S. (1999). Correlation of 16S ribosomal DNA signature sequences with temperature-dependent growth rates of mesophilic and psychrotolerant strains of the *Bacillus cereus* group. J. Bacteriol. *181*, 2624–2630.

Sakamoto, J.J., Sasaki, M., and Tsuchido, T. (2001). Purification and characterization of a *Bacillus subtilis* 168 nuclease, YokF, involved in chromosomal DNA degradation and cell death caused by thermal shock treatments. J. Biol. Chem. *276*, 47046–47051.

Salvetti, S., Celandroni, F., Ceragioli, M., Senesi, S., and Ghelardi, E. (2009). Identification of non-flagellar genes involved in swarm cell differentiation using a *Bacillus thuringiensis* mini-Tn10 mutant library. Microbiology *155*, 912–921.

van Schaik, W., Tempelaars, M.H., Wouters, J.A., de Vos, W.M., and Abee, T. (2004). The alternative sigma factor sigmaB of *Bacillus cereus*: response to stress and role in heat adaptation. J. Bacteriol. *186*, 316–325.

Schoeni, J.L., and Wong, A.C. (2005). *Bacillus cereus* food poisoning and its toxins. J. Food Prot. *68*, 636–648.

Schujman, G.E., Paoletti, L., Grossman, A.D., and de Mendoza, D. (2003). FapR, a bacterial transcription factor involved in global regulation of membrane lipid biosynthesis. Dev. Cell *4*, 663–672.

Schujman, G.E., Guerin, M., Buschiazzo, A., Schaeffer, F., Llarrull, L.I., Reh, G., Vila, A.J., Alzari, P.M., and de Mendoza, D. (2006). Structural basis of lipid biosynthesis regulation in Gram-positive bacteria. EMBO J. *25*, 4074–4083.

Schumann, W. (2009). Temperature sensors of Eubacteria. In Advances in Applied Microbiology, Laskin, A.I., ed. (Elsevier Academic Press, San Diego), pp. 213–256.

Setlow, B., Atluri, S., Kitchel, R., Koziol-Dube, K., and Setlow, P. (2006). Role of dipicolinic acid in resistance and stability of spores of *Bacillus subtilis* with or without DNA-protective alpha/beta-type small acid-soluble proteins. J. Bacteriol. *188*, 3740–3747.

Shin, J.H., and Price, C.W. (2007). The SsrA-SmpB ribosome rescue system is important for growth of *Bacillus subtilis* at low and high temperatures. J. Bacteriol. *189*, 3729–3737.

Sorokin, A., Candelon, B., Guilloux, K., Galleron, N., Wackerow-Kouzova, N., Ehrlich, S.D., Bourguet, D., and Sanchis, V. (2006). Multiple-locus sequence typing analysis of *Bacillus cereus* and *Bacillus thuringiensis* reveals separate clustering and a distinct population structure of psychrotrophic strains. Appl. Environ. Microbiol. *72*, 1569–1578.

Stenfors, L.P., and Granum, P.E. (2001). Psychrotolerant species from the *Bacillus cereus* group are not necessarily *Bacillus weihenstephanensis*. FEMS Microbiol. Lett. *197*, 223–228.

Stenfors, L.P., Mayr, R., Scherer, S., and Granum, P.E. (2002). Pathogenic potential of fifty *Bacillus weihenstephanensis* strains. FEMS Microbiol. Lett. *215*, 47–51.

Stenfors Arnesen, L.P., Fagerlund, A., and Granum, P.E. (2008). From soil to gut: *Bacillus cereus* and its food poisoning toxins. FEMS Microbiol. Rev. *32*, 579–606.

Stephenson, K., and Lewis, R.J. (2005). Molecular insights into the initiation of sporulation in Gram-positive bacteria: new technologies for an old phenomenon. FEMS Microbiol. Rev. *29*, 281–301.

von Stetten, F., Mayr, R., and Scherer, S. (1999). Climatic influence on mesophilic *Bacillus cereus* and psychrotolerant *Bacillus weihenstephanensis* populations in tropical, temperate and alpine soil. Environ. Microbiol. *1*, 503–515.

Swiecicka, I. (2008). Natural occurrence of *Bacillus thuringiensis* and *Bacillus cereus* in eukaryotic organisms: a case for symbiosis. Biocontrol Sci. Technol. *18*, 221–239.

Tatusov, R.L., Galperin, M.Y., Natale, D.A., and Koonin, E.V. (2000). The COG database: a tool for genome-scale analysis of protein functions and evolution. Nucleic Acids Res. *28*, 33–36.

Thieringer, H.A., Jones, P.G., and Inouye, M. (1998). Cold shock and adaptation. Bioessays *20*, 49–57.

Tourasse, N.J., Helgason, E., Klevan, A., Sylvestre, P., Moya, M., Haustant, M., Okstad, O.A., Fouet, A., Mock, M., and Kolsto, A.B. (2011). Extended and global phylogenetic view of the *Bacillus cereus* group population by combination of MLST, AFLP, and MLEE genotyping data. Food Microbiol. *28*, 236–244.

Trubetskoy, D., Proux, F., Allemand, F., Dreyfus, M., and Iost, I. (2009). SrmB, a DEAD-box helicase involved in *Escherichia coli* ribosome assembly, is specifically targeted to 23S rRNA *in vivo*. Nucleic Acids Res. *37*, 6540–6549.

VanBogelen, R.A., and Neidhardt, F.C. (1990). Ribosomes as sensors of heat and cold shock in *Escherichia coli*. Proc. Natl. Acad. Sci. U.S.A. *87*, 5589–5593.

Weber, M.H., and Marahiel, M.A. (2002). Coping with the cold: the cold shock response in the Gram-positive soil bacterium *Bacillus subtilis*. Phil. Trans. R. Soc. Lond. B Biol. Sci. *357*, 895–907.

Weber, M.H., Klein, W., Muller, L., Niess, U.M., and Marahiel, M.A. (2001). Role of the *Bacillus subtilis* fatty acid desaturase in membrane adaptation during cold shock. Mol. Microbiol. *39*, 1321–1329.

Wiegeshoff, F., Beckering, C.L., Debarbouille, M., and Marahiel, M.A. (2006). Sigma L is important for cold shock adaptation of *Bacillus subtilis*. J. Bacteriol. *188*, 3130–3133.

Xia, B., Ke, H.P., Jiang, W., and Inouye, M. (2002). The cold box stem-loop proximal to the 5′-end of the *Escherichia coli cspA* gene stabilizes its mRNA at low temperature. J. Biol. Chem. *277*, 6005–6011.

Zhang, Y.M., and Rock, C.O. (2008). Membrane lipid homeostasis in bacteria. Nat. Rev. Microbiol. *6*, 222–233.

Zhang, Y.M., and Rock, C.O. (2009). Transcriptional regulation in bacterial membrane lipid synthesis. J. Lipid Res. *50*, S115–119.

Stress Responses in *Salmonella*

6

Suzanne Humphrey, Tom J. Humphrey and Mark A. Jepson

Abstract

Salmonella enterica strains are the causative agents of a spectrum of diseases, including enteric fever and self-limiting gastroenteritis, and remain significant food-borne pathogens throughout both the developed and developing worlds. The ability to actively invade and reside within gut epithelia and macrophages is an important process in the establishment of *Salmonella* infection, generating localized inflammatory responses and facilitating systemic spread of the pathogen within the host. Many environments, including food matrices, the external environment and conditions within the host, present a range of stressful challenges that *Salmonella* must overcome in order to survive and establish infection. *Salmonella* strains utilize a diverse range of stress response strategies – including expression of alternative RNA polymerase sigma factors, uptake of compatible solutes, increased expression of genes encoding uptake or efflux pumps, and production of proteins with roles in protecting and repairing stress-induced damage – in order to facilitate their survival in suboptimal and stressful growth environments. Additionally, the ability of *Salmonella* to undergo morphological changes during stress exposure and rapidly recover from stress conditions commonly encountered within food matrices represents a pertinent issue for food processing and public health.

Introduction

Pathology and epidemiology of *Salmonella* infection

Salmonella are Gram-negative, facultatively anaerobic, rod-shaped γ-proteobacteria belonging to the family *Enterobacteriaceae*. The genus *Salmonella* comprises more than 2463 distinct serotypes, of which almost 60% are classified as *Salmonella enterica* subspecies *enterica* (Brenner *et al.*, 2000). *S. enterica* subsp. *enterica* represents the most clinically significant group of *Salmonella*, accountable for almost all cases of salmonellosis in mammals and birds (Heyndrickx *et al.*, 2005). Faecal–oral transmission of *Salmonella* is most often mediated through consumption of contaminated food or water, with transmission generally occurring in conditions associated with poor hygiene and sanitation (Connor and Schwartz, 2005; Kanungo *et al.*, 2008; Meltzer and Schwartz, 2010).

S. enterica are the causative agents of a range of disease syndromes, including systemic enteric fever and self-limiting gastroenteritis (salmonellosis). While many *S. enterica* infections are zoonotic in nature, arising from *S. enterica* with broad host ranges, *S. enterica* subsp. *enterica* serovars Typhi (*S.* Typhi) and Paratyphi (*S.* Paratyphi) are host-restricted pathogens, exclusively causing disease in humans (Meltzer and Schwartz, 2010). Typhoid and Paratyphoid fevers typically present as a high fever accompanied by abdominal pain and discomfort (Siddiqui *et al.*, 2006; Rafiq *et al.*, 2009; Meltzer and Schwartz, 2010), while a range of non-specific symptoms such as headache, vomiting, transient diarrhoea, coughing, weakness, and hepatic and/or splenic enlargement may also be noted (Siddiqui *et al.*, 2006; Rafiq *et al.*, 2009). Endemic *S.* Typhi and *S.* Paratyphi infection is a continuing problem in a number of South Asian and South-East Asian countries (Crump *et al.*, 2004; Siddiqui *et al.*, 2006), with mortality rates standing at just under a third of cases in some endemic regions (Connor and Schwartz, 2005). Notably, while sporadic outbreaks of enteric fever do occur in industrialized nations, these are generally associated with travel to endemic regions and usually exhibit lower mortality rates (Connor and Schwartz, 2005; Meltzer and Schwartz, 2010).

In contrast, non-typhoidal (NT) *S. enterica*

continue to be widespread food-borne pathogens throughout both the developing and developed worlds, as exemplified by their identification as the causative agents of disease in 131,468 laboratory-confirmed cases across the European Union in 2008 (Anon, 2010). NT salmonellosis typically presents as a self-limiting gastroenteritis characterized by diarrhoea, abdominal cramps and fever, 12–72 hours following ingestion of the causative bacteria. A range of foodstuffs have been implicated as vehicles for NT *S. enterica* transmission, including eggs (Webpage 1), beef (Dechet *et al.*, 2006), cheese (Villar *et al.*, 1999), peanut butter (Webpage 2), ice cream (Hennessy *et al.*, 1996), tomatoes (Webpage 3), basil (Pezzoli *et al.*, 2008) and orange juice (Jain *et al.*, 2009); demonstrating the diverse capacity for *Salmonella* transmission in the food chain.

Salmonella invasion and intracellular proliferation

Following ingestion, *Salmonella* pass through the host stomach and enter the small intestine. Intimate attachment of *Salmonella* at the apical surface of intestinal epithelial cells stimulates the rapid translocation of a battery of bacterial effector proteins – including SipA, SipC, SopE, SopE2, SopB/SigD and SptP – directly into the host cell via the type III secretion system (T3SS) encoded within *Salmonella* Pathogenicity Island (SPI)-1 (Raffatellu *et al.*, 2005; Schlumberger *et al.*, 2005; McGhie *et al.*, 2009). Effectors SipA and SipC are actin-binding proteins with roles in localizing and driving actin polymerization in close proximity to the adhered bacterium and in assisting invasion (Hayward and Koronakis, 1999; Zhou *et al.*, 1999; McGhie *et al.*, 2001; Higashide *et al.*, 2002; Chang *et al.*, 2005; Perrett *et al.*, 2009), while SopE and SopE2 are guanine nucleotide exchange factors (GEFs) that directly activate host Rac1 and Cdc42 Rho GTPases (Hardt *et al.*, 1998; Rudolph *et al.*, 1999; Stender *et al.*, 2000; Friebel *et al.*, 2001), modulating actin cytoskeletal assembly through the Arp2/3 complex and stimulating formation of the characteristic membrane ruffles which facilitate *Salmonella* uptake into membrane-bound vacuoles (Hardt *et al.*, 1998; Stender *et al.*, 2000) (Fig. 6.1). Similarly, SopB/SigD is a phosphatidylinositol phosphatase that contributes to actin remodelling through indirect activation of the host RhoG GTPase (Zhou *et al.*, 2001; Patel and Galan, 2006). Following *Salmonella* uptake, normal host cell architecture is restored through the activity of the GTPase activating protein (GAP) SptP (Fu and Galan,

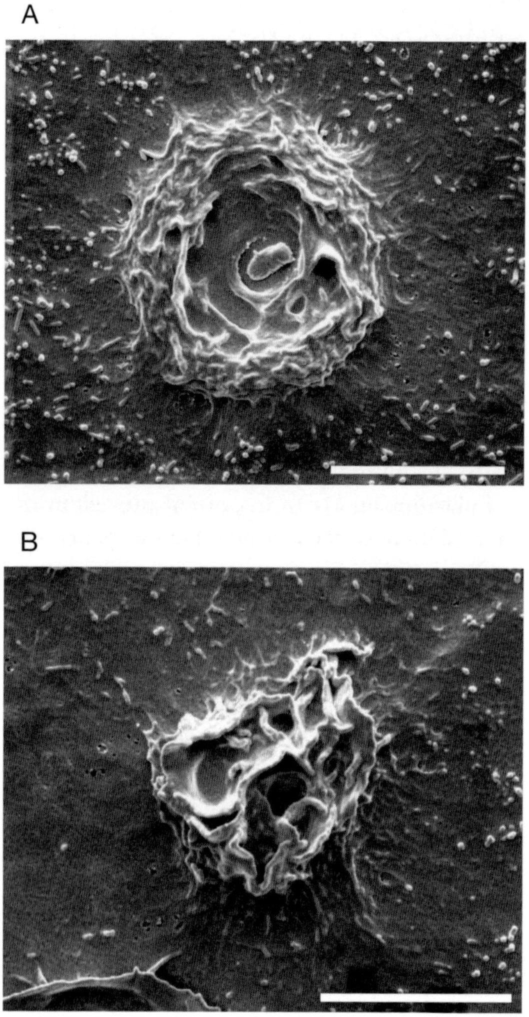

Figure 6.1 SPI-1-mediated host cytoskeletal rearrangements during *Salmonella* infection. (A) This scanning electron micrograph shows an *S.* Typhimurium bacterium inducing the formation of a characteristic membrane ruffle on the apical surface of a Madin–Darby canine kidney (MDCK) strain I epithelial cell. (B) Ruffle closure around invading bacteria results in engulfment in a process similar to macropinocytosis. Scale bars represent 5 μM.

1999). In addition to their roles in stimulating actin remodelling, SipA, SopB/SigD, SopE and SopE2 have been implicated in mediating gut inflammation by compromising the integrity of the intestinal barrier through tight junction modulation (Boyle *et al.*, 2006) and by stimulating production of pro-inflammatory mediators (Bruno *et al.*, 2009; Muller *et al.*, 2009). Together, these responses result in increased transepithelial permeability, polymorphonuclear leucocyte (PMN) influx and fluid accumulation in the intestinal lumen (Jepson *et al.*, 1995; Galyov *et al.*, 1997; Lee *et al.*, 2000), culminating in the characteristic diarrhoeal symptoms associated with gastrointestinal infection.

Following uptake into the host cell, *Salmonella* reside within a spacious phagosome that rapidly matures into a distinct membrane-bound compartment known as the *Salmonella*-containing vacuole (SCV) (Steele-Mortimer *et al.*, 1999). Acidification of the SCV, combined with limited divalent cation and phosphate concentrations, stimulates expression of a second T3SS system encoded within *Salmonella* Pathogenicity Island (SPI)-2 (Cirillo *et al.*, 1998; Deiwick *et al.*, 1999; Garmendia *et al.*, 2003), which *Salmonella* utilizes to facilitate the translocation of at least 20 effector proteins across the SCV membrane and into the host cell cytosol (reviewed in Haraga *et al.*, 2008; McGhie *et al.*, 2009). These effector proteins have roles in manipulating host cell trafficking pathways and adapting the SCV into a favourable site for *Salmonella* proliferation (Uchiya *et al.*, 1999; Beuzon *et al.*, 2000; Boucrot *et al.*, 2005). The intracellular lifestyle of *Salmonella* is a critical virulence determinant for effective intracellular proliferation and systemic spread of the organism within epithelial cells and non-activated macrophages (Fields *et al.*, 1986; Cirillo *et al.*, 1998). Notably, in addition to SPI-1 induced invasion of epithelial cells, the ability of *Salmonella* to translocate across intestinal antigen-sampling M cells located within Peyer's patches enables the bacteria to establish contact with underlying macrophages (Clark *et al.*, 1994, 1996; Sansonetti and Phalipon, 1999; Martinez-Argudo and Jepson, 2008), mediating dissemination of *Salmonella* to other anatomical sites, such as the liver and spleen.

Stress conditions encountered during the *Salmonella* life cycle

Salmonella are truly versatile, multiniche organisms, capable of survival in the external environment, in food matrices, and within a range of cell types in both warm and cold-blooded hosts. *Salmonella* may be exposed to a multitude of stress conditions, including desiccation, extremes of temperature, and a spectrum of oxygen concentrations, osmotic strengths and pH conditions in environments such as soil or on surfaces. Similarly, owing to their transmission vehicles of food and water, *Salmonella* must be able to appropriately respond to exposure to conditions commonly encountered during food production and storage, including, for example, osmotic stress in low water activity (a_w) and desiccated foodstuffs, exposure to organic and inorganic acidifiers, the presence of competing bacteria, antimicrobial peptides and low-temperature storage (Leistner and Gorris, 1995).

Importantly, many similar stress conditions may also be encountered within the host. Immediately upon infection of warm-blooded hosts, *Salmonella* are exposed to a sudden temperature upshift from that of the external environment, for instance 37°C in humans and 42°C in chickens. Following ingestion, *Salmonella* must survive the activity of antimicrobial peptides in the saliva and the low pH environment of the stomach before passing into the small intestine. Here, *Salmonella* must endure hyperosmotic and outer membrane stresses owing to the action of bile and antimicrobial peptides, coupled with limited oxygen availability and the need to compete with the normal flora in the intestine in order to initiate invasion of intestinal enterocytes and M cells. Further, following penetration of the epithelial cell barrier, *Salmonella* must sense and respond to the intracellular environment within epithelial cells, neutrophils and non-activated macrophages; this includes the low pH environment of the phagolysosome, nutrient limitation and avoidance of NADPH oxidase-mediated killing.

This review aims to give an overview of the main stress response mechanisms employed by *Salmonella* for survival in nutrient-limited conditions and during osmotic and acid stress exposure. Further, the heat-shock and cold-shock mechanisms for enduring challenge with extremes of temperature will be described, as well as the extracytoplasmic, oxidative and SOS stress responses, respectively, to outer membrane stress and DNA-damaging conditions. These responses are summarized briefly in Table 6.1. Finally, the impact of the ability of *Salmonella* to exhibit morphological plasticity under stress, coupled with the ability to recover from conditions relevant to food preservation will be discussed.

Salmonella stress responses

The general (σ^S) stress response

In *Salmonella* and other bacteria such as *Escherichia coli* and *Vibrio cholerae*, the general stress response is mediated by the alternative RNA polymerase sigma factor ($\sigma^{S/38}$) encoded by *rpoS* (Jones *et al.*, 2006; Dong and Schellhorn, 2010). Alternative sigma factors are global regulators that alter the affinity of RNA polymerase for a specific range of promoters and, upon exposure to stress, some alternative sigma factors can associate with the core RNA polymerase and drive expression of particular subsets of genes whose products play important roles in maintaining homeostasis and reducing damage associated with stress exposure (Kazmierczak *et al.*, 2005; Dong and Schellhorn, 2010).

Regulation of σ^S expression is complex, with control exerted at the levels of transcription, translation and proteolysis (Lange and Hengge-Aronis, 1994; Hirsch and Elliott, 2005). In *S. enterica*, transcription of *rpoS* is repressed

Table 6.1 Summary of *Salmonella* stress responses

Stress response	Inducing signal(s)	Key regulator(s)	Key effector(s)	Comments
General stress response	C, N_2 or PO_4 starvation; stationary phase growth	σ^S	StiA-C; KatE; SodCII; OtsA; SpvR; SpvA-D	Required for full virulence in mice
Osmotic stress response	Hyperosmotic stress	σ^S; OmpR	OmpC; OtsA and OtsB; ProU; ProP; Kdp	Important for survival in food matrices
Acid tolerance response	Low pH	σ^S, Fur, PhoP, OmpR	OmpC; DnaK; GroEL; CadA and CadB; Atp	May induce cross-protective responses against osmotic and temperature stress. Important for virulence
Heat-shock response	High temperature (>40°C); intramacrophage growth	σ^H	DnaK/DnaJ; GroEL/GroES; GrpE; Lon; ClpXP	Required for virulence in mice. Negatively regulates SPI-1 gene expression
Cold-shock response	Low temperature exposure (<10°C)	Unknown	CspA-E and CspH	Important for survival in food matrices during refrigeration
Extracytoplasmic stress responses	Outer membrane stress; misfolded proteins; CAMPs; hyperosmotic stress	σ^E, CpxAR, PspA-PspF	HtrA; SurA; FkpA; Skp; PspA	Required for resistance to CAMPs and for virulence in mice. Roles in maintaining the PMF
Oxidative stress response	Superoxide; H_2O_2, reactive O_2 and N_2 species	σ^S; SoxRS, OxyR, SsrAB	SodA; KatE and KatG; AhpCF; TsaA; SodCII; SpiC	Some functional redundancy between effectors. Critical for intracellular survival within macrophages
SOS response	DNA damage	LexA, RecA	SulA	Loss of *recA* or *recBCD* results in severe attenuation in mice. Induction may induce a filamentous phenotype

CAMPs, cationic antimicrobial peptides; PMF, proton motive force.

during exponential growth by binding of the DNA-binding protein, Fis, at the *rpoS* promoter (Hirsch and Elliot, 2005). Following the onset of stationary phase, reduced cellular concentrations of Fis serve to relieve Fis-mediated repression of *rpoS* transcription, enabling expression of σ^S and σ^S-regulated genes (Hirsch and Elliot, 2005). Additional factors reported to influence transcriptional regulation of *rpoS* in *E. coli* include cAMP-CRP (Lange and Hengge-Aronis, 1994), ppGpp (Lange *et al.*, 1995) and the cellular ratio of NADH:NAD^+ (Sevcik *et al.*, 2001). While transcriptional regulation of *rpoS* is important, the principal mechanisms governing regulation of σ^S expression are exerted at the post-transcriptional level (Lange and Hengge-Aronis, 1994). In *S. enterica,* translational regulation of σ^S expression involves the gene product of *hfq,* encoding host factor I (HF-I) (Brown and Elliott, 1996). HF-1 is a ribosomal-associated RNA-binding protein and *S.* Typhimurium *hfq* mutants have been shown to exhibit substantially reduced cellular levels of σ^S as a consequence of diminished efficiency in *rpoS* translation (Brown and Elliot, 1996). Other factors, including the histone-like proteins HU and H-NS (Yamashino *et al.*, 1995; Balandina *et al.*, 2001; Mangan *et al.*, 2011), and the small regulatory RNA molecules DsrA, OxyS and RprA (Sledjeski *et al.*, 1996; Zhang *et al.*, 1998; Majdalani *et al.*, 2001), have also been reported to have roles in stabilizing and regulating translation of *rpoS* mRNA in *Salmonella* and *E. coli.* Finally, further control over the cellular level of σ^S is mediated by proteolysis of the protein under conditions inappropriate for its expression. σ^S degradation is principally mediated through the action of the ATP-dependent ClpXP protease – where loss of either *clpP* or *clpX* results in enhanced stability of σ^S (Schweder *et al.*, 1996) – and a response regulator known as MviA (RssB in *E. coli*) (Moreno *et al.*, 2000). In this system, phosphorylated MviA sequesters σ^S and targets the MviA–σ^S complex to ClpXP for degradation of σ^S and recycling of MviA (Moreno *et al.*, 2000), regulating the cellular levels of σ^S available for driving downstream expression of σ^S-regulated genes.

During *Salmonella* growth in stationary phase and in response to starvation conditions where carbon, phosphate and nitrogen are limited, σ^S regulates the expression of at least 60 genes (O'Neal *et al.*, 1994; Hirsch and Elliott, 2005). Some examples of σ^S-regulated genes include the virulence plasmid genes *spvRABCD* (Heiskanen *et al.*, 1994; Kowarz *et al.*, 1994; Chen *et al.*, 1995); *katE,* encoding HPII catalase (Chen *et al.*, 1996); *otsA,* encoding trehalose-6-phosphate synthase (Fang *et al.*, 1996); *poxB and ogt,* encoding a pyruvate oxidase and O^6-methylguanine DNA methyltransferase, respectively (Ibanez-Ruiz *et al.*, 2000); the multiple-nutrient starvation-inducible loci *stiA, stiB* and *stiC* (O'Neal *et al.*, 1994); and at least nine genes of unknown function, *yciF, yehY, yhjY, yncC, yigB, yahO, ygaU, ycgB* and *yeaG* (Ibanez-Ruiz *et al.*, 2000). In addition, σ^S is involved in the response of *Salmonella* to a number of other physical and physiochemical stress factors, including hyperosmotic stress at high and low temperatures (McMeechan *et al.*, 2007), low pH stress (Fang *et al.*, 1992; Lee *et al.*, 1994), oxidative stress and DNA damage (Fang *et al.*, 1992; Lee *et al.*, 1995). Thus, σ^S is important for *Salmonella* survival in a number of environments including those within food matrices. McMeechan *et al.* (2007) demonstrated that *rpoS* is important for long-term *Salmonella* survival in nutritionally limited environments, such as saline, at high (37°C) and low (4.5°C) temperatures. This study also showed that *rpoS* is particularly important for prolonged survival of *S.* Typhimurium in high osmolarity conditions, where Δ*rpoS* mutants cultured in 1.0 M NaCl at 4.5°C decreased to undetectable levels after 37 days; 42 days earlier than the wild-type control (McMeechan *et al.*, 2007), illustrating the importance of σ^S-mediated stress responses to nutritionally limited conditions that may be encountered in the external environment and during food production and storage.

Interestingly, while σ^S is necessary for facilitating stress resistance against a number of environmental conditions, the requirement of σ^S for virulence varies between bacterial species (Dong and Schellhorn, 2010). In addition to its roles in moderating the response to starvation or exposure to a range of physical and physiochemical stresses, σ^S has also been shown to be critical for *Salmonella* virulence (Fang *et al.*, 1992; Nickerson and Curtiss, 1997). A number of virulence genes, including *spvR, spvABCD* and some chromosomally encoded virulence factors – such as *yedI, sodCII* and the genes involved in curli fimbriae synthesis – are known to be regulated by σ^S (Fang *et al.*, 1992; Kowarz *et al.*, 1994; ElGedaily *et al.*, 1997; Romling *et al.*, 1998; Erickson and Detweiler, 2006; Dong and Schellhorn, 2010). Further, *Salmonella* Δ*rpoS* mutants exhibit significantly reduced virulence compared to wild-type *Salmonella* in the murine oral infection model (Fang *et al.*, 1992), while their ability to persist in the liver and spleen is also significantly compromised (Kowarz *et al.*, 1994; Coynault *et al.*, 1996). This is, at least in part, probably owing to the significantly compromised ability of *rpoS* mutants to survive the gastric environment (Bearson *et al.*, 2006), coupled with the importance of σ^S for mediating resistance against a range of host stresses, including cationic antimicrobial peptides (Bader *et al.*, 2003). Accordingly, σ^S represents an important component of the defensive arsenal that *Salmonella* utilizes to withstand a wide range of stress conditions in the environment, within food matrices and during infection of the host.

Salmonella responses to osmotic stress

The bacterial cell wall acts as a semi-rigid barrier, limiting the expansion of the cytoplasm and cytoplasmic membrane following the influx of water into the cell. Accordingly, continued water influx into the bacterial cell generates increasing internal pressure which is exerted onto the cell wall by the cytoplasmic membrane (Csonka, 1989). This is known as turgor pressure and is important to the cell for providing the mechanical force necessary for driving cell elongation (Csonka, 1989; Gutierrez *et al.*, 1995). Hyperosmotic shock, caused by exposure to environments containing high solute concentrations – such as those encountered in low-a_w foodstuffs or the intestinal lumen – results in a rapid osmotic efflux of water out of the bacterial cell, causing a considerable reduction in cell volume and loss of normal turgor pressure (Csonka, 1989). Thus, in order to survive and cause infection, *Salmonella* must be able to adapt to and overcome hyperosmotic stresses encountered in foodstuffs and within the host.

Upon exposure to hyperosmotic stress, water is lost rapidly from the bacterial cell, resulting in shrinkage (Delamarche *et al.*, 1999). This loss of turgor pressure and the sudden relative increase in internal ion concentrations within the bacterial cytoplasm serve as inducing signals for the activation of a number of systems for restoring normal turgor pressure (Balaji *et al.*, 2005). Initially, uptake of potassium (K^+) and the counter-ion glutamate is implemented as the primary method for restoring normal turgor pressure to the bacterial cell (Botsford *et al.*, 1994; Gutierrez *et al.*, 1995; Capozzi *et al.*, 2009). However, under conditions of significant hyperosmolarity, excessive uptake of K^+ can result in high internal concentrations of K^+ and glutamate, which can begin to adversely affect enzymatic activities within the cell (Gutierrez *et al.*, 1995). Accordingly, *Salmonella* induce uptake of compatible solutes – such as glycine betaine and proline – from the environment or induce their *de novo* synthesis, as in the case of trehalose, in order to stabilize the osmotic pressure inside the cell (Cairney *et al.*, 1985a,b; Gutierrez *et al.*, 1995; Wood, 1999; Howells *et al.*, 2002). Compatible solutes are typically small organic end-point metabolites, usually with neutral or zwitterionic charge, that are soluble to high concentrations within the bacterial cytoplasm, increasing the inner cytoplasmic pressure without adversely affecting the activity of cytoplasmic enzymes (Gutierrez *et al.*, 1995; Capozzi *et al.*, 2009). Uptake of compatible solutes is regulated by membrane-bound transporters which display varying affinities for their substrate depending on the external osmotic pressure (Gutierrez *et al.*, 1995). *Salmonella* utilize several compatible solutes, many of which are freely available in food matrices, including proline and glycine betaine, which are imported via the proline/glycine betaine transporters ProP and ProU (Cairney *et al.*, 1985a,b; Dunlap and Csonka, 1985). Further, in circumstances of osmotic stress where glycine betaine and proline are unavailable or at insufficient concentrations, the compatible solute trehalose can be synthesized *de novo* by the products of the *otsA* and *otsB* genes (Howells *et al.*, 2002) to raise the cytoplasmic solute concentration and restore turgor pressure.

Following osmotic shock with 300 mM NaCl, a range of osmoregulated genes are rapidly expressed in *S.* Typhimurium. Balaji *et al.*

(2005) demonstrated that four minutes following osmotic shock, *S.* Typhimurium induced expression of the *proU* operon (encoding an inducible glycine betaine transporter; Cairney *et al.*, 1985a), while expression of *proP* (encoding the proline/glycine betaine transporter; Cairney *et al.*, 1985b) occurred after 5–6 minutes. Expression of the *kdp* operon, encoding the high-affinity potassium-transporting ATPase (Laimins *et al.*, 1981), was also induced 8–9 minutes after exposure to 300 mM NaCl, while induction of the gene encoding the trehalose-6-phosphate phosphatase involved in trehalose biosynthesis (*ostB*) occurred 10–12 minutes after osmotic shock (Balaji *et al.*, 2005). Together, the induction profiles of these genes suggest that a combined response to osmotic stress is elicited in *Salmonella*, with a number of mechanisms cooperating to re-establish normal turgor pressure under hyperosmotic conditions.

The *Salmonella* acid tolerance response

While *Salmonella* preferentially grow at neutral pH, the multiniche lifestyle of these organisms demands that they display a comprehensive ability to tolerate exposure to extreme low pH conditions, such as those encountered in the host stomach, within the macrophage phagolysosomal compartment and in acidic conditions in the external environment; including within food matrices (Bearson *et al.*, 1997, 1998). In *Salmonella*, internal housekeeping pH homeostasis mechanisms, such as the Na^+/H^+ and K^+/H^+ antiporters, are capable of maintaining the intracellular pH within satisfactory levels in moderately acidic environments, thus facilitating survival in mild to moderate acid conditions with no requirement for protein synthesis (Macnab and Castle, 1987; Foster, 1991). In contrast, in extreme acid environments extracellular protons diffuse across the bacterial membrane along the concentration gradient more rapidly than they can be extruded by normal physiological homeostasis mechanisms, resulting in acidification of the cytoplasm to damaging levels (Bearson *et al.*, 1998). Consequently, *Salmonella* have evolved a complex acid stress response comprising several mechanisms that facilitate the survival of both exponential and stationary phase bacteria during periods of extreme pH stress. Some of these inducible factors include amino acid decarboxylases (Park *et al.*, 1996; Bearson *et al.*, 1997; Kim *et al.*, 2001) and a collection of at least 51 acid shock proteins (ASP) whose expression is controlled by RpoS, Fur, PhoP and OmpR in response to organic and inorganic acid conditions (Bearson *et al.*, 1997, 1998; Bang *et al.*, 2000).

The acid stress response is an extremely complex mechanism of stress resistance and comprises at least three distinct, yet complementary, acid tolerance response (ATR) systems that can be utilized in response to acid stress depending on the growth phase of the bacteria (Lee *et al.*, 1994; Greenacre and Brocklehurst, 2006). The first of these systems is a two-phase response, known as the logarithmic-phase ATR, which is employed by exponentially growing *Salmonella* during acid exposure (Foster, 1991; Lee *et al.*, 1994). While each of the two parts of the logarithmic-phase ATR can independently provide some protection against acid shock down to around pH 3.0, both are utilized in combination for optimal survival in severe acid conditions (Greenacre and Brocklehurst, 2006). The first stage of the log-phase ATR consists of an adaptation phase, termed as 'pre-shock', in which *Salmonella* exposed to mildly acidic conditions (pH 5.8) induce production of at least 12 ASPs with roles in priming the bacterium for the subsequent 'acid shock' stage of the ATR (Foster, 1991). Acid shock occurs when *Salmonella* are transferred from mild (pH 5.8) to moderate/severe (pH <4.5) acidity and induces the production of at least 37 ASPs that facilitate *Salmonella* survival in highly acidic (pH ≤ 3.0) environments (Foster, 1991). Proteins produced in response to acid shock have many roles in protection and maintenance of intracellular homeostasis and include chaperone proteins – such as DnaK and GroEL (Foster, 1991) – which function to prevent protein aggregation or misfolding in the cytoplasm, membrane porins such as OmpC (Bang *et al.*, 2000), amino acid decarboxylases such as the lysine decarboxylase encoded by *cadA* (Park *et al.*, 1996; Kim *et al.*, 2001) and inducible proton pump translocators such as the Mg^{2+}-dependent proton translocating ATPase encoded by genes in the *atp* operon (Foster, 1991; Foster and Hall, 1991). Crucially, following a period of adaptation at mild pH, survival of *Salmonella* at severe pH is significantly (100- to 1000-fold) increased from that of non-adapted *Salmonella* (Foster, 1991; Bearson *et al.*, 1998).

Two further acid stress responses have been described for *Salmonella* in stationary phase; a pH-dependent system known as the stationary-phase acid tolerance response (ATR) and an alternative pH-independent system that requires the general stress response stationary phase sigma factor, σ^S (Lee *et al.*, 1994). The former stationary-phase ATR is induced between pH 4.3–5.0 and confers enhanced survival at extreme low pH (pH 3.0), where acid adapted cells are around 1000-fold more resistant to acid stress than their non-adapted counterparts (Lee *et al.*, 1994). Following adaptation at pH 4.3, 15 acid shock proteins (ASP) are produced as part of the stationary-phase ATR, four of which have also been identified as log-phase ASPs (Lee *et al.*, 1994). In contrast to the log-phase ATR, production of ASPs in the stationary-phase ATR requires prolonged (< 120 minutes) exposure to moderate acid conditions, with continued protein synthesis occurring during exposure to severe low pH, suggesting that this mechanism is appropriately tailored for facilitating survival of non-growing *Salmonella* during gradual transitions into severe acid environments, such as food matrices, the external environment and the host stomach (Lee *et al.*, 1994).

Finally, σ^S has been proposed to play a role in mediating a third, pH-independent acid stress response in *Salmonella* (Fang *et al.*, 1992; Lee *et al.*, 1994, 1995; Rychlik and Barrow, 2005). Expression of *rpoS* is strongly induced during growth to stationary phase (Hirsch and Elliott, 2005), with σ^S mediating the general stress response to nutrient limitation and other stress conditions commonly encountered by stationary phase bacteria (Fang *et al.*, 1992; McMeechan *et al.*, 2007). Stationary phase cells naturally exhibit around 1000-fold higher acid resistance than logarithmic phase cells exposed to low pH conditions (pH 3.0 for 1 hour) (Lee *et al.*, 1994). Importantly, the role of σ^S in stationary phase acid resistance appears to be pH independent, as the pH-inducible ATR continues to confer enhanced acid survival in *rpoS* mutants, while the total levels of stationary-phase acid resistance are decreased by up to 100-fold from that of the wild type (Lee *et al.*, 1994).

The ATR of *Salmonella* is not only important for survival of acid stress, but has also been shown to induce cross-resistance against a variety of other stress conditions. Exposure of *Salmonella* to organic acids can offer protection against subsequent challenge with inorganic acids such as HCl (Greenacre *et al.*, 2003) and the ATR induced by exposure to acetic acid also exhibits protection against hyperosmotic stress conditions with salts such as NaCl and KCl (Greenacre and Brocklehurst, 2006). In addition, other work has indicated that acid-adapted *Salmonella* cells elicit cross-protection against low temperature stress (Xu *et al.*, 2008). Thus, the ATR is critically important for *Salmonella* survival and virulence, while the phenomenon of cross-protection also has serious implications for food preservation and food-borne disease.

Salmonella responses to high- and low-temperature stress

Heat shock

Exposure to temperatures above 40°C, for instance during food processing treatments or during infection of avian hosts where body temperature is around 42°C, can result in the accumulation of misfolded proteins in the bacterial cytoplasm and periplasmic space (Morimoto, 1998; Takaya *et al.*, 2004). In *Salmonella* and *E. coli,* the presence of excessive unfolded or misfolded proteins in the bacterial cytoplasm activates the heat-shock $\sigma^{H/32}$ regulon (Gruber and Gross, 2003), driving expression of a repertoire of heat shock proteins (HSPs) with chaperone, repair, and degradative functions that cooperate to prevent protein aggregation within the stressed cell (Straus *et al.*, 1988; Hayes and Dice, 1996; Dougan *et al.*, 2002). Members of the heat-shock response include the chaperone proteins DnaK and GroEL, co-chaperones DnaJ, GrpE, and GroES, and proteases Lon and ClpXP (Hayes and Dice, 1996; Dougan *et al.*, 2002).

The HSP chaperone GroEL and its co-chaperone GroES, are responsible for mediating the folding of native-state proteins in the bacterial cytoplasm under both normal conditions and during stress exposure, and functional copies of GroEL and GroES have been shown to be required for bacterial viability at all temperatures (Fayet *et al.*, 1989). Similarly, the HSP DnaK represents an integral component of the chaperone machinery induced by the heat shock response and cooperates with co-chaperones DnaJ and GrpE to regulate a number of cellular functions, including negative regulation of the heat shock response through reversible inhibition of σ^H (Gamer *et al.*, 1996; Tomoyasu *et al.*, 1998). A critical function of this chaperone machinery is to assist in protein folding and to prevent aggregation of damaged or misfolded proteins within the cytoplasm by stabilizing them in an appropriate state for proteolysis (Hartl, 1996; Jubete *et al.*, 1996). In addition to production of chaperone proteins, several heat shock proteins, including the ATP-dependent Lon and ClpXP proteases, are also known to contribute to cell viability during stress by degrading misfolded or damaged proteins in the cytoplasm (Downs *et al.*, 1986; Thomsen *et al.*, 2002).

Interestingly, a number of heat shock proteins are known to have important roles in virulence, most likely regulating the response of *Salmonella* to stresses encountered within the host and repressing expression of virulence genes where conditions are unsuitable for invasion of host cells (Buchmeier and Heffron, 1990; Takaya *et al.*, 2004; Matsui *et al.*, 2008). The DnaK/DnaJ chaperone machinery is a critical factor for virulence, as DnaK/DnaJ-depleted *S.* Typhimurium are defective for invasion in epithelial cells and are unable to colonize or cause systemic disease in BALB/c mice (Takaya *et al.*, 2004). Further, σ^H-induced Lon protease and ClpXP protease have been shown to negatively regulate expression of the genes required for invasion of intestinal epithelial cells (Takaya *et al.*, 2002; Takaya *et al.*, 2005; Kage *et al.*, 2008; Matsui *et al.*, 2008), highlighting the importance of this response for *Salmonella* virulence and survival in a multitude of environments.

Cold shock

NT *S. enterica* infections have been associated with a variety of foodstuffs, including meat, cheese, fruit juice and salad vegetables (Davies *et al.*, 1996; Cook *et al.*, 1998; Villar *et al.*, 1999; Cummings *et al.*, 2001; Horby *et al.*, 2003; Greene *et al.*, 2008; Nygard *et al.*, 2008; Jain *et al.*, 2009). Following processing and prior to consumption, many of these foodstuffs are stored at refrigeration temperatures (recommended range 4–7°C) in order to restrict microbial growth. Low-temperature environments present a number of challenges for mesophilic bacteria such as *Salmonella*, as cold temperatures result in diminished enzyme activity, loss of substrate affinity in bacterial transport systems, decreased stability and efficiency of mRNA translation and increased rigidity of bacterial membranes (Graumann and Marahiel, 1996). Thus, in order to survive in foodstuffs during low-temperature storage, *Salmonella* must utilize protective mechanisms known collectively as the cold-shock response.

Studies of a number of bacteria, including *Salmonella*, have elucidated that changes occur in the global cellular protein expression profile under conditions of low temperature stress, coupled with alterations in cellular physiology (Graumann and Marahiel, 1996; Jeffreys *et al.*, 1998; Horton *et al.*, 2000). Expression of cold acclimatisation proteins (CAPs) and cold shock proteins (CSPs) is induced following a decrease in temperature, with cold shock proteins highly expressed in *S. enterica* at temperatures ≤ 10°C (Gaumann and Marahiel, 1996; Jeffreys *et al.*, 1998). CSPs are highly conserved, acidic, low molecular weight (around 7 kDa) proteins that have protective roles in maintaining cellular physiology during low temperature exposure (Graumann and Marahiel, 1996; Jeffreys *et al.*, 1998; Horn *et al.*, 2007). Following transfer from 37°C, incubation of *S. enterica* at 5–10°C stimulates increased expression of the major cold shock protein CspA (also known as CS7.4 in *E. coli*) from the gene *cspA* (Jeffreys *et al.*, 1998; Horton *et al.*, 2000). Interestingly, *Salmonella* have been reported to exhibit low-level expression of CspA at 37°C (Jeffreys *et al.*, 1998). This may suggest that mRNA transcripts of *cspA* exhibit increased stability at lower temperatures – a mechanism thought to regulate expression of the CspA homologue in *E. coli* (Goldenberg *et al.*, 1996) – thus implicating post-translational regulation as having a major role in controlling production of these proteins following a temperature down-shift (Graumann and Marahiel, 1996). Additionally, induction of the cold shock response has been observed in *E. coli* under conditions where ribosomal function has been impaired, such as cold-sensitive mutations in ribosomal factors or the presence of the antibiotic chloramphenicol (Jiang *et al.*, 1993; Jones and Inouye, 1996), lending further support for ribosomal involvement in regulating induction of the cold shock response.

Six homologous cold-shock proteins, CspA–E and CspH, have been identified in *Salmonella* and although the precise roles of these proteins remain unclear, they are now often classified as RNA chaperones with DNA-binding ability (Morgan *et al.*, 2009). Preferential expression of CSPs at low temperature implies that CSP mRNA transcripts are capable of forming a stable translation initiation complex with the ribosome at low temperatures in the absence of other stabilizing factors (Thieringer *et al.*, 1998). Once translated, CSPs can go on to behave as moderators of a number of cellular functions, including transcriptional activation and enhancement (La Teana *et al.*, 1991; Jones *et al.*, 1992; Jones and Inouye, 1996) and stabilization of the ribosomal translation initiation complex by prevention of mRNA secondary structure formation (Jiang *et al.*, 1997; Morgan *et al.*, 2009). Further, it is probable that some cold shock proteins are also capable of behaving as chaperones that enhance stability of the RNA polymerase/DNA open complex during transcription (Graumann and Marahiel, 1996) and may also enhance DNA replication at low temperatures (Thieringer *et al.*, 1998).

Finally, exposure to low temperature can result in additional physiological changes in bacteria, such as increased membrane rigidity. In *Salmonella*, low-temperature adaptation is associated with desaturation of the membrane in order to increase fluidity (Alvarez-Ordonez *et al.*, 2008). Desaturation of the membrane is achieved through the incorporation of phospholipids with unsaturated hydrocarbon chains into the bilayer in exchange for those containing saturated fatty acids (Jeffreys *et al.*, 1998; Alvarez-Ordonez *et al.*, 2008, 2010). Phospholipids with unsaturated hydrocarbon chains have lower melting points, lending enhanced flexibility and fluidity to the bacterial membrane at low temperatures (Thieringer *et al.*, 1998) and enhancing survival in cold environments.

Extracytoplasmic stress responses

The Gram-negative bacterial outer membrane represents the primary interface for contact between the bacterium and its environment and thus must be capable of withstanding exposure to a number of stress conditions, including components of host innate immunity such as antimicrobial peptides and the action of bile. Accordingly, enteric bacteria employ a number of stress response systems collectively known as envelope or extracytoplasmic stress responses (ESRs) to sense stress and mediate resistance against conditions detrimental to the cell envelope and its components (Rowley *et al.*, 2006).

The alternative sigma factor, σ^E

σ^E, encoded by *rpoE,* is an alternative sigma factor similar to the general stress response and heat-shock sigma factors, σ^S and σ^H respectively, that drives the expression of a subset of genes with roles in responding to outer membrane damaging conditions (Rowley *et al.*, 2006). In *Salmonella* and other enteric bacteria, exposure to conditions that induce protein misfolding – such as carbon starvation, extremes of pH, hyperosmotic shock, high- or low-temperature shock, oxidative stress, outer membrane protein overexpression, antimicrobial peptides and infection of host cells – activate the σ^E-mediated ESR (Humphreys *et al.*, 1999; Rowley *et al.*, 2006; McMeechan *et al.*, 2007). Under conditions inappropriate for its activity, σ^E is regulated by sequestration to its membrane-bound anti-sigma factor, RseA (Missiakas *et al.*, 1997). Induction of the σ^E response occurs following the interaction of misfolded proteins in the periplasm with the PDZ domain of the inner membrane-bound DegS/HtrA protease, stimulating the sequential proteolysis of the anti-sigma factor RseA by HtrA and RseP/YaeL (Ades *et al.*, 1999; Rowley *et al.*, 2006). This action releases the RseA–σ^E complex into the cytoplasm, where its subsequent proteolysis by ClpXP protease liberates the σ^E subunit from its anti-sigma factor, enabling σ^E to bind to the core RNA polymerase subunit and drive σ^E-dependent gene expression (Flynn *et al.*, 2004). Some important σ^E-regulated factors include the periplasmic serine protease and chaperone protein HtrA, peptidyl-prolyl isomerases SurA and FkpA with roles in protein folding, and the chaperone Skp (Erickson and Gross, 1989; Strauch *et al.*, 1989; Missiakas *et al.*, 1996; Rowley *et al.*, 2011); all of which are located and active within the periplasmic space. Further, σ^E has also been shown to exhibit some functional overlap with the alternative sigma factors σ^S and σ^H, cooperatively inducing maximal expression of a range of genes involved in the response of *Salmonella* to a number of stress conditions (Bang *et al.*, 2005; Kenyon *et al.*, 2010). Importantly, σ^E and σ^E-regulated effectors have also been shown to be critical components for *Salmonella* virulence (Humphreys *et al.*, 1999), with Δ*rpoE*, Δ*htrA*, Δ*surA* and Δ*skp* mutants exhibiting significant attenuation in mice (Johnson *et al.*, 1991; Chatfield *et al.*, 1992; Humphreys *et al.*, 1999; Sydenham *et al.*, 2000; Rowley *et al.*, 2011). This is probably due to the reduced ability of strains deficient in σ^E to withstand exposure to reactive oxygen species and other antimicrobial host factors (Humphreys *et al.*, 1999; Kenyon *et al.*, 2002; Testerman *et al.*, 2002).

The CpxAR two-component regulatory system

The two-component signal transduction system CpxAR represents another important ESR for a number of Gram-negative bacteria (Rowley *et al.*, 2006) and has been shown to be primarily involved in regulating the expression of virulence-associated macromolecular surface structures, such as pili and T3SSs (Hung *et al.*, 2001; MacRitchie *et al.*, 2008; Vogt *et al.*, 2010). In the CpxAR response to envelope stress, phosphorylation of the CpxA sensor kinase activates the CpxR response regulator, enabling its DNA-binding activity and driving expression or repression of genes within the Cpx regulon (De Wulf *et al.*, 2002). Interestingly, there is a degree of overlap among the σ^E and Cpx regulons, with *htrA* expression regulated by both σ^E and CpxAR (Danese and Silhavy, 1997). Finally, up-regulated expression of CpxP by activated CpxR serves to negatively regulate the CpxAR system by preventing autophosphorylation of CpxA through CpxP binding (DiGiuseppe and Silhavy, 2003).

Similarly to that observed with σ^E, the CpxAR system has been shown to contribute to *Salmonella* virulence. The presence of the periplasmic sensor kinase, CpxA, is critical for *Salmonella* invasion via the T3SS encoded on SPI-1 (Nakayama *et al.*, 2003). Interestingly, while expression of the SPI-1 master regulator, *hilA*, is unchanged from parent strain levels at mildly acidic pH in Δ*cpxR* mutants, its expression is significantly reduced under identical conditions in Δ*cpxA* mutants, suggesting a specific role for *cpxA* in stimulating *hilA* expression at pH 6 (Nakayama *et al.*, 2003). Further, the CpxA sensor kinase and CpxR response regulator components of the CpxAR system appear to be capable of operating independently, as demonstrated by a considerable loss of virulence in *Salmonella* Δ*cpxA* mutants compared with retention of wild-type virulence levels in Δ*cpxR* mutants in the murine infection model (Humphreys *et al.*, 2004).

The phage shock response

Since its identification in *E. coli* in the early 1990s (Brissette *et al.*, 1991), the phage shock protein response (PSP) *psp* operon has been identified in a number of Gram-negative bacteria, including *Salmonella* and *Yersinia* species (Darwin, 2005). In comparison with the other systems, this ESR remains relatively mysterious and its role in *Salmonella* physiology and pathogenesis remains poorly characterized. The PSP ESR is induced by a number of stimuli – including filamentous phage infection, osmotic stress, extremes of temperature and mislocalization of secretin proteins – all of which may result in dissipation of the proton motive force (Brissette *et al.*, 1991; Darwin, 2005; Rowley *et al.*, 2006). Indeed, the product of *pspA* is thought to assist in maintaining the proton motive force through association with the cytoplasmic membrane upon induction of the PSP response (Yamaguchi *et al.*, 2010). While relatively little is known about this ESR in *Salmonella*, a putative role has been proposed for the PSP response in *Salmonella* virulence owing to the high levels of *psp* gene expression during intracellular growth within macrophages and epithelial cells (Eriksson *et al.*, 2003; Hautefort *et al.*, 2008).

The Rcs phosphorelay system

The Rcs phosphorelay system is an atypical two-component signal transduction system encoded by a number of enterobacterial pathogens, including *Salmonella* (Huang *et al.*, 2006), which responds to outer membrane perturbations. Conditions reported to induce the Rcs system include osmotic shock (Sledjeski and Gottesman, 1996), desiccation (Ophir and Gutnick, 1994), growth in the presence of glucose or zinc at 20°C (Hagiwara *et al.*, 2003), overproduction of the membrane-associated protein DjlA (Clarke *et al.*, 1997), growth on solid surfaces (Ferrieres and Clarke, 2003) and exposure to antimicrobial peptides such as polymyxin B (Bader *et al.*, 2003; Farris *et al.*, 2010). Outer membrane disruption induces phosphorylation of the inner membrane-bound sensor kinase, RcsC, stimulating sequential activation of the histidine phosphotransfer intermediate protein, RcsD, and the response regulator, RcsB, via a histidine-to-aspartate phosphorylation cascade (Majdalani *et al.*, 2005). Certain conditions, such as exposure to cationic antimicrobial peptides, stimulate RcsCDB activation through the action of the outer membrane lipoprotein, RcsF (Farris *et al.*, 2010). In addition, transcriptional activation of certain genes by RcsB also requires the co-activating activity of the auxiliary protein, RcsA (Stout *et al.*, 1991). In *Salmonella*, the Rcs phosphorelay system contributes to virulence in the murine model of typhoid fever at the late stage of infection, with Δ*rcsC* mutants exhibiting attenuation compared with the wild type (Detweiler *et al.*, 2003). In this study, Δ*rcsC* mutants showed a marked decrease in their ability to persist in murine tissues at 11 days post-infection, while the wild type had established lethal infection within the same interval (Detweiler *et al.*, 2003). This phenotype may be due to the loss of Rcs-mediated activation of *ydeI* in these mutants, therefore increasing their sensitivity to the antibacterial activity of CAMPs during systemic infection of the host (Erickson and Detweiler, 2006).

Additional ESRs in *Salmonella*

Finally, two additional ESR systems have also been described; however, their precise roles in responding to outer membrane stress remain poorly characterized in *Salmonella*. The first of these systems is a two-component signal transduction system, BaeSR, which was first described in *E. coli* (Raffa and Raivio, 2002). Phosphorylation of the inner membrane-bound sensor histidine kinase, BaeS, activates the response regulator, BaeR, enabling its DNA-binding activity and driving expression of members of the BaeSR regulon. This includes the *mdtABCD* and *acrD* genes, which encode two multidrug efflux pumps, conferring enhanced resistance to antibiotics including β-lactams, novobicin and deoxycholate (Baranova and Nikaido, 2002; Nishino *et al.*, 2007). In addition, MdtABC and AcrD have also been shown to have roles in mediating resistance to metals, such as copper and zinc, in *Salmonella* (Nishino *et al.*, 2007). Notably, while this system has been shown to be important for antibiotic and metal stress resistance, the role of this system in *Salmonella* virulence remains to be fully characterized.

The second putative ESR concerns the release of outer membrane vesicles (OMV) into the environment following exposure of Gram-negative bacteria to stress. This putative ESR is proposed to function by preventing accumulation of misfolded proteins in the periplasm through selective release of OMVs containing unwanted material into the environment, the rate of which is regulated by the level of protein accumulation in the periplasm (McBroom and Kuehn, 2007). Importantly, although this response has been observed in *Salmonella* (McBroom and Kuehn, 2007), relatively little is truly known about this ESR and thus fuller characterization is required to establish the role of this response in *Salmonella* virulence.

The oxidative stress response

Salmonella may encounter oxidative stress conditions in both the external environment and within hosts. In food processing environments, chlorine is routinely used as an oxidative disinfecting agent for the sanitation of food surfaces and processing facilities, while in host cells *Salmonella* are exposed to an arsenal of antimicrobial oxygen radicals – including superoxide (O_2^-), hydrogen peroxide (H_2O_2) and other reactive oxygen and nitrogen species (ROS or RNS, respectively) – produced by phagocyte NADPH oxidase, inducible nitric oxide synthase (iNOS) and prostaglandin (PG)-endoperoxide synthase 2 (Rosenberger and Finlay,

2002; Withanage *et al.*, 2005; Shi *et al.*, 2009), which may cause oxidative damage to bacterial lipids, proteins, haem proteins, thiols and DNA (Vazquez-Torres and Fang, 2001). Oxidative stress-mediated bacterial killing is an important facet of host innate immunity, as exemplified by the increased susceptibility of both human and murine hosts with defects in superoxide production – such as in chronic granulomatous disease (CGD) – to bacterial and fungal infections (Mouy *et al.*, 1989; Mastroeni *et al.*, 2000).

Owing to its replicative niche within macrophages, *Salmonella* has evolved a range of mechanisms to facilitate circumvention of oxidative stress-mediated killing. The presence of superoxide and H_2O_2 within the cell serves as an inducing stimulus for several transcriptional regulators, such as SoxRS and OxyR, driving the expression of a number of genes encoding detoxifying enzymes, including *micF, sodA, katG* and *ahpCF* (Pomposiello and Demple, 2000; Zheng *et al.*, 2001). Following penetration of the *Salmonella* outer membrane, superoxide anions are dismutated to molecular oxygen and H_2O_2 through the action of two periplasmic copper, zinc-cofactored superoxide dismutases (SODs) (SodCI and SodCII) and two cytosolic manganese- and iron-cofactored SODs (SodA and SodB, respectively) (Vazquez-Torres and Fang, 2001). While loss of SodA fails to substantially alter *Salmonella* virulence in the murine model (Tsolis *et al.*, 1995), loss of either or both SodCI and SodCII results in significant attenuation of virulence in C3H/HeN mice infected via the intraperitoneal route (Fang *et al.*, 1999), implicating an important role for superoxide detoxification by the Cu, Zn-cofactored SOD enzymes during systemic infection. In addition, *Salmonella* also harbour three catalase enzymes, encoded by *katE, katG* and *katN* (Loewen, 1984; Loewen and Stauffer, 1990; Robbe-Saule *et al.*, 2001), with roles in detoxifying H_2O_2 to water and molecular oxygen, and two alkyl hydroperoxide reductases, AhpC and TsaA, which convert H_2O_2 to water (Hebrard *et al.*, 2009). Interestingly, these enzymes demonstrate relatively high levels of functional redundancy in *Salmonella*, as neither Δ*katE*Δ*katG*Δ*katN* nor Δ*ahpCF*Δ*tsaA* mutants exhibit defects in degrading exogenous H_2O_2 (Hebrard *et al.*, 2009). Conversely, Δ*katE*Δ*katG*Δ*katN*Δ*ahpCF*Δ*tsaA* mutants display significantly enhanced sensitivity to exogenous H_2O_2, are defective for replication in macrophages and exhibit significantly reduced virulence in mice, suggesting that while a considerable degree of functional overlap is afforded by these systems, at least one functional mechanism for detoxification of H_2O_2 is required for *Salmonella* virulence (Hebrard *et al.*, 2009).

Reactive oxygen species may cause considerable damage to DNA (Imlay and Linn, 1988); thus *Salmonella* have evolved a number of mechanisms for mediating repair in response to the oxidative burst of host phagocytes. In enteric bacteria such as *Salmonella*, DNA repair is facilitated by the products of the genes *recA* and *recBCD,* which behave as recombinase enzymes, mediating recombination between homologous DNA molecules and allowing for repair (reviewed by Kowalczykowski, 2000). In addition, RecA also plays a critical role in the bacterial SOS response by behaving as a co-protease for the autocatalytic degradation of the LexA repressor, stimulating downstream expression of a network of LexA-repressed genes involved in mounting a response to DNA damaging conditions (Little, 1984). The RecA and RecBCD recombinational repair pathways are critical for *Salmonella* virulence, as mutations in *recA* and *recBC* result in significant attenuation in mice and an inability to proliferate in macrophages competent for superoxide production (Buchmeier *et al.*, 1993; Cano *et al.*, 2002). Indeed, these mutations are significantly more detrimental to *Salmonella* virulence than loss of the HPI or HPII catalase enzymes encoded by *katE* and *katG,* suggesting that the ability to repair DNA damage is more critical to *Salmonella* virulence than the ability to produce certain detoxifying agents (Buchmeier *et al.*, 1995).

Finally, *Salmonella* utilize the T3SS encoded on *Salmonella* pathogenicity island (SPI)-2 to manipulate host cell trafficking pathways following infection in order to circumvent exposure to ROS or RNS. The SPI-2 effector protein, SpiC, is important for virulence in systemic *Salmonella* infection and has been shown to prevent fusion of *Salmonella* containing vacuoles (SCVs) with lysosomes and endosomes in macrophages (Uchiya *et al.*, 1999), thus avoiding exposure to oxidative species. Further, SPI-2 has been shown to prevent assembly of the NADPH oxidase complex in macrophages by actively excluding or removing the membrane component of NADPH oxidase, cytochrome b_{558}, thus preventing superoxide production and enhancing intra-macrophage proliferation and survival (Gallois *et al.*, 2001). Similarly, SPI-2 has also been shown to uncouple co-localization of iNOS with SCVs, preventing peroxynitrite delivery into SCVs and enabling avoidance of damaging RNS during intracellular growth within macrophages (Chakravortty *et al.*, 2002). Together, these features of SPI-2 and general *Salmonella* physiology are critical for facilitating efficient intracellular survival and proliferation during the systemic phase of infection.

SOS stress response to DNA damage

Initially described in *E. coli*, the SOS stress response has subsequently been identified in a number of Gram-positive and Gram-negative organisms, including *Salmonella* (Erill *et al.*, 2007). Directly regulated by the products of *lexA* and *recA*, the SOS response encompasses a network of at least 30 genes that together form a coordinated response against a range of stress conditions that may induce DNA damage (Quillardet *et al.*, 2003), including oxidative stress, UV radiation, β-lactam antibiotics and exposure to high or low pH stress (Quillardet *et al.*, 2003; Erill *et al.*, 2007). In addition, an essential role for the SOS response in virulence has been implicated for uropathogenic *E. coli* (UPEC) in the murine cystitis model (Li *et al.*, 2010), while in *Salmonella* the SOS-regulated gene *sulA* has been shown to be up-regulated during intracellular growth within macrophages (Eriksson *et al.*, 2003), suggesting a putative role in virulence.

The SOS stress response is negatively regulated by the binding of LexA protein dimers at SOS boxes – 16–20 bp palindromic sequences – present in the promoter sequences of SOS response-regulated genes (Brent and Ptashne, 1981; Erill *et al.*, 2007). SOS boxes are typically located within or in close proximity to the RNA polymerase binding site and thus the binding of LexA physically prevents their transcription (Brent and Ptashne, 1981; Quinones *et al.*, 2006; Erill *et al.*, 2007). Exposure to conditions causing DNA damage often results in stalling of the replication fork, which generates single-stranded (ssDNA) DNA fragments. Unspecific binding of RecA to ssDNA fragments in the presence of ATP triggers polymerization of RecA into a helical filament-like nucleoprotein complex (Roca and Singleton, 2003). Activated RecA-ssDNA triggers the self-catalytic properties of LexA, stimulating cleavage of LexA and providing relief from LexA-mediated transcriptional repression of SOS genes (Little, 1984). Some of the products of genes induced during the SOS response have roles in stabilization of the stalled replication fork, while others encode proteins that are involved in actively repairing DNA damage or facilitating translesion DNA synthesis around the damaged region (Erill *et al.*, 2007). In addition, the product of the SOS-induced gene *sulA* is of significant importance following DNA damage. Through sequestration of the FtsZ cell division protein, SulA inhibits septum formation and subsequent cell division (Trusca *et al.*, 1998; Justice *et al.*, 2000; Cordell *et al.*, 2003), giving the cell time to address and repair DNA damage, and minimizing the risk of passing mutations on to daughter cells. Notably, bacterial growth can continue in the absence of septation and the formation of elongated filamentous bacterial cells has been described as a characteristic phenotype associated with induction of the SOS response (Justice *et al.*, 2008). Finally, following repair of damaged DNA, LexA-mediated SOS gene repression is restored and cell division resumes following SulA degradation by the cytoplasmic Lon protease (Ishii *et al.*, 2000; Justice *et al.*, 2008).

Morphological changes in stressed *Salmonella*

The bacterial Z-ring acts as a scaffolding structure for the recruitment of proteins involved in mediating cytokinesis to the midcell (Pichoff and Lutkenhaus, 2002; Dajkovic *et al.*, 2010). The MinCDE system controls the positioning of Z-ring formation at the bacterial midcell by specifically inactivating potential division sites at the cell poles (de Boer *et al.*, 1989, 1992). Polymerization of the bacterial tubulin homologue, FtsZ, into a ring structure around the circumference of the cell induces formation of the Z-ring at the putative division site (Bi and Lutkenhaus, 1991). Recruitment of membrane-associated proteins ZipA and FtsA to the immature Z-ring complex through binding of FtsZ enables the ring structure to be tethered in close proximity to the cytoplasmic membrane (Liu *et al.*, 1999; Pichoff and Lutkenhaus, 2005). Following assembly of the divisome by recruitment of proteins involved in subsequent stages of cell division to the complex, including FtsI, FtsK, FtsL, FtsN, FtsQ and FtsW (Rothfield *et al.*, 1999), constriction of the Z-ring in rod-shaped bacteria generates two daughter cells, each half the length of the original parent cell.

In *Salmonella,* normal cell division produces rod-shaped daughter cells of around 1–2 μm in length. However, under conditions suboptimal for growth, such as intracellular growth within macrophages or in response to conditions encountered in food matrices, *Salmonella* have been reported to adopt a filamentous morphology (Phillips *et al.*, 1998; Mattick *et al.*, 2000a; Rosenberger and Finlay, 2002; Mattick *et al.*, 2003a,b; Hazeleger *et al.*, 2006; Humphrey *et al.*, 2011a). Filaments are formed when bacteria continue to grow (increasing biomass) in the absence of septation, producing a single elongated cell with multiple evenly spaced chromosomal copies distributed along the length of the filament, which may be up to 200 μm long (Mattick *et al.*, 2000a, 2003a; Hazeleger *et al.*, 2006; Justice *et al.*, 2008). *S.* Enteritidis and *S.* Typhimurium have been shown to produce filaments in response to hyperosmotic conditions representative of those encountered in low a_w foods (Mattick *et al.*, 2000a; Humphrey *et al.*, 2011a). Further, Mattick *et al.* (2000a) demonstrated that the filamentous phenotype appeared to confer enhanced resistance to reduced a_w (0.98–0.93 rvp) stress over that of non-filamentatous *Salmonella*. In a culture exposed to hyperosmotic stress, some filamentous *Salmonella* were observed to be viable by fluorescent labelling, while no viable non-filamentous *Salmonella* were detected. The filamentation response to suboptimal growth conditions is not exclusive to *Salmonella,* with many other bacteria, including *E. coli, Mycobacterium tuberculosis, Listeria monocytogenes* and *Legionella pneumophila,* also exhibiting filament formation in a range of environments, as described in Table 6.2. A critical role for filamentation in pathogenesis has been described for uropathogenic *E. coli* (UPEC), where filamentation serves to prevent phagocytosis and destruction of the bacteria by host PMNs (Justice *et al.*, 2004). Similarly, a putative role for

Table 6.2 Filamentous growth is observed in a number of bacterial species (adapted from Justice *et al.*, 2008)

Organism	Disease condition	Filament-inducing condition(s)	Source
Salmonella enterica	Enteric fever; salmonellosis	Intracellular growth within macrophages; low a_w; low temperature; low pH; *his* mutation; SulA overexpression; oxidative stress	Gibert and Casadesus (1990), Phillips *et al*. (1998), Mattick *et al*. (2000a, 2003a,b), Rosenberger and Finlay (2002), Mattick *et al*. (2003a,b), Henry *et al*. (2005), Hazeleger *et al*. (2006), Hebrard *et al*. (2009), Humphrey *et al*. (2011a,b)
Uropathogenic *Escherichia coli*	Urinary tract infections	Murine bladder; human urine	Justice *et al*. (2006)
Listeria monocytogenes	Listeriosis	Alkaline stress; low a_w	Hazeleger *et al*. (2006), Giotis *et al*. (2007)
Legionella pneumophila	Legionnaires' disease	*In vitro* biofilms; intracellular growth within Vero cells	Ogawa *et al*. (2001), Piao *et al*. (2006)
Proteus mirabilis	Urinary tract infections	Murine bladder	Allison *et al*. (1992)
Mycobacterium tuberculosis	Tuberculosis	Intracellular growth within macrophages	Chauhan *et al*. (2006)

filamentation in virulence has also been described for the respiratory pathogen *L. pneumophila*, since biofilms formed in the lungs at physiological temperature consist of filamentous cells, while those formed at lower temperatures comprise only non-filamentous bacteria (Piao *et al.*, 2006). Enhanced survival through filamentation may reflect the up-regulation of stress response systems associated with filamentation, such as the alternative sigma factor σ^S and the SOS stress response. Humphrey *et al.* (1998) demonstrated that an *S.* Enteritidis strain shown to be defective in filament formation at low temperature (Phillips *et al.*, 1998) was in fact an *rpoS* mutant, suggesting a putative role for σ^S in filament formation at low temperature. Notably, however, other studies have indicated that filamentation of *Salmonella* in response to low a_w stress is independent of *rpoS* expression (Mattick *et al.*, 2000a), leaving the role of σ^S in mediating filamentation unclear. Similarly, SOS stress response genes (*sulA*, *recA* and *lexA*) have been shown to be up-regulated in *Salmonella* exposed to nitric oxide radicals during intracellular growth within macrophages *in vitro* (Eriksson *et al.*, 2003) and in uropathogenic *E. coli* during *in vivo* murine bladder infection (Justice *et al.*, 2004, 2006), which are conditions reported to induce filamentous growth (Rosenberger and Finlay, 2002; Justice *et al.*, 2006).

Although the precise mechanism that mediates filamentation remains unknown, the cell division inhibitor protein SulA, an integral mediator of the SOS stress response, has been hypothesized to play a central role in triggering *in vivo* filamentation during UPEC infection (Justice *et al.*, 2006). SulA inhibits FtsZ polymer formation by sequestration of FtsZ monomers through direct binding at the T7 loop surface (Cordell *et al.*, 2003; Dajkovic *et al.*, 2008), thereby destabilizing and promoting the disassembly of existing Z-rings and preventing the formation of new Z-rings (Dajkovic *et al.*, 2008). Indeed, overproduction of SulA in *S.* Typhimurium induces filament formation (Humphrey *et al.*, 2011b) (Fig. 6.2). However, SulA-independent mechanisms of filamentation have also been described in *Salmonella* that are associated with disruptions in the histidine biosynthetic pathway, where auxotrophic *hisG* mutants display filamentous growth within human melanoma cells (Henry *et al.*, 2005) and overexpression of *his* in regulatory mutants induces filamentous growth in a range of media (Gibert and Casadesus, 1990).

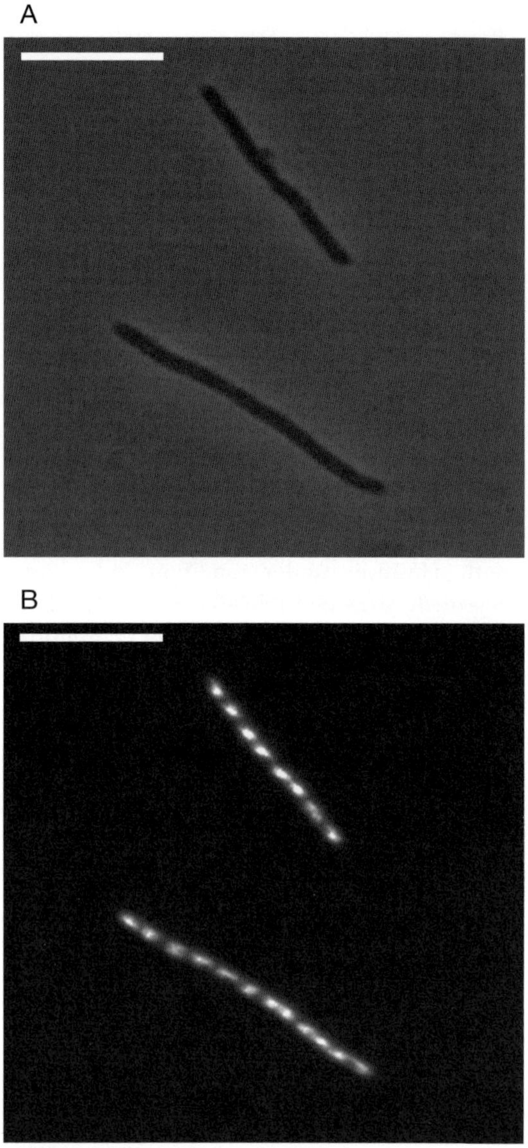

Figure 6.2 Overproduction of SulA in *S.* Typhimurium produces filaments with multiple chromosomal copies distributed along their lengths. (A) Expression of SulA from an arabinose-inducible plasmid vector (pSH02) produced filamentous *Salmonella* cells (phase contrast image). (B) DAPI staining revealed the presence of evenly spaced double-stranded DNA-rich areas distributed along the length of each filamentous cell. Scale bars represent 10 µM.

The ability of food-borne pathogens to survive stresses present in foodstuffs has implications for food production and storage, while the presence of filamentous bacteria in foodstuffs presents an important surreptitious threat to public health. Contamination of foodstuffs with filamentous *Salmonella* has extremely important implications for food monitoring as the traditional colony-forming unit (CFU) viable plate count assay cannot distinguish between filamentous and bacillary bacteria, potentially leading to significant underestimation of the extent of microbial contamination in foodstuffs (Mattick *et al.*, 2000a). Indeed, increasing biomass in the absence of increasing viable counts has been reported for *Salmonella* cultured in marginal conditions, such as low a_w (Mattick *et al.*, 2003b), suggesting that these organisms are capable of growth in the absence of septation under conditions widely considered to be bacteriostatic and highlighting the inefficiency of traditional plating techniques for detection of filamentous cells. Further, stress-induced *Salmonella* filaments have been shown to undergo rapid septation following reversion to conditions favourable for bacterial growth, with each filament generating many daughter cells (Phillips *et al.*, 1998; Mattick *et al.*, 2000a; Humphrey *et al.*, 2011a). This scenario has important epidemiological implications for food-borne salmonellosis, as division of filamentous cells following consumption of contaminated foodstuffs may enable the infectious dose to be exceeded, culminating in disease.

Recovery of *Salmonella* from stress: concerns for food safety

In order to successfully establish infection, *Salmonella* must tolerate exposure to multiple simultaneous and/or repeated stress conditions – including extremes of pH, temperature, oxygen concentration, pressure, osmotic stress and exposure to a range of antimicrobial, preservative and sanitizing agents – whilst also retaining the ability to express the SPI-1 effector proteins and T3SS when conditions are appropriate for invasion of host cells. The extensive stress response repertoire of *Salmonella* is an issue of critical importance to the food industry, as food preservation techniques rely on producing environments deleterious for bacterial growth and survival in order to minimize spoilage and prevent disease transmission. Many studies have examined the effects of *Salmonella* exposure to severe types of stress that may be encountered in foodstuffs or during processing, including high (>2.0 M) NaCl concentration, high temperature (54–60°C) and low pH (pH ≤ 3.0) (Lin *et al.*, 1995; Mattick *et al.*, 2000a,b; Fratamico, 2003). However, owing to the necessity for balance between preservative effects and food quality – particularly with increasing consumer demand for fresher, less processed foods – bacteria in foodstuffs may also be exposed to more moderate stresses that can cause sublethal damage to contaminating bacterial cells. Sublethally injured bacteria in food matrices are of critical importance to public health, as a number of *S.* Typhimurium and *S.* Enteritidis strains have demonstrated the ability to rapidly recover from osmotic and/or low pH stress-induced membrane damage, with active repair of membrane integrity, following a return to conditions optimal for growth (Humphrey *et al.*, 2011a; L. Clark, T. Humphrey and M. Jepson, unpublished).

The SPI-1 T3SS facilitates an important part of *Salmonella* disease establishment: invasion of host enterocytes (McGhie *et al.*, 2009). Expression of the SPI-1 virulence genes is tightly regulated in response to a variety of environmental factors, such as growth phase, osmolarity, pH, and oxygen concentration (Ernst *et al.*, 1990; Galan and Curtiss, 1990; Lee and Falkow, 1990; Bajaj *et al.*, 1996). Many conditions typical of food preservation, such as elevated (>300 mM) NaCl, acidic pH and low temperature, have been shown to abrogate expression of the SPI-1 virulence genes in *Salmonella* (Behlau and Miller, 1993; Bajaj *et al.*, 1996; Leclerc *et al.*, 1998), suggesting that *Salmonella* are unlikely to express these virulence factors in foodstuffs. Importantly, however, recovery of SPI-1 gene expression and associated adhesion to and invasion of epithelial cells has been observed for a number of *S.* Typhimurium and *S.* Enteritidis strains following removal from conditions including 685 mM NaCl, pH 3.8 and 6°C (Humphrey *et al.*, 2011a; L. Clark, T. Humphrey and M. Jepson, unpublished). In these studies, *Salmonella* exposed to food-related stress conditions for 8 days were subcultured into standard formulation Luria-Bertani (LB) medium and single cell analysis revealed recovery of *prgH* expression to normal maximal levels within 4–6 hours (Humphrey *et al.*, 2011a). In humans, gastric emptying of at least 50% of an ingested meal occurs 2–4 hours after consumption (Camilleri *et al.*, 1989) and the composition of LB medium reflects the nutrient-rich environment of the intestinal lumen (Bustamante *et al.*, 2008). Thus, it is possible that the timescale of SPI-1 virulence gene expression and invasiveness recovery observed following a return to nutrient-rich conditions might correlate with the time required for digestion of contaminated foodstuffs and transit to the appropriate anatomical location for SPI-1-mediated infection in the gut. In addition, translocation of *Salmonella* across model M-cells is known to occur independently of SPI-1 and SPI-2 (Martinez-Argudo and Jepson, 2008) suggesting that alternative mechanisms of *Salmonella* uptake are also probably important for establishing infection. Together, these examples serve to highlight the plasticity of *Salmonella* gene expression in response to different environments and indicate that current food preservation methods may not be fully sufficient to prevent food-borne *Salmonella* transmission.

Conclusions and perspectives

Salmonella has evolved an extensive repertoire of stress response mechanisms, several of which exhibit considerable overlap, in response to a variety of insults that the organism may encounter during infection of the host or in the external non-host environment.

Of all the stress conditions that *Salmonella* may encounter during its life cycle, low pH is among the most frequent. Exposure to acidic pH may occur in the external environment or following contamination of foodstuffs, such as fruit juices and mayonnaise, and is one of the first stress conditions to be encountered in the stomach of mammalian hosts or in the crop of birds following infection. *Salmonella* present in the environment is widely accepted to exist in a non-replicating state similar to that of stationary phase, and given the enhanced σ^S-mediated acid resistance of stationary phase *Salmonella* (Lee *et al.*, 1994), this feature may serve to improve their survival following exposure to the acid conditions of the host stomach or crop. In addition, the presence of short-chain fatty acids (SCFA) in the host crop or small intestine may induce the stationary phase acid tolerance response of infecting *Salmonella*, further enhancing their natural resistance to acid stress and inducing cross-protection against a variety of additional conditions, such as organic and inorganic acids (Baik *et al.*, 1996; Greenacre *et al.*, 2003), NaCl and KCl salts (Lee *et al.*, 1995; Greenacre and Brocklehurst, 2006), hydrogen peroxide (Lee *et al.*, 1995) and extremes of temperature (Lee *et al.*, 1995; Xu *et al.*, 2008).

The *Salmonella* ATRs induce expression of at least 51 acid-shock proteins, one of which is the general stress response mediator, σ^S (Lee *et al.*, 1995). Enhanced thermotolerance, salt and H_2O_2 resistance following acid stress requires the presence of σ^S (Lee *et al.*, 1995), implicating a critical role for this sigma factor in *Salmonella* resistance to a variety of stress conditions. Indeed, regulation of a variety of stress response genes derives from the activity of σ^S, including the HPII catalase encoded by *katE* and the product of *otsA* which is involved in trehalose synthesis during osmotic stress (Chen *et al.*, 1996; Fang *et al.*, 1996). The alternative sigma factors σ^S, σ^E and σ^H also display cooperative interactions in response to stress exposure, with σ^E capable of regulating expression of genes encoded within the σ^E-, σ^S- and σ^H-regulons (Bang *et al.*, 2005). Additional overlapping roles have also been described for regulation of these alternative sigma factors, such as σ^E-mediated activation of *rpoH* expression (Erickson and Gross, 1989; Wang and Kaguni, 1989) and stimulation of *hfq* expression, which is required for efficient translation of *rpoS* mRNA transcripts (Brown and Elliott, 1996), by σ^H (Tsui *et al.*, 1996). Interestingly, homologues of the genes encoding σ^S, σ^H, and σ^E (*rpoS, rpoH* and *rpoE*, respectively) are present in a number of Gram-negative bacterial species (Erickson and Gross, 1989; Nakahigashi *et al.*, 1995; Humphreys *et al.*, 1999; Craig *et al.*, 2002; Kovacikova and Skorupski, 2002; Chiang and Schellhorn, 2010; Dong and Schellhorn, 2010), implying that evolution of σ^S-, σ^H- and σ^E-mediated stress responses in a number of organisms may have descended from a common ancestor.

Salmonella also possess a range of other physiological and genetic mechanisms for abrogating damage associated with stress exposure. These include inducible amino acid carboxylases, uptake and synthesis of compatible solutes, RecA-, SulA- and LexA-mediated DNA repair systems and interfering with host cell trafficking pathways in order to circumvent exposure to damaging oxygen and nitrogen species (Gutierrez *et al.*, 1995; Park *et al.*, 1996; Uchiya *et al.*, 1999; Gallois *et al.*, 2001; Chakravortty *et al.*, 2002). Together with σ^S-, σ^H- and σ^E-regulated responses, these approaches provide *Salmonella* with an arsenal of mechanisms for maintaining cellular homeostasis and repairing damage following stress exposure. The resilience of *Salmonella* cells, coupled with their ability to alter their cellular shape, in the presence of potentially lethal stress conditions carries important epidemiological and public health concerns for the food production industry. Further, the ability of *Salmonella* to quickly repair membrane damage and resume invasiveness in epithelial cells following transfer from sublethal stress conditions typical of food matrices to nutrient-rich conditions suggests that traditional approaches to food preservation may not be fully sufficient to prevent salmonellosis (Humphrey *et al.*, 2011a).

Most work examining the cellular and genetic responses of *Salmonella* has involved the use of a limited number of strains, often *S.* Typhimurium strains SL1344 and 14028, which is subsequently extrapolated to other *Salmonella* strains as the accepted dogma. Importantly, these commonly studied strains are not typically associated with outbreaks of disease, and relatively little attention has been paid to stress responses in alternative serovars that are common causes of food-borne infections. Recent work has highlighted extensive heterogeneity in expression of the SPI-1 virulence gene *prgH*, generating heterogeneity in the relative levels of invasion in epithelial cells between different *S.* Typhimurium strains (Clark *et al.*, 2011). *Salmonella* do not display a uniform ability to tolerate, adapt to and recover from stress exposure; exemplified in a recent study in which two *S.* Typhimurium Definitive phage type 104 (DT104) strains demonstrated an enhanced ability to recover growth, SPI-1 virulence gene expression and invasion of epithelial cells compared to that of non-DT104 *S.* Typhimurium strains following removal from prolonged exposure to combined 685 mM NaCl and pH 3.8 stress at 6°C (Humphrey *et al.*, 2011a). It is possible that if virulence genes are subject to heterogeneous expression in different *Salmonella* strains, perhaps stress response genes are also differentially expressed. This scenario could go some way to explaining the success of some outbreak strains over others and highlights the importance of continued research to establish the reasons why some strains tolerate environmental and host stress more effectively than others.

While much has been elucidated about the stress response repertoire of *Salmonella* over the past few decades, several aspects of this organism's response to stress remain poorly characterized. Some insights into the interplay between the different stress responses, coupled with the role of stress response factors in regulating SPI-1 and SPI-2 virulence gene expression, have been gained through recent studies (Bang *et al.*, 2005; Matsui *et al.*, 2008). However, given the range of stress conditions to which *Salmonella* may be exposed during its life cycle, future work is required to enable fuller characterization of the interaction between stress response and virulence gene expression in order to establish how this pathogen coordinates its response to potentially damaging conditions in its environment while also expressing factors necessary for establishing infection. Ultimately, the extensive stress response of *Salmonella* has enabled this organism to effectively adapt to a wide variety of niches within host and non-host environments, thus preventing its eradication from the food chain and sustaining the threat of food-borne salmonellosis to public health.

Web resources

Web page 1: www.hpa.org.uk/NewsCentre/NationalPressReleases/2010PressReleases/100914cooperationcurtailedsalmonellaoutbreak

Web page 2: www.cdc.gov/mmwr/preview/mmwrhtml/mm58e0129a1.htm

Web page 3: www.cdc.gov/ncidod/dbmd/diseaseinfo/salmonellosis_2006/110306_outbreak_notice.htm

References

Ades, S.E., Connolly, L.E., Alba, B.M., and Gross, C.A. (1999). The *Escherichia coli* sigma(E)-dependent extracytoplasmic stress response is controlled by the regulated proteolysis of an anti-sigma factor. Genes Dev. *13*, 2449–2461.

Allison, C., Coleman, N., Jones, P.L., and Hughes, C. (1992). Ability of *Proteus mirabilis* to invade human urothelial cells is coupled to motility and swarming differentiation. Infect. Immun. *60*, 4740–4746.

Alvarez-Ordonez, A., Fernandez, A., Lopez, M., Arenas, R., and Bernardo, A. (2008). Modifications in membrane fatty acid composition of *Salmonella typhimurium* in response to growth conditions and their effect on heat resistance. Int. J. Food Microbiol. *123*, 212–219.

Alvarez-Ordonez, A., Halisch, J., and Prieto, M. (2010). Changes in Fourier transform infrared spectra of *Salmonella enterica* serovars Typhimurium and Enteritidis after adaptation to stressful growth conditions. Int. J. Food Microbiol. *142*, 97–105.

Anon (2010). The community summary report on trends and sources of zoonoses, zoonotic agents and food-borne outbreaks in the European Union in 2008. EFSA J. *8*, 1496.

Bader, M.W., Navarre, W.W., Shiau, W., Nikaido, H., Frye, J.G., McClelland, M., Fang, F.C., and Miller, S.I. (2003). Regulation of *Salmonella typhimurium* virulence gene expression by cationic antimicrobial peptides. Mol. Microbiol. *50*, 219–230.

Baik, H.S., Bearson, S., Dunbar, S., and Foster, J.W. (1996). The acid tolerance response of *Salmonella typhimurium* provides protection against organic acids. Microbiology *142*, 3195–3200.

Bajaj, V., Lucas, R.L., Hwang, C., and Lee, C.A. (1996). Co-ordinate regulation of *Salmonella typhimurium* invasion genes by environmental and regulatory factors is mediated by control of *hilA* expression. Mol. Microbiol. *22*, 703–714.

Balaji, B., O'Connor, K., Lucas, J.R., Anderson, J.M., and Csonka, L.N. (2005). Timing of induction of osmotically controlled genes in *Salmonella enterica* serovar Typhimurium, determined with quantitative real-time reverse transcription-PCR. Appl. Environ. Microbiol. *71*, 8273–8283.

Balandina, A., Claret, L., Hengge-Aronis, R., and Rouviere-Yaniv, J. (2001). The *Escherichia coli* histone-like protein HU regulates *rpoS* translation. Mol. Microbiol. *39*, 1069–1079.

Bang, I.S., Kim, B.H., Foster, J.W., and Park, Y.K. (2000). OmpR regulates the stationary-phase acid tolerance response of *Salmonella enterica* serovar Typhimurium. J. Bacteriol. *182*, 2245–2252.

Bang, I.S., Frye, J.G., McClelland, M., Velayudhan, J., and Fang, F.C. (2005). Alternative sigma factor interactions in *Salmonella*: sigma and sigma promote antioxidant defences by enhancing sigma levels. Mol. Microbiol. *56*, 811–823.

Baranova, N., and Nikaido, H. (2002). The BaeSR two-component regulatory system activates transcription of the *yegMNOB* (*mdtABCD*) transporter gene cluster in *Escherichia coli* and increases its resistance to novobiocin and deoxycholate. J. Bacteriol. *184*, 4168–4176.

Bearson, B.L., Wilson, L., and Foster, J.W. (1998). A low pH-inducible, PhoPQ-dependent acid tolerance response protects *Salmonella typhimurium* against inorganic acid stress. J. Bacteriol. *180*, 2409–2417.

Bearson, S., Bearson, B., and Foster, J.W. (1997). Acid stress responses in enterobacteria. FEMS Microbiol. Lett. *147*, 173–180.

Bearson, S.M.D., Bearson, B.L., and Rasmussen, M.A. (2006). Identification of *Salmonella enterica* serovar Typhimurium genes important for survival in the swine gastric environment. Appl. Environ. Microbiol. *72*, 2829–2836.

Behlau, I., and Miller, S.I. (1993). A PhoP-repressed gene promotes *Salmonella typhimurium* invasion of epithelial cells. J. Bacteriol. *175*, 4475–4484.

Beuzon, C.R., Meresse, S., Unsworth, K.E., Ruiz-Albert, J., Garvis, S., Waterman, S.R., Ryder, T.A., Boucrot, E., and Holden, D.W. (2000). *Salmonella* maintains the integrity of its intracellular vacuole through the action of SifA. EMBO J. *19*, 3235–3249.

Bi, E.F., and Lutkenhaus, J. (1991). FtsZ ring structure associated with division in *Escherichia coli*. Nature *354*, 161–164.

de Boer, P.A., Crossley, R.E., and Rothfield, L.I. (1989). A division inhibitor and a topological specificity factor coded for by the minicell locus determine proper placement of the division septum in *E. coli*. Cell *56*, 641–649.

de Boer, P.A., Crossley, R.E., and Rothfield, L.I. (1992). Roles of MinC and MinD in the site-specific septation block mediated by the MinCDE system of *Escherichia coli*. J. Bacteriol. *174*, 63–70.

Botsford, J.L., Alvarez, M., Hernandez, R., and Nichols, R. (1994). Accumulation of glutamate by *Salmonella typhimurium* in response to osmotic stress. Appl. Environ. Microbiol. *60*, 2568–2574.

Boucrot, E., Henry, T., Borg, J.P., Gorvel, J.P., and Meresse, S. (2005). The intracellular fate of *Salmonella* depends on the recruitment of kinesin. Science *308*, 1174–1178.

Boyle, E.C., Brown, N.F., and Finlay, B.B. (2006). *Salmonella enterica* serovar Typhimurium effectors SopB, SopE, SopE2 and SipA disrupt tight junction structure and function. Cell. Microbiol. *8*, 1946–1957.

Brenner, F.W., Villar, R.G., Angulo, F.J., Tauxe, R., and Swaminathan, B. (2000). *Salmonella* nomenclature – Guest commentary. J. Clin. Microbiol. *38*, 2465–2467.

Brent, R., and Ptashne, M. (1981). Mechanism of action of the *lexA* gene product. Proc. Natl. Acad. Sci. U.S.A. *78*, 4204–4208.

Brissette, J.L., Weiner, L., Ripmaster, T.L., and Model, P. (1991). Characterization and sequence of the *Escherichia coli* stress-induced *psp* operon. J. Mol. Biol. *220*, 35–48.

Brown, L., and Elliott, T. (1996). Efficient translation of the RpoS sigma factor in *Salmonella typhimurium* requires host factor I, an RNA-binding protein encoded by the *hfq* gene. J. Bacteriol. *178*, 3763–3770.

Bruno, V.M., Hannemann, S., Lara-Tejero, M., Flavell, R.A., Kleinstein, S.H., and Galan, J.E. (2009). *Salmonella*

Typhimurium type III secretion effectors stimulate innate immune responses in cultured epithelial cells. PLoS Pathog. *5*, 11.

Buchmeier, N.A., and Heffron, F. (1990). Induction of *Salmonella* stress proteins upon infection of macrophages. Science *248*, 730–732.

Buchmeier, N.A., Lipps, C.J., So, M.Y., and Heffron, F. (1993). Recombination-deficient mutants of *Salmonella typhimurium* are avirulent and sensitive to the oxidative burst of macrophages. Mol. Microbiol. *7*, 933–936.

Buchmeier, N.A., Libby, S.J., Xu, Y., Loewen, P.C., Switala, J., Guiney, D.G., and Fang, F.C. (1995). DNA repair is more important than catalase for *Salmonella* virulence in mice. J. Clin. Invest. *95*, 1047–1053.

Bustamante, V.H., Martinez, L.C., Santana, F.J., Knodler, L.A., Steele-Mortimer, O., and Puente, J.L. (2008). HilD-mediated transcriptional cross-talk between SPI-1 and SPI-2. Proc. Natl. Acad. Sci. U.S.A. *105*, 14591–14596.

Cairney, J., Booth, I.R., and Higgins, C.F. (1985a). Osmoregulation of gene expression in *Salmonella typhimurium*: *proU* encodes an osmotically induced betaine transport system. J. Bacteriol. *164*, 1224–1232.

Cairney, J., Booth, I.R., and Higgins, C.F. (1985b). *Salmonella typhimurium proP* gene encodes a transport system for the osmoprotectant betaine. J. Bacteriol. *164*, 1218–1223.

Camilleri, M., Colemont, L.J., Phillips, S.F., Brown, M.L., Thomforde, G.M., Chapman, N., and Zinsmeister, A.R. (1989). Human gastric emptying and colonic filling of solids characterized by a new method. Am. J. Physiol. *257*, G284–290.

Cano, D.A., Pucciarelli, M.G., Garcia-del Portillo, F., and Casadesus, J. (2002). Role of the RecBCD recombination pathway in *Salmonella* virulence. J. Bacteriol. *184*, 592–595.

Capozzi, V., Fiocco, D., Amodio, M.L., Gallone, A., and Spano, G. (2009). Bacterial stressors in minimally processed food. Int. J. Mol. Sci. *10*, 3076–3105.

Chakravortty, D., Hansen-Wester, I., and Hensel, M. (2002). *Salmonella* pathogenicity island 2 mediates protection of intracellular *Salmonella* from reactive nitrogen intermediates. J. Exp. Med. *195*, 1155–1166.

Chang, J., Chen, J., and Zhou, D. (2005). Delineation and characterization of the actin nucleation and effector translocation activities of *Salmonella* SipC. Mol. Microbiol. *55*, 1379–1389.

Chatfield, S.N., Strahan, K., Pickard, D., Charles, I.G., Hormaeche, C.E., and Dougan, G. (1992). Evaluation of *Salmonella typhimurium* strains harbouring defined mutations in *htrA* and *aroA* in the murine salmonellosis model. Microb. Pathog. *12*, 145–151.

Chauhan, A., Madiraju, M., Fol, M., Lofton, H., Maloney, E., Reynolds, R., and Rajagopalan, M. (2006). *Mycobacterium tuberculosis* cells growing in macrophages are filamentous and deficient in FtsZ rings. J. Bacteriol. *188*, 1856–1865.

Chen, C.Y., Buchmeier, N.A., Libby, S., Fang, F.C., Krause, M., and Guiney, D.G. (1995). Central regulatory role for the RpoS sigma factor in expression of *Salmonella dublin* plasmid virulence genes. J. Bacteriol. *177*, 5303–5309.

Chen, C.Y., Eckmann, L., Libby, S.J., Fang, F.C., Okamoto, S., Kagnoff, M.F., Fierer, J., and Guiney, D.G. (1996). Expression of *Salmonella typhimurium rpoS* and *rpoS*-dependent genes in the intracellular environment of eukaryotic cells. Infect. Immun. *64*, 4739–4743.

Chen, M.H., Takeda, S., Yamada, H., Ishii, Y., Yamashino, T., and Mizuno, T. (2001). Characterization of the RcsC-->YojN-->RcsB phosphorelay signaling pathway involved in capsular synthesis in *Escherichia coli*. Biosci. Biotechnol. Biochem. *65*, 2364–2367.

Chiang, S.M., and Schellhorn, H.E. (2010). Evolution of the RpoS regulon: origin of RpoS and the conservation of RpoS-dependent regulation in bacteria. J. Mol. Evol. *70*, 557–571.

Cirillo, D.M., Valdivia, R.H., Monack, D.M., and Falkow, S. (1998). Macrophage-dependent induction of the *Salmonella* pathogenicity island 2 type III secretion system and its role in intracellular survival. Mol. Microbiol. *30*, 175–188.

Clarke, D.J., Holland, I.B., and Jacq, A. (1997). Point mutations in the transmembrane domain of DjlA, a membrane-linked DnaJ-like protein, abolish its function in promoting colanic acid production via the Rcs signal transduction pathway. Mol. Microbiol. *25*, 933–944.

Clark, L., Perrett, C.A., Malt, L., Harward, C., Humphrey, S., Jepson, K.A., Martinez Argudo, I., Carney, L.J., La Ragione, R.M., Humphrey, T., *et al.* (2011). Differences in *Salmonella enterica* serovar Typhimurium strain invasiveness are associated with heterogeneity in SPI-1 gene expression. Microbiology *157*, 2072–2083.

Clark, M.A., Jepson, M.A., Simmons, N.L., and Hirst, B.H. (1994). Preferential interaction of *Salmonella typhimurium* with mouse Peyers Patch M cells. Res. Microbiol. *145*, 543–552.

Clark, M.A., Reed, K.A., Lodge, J., Stephen, J., Hirst, B.H., and Jepson, M.A. (1996). Invasion of murine intestinal M cells by *Salmonella typhimurium inv* mutants severely deficient for invasion of cultured cells. Infect. Immun. *64*, 4363–4368.

Connor, B.A., and Schwartz, E. (2005). Typhoid and paratyphoid fever in travellers. Lancet Infect. Dis. *5*, 623–628.

Cook, K.A., Dobbs, T.E., Hlady, W.G., Wells, J.G., Barrett, T.J., Puhr, N.D., Lancette, G.A., Bodager, D.W., Toth, B.L., Genese, C.A., *et al.* (1998). Outbreak of *Salmonella* serotype Hartford infections associated with unpasteurized orange juice. JAMA *280*, 1504–1509.

Cordell, S.C., Robinson, E.J.H., and Lowe, J. (2003). Crystal structure of the SOS cell division inhibitor SulA and in complex with FtsZ. Proc. Natl. Acad. Sci. U.S.A. *100*, 7889–7894.

Coynault, C., RobbeSaule, V., and Norel, F. (1996). Virulence and vaccine potential of *Salmonella typhimurium* mutants deficient in the expression of the RpoS (sigma (S)) regulon. Mol. Microbiol. *22*, 149–160.

Craig, J.E., Nobbs, A., and High, N.J. (2002). The extracytoplasmic sigma factor, final sigma(E), is required for intracellular survival of nontypeable *Haemophilus influenzae* in J774 macrophages. Infect. Immun. *70*, 708–715.

Crump, J.A., Luby, S.P., and Mintz, E.D. (2004). The global burden of typhoid fever. Bull. World Health Organ. *82*, 346–353.

Csonka, L.N. (1989). Physiological and genetic responses of bacteria to osmotic stress. Microbiol. Rev. *53*, 121–147.

Cummings, K., Barrett, E., Mohle-Boetani, J.C., Brooks, J.T., Farrar, J., Hunt, T., Fiore, A., Komatsu, K., Werner, S.B., and Slutsker, L. (2001). A multistate outbreak of *Salmonella enterica* serotype baildon associated with domestic raw tomatoes. Emerg. Infect. Dis. *7*, 1046–1048.

Dajkovic, A., Mukherjee, A., and Lutkenhaus, J. (2008). Investigation of regulation of FtsZ assembly by SulA and development of a model for FtsZ polymerization. J. Bacteriol. *190*, 2513–2526.

Dajkovic, A., Pichoff, S., Lutkenhaus, J., and Wirtz, D. (2010). Cross-linking FtsZ polymers into coherent Z rings. Mol. Microbiol. *78*, 651–668.

Danese, P.N., and Silhavy, T.J. (1997). The sigma(E) and the Cpx signal transduction systems control the synthesis of periplasmic protein-folding enzymes in *Escherichia coli*. Genes Dev. *11*, 1183–1193.

Darwin, A.J. (2005). The phage-shock-protein response. Mol. Microbiol. *57*, 621–628.

Davies, A., O'Neill, P., Towers, L., and Cooke, M. (1996). An outbreak of *Salmonella typhimurium* DT104 food poisoning associated with eating beef. Commun. Dis. Rep. CDR Rev. *6*, R159–162.

De Wulf, P., McGuire, A.M., Liu, X., and Lin, E.C. (2002). Genome-wide profiling of promoter recognition by the two-component response regulator CpxR-P in *Escherichia coli*. J. Biol. Chem. *277*, 26652–26661.

Dechet, A.M., Scallan, E., Gensheimer, K., Hoekstra, R., Gunderman-King, J., Lockett, J., Wrigley, D., Chege, W., Sobel, J., and Multistate Working, G. (2006). Outbreak of multidrug-resistant *Salmonella enterica* serotype Typhimurium definitive type 104 infection linked to commercial ground beef, northeastern United States, 2003–2004. Clin. Infect. Dis. *42*, 747–752.

Deiwick, J., Nikolaus, T., Erdogan, S., and Hensel, M. (1999). Environmental regulation of *Salmonella* pathogenicity island 2 gene expression. Mol. Microbiol. *31*, 1759–1773.

Delamarche, C., Thomas, D., Rolland, J.P., Froger, A., Gouranton, J., Svelto, M., Agre, P., and Calamita, G. (1999). Visualization of AqpZ-mediated water permeability in *Escherichia coli* by cryoelectron microscopy. J. Bacteriol. *181*, 4193–4197.

Detweiler, C.S., Monack, D.M., Brodsky, I.E., Mathew, H., and Falkow, S. (2003). *virK*, *somA* and *rcsC* are important for systemic *Salmonella enterica* serovar Typhimurium infection and cationic peptide resistance. Mol. Microbiol. *48*, 385–400.

DiGiuseppe, P.A., and Silhavy, T.J. (2003). Signal detection and target gene induction by the CpxRA two-component system. J. Bacteriol. *185*, 2432–2440.

Dong, T., and Schellhorn, H.E. (2010). Role of RpoS in virulence of pathogens. Infect. Immun. *78*, 887–897.

Dougan, D.A., Mogk, A., and Bukau, B. (2002). Protein folding and degradation in bacteria: to degrade or not to degrade? That is the question. Cell. Mol. Life Sci. *59*, 1607–1616.

Downs, D., Waxman, L., Goldberg, A.L., and Roth, J. (1986). Isolation and characterization of *lon* mutants in *Salmonella typhimurium*. J. Bacteriol. *165*, 193–197.

Dunlap, V.J., and Csonka, L.N. (1985). Osmotic regulation of L-proline transport in *Salmonella typhimurium*. J. Bacteriol. *163*, 296–304.

ElGedaily, A., Paesold, G., Chen, C.Y., Guiney, D.G., and Krause, M. (1997). Plasmid virulence gene expression induced by short-chain fatty acids in *Salmonella dublin*: Identification of *rpoS*-dependent and *rpoS*-independent mechanisms. J. Bacteriol. *179*, 1409–1412.

Erickson, J.W., and Gross, C.A. (1989). Identification of the sigma E subunit of *Escherichia coli* RNA polymerase: a second alternate sigma factor involved in high-temperature gene expression. Genes Dev. *3*, 1462–1471.

Erickson, K.D., and Detweiler, C.S. (2006). The Rcs phosphorelay system is specific to enteric pathogens/commensals and activates *ydeI*, a gene important for persistent *Salmonella* infection of mice. Mol. Microbiol. *62*, 883–894.

Eriksson, S., Lucchini, S., Thompson, A., Rhen, M., and Hinton, J.C. (2003). Unravelling the biology of macrophage infection by gene expression profiling of intracellular *Salmonella enterica*. Mol. Microbiol. *47*, 103–118.

Erill, I., Campoy, S., and Barbe, J. (2007). Aeons of distress: an evolutionary perspective on the bacterial SOS response. FEMS Microbiol. Rev. *31*, 637–656.

Ernst, R.K., Dombroski, D.M., and Merrick, J.M. (1990). Anaerobiosis, type-1 fimbriae, and growth-phase are factors that affect invasion of HEP-2 cells by *Salmonella* Typhimurium. Infect. Immun. *58*, 2014–2016.

Fang, F.C., Libby, S.J., Buchmeier, N.A., Loewen, P.C., Switala, J., Harwood, J., and Guiney, D.G. (1992). The alternative sigma-factor KatF (RpoS) regulates *Salmonella* virulence. Proc. Natl. Acad. Sci. U.S.A. *89*, 11978–11982.

Fang, F.C., Chen, C.Y., Guiney, D.G., and Xu, Y. (1996). Identification of sigma S-regulated genes in *Salmonella typhimurium*: complementary regulatory interactions between sigma S and cyclic AMP receptor protein. J. Bacteriol. *178*, 5112–5120.

Fang, F.C., DeGroote, M.A., Foster, J.W., Baumler, A.J., Ochsner, U., Testerman, T., Bearson, S., Giard, J.C., Xu, Y.S., Campbell, G., *et al.* (1999). Virulent *Salmonella typhimurium* has two periplasmic Cu, Zn-superoxide dismutases. Proc. Natl. Acad. Sci. U.S.A. *96*, 7502–7507.

Farris, C., Sanowar, S., Bader, M.W., Pfuetzner, R., and Miller, S.I. (2010). Antimicrobial peptides activate the

Rcs regulon through the outer membrane lipoprotein RcsF. J. Bacteriol. *192*, 4894–4903.

Fayet, O., Ziegelhoffer, T., and Georgopoulos, C. (1989). The *groES* and *groEL* heat shock gene products of *Escherichia coli* are essential for bacterial growth at all temperatures. J. Bacteriol. *171*, 1379–1385.

Ferrieres, L., and Clarke, D.J. (2003). The RcsC sensor kinase is required for normal biofilm formation in *Escherichia coli* K-12 and controls the expression of a regulon in response to growth on a solid surface. Mol. Microbiol. *50*, 1665–1682.

Fields, P.I., Swanson, R.V., Haidaris, C.G., and Heffron, F. (1986). Mutants of *Salmonella typhimurium* that cannot survive within the macrophage are avirulent. Proc. Natl. Acad. Sci. U.S.A. *83*, 5189–5193.

Flynn, J.M., Levchenko, I., Sauer, R.T., and Baker, T.A. (2004). Modulating substrate choice: the SspB adaptor delivers a regulator of the extracytoplasmic-stress response to the AAA+ protease ClpXP for degradation. Genes Dev. *18*, 2292–2301.

Foster, J.W. (1991). *Salmonella* acid shock proteins are required for the adaptive acid tolerance response. J. Bacteriol. *173*, 6896–6902.

Foster, J.W., and Hall, H.K. (1991). Inducible pH homeostasis and the acid tolerance response of *Salmonella typhimurium*. J. Bacteriol. *173*, 5129–5135.

Fratamico, P.M. (2003). Tolerance to stress and ability of acid-adapted and non-acid-adapted *Salmonella enterica* serovar Typhimurium DT104 to invade and survive in mammalian cells *in vitro*. J. Food Prot. *66*, 1115–1125.

Friebel, A., Ilchmann, H., Aepfelbacher, M., Ehrbar, K., Machleidt, W., and Hardt, W.D. (2001). SopE and SopE2 from *Salmonella typhimurium* activate different sets of RhoGTPases of the host cell. J. Biol. Chem. *276*, 34035–34040.

Fu, Y., and Galan, J.E. (1999). A *Salmonella* protein antagonizes Rac-1 and Cdc42 to mediate host-cell recovery after bacterial invasion. Nature *401*, 293–297.

Galan, J.E., and Curtiss, R. (1990). Expression of *Salmonella typhimurium* genes required for invasion is regulated by changes in DNA supercoiling. Infect. Immun. *58*, 1879–1885.

Gallois, A., Klein, J.R., Allen, L.A., Jones, B.D., and Nauseef, W.M. (2001). *Salmonella* pathogenicity island 2-encoded type III secretion system mediates exclusion of NADPH oxidase assembly from the phagosomal membrane. J. Immunol. *166*, 5741–5748.

Galyov, E.E., Wood, M.W., Rosqvist, R., Mullan, P.B., Watson, P.R., Hedges, S., and Wallis, T.S. (1997). A secreted effector protein of *Salmonella dublin* is translocated into eukaryotic cells and mediates inflammation and fluid secretion in infected ileal mucosa. Mol. Microbiol. *25*, 903–912.

Gamer, J., Multhaup, G., Tomoyasu, T., McCarty, J.S., Rudiger, S., Schonfeld, H.J., Schirra, C., Bujard, H., and Bukau, B. (1996). A cycle of binding and release of the DnaK, DnaJ and GrpE chaperones regulates activity of the *Escherichia coli* heat shock transcription factor sigma32. EMBO J. *15*, 607–617.

Garmendia, J., Beuzon, C.R., Ruiz-Albert, J., and Holden, D.W. (2003). The roles of SsrA-SsrB and OmpR-EnvZ in the regulation of genes encoding the *Salmonella typhimurium* SPI-2 type III secretion system. Microbiology *149*, 2385–2396.

Gibert, I., and Casadesus, J. (1990). SulA-Independent Division Inhibition in His-Constitutive Strains of *Salmonella* Typhimurium. FEMS Microbiol. Lett. *69*, 205–210.

Giotis, E.S., Blair, I.S., and McDowell, D.A. (2007). Morphological changes in *Listeria monocytogenes* subjected to sublethal alkaline stress. Int. J. Food Microbiol. *120*, 250–258.

Goldenberg, D., Azar, I., and Oppenheim, A.B. (1996). Differential mRNA stability of the *cspA* gene in the cold-shock response of *Escherichia coli*. Mol. Microbiol. *19*, 241–248.

Graumann, P., and Marahiel, M.A. (1996). Some like it cold: response of microorganisms to cold shock. Arch. Microbiol. *166*, 293–300.

Greenacre, E.J., and Brocklehurst, T.F. (2006). The acetic acid tolerance response induces cross-protection to salt stress in *Salmonella typhimurium*. Int. J. Food Microbiol. *112*, 62–65.

Greenacre, E.J., Brocklehurst, T.F., Waspe, C.R., Wilson, D.R., and Wilson, P.D.G. (2003). *Salmonella enterica* serovar Typhimurium and *Listeria monocytogenes* acid tolerance response induced by organic acids at 20 degrees C: Optimization and modeling. Appl. Environ. Microbiol. *69*, 3945–3951.

Greene, S.K., Daly, E.R., Talbot, E.A., Demma, L.J., Holzbauer, S., Patel, N.J., Hill, T.A., Walderhaug, M.O., Hoekstra, R.M., Lynch, M.F., *et al.* (2008). Recurrent multistate outbreak of *Salmonella* Newport associated with tomatoes from contaminated fields, 2005. Epidemiol. Infect. *136*, 157–165.

Gruber, T.M., and Gross, C.A. (2003). Multiple sigma subunits and the partitioning of bacterial transcription space. Annu. Rev. Microbiol. *57*, 441–466.

Gutierrez, C., Abee, T., and Booth, I.R. (1995). Physiology of the osmotic stress response in microorganisms. Int. J. Food Microbiol. *28*, 233–244.

Hagiwara, D., Sugiura, M., Oshima, T., Mori, H., Aiba, H., Yamashino, T., and Mizuno, T. (2003). Genome-wide analyses revealing a signaling network of the RcsC-YojN-RcsB phosphorelay system in *Escherichia coli*. J. Bacteriol. *185*, 5735–5746.

Haraga, A., Ohlson, M.B., and Miller, S.I. (2008). Salmonellae interplay with host cells. Nat. Rev. Microbiol. *6*, 53–66.

Hardt, W.D., Chen, L.M., Schuebel, K.E., Bustelo, X.R., and Galan, J.E. (1998). *S.* Typhimurium encodes an activator of Rho GTPases that induces membrane ruffling and nuclear responses in host cells. Cell *93*, 815–826.

Hartl, F.U. (1996). Molecular chaperones in cellular protein folding. Nature *381*, 571–579.

Hautefort, I., Thompson, A., Eriksson-Ygberg, S., Parker, M.L., Lucchini, S., Danino, V., Bongaerts, R.J., Ahmad, N., Rhen, M., and Hinton, J.C. (2008). During infection of epithelial cells *Salmonella enterica* serovar Typhimurium undergoes a time-dependent transcriptional adaptation that results in simultaneous expression of three type 3 secretion systems. Cell. Microbiol. *10*, 958–984.

Hayes, S.A., and Dice, J.F. (1996). Roles of molecular chaperones in protein degradation. J. Cell Biol. *132*, 255–258.

Hayward, R.D., and Koronakis, V. (1999). Direct nucleation and bundling of actin by the SipC protein of invasive *Salmonella*. EMBO J. *18*, 4926–4934.

Hazeleger, W.C., Dalvoorde, M., and Beumer, R.R. (2006). Fluorescence microscopy of NaCl-stressed, elongated *Salmonella* and *Listeria* cells reveals the presence of septa in filaments. Int. J. Food Microbiol. *112*, 288–290.

Hebrard, M., Viala, J.P.M., Meresse, P., Barras, F., and Aussel, L. (2009). Redundant Hydrogen Peroxide Scavengers Contribute to *Salmonella* Virulence and Oxidative Stress Resistance. J. Bacteriol. *191*, 4605–4614.

Heiskanen, P., Taira, S., and Rhen, M. (1994). Role of *rpoS* in the regulation of *Salmonella* plasmid virulence (*spv*) genes. FEMS Microbiol. Lett. *123*, 125–130.

Hennessy, T.W., Hedberg, C.W., Slutsker, L., White, K.E., Besser-Wiek, J.M., Moen, M.E., Feldman, J., Coleman, W.W., Edmonson, L.M., MacDonald, K.L., *et al.* (1996). A national outbreak of *Salmonella enteritidis* infections from ice cream. The Investigation Team. N. Engl. J. Med. *334*, 1281–1286.

Henry, T., Garcia-del Portillo, F., and Gorvel, J.P. (2005). Identification of *Salmonella* functions critical for bacterial cell division within eukaryotic cells. Mol. Microbiol. *56*, 252–267.

Heyndrickx, M., Pasmans, F., Ducatelle, R., Decostere, A., and Haesebrouck, F. (2005). Changes in *Salmonella* nomenclature: The need for clarification. Vet. J. *170*, 275–277.

Higashide, W., Dai, S., Hombs, V.P., and Zhou, D. (2002). Involvement of SipA in modulating actin dynamics during *Salmonella* invasion into cultured epithelial cells. Cell. Microbiol. *4*, 357–365.

Hirsch, M., and Elliott, T. (2005). Fis regulates transcriptional induction of RpoS in *Salmonella enterica*. J. Bacteriol. *187*, 1568–1580.

Horby, P.W., O'Brien, S.J., Adak, G.K., Graham, C., Hawker, J.I., Hunter, P., Lane, C., Lawson, A.J., Mitchell, R.T., Reacher, M.H., *et al.* (2003). A national outbreak of multi-resistant *Salmonella enterica* serovar Typhimurium definitive phage type (DT) 104 associated with consumption of lettuce. Epidemiol. Infect. *130*, 169–178.

Horn, G., Hofweber, R., Kremer, W., and Kalbitzer, H.R. (2007). Structure and function of bacterial cold shock proteins. Cell. Mol. Life Sci. *64*, 1457–1470.

Horton, A.J., Hak, K.M., Steffan, R.J., Foster, J.W., and Bej, A.K. (2000). Adaptive response to cold temperatures and characterization of *cspA* in *Salmonella typhimurium* LT2. Antonie Van Leeuwenhoek *77*, 13–20.

Howells, A.M., Bullifent, H.L., Dhaliwal, K., Griffin, K., Garcia de Castro, A., Frith, G., Tunnacliffe, A., and Titball, R.W. (2002). Role of trehalose biosynthesis in environmental survival and virulence of *Salmonella enterica* serovar *typhimurium*. Res. Microbiol. *153*, 281–287.

Huang, Y.H., Ferrieres, L., and Clarke, D.J. (2006). The role of the Rcs phosphorelay in *Enterobacteriaceae*. Res. Microbiol. *157*, 206–212.

Humphrey, S., Clark, L.F., Humphrey, T.J., and Jepson, M.A. (2011a). Enhanced recovery of *Salmonella* Typhimurium DT104 from exposure to stress at low temperature. Microbiology *157*, 1103–1114.

Humphrey, S., Macvicar, T., Stevenson, A., Roberts, M., Humphrey, T.J., and Jepson, M.A. (2011b). SulA-induced filamentation in *Salmonella enterica* serovar Typhimurium: effects on SPI-1 expression and epithelial infection. J. Appl. Microbiol. *111*, 185–196.

Humphrey, T.J., Williams, A., McAlpine, K., Jorgensen, F., and O'Byrne, C. (1998). Pathogenicity in isolates of *Salmonella enterica* serotype *enteritidis* PT4 which differ in RpoS expression: effects of growth phase and low temperature. Epidemiol. Infect. *121*, 295–301.

Humphreys, S., Stevenson, A., Bacon, A., Weinhardt, A.B., and Roberts, M. (1999). The alternative sigma factor, sigmaE, is critically important for the virulence of *Salmonella typhimurium*. Infect. Immun. *67*, 1560–1568.

Humphreys, S., Rowley, G., Stevenson, A., Anjum, M.F., Woodward, M.J., Gilbert, S., Kormanec, J., and Roberts, M. (2004). Role of the two-component regulator CpxAR in the virulence of *Salmonella enterica* serotype Typhimurium. Infect. Immun. *72*, 4654–4661.

Hung, D.L., Raivio, T.L., Jones, C.H., Silhavy, T.J., and Hultgren, S.J. (2001). Cpx signaling pathway monitors biogenesis and affects assembly and expression of P pili. EMBO J. *20*, 1508–1518.

Ibanez-Ruiz, M., Robbe-Saule, V., Hermant, D., Labrude, S., and Norel, F. (2000). Identification of RpoS (sigma(S))-regulated genes in *Salmonella enterica* serovar *typhimurium*. J. Bacteriol. *182*, 5749–5756.

Imlay, J.A., and Linn, S. (1988). DNA damage and oxygen radical toxicity. Science *240*, 1302–1309.

Ishii, Y., Sonezaki, S., Iwasaki, Y., Miyata, Y., Akita, K., Kato, Y., and Amano, F. (2000). Regulatory role of C-terminal residues of SulA in its degradation by Lon protease in *Escherichia coli*. J. Biochem. *127*, 837–844.

Jain, S., Bidol, S.A., Austin, J.L., Berl, E., Elson, F., LeMaile-Williams, M., Deasy, M., Moll, M.E., Rea, V., Vojdani, J.D., *et al.* (2009). Multistate outbreak of *Salmonella* Typhimurium and Saintpaul infections associated with unpasteurized orange juice – United States, 2005. Clin. Infect. Dis. *48*, 1065–1071.

Jeffreys, A.G., Hak, K.M., Steffan, R.J., Foster, J.W., and Bej, A.K. (1998). Growth, survival and characterization of *cspA* in *Salmonella enteritidis* following cold shock. Curr. Microbiol. *36*, 29–35.

Jepson, M.A., Collares-Buzato, C.B., Clark, M.A., Hirst, B.H., and Simmons, N.L. (1995). Rapid disruption of epithelial barrier function by *Salmonella typhimurium* is associated with structural modification of intercellular junctions. Infect. Immun. *63*, 356–359.

Jiang, W., Jones, P., and Inouye, M. (1993). Chloramphenicol induces the transcription of the major cold shock gene of *Escherichia coli, cspA*. J. Bacteriol. *175*, 5824–5828.

Jiang, W., Hou, Y., and Inouye, M. (1997). CspA, the major cold-shock protein of *Escherichia coli,* is an RNA chaperone. J. Biol. Chem. *272*, 196–202.

Johnson, K., Charles, I., Dougan, G., Pickard, D., O'Gaora, P., Costa, G., Ali, T., Miller, I., and Hormaeche, C. (1991). The role of a stress-response protein in *Salmonella typhimurium* virulence. Mol. Microbiol. *5*, 401–407.

Jones, A.M., Goodwill, A., and Elliott, T. (2006). Limited role for the DsrA and RprA regulatory RNAs in *rpoS* regulation in *Salmonella enterica*. J. Bacteriol. *188*, 5077–5088.

Jones, P.G., and Inouye, M. (1996). RbfA, a 30S ribosomal binding factor, is a cold-shock protein whose absence triggers the cold-shock response. Mol. Microbiol. *21*, 1207–1218.

Jones, P.G., Krah, R., Tafuri, S.R., and Wolffe, A.P. (1992). DNA gyrase, CS7.4, and the cold shock response in *Escherichia coli*. J. Bacteriol. *174*, 5798–5802.

Jubete, Y., Maurizi, M.R., and Gottesman, S. (1996). Role of the heat shock protein DnaJ in the *lon*-dependent degradation of naturally unstable proteins. J. Biol. Chem. *271*, 30798–30803.

Justice, S.S., Garcia-Lara, J., and Rothfield, L.I. (2000). Cell division inhibitors SulA and MinC/MinD block septum formation at different steps in the assembly of the *Escherichia coli* division machinery. Mol. Microbiol. *37*, 410–423.

Justice, S.S., Hung, C., Theriot, J.A., Fletcher, D.A., Anderson, G.G., Footer, M.J., and Hultgren, S.J. (2004). Differentiation and developmental pathways of uropathogenic *Escherichia coli* in urinary tract pathogenesis. Proc. Natl. Acad. Sci. U.S.A. *101*, 1333–1338.

Justice, S.S., Hunstad, D.A., Seed, P.C., and Hultgren, S.J. (2006). Filamentation by *Escherichia coli* subverts innate defenses during urinary tract infection. Proc. Natl. Acad. Sci. U.S.A. *103*, 19884–19889.

Justice, S.S., Hunstad, D.A., Cegelski, L., and Hultgren, S.J. (2008). Morphological plasticity as a bacterial survival strategy. Nature Reviews Microbiology *6*, 162–168.

Kage, H., Takaya, A., Ohya, M., and Yamamoto, T. (2008). Coordinated regulation of expression of *Salmonella* pathogenicity island 1 and flagellar type III secretion systems by ATP-dependent ClpXP protease. J. Bacteriol. *190*, 2470–2478.

Kanungo, S., Dutta, S., and Sur, D. (2008). Epidemiology of typhoid and paratyphoid fever in India. J. Infect. Dev. Ctries. *2*, 454–460.

Kazmierczak, M.J., Wiedmann, M., and Boor, K.J. (2005). Alternative sigma factors and their roles in bacterial virulence. Microbiol. Mol. Biol. Rev. *69*, 527–543.

Kenyon, W.J., Sayers, D.G., Humphreys, S., Roberts, M., and Spector, M.P. (2002). The starvation-stress response of *Salmonella enterica* serovar Typhimurium requires sigma(E)-, but not CpxR-regulated extracytoplasmic functions. Microbiology *148*, 113–122.

Kenyon, W.J., Humphreys, S., Roberts, M., and Spector, M.P. (2010). Periplasmic peptidyl-prolyl isomerases SurA and FkpA play an important role in the starvation-stress response (SSR) of *Salmonella enterica* serovar Typhimurium. Antonie Van Leeuwenhoek International *98*, 51–63.

Kim, B.H., Lee, H.J., Lee, I.S., Bang, S.H., Kim, J., and Park, Y.K. (2001). Cloning and sequencing analysis of *cadC* encoding transcriptional activator CadC from *Salmonella typhimurium*. J. Microbiol. *39*, 109–115.

Kovacikova, G., and Skorupski, K. (2002). The alternative sigma factor sigma(E) plays an important role in intestinal survival and virulence in *Vibrio cholerae*. Infect. Immun. *70*, 5355–5362.

Kowalczykowski, S.C. (2000). Initiation of genetic recombination and recombination-dependent replication. Trends Biochem. Sci. *25*, 156–165.

Kowarz, L., Coynault, C., Robbesaule, V., and Norel, F. (1994). The *Salmonella typhimurium* Katf (Rpos) Gene – Cloning, Nucleotide-Sequence, and Regulation of Spvr and Spvabcd Virulence Plasmid Genes. J. Bacteriol. *176*, 6852–6860.

La Teana, A., Brandi, A., Falconi, M., Spurio, R., Pon, C.L., and Gualerzi, C.O. (1991). Identification of a cold shock transcriptional enhancer of the *Escherichia coli* gene encoding nucleoid protein H-NS. Proc. Natl. Acad. Sci. U.S.A. *88*, 10907–10911.

Laimins, L.A., Rhoads, D.B., and Epstein, W. (1981). Osmotic control of *kdp* operon expression in *Escherichia coli*. Proc. Natl. Acad. Sci. U.S.A. *78*, 464–468.

Lange, R., and Hengge-Aronis, R. (1994). The cellular concentration of the sigma S subunit of RNA polymerase in *Escherichia coli* is controlled at the levels of transcription, translation, and protein stability. Genes Dev. *8*, 1600–1612.

Lange, R., Fischer, D., and Hengge-Aronis, R. (1995). Identification of transcriptional start sites and the role of ppGpp in the expression of *rpoS*, the structural gene for the sigma S subunit of RNA polymerase in *Escherichia coli*. J. Bacteriol. *177*, 4676–4680.

Leclerc, G.J., Tartera, C., and Metcalf, E.S. (1998). Environmental regulation of *Salmonella typhi* invasion-defective mutants. Infect. Immun. *66*, 682–691.

Lee, C.A., and Falkow, S. (1990). The ability of *Salmonella* to enter mammalian cells is affected by bacterial growth state Proc. Natl. Acad. Sci. U.S.A. *87*, 4304–4308.

Lee, C.A., Silva, M., Siber, A.M., Kelly, A.J., Galyov, E., and McCormick, B.A. (2000). A secreted *Salmonella* protein induces a proinflammatory response in epithelial cells, which promotes neutrophil migration. Proc. Natl. Acad. Sci. U.S.A. *97*, 12283–12288.

Lee, I.S., Slonczewski, J.L., and Foster, J.W. (1994). A low-pH-inducible, stationary-phase acid tolerance response in *Salmonella typhimurium*. J. Bacteriol. *176*, 1422–1426.

Lee, I.S., Lin, J.S., Hall, H.K., Bearson, B., and Foster, J.W. (1995). The stationary-phase sigma-factor sigma(S) (Rpos) is required for a sustained acid tolerance response in virulent *Salmonella typhimurium*. Mol. Microbiol. *17*, 155–167.

Leistner, L., and Gorris, L.G.M. (1995). Food preservation by hurdle technology. Trends Food Sci. Technol. *6*, 41–46.

Li, B.R., Smith, P., Horvath, D.J., Romesberg, F.E., and Justice, S.S. (2010). SOS regulatory elements are essential for UPEC pathogenesis. Microbes Inf. *12*, 662–668.

Lin, J., Lee, I.S., Frey, J., Slonczewski, J.L., and Foster, J.W. (1995). Comparative analysis of extreme acid survival in *Salmonella typhimurium, Shigella flexneri,* and *Escherichia coli*. J. Bacteriol. *177*, 4097–4104.

Little, J.W. (1984). Autodigestion of *lexA* and phage lambda repressors. Proc. Natl. Acad. Sci. U.S.A. *81*, 1375–1379.

Liu, Z., Mukherjee, A., and Lutkenhaus, J. (1999). Recruitment of ZipA to the division site by interaction with FtsZ. Mol. Microbiol. *31*, 1853–1861.

Loewen, P.C. (1984). Isolation of catalase-deficient *Escherichia coli* mutants and genetic mapping of *katE*, a locus that affects catalase activity. J. Bacteriol. *157*, 622–626.

Loewen, P.C., and Stauffer, G.V. (1990). Nucleotide sequence of *katG* of *Salmonella typhimurium* LT2 and characterization of its product, hydroperoxidase I. Mol. Gen. Genet. *224*, 147–151.

McBroom, A.J., and Kuehn, M.J. (2007). Release of outer membrane vesicles by Gram-negative bacteria is a novel envelope stress response. Mol. Microbiol. *63*, 545–558.

McGhie, E.J., Hayward, R.D., and Koronakis, V. (2001). Cooperation between actin-binding proteins of invasive *Salmonella*: SipA potentiates SipC nucleation and bundling of actin. EMBO J. *20*, 2131–2139.

McGhie, E.J., Brawn, L.C., Hume, P.J., Humphreys, D., and Koronakis, V. (2009). *Salmonella* takes control: effector-driven manipulation of the host. Curr. Opin. Microbiol. *12*, 117–124.

McMeechan, A., Roberts, M., Cogan, T.A., Jorgensen, F., Stevenson, A., Lewis, C., Rowley, G., and Humphrey, T.J. (2007). Role of the alternative sigma factors sigma(E) and sigma(S) in survival of *Salmonella enterica* serovar Typhimurium during starvation, refrigeration and osmotic shock. Microbiology *153*, 263–269.

Macnab, R.M., and Castle, A.M. (1987). A variable stoichiometry model for pH homeostasis in bacteria. Biophys. J. *52*, 637–647.

MacRitchie, D.M., Ward, J.D., Nevesinjac, A.Z., and Raivio, T.L. (2008). Activation of the Cpx envelope stress response down-regulates expression of several locus of enterocyte effacement-encoded genes in enteropathogenic *Escherichia coli*. Infect. Immun. *76*, 1465–1475.

Majdalani, N., Chen, S., Murrow, J., St John, K., and Gottesman, S. (2001). Regulation of RpoS by a novel small RNA: the characterization of RprA. Mol. Microbiol. *39*, 1382–1394.

Majdalani, N., Heck, M., Stout, V., and Gottesman, S. (2005). Role of RcsF in signaling to the Rcs phosphorelay pathway in *Escherichia coli*. J. Bacteriol. *187*, 6770–6778.

Mangan, M.W., Lucchini, S., Ó Cróinín, T., Fitzgerald, S., Hinton, J.C., and Dorman, C.J. (2011). The nucleoid-associated protein HU controls three regulons that coordinate virulence, response to stress and general physiology in *Salmonella enterica* serovar Typhimurium. Microbiology *157*, 1075–1087.

Martinez-Argudo, I., and Jepson, M.A. (2008). *Salmonella* translocates across an *in vitro* M cell model independently of SPI-1 and SPI-2. Microbiology *154*, 3887–3894.

Mastroeni, P., Vazquez-Torres, A., Fang, F.C., Xu, Y., Khan, S., Hormaeche, C.E., and Dougan, G. (2000). Antimicrobial actions of the NADPH phagocyte oxidase and inducible nitric oxide synthase in experimental salmonellosis. II. Effects on microbial proliferation and host survival *in vivo*. J. Exp. Med. *192*, 237–248.

Matsui, M., Takaya, A., and Yamamoto, T. (2008). Sigma(32)-Mediated negative regulation of *Salmonella* pathogenicity island 1 expression. J. Bacteriol. *190*, 6636–6645.

Mattick, K.L., Jorgensen, F., Legan, J.D., Cole, M.B., Porter, J., Lappin-Scott, H.M., and Humphrey, T.J. (2000a). Survival and filamentation of *Salmonella enterica* serovar Enteritidis PT4 and *Salmonella enterica* serovar Typhimurium DT104 at low water activity. Appl. Environ. Microbiol. *66*, 1274–1279.

Mattick, K.L., Jorgensen, F., Legan, J.D., Lappin-Scott, H.M., and Humphrey, T.J. (2000b). Habituation of *Salmonella* spp. at reduced water activity and its effect on heat tolerance. Appl. Environ. Microbiol. *66*, 4921.

Mattick, K.L., Phillips, L.E., Jorgensen, F., Lappin-Scott, H.M., and Humphrey, T.J. (2003a). Filament formation by *Salmonella* spp. inoculated into liquid food matrices at refrigeration temperatures, and growth patterns when warmed. J. Food. Prot. *66*, 215–219.

Mattick, K.L., Rowbury, R.J., and Humphrey, T.J. (2003b). Morphological changes to *Escherichia coli* O157:H7, commensal *E. coli* and *Salmonella spp* in response to marginal growth conditions, with special reference to mildly stressing temperatures. Sci. Prog. *86*, 103–113.

Meltzer, E., and Schwartz, E. (2010). Enteric fever: a travel medicine oriented view. Curr. Opin. Infect. Dis. *23*, 432–437.

Missiakas, D., Betton, J.M., and Raina, S. (1996). New components of protein folding in extracytoplasmic compartments of *Escherichia coli* SurA, FkpA and Skp/OmpH. Mol. Microbiol. *21*, 871–884.

Missiakas, D., Mayer, M.P., Lemaire, M., Georgopoulos, C., and Raina, S. (1997). Modulation of the *Escherichia coli* sigmaE (RpoE) heat-shock transcription-factor activity by the RseA, RseB and RseC proteins. Mol. Microbiol. *24*, 355–371.

Moreno, M., Audia, J.P., Bearson, S.M., Webb, C., and Foster, J.W. (2000). Regulation of sigma S degradation in *Salmonella enterica* var *typhimurium*: *in vivo* interactions between sigma S, the response regulator MviA(RssB) and ClpX. J. Mol. Microbiol. Biotechnol *2*, 245–254.

Morgan, H.P., Wear, M.A., McNae, I., Gallagher, M.P., and Walkinshaw, M.D. (2009). Crystallization and X-ray structure of cold-shock protein E from *Salmonella typhimurium*. Acta Crystallogr. Sect. F Struct. Biol. Cryst. Commun. *65*, 1240–1245.

Morimoto, R.I. (1998). Regulation of the heat shock transcriptional response: cross talk between a family of heat shock factors, molecular chaperones, and negative regulators. Genes Dev. *12*, 3788–3796.

Mouy, R., Fischer, A., Vilmer, E., Seger, R., and Griscelli, C. (1989). Incidence, severity, and prevention of infections in chronic granulomatous disease. J. Pediatr. *114*, 555–560.

Muller, A.J., Hoffmann, C., Galle, M., Van Den Broeke, A., Heikenwalder, M., Falter, L., Misselwitz, B., Kremer, M., Beyaert, R., and Hardt, W.-D. (2009). The *S.* Typhimurium effector SopE induces caspase-1 activation in stromal cells to initiate gut inflammation. Cell Host Microbe *6*, 125–136.

Nakahigashi, K., Yanagi, H., and Yura, T. (1995). Isolation and sequence analysis of *rpoH* genes encoding sigma 32 homologs from Gram-negative bacteria: conserved mRNA and protein segments for heat shock regulation. Nucleic Acids Res. *23*, 4383–4390.

Nakayama, S., Kushiro, A., Asahara, T., Tanaka, R., Hu, L., Kopecko, D.J., and Watanabe, H. (2003). Activation of *hilA* expression at low pH requires the signal sensor CpxA, but not the cognate response regulator CpxR, in *Salmonella enterica* serovar Typhimurium. Microbiology *149*, 2809–2817.

Nickerson, C.A., and Curtiss, R. (1997). Role of sigma factor RpoS in initial stages of *Salmonella typhimurium* infection. Infect. Immun. *65*, 1814–1823.

Nishino, K., Nikaido, E., and Yamaguchi, A. (2007). Regulation of multidrug efflux systems involved in multidrug and metal resistance of *Salmonella enterica* serovar Typhimurium. J. Bacteriol. *189*, 9066–9075.

Nygard, K., Lassen, J., Vold, L., Andersson, Y., Fisher, I., Lofdahl, S., Threlfall, J., Luzzi, I., Peters, T., Hampton, M., *et al.* (2008). Outbreak of *Salmonella* Thompson infections linked to imported rucola lettuce. Food-borne Pathog. Dis. *5*, 165–173.

O'Neal, C.R., Gabriel, W.M., Turk, A.K., Libby, S.J., Fang, F.C., and Spector, M.P. (1994). RpoS is necessary for both the positive and negative regulation of starvation survival genes during phosphate, carbon, and nitrogen starvation in *Salmonella* Typhimurium. J. Bacteriol. *176*, 4610–4616.

Ogawa, M., Takade, A., Miyamoto, H., Taniguchi, H., and Yoshida, S. (2001). Morphological variety of intracellular microcolonies of *Legionella* species in Vero cells. Microbiol. Immunol. *45*, 557–562.

Ophir, T., and Gutnick, D.L. (1994). A role for exopolysaccharides in the protection of microorganisms from desiccation. Appl. Environ. Microbiol. *60*, 740–745.

Park, Y.K., Bearson, B., Bang, S.H., Bang, I.S., and Foster, J.W. (1996). Internal pH crisis, lysine decarboxylase and the acid tolerance response of *Salmonella typhimurium*. Mol. Microbiol. *20*, 605–611.

Patel, J.C., and Galan, J.E. (2006). Differential activation and function of Rho GTPases during *Salmonella*-host cell interactions. J. Cell Biol. *175*, 453–463.

Perrett, C.A., and Jepson, M.A. (2009) Regulation of *Salmonella*-induced membrane ruffling by SipA differs in strains lacking other effectors. Cell. Microbiol. *11*, 475–487.

Pezzoli, L., Elson, R., Little, C.L., Yip, H., Fisher, I., Yishai, R., Anis, E., Valinsky, L., Biggerstaff, M., Patel, N., *et al.* (2008). Packed with *Salmonella* – investigation of an international outbreak of *Salmonella* Senftenberg infection linked to contamination of prepacked basil in 2007. Food-borne Pathog. Dis. *5*, 661–668.

Phillips, L.E., Humphrey, T.J., and Lappin-Scott, H.M. (1998). Chilling invokes different morphologies in two *Salmonella enteritidis* PT4 strains. J. Appl. Microbiol. *84*, 820–826.

Piao, Z., Sze, C.C., Barysheva, O., Iida, K., and Yoshida, S. (2006). Temperature-regulated formation of mycelial mat-like biofilms by *Legionella pneumophila*. Appl. Environ. Microbiol. *72*, 1613–1622.

Pichoff, S., and Lutkenhaus, J. (2002). Unique and overlapping roles for ZipA and FtsA in septal ring assembly in *Escherichia coli*. EMBO J. *21*, 685–693.

Pichoff, S., and Lutkenhaus, J. (2005). Tethering the Z ring to the membrane through a conserved membrane targeting sequence in FtsA. Mol. Microbiol. *55*, 1722–1734.

Pomposiello, P.J., and Demple, B. (2000). Identification of SoxS-regulated genes in *Salmonella enterica* serovar *typhimurium*. J. Bacteriol. *182*, 23–29.

Quillardet, P., Rouffaud, M.A., and Bouige, P. (2003). DNA array analysis of gene expression in response to UV irradiation in *Escherichia coli*. Res. Microbiol. *154*, 559–572.

Quinones, M., Kimsey, H.H., Ross, W., Gourse, R.L., and Waldor, M.K. (2006). LexA represses CTXphi transcription by blocking access of the alpha C-terminal domain of RNA polymerase to promoter DNA. J. Biol. Chem. *281*, 39407–39412.

Raffa, R.G., and Raivio, T.L. (2002). A third envelope stress signal transduction pathway in *Escherichia coli*. Mol. Microbiol. *45*, 1599–1611.

Raffatellu, M., Wilson, R.P., Chessa, D., Andrews-Polymenis, H., Tran, Q.T., Lawhon, S., Khare, S., Adams, L.G., and Baumler, A.J. (2005). SipA, SopA, SopB, SopD, and SopE2 contribute to *Salmonella enterica* serotype Typhimurium invasion of epithelial cells. Infect. Immun. *73*, 146–154.

Rafiq, H., Zia, R., and and Naeem, S. (2009). Typhoid fever – continues as a major threat in children. Biomedica *25*, 1–2.

Robbe-Saule, V., Coynault, C., Ibanez-Ruiz, M., Hermant, D., and Norel, F. (2001). Identification of a non-haem catalase in *Salmonella* and its regulation by RpoS (sigmaS). Mol. Microbiol. *39*, 1533–1545.

Roca, A.I., and Singleton, S.F. (2003). Direct evaluation of a mechanism for activation of the RecA nucleoprotein filament. J. Am. Chem. Soc. *125*, 15366–15375.

Romling, U., Bian, Z., Hammar, M., Sierralta, W.D., and Normark, S. (1998). Curli fibers are highly conserved between *Salmonella typhimurium* and *Escherichia coli* with respect to operon structure and regulation. J. Bacteriol. *180*, 722–731.

Rosenberger, C.M., and Finlay, B.B. (2002). Macrophages inhibit *Salmonella typhimurium* replication through MEK/ERK kinase and phagocyte NADPH oxidase activities. J. Biol. Chem. *277*, 18753–18762.

Rothfield, L., Justice, S., and Garcia-Lara, J. (1999). Bacterial cell division. Annu. Rev. Genet. *33*, 423–448.

Rowley, G., Spector, M., Kormanec, J., and Roberts, M. (2006). Pushing the envelope: extracytoplasmic stress responses in bacterial pathogens. Nat. Rev. Microbiol. *4*, 383–394.

Rowley, G., Skovierova, H., Stevenson, A., Rezuchova, B., Homerova, D., Lewis, C., Sherry, A., Kormanec, J., and Roberts, M. (2011). The periplasmic chaperone Skp is required for successful *Salmonella typhimurium* infection in a murine typhoid model. Microbiology *157*, 848–858.

Rudolph, M.G., Weise, C., Mirold, S., Hillenbrand, B., Bader, B., Wittinghofer, A., and Hardt, W.D. (1999). Biochemical analysis of SopE from *Salmonella typhimurium*, a highly efficient guanosine nucleotide exchange factor for RhoGTPases. J. Biol. Chem. *274*, 30501–30509.

Rychlik, I., and Barrow, P.A. (2005). *Salmonella* stress management and its relevance to behaviour during intestinal colonisation and infection. FEMS Microbiol. Rev. *29*, 1021–1040.

Sansonetti, P.J., and Phalipon, A. (1999). M cells as ports of entry for enteroinvasive pathogens: mechanisms of interaction, consequences for the disease process. Semin. Immunol. *11*, 193–203.

Schlumberger, M.C., Muller, A.J., Ehrbar, K., Winnen, B., Duss, I., Stecher, B., and Hardt, W.D. (2005). Real-time imaging of type III secretion: *Salmonella* SipA injection into host cells. Proc. Natl. Acad. Sci. U.S.A. *102*, 12548–12553.

Schweder, T., Lee, K.H., Lomovskaya, O., and Matin, A. (1996). Regulation of *Escherichia coli* starvation sigma factor (sigma s) by ClpXP protease. J. Bacteriol. *178*, 470–476.

Sevcik, M., Sebkova, A., Volf, J., and Rychlik, I. (2001). Transcription of *arcA* and *rpoS* during growth of *Salmonella typhimurium* under aerobic and microaerobic conditions. Microbiology *147*, 701–708.

Shi, L., Chowdhury, S.M., Smallwood, H.S., Yoon, H., Mottaz-Brewer, H.M., Norbeck, A.D., McDermott, J.E., Clauss, T.R.W., Heffron, F., Smith, R.D., *et al.* (2009). Proteomic investigation of the time course responses of RAW 264.7 macrophages to infection with *Salmonella enterica*. Infect. Immun. *77*, 3227–3233.

Siddiqui, F.J., Rabbani, F., Hasan, R., Nizami, S.Q., and Bhutta, Z.A. (2006). Typhoid fever in children: some epidemiological considerations from Karachi, Pakistan. Int. J. Infect. Dis. *10*, 215–222.

Sledjeski, D.D., and Gottesman, S. (1996). Osmotic shock induction of capsule synthesis in *Escherichia coli* K-12. J. Bacteriol. *178*, 1204–1206.

Sledjeski, D.D., Gupta, A., and Gottesman, S. (1996). The small RNA, DsrA, is essential for the low temperature expression of RpoS during exponential growth in *Escherichia coli*. EMBO J. *15*, 3993–4000.

Steele-Mortimer, O., Meresse, S., Gorvel, J.P., Toh, B.H., and Finlay, B.B. (1999). Biogenesis of *Salmonella typhimurium*-containing vacuoles in epithelial cells involves interactions with the early endocytic pathway. Cell. Microbiol. *1*, 33–49.

Stender, S., Friebel, A., Linder, S., Rohde, M., Mirold, S., and Hardt, W.D. (2000). Identification of SopE2 from *Salmonella typhimurium*, a conserved guanine nucleotide exchange factor for Cdc42 of the host cell. Mol. Microbiol. *36*, 1206–1221.

Stout, V., Torres-Cabassa, A., Maurizi, M.R., Gutnick, D., and Gottesman, S. (1991). RcsA, an unstable positive regulator of capsular polysaccharide synthesis. J. Bacteriol. *173*, 1738–1747.

Strauch, K.L., Johnson, K., and Beckwith, J. (1989). Characterization of *degP*, a gene required for proteolysis in the cell envelope and essential for growth of *Escherichia coli* at high temperature. J. Bacteriol. *171*, 2689–2696.

Straus, D.B., Walter, W.A., and Gross, C.A. (1988). *Escherichia coli* heat shock gene mutants are defective in proteolysis. Genes Dev. *2*, 1851–1858.

Sydenham, M., Douce, G., Bowe, F., Ahmed, S., Chatfield, S., and Dougan, G. (2000). *Salmonella enterica* serovar Typhimurium *surA* mutants are attenuated and effective live oral vaccines. Infect. Immun. *68*, 1109–1115.

Takaya, A., Tomoyasu, T., Tokumitsu, A., Morioka, M., and Yamamoto, T. (2002). The ATP-dependent Lon protease of *Salmonella enterica* serovar Typhimurium regulates invasion and expression of genes carried on *Salmonella* pathogenicity island 1. J. Bacteriol. *184*, 224–232.

Takaya, A., Tomoyasu, T., Matsui, H., and Yamamoto, T. (2004). The DnaK/DnaJ chaperone machinery of *Salmonella enterica* serovar Typhimurium is essential for invasion of epithelial cells and survival within macrophages, leading to systemic infection. Infect. Immun. *72*, 1364–1373.

Takaya, A., Kubota, Y., Isogai, E., and Yamamoto, T. (2005). Degradation of the HilC and HilD regulator proteins by ATP-dependent Lon protease leads to down-regulation of *Salmonella* pathogenicity island 1 gene expression. Mol. Microbiol. *55*, 839–852.

Testerman, T.L., Vazquez-Torres, A., Xu, Y., Jones-Carson, J., Libby, S.J., and Fang, F.C. (2002). The alternative sigma factor sigmaE controls antioxidant defences required for *Salmonella* virulence and stationary-phase survival. Mol. Microbiol. *43*, 771–782.

Thieringer, H.A., Jones, P.G., and Inouye, M. (1998). Cold shock and adaptation. Bioessays *20*, 49–57.

Thomsen, L.E., Olsen, J.E., Foster, J.W., and Ingmer, H. (2002). ClpP is involved in the stress response and degradation of misfolded proteins in *Salmonella enterica* serovar Typhimurium. Microbiology *148*, 2727–2733.

Tomoyasu, T., Ogura, T., Tatsuta, T., and Bukau, B. (1998). Levels of DnaK and DnaJ provide tight control of heat shock gene expression and protein repair in *Escherichia coli*. Mol. Microbiol. *30*, 567–581.

Trusca, D., Scott, S., Thompson, C., and Bramhill, D. (1998). Bacterial SOS checkpoint protein SulA inhibits polymerization of purified FtsZ cell division protein. J. Bacteriol. *180*, 3946–3953.

Tsolis, R.M., Baumler, A.J., and Heffron, F. (1995). Role of *Salmonella typhimurium* Mn-Superoxide dismutase (SodA) in protection against early killing by J774 macrophages. Infect. Immun. *63*, 1739–1744.

Tsui, H.C., Feng, G., and Winkler, M.E. (1996). Transcription of the *mutL* repair, *miaA* tRNA modification, *hfq* pleiotropic regulator, and *hflA* region protease genes of *Escherichia coli* K-12 from clustered Esigma32-specific promoters during heat shock. J. Bacteriol. *178*, 5719–5731.

Uchiya, K., Barbieri, M.A., Funato, K., Shah, A.H., Stahl, P.D., and Groisman, E.A. (1999). A *Salmonella* virulence protein that inhibits cellular trafficking. EMBO J. *18*, 3924–3933.

Vazquez-Torres, A., and Fang, F.C. (2001). *Salmonella* evasion of the NADPH phagocyte oxidase. Microbes Infect. 3, 1313–1320.

Villar, R.G., Macek, M.D., Simons, S., Hayes, P.S., Goldoft, M.J., Lewis, J.H., Rowan, L.L., Hursh, D., Patnode, M., and Mead, P.S. (1999). Investigation of multidrug-resistant *Salmonella* serotype Typhimurium DT104 infections linked to raw-milk cheese in Washington State. JAMA-J. Am. Med. Assoc. *281*, 1811–1816.

Vogt, S.L., Nevesinjac, A.Z., Humphries, R.M., Donnenberg, M.S., Armstrong, G.D., and Raivio, T.L. (2010). The Cpx envelope stress response both facilitates and inhibits elaboration of the enteropathogenic *Escherichia coli* bundle-forming pilus. Mol. Microbiol. *76*, 1095–1110.

Wang, Q.P., and Kaguni, J.M. (1989). A novel sigma factor is involved in expression of the *rpoH* gene of *Escherichia coli*. J. Bacteriol. *171*, 4248–4253.

Withanage, G.S.K., Mastroeni, P., Brooks, H.J., Maskell, D.J., and McConnell, I. (2005). Oxidative and nitrosative responses of the chicken macrophage cell line MQ-NCSU to experimental *Salmonella* infection. Br. Poult. Sci. *46*, 261–267.

Wood, J.M. (1999). Osmosensing by bacteria: signals and membrane-based sensors. Microbiol. Mol. Biol. Rev. *63*, 230–262.

Xu, H., Lee, H.Y., and Ahn, J. (2008). Cross-protective effect of acid-adapted *Salmonella enterica* on resistance to lethal acid and cold stress conditions. Lett. Appl. Microbiol. *47*, 290–297.

Yamaguchi, S., Gueguen, E., Horstman, N.K., and Darwin, A.J. (2010). Membrane association of PspA depends on activation of the phage-shock-protein response in *Yersinia enterocolitica*. Mol. Microbiol. *78*, 429–443.

Yamashino, T., Ueguchi, C., and Mizuno, T. (1995). Quantitative control of the stationary phase-specific sigma factor, sigma S, in *Escherichia coli*: involvement of the nucleoid protein H-NS. EMBO J. *14*, 594–602.

Zhang, A., Altuvia, S., Tiwari, A., Argaman, L., Hengge-Aronis, R., and Storz, G. (1998). The OxyS regulatory RNA represses *rpoS* translation and binds the Hfq (HF-I) protein. EMBO J. *17*, 6061–6068.

Zheng, M., Wang, X., Templeton, L.J., Smulski, D.R., LaRossa, R.A., and Storz, G. (2001). DNA microarray-mediated transcriptional profiling of the *Escherichia coli* response to hydrogen peroxide. J. Bacteriol. *183*, 4562–4570.

Zhou, D., Mooseker, M.S., and Galan, J.E. (1999). Role of the *S. typhimurium* actin-binding protein SipA in bacterial internalization. Science *283*, 2092–2095.

Zhou, D., Chen, L.M., Hernandez, L., Shears, S.B., and Galan, J.E. (2001). A *Salmonella* inositol polyphosphatase acts in conjunction with other bacterial effectors to promote host cell actin cytoskeleton rearrangements and bacterial internalization. Mol. Microbiol. *39*, 248–259.

Stress Response in the Pathogenic *Yersinia* Species

7

N. Kaye Horstman and Andrew J. Darwin

Abstract

Pathogenic *Yersinia* species have long been studied as important causes of human disease and as model organisms to understand widely conserved mechanisms of bacterial virulence. Like all bacteria, these pathogens must respond to a variety of potentially damaging conditions to ensure their survival. This chapter begins by introducing the pathogenic *Yersinia* and the aspects of their lifestyles that are likely to require successful response to stress. The emphasis is primarily on conditions relevant to pathogenesis. Then, some genome-wide transcription and gene function studies that have identified or implicated stress response mechanisms are summarized. Next, more focused analyses of response to increased and decreased temperature, encounter with macrophages, and macrophage-like conditions are covered in more detail. Finally, the so-called extracytoplasmic stress responses (ESRs) that are activated by changes to the cell envelope will be described. Several of these ESRs have been directly associated with the infectious process in *Yersinia*. Inactivation of one, the phage shock protein (Psp) system, completely attenuates *Y. enterocolitica*. As a result, the Psp system has become the most extensively studied *Yersinia* stress response. Therefore, the final section specifically describes the regulation and function of this critical stress response system.

Introduction

Yersinia species overview

Bacteria in the *Yersinia* genus have been found in a variety of environments and can be commensal, environmental, or pathogenic. These bacteria are Gram-negative facultatively anaerobic bacilli and belong to the Enterobacteriaceae family. As of 2005, *Bergey's Manual* listed 11 species in the *Yersinia* genus (Bottone *et al.*, 2005). Three of these species, *Yersinia pestis, Y. pseudotuberculosis,* and *Y. enterocolitica,* are well established as causative agents of human disease (Carniel and Mollaret, 1990). These three species exhibit different disease characteristics and exist in different reservoirs. Although the diseases differ, these pathogens all demonstrate lymph tropism and the ability to evade killing by macrophages (Zhang and Bliska, 2005; Bergsbaken and Cookson, 2009). A common molecular feature of these three species is the presence of an approximately 70 kb virulence plasmid that is lacking in the non-pathogenic species (Gemski *et al.*, 1980; Ben-Gurion and Shafferman, 1981; Portnoy *et al.*, 1981; Laroche *et al.*, 1984; Sulakvelidze, 2000). This plasmid, denoted pYV (or pCD1 in *Y. pestis*), is essential for virulence (reviewed by Cornelis *et al.*, 1998). pYV encodes the Ysc type 3 secretion system along with the cytotoxic proteins it exports (the Yop proteins). The Yops function to inhibit phagocytosis by the intoxicated host cell and to interfere with important signalling pathways of the innate immune response (recently reviewed by Trosky *et al.*, 2008).

Along with species pathogenic to humans, the *Yersinia* genus contains a species, *Y. ruckeri*, which causes disease in the salmonid family of fish (most commonly rainbow trout). This fish pathogen, causing enteric red mouth disease, is the least similar to the other *Yersinia* species, leading to question the appropriateness of its classification (reviewed by Fernandez *et al.*, 2007). The remaining *Yersinia* species are generally considered environmental and non-pathogenic. These species are commonly isolated from pets, wild animals, water sources, vegetables, and meat products (reviewed by Sulakvelidze, 2000). That being said, many have also been isolated from healthy and ill humans, leading to the question of whether they could be opportunistic pathogens (Sulakvelidze, 2000; Carniel *et al.*, 2006; Souza *et al.*, 2010). Regardless, as of yet, none have been shown to cause disease.

Recent studies have proposed the existence of three new environmental *Yersinia* species based on 16S rRNA sequencing as well as other genotypic and phenotypic analyses. *Y. aleksiciae* has been isolated from a variety of warm-blooded animals including humans (Sprague and Neubauer, 2005); *Y. massiliensis* was isolated from water sources in France (Merhej *et al.*, 2008); and *Y. similis* was isolated from a rabbit in Germany (Sprague *et al.*, 2008). The molecular typing analyses currently available should help distinguish and better characterize the taxonomy for this genus (Souza *et al.*, 2010).

Overall, the *Yersinia* species are varied in molecular nature, in habitat, and in their effects on hosts. Despite the growing number of species in this genus, most studies have examined the three human pathogens. Accordingly, this review focuses on *Y. pestis*, *Y. pseudotuberculosis*, and *Y. enterocolitica*. However, detailed knowledge of their stress responses is notably lacking. The small number of studies may be a result of the heavier focus on virulence mechanisms. Another possible explanation is the reliance on extrapolation from the closely related model bacterium, *E. coli*. Here we present what is known of stress response in *Yersinia* and hopefully underscore the importance of studying responses in this genus directly. As studies of stress responses in *Yersinia* species have been confined to those implicated in pathogenesis, the lifestyles, virulence mechanisms, and associated diseases for these species will be introduced first.

Yersinia pestis

Y. pestis is best known as the causative agent of bubonic and pneumonic plague. Key features of its pathogenic lifestyle are summarized in Fig. 7.1. *Y. pestis* can reside in various rodent species (Butler, 2009; Smith *et al.*, 2010) and transmission to domestic pets and humans occurs via fleas (McElroy *et al.*, 2010). After a flea bites an infected individual, the bacteria reside in the flea midgut (Bacot and Martin, 1914; Hinnebusch, 1997). From there, the bacteria can be regurgitated during the flea's next blood meal, transmitting to the flea's new victim. Once inside the new host, the bacteria migrate to the lymph nodes where they can be taken up by macrophages. Importantly, *Y. pestis* is able to resist killing by the macrophage (e.g. Cavanaugh and Randall, 1959; Straley and Harmon, 1984). Swelling of the infected lymph nodes causes the infamous buboes. When the bacteria exit the macrophages, they can multiply extracellularly, travel to other organs via the bloodstream, and ultimately cause severe septicaemia. This method of flea-borne transmission causes bubonic plague. If *Y. pestis* spreads to the lungs, it can lead to the more severe pneumonic plague (Fig. 7.1), which is characterized by human-to-human transmission and is highly contagious (disease progression reviewed by Zhou *et al.*, 2006b).

The rodent reservoir and flea transmission of *Y. pestis* are of great importance to its continued cause of human disease (Wimsatt and Biggins, 2009). For example, studies in North America suggest that the occurrence of plague in humans relates not to host susceptibility, but to overlap between human and rodent populations (Maher *et al.*, 2010). Additionally, the effects of climate change on existing rodent populations may affect the prevalence of plague reservoirs (Kausrud *et al.*, 2010). The continued surveillance of rodent and flea populations is important, especially as

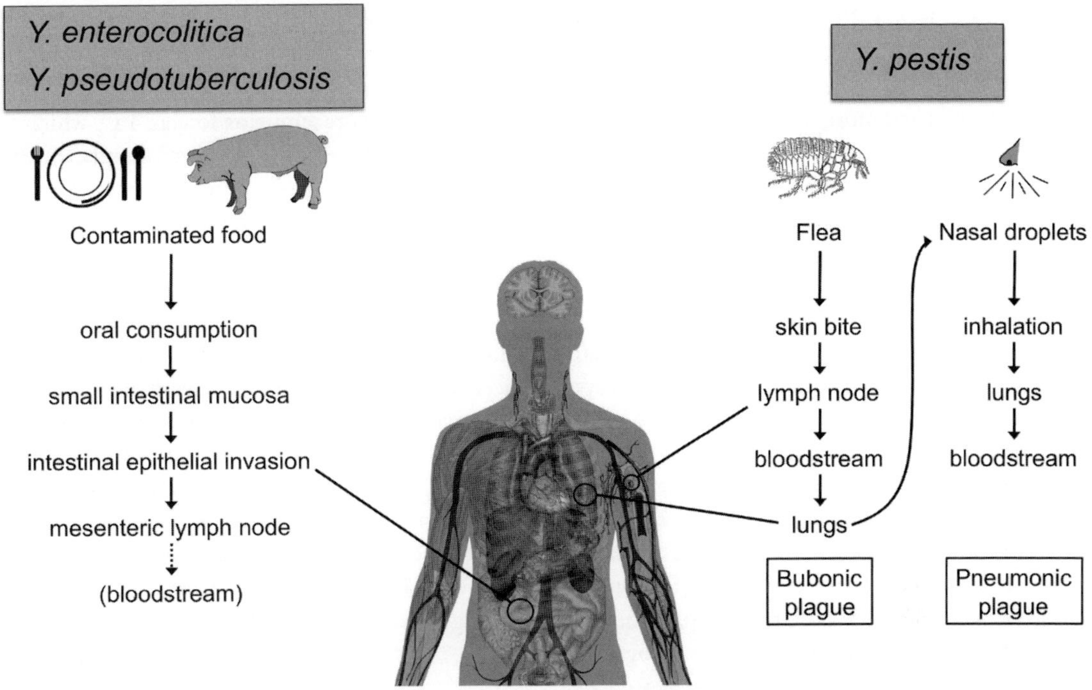

Figure 7.1 The human pathogenic *Yersinia* species: routes of entry and infection sites. The enteropathogenic species *Y. pseudotuberculosis* and *Y. enterocolitica* are ingested from contaminated food, with pork products being a frequently cited source of *Y. enterocolitica* infection in particular. The organisms invade across the small intestinal epithelium and can drain to the mesenteric lymph nodes. Spread to the bloodstream is a rare occurrence generally limited to individuals with underlying health conditions. In bubonic plague, during a flea bite *Y. pestis* is regurgitated from the flea midgut through the skin, eventually reaching a proximal lymph node. The bacteria can then spread to the bloodstream and become systemic. Bacteria that subsequently reach the lung can be spread from human to human in nasal droplets leading to the more severe pneumonic plague.

the bacteria do not always cause disease in rodent species.

Y. pseudotuberculosis and *Y. enterocolitica*

Both *Y. pseudotuberculosis* and *Y. enterocolitica* cause gastroenteritis (Fig. 7.1). Transmission most commonly occurs through the ingestion of contaminated food and water. Upon reaching the intestine, bacteria are able to invade host tissue via M cells in Peyer's patches (Clark *et al.*, 1998). There, the bacteria replicate extracellularly and are able to evade phagocytosis by macrophages owing to the factors encoded on the virulence plasmid, pYV. Once in Peyer's Patches, bacteria can drain into the lymphatic system, leading to infection of the mesenteric lymph nodes. Occasionally, *Y. enterocolitica* can progress further and invade the liver and spleen (reviewed by Cornelis *et al.*, 1987; Bottone, 1999). Although the progression of infection is similar for these two species (Fig. 7.1), the disease symptoms that arise can differ. *Y. pseudotuberculosis* most commonly infects lymph nodes, where the bacteria elicit a strong macrophage response, leading to the formation of tuberculosis-like granulomas. In general, *Y. enterocolitica* infection causes more acute illness than *Y. pseudotuberculosis* (Carniel *et al.*, 2006).

Y. pseudotuberculosis and *Y. enterocolitica* can both survive in environmental habitats, such as soil and water supplies, as well as in animal reservoirs. *Y. pseudotuberculosis* infects a variety of mammals, domestic and wild, as well as humans. Generally thought to be asymptomatic, disease in animals can occur after a stress, like cold and humid weather or starvation (Carniel and Mollaret, 1990). This leads to the dissemination of bacteria into the environment via animal faeces. Interestingly, when the bacteria do cause disease, it can do so more severely in animals than in humans. Because colder temperatures can lead to disease in infected animals and because it can itself survive cold temperatures, *Y. pseudotuberculosis* is widely distributed in countries with colder climates (Carniel and Mollaret, 1990). The 21 different serotypes of *Y. pseudotuberculosis* exhibit different geographic distribution and correlate with disease severity. For example, cases in Europe are generally restricted to enteritis, while those in Asia more frequently exhibit systemic symptoms (*Y. pseudotuberculosis* serotypes and epidemiology were reviewed in detail by Carniel *et al.*, 2006).

Y. enterocolitica is more heterogeneous than the other pathogenic *Yersinia*. Grouping of *Y. enterocolitica* strains occurs by both biogroup, according to biochemical properties, and by serogroup, according to O-antigen reactivity (reviewed by Cornelis *et al.*, 1987). As for *Y. pseudotuberculosis*, the different groups of *Y. enterocolitica* associate with different disease severity and geographic distribution (reviewed by Bottone, 1999; Carniel *et al.*, 2006). Interestingly, not all bioserogroups of *Y. enterocolitica* cause disease – the 1A biogroup lacks the pYV and is thus considered non-pathogenic. Pigs are considered an important reservoir for *Y. enterocolitica*, as the strains from pig samples correlate with those causing disease in humans (Fredriksson-Ahomaa *et al.*, 2006). Pathogenic *Y. enterocolitica* strains have also been uncovered in environmental water supplies, including cooler lakes and streams in the United States (Harvey *et al.*, 1976) and water sources and sewage in Brazil (Falcao *et al.*, 2004). Owing to its ability to survive cold temperatures, *Y. enterocolitica* multiplication during refrigeration is of particular concern (Schiemann, 1987).

Stress conditions encountered by *Yersinia* species during the infection cycle

In order for the pathogenic *Yersinia* to progress through the disease cycles described above, the bacteria must survive many potentially stressful conditions. All three pathogenic species must respond to an increase in temperature upon transition from infectious vector (environment, food, or flea) to mammalian host. *Y. enterocolitica* can also adapt to allow growth to continue at temperatures reaching as low as 4°C, which is an obvious concern for the food supply. In the mammalian host, other sources of stress include changes in osmolarity, pH, exposure to bile salts, antimicrobial peptides and macrophages. Most, if not all, of these environmental insults can cause potentially catastrophic physical changes to the interface between bacterium and host: the bacterial cell envelope. Therefore, specific responses to cell envelope damage are of particular importance during host infection. Owing to the strong focus on studying *Yersinia* pathogenesis, and owing to the variety of insults the bacteria face during pathogenesis, stress response studies have been relatively limited to those stresses important during infection.

Global studies that have identified *Yersinia* stress responses

Only a very limited amount of detailed mechanistic work has been done on specific stress responses in *Yersinia*. Rather, the majority of stress-related *Yersinia* studies have been limited to microarray surveys of genome wide transcriptional changes occurring in situations the bacteria might encounter during host infection. Similarly, global identification of genes required for host infection, or induced during it, has been another method by which putative *Yersinia* stress responses have been uncovered. In most cases, studies of stress responses in other bacteria have been used to infer the function of homologous *Yersinia* genes identified by these approaches. In this section, a selection of these global surveys is discussed briefly.

Microarray studies

A significant number of *Y. pestis* microarray studies have examined changes in gene expression in response to altered environmental conditions. This has created a wealth of data with regards to the genes expressed in particular conditions. However, to date there has not been much detailed follow-up work on determining the function of these genes and gene products. Therefore, as mentioned above, the interpretation of data from these studies relies heavily upon gene homology with other bacterial species. In this way, although many *Yersinia* genes are known to respond to a particular change in the environment, their specific roles in *Yersinia* during those conditions remain generally undetermined.

Microarray studies have been done with *Y. pestis* looking for changes in gene expression upon change in temperature. A change in temperature from 26°C to 37°C led to up-regulation of a variety of genes associated with stress response (Motin *et al.*, 2004). For example, the genes encoding the periplasmic serine protease HtrA, the universal heat shock chaperonins GroEL and GroES, the extracytoplasmic sigma factor σ^E (RpoE), and PspA from the phage shock protein (Psp) system were among those up-regulated at 37°C. Another study examined changes in expression during temperature shock treatments, heat shock at 45°C and cold shock at 10°C (Han *et al.*, 2005). Among the up-regulated genes during heat shock were those encoding known heat shock proteases (Lon, Clp, HslVU, HtrA, and DegQ) as well as those encoding chaperones (GroEL, DnaK, DnaJ, GrpE, HtpG, IbpA, IbpB). Cold shock led to the up-regulation of genes encoding known cold shock responders, CspA1, CspA2, and RNA helicases. More detailed information for heat and cold responses is discussed in 'Response to temperature change', below.

Arrays have also been used to examine changes in gene expression when *Y. pestis* enters macrophages. These studies have either been *in vivo* intracellular assays or have used *in vitro* conditions that mimic those likely to occur within a macrophage phagolysosome. Conditions that mimicked a macrophage environment included low magnesium, treatment with the cationic peptide Polymixin B, and treatment with hydrogen peroxide (Zhou *et al.*, 2006a). While the expression of a variety of stress-related genes did change upon these conditions, collectively the changes were not synonymous. For example, under low magnesium genes many heat shock proteins were down-regulated, while they were up-regulated during Polymixin B treatment. This difference in response to the variety of stresses might explain why the number of genes whose expression changed when residing in a macrophage decreased compared to the *in vitro* conditions. Interestingly, the *in vivo* studies of *Y. pestis* gene expression during infection of macrophages mainly showed an up-regulation of genes associated with detoxifying reactive oxygen species (superoxide dismutases and a catalase; Fukuto *et al.*, 2010). A very recent study used proteomics to determine *Y. pestis* proteins that were more abundant when the bacteria were in mouse macrophages compared to *in vitro* culture (Ponnusamy *et al.*, 2011). Nine proteins were uncovered that highlighted the importance of stress responses during survival in macrophages – superoxide dismutase SodA, inorganic pyrophosphatase, the glycine radical cofactor GrcA, chaperone DnaK, protease HtrA, transcriptional regulator H-NS, the urease subunit UreA, and the tellurite resistance proteins TerD and TerE.

Owing to the wealth of information from various microarrays, a composite study was undertaken to examine the array results in *Y. pestis* as a whole. Examining gene expression profiles *en masse*, this study proposed 39 stress-responsive operons (Han *et al.*, 2007). Of these, only four were specifically classified as being involved in stress response and adaptation based on homology (*dnaKJ*, *hslUV*, *katY-cybCB* and the *pspA* operon). The list also included several operons of unknown function. Perhaps examination of these operons is a good place to begin looking for novel stress responses in *Yersinia*. In fact, Jim Bliska's group have shown that deletion of one of these operons (y2313–y2315–y2316 and *orfX* immediately downstream) in the *Y. pestis* KIM6$^+$ strain caused several phenotypes including reduced intracellular replication and reduced translocation of the YopJ protein into host cells (Fukuto *et al.*, 2010). The authors hypothesized a role for this operon in regulating cell envelope characteristics in the intracellular environment.

Genome-wide virulence screens

The use of screens to uncover virulence factors important for *Yersinia* pathogenesis has also confirmed the importance of stress responses in virulence (for an extensive review of these techniques in *Yersinia* see Darwin, 2005a). One such screening technique, *in vivo* expression technology (IVET), determines genes that are expressed during infection but not during laboratory culture (Mahan *et al.*, 1993). This technique was employed in *Y. enterocolitica* to determine genes that were expressed in Peyer's patches of mice (Young and Miller, 1997). This study uncovered several chromosomal loci predicted to be involved in stress response as determined by homology to known stress responders in other bacteria – *rpoE* encoding an alternate sigma factor, *clpX* encoding a protease cofactor, and *acrR* encoding a repressor of production of the AcrAB efflux pump. Interestingly, these genes were not identified in a later IVET screen to identify *Y. enterocolitica* genes expressed in the mouse spleen (Gort and Miller, 2000). However, one common feature of these two IVET screens was the identification of genes likely to be involved in the biosynthesis and/or modification of cell envelope components.

A second technique, signature-tagged mutagenesis (STM), involves screening transposon-insertion mutants for decreased survival in a mouse infection (Hensel *et al.*, 1995). This technique has been employed in all three pathogenic *Yersinia* species. Intraperitoneal infection of mice by *Y. enterocolitica* showed the requirement of the chaperone DnaJ and a member of the phage shock protein response, PspC (Darwin and Miller, 1999). However, STM in *Y. pseudotuberculosis* did not reveal obvious stress response genes (Karlyshev *et al.*, 2001; Mecsas *et al.*, 2001). STM in *Y. pestis* revealed that the *pcm* locus was required for survival during a mouse infection (Flashner *et al.*, 2004; Tidhar *et al.*, 2009). The Pcm enzyme has been found to be important during heat shock in *E. coli* and has a proposed function in recognizing damaged L-isoaspartyl residues (Li and Clarke, 1992). Intriguingly, the *Y. enterocolitica* STM screen isolated a transposon insertion in the *nlpD* gene, which is located immediately downstream of *pcm* (Darwin and Miller, 1999).

Overall, these genome-wide studies have provided strong support for the contention that conditions encountered during mammalian host infection led to the activation of various stress responses in the pathogenic *Yersinia*. Although more follow-up work will be needed to test the importance of many of the individual genes identified, it is already clear that host infection requires the ability to mount a successful response to stress. In the following sections some of the work focused on specific stress responses/conditions in *Yersinia* is described in more depth.

Response to temperature change

All three pathogenic *Yersinia* species experience changes in temperature as a result of their lifestyles: *Y. pestis* experiences the different body temperatures of fleas and mammals; *Y. pseudotuberculosis* and *Y. enterocolitica* both can experience colder temperatures in the environment and during refrigeration, as well as elevated temperature in the mammalian body. As a result, many studies have examined temperature regulation in *Yersinia* species. Perhaps the most studied example of changes in gene expression as a result of temperature is the expression of pYV-encoded genes (reviewed by Cornelis *et al.*, 1998). Upon mammalian infection, the change in temperature to 37°C triggers pYV gene expression, leading to production of the critical Ysc-Yop type 3 secretion system. Other temperature-regulated phenomena include, but are not limited to, motility and O-antigen biosynthesis in *Y. enterocolitica* (e.g. Kapatral and Minnich, 1995; Bengoechea *et al.*, 2002).

Increased temperature

Although many studies have examined temperature regulation in *Yersinia*, significantly fewer have been focused on heat shock. Heat shock is one of the quintessential stress responses in bacteria, generally defined as a shift in temperature to 10°C above normal growth temperature (Lindquist, 1986). In *Yersinia* there are genes homologous to known heat shock genes, but this response has not been well characterized. In fact, the pathogenic *Yersinia* are often grown at 26°C in the laboratory, but can also be propagated successfully at 37°C. Therefore, although a shift from 26°C to 37°C meets the 10°C difference threshold of classic heat shock, it can be argued that 37°C is still a normal growth temperature for *Yersinia*. Nevertheless, as described above, transition of *Y. pestis* from 26°C to 37°C does lead to up-regulation of genes traditionally associated with heat shock. These include genes encoding the proteases HtrA, Lon, and Clp, as well as the chaperones GroEL, GroES, DnaK, and DnaJ. Much of what is known, or assumed, about heat shock in *Yersinia* has been extrapolated from studies in *E. coli*. One paper did find that heat shock of *Y. enterocolitica* led to increased survival during subsequent prolonged increased temperature (McMahon *et al.*, 2000).

One chaperone known to be involved in *Y. enterocolitica* heat shock is Hsp60/GroEL (Yamaguchi *et al.*, 1990). Based on the homology of *hsp60* genes in bacterial species, amplification of *hsp60* from samples can lead to identification of an enteric pathogen (Wong and Chow, 2002). Literature searches also reveal that *Y. enterocolitica* Hsp60 has been implicated in the induction of colitis in mice (Yagita *et al.*, 1999; Sukegawa *et al.*, 2000). While this information is interesting and informative, most heat shock studies in *Yersinia* species are related to roles of induced genes in pathogenesis rather than any mechanism they might have in stress tolerance.

Decreased temperature

As for high temperatures, there are two types of response to cold temperatures – either adaption for prolonged exposure or shock response for a brief exposure to 10°C below normal temperature. Response to cold temperature has been particularly important in studies of *Y. enterocolitica* and *Y. pseudotuberculosis* because both species can survive at lower temperatures, and *Y. enterocolitica* can actually grow at normal food refrigeration temperatures (Eiss, 1975). Studies have attempted to determine the requirements for *Yersinia* to survive at temperatures close to refrigeration. One study examined the long-term changes in *Y. enterocolitica* gene expression during growth at low temperatures (Bresolin *et al.*, 2006). This work found that of the 42 genes with enhanced promoter activity at 10°C versus 30°C, none was required for growth at 10°C, and none overlapped with genes generally considered cold shock genes in *E. coli*. Another study examining the genes required for *Y. enterocolitica* during cold (5°C) adaptation uncovered *pnp* (Goverde *et al.*, 1998). Interestingly, this gene, encoding a nuclease, is up-regulated in response to cold shock in *E. coli* (Takata *et al.*, 1985; Jones *et al.*, 1987). Collectively, these results emphasize the distinction between adaptation and shock while also suggesting some interplay between them (see below).

There have been relatively few studies focused on cold shock response in psychrotrophic species such as *Yersinia* (Palonen *et al.*, 2010). However, it is known that the enteropathogenic *Yersinia* contain a unique gene tandem duplication of *cspA*, encoding the major cold shock protein (Neuhaus *et al.*, 1999). CspA is an RNA chaperone that affects translation and transcription at low temperatures. This gene duplication occurs in *Y. enterocolitica* and *Y. pseudotuberculosis* as well as the fish pathogen *Y. ruckeri*, but not in the other *Yersinia* species. Intriguingly, both copies of the gene are transcribed and translated, perhaps suggesting that they may play a role in the capacity to respond to cold temperatures. Further work showed that in order for *Y. enterocolitica* to restart growth after cold shock, the *cspA* mRNAs needed to be degraded (Neuhaus *et al.*, 2000). This mRNA degradation was found to be dependent on the nuclease Pnp (Neuhaus *et al.*, 2000). This finding might suggest why *pnp* would be up-regulated during cold shock and required for growth at low temperatures: after the initial cold shock response to a drop in temperature, the polynuclease might be required to degrade cold shock response mRNAs in order to allow the bacteria to subsequently grow.

Another study found the osmolyte uptake system encoded by *proP* and *proU* to be important for *Y. enterocolitica* growth when under both osmolyte and cold shocks (Annamalai and Venkitanarayanan, 2009). The importance of this system is due to its role in the uptake of the osmolyte betaine, helping to permit intracellular water retention. The reason for this requirement during a double shock treatment, but not during single shock treatments, is still unclear.

A more detailed review about cold shock and cold adaptation in the enteropathogenic *Yersinia* species is available (Palonen *et al.*, 2010). However, those authors also noted the limited availability of studies on cold shock response in *Yersinia*.

Conditions faced during infection of macrophages

The Yop effectors of the Ysc type 3 secretion system are known to have anti-phagocytic roles during *Yersinia* infection (e.g. Cornelis, 2002; Grosdent *et al.*, 2002). Regardless, some phagocytosis does occur and there is evidence for *Y. pestis and Y. pseudotuberculosis* survival and replication in macrophages, even leading to the proposal that the macrophage provides an environment where the bacteria will be safe from the more efficient killers, the neutrophils (Pujol and Bliska, 2005). In order to survive and replicate in a phagosome, the bacteria must resist the harsh environment and an arsenal of antimicrobial molecules. The phagosome environment is known to be acidic and contain both reactive oxygen and nitrogen species, all affecting bacterial proteins and lipids; it contains scavengers to deplete necessary nutrients like iron; it contains antimicrobial peptides that disrupt the bacterial cell envelope as well as a collection of proteases and hydrolases (reviewed by Flannagan *et al.*, 2009).

A host of stress response factors have been implicated in *Yersinia* survival in macrophages. However, well-defined stress response systems have not been elucidated. Based on our knowledge of the phagosomal environment and the microarray studies described above, responses to oxidative stress and resistance to antimicrobial peptides will probably be involved in *Yersinia* intracellular survival. Here we review some of the most cited proteins that play a role in macrophage survival – PhoPQ, OmpR/EnvZ, and HtrA.

PhoPQ

PhoPQ is a classical two-component system that is 'off' in the presence of divalent cations (such as Mg^{2+} and Ca^{2+}) and activated by cationic peptides and acidic pH (Garcia Vescovi *et al.*, 1996; reviewed by Prost and Miller, 2008). PhoQ is the histidine kinase sensor protein and forms a dimer in the cytoplasmic membrane. PhoP is the phosphorylation-dependent DNA-binding response regulator. The classic stimulus for the PhoPQ system is low Mg^{2+}. In the presence of high Mg^{2+}, PhoQ binds the cations in the periplasm, which then promotes the dephosphorylation of PhoP and repression of target gene expression. Conversely, in low-Mg^{2+} conditions, PhoQ autokinase activity dominates, leading to PhoP phosphorylation and gene activation (Groisman, 2001). However, it is the presence of antimicrobial peptides and low pH that are probably responsible for PhoP activation in macrophages (Prost and Miller, 2008). In these cases, PhoQ activation occurs through binding of cationic antimicrobial peptides and possible destructuring of PhoQ at low pH (Prost and Miller, 2008). This two-component system was found to play a role in *Salmonella enterica* serovar Typhimurium virulence (Miller *et al.*, 1989; Miller and Mekalanos, 1990), leading to further study of its role in other species. Among the genes found to be regulated by PhoPQ in *S. enterica* are those encoding a magnesium transporter (*mgtABC*), an iron-responsive two component system (*pmrAB*), components of the Inv-Spa type 3 secretion system (*prgHIJK*), and those involved in membrane lipid synthesis (reviewed by Groisman, 2001).

Mutating *phoP* in *Y. pestis* was found to decrease bacterial survival in macrophages and possibly to decrease virulence (Oyston *et al.*, 2000; Bozue *et al.*, 2011). When *Y. pestis* is inside a macrophage, the phagosome still fuses with the lysosome. However, the bacteria are able to maintain an expanded vacuole, mediated by PhoP, which may be important for its survival (Grabenstein *et al.*, 2006). The phenotype of PhoP playing a role in intracellular survival was recapitulated in *Y. pseudotuberculosis*, although the *phoP* mutant was less sensitive to killing by macrophages than the *Y. pestis* mutant (Grabenstein *et al.*, 2004; Bozue *et al.*, 2011). Interestingly, one strain of *Y. pseudotuberculosis* is naturally mutated with a non-functional *phoP* gene and is also naturally defective in replicating in murine macrophages (Grabenstein *et al.*, 2004).

PhoP was also found to play a role in *Y. pestis* and *Y. pseudotuberculosis* resistance to antimicrobial peptides (Rebeil *et al.*, 2004; O'Loughlin *et al.*, 2010; Bozue *et al.*, 2011), as well as serine proteases and general survival near neutrophils (O'Loughlin *et al.*, 2010). The resistance of *Y. pestis* to antimicrobial peptides is related to its ability to modify its lipid structure, which requires functional PhoP (Hitchen *et al.*, 2002). Among the genes regulated in *Y. pestis* by PhoP are *ugb* and the *pgbP* operon (Winfield *et al.*, 2005; Perez *et al.*, 2009). These genes are important for their role in modifying lipid A of the LPS. It is the regulation of these genes that leads to the resistance to antimicrobial peptides (Grabenstein *et al.*, 2006; O'Loughlin *et al.*, 2010). Another study examined the *Y. pestis* transcriptional changes in response to magnesium with or without *phoP*, finding that PhoP was somehow involved in up-regulating several genes involved in oxidative stress, heat shock proteins, and those that prevent DNA damage, while negatively regulating cold shock and phage shock proteins (Zhou *et al.*, 2005). These results hint at the interplay between the various responses to stresses.

OmpR/EnvZ

OmpR and EnvZ are another two-component regulatory system involved in responding to stress. OmpR is considered the regulator of porin production – porins form regulated channels in the bacterial outer membrane (De la Cruz and Calva, 2010). EnvZ serves as the sensor of changes in pH, temperature, osmolarity, nutrients, and toxins.

Deletion of *ompR* in a *Y. enterocolitica* strain of the O:9 serogroup led to increased sensitivity to osmotic stress, oxidative stress, and low pH, along with a failure to survive in macrophages (Brzostek *et al.*, 2003). Similarly, deletion of *ompR* in a strain from the O:8 serogroup led to reduced survival during osmotic stress and high temperature, as well as attenuation during a mouse infection (Dorrell *et al.*, 1998). OmpR is also required for *Y. pseudotuberculosis* survival at acidic pH, due to its ability to mediate the expression of urease genes (Hu *et al.*, 2009). Urease is proposed to aid in acid survival through its degradation of urea, leading to an ammonia byproduct that raises the surrounding pH (Stingl and De Reuse, 2005). Similarly, *Y. enterocolitica* is also known to require urease activity in order to survive the acidic pH in the stomach (De Koning-Ward and Robins-Browne, 1995; Young *et al.*, 1996). Finally, an *ompR* mutant of *Y. pestis* is more sensitive to osmotic, oxidative, and temperature stress; additionally, more bacteria are taken up by macrophages (Gao *et al.*, 2011). However, these phenotypes did not contribute to an effect on *Y. pestis* virulence during a mouse infection.

HtrA

HtrA, which has also been known as GsrA in *Y. enterocolitica* (see below) and DegP in *E. coli*, is a periplasmic serine protease. This protease was unique upon its discovery, leading to its name being given to this novel class of oligomeric serine proteases (Meltzer *et al.*, 2009). Two functions of HtrA are to degrade unfolded proteins and to serve as a chaperone in the periplasm (Ingmer and Brondsted, 2009).

Disruption of *htrA* in *Y. enterocolitica* led to increased sensitivity to elevated temperature, oxidative, and osmotic stresses; decreased survival in macrophages; and attenuation in mice (Li *et al.*, 1996; Yamamoto *et al.*, 1996; Pallen and Wren, 1997; Yamamoto *et al.*, 1997). The pleiotropic nature of the mutant phenotypes led to it being called a global-stress requirement (*gsrA*) gene (Yamamoto *et al.*, 1996). In *Y. pestis*, disruption of *htrA* only slightly increased sensitivity to oxidative stress and also led to only slight attenuation in a mouse infection (Williams *et al.*, 2000).

Expression of *htrA* is regulated by both the alternative sigma factor RpoE and by the CpxAR two-component system (Heusipp *et al.*, 2004). Interestingly, although this is true in both *Y. enterocolitica* and *E. coli*, the same stress can lead to *htrA* activation through different mediators in the two species (Heusipp *et al.*, 2004; Meltzer *et al.*, 2009). In *Y. enterocolitica*, envelope stress will lead to *htrA* activation through the Cpx system. The control of *htrA* expression by these factors firmly places it within a classic extracytoplasmic stress response. This particularly important category of stress response will be the focus of the remainder of this chapter.

Extracytoplasmic stress responses

The extracytoplasmic stress responses (ESRs) are activated by potentially detrimental changes affecting the bacterial cell envelope. This part of the bacterial cell is the interface between host and pathogen. Its integrity is critical for bacterial survival because it acts as a permeability barrier, facilitates the establishment of the energy-generating proton motive force, and is home to important structures including a variety of protein export machineries. However, the integrity of the cell envelope is threatened by environmental insults such as heat, high osmolarity, and organic solvents, which can increase envelope protein misfolding and/or mislocalization or directly alter the physical properties of the membranes. Some antibiotics and antimicrobial peptides also attack the cell envelope directly. It is already clear that ESRs, which are activated by these stimuli, have an important role to play during mammalian infection by a variety of bacterial pathogens (reviewed by Raivio, 2005; Rowley *et al.*, 2006). The *Yersinia* species are no exception to this, as described below.

Rcs

The Rcs phosphorelay system is not always included in lists of Gram-negative bacterial ESRs. However, some of its features mean that it can perhaps be considered as a member. This system, discovered in *E. coli*, is conserved throughout the Enterobacteriaceae (Huang *et al.*, 2006). Originally the Rcs system genes were found as regulators of the *cps* operon, which is involved in capsular biosynthesis (Gottesman *et al.*, 1985). Based on work in *E. coli*, the Rcs components comprise a complex phosphorelay system. An inducing signal is sensed by RcsF at the outer membrane, or by RcsC at the cytoplasmic membrane. RscC is then autophosphorylated, ultimately transferring the phosphate to RcsD, which is also located at the cytoplasmic membrane. The phosphate is subsequently transferred to RcsB, which activates transcription of response genes as a homodimer, or as part of a heterodimer with the short-lived RcsA (more detailed reviews of the system have been written by Majdalani and Gottesman, 2005; Huang *et al.*, 2006; MacRitchie *et al.*, 2008). The Rcs system is activated by a variety of stresses but most commonly by those that cause changes in outer membrane structures or integrity, including osmotic shock, desiccation, and overproduction of envelope proteins (MacRitchie *et al.*, 2008). Genes regulated by the system in several Gram-negative species include those that play roles in biofilm formation, flagella, lipopolysaccharide and peptidoglycan biosynthesis (Majdalani and Gottesman, 2005; Huang *et al.*, 2006).

The *rcsB* gene of *Y. enterocolitica* was found to be necessary for the secretion of Ysp proteins by the chromosomally encoded Ysa type 3 secretion system (Venecia and Young, 2005). The mechanism involves RcsB-dependent expression of the *ysa* genes, encoding the structural components of the secretion system, and the *ysp* genes, encoding the exported substrates (Venecia and Young, 2005; Walker and Miller, 2009). However, the effect of RcsB on the *ysp* genes might be indirectly mediated through reduced levels of YsaE/SycB, which are themselves involved in controlling *ysp* gene expression (Walker and Miller, 2009). The Ysa-Ysp system is present only in the highly pathogenic 1B biogroup of *Y. enterocolitica* (Haller *et al.*, 2000; Foultier *et al.*, 2002). The genes for this system and its secreted factors are expressed at neutral pH, 26°C, and high salt concentrations, suggesting this system might be induced during the early stages of infection (Haller *et al.*, 2000; Venecia and Young, 2005; Walker and Miller, 2009). Furthermore, mouse studies indicate that the Ysa-Ysp system is needed for effective colonization of the terminal ileum (Matsumoto and Young, 2006). Therefore, the Rcs system apparently represents another stress response involved in *Yersinia* pathogenesis.

rcs genes are also present in the other pathogenic *Yersinia* species. However, although *Y. pestis* strains contain intact *rcsB* and *rcsC* genes, genome sequence analysis has revealed that *rcsD* is a pseudogene in this species (Song *et al.*, 2004). It is unknown how the loss of this intermediate phosphotransfer protein affects the Rcs system or whether it plays a role in the inability of *Y. pestis* to infect the intestine (Hinchliffe *et al.*, 2008).

The Rcs components have been most frequently examined in *Y. pestis* and *Y. pseudotuberculosis* for their roles in biofilm formation. Both species can form biofilms (although not all strains of *Y. pseudotuberculosis* possess this ability), but only *Y. pestis* can form a specialized biofilm in the flea midgut (Erickson *et al.*, 2006). This difference in biofilm formation correlates with differences in the *rcs* genes, making the Rcs components attractive candidates for causing a major difference in pathogenesis. *rcsB* was found to be a negative regulator of biofilm formation in *Y. pestis*, while replacing *rcsA* from *Y. pestis* with that from *Y. pseudotuberculosis* led to the loss of biofilm formation in fleas (Sun *et al.*, 2008). Additionally, complementing an *rcsC* deletion in *E. coli* with *Y. pestis rcsC* does not restore biofilm formation (Huang *et al.*, 2009). In that case, biofilm formation was attributed to differences in the periplasmic domain (or input domain) of RscC.

Microarrays have compared wild-type, *rscC*, and *rcsD* null mutants in strains of *Y. pseudotuberculosis* as well as the effects of overproducing either RcsB or the membrane chaperone DjlA (Hinchliffe *et al.*, 2008). Overproduction of RcsB or DjlA induces the Rcs system in *E. coli* (Majdalani and Gottesman, 2005). The overall trend of the data suggested that the *Y. pseudotuberculosis rcs* genes play roles in adhesion, biofilm formation, motility, and resistance to bile salts (Hinchliffe *et al.*, 2008). Most genes affected by *rcsB* overexpression were involved in cell envelope modifications or encoded proteins located in the cell envelope. The potential phenotypes suggested by the microarrays were then tested in several strains of both *Y. pseudotuberculosis* and *Y. enterocolitica*. Interestingly, the phenotypes of *rcs* mutants (motility, resistance to bile salts, and tolerance to high NaCl concentrations) were not always consistent across the different strains of the species (Hinchliffe *et al.*, 2008). Overall, these results strongly support a function of the Rcs system in managing cell envelope changes and suggest that its specific effects on bacterial physiology depends upon other characteristics of the individual *Yersinia* strains.

BaeSR

The genes for the two-component system BaeSR were originally found in a screen for genes that could compensate for deletions of *envZ* and *phoR*, both encoding two-component sensors, in *E. coli* (Nagasawa *et al.*, 1993). A 2002 paper classified BaeSR as involved in envelope stress due to its induction by spheroplasting, indole, and overproduction of PapG (a P pilus subunit from uropathogenic *E. coli*; Raffa and Raivio, 2002). Around the same time, overproduction of BaeR was shown to increase drug resistance by inducing expression of genes encoding a drug efflux pump, *mdtABCD*, which are located upstream of, and co-transcribed with, *baeSR* (Baranova and Nikaido, 2002; Nagakubo *et al.*, 2002). Very recent work suggests that *E. coli* BaeS senses a result of oxidative stress, perhaps the oxidation of a membrane component, or compounds that can be effluxed by the Mdt pump (Leblanc *et al.*, 2011). Therefore, the major function of the Bae response was hypothesized to be up-regulation of efflux pump expression in response to specific envelope damaging agents (Leblanc *et al.*, 2011).

To our knowledge, so far only one paper has been published with information about the BaeSR system in *Yersinia* species. This study surveyed the involvement of various membrane stress responses in Ysc-Yop type 3 secretion by *Y. pseudotuberculosis* (Carlsson *et al.*, 2007a). However, the authors found that deletion of *baeSR* did not hamper Yop secretion. Thus, any cell envelope stress that might be caused by the production of this type 3 secretion system presumably does not trigger the BaeSR system.

Cpx

The Cpx system includes another two-component system that responds to envelope stress. CpxP is a periplasmic protein that interacts with misfolded envelope proteins during envelope stress, leading to its dissociation from the periplasmic domain of the transmembrane sensor protein CpxA. In turn, CpxA then phosphorylates the response regulator CpxR, leading to regulation of target gene expression (for reviews see Rowley *et al.*, 2006; MacRitchie *et al.*, 2008). Stresses that induce this system in *E. coli* include elevated pH, alterations in the cell envelope, P pilus subunits, and, the classical stimulus, overproduction of the outer membrane lipoprotein NlpE/CutF. In *E. coli*, 41 genes have been found to be responsive to Cpx, most of which are involved in envelope protein folding or degradation, such as the gene encoding the periplasmic protease DegP/HtrA (MacRitchie *et al.*, 2008). The Cpx system has been associated with virulence in several species. Cpx affects the regulation of pili in *E. coli* (Hung *et al.*, 2001; Hernday *et al.*, 2004; Nevesinjac and Raivio, 2005); regulation of type 3 secretion genes in *Shigella sonnei* (Mitobe *et al.*, 2005); regulation of the Icm-Dot type 4 secretion system in *Legionella pneumophila* (Gal-Mor and Segal, 2003); and a *cpxA* deletion attenuates *S.* Typhimurium (Humphreys *et al.*, 2004).

In *Yersinia,* Cpx has been evaluated for its involvement in virulence. Interestingly, there do appear to be some differences between *Y. enterocolitica* and *Y. pseudotuberculosis.* Deletion of *cpxR* in *Y. enterocolitica* did not affect virulence or host cell invasion (Heusipp *et al.*, 2004; Ronnebaumer *et al.*, 2009). However, overproduction of either CpxR or CpxA caused a growth defect and *cpxA* could not be deleted, suggesting that an appropriate level of Cpx system production might be required for the growth of *Y. enterocolitica* (Ronnebaumer *et al.*, 2009). In *Y. pseudotuberculosis,* however, *cpxA* could be deleted, leading to a defect in the production and secretion of the Yop effector proteins. This defect was probably due to reduced production of structural components of the Ysc type 3 secretion system (Carlsson *et al.*, 2007a). Furthermore, *cpxA* is needed for *Y. pseudotuberculosis* association with mammalian cells through its influence on the production of adhesins, including the well-characterized Invasin protein (Carlsson *et al.*, 2007b). Conversely, deletion of *cpxR* increased Invasin production and association with host cells (Carlsson *et al.*, 2007b). The authors proposed that phosphorylated CpxR negatively regulates the Invasin gene (*inv*) and the gene encoding its direct transcriptional activator, RovA. Thus, deleting CpxA, which can cause hyper CpxR phosphorylation in normally non-Cpx-inducing conditions, reduces *inv* and *rovA* expression, whereas deletion of *cpxR* increases it. This means that in *Y. pseudotuberculosis,* extracytoplasmic stress sensing by CpxAR modulates bacteria–host cell interaction. Perhaps differences in the Cpx virulence-related requirements for the two enteropathogenic *Yersinia* species relates to their different interactions with macrophages.

RpoE/σ^E

rpoE encodes the alternative sigma factor σ^E, which is a member of the extracytoplasmic function sigma factors found in both Gram-negative and Gram-positive bacteria. In some species, such as *E. coli,* the *rpoE* gene is essential (De Las Penas *et al.*, 1997). Stresses that have been found to induce the RpoE response in several Gram-negative species include heat shock, cold shock, oxidative and osmotic stresses, starvation, and biofilm formation, all of which may lead to the accumulation of misfolded envelope proteins (reviewed by Rowley *et al.*, 2006). Furthermore, depletion of *rpoE* in *E. coli* leads to a general compromise in cell envelope integrity (Hayden and Ades, 2008). σ^E has been shown to play a role in the pathogenesis of *S.* Typhimurium, *Pseudomonas aeruginosa, Vibrio cholerae,* and *Haemophilus influenzae* (Rowley *et al.*, 2006) and has been linked to protecting pathogenic bacteria from lethal host defences (Raivio, 2005). Regulation of the σ^E response is complex. In brief, the membrane-bound RseA protein sequesters inactive σ^E at the membrane. Upon an inducing trigger, RseA is cleaved and σ^E is released. σ^E can then bind RNA polymerase and activate transcription of target promoters (for more detailed reviews of the regulation mechanism see Alba and Gross, 2004; Ades, 2008). Amongst the considerable number of genes activated by σ^E in *E. coli* are *rpoH,* encoding the heat-shock sigma factor σ^{32}, *htrA,* and the *rpoE* operon itself (e.g. Rhodius *et al.*, 2006).

The RpoE system has been investigated in a few *Yersinia* studies. As in *E. coli, rpoE* is essential in *Y. enterocolitica* (Heusipp *et al.*, 2003; Seo *et al.*, 2007). It was reported that *Y. enterocolitica rpoE* expression is not induced by heat shock, ethanol or by overproduction of Ail (a homologue of OmpX, which does induce the system in *E. coli*). This is in contrast to the situation in *E. coli.* However, as in *E. coli, Y. enterocolitica rpoE* expression was induced by high salt concentration (Heusipp *et al.*, 2003) and by overproduction of the *Y. enterocolitica* OmpX orthologue (Maxson and Darwin, 2004). Additionally, the σ^E response was induced in Peyer's patches during *Y. enterocolitica* infection of mice (Young and Miller, 1997; Heusipp *et al.*, 2003). In *Y. pseudotuberculosis,* σ^E was reported to positively influence production and secretion of the Yop proteins and, as in *Y. enterocolitica, rpoE* was also found to be essential (Carlsson *et al.*, 2007a). However, the mechanism for the effect of σ^E on the *Y. pseudotuberculosis* Ysc-Yop system has not been reported.

The phage shock-protein system

Introduction to the Psp system

The phage shock protein (Psp) system is a highly specific extracytoplasmic stress response (for other Psp system reviews see Model *et al.*, 1997; Darwin, 2005b; Joly *et al.*, 2010). Peter Model and his colleagues originally discovered this system in *E. coli* when it was infected by a filamentous phage, which is how the system received its 'phage shock' name (Brissette *et al.*, 1990). Found to comprise a regulatory system that was induced by the f1 phage protein IV (Brissette *et al.*, 1991), subsequent genomic analysis has found that the *psp* genes are widely conserved (Darwin, 2005b; Huvet *et al.*, 2010). The Psp system can also be induced by heat shock, osmotic shock, proton ionophore treatment, ethanol, and by proteins trapped in the cytoplasmic membrane translocation machinery (Brissette *et al.*, 1990; Kleerebezem and Tommassen, 1993; Weiner and Model, 1994). However, the mislocalization of the f1 phage protein IV or one of its homologues, known as secretins, is one of the most potent and remarkably specific inducers of the Psp response (e.g. Lloyd *et al.*, 2004; Seo *et al.*, 2007). Based on the characteristics of all these inducing stimuli, the overwhelming hypothesis is that Psp system induction and function relate to disruption of cytoplasmic membrane integrity. To date, the Psp system is by far the most extensively studied *Yersinia* stress response because in *Y. enterocolitica* it was found to be essential for virulence (Darwin and Miller, 1999, 2001). This provided strong motivation to further study the function of the Psp system in *Y. enterocolitica* virulence and physiology.

Defining the core components of the *Y. enterocolitica* Psp system

The *Y. enterocolitica* Psp system contains eight genes: the major *psp* locus, consisting of the *pspA* operon (*pspABCDycjXF*) and the divergently transcribed *pspF* gene (Darwin and Miller, 2001), and *pspG* at a distinct chromosomal locus (Green and Darwin, 2004). The *pspD* and *ycjXF* genes are enigmatic because their disruption does not detectably affect stress tolerance or regulation of the remaining *psp* genes. Similarly, a *pspG* null mutation does not affect *psp* gene regulation and only impacts stress tolerance if the *pspA* operon is also disrupted (Green and Darwin, 2004). In contrast, individual deletion of any one of the *pspF, A, B* or *C* genes causes a robust regulatory and/or physiological phenotype. Therefore, we have come to consider these four well-conserved genes as encoding the core components of the Psp system in *Y. enterocolitica* (Fig. 7.2). The functions of these core components are elaborated below.

The core components of the Psp system form a regulatory signal transduction system

The Psp system is tightly regulated such that *psp* gene expression is highly induced only during the specific conditions listed above. This regulation is mediated by the PspABC proteins, which together form a signal transduction system that controls the activity of PspF (Fig. 7.2). PspF is a DNA-binding transcriptional activator, which activates expression from the σ^{54}-dependent promoters upstream of *pspA* and *pspG* (Jovanovic *et al.*, 1996; Green and Darwin, 2004). According to current understanding, in non-inducing conditions PspA inhibits PspF by binding to it in the cytoplasm (Dworkin *et al.*, 2000; Elderkin *et al.*, 2002; Yamaguchi *et al.*, 2010). PspB and PspC are integral cytoplasmic membrane proteins required for stress-dependent induction of the Psp system (Weiner *et al.*, 1991; Kleerebezem *et al.*, 1996; Maxson and Darwin, 2006; Gueguen *et al.*, 2009). Upon an inducing stimulus, PspB and/or PspC recognize stress at the cytoplasmic membrane leading to their sequestration of PspA by direct interaction (Yamaguchi *et al.*, 2010). This frees PspF to activate the *pspA* and *pspG* promoters. Cytoplasmic membrane stress is then presumably alleviated by the elevation in Psp protein levels (see below). Unlike many of the stress response systems discussed above, the Psp system has a very limited regulatory response – the *pspA* and *pspG* promoters apparently make up the entire PspF regulon in *Y. enterocolitica, E. coli* and *S. enterica* (Lloyd *et al.*, 2004; Seo *et al.*, 2007).

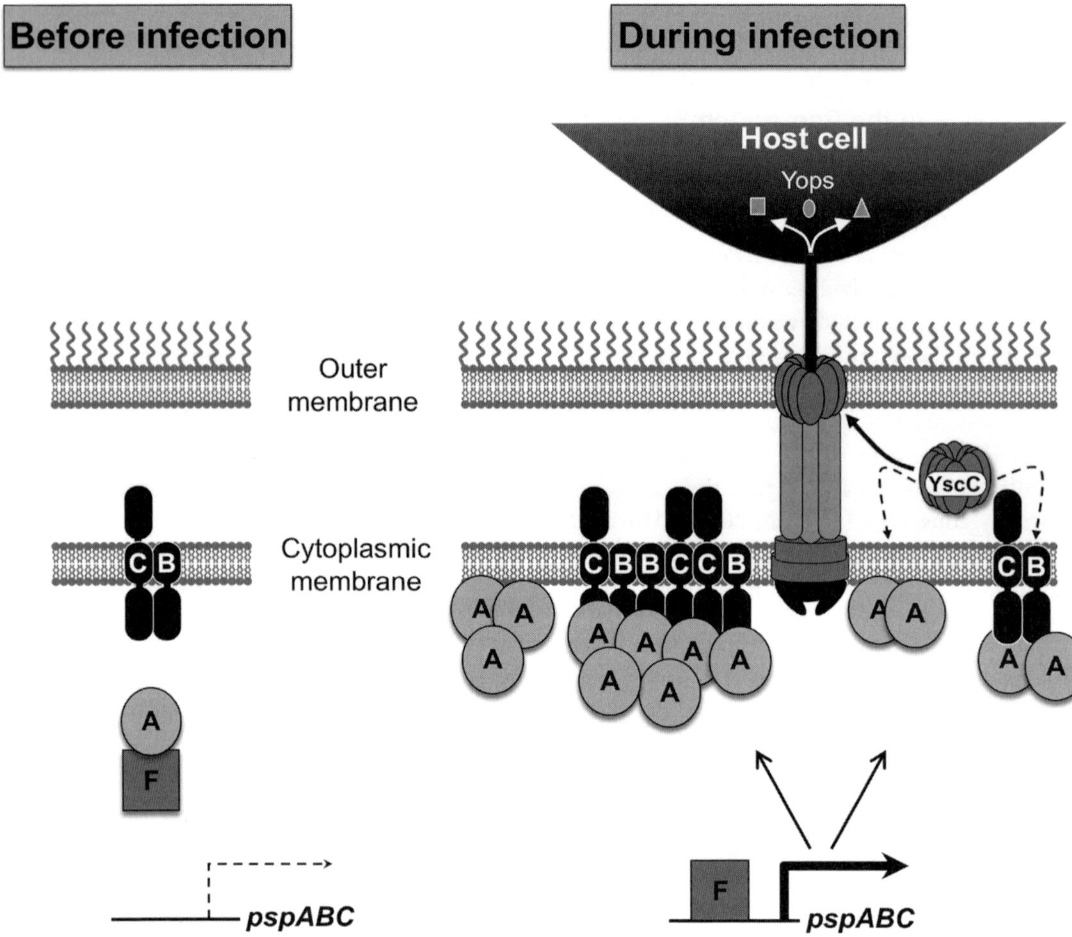

Figure 7.2 The *Y. enterocolitica* Psp response. Before infection, PspA binds to PspF in the cytoplasm and prevents it from activating *pspA* operon expression. The resulting low basal levels of PspA, B and C are poised as a stress-responsive regulatory switch. After entry into the host, the Ysc-Yop type 3 secretion system is produced with some of the YscC secretin multimer mislocalizing to the cytoplasmic membrane. PspB and/or PspC detect this stress and sequester PspA. This frees PspF to activate the *pspA* promoter, causing a large increase in PspABC concentration. PspBC then prevent or reverse secretin-induced damage to the cytoplasmic membrane by an unknown mechanism. The elevated PspA concentration might allow it to contact the membrane directly for its own physiological function (Yamaguchi *et al.*, 2010). Products of the distal *pspA* operon genes (PspD-YcjXF) and the unlinked *pspG* gene are not shown in the model because individual loss of these genes does not cause a detectable phenotype.

A brief word about the Psp system and proton motive force

A drop in PMF has long been hypothesized to be an inducing signal of the Psp system (Model *et al.*, 1997; Darwin, 2005b; Joly *et al.*, 2010). However, a lot of evidence actually argues against this. Numerous studies have in fact shown that the PMF does not decrease significantly when the Psp system is induced, unless the Psp system is defective (Kleerebezem *et al.*, 1996; Guilvout *et al.*, 2006; Jovanovic *et al.*, 2006; Engl *et al.*, 2009; Jovanovic *et al.*, 2009, 2010). Therefore, induction of the Psp system clearly occurs without a detectable drop in PMF, a conclusion that has been formalized recently (Engl *et al.*, 2011). A drop in PMF might simply be a secondary consequence of impaired cytoplasmic membrane integrity that occurs when a Psp-inducing stress is encountered but there is no functional Psp system to respond to it (i.e. in a *psp* null strain). Nevertheless, tentative links between PMF and Psp system induction remain. The proton ionophore carbonyl cyanide *m*-chlorophenyl hydrazone induces the Psp response in *E. coli* and *S. enterica* (e.g. Brissette *et al.*, 1990; Becker *et al.*, 2005). Null mutations affecting the FoF1ATPase, which interconverts PMF and ATP, induce the Psp system in *S. enterica* and *Y. enterocolitica* (Maxson and Darwin, 2004; Becker *et al.*, 2005). Perhaps a drop in PMF can be an inducing stimulus but the magnitude of the decrease in a Psp^{+} cell is too low to reliably detect. Regardless, for now the precise nature of the Psp-inducing stimulus remains unclear. All we can conclude is that many of the known inducing conditions have the potential to compromise cytoplasmic membrane integrity and that the signal(s) is presumably somehow related to this.

Induction of the *Y. enterocolitica* Psp system by secretins

As outlined above, a common theme of Psp-inducing events might be disruption of cytoplasmic membrane integrity, as can be achieved by a number of environmental shocks in *E. coli* such as ethanol, heat and high osmolarity (reviewed in Model *et al.*, 1997). Consistent with this, a random screen uncovered three cytoplasmic membrane proteins of unknown function (AmpE, YggT, and YE0566) that induced the *Y. enterocolitica* Psp system when they were overproduced (Maxson and Darwin, 2004). This random screen in *Y. enterocolitica* also revealed that overproduction of four chromosomally encoded secretins induced the Psp system (Maxson and Darwin, 2004), as does production of the pYV-encoded YscC secretin (Darwin and Miller, 2001). Secretins are outer membrane channel components of filamentous phage, type IV pili, and type 2 and type 3 secretion systems (reviewed by Genin and Boucher, 1994; Korotkov *et al.*, 2011). However, there is a significant amount of data showing that secretins can mislocalize to the cytoplasmic membrane (e.g. Koster *et al.*, 1997; Burghout *et al.*, 2004; Guilvout *et al.*, 2006). Secretin mislocalization can be exacerbated by overproduction relative to factors, such as pilotin proteins, involved in their trafficking to the outer membrane (reviewed by Korotkov *et al.*, 2011). Importantly, data strongly suggest that it is mislocalization of a secretin to the cytoplasmic membrane that activates the Psp response (Hardie *et al.*, 1996; Darwin and Miller, 2001; Guilvout *et al.*, 2006). Therefore, induction by secretins is fully consistent with cytoplasmic membrane integrity problems being the Psp inducing stimulus.

Our laboratory has focused on secretin mislocalization as a model for Psp-induction for three main reasons. First, secretin mislocalization is an exquisitely specific Psp-inducing stimulus. Studies in *Y. enterocolitica*, and also in *E. coli* and *S. enterica*, have revealed that secretin overproduction induces *psp* gene expression without significantly affecting the transcription of any other genes, including other ESRs (Lloyd *et al.*, 2004; Maxson and Darwin, 2004; Seo *et al.*, 2007). Second, *Yersinia* secretin-containing export systems have obvious significance. The virulence plasmid of all three pathogenic species encodes the Ysc-Yop type 3 secretion system, which is essential for virulence (reviewed by Cornelis, 2002). Highly pathogenic *Y. enterocolitica* 1B strains also have the chromosomal Ysa-Ysp type 3 secretion system (Haller *et al.*, 2000) and two type 2 secretion systems (Iwobi *et al.*, 2003). Third, the link between secretins and the Psp system in *Y. enterocolitica* almost certainly underlies the physiological requirement of the Psp system for virulence, as explained below.

The *Y. enterocolitica* Psp system mediates survival during secretin mislocalization stress

Originally, the only known link between secretins and the Psp system was that secretins induced *psp* gene expression (Model *et al.*, 1997). However, after the discovery that a *Y. enterocolitica pspC* null mutant was attenuated to the same level as a strain lacking the virulence plasmid, the link between secretins and the Psp system became more significant (Darwin and Miller, 1999). Subsequent work demonstrated that the *pspC* mutant was attenuated because it could not survive when it produced the Ysc type 3 secretion system (Darwin and Miller, 2001; Fig. 7.2). In fact, production of only the YscC secretin component was sufficient to prevent growth and this was exacerbated under conditions that maximized YscC mislocalization (Darwin and Miller, 2001). Therefore, it is reasonable to assume that when a *Y. enterocolitica pspC* null mutant is introduced into mice the production of its Ysc type 3 secretion system produces a lethal stress, caused by YscC mislocalization, which results in clearance of the bacteria from the animal. Later work has demonstrated that the sensitivity of *psp* null mutants to secretin production is also observed in *E. coli* and *S. enterica* (Seo *et al.*, 2007, 2009). Furthermore, a random transposon mutant screen in *Y. enterocolitica* suggested that only *psp* null mutations cause highly specific sensitivity to secretin mislocalization, which further emphasizes a specific secretin-Psp system relationship.

PspB and PspC are the mediators of secretin stress tolerance

Most of the Psp studies in *E. coli* have focused on PspA as the major mediator of the poorly characterized physiological output of the stress response. This is for good reason. PspA is the most abundant Psp protein when the system is induced and perhaps one of the most abundant proteins in the cell (e.g. Brissette *et al.*, 1990). PspA has been linked to aiding survival in stationary phase at alkaline pH (Weiner and Model, 1994) and to maintenance of the proton motive force by both *in vivo* and *in vitro* experimentation (Kleerebezem *et al.*, 1996; Kobayashi *et al.*, 2007). It has been hypothesized that PspA–membrane interactions might somehow help to stabilize membrane integrity to prevent permeability to protons (Kobayashi *et al.*, 2007; Standar *et al.*, 2008). However, the visualization of highly discretely localized GFP-PspA protein might argue for a different mechanism (Engl *et al.*, 2009).

The evidence suggesting a stress-tolerance role for PspA is both considerable and convincing. However, it cannot explain why *Y. enterocolitica psp* null strains are sensitive to secretin mislocalization and avirulent in mice. This is because a *pspA* in frame deletion mutant is completely insensitive to secretin overproduction (Darwin and Miller, 2001). Surprisingly, we have shown that PspB and PspC do not just function as positive regulators of the Psp system, but are also the components required to tolerate secretin stress (Maxson and Darwin, 2006; Gueguen *et al.*, 2009; N. K. Horstman and A. J. Darwin, unpublished). Therefore, PspA, PspB and PspC are all dual function proteins involved in both regulating *psp* gene expression and in mediating a physiological output. This makes sense because the level of each of these proteins increases when an inducing stress is encountered, which indicates that the cell needs more of them all when under stress, and not just more PspA (Fig. 7.2).

Some of the unanswered questions for the *Y. enterocolitica* Psp system

There is still much to learn about the Psp response. Indeed, the two most fundamental questions remain unanswered: what is the precise nature of the inducing stimulus and what is the mechanism(s) of the physiological output of the system. Beyond that, the individual physiological effector functions of PspA, PspB and PspC (and the other PspF regulon proteins for which phenotypes are currently unidentified) have yet to be defined. If PspA is the most abundant Psp protein, and the presumed 'master effector' of the system, why is it dispensable for secretin-stress tolerance? One possibility is that different problems at the cytoplasmic membrane might trigger the Psp response and they might require different effectors to counter them. Another possibility is that all inducers cause a similar stress, but secretins do so to a much more severe degree. In this 'threshold' model, PspA might be sufficient to counter low levels of membrane stress whereas more acute membrane damage demands the functions of PspBC. In considering these possibilities it might be relevant to think about the random screen to identify overproduction inducers of the Psp response in *Y. enterocolitica* (Maxson and Darwin, 2004). Of the proteins that were found to induce the Psp system in *Y. enterocolitica*, only the production of secretins caused a growth defect that was specific to a *psp* null strain. Perhaps secretins and the other inducers have different effects on the cell, or cause a similar effect to vastly different degrees. Regardless, in *Y. enterocolitica* our ultimate goal is to understand how the Psp system permits bacterial survival during host infection. To this end, the most pressing question to answer is how two small cytoplasmic membrane proteins (PspBC) can prevent or reverse potential damage caused by the insertion of a massive secretin oligomer into the incorrect membrane.

Discussion and future directions

Yersinia species have been studied mainly for their roles in human disease, which has meant that most work has been devoted to elucidating virulence mechanisms of the human pathogens, *Y. pestis*, *Y. pseudotuberculosis*, and *Y. enterocolitica*. Nevertheless, during the disease process the bacteria inevitably encounter various potentially stressful conditions. Therefore, by studying pathogenesis, putative *Yersinia* stress responses have been discovered. However, what is perhaps most obviously lacking is post-identification detailed analysis of the majority of these systems in *Yersinia* directly.

Microarray studies of gene expression changes in conditions that mimic an aspect the host environment have produced a wealth of information about genes that respond to stress, especially in *Y. pestis*. However, classification of these as 'stress response' genes relies heavily upon our knowledge of homologous genes studied in *E. coli* and/or other species. More work is needed to validate these assumptions. Follow-up studies should address the significance of the data from the individual studies and how it might all fit together. For example, while determining which genes respond to a specific environmental change is useful, it does not conclusively demonstrate that these genes are important for survival in that condition and/or during a human infection. Furthermore, during host infection the bacteria will encounter multiple conditions simultaneously rather than the individual stresses used in the majority of these studies.

Identification of *Yersinia* genes required or induced in animal infection models has also demonstrated the importance of stress response genes. However, these studies do not reveal the specific condition(s) during infection that necessitates those genes. For example, studies of the Psp system have shown that its requirement during infection is almost certainly due to its role in relieving envelope stress caused by production of the Ysc type 3 secretion system (and possibly other secretin-containing systems). Perhaps some other *Yersinia* stress responses have similarly specific roles during the infection process.

One particularly interesting area to consider for future research is the overlap/interplay of different stress responses, most notably during host infection. For example, many stress responses can be simultaneously induced by common environmental triggers likely to be encountered in the host, such as changes in temperature, pH, osmolarity, exposure to membrane damaging agents and so on. Furthermore, some stress responses may share some common downstream targets. Indeed, this is already known to be the case for the Cpx and RpoE responses in *E. coli*. Do any *Yersinia* stress response systems have mechanistic interaction, where activation of one response leads to activation or repression of another? Do different responses repair different consequences of the same stresses? Of course, some of these questions might be difficult to address, especially in the context of host infection. Nevertheless, they represent an interesting and important challenge for the future.

In summary, global approaches have provided an excellent starting point for uncovering the potential importance of *Yersinia* stress responses during infection. However, with some limited exceptions (most notably, the Psp system), much has been assumed about similarities between *Yersinia* and *E. coli* stress responses, which might not always be safe. Continuing to assume identical regulation, function, and components of these responses should probably be avoided without experimental validation. Overall it is fair to say that much work remains to determine the importance and mechanisms of action of the individual putative stress responses present in *Yersinia* species.

Acknowledgements

We thank Josué Flores-Kim, Catherine Potenski, and Jin Seo for critically reviewing a draft version of the manuscript. Work in our laboratory is supported in part by Award Number R01AI052148 from the National Institute of Allergy and Infectious Diseases (NIAID). The content of this article is solely the responsibility of the authors and does not necessarily represent the official views of the NIAID or the National Institutes of Health. AJD holds an Investigators in Pathogenesis of Infectious Disease Award from the Burroughs Wellcome Fund.

Web resources

Sanger Institute: http://www.sanger.ac.uk/resources/

Sanger Institute page with information and analysis tools. Go to downloads/bacteria/*Yersinia*.html, for *Yersinia* genome sequence projects.

JCVI: http://gsc.jcvi.org/projects/msc/*Yersinia*

J. Craig Venter Institute page for its *Yersinia* genome sequence projects.

PATRIC: http://www.patricbrc.org/portal/portal/patric/Taxon?cType=taxon&cId=629

The Pathogen Resource Integration Center provides *Yersinia* genome information and analysis tools.

NCBI: http://www.ncbi.nlm.nih.gov/genomes

The National Center for Biotechnology Information provides information about all complete and ongoing *Yersinia* genome sequence projects as well as genome analysis tools.

References

Ades, S.E. (2008). Regulation by destruction: design of the sigmaE envelope stress response. Curr. Opin. Microbiol. *11*, 535–540.

Alba, B.M., and Gross, C.A. (2004). Regulation of the *Escherichia coli* sigma-dependent envelope stress response. Mol. Microbiol. *52*, 613–619.

Annamalai, T., and Venkitanarayanan, K. (2009). Role of *proP* and *proU* in betaine uptake by *Yersinia enterocolitica* under cold and osmotic stress conditions. Appl. Environ. Microbiol. *75*, 1471–1477.

Bacot, A.W., and Martin, C.J. (1914). LXVII. Observations on the mechanism of the transmission of plague by fleas. J. Hyg. (Lond.) *13*, 423–439.

Baranova, N., and Nikaido, H. (2002). The BaeSR two-component regulatory system activates transcription of the *yegMNOB* (*mdtABCD*) transporter gene cluster in *Escherichia coli* and increases its resistance to novobiocin and deoxycholate. J. Bacteriol. *184*, 4168–4176.

Becker, L.A., Bang, I.S., Crouch, M.L., and Fang, F.C. (2005). Compensatory role of PspA, a member of the phage shock protein operon, in *rpoE* mutant *Salmonella enterica* serovar Typhimurium. Mol. Microbiol. *56*, 1004–1016.

Ben-Gurion, R., and Shafferman, A. (1981). Essential virulence determinants of different *Yersinia* species are carried on a common plasmid. Plasmid *5*, 183–187.

Bengoechea, J.A., Zhang, L., Toivanen, P., and Skurnik, M. (2002). Regulatory network of lipopolysaccharide O-antigen biosynthesis in *Yersinia enterocolitica* includes cell envelope-dependent signals. Mol. Microbiol. *44*, 1045–1062.

Bergsbaken, T., and Cookson, B.T. (2009). Innate immune response during *Yersinia* infection: critical modulation of cell death mechanisms through phagocyte activation. J. Leukoc. Biol. *86*, 1153–1158.

Bottone, E.J. (1999). *Yersinia enterocolitica*: overview and epidemiologic correlates. Microbes Infect. *1*, 323–333.

Bottone, E.J., Bercovier, H., and Mollaret, H.H. (2005). *Yersinia*. In Bergey's Manual of Systematic Bacteriology, volume 2, Garrity, G.M., ed. (Springer, New York, USA), pp. 838–848.

Bozue, J., Mou, S., Moody, K.L., Cote, C.K., Trevino, S., Fritz, D., and Worsham, P. (2011). The role of the *phoPQ* operon in the pathogenesis of the fully virulent CO92 strain of *Yersinia pestis* and the IP32953 strain of *Yersinia pseudotuberculosis*. Microb. Pathog. *50*, 314–321.

Bresolin, G., Neuhaus, K., Scherer, S., and Fuchs, T.M. (2006). Transcriptional analysis of long-term adaptation of *Yersinia enterocolitica* to low-temperature growth. J. Bacteriol. *188*, 2945–2958.

Brissette, J.L., Russel, M., Weiner, L., and Model, P. (1990). Phage shock protein, a stress protein of *Escherichia coli*. Proc. Natl. Acad. Sci. U.S.A. *87*, 862–866.

Brissette, J.L., Weiner, L., Ripmaster, T.L., and Model, P. (1991). Characterization and sequence of the *Escherichia coli* stress-induced *psp* operon. J. Mol. Biol. *220*, 35–48.

Brzostek, K., Raczkowska, A., and Zasada, A. (2003). The osmotic regulator OmpR is involved in the response of *Yersinia enterocolitica* O:9 to environmental stresses and survival within macrophages. FEMS Microbiol. Lett. *228*, 265–271.

Burghout, P., Beckers, F., de Wit, E., van Boxtel, R., Cornelis, G.R., Tommassen, J., and Koster, M. (2004). Role of the pilot protein YscW in the biogenesis of the YscC secretin in *Yersinia enterocolitica*. J. Bacteriol. *186*, 5366–5375.

Butler, T. (2009). Plague into the 21st century. Clin. Infect. Dis. *49*, 736–742.

Carlsson, K.E., Liu, J., Edqvist, P.J., and Francis, M.S. (2007a). Extracytoplasmic stress-responsive pathways modulate type III secretion in *Yersinia pseudotuberculosis*. Infect. Immun. *75*, 3913–3924.

Carlsson, K.E., Liu, J., Edqvist, P.J., and Francis, M.S. (2007b). Influence of the Cpx extracytoplasmic-stress-responsive pathway on *Yersinia* sp.-eukaryotic cell contact. Infect. Immun. *75*, 4386–4399.

Carniel, E., and Mollaret, H.H. (1990). Yersiniosis. Comp. Immun. Microbiol. infect. Dis. *13*, 51–58.

Carniel, E., Autenrieth, I., Cornelis, G., Fukushima, H., Guinet, F., Isberg, R., Pham, J., Prentice, M.B., Simonet, M., Skurnik, M., *et al.* (2006). *Y. enterocolitica* and *Y. pseudotuberculosis*. In The Prokaryotes, Dworkin, M., Falkow, S., Rosenberg, E., Schleifer, K., and Stackebrandt, E., eds (Springer, New York), pp. 270–398.

Cavanaugh, D.C., and Randall, R. (1959). The role of multiplication of *Pasteurella pestis* in mononuclear phagocytes in the pathogenesis of flea-borne plague. J. Immunol. *83*, 348–363.

Clark, M.A., Hirst, B.H., and Jepson, M.A. (1998). M-cell surface beta1 integrin expression and invasin-mediated targeting of *Yersinia pseudotuberculosis* to mouse Peyer's patch M cells. Infect. Immun. *66*, 1237–1243.

Cornelis, G., Laroche, Y., Balligand, G., Sory, M.-P., and Wauters, G. (1987). *Yersinia enterocolitica*, a primary model for bacterial invasiveness. Rev. Infect. Dis. *9*, 64–87.

Cornelis, G.R. (2002). The *Yersinia* Ysc–Yop 'Type III' weaponry. Nat. Rev. Mol. Cell. Biol. 3, 742–754.

Cornelis, G.R., Boland, A., Boyd, A.P., Geuijen, C., Iriarte, M., Neyt, C., Sory, M.P., and Stainier, I. (1998). The virulence plasmid of *Yersinia*, an antihost genome. Microbiol. Mol. Biol. Rev. *62*, 1315–1352.

Darwin, A.J. (2005a). Genome-wide screens to identify genes of human pathogenic *Yersinia* species that are expressed during host infection. Curr. Issues Mol. Biol. *7*, 135–149.

Darwin, A.J. (2005b). The phage-shock-protein response. Mol. Microbiol. *57*, 621–628.

Darwin, A.J., and Miller, V.L. (1999). Identification of *Yersinia enterocolitica* genes affecting survival in an animal host using signature-tagged transposon mutagenesis. Mol. Microbiol. *32*, 51–62.

Darwin, A.J., and Miller, V.L. (2001). The *psp* locus of *Yersinia enterocolitica* is required for virulence and for growth *in vitro* when the Ysc type III secretion system is produced. Mol. Microbiol. *39*, 429–444.

De Koning-Ward, T.F., and Robins-Browne, R.M. (1995). Contribution of urease to acid tolerance in *Yersinia enterocolitica*. Infect. Immun. *63*, 3790–3795.

De la Cruz, M.A., and Calva, E. (2010). The complexities of porin genetic regulation. J. Mol. Microbiol. Biotechnol. *18*, 24–36.

De Las Penas, A., Connolly, L., and Gross, C.A. (1997). SigmaE is an essential sigma factor in *Escherichia coli*. J. Bacteriol. *179*, 6862–6864.

Dorrell, N., Li, S.R., Everest, P.H., Dougan, G., and Wren, B.W. (1998). Construction and characterisation of a *Yersinia enterocolitica* O:8 *ompR* mutant. FEMS Microbiol. Lett. *165*, 145–151.

Dworkin, J., Jovanovic, G., and Model, P. (2000). The PspA protein of *Escherichia coli* is a negative regulator of σ^{54}-dependent transcription. J. Bacteriol. *182*, 311–319.

Eiss, J. (1975). Selective culturing of *Yersinia enterocolitica* at a low temperature. Scand. J. Infect. Dis. *7*, 249–251.

Elderkin, S., Jones, S., Schumacher, J., Studholme, D., and Buck, M. (2002). Mechanism of action of the *Escherichia coli* phage shock protein PspA in repression of the AAA family transcription factor PspF. J. Mol. Biol. *320*, 23–37.

Engl, C., Jovanovic, G., Lloyd, L.J., Murray, H., Spitaler, M., Ying, L., Errington, J., and Buck, M. (2009). *In vivo* localizations of membrane stress controllers PspA and PspG in *Escherichia coli*. Mol. Microbiol. *73*, 382–396.

Engl, C., Beek, A.T., Bekker, M., de Mattos, J.T., Jovanovic, G., and Buck, M. (2011). Dissipation of proton motive force is not sufficient to induce the phage shock protein response in *Escherichia coli*. Curr. Microbiol. *62*, 1374–1385.

Erickson, D.L., Jarrett, C.O., Wren, B.W., and Hinnebusch, B.J. (2006). Serotype differences and lack of biofilm formation characterize *Yersinia pseudotuberculosis* infection of the *Xenopsylla cheopis* flea vector of *Yersinia pestis*. J. Bacteriol. *188*, 1113–1119.

Falcao, J.P., Brocchi, M., Proenca-Modena, J.L., Acrani, G.O., Correa, E.F., and Falcao, D.P. (2004). Virulence characteristics and epidemiology of *Yersinia enterocolitica* and yersiniae other than *Y. pseudotuberculosis* and *Y. pestis* isolated from water and sewage. J. Appl. Microbiol. *96*, 1230–1236.

Fernandez, L., Mendez, J., and Guijarro, J.A. (2007). Molecular virulence mechanisms of the fish pathogen *Yersinia ruckeri*. Vet. Microbiol. *125*, 1–10.

Flannagan, R.S., Cosio, G., and Grinstein, S. (2009). Antimicrobial mechanisms of phagocytes and bacterial evasion strategies. Nat. Rev. Microbiol. *7*, 355–366.

Flashner, Y., Mamroud, E., Tidhar, A., Ber, R., Aftalion, M., Gur, D., Lazar, S., Zvi, A., Bino, T., Ariel, N., *et al.* (2004). Generation of *Yersinia pestis* attenuated strains by signature-tagged mutagenesis in search of novel vaccine candidates. Infect. Immun. 72, 908–915.

Foultier, B., Troisfontaines, P., Muller, S., Opperdoes, F.R., and Cornelis, G.R. (2002). Characterization of the *ysa* pathogenicity locus in the chromosome of *Yersinia enterocolitica* and phylogeny analysis of type III secretion systems. J. Mol. Evol. *55*, 37–51.

Fredriksson-Ahomaa, M., Stolle, A., and Korkeala, H. (2006). Molecular epidemiology of *Yersinia enterocolitica* infections. FEMS Immunol. Med. Microbiol. *47*, 315–329.

Fukuto, H.S., Svetlanov, A., Palmer, L.E., Karzai, A.W., and Bliska, J.B. (2010). Global gene expression profiling of *Yersinia pestis* replicating inside macrophages reveals the roles of a putative stress-induced operon in regulating type III secretion and intracellular cell division. Infect. Immun. *78*, 3700–3715.

Gal-Mor, O., and Segal, G. (2003). Identification of CpxR as a positive regulator of *icm* and *dot* virulence genes of *Legionella pneumophila*. J. Bacteriol. *185*, 4908–4919.

Gao, H., Zhang, Y., Han, Y., Yang, L., Liu, X., Guo, Z., Tan, Y., Huang, X., Zhou, D., and Yang, R. (2011). Phenotypic and transcriptional analysis of the osmotic regulator OmpR in *Yersinia pestis*. BMC Microbiol. *11*, 39.

Garcia Vescovi, E., Soncini, F.C., and Groisman, E.A. (1996). Mg^{2+} as an extracellular signal: environmental regulation of *Salmonella* virulence. Cell *84*, 165–174.

Gemski, P., Lazere, J.R., Casey, T., and Wohlhieter, J.A. (1980). Presence of a virulence-associated plasmid in *Yersinia pseudotuberculosis*. Infect. Immun. *28*, 1044–1047.

Genin, S., and Boucher, C.A. (1994). A superfamily of proteins involved in different secretion pathways in Gram-negative bacteria: modular structure and specificity of the N-terminal domain. Mol. Gen. Genet. *243*, 112–118.

Gort, A.S., and Miller, V.L. (2000). Identification and characterization of *Yersinia enterocolitica* genes induced during systemic infection. Infect. Immun. *68*, 6633–6642.

Gottesman, S., Trisler, P., and Torres-Cabassa, A. (1985). Regulation of capsular polysaccharide synthesis in *Escherichia coli* K-12: characterization of three regulatory genes. J. Bacteriol. *162*, 1111–1119.

Goverde, R.L., Huis in't Veld, J.H., Kusters, J.G., and Mooi, F.R. (1998). The psychrotrophic bacterium *Yersinia enterocolitica* requires expression of *pnp*, the gene for polynucleotide phosphorylase, for growth at low temperature (5°C). Mol. Microbiol. *28*, 555–569.

Grabenstein, J.P., Marceau, M., Pujol, C., Simonet, M., and Bliska, J.B. (2004). The response regulator PhoP of *Yersinia pseudotuberculosis* is important for replication in macrophages and for virulence. Infect. Immun. *72*, 4973–4984.

Grabenstein, J.P., Fukuto, H.S., Palmer, L.E., and Bliska, J.B. (2006). Characterization of phagosome trafficking and identification of PhoP-regulated genes important for survival of *Yersinia pestis* in macrophages. Infect. Immun. *74*, 3727–3741.

Green, R.C., and Darwin, A.J. (2004). PspG, a new member of the *Yersinia enterocolitica* phage shock protein regulon. J. Bacteriol. *186*, 4910–4920.

Groisman, E.A. (2001). The pleiotropic two-component regulatory system PhoP-PhoQ. J. Bacteriol. *183*, 1835–1842.

Grosdent, N., Maridonneau-Parini, I., Sory, M.P., and Cornelis, G.R. (2002). Role of Yops and adhesins in resistance of *Yersinia enterocolitica* to phagocytosis. Infect. Immun. *70*, 4165–4176.

Gueguen, E., Savitzky, D.C., and Darwin, A.J. (2009). Analysis of the *Yersinia enterocolitica* PspBC proteins defines functional domains, essential amino acids and new roles within the phage-shock-protein response. Mol. Microbiol. *74*, 619–633.

Guilvout, I., Chami, M., Engel, A., Pugsley, A.P., and Bayan, N. (2006). Bacterial outer membrane secretin PulD assembles and inserts into the inner membrane in the absence of its pilotin. EMBO J. *25*, 5241–5249.

Haller, J.C., Carlson, S., Pederson, K.J., and Pierson, D.E. (2000). A chromosomally encoded type III secretion pathway in *Yersinia enterocolitica* is important in virulence. Mol. Microbiol. *36*, 1436–1446.

Han, Y., Zhou, D., Pang, X., Zhang, L., Song, Y., Tong, Z., Bao, J., Dai, E., Wang, J., Guo, Z., *et al.* (2005). DNA microarray analysis of the heat- and cold-shock stimulons in *Yersinia pestis*. Microbes Infect. *7*, 335–348.

Han, Y., Qiu, J., Guo, Z., Gao, H., Song, Y., Zhou, D., and Yang, R. (2007). Comparative transcriptomics in *Yersinia pestis*: a global view of environmental modulation of gene expression. BMC Microbiol. *7*, 96.

Hardie, K.R., Lory, S., and Pugsley, A.P. (1996). Insertion of an outer membrane protein in *Escherichia coli* requires a chaperone-like protein. EMBO J. *15*, 978–988.

Harvey, S., Greenwood, J.R., Pickett, M.J., and Mah, R.A. (1976). Recovery of *Yersinia enterocolitica* from streams and lakes of California. Appl. Environ. Microbiol. *32*, 352–354.

Hayden, J.D., and Ades, S.E. (2008). The extracytoplasmic stress factor, sigmaE, is required to maintain cell envelope integrity in *Escherichia coli*. PLoS One *3*, e1573.

Hensel, M., Shea, J.E., Gleeson, C., Jones, M.D., Dalton, E., and Holden, D.W. (1995). Simultaneous identification of bacterial virulence genes by negative selection. Science *269*, 400–403.

Hernday, A.D., Braaten, B.A., Broitman-Maduro, G., Engelberts, P., and Low, D.A. (2004). Regulation of the *pap* epigenetic switch by CpxAR: phosphorylated CpxR inhibits transition to the phase ON state by competition with Lrp. Mol. Cell *16*, 537–547.

Heusipp, G., Schmidt, M.A., and Miller, V.L. (2003). Identification of *rpoE* and *nadB* as host responsive elements of *Yersinia enterocolitica*. FEMS Microbiol. Lett. *226*, 291–298.

Heusipp, G., Nelson, K.M., Schmidt, M.A., and Miller, V.L. (2004). Regulation of *htrA* expression in *Yersinia enterocolitica*. FEMS Microbiol. Lett. *231*, 227–235.

Hinchliffe, S.J., Howard, S.L., Huang, Y.H., Clarke, D.J., and Wren, B.W. (2008). The importance of the Rcs phosphorelay in the survival and pathogenesis of the enteropathogenic yersiniae. Microbiology *154*, 1117–1131.

Hinnebusch, B.J. (1997). Bubonic plague: a molecular genetic case history of the emergence of an infectious disease. J. Mol. Med. *75*, 645–652.

Hitchen, P.G., Prior, J.L., Oyston, P.C., Panico, M., Wren, B.W., Titball, R.W., Morris, H.R., and Dell, A. (2002). Structural characterization of lipo-oligosaccharide (LOS) from *Yersinia pestis*: regulation of LOS structure by the PhoPQ system. Mol. Microbiol. *44*, 1637–1650.

Hu, Y., Lu, P., Wang, Y., Ding, L., Atkinson, S., and Chen, S. (2009). OmpR positively regulates urease expression to enhance acid survival of *Yersinia pseudotuberculosis*. Microbiology *155*, 2522–2531.

Huang, Y.H., Ferrieres, L., and Clarke, D.J. (2006). The role of the Rcs phosphorelay in Enterobacteriaceae. Res. Microbiol. *157*, 206–212.

Huang, Y.H., Ferrieres, L., and Clarke, D.J. (2009). Comparative functional analysis of the RcsC sensor kinase from different Enterobacteriaceae. FEMS Microbiol. Lett. *293*, 248–254.

Humphreys, S., Rowley, G., Stevenson, A., Anjum, M.F., Woodward, M.J., Gilbert, S., Kormanec, J., and Roberts, M. (2004). Role of the two-component regulator CpxAR in the virulence of *Salmonella enterica* serotype Typhimurium. Infect. Immun. *72*, 4654–4661.

Hung, D.L., Raivio, T.L., Jones, C.H., Silhavy, T.J., and Hultgren, S.J. (2001). Cpx signaling pathway monitors biogenesis and affects assembly and expression of P pili. EMBO J. *20*, 1508–1518.

Huvet, M., Toni, T., Sheng, X., Thorne, T., Jovanovic, G., Engl, C., Buck, M., Pinney, J.W., and Stumpf, M.P. (2010). The evolution of the Phage shock protein (Psp) response system: interplay between protein function, genomic organization and system function. Mol. Biol. Evol. *28*, 1141–1155.

Ingmer, H., and Brondsted, L. (2009). Proteases in bacterial pathogenesis. Res. Microbiol. *160*, 704–710.

Iwobi, A., Heesemann, J., Garcia, E., Igwe, E., Noelting, C., and Rakin, A. (2003). Novel virulence-associated type II secretion system unique to high-pathogenicity *Yersinia enterocolitica*. Infect. Immun. *71*, 1872–1879.

Joly, N., Engl, C., Jovanovic, G., Huvet, M., Toni, T., Sheng, X., Stumpf, M.P.H., and Buck, M. (2010). Managing membrane stress: the phage shock protein (Psp) response, from molecular mechanisms to physiology. FEMS Microbiol. Rev. *34*, 797–827.

Jones, P.G., VanBogelen, R.A., and Neidhardt, F.C. (1987). Induction of proteins in response to low temperature in *Escherichia coli*. J. Bacteriol. *169*, 2092–2095.

Jovanovic, G., Weiner, L., and Model, P. (1996). Identification, nucleotide sequence, and characterization of PspF, the transcriptional activator of the *Escherichia coli* stress-induced *psp* operon. J. Bacteriol. *178*, 1936–1945.

Jovanovic, G., Lloyd, L.J., Stumpf, M.P., Mayhew, A.J., and Buck, M. (2006). Induction and function of the phage shock protein extracytoplasmic stress response in *Escherichia coli*. J. Biol. Chem. *281*, 21147–21161.

Jovanovic, G., Engl, C., and Buck, M. (2009). Physical, functional and conditional interactions between ArcAB and phage shock proteins upon secretin-induced stress in *Escherichia coli*. Mol. Microbiol. *74*, 16–28.

Jovanovic, G., Engl, C., Mayhew, A.J., Burrows, P.C., and Buck, M. (2010). Properties of the phage-shock-protein (Psp) regulatory complex that govern signal transduction and induction of the Psp response in *Escherichia coli*. Microbiology *156*, 2920–2932.

Kapatral, V., and Minnich, S.A. (1995). Co-ordinate, temperature-sensitive regulation of the three *Yersinia enterocolitica* flagellin genes. Mol. Microbiol. *17*, 49–56.

Karlyshev, A.V., Oyston, P.C., Williams, K., Clark, G.C., Titball, R.W., Winzeler, E.A., and Wren, B.W. (2001). Application of high-density array-based signature-tagged mutagenesis to discover novel *Yersinia* virulence-associated genes. Infect. Immun. *69*, 7810–7819.

Kausrud, K.L., Begon, M., Ari, T.B., Viljugrein, H., Esper, J., Buntgen, U., Leirs, H., Junge, C., Yang, B., Yang, M., *et al.* (2010). Modeling the epidemiological history of plague in Central Asia: palaeoclimatic forcing on a disease system over the past millennium. BMC Biol. *8*, 112.

Kleerebezem, M., and Tommassen, J. (1993). Expression of the *pspA* gene stimulates efficient protein export in *Escherichia coli*. Mol. Microbiol. *7*, 947–956.

Kleerebezem, M., Crielaard, W., and Tommassen, J. (1996). Involvement of stress protein PspA (phage shock protein A) of *Escherichia coli* in maintenance of the protonmotive force under stress conditions. EMBO J. *15*, 162–171.

Kobayashi, R., Suzuki, T., and Yoshida, M. (2007). *Escherichia coli* phage-shock protein A (PspA) binds to membrane phospholipids and repairs proton leakage of the damaged membranes. Mol. Microbiol. *66*, 100–109.

Korotkov, K.V., Gonen, T., and Hol, W.G. (2011). Secretins: dynamic channels for protein transport across membranes. Trends Biochem. Sci. *36*, 433–443.

Koster, M., Bitter, W., de Cock, H., Allaoui, A., Cornelis, G.R., and Tommassen, J. (1997). The outer membrane component, YscC, of the Yop secretion machinery of *Yersinia enterocolitica* forms a ring-shaped multimeric complex. Mol. Microbiol. *26*, 789–797.

Laroche, Y., van Bouchaute, M., and Cornelis, G. (1984). A restriction map of virulence plasmid pVYE439–80 from a serogroup 9 *Yersinia enterocolitica* strain. Plasmid *12*, 67–70.

Leblanc, S.K., Oates, C.W., and Raivio, T.L. (2011). Characterization of the induction and cellular role of the BaeSR two-component envelope stress response of *Escherichia coli*. J. Bacteriol. *193*, 3367–3375.

Li, C., and Clarke, S. (1992). A protein methyltransferase specific for altered aspartyl residues is important in *Escherichia coli* stationary-phase survival and heat-shock resistance. Proc. Natl. Acad. Sci. U.S.A. *89*, 9885–9889.

Li, S.R., Dorrell, N., Everest, P.H., Dougan, G., and Wren, B.W. (1996). Construction and characterization of a *Yersinia enterocolitica* O:8 high-temperature requirement (*htrA*) isogenic mutant. Infect. Immun. *64*, 2088–2094.

Lindquist, S. (1986). The heat-shock response. Annu. Rev. Biochem. *55*, 1151–1191.

Lloyd, L.J., Jones, S.E., Jovanovic, G., Gyaneshwar, P., Rolfe, M.D., Thompson, A., Hinton, J.C., and Buck, M. (2004). Identification of a new member of the phage shock protein response in *Escherichia coli*, the phage shock protein G (PspG). J. Biol. Chem. *279*, 55707–55714.

McElroy, K.M., Blagburn, B.L., Breitschwerdt, E.B., Mead, P.S., and McQuiston, J.H. (2010). Flea-associated zoonotic diseases of cats in the USA: bartonellosis, flea-borne rickettsioses, and plague. Trends Parasitol. *26*, 197–204.

McMahon, C.M., Byrne, C.M., Sheridan, J.J., McDowell, D.A., Blair, I.S., and Hegarty, T. (2000). The effect of culture growth phase on induction of the heat shock response in *Yersinia enterocolitica* and *Listeria monocytogenes*. J. Appl. Microbiol. *89*, 198–206.

MacRitchie, D.M., Buelow, D.R., Price, N.L., and Raivio, T.L. (2008). Two-component signaling and Gram-negative envelope stress response systems. Adv. Exp. Med. Biol. *631*, 80–110.

Mahan, M.J., Slauch, J.M., and Mekalanos, J.J. (1993). Selection of bacterial virulence genes that are specifically induced in host tissues. Science *259*, 686–688.

Maher, S.P., Ellis, C., Gage, K.L., Enscore, R.E., and Peterson, A.T. (2010). Range-wide determinants of plague distribution in North America. Am. J. Trop. Med. Hyg. *83*, 736–742.

Majdalani, N., and Gottesman, S. (2005). The Rcs phosphorelay: a complex signal transduction system. Annu. Rev. Microbiol. *59*, 379–405.

Matsumoto, H., and Young, G.M. (2006). Proteomic and functional analysis of the suite of Ysp proteins exported by the Ysa type III secretion system of *Yersinia enterocolitica* Biovar 1B. Mol. Microbiol. *59*, 689–706.

Maxson, M.E., and Darwin, A.J. (2004). Identification of inducers of the *Yersinia enterocolitica* phage shock protein system and comparison to the regulation of the RpoE and Cpx extracytoplasmic stress responses. J. Bacteriol. *186*, 4199–4208.

Maxson, M.E., and Darwin, A.J. (2006). PspB and PspC of *Yersinia enterocolitica* are dual function proteins: regulators and effectors of the phage-shock-protein response. Mol. Microbiol. *59*, 1610–1623.

Mecsas, J., Bilis, I., and Falkow, S. (2001). Identification of attenuated *Yersinia pseudotuberculosis* strains and characterization of an orogastric infection in BALB/c mice on day 5 postinfection by signature-tagged mutagenesis. Infect. Immun. *69*, 2779–2787.

Meltzer, M., Hasenbein, S., Mamant, N., Merdanovic, M., Poepsel, S., Hauske, P., Kaiser, M., Huber, R., Krojer, T., Clausen, T., *et al.* (2009). Structure, function and regulation of the conserved serine proteases DegP and DegS of *Escherichia coli*. Res. Microbiol. *160*, 660–666.

Merhej, V., Adekambi, T., Pagnier, I., Raoult, D., and Drancourt, M. (2008). *Yersinia massiliensis* sp. nov., isolated from fresh water. Int. J. Syst. Evol. Microbiol. *58*, 779–784.

Miller, S.I., and Mekalanos, J.J. (1990). Constitutive expression of the *phoP* regulon attenuates *Salmonella* virulence and survival within macrophages. J. Bacteriol. *172*, 2485–2490.

Miller, S.I., Kukral, A.M., and Mekalanos, J.J. (1989). A two-component regulatory system (*phoP phoQ*) controls *Salmonella typhimurium* virulence. Proc. Natl. Acad. Sci. U.S.A. *86*, 5054–5058.

Mitobe, J., Arakawa, E., and Watanabe, H. (2005). A sensor of the two-component system CpxA affects expression of the type III secretion system through posttranscriptional processing of InvE. J. Bacteriol. *187*, 107–113.

Model, P., Jovanovic, G., and Dworkin, J. (1997). The *Escherichia coli* phage-shock-protein (*psp*) operon. Mol. Microbiol. *24*, 255–261.

Motin, V.L., Georgescu, A.M., Fitch, J.P., Gu, P.P., Nelson, D.O., Mabery, S.L., Garnham, J.B., Sokhansanj, B.A., Ott, L.L., Coleman, M.A., *et al.* (2004). Temporal global changes in gene expression during temperature transition in *Yersinia pestis*. J. Bacteriol. *186*, 6298–6305.

Nagakubo, S., Nishino, K., Hirata, T., and Yamaguchi, A. (2002). The putative response regulator BaeR stimulates multidrug resistance of *Escherichia coli* via a novel multidrug exporter system, MdtABC. J. Bacteriol. *184*, 4161–4167.

Nagasawa, S., Ishige, K., and Mizuno, T. (1993). Novel members of the two-component signal transduction genes in *Escherichia coli*. J. Biochem. *114*, 350–357.

Neuhaus, K., Francis, K.P., Rapposch, S., Gorg, A., and Scherer, S. (1999). Pathogenic *Yersinia* species carry a novel, cold-inducible major cold shock protein tandem gene duplication producing both bicistronic and monocistronic mRNA. J. Bacteriol. *181*, 6449–6455.

Neuhaus, K., Rapposch, S., Francis, K.P., and Scherer, S. (2000). Restart of exponential growth of cold-shocked *Yersinia enterocolitica* occurs after down-regulation of *cspA1/A2* mRNA. J. Bacteriol. *182*, 3285–3288.

Nevesinjac, A.Z., and Raivio, T.L. (2005). The Cpx envelope stress response affects expression of the type IV bundle-forming pili of enteropathogenic *Escherichia coli*. J. Bacteriol. *187*, 672–686.

O'Loughlin, J.L., Spinner, J.L., Minnich, S.A., and Kobayashi, S.D. (2010). *Yersinia pestis* two-component gene regulatory systems promote survival in human neutrophils. Infect. Immun. *78*, 773–782.

Oyston, P.C., Dorrell, N., Williams, K., Li, S.R., Green, M., Titball, R.W., and Wren, B.W. (2000). The response regulator PhoP is important for survival under conditions of macrophage-induced stress and virulence in *Yersinia pestis*. Infect. Immun. *68*, 3419–3425.

Pallen, M.J., and Wren, B.W. (1997). The HtrA family of serine proteases. Mol. Microbiol. *26*, 209–221.

Palonen, E., Lindstrom, M., and Korkeala, H. (2010). Adaptation of enteropathogenic *Yersinia* to low growth temperature. Crit. Rev. Microbiol. *36*, 54–67.

Perez, J.C., Shin, D., Zwir, I., Latifi, T., Hadley, T.J., and Groisman, E.A. (2009). Evolution of a bacterial regulon controlling virulence and Mg^{2+} homeostasis. PLoS Genet. *5*, e1000428.

Ponnusamy, D., Hartson, S.D., and Clinkenbeard, K.D. (2011). Intracellular *Yersinia pestis* expresses general stress response and tellurite resistance proteins in mouse macrophages. Vet. Microbiol. *150*, 146–151.

Portnoy, D.A., Moseley, S.L., and Falkow, S. (1981). Characterization of plasmids and plasmid-associated determinants of *Yersinia enterocolitica* pathogenesis. Infect. Immun. *31*, 775–782.

Prost, L.R., and Miller, S.I. (2008). The Salmonellae PhoQ sensor: mechanisms of detection of phagosome signals. Cell. Microbiol. *10*, 576–582.

Pujol, C., and Bliska, J.B. (2005). Turning *Yersinia* pathogenesis outside in: subversion of macrophage function by intracellular yersiniae. Clin. Immunol. *114*, 216–226.

Raffa, R.G., and Raivio, T.L. (2002). A third envelope stress signal transduction pathway in *Escherichia coli*. Mol. Microbiol. *45*, 1599–1611.

Raivio, T.L. (2005). Envelope stress responses and Gram-negative bacterial pathogenesis. Mol. Microbiol. *56*, 1119–1128.

Rebeil, R., Ernst, R.K., Gowen, B.B., Miller, S.I., and Hinnebusch, B.J. (2004). Variation in lipid A structure in the pathogenic yersiniae. Mol. Microbiol. *52*, 1363–1373.

Rhodius, V.A., Suh, W.C., Nonaka, G., West, J., and Gross, C.A. (2006). Conserved and variable functions of the sigmaE stress response in related genomes. PLoS Biol. *4*, e2.

Ronnebaumer, K., Sander, G., Shutinoski, B., Schmidt, M.A., and Heusipp, G. (2009). Controlled activation of the Cpx system is essential for growth of *Yersinia enterocolitica*. FEMS Microbiol. Lett. *296*, 274–281.

Rowley, G., Spector, M., Kormanec, J., and Roberts, M. (2006). Pushing the envelope: extracytoplasmic stress responses in bacterial pathogens. Nat. Rev. Microbiol. *4,* 383–394.

Schiemann, D.A. (1987). *Yersinia enterocolitica* in milk and dairy products. J. Dairy Sci. *70,* 383–391.

Seo, J., Savitzky, D.C., Ford, E., and Darwin, A.J. (2007). Global analysis of tolerance to secretin-induced stress in *Yersinia enterocolitica* suggests that the phage-shock-protein system may be a remarkably self-contained stress response. Mol. Microbiol. *65,* 714–727.

Seo, J., Brencic, A., and Darwin, A.J. (2009). Analysis of secretin-induced stress in *Pseudomonas aeruginosa* suggests prevention rather than response and identifies a novel protein involved in secretin function. J. Bacteriol. *191,* 898–908.

Smith, C.R., Tucker, J.R., Wilson, B.A., and Clover, J.R. (2010). Plague studies in California: a review of long-term disease activity, flea–host relationships and plague ecology in the coniferous forests of the Southern Cascades and northern Sierra Nevada mountains. J. Vector. Ecol. *35,* 1–12.

Song, Y., Tong, Z., Wang, J., Wang, L., Guo, Z., Han, Y., Zhang, J., Pei, D., Zhou, D., Qin, H., *et al.* (2004). Complete genome sequence of *Yersinia pestis* strain 91001, an isolate avirulent to humans. DNA Res. *11,* 179–197.

Souza, R.A., Pitondo-Silva, A., Falcao, D.P., and Falcao, J.P. (2010). Evaluation of four molecular typing methodologies as tools for determining taxonomy relations and for identifying species among *Yersinia* isolates. J. Microbiol. Methods. *82,* 141–150.

Sprague, L.D., and Neubauer, H. (2005). *Yersinia aleksiciae* sp. nov. Int. J. Syst. Evol. Microbiol. *55,* 831–835.

Sprague, L.D., Scholz, H.C., Amann, S., Busse, H.J., and Neubauer, H. (2008). *Yersinia similis* sp. nov. Int. J. Syst. Evol. Microbiol. *58,* 952–958.

Standar, K., Mehner, D., Osadnik, H., Berthelmann, F., Hause, G., Lunsdorf, H., and Bruser, T. (2008). PspA can form large scaffolds in *Escherichia coli.* FEBS Lett. *582,* 3585–3589.

Stingl, K., and De Reuse, H. (2005). Staying alive overdosed: how does *Helicobacter pylori* control urease activity? Int. J. Med. Microbiol. *295,* 307–315.

Straley, S.C., and Harmon, P.A. (1984). *Yersinia pestis* grows within phagolysosomes in mouse peritoneal macrophages. Infect. Immun. *45,* 655–659.

Sukegawa, Y., Kamiya, S., Yagita, A., Sugamata, M., and Atomi, Y. (2000). Induction of autoimmune colitis by *Yersinia enterocolitica* 60-kilodalton heat-shock protein. Scand. J. Gastroenterol. *35,* 1188–1193.

Sulakvelidze, A. (2000). Yersiniae other than *Y. enterocolitica, Y. pseudotuberculosis,* and *Y. pestis*: the ignored species. Microbes. Infect. *2,* 497–513.

Sun, Y.C., Hinnebusch, B.J., and Darby, C. (2008). Experimental evidence for negative selection in the evolution of a *Yersinia pestis* pseudogene. Proc. Natl. Acad. Sci. U.S.A. *105,* 8097–8101.

Takata, R., Mukai, T., and Hori, K. (1985). Attenuation and processing of RNA from the *rpsO-pnp* transcription unit of *Escherichia coli.* Nucleic Acids Res. *13,* 7289–7297.

Tidhar, A., Flashner, Y., Cohen, S., Levi, Y., Zauberman, A., Gur, D., Aftalion, M., Elhanany, E., Zvi, A., Shafferman, A., *et al.* (2009). The NlpD lipoprotein is a novel *Yersinia pestis* virulence factor essential for the development of plague. PLoS One *4,* e7023.

Trosky, J.E., Liverman, A.D.B., and Orth, K. (2008). *Yersinia* outer proteins: Yops. Cell. Microbiol. *10,* 557–565.

Venecia, K., and Young, G.M. (2005). Environmental regulation and virulence attributes of the Ysa type III secretion system of *Yersinia enterocolitica* biovar 1B. Infect. Immun. *73,* 5961–5977.

Walker, K.A., and Miller, V.L. (2009). Synchronous gene expression of the *Yersinia enterocolitica* Ysa type III secretion system and its effectors. J. Bacteriol. *191,* 1816–1826.

Weiner, L., and Model, P. (1994). Role of an *Escherichia coli* stress-response operon in stationary-phase survival. Proc. Natl. Acad. Sci. U.S.A. *91,* 2191–2195.

Weiner, L., Brissette, J.L., and Model, P. (1991). Stress-induced expression of the *Escherichia coli* phage shock protein operon is dependent on σ^{54} and modulated by positive and negative feedback mechanisms. Genes Dev. *5,* 1912–1923.

Williams, K., Oyston, P.C., Dorrell, N., Li, S., Titball, R.W., and Wren, B.W. (2000). Investigation into the role of the serine protease HtrA in *Yersinia pestis* pathogenesis. FEMS Microbiol. Lett. *186,* 281–286.

Wimsatt, J., and Biggins, D.E. (2009). A review of plague persistence with special emphasis on fleas. J. Vector Borne Dis. *46,* 85–99.

Winfield, M.D., Latifi, T., and Groisman, E.A. (2005). Transcriptional regulation of the 4-amino-4-deoxy-L-arabinose biosynthetic genes in *Yersinia pestis.* J. Biol. Chem. *280,* 14765–14772.

Wong, R.S., and Chow, A.W. (2002). Identification of enteric pathogens by heat shock protein 60 kDa (HSP60) gene sequences. FEMS Microbiol. Lett. *206,* 107–113.

Yagita, A., Sukegawa, Y., Maruyama, S., Sato, N., Atomi, Y., Yamaguchi, H., Kamiya, S., Ihara, T., and Sugamata, M. (1999). Mouse colitis induced by *Escherichia coli* producing *Yersinia enterocolitica* 60-kilodalton heat-shock protein: light and electron microscope study. Dig. Dis. Sci. *44,* 445–451.

Yamaguchi, H., Yamamoto, T., Taguchi, H., and Ogata, S. (1990). *Yersinia enterocolitica* immunodominant 60 kDa antigen, common to a broad range of bacteria, is a heat-shock protein. J. Gen. Microbiol. *136,* 1091–1097.

Yamaguchi, S., Gueguen, E., Horstman, N.K., and Darwin, A.J. (2010). Membrane association of PspA depends on activation of the phage-shock-protein response in *Yersinia enterocolitica.* Mol. Microbiol. *78,* 429–443.

Yamamoto, T., Hanawa, T., Ogata, S., and Kamiya, S. (1996). Identification and characterization of the *Yersinia enterocolitica gsrA* gene, which protectively responds to intracellular stress induced by macrophage phagocytosis and to extracellular environmental stress. Infect. Immun. *64,* 2980–2987.

Yamamoto, T., Hanawa, T., Ogata, S., and Kamiya, S. (1997). The *Yersinia enterocolitica* GsrA stress protein, involved in intracellular survival, is induced by macrophage phagocytosis. Infect. Immun. *65*, 2190–2196.

Young, G.M., and Miller, V.L. (1997). Identification of novel chromosomal loci affecting *Yersinia enterocolitica* pathogenesis. Mol. Microbiol. *25*, 319–328.

Young, G.M., Amid, D., and Miller, V.L. (1996). A bifunctional urease enhances survival of pathogenic *Yersinia enterocolitica* and *Morganella morganii* at low pH. J. Bacteriol. *178*, 6487–6495.

Zhang, Y., and Bliska, J.B. (2005). Role of macrophage apoptosis in the pathogenesis of *Yersinia*. Curr. Top. Microbiol. Immunol. *289*, 151–173.

Zhou, D., Han, Y., Qin, L., Chen, Z., Qiu, J., Song, Y., Li, B., Wang, J., Guo, Z., Du, Z., *et al.* (2005). Transcriptome analysis of the Mg^{2+}-responsive PhoP regulator in *Yersinia pestis*. FEMS Microbiol. Lett. *250*, 85–95.

Zhou, D., Han, Y., Qiu, J., Qin, L., Guo, Z., Wang, X., Song, Y., Tan, Y., Du, Z., and Yang, R. (2006a). Genome-wide transcriptional response of *Yersinia pestis* to stressful conditions simulating phagolysosomal environments. Microbes Infect. *8*, 2669–2678.

Zhou, D., Han, Y., and Yang, R. (2006b). Molecular and physiological insights into plague transmission, virulence and etiology. Microbes Infect. *8*, 273–284.

Adaptations to Environmental Changes: Stress Response Mechanisms among *Vibrio* Species

8

W. Brian Whitaker and E. Fidelma Boyd

Abstract

Members of the genus *Vibrio* are Gram-negative ubiquitous marine bacteria. They can be isolated directly from the water column but are perhaps most known for their association with eukaryotic organisms. In their association with eukaryotic hosts, be it pathogenic or symbiotic, these bacteria must respond to a variety of stress conditions present within the host environment. Often times, these stress response systems are vitally important for the vibrios to successfully establish in the host. Here, we will discuss the systems used by the three main human pathogens of the genus, *V. cholerae, V. parahaemolyticus,* and *V. vulnificus* as well as briefly discussing the stress response systems of *V. fischeri, V. splendidus,* and *V. anguillarum*, all of which form close associations with marine organisms.

The genus *Vibrio*

The genus *Vibrio* comprises almost exclusively marine bacteria. The name of the genus is derived from Greek meaning 'curved rod', which accurately describes the comma-shaped morphology of the members of this group. Members of this genus are ubiquitous in the marine environment where they can be isolated directly from the water column or often in association with marine organisms. Perhaps the best-known representatives of this genus are the human pathogens *V. cholerae, V. parahaemolyticus,* and *V. vulnificus*. These species have been the subject of much study as they can pose a significant health risk in areas around the world where they take hold. The genus is not restricted to human pathogens, many members of this group are associated with fish, crustaceans and molluscs, often forming either a parasitic or symbiotic relationship with marine organisms.

The lifestyles of many *Vibrio* species find them encountering a diverse set of environmental conditions. In the marine and estuarine environment, the bacteria must overcome changing temperatures and salinities as well as nutrient limitations. In an era of global climate change, the overall increase in ocean temperatures as well as the acidification of these waters, also impacts these species in terms of their numbers, distribution and function. When these bacteria infect the host, typically through the gastrointestinal tract, they are faced with a new set of stress conditions, including, but not limited to, low-pH conditions, low-salinity conditions, the presence of bile as well as challenges from the host immune system. Because of this, many *Vibrio* species have developed a number of stress response systems that allow them to traverse these two vastly different environments. Here, we will detail the main stress response systems employed by the key members of this genus with an emphasis on how these systems play a role in colonizing or interacting with their eukaryotic hosts.

Vibrio cholerae

Perhaps the best-known of the pathogenic *Vibrio* species is the human intestinal pathogen *V. cholerae,* which causes the disease cholera. *Vibrio cholerae* is typically found in freshwater environments and is transmitted via the faecal–oral route upon consumption of contaminated drinking water. There are approximately 3–5 million cases of cholera a year, which lead to approximately

100,000–130,000 deaths. Even though treatment of this disease is relatively simple, administering oral rehydration salts, cholera remains a significant disease in parts of the world where medical care is poor or non-existent. Aside from being the best-known member of this genus, *V. cholerae* is also the most extensively studied. It was first determined that the disease cholera was spread through contaminated drinking water in the mid-1800s by John Snow (Snow, 1855). Later, in 1883, Robert Koch isolated the bacteria from clinical stool samples of patients suffering from cholera, providing evidence that this disease was of bacterial origin (Koch, 1884).

Vibrio cholerae is endemic to the Indian subcontinent and has been described there for centuries, and outbreaks of the disease may even be traced to ancient Greek and Sanskrit writings. Onset of disease takes approximately 18 hours to 5 days from the time a patient becomes infected with the bacterium to when symptoms of the disease present themselves. Symptoms of cholera are typically explosive watery diarrhoea and vomiting. The diarrhoea associated with cholera is often painless, voluminous, and resembles rice water. Likewise, the vomitus is typically clear and watery. The rate of diarrhoea may be so severe that is it not uncommon for patients to lose up to 1 l per hour. This high rate of water loss leads to severe dehydration in patients, and the symptoms that often accompany dehydration, such as low peripheral pulse, extremely low to undetectable blood pressure, poor skin turgor, and sunken eyes. Mortality associated with cholera often occurs within 24 hours but in some cases where water loss is severe, death may occur within hours of the onset of symptoms. Without treatment, mortality rates for patients presenting with cholera can be as high as 50%. However, current treatment methods, such as the administering of oral rehydration salts, have proved to be very effective in preventing mortality in cholera patients and are also relatively simple and easy to administer.

The pathogenesis of pandemic strains of *V. cholerae* is intimately related to their ability to acquire particular genetic elements via horizontal gene transfer. The genes that encode for the cholera toxin subunits A and B are encoded on the horizontally acquired CTX phage (Waldor and Mekalanos, 1996). The cholera toxin is an oligomeric complex comprises six subunits, one A subunit and five B subunits. The B subunits form a five-membered ring that contains the single A subunit. In the intestine, the toxin is secreted by *V. cholerae* and subsequently binds to the GM1 gangliosides, which are present on the surface of the intestine epithelial cells (Holmgren *et al.*, 1975). The toxin is endocytosed and after entry into the host cells, the A subunit detaches from the toxin complex and catalyses the ADP-rybosylation of $G\alpha_s$ subunit of the G-protein coupled receptors. This causes constitutive production of cAMP, which in turns leads to a massive efflux of electrolytes and H_2O from the intestinal cells, and thus the profuse watery diarrhoea associated with cholera (Holmgren, 1981). The toxin coregulated pilus (TCP) is encoded by a cluster of 15 genes located on the mobile genetic element known as *Vibrio* pathogenicity island-1 (VPI-1) (Karaolis *et al.*, 1998). TCP is a type IV pilus that has been shown to be an essential colonization factor, as TCP strains are unable to colonize the infant mouse (Taylor *et al.*, 1987). Additionally, TCP serves as a receptor for the CTXφ (Waldor and Mekalanos, 1996). Also encoded on this island are four accessory colonization factors. Encoded on the Vibrio pathogenicity island-2 (VPI-2) is the gene for sialidase, which is an enzyme that cleaves sialic acid from the GM1 ganglioside, which is necessary for the binding of cholera toxin to the gangliosides (Jermyn and Boyd, 2002, 2005). These horizontally acquired elements are typically not present in non-O1 non-O139 strains of *V. cholerae* (Jermyn and Boyd, 2002, 2005).

A unique *Vibrio* extracellular stress response system: ToxRS

Bacteria have evolved a number of ways in which to sense changes in the extracellular environment and respond to these changes accordingly. This is especially true for the human *Vibrio* pathogens, which often must alternate between the vastly different marine and human host environments. The prototypical example of a two component regulator in bacteria is the OmpR/EnvZ system of *E. coli*. The EnvZ protein contains a periplasmic N-terminal domain, two transmembrane domains, and a cytoplasmic C-terminal domain

(Stock *et al.*, 2000; Cai and Inouye, 2002). It is thought that the periplasmic domain of this protein responds to changing osmolarity conditions, which leads to the autophosphorylation of a conserved histidine residue in the cytoplasmic kinase domain of the protein. The phosphate group is subsequently transferred to an aspartate residue found on the response regulator OmpR (Stock *et al.*, 2000; Cai and Inouye, 2002). Once phosphorylated, OmpR can bind DNA and serves as a transcription factor that modulates the expression of several outer membrane porins (Stock *et al.*, 2000; Cai and Inouye, 2002).

ToxRS is a two-component regulatory unit that has been well characterized in *V. cholerae*. This system is unique to members of the Vibrionaceae and differs structurally from EnvZ/OmpR-type two-component regulators. In the ToxRS system, ToxR serves as both the sensor and response regulator (Miller and Mekalanos, 1984). The ToxR protein contains a C-terminal periplasmic domain, a single transmembrane domain, and a DNA binding N-terminal cytoplasmic domain. The periplasmic domain acts as the sensor, and has been shown to respond to changes in pH, temperature, amino acids, and osmolarity (Miller *et al.*, 1987; Miller and Mekalanos, 1988). In response to changes in environmental conditions, it is thought that ToxR forms homodimers (Dziejman and Mekalanos, 1994; Ottemann and Mekalanos, 1995). This dimerization leads to a conformational change in the cytoplasmic portion of the protein, subsequently allowing it to bind DNA. The *toxR* gene is transcribed in an operon with a second gene, known as *toxS*. The ToxS protein contains both a periplasmic and a transmembrane domain and it has been shown to have the ability to interact with the periplasmic domain of ToxR (DiRita and Mekalanos, 1991). Additionally, the presence of ToxS is necessary for the expression of the ToxR-regulon as it has been shown that genes in the ToxR-regulon are not expressed in a Δ*toxS* deletion mutant (DiRita and Mekalanos, 1991). It is proposed that ToxS stabilizes the homodimerization of ToxR, which is necessary for ToxR binding to its cognate promoter regions (DiRita and Mekalanos, 1991). The ToxR regulon in *V. cholerae* comprises two main branches, the ToxT-dependent branch and the ToxT-independent branch. In the ToxT-dependent branch of the regulon, ToxR acts as a co-activator of the *toxT* gene, which encodes a transcription factor that directly regulates the expression of the cholera toxin (the main cause of the explosive diarrhoea), as well as the toxin co-regulated pilus (TCP), which is an essential intestinal colonization factor (DiRita *et al.*, 1991).

Aside from its role as an important regulator of virulence, the ToxRS system has also been shown to be important for survival under a variety of environmental conditions. The role of stress response regulator is attributed to the ToxT-independent branch of the ToxR regulon. Here, ToxR directly regulates the expression of two outer membrane porins, OmpU and OmpT, in an inverse fashion. ToxR activation has been shown to lead to an increase in the expression of OmpU, while at the same time repressing the expression of OmpT. A *toxR*-deletion mutant of *V. cholerae* has been demonstrated to be deficient in mounting an adaptive acid tolerance response, and is hypersensitive to bile and other anionic detergents and will be discussed at greater length later (Merrell and Camilli, 1999, 2002; Provenzano *et al.*, 2000; Provenzano and Klose, 2000; Merrell *et al.*, 2002). Additionally, it has been shown that the lack of OmpU present in the *V. cholerae toxR*-deletion mutant is responsible for the sensitivity exhibited by this mutant in the face of the previously mentioned stresses (Provenzano and Klose, 2000; Merrell *et al.*, 2001).

Transcriptional regulation in bacteria: sigma factors

An additional method by which bacteria regulate gene expression is through the use of alternate sigma factors. Bacterial sigma factors are proteins that bind reversibly to the core RNA polymerase forming what is known as the holoenzyme. The bound sigma factor allows the holoenzyme to recognize promoter sequences that are specific to the particular bound sigma factor. Once transcription has been initiated, the sigma factor dissociates from the holoenzyme complex and the core RNA polymerase continues with transcription elongation (Helmann and Chamberlin, 1988; Kazmierczak *et al.*, 2005). In *E. coli*, the majority of gene expression is controlled by RpoD (also known as σ^{70}), which is known as the primary sigma factor. Alternative sigma factors are used to direct transcription under specific environmental conditions, allowing the bacteria to quickly alter their transcription depending upon different circumstances. The number of alternate sigma factors can vary by species.

RpoS is found in both Gram-negative and Gram-positive bacteria. In the enterobacteria *E. coli* and *Salmonella enterica*, RpoS is described as the stationary or stress response sigma factor. RpoS is essential for transcription of genes necessary for cell viability in stationary phase and under differing stress conditions (Lange and Hengge-Aronis, 1991; Fang *et al.*, 1992; Kazmierczak *et al.*, 2005). In *E. coli* RpoS has also been implicated in the acid tolerance and osmotic tolerance responses (Arnold *et al.*, 2001; Culham *et al.*, 2001; Kazmierczak *et al.*, 2005). In *S. enterica* serovar Typhimurium, RpoS is a major regulator of virulence, through its regulation of the *spv* gene cluster (Fang *et al.*, 1992; Heiskanen *et al.*, 1994). In *V. cholerae*, a *rpoS*-deletion mutant was shown to be deficient for survival in the face of hydrogen peroxide stress, hyperosmolarity stress, and carbon starvation (Yildiz and Schoolnik, 1998). Studies have shown that RpoS is necessary for *in vivo* survival of *V. cholerae* (Yildiz and Schoolnik, 1998; Merrell *et al.*, 2000). This was further examined by Schoolnik and colleagues, who demonstrated that RpoS was involved in the down-regulation of virulence gene expression *in vitro* and *in vivo*. Interestingly they found that *V. cholerae* isolated from ileal loop fluid and mucus showed significant RpoS-dependent up-regulation of many chemotaxis and motility genes. In addition, they found that in stationary phase cultures, RpoS was also required for up-regulation of chemotaxis and motility genes, production of flagella, and swarming movement (Nielsen *et al.*, 2006).

Acid stress and the adaptive acid tolerance response

Apart from its role in the digestive process, the low pH of the human stomach also serves as an innate physiological barrier against enteric pathogens which must first pass through the stomach in order to reach their preferred site of colonization in the intestine. Additionally, even though the small and large intestine are relatively neutral in pH, they do contain varying concentrations of fatty acids, which are the fermentative by-products of commensal microflora. Therefore, enteric pathogens must be able to not only survive the low, inorganic pH challenge presented during passage through the host stomach, but must also overcome the organic acid challenge presented in the host intestinal tract. *Vibrio cholerae*, like all *Vibrio*, is a neutrophile, and is sensitive to even mild pH challenges. This sensitivity is translated into the large infectious dose this organism requires to effectively colonize the human host. Studies have demonstrated that human volunteers required an oral dose of up to 10^{11} CFUs of *V. cholerae* to induce diarrhoea (Cash *et al.*, 1974; Reidl and Klose, 2002). However, when $NaHCO_3$ was administered concomitantly with an oral dose of *V. cholerae*, only 10^4 CFUs were required to cause diarrhoea in approximately 90% of the volunteers (Cash *et al.*, 1974; Reidl and Klose, 2002). Regardless of its sensitivities to acidic conditions, studies have indeed shown that *V. cholerae*, like *E. coli* and *S.* Typhimurium, possess systems that allow it to withstand inorganic and organic acid stress under certain conditions. Many enteric bacteria, and *V. cholerae* is no exception, are capable of a phenomenon known as the adaptive acid tolerance response (Foster, 1999; Merrell and Camilli, 2002). When bacteria are subjected to mild, non-lethal pH stress, they begin expressing systems necessary for survival in the face of lower, more lethal pH stress. This primes the cells for when they are subjected to what would typically be considered lethal pH stress. When *V. cholerae* is grown to mid-exponential phase and resuspended in LB broth adjusted to pH 7.0 for 1 hour, and then shifted to LB broth adjusted to pH 4.5 (HCl), there was a 98% reduction in CFUs after only 15 minutes. By 30 minutes, all of the initial population had been killed (Merrell and Camilli, 1999). Alternatively, when exponentially grown cells were transferred to LB broth adjusted to pH 5.7 (HCl) for 1 hour, and then transferred again to LB broth pH 4.5 (HCl), *V. cholerae* exhibited 100% survival, even after 60 minutes at pH 4.5 (Merrell and Camilli, 1999). This indicates that this organism is capable of mounting an adapted acid tolerance response that can drastically alter its ability to survive low pH environments (Merrell and Camilli, 1999). When *V. cholerae* cells were transferred to LB broth pH 5.7 plus the addition of chloramphenicol, they failed to mount an adaptive response and were unable to survive subsequent acid shock in pH 4.5 (Merrell and Camilli, 1999). This further indicates that the adapted acid tolerance response in *V. cholerae* is reliant upon the induction and synthesis of stress response proteins or, in this particular case, acid shock proteins.

Molecular mechanisms of acid tolerance: *cadBA* and *cadC*

Amino acid decarboxylation systems are common inducible acid stress response systems found throughout enteric bacteria. These operons typically comprise two genes, one encoding the decarboxylase enzyme and another encoding a substrate/product antiporter. There are also typically response elements associated with these operons that regulate expression in response to the presence of the particular amino acid substrate or changing extracellular pH conditions (Merrell and Camilli, 2002; Foster, 2004) (Fig. 8.1). In *V. cholerae,* the amino acid decarboxylase system responsible for responding to acid stress conditions is the lysine decarboxylase enzyme (Merrell and Camilli, 2002). The enzyme lysine decarboxylase is encoded by the gene *cadA* and is expressed in an operon with the lysine/cadaverine antiporter, encoded by the gene *cadB*. As the name implies, lysine decarboxylase facilitates the conversion of lysine into cadaverine + carbon dioxide. This reaction consumes an H^+ in the cytoplasm, and thus can be used to increase the internal pH of a cell during times of acid stress (Merrell and Camilli, 1999, 2002) (Fig. 8.1). The *cadA* gene in *V. cholerae* was first identified by Merrell and Camilli using IVET (*in vivo* expression technology) in conjunction with the rabbit

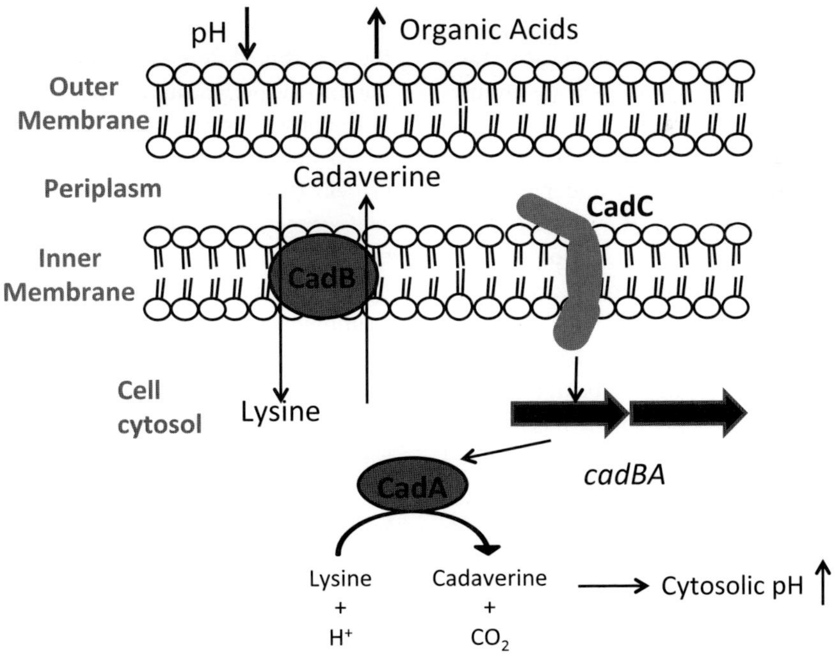

Figure 8.1 The lysine decarboxylase pathway in *V. cholerae* is important for survival under acidic conditions. CadC is a member of the OmpR-family DNA binding proteins and is localized to the inner membrane of the cell. When the pH of the extracellular environment decreases, or the concentration of extracellular organic acids increase, CadC increases the expression of *cadBA.* CadB is a lysine/cadaverine antiporter. CadA is the enzyme lysine decarboxylase, which converts lysine into cadaverine and carbon dioxide, and in doing so, consumes a proton in the cytoplasm. Cells employ this system in order to maintain the proper cytosolic pH in the face of acidic conditions.

ileal loop model and the suckling mouse model, in which it was found that the *cadA* gene showed an increase in expression in both *in vivo* models tested (Merrell and Camilli, 1999). The *V. cholerae* CadA was shown to share 73% and 75% amino acid identity with the those found in *E. coli* and *S.* Typhimurium respectively. Subsequent bioinformatic analysis revealed the presence of the *cadB* and *cadC* genes in the *V. cholerae* genome, which also showed similar homology and arrangement to those found in *E. coli* and *S. typhymurium*. When the *cadA* gene of *V. cholerae* is deleted via insertion mutagenesis, the *cadA* deletion strain (Δ*cadA*) exhibited no difference in the non-adapted acid response when compared with the isogenic wild-type strain. However, the Δ*cadA* was unable to mount an effective adapted acid tolerance response when this strain was subjected to mild, non-lethal inorganic (pH 5.7) and organic (pH 5.7 plus 6.5 mM acetic acid, 1.9 mM butyric acid, and 2.8 mM propionic acid) for 1 hour and then transferred to lethal inorganic acid (pH 4.5) or lethal organic acid (pH 4.5 plus 8.7 mM acetic acid, 2.5 mM butyric acid, and 3.7 mM propionic acid). In inorganic acid stress, the Δ*cadA* strain exhibited less than 10% survival after 60 minutes in lethal acid stress conditions, compared with the near 100% survival of the isogenic wild-type strain that was also pre-adapted under mild pH stress conditions. Likewise, the Δ*cadA* strain exhibited an approximate 90% reduction in viable cells after 30 minutes in lethal organic acid conditions, and no viable cells were able to be recovered after 60 minutes. Again this differs from the isogenic wild-type strain, which has been shown to be able to mount an affective adaptive acid tolerance response to organic acid stress (Merrell and Camilli, 1999). Bioinformatic searches have also revealed that the *V. cholerae* genome also encodes an ornithine decarboxylase gene, and it has been shown *in vitro* that *V. cholerae* does have the ability to convert ornithine to putrescine. Additionally, it does not appear that this organism contains any other amino acid decarboxylase systems. When the gene encoding ornithine decarboxylase (*speF*) is deleted, there is no apparent defect in this strain to mount an adaptive acid tolerance response, revealing that the lysine decarboxylase system is the sole amino acid decarboxylase system in this organism responsible for acid stress response (Merrell and Camilli, 1999). Overall, it appears that the CadBA system of *V. cholerae* plays a very important role in this organism, allowing it to effectively mount an adaptive acid tolerance response under both inorganic and organic stress conditions.

In *E. coli*, the lysine decarboxylase operon (*cadBA*) is regulated by an upstream element known as CadC. Genomic analysis reveals that *V. cholerae* contains *cadC* homologue, which encodes a protein that shares a 26% amino acid identity with the CadC of *E. coli* (Merrell and Camilli, 2000). The *cadC* gene of *V. cholerae* is upstream of the *cadB* gene, and is separated from the *cadB* gene by 699 base pairs, which encodes a hypothetical protein. The N-terminal portion of CadC is homologous to the N-terminal domain of the ToxR regulator. This region in the ToxR protein is cytoplasmic and contains an OmpR-like DNA binding domain. Further sequence analysis revealed that CadC also contains a transmembrane domain. Taken together CadC appears to localize to the inner membrane of the cell, with the C-terminal domain of the protein extending out into the periplasm and the N-terminal domain of the protein localizing to the cytoplasm, where it can interact with DNA, much like the ToxR protein (Merrell and Camilli, 2000). When the *cadC* gene is deleted in *V. cholerae,* the strain losses the ability to decarboxylate lysine, much like the Δ*cadA* strain. When *cadA* transcript levels were examined in both the isogenic wild-type strain and Δ*cadC* strains that were exposed to low pH, the wild-type strain was able to induce the expression of *cadA* to high levels. This induction of *cadA* was absent in the Δ*cadC* strain even when the cells were exposed to low pH stress, although there were low basal levels of *cadA* present regardless of external pH. Like the Δ*cadA* strain, the Δ*cadC* was also unable to mount an effective adaptive acid tolerance response in the face of both inorganic and organic acid stress, which is indicative of its role as a positive regulator of *cadA* expression (Merrell and Camilli, 2000). Interestingly, the loss of viability rate of the acid adapted Δ*cadC* strain was not as severe as the loss of viability rate of unadapted wild-type cells (Merrell and Camilli, 2000). This indicates two things:

1. Even though the Δ*cadC* cells are killed off at a much faster rate than adapted wild-type cells, they are still able to mount a small adaptive acid tolerance response since they do not die as fast as unadapted wild-type cells.
2. This response is independent of CadC regulation, indicating that the adapted acid tolerance response is the result of global changes in the *V. cholerae* cell. Interestingly, the Δ*cadC* deletion strain showed no defect in colonization when it was co-infected with the isogenic wild-type strain in the suckling mouse model, in much the same way the Δ*cadA* strain showed no detectable difference in colonization (Merrell and Camilli, 2000).

Additional mechanisms of acid tolerance

In addition to studying the role of lysine decarboxylase in *V. cholerae* acid tolerance response, Merrell and Camilli also examined if the global regulator ToxRS played a key role in mounting an effective stress response in the face of low pH stress (Merrell and Camilli, 1999). As stated previously, it is well known that ToxRS responds to external stimuli, including low pH, and in turn, regulates the expression of virulence genes via the ToxT-dependent pathway. For this reason, it was hypothesized that the ToxRS might also play a role in regulating stress response genes in the face of low pH as well. A *V. cholerae* Δ*toxRS* mutant was able to mount an adaptive acid tolerance response when it was pre-adapted in media that was adjusted to pH 5.7 before being subjected to lethal inorganic acid stress (pH 4.5). However, this same mutant was unable to mount an effective adaptive acid tolerance response when pre-adapted in pH 5.7 plus organic acids and then subjected to lethal organic acid stress. In an attempt to close in on the mechanism of ToxRS mediated organic acid resistance, the same experiments were repeated with a *toxT* deletion mutant. This mutant behaved as the isogenic wild-type strain in the face of organic acid challenge, indicating that the ToxRS system mediates adaptive acid tolerance through the regulation of genes in the ToxT-independent pathway (Merrell and Camilli, 1999). As stated earlier, it has been well established that ToxR directly acts as a positive regulator of the *ompU* gene and as a negative regulator of the *ompT* gene (Fig. 8.2). When the *ompU* gene is ectopically expressed *in trans* in a Δ*toxRS* background, the ability of the mutant strain to effectively mount an adaptive organic acid tolerance response is restored back to wild-type levels of survival. Interestingly, outer membrane profiles of wild-type cells grown in pH 7, 5.7, and 4.5 reveal that OmpU levels appear to be constant regardless of external pH, whereas in the Δ*toxRS* strain, OmpU is undetectable regardless of culture conditions. This could imply that OmpU is not an acid shock protein *per se*, but rather its presence in the outer membrane alone is required for survival under organic acid conditions (Merrell *et al.*, 2001).

Outer membrane stress response systems

During the pathogenic cycle of *V. cholerae*, it is faced with a variety of innate stresses as it progresses through the human host. The first is the acidity of the stomach, and *V. cholerae* possess the appropriate stress response systems that allow enough cells to survive this condition, as discussed in the previous section. Next, *V. cholerae* reaches its preferred site of colonization within the human gastrointestinal tract, the small intestine. Here it is faced with a variety of stress conditions, such as the presence of bile, whose concentration peaks in the small intestine, as well as organic acids, which are the metabolic byproducts of the gut microflora. Bile exerts a very power antimicrobial effect, so much so that while the large intestine (which contains very little bile) can support as many as 10^{14} bacteria cells/ml, the small intestine is relatively low in microbial life, with only 10^5 bacterial cells/ml (Hofmann and Eckmann, 2006). In the small intestine, bile concentrations can reach as high as 10 mM. To further prove the importance of bile as not only a digestive aid, but also as an antimicrobial compound, bile duct ligations and liver cirrhosis have both been shown to lead to a decrease in bile output and subsequently, bacterial overgrowth in the small intestine (Hofmann and Eckmann, 2006).

Bile itself is an emulsifying agent that is used to assist in the digestion and subsequent absorption of lipids in the small intestine. The functional

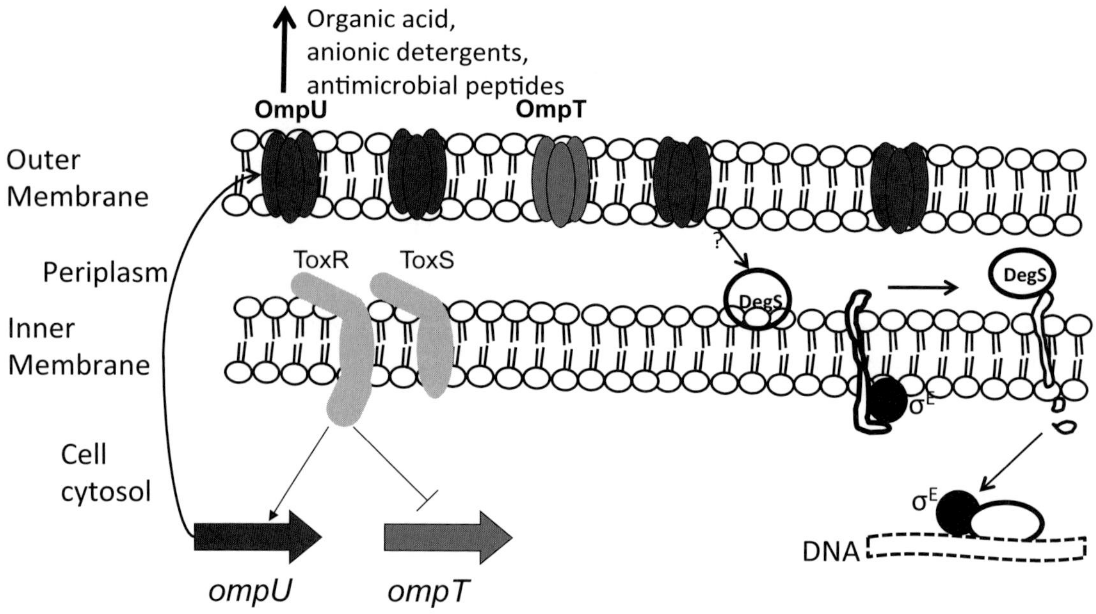

Figure 8.2 ToxRS is an important regulator of the outer membrane stress response. While best known as a virulence regulator, it also plays a role in the regulation of outer membrane proteins that are important in stress response. Outer membrane protein U (OmpU), which is regulated by ToxRS plays a role in stress response to organic acids, anionic detergents, or antimicrobial peptides. It is believed that OmpU may act as a signal that activates σ^E, which is the extracytoplasmic sigma factor responsible for mediated stresses involved with the outer membrane.

portion of the bile acid is the anion, which contains both hydrophilic and hydrophobic domains. It is believed that bile exerts its bactericidal effects by inducing membrane damage. Studies have demonstrated that bile has the ability to lysis erythrocytes. Given that this cell type lacks any type of uptake or metabolism mechanisms for bile, the only way haemolysis can occur is through damage of the cellular membrane (Begley *et al.*, 2005; Billington *et al.*, 1979). There is also evidence that the presence of bile may have a deleterious effect on bacterial proteins, as bile has been shown to increase the expression of a number of molecular chaperones, which are necessary to ensure proper folding of proteins (Begley *et al.*, 2005).

One of the most important means *V. cholerae* possesses that allows it to sense extracellular bile and respond to it accordingly, is the ToxRS system discusses previously. Studies have demonstrated that the ToxRS system responds to external bile concentration and increases the expression of the outer membrane protein OmpU. There has been conflicting reports on whether or not bile plays a role in the activation of virulence genes in *V. cholerae*. Gupta and Chowdhury (1997) reported that the presence of external bile did not increase the expression of ToxT, nor the expression of CT or TCP. However, Hung and Mekalanos (2005) reported that bile can stimulate CT production by activating ToxR, which in this case directly activated the expression of the cholera toxin genes.

The precise role of ToxRS in responding to bile stress has been extensively studied in *V. cholerae*. Provenzano and colleagues demonstrated that in both classical and El Tor strains of *V. cholerae*, a *toxR* deletion mutant exhibited a minimum bactericidal concentration (MBC) in bile that was half (3.75 and 7.5 respectively) of the MBC of the isogenic wild-type strains (7.5 and 15.0 respectively) (Provenzano *et al.*, 2000). Additionally, *toxR* deletion mutants also exhibited substantially less minimum bactericidal concentrations for deoxycholate (which is an individual component of bile) as well as in the presence of sodium dodecyl sulfate (SDS). Like bile, SDS is classified as an anionic detergent and most probably exerts a similar bactericidal mechanism against cells. Interestingly, both the *toxR* deletion strains and the isogenic wild-type strains showed no difference in the minimum bactericidal concentration when subjected to Triton X-100 stress, which is a non-ionic detergent (Provenzano *et al.*, 2000). This would indicate that the mechanism by which ToxRS confers resistance to *V. cholerae* may be specific to anionic detergents, and other stresses that act in a similar manner. That same study also examined the outer membrane profiles of *toxR* deletion mutants and the isogenic wild-type strains that were grown under standard conditions and also in the presence of sublethal bile concentrations. They found that the wild-type strains showed a marked increase in OmpU expression when grown in the presence of bile when compared with cells grown under standard conditions (LB broth). However, the *toxR* deletion strains showed no appreciable OmpU protein in the outer membrane regardless of growth conditions, which was to be expected since it is well known that ToxR controls the regulation of OmpU (Provenzano *et al.*, 2000). Taken together, it was hypothesized that presence or absence of OmpU determined a strain's sensitivity to external anionic detergents, much in the same way the ability to produce OmpU played a role in a strain's ability to mount an organic acid adapted response. The importance of OmpU was further confirmed in a follow up study (Provenzano and Klose, 2000), that confirmed the presence of OmpU and not OmpT was important for survival in the presence of bile. However, the exact mechanism for OmpU's role as a stress response protein remained elusive at the time.

Since the initial acid and bile studies, the role of OmpU in *V. cholerae* stress response and pathogenesis has been studied even further. Mathur and Waldor (2004) also examined the role of ToxR and OmpU in response to the presence of antimicrobial peptides. Antimicrobial peptides are part of the body's innate defence system against invading bacteria and are commonly found in the small intestine. BPI (bactericidal/permeability-increasing) is once such peptide that is typically found in neutrophil granules and on the neutrophil surface, but may also be found decorated on the surface of intestinal epithelial cells as well (Weiss *et al.*, 1978; Canny *et al.*, 2002; Mathur and Waldor, 2004). This peptide has affinity for the lipid A domain of lipopolysaccharide of Gram-negative bacteria and the protein itself contains several functional domains that exert different antimicrobial effects. For the purposes of the aforementioned study, P2, which is a synthetic peptide that contains a region of BPI, was used. P2 has been shown in *E. coli* to increase outer membrane permeability and also inhibit O_2 consumption. It was demonstrated that P2 was able to kill wild-type *V. cholerae* in a dose-dependent manner. Additionally, it was found that ToxR, and not ToxT, contributed resistance to P2, as a *toxR* deletion mutant was found to be nearly 100 times more sensitive to P2 than the isogenic wild type (Mathur and Waldor, 2004). A *toxT* deletion mutant was not significantly more sensitive to P2 than the isogenic wild type, indicating that ToxR-mediated resistance to P2 was independent of the ToxT pathway; leaving OmpU as the probable ToxR regulated stress response protein that enhances sensitivity to P2. Not surprisingly, when the *ompU* gene was deleted, and this strain subjected to the presence of P2, there was a near 1000-fold increase in sensitivity compared with the isogenic wild-type strain (Mathur and Waldor, 2004).

Additionally, the *ompU* deletion strain was also very sensitive to the antibiotic polymyxin B sulfate (Mathur and Waldor, 2004). Polymyxin B contains a hydrophilic and hydrophobic domain and is considered to fall under the category of cationic detergents. The hydrophilic domain of this peptide binds to Gram-negative LPS, after which the hydrophobic domain disrupts both the outer and inner membranes. As such, this antibiotic behaves much like a detergent, in that it causes disruption and increased permeability in bacterial membranes. Initially, this would appear to confirm earlier thoughts that the lack of OmpU in the outer membrane exhibited by *V. cholerae toxR* deletion mutants may result in either a destabilized membrane or one that is more permeable than the isogenic wild-type strains (Merrell *et al.*, 2001). However, further experimentation by separate studies indicated that this is not the mechanism by OmpU confers resistance to various membrane stress conditions (Mathur and Waldor, 2004). In the same study, they examined the effect of the deletion of the *omoU* gene had on the *V. cholerae* outer membrane (Mathur and Waldor, 2004). Here the *ompU* deletion mutant as well as the isogenic wild-type strain were grown separately under standard growth conditions that included the addition of nitrocefin, which is the chromogenic substrate for the enzyme β-lactamase (Mathur and Waldor, 2004). β-Lactamase is a periplasmic enzyme that hydrolyses the lactam class of antibiotics, and also in this case, the substrate nitrocefin. Since this enzyme is located in the cell's periplasm, hydrolysis of nitrocefin would only occur if the bacterial outer membrane is damaged or demonstrates increased permeability, which would result in leakage of periplasmic enzymes into the extracellular milieu. Both the *ompU* deletion mutant and the isogenic wild-type strain showed limited hydrolysis of nitrocefin when grown under standard conditions (Weiss *et al.*, 1978; Canny *et al.*, 2002; Mathur and Waldor, 2004). Similarly, when both strains were exposed to P2, they exhibited similar rates of nitrocefin hydrolysis, indicating that while P2 probably causes perturbations of the outer membrane, this is probably independent of the presence of OmpU and that this outer membrane protein must act in different way to confer resistance to P2-mediated stress (Weiss *et al.*, 1978; Canny *et al.*, 2002; Mathur and Waldor, 2004). This was also confirmed in a separate study by Provenzano and Klose, in which they too examined leakage of β-lactamase into culture supernatant and found that there was no difference between wild-type *V. cholerae* and an *ompU* deletion mutant (Provenzano *et al.*, 2001). Additionally, they observed no alterations in LPS structure in the *ompU* mutant strain, further indicating that the lack of OmpU does not in fact lead to alteration of the outer membrane structure (Provenzano *et al.*, 2001).

One of the main ways in which bacteria respond to outer membrane stress and alter their gene expression accordingly is through the alternate sigma factor E, encoded by the *rpoE* gene. Sigma factor E belongs to the sigma-70 family of sigma factors and is part of the extracytoplasmic function sigma factors. These sigma factors, under non-stress conditions, are inactivated by their cognate anti-sigma factor. The anti-sigma factor localizes to the inner membrane, where it holds the sigma factor until times of membrane stress. In *E. coli*, under conditions that lead to stress of the outer membrane (detergents, antimicrobial peptides, etc.) or the misfolding of proteins in the periplasm, sigma E is proteolytically cleaved from its anti-sigma factor and then directs the transcription of the appropriate response genes (Bashyam and Hasnain, 2004). It has been demonstrated that certain outer membrane proteins can act as upstream signals that modulate RpoE activity. The outer membrane proteins OmpC and OmpF have been shown to contain a conserved C-terminus consisting of a Y-Q-F motif (later modified to Y-X-F bases on sequences in other species). This C-terminal motif interacts with a conserved PDZ domain of the periplasmic protease DegS, which alters the conformation of the protein to reveal the catalytic site. Activated DegS then cleaves RpoE from its cognate anti-sigma factor, allowing it to properly function (Walsh *et al.*, 2003; Mathur *et al.*, 2007) (Fig. 8.2).

RpoE has been shown to play an important role in outer membrane stress response and resistance to antimicrobial peptides in *V. cholerae.* When the *rpoE* gene is deleted in *V. cholerae*, this strain exhibits sensitivity to the synthetic antimicrobial peptide P2, much like what is observed in *ompU* deletion mutants (Walsh *et al.*, 2003; Mathur *et al.*, 2007). Interestingly like OmpC and OmpF, the OmpU protein of *V. cholerae* contains a Y-D-F, C-terminal motif, and thus may serve as a signal for RpoE activation. Indeed studies have shown that complementation of *ompU* deletion mutants with RpoE relieves the deletion mutant's sensitivity to antimicrobial peptides, indicating that OmpU's effect on outer membrane stress response acts upstream to RpoE activation (Walsh *et al.*, 2003; Mathur *et al.*, 2007) (Fig. 8.2). Thus, the loss of OmpU from the *V. cholerae* outer membrane appears to prevent activation of RpoE, which is needed to induce the transcription of the necessary genetic response elements, instead of OmpU playing a role in maintaining outer membrane integrity or permeability.

Iron limitation and scavenging

Iron is essential for nearly all living organisms throughout all three major domains in life. Organisms must acquire the necessary amounts of iron from their environment in order to utilize this element in various physiological processes. Once acquired, iron is incorporated into proteins by organisms. This allows organisms to control the environment surrounding the iron ion, which in turn allows the organism to vary the redox potential, spin state, or geometry of the iron, depending on its intended biological function. Pathogenic microorganisms are unique in that many occupy multiple niches in their life cycles; that of the natural environment and that of their host organism. Inside the host, pathogens can acquire all of the necessary carbon sources (from host tissues and fluids), nitrogen (amino acids, nucleotides) as well as phosphate and other inorganic ions. However, iron is extremely limited within the host, much more so than in the pathogens non-host environment. In fact, while in external aerobic environments, ferric iron has a solubility of approximately 1×10^{-9} to 1×10^{-18} M (at pH 7) (Butler, 2003), whereas the concentration of soluble iron in the mammalian cell is approximately 1×10^{-24} M at pH 7 (Butler, 2003). This increased scarcity of iron within the eukaryotic host does not happen by accident. Any free iron is purposely sequestered by the host in the hopes of limiting this important nutrient in order to prevent colonization by potential pathogens. This phenomenon is often referred to as nutritional immunity and is a main part of the eukaryotic host's innate defence against colonization by pathogens.

One important compound used by eukaryotic hosts to limit the amount of free iron available to pathogens in through the use of transferrin. Transferrin is a type of glycoprotein that is produced by eukaryotes to sequester free iron. It is typically found in plasma and mucosal layers within eukaryotic organisms. Transferrin has the ability to reversibly bind up to two free Fe^{3+} ions, which it can then deliver to host cells via cell surface receptors. Another type of eukaryotic glycoprotein important in iron sequestering is lactoferrin. Lactoferrin is typically found in mammalian secretions, such as milk, mucous, tears, etc. Similarly to transferrin, it exerts its antimicrobial function via the uptake of free iron, thereby preventing its uptake by pathogenic bacteria. There is also evidence that lactoferrin is secreted by neutrophils in response to pathogen colonization. Lactoferrin is an important component of breast milk and play an important role in establishing the innate immune system in infants as well as promoting the establishment of beneficial gut microbiota (Ratledge and Dover, 2000; Andrews *et al.*, 2003; Butler, 2003). Ferritin is ubiquitous to both eukaryotes and bacteria alike. Ferritin is a large globular protein with a weight of 450 kDa and exists as a complex of 24 subunits. A single ferritin complex has the ability to bind 4500 Fe^{3+} ions, making it an ideal storage unit for iron. Ferritin is an intracellular protein, and this allows eukaryotes to store iron inside of cells, making it less available to colonizing bacteria (Ratledge and Dover, 2000; Andrews *et al.*, 2003; Butler, 2003).

Extracellular iron as a signal for genetic regulation

Given the relative importance of iron to pathogenic bacterial species and the lengths these organisms go to capture free iron and transport it into the cell, it would stand to reason that the abundance, or absence, of free soluble iron would act as a strong environmental regulator signal for these bacteria. One such method used to detect, and respond the presence of iron is the ferric uptake regulator (Fur). In *V. cholerae*, the ferric uptake regulator is a 150 amino acid protein encoded by the *fur* gene which shares a 76% amino acid identity with the Fur protein of *E. coli*. The 17 kDa Fur protein contains two functional domains: a DNA binding domain in the amino terminus with a helix–turn–helix motif and two metal binding sites in the carboxy terminus. It has been determined in *V. cholerae* that in a number of strains the Fur protein is typically expressed at very high levels (>5000 molecules/cell) (Watnick *et al.*, 1997). The DNA binding domain of the Fur protein has a very weak affinity for DNA and under iron limiting conditions, i.e. when all available iron is complexed with haem, cytochrome, etc. and no free iron exists in the cytoplasm, the Fur protein is unbound to iron and inactive. However, when iron is plentiful and there is free iron in the cytoplasm, the Fur protein will bind the excess iron, which effectively increases its binding affinity for DNA approximately 1000-fold. Once bound with iron, the Fur protein binds a sequence of DNA known as the fur-box (Smith *et al.*, 1996). Evidence has shown that the *V. cholera* fur-box does not appear to the deviate from the canonical *E. coli* fur-box sequence (GATAATGAT[A/T]ATCATTATC) (Mey *et al.*, 2005).

Iron acquisition

Under iron-limited conditions, *V. cholerae* is forced to scavenge iron from the extracellular environment, which it does through several different avenues. One such way of acquiring iron is through the use of siderophores (Greek for 'iron carriers'). Siderophores are iron chelating compounds that are synthesized and secreted out of bacterial cells in order to sequester iron. The usage of siderophores for iron scavenging is not confined solely to microorganisms, and this far over 500 different siderophores have been identified and isolated from bacteria, yeast, and fungi (Ratledge and Dover, 2000). Siderophores have been demonstrated to have a very high binding affinity to Fe^{3+}, with association constants better than 10^{20} and upwards of 10^{50} being reported for many of these compounds. Conversely, siderophores have been shown to have a much lower affinity for Fe^{2+} as well as other divalent and trivalent cations (Ratledge and Dover, 2000).

Vibrio cholerae has been demonstrated to synthesize the siderophore vibriobactin. Vibriobactin comprises a norspermidine backbone that contains three dihydroxybenzoate moieties (Fig. 8.3A). Two of the dihydroxybenzoate groups are attached to the norspermidine via cyclized threonine bridges, while the remaining dihydoxybenzoate group is attached directly to the norspermidine backbone. Synthesis of the dihydoxybenzoate groups requires the *vibABC* operon and assembly of the siderophore is accomplished by a non-ribosomal peptide synthesis-type mechanism, which is encoded by the *vibBDEFH* operon (Wyckoff *et al.*, 1997, 2007).

In order to acquire the necessary amount of iron for growth within the host, pathogenic bacteria will often secrete siderophores into the extracellular milieu to chelate free iron or strip away iron bound to host compounds. Once the siderophores are bound to Fe^{3+}, pathogens must have a way to re-assimilate the siderophores back into their cytoplasm where they can

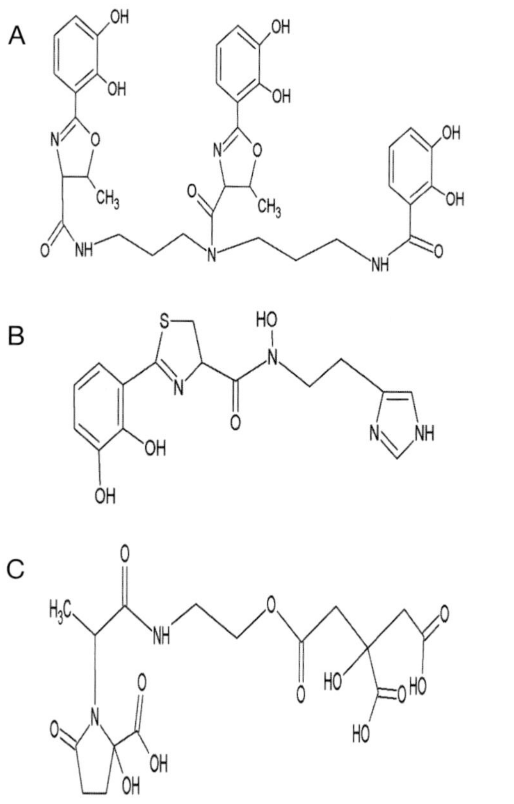

Figure 8.3 *Vibrio* spp. siderophore structure. (A) Vibriobactin produced by *V. cholerae*. (B) Anguibactin produced by *V. anguillarum*. (C) Vibrioferrin produced by *V. parahaemolyticus*.

make use of the scavenged iron. Ferri-siderophore uptake is similar throughout the different pathogenic bacteria, though each species produces different outer membrane receptors that are specific to the type of siderophore that particular species synthesizes. The ViuA protein is the receptor for ferri-vibriobactin and is encoded by the gene *viuA*. *V. cholerae* can also utilize the siderophore enterobactin, which is produced in a variety of Gram-negative bacteria, including *E. coli* and *S. typhymurium*. *V. cholerae* produces two proteins, IrgA and VctA, both of which can serve as receptors for ferri-enterobactin (Fig. 8.4). Beyond the ability to synthesize and uptake siderophores, *V. cholerae* also has the ability to scavenge iron directly from host haem and ferrichrome. *V. cholerae* produces three haem receptors, HutA, HutR, and HasR and one ferrichrome receptor, FhuA (Wyckoff *et al.*, 2007).

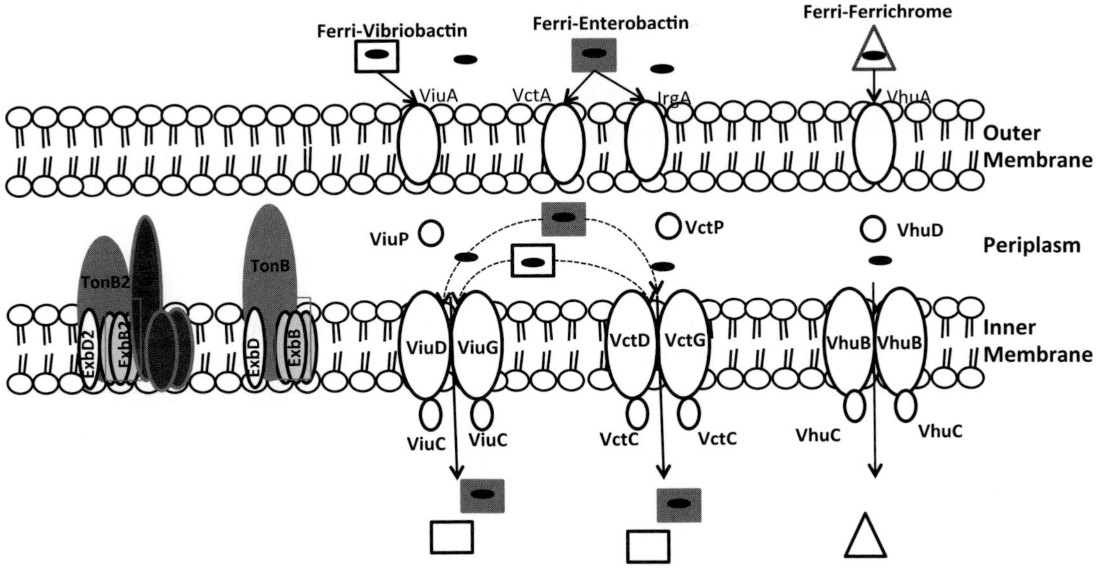

Figure 8.4 Siderophore uptake in *V. cholerae. V. cholerae* contains receptors for the siderophore vibriobactin, enterobactin, and ferrichrome. Once the ferri-siderophore binds to the appropriate receptor, it is translocated into the periplasm. The energy for this system is provided by either the TonB1 or TonB2 systems. Once in the periplasm, the ferri-siderophore is once again translocated into the cytoplasm via its cognate ABC transporter.

In Gram-negative bacteria, transport of the receptor bound siderophore across the outer membrane into the periplasm is accomplished using the TonB-ExbB-ExbD transport system (Ratledge and Dover, 2000; Andrews *et al.*, 2003; Krewulak and Vogel, 2011). TonB is a periplasmic protein that interacts with both the outer membrane and cytoplasmic membrane equally. The N-terminus of the TonB protein is anchored to the cytoplasmic membrane where it forms a complex with ExbB and ExbD, which are two proteins that reside in the cytoplasmic membrane. The C-terminus domain of the TonB protein has been demonstrated to interact directly with the periplasmic face of the outer membrane siderophore receptor (Ratledge and Dover, 2000; Andrews *et al.*, 2003; Krewulak and Vogel, 2011). Translocation of the ferri-siderophore complex involves two steps. The ferri-siderophore binds to the siderophore binding domain of the outer membrane siderophore receptor. Second, activation of the TonB protein causes a conformation change in the ligand bound outer membrane receptor, which allows for the translocation of the ferri-siderophore into the periplasm. The ExbB and ExbD proteins are believed to function by transferring the energy generated by the proton motive force conserved in the cytoplasmic membrane to the TonB protein, which in turn, allows the TonB to alter the conformation of the outer membrane receptor (Ratledge and Dover, 2000; Andrews *et al.*, 2003; Krewulak and Vogel, 2011). *V. cholerae,* as well as many other *Vibrio* spp. are unique in that they contain two or more TonB systems. *V. cholerae* contains two such systems, TonB1, which is most similar to the system found in *E. coli,* and TonB2. The TonB2 system contains a third novel protein, TtpC, which is believed to aid TonB2 to span the periplasm (Seliger *et al.*, 2001; Wyckoff *et al.*, 2007). In *V. cholerae,* the TonB1 system functions by supplying the energy to power haem uptake, while the TonB2 system powers siderophore uptake (Seliger *et al.*, 2001; Wyckoff *et al.*, 2007; Kuehl and Crosa, 2010). Both TonB systems appear to be important *in vivo*, as both deletion mutants exhibit a defect in the infant mouse model, which is exacerbated in a strain with TonB1 and TonB2 deleted (Seliger *et al.*, 2001).

Vibrio parahaemolyticus

Vibrio parahaemolyticus is a moderate halophile and is typically isolated from coastal and marine waters throughout much of the world. This organism is of significant medical importance as it is the leading cause of seafood-associated gastroenteritis in both the United States and much of Asia (Daniels *et al.*, 2000; Su and Liu, 2007). *Vibrio parahaemolyticus* has been isolated in the environment from the water column, sediment and quite commonly in association with crustaceans, molluscs, fish and planktonic copepods (Krantz *et al.*, 1969; Kaneko and Colwell, 1973; Colwell *et al.*, 1977). In these habitats, the organism can experience fluctuations in both salinity and temperature. Indeed, it has been shown that *V. parahaemolyticus* levels in marine and estuarine waters are linearly dependent on both salinity concentrations and water temperature. During winter months, *V. parahaemolyticus* is rarely isolated from the water column and is typically only found in small numbers in sediment (Kaneko and Colwell, 1973; Kaper, 1981). When the water temperature reaches above 15°C in warmer months, *V. parahaemolyticus* is released from the sediment and can be isolated in the water column and in association with marine organisms (Kaneko and Colwell, 1973; Kaper, 1981; DePaola *et al.*, 1990). It is typically during the warmer months of the year where *V. parahaemolyticus* infections are at the highest.

Vibrio parahaemolyticus was first identified as the causative agent of a food-borne illness outbreak in Japan in 1950. In this incidence, 272 individuals became ill and 20 died after the consumption of shirasu, which is lightly boiled juvenile sardines (Fujino, 1953). From 1950 until the mid-1990s, many different strains and serotypes of *V. parahaemolyticus* were found to be the aetiological agents of sporadic illness associated with the consumption of seafood in several countries, including the United States (Daniels *et al.*, 2000); however, these strains appeared to be endemic to particular regions. The mid-1990s saw the emergence of the O3:K6 serotype. This strain was responsible for an increase of *V. parahaemolyticus* hospitalizations in Calcutta in February of 1996. In the months that followed it is estimated that this serotype of *V. parahaemolyticus* was responsible for up to 80% of hospitalizations in Calcutta that resulted from *V. parahaemolyticus* infections (Okuda *et al.*, 1997; Nair *et al.*, 2007).

Infection by this bacterium is typically characterized by three types of illness: acute inflammatory gastroenteritis, and to a much lesser extent wound infections, and septicaemia (Morris and Black, 1985). Inflammatory gastroenteritis is the most common disease associated with *V. parahaemolyticus* infection, with symptoms that include diarrhoea with abdominal cramping, nausea, vomiting, and mild fevers. Gastroenteritis caused by *V. parahaemolyticus* is self-limiting and lasts approximately 3 days in patients that are immunocompetent. Diarrhoea caused by *V. parahaemolyticus* is distinct from the profuse rice water diarrhoea associated with the related species *V. cholerae*. While there is a wealth of knowledge regarding the synergistic interaction of the major virulence genes in *V. cholerae* (neuraminidase, the toxin co-regulated pilus, and the cholera toxin) much less is known about the virulence factors possessed by *V. parahaemolyticus* and how they contribute to pathogenesis.

Osmotic stress response

Of the three main human *Vibrio* pathogens, *V. parahaemolyticus* is perhaps the most osmotically tolerant (compared with *V. cholerae,* and *V. vulnificus)* (Naughton *et al.*, 2009). Under the standard laboratory conditions, this organism grows best in LB broth containing 3% NaCl, although *V. parahaemolyticus* can grow in upwards of 12% NaCl (Whitaker and Boyd, unpublished data) when grown in rich media. In fact, because *V. parahaemolyticus* and *V. vulnificus* are difficult to differentiate when grown on the *Vibrio* selective media TCBS, the accepted differential test is to then grow in 8% NaCl: *V. parahaemolyticus* will grow and also has an absolute requirement for salt, and will not grow in media that contains less than 0.05% NaCl.

High osmotic conditions can be detrimental to bacterial cells, even those that are considered halophilic, like *V. parahaemolyticus*. Bacteria possess two systems for dealing with high external salt conditions. The first system is typically the uptake of K^+ ions to compensate for high external osmotic concentration. This strategy is common among extreme halophile Archaea and a few select halophilic Bacteria. While many species of bacteria can utilize this system as a short-term response, the long-term presence of K^+ is detrimental to cell and therefore this method of osmotolerance is transient in most bacteria (Santos and da Costa, 2002; Empadinhas and da Costa, 2008; Naughton *et al.*, 2009). Another system employed by Bacteria and even some eukaryotes to withstand osmotic stress is the use of compatible solutes. Under conditions of high osmolality, compatible solutes, which are low molecular weight compounds, are accumulated in the cytoplasm of the cell. These compatible solutes can either be transported into the cell from the extracellular environment or synthesized *de novo* (Empadinhas and da Costa, 2008; Wood, 2011).

Two common compatible solutes utilized by bacteria are glycine-betaine and ectoine. Glycine-betaine, also known as *N,N,N*-trimethylglycine, is perhaps the most commonly used compatible solute regardless of the domain of life, and prokaryotes are no exception to this. In most prokaryotes, glycine-betaine is synthesized via the conversion of choline to betaine aldehyde and subsequently to glycine betaine. The genes needed for the synthesis of glycine betaine are encoded by the *betAB* (Lamark *et al.*, 1991). Ectoine, also known as 1,4,5,6-tetrahydro-2-methyl pyrimidine-4-carboxylate, is another compatible solute commonly used by halophilic prokaryotic organisms. Thus far, in all organisms examined the pathway for ectoine synthesis is always encoded by the *ectABC* genes. In this pathway, aspartokinase converts aspartate to β-aspartyl phosphate. β-aspartyl phosphate is then converted to L-aspartate-β-semialdehyde by the enzyme Aspartate semialdehyde dehydrogenase. L-aspartate-β-semialdehyde is then shuttled to the ectoine synthesis pathway where it is first converted to L-2,4 diaminobutyric acid by the enzyme L-2,4-diaminobutyric acid acetyltransferase, which is encoded by the *ectA* gene. L-2,4-Diaminobutyric acid is then converted to N^{γ}-acetyl-l-2,4-diaminobutyric acid by diaminobutyrate-pyruvate aminotransferase, which is encoded by the *ectB* gene. Lastly, ectoine synthase, which is encoded by the *ectC* gene, converts N^{γ}-acetyl-l-2,4-diaminobutyric acid to ectoine (Pflughoeft *et al.*, 2003).

Aside from pathways which encode the necessary enzymes to synthesize compatible solutes, many prokaryotes, including halophiles and non-halophiles, have the ability to take up the fully formed compatible solutes directly from the environment. This is commonly accomplished via the use of specialized transports such as BCCT (betaine/carnitine/choline transporters) (Ziegler *et al.*, 2010) and ProU compatible solute transport systems, which have been shown to import a variety of compatible solutes and compatible solute precursors (Fig. 8.5).

As previously stated, *V. parahaemolyticus* is a moderate halophile, and as such, has been demonstrated to encode both a number of compatible solute synthesis systems and compatible solute transporters. Bioinformatic analysis of the *V. parahaemolyticus* RIMD 2210633 genome has revealed the presence of two putative compatible solute synthesis systems (glylcine-betaine and ectoine) as well as six putative compatible transport systems (Naughton *et al.*, 2009) (Fig. 8.5). Additionally, the compatible solute synthesis systems appear to be universal within the species as over 40 separate *V. parahaemolyticus* strains tested positive for the presence for both glycine-betaine and ectoine synthesis systems. By comparison, *V. parahaemolyticus* contains four times as many compatible solute systems as *V. cholerae,* and approximately two times as many systems as *V. vulnificus* (Naughton *et al.*, 2009). Additionally, *V. parahaemolyticus* contains the same number of compatible solute synthesis/transport systems as the closely related *V. alginolyticus* and *V. harveyi,* although the systems in *V. harveyi* are arranged differently in the genome (Naughton *et al.*, 2009). This large amount of compatible solute systems does appear to convey a competitive advantage to *V. parahaemolyticus* when grown under a variety of salt as well as temperature conditions. In LB broth, which is rich media that contains compatible solutes and their precursors, supplemented with 6% NaCl, *V. vulnificus* cells experienced a 5-hour lag phase and *V. cholerae* experienced a 9-hour lag phase whereas *V. parahaemolyticus,* however, did not exhibit a lag phase under the same conditions. When cultured in LB broth supplemented with 9% NaCl, *V. cholerae,* and *V. vulnificus* were unable to grow, whereas *V. parahaemolyticus* was able to

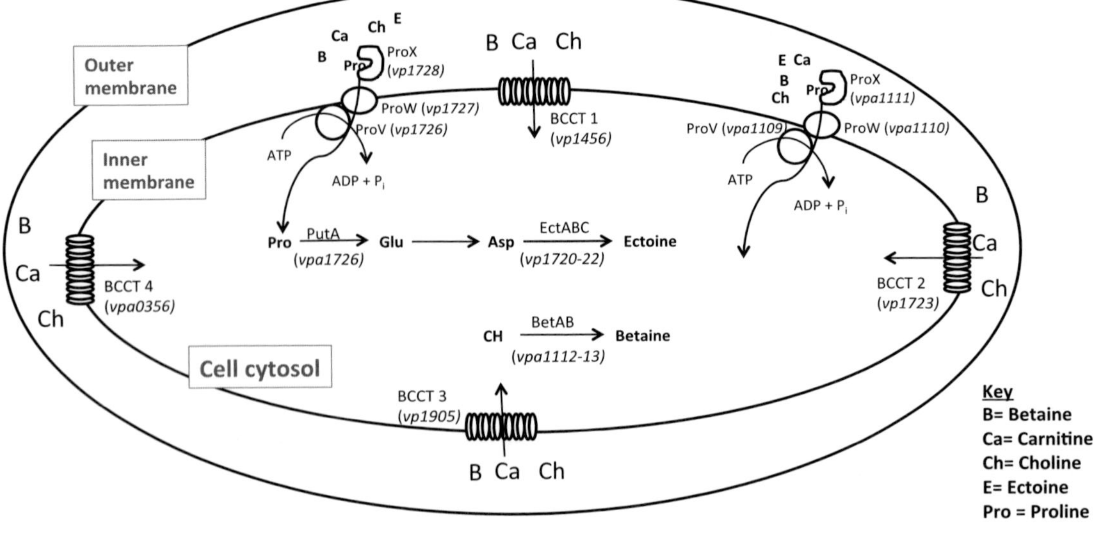

Figure 8.5 Compatible solute transport and synthesis in *V. parahaemolyticus*. *V. parahaemolyticus* contains a large number of osmotolerance systems, including four BCCT transporters, two ProU ABC transporters, and two compatible solute synthesis systems (ectoine and glycine-betaine). The six transport systems can potentially directly uptake compatible solutes from the extracellular environment. If these solutes are absent in the environment, the transporters can also function via the uptake of compatible solute precursors, which the cell uses to synthesis either ectoine or glycine-betaine *de novo*.

grow, albeit it experienced approximately a 5-hour lag phase (Naughton *et al.*, 2009).

Acid tolerance

Like its relative *V. cholerae, V. parahaemolyticus* is considered to be neutrophilic and is sensitive to changes in external pH conditions. Like *V. cholerae, V. parahaemolyticus* has also been demonstrated to be able to mount an effective acid tolerance response when cells are pre-adapted in mild, non-lethal pH conditions. Wong *et al.* (1998) determined that as little as 30 minutes in pH 5.0 was sufficient to elicit an adapted response to lethal pH 4.4 stress. Survival was greatest when an adaptation gradient was used, where *V. parahaemolyticus* cells were grown to exponential phase, transferred to pH 5.8 for 30 minutes, transferred next to pH 5.0 for 30 minutes and finally transferred to pH 4.4. When this was carried out, there was great than 100% survival for cells, whereas cells that were only subjected to a single-phase acid adaptation yielded approximately 75% in pH 4.4 after 60 minutes (Wong *et al.*, 1998). *V. parahaemolyticus* has also been demonstrated to mount an effective adapted response to organic acid as well. When cells were grown exponentially in LB (pH 7) supplemented with 3% NaCl and then shifted to the same media that was adjusted to pH 5.5 and also contained 1 mM acetic acid, there was near 100% survival of *V. parahaemolyticus* cells when they were subsequently transferred to media that was adjusted to pH 4 and also contained 4 mM acetic acid. Un-adapted cells, however, exhibited an approximate 5-log reduction in viability when shifted to media that was pH 4 that also contained 4 mM acetic acid (Whitaker *et al.*, 2010). When protein profiles of *V. parahaemolyticus* acid adapted cells (pH 5.8 for 30 minutes and pH 5.0 for 30 minutes), twenty proteins showed enhanced expression while 26 proteins showed a decrease in expression, when compared with un-adapted cells. This indicates that again, like *V. cholerae*, acid adaptation results in pleitropic protein expression alterations and that the induction of acid shock proteins is necessary for the cells to mount an effect response to acid stress (Chiang *et al.*, 2005).

Interestingly, acid adaptation also seems to

have an effect on the pathogenicity of this organism. When acid adapted and non-acid-adapted cells were used to infect suckling mice, the intestinal weights of mice that had been infected with acid adapted cells were significantly larger than the intestines of mice that had received non-acid-adapted inoculum. There was however, no difference in the recoverable number of vibrios from the intestines after the mice had been sacrificed nor was there is a difference in the time to death observed in the mice regardless if they had been given the adapted or non-adapted inoculums (Wong *et al.*, 1998). The cytotoxicity of *V. parahaemolyticus* has also been shown to be affected by acid adaptation. Cells that were transferred to media adjusted to pH 5.0 for 30 minutes exhibited an approximately 2-fold increase in their cytotoxicity towards the intestinal epithelial cell line CaCo-2 (Whitaker *et al.*, 2010).

Mechanisms of acid tolerance

HPLC analysis of acid adapted *V. parahaemolyticus* culture supernatant revealed large amounts of cadaverine, indicating that this organism was actively converting lysine into cadaverine in the presence of mild acid. This was further bolstered by genomic analysis, which revealed the presence of the *cadBAC* operon (Tanaka *et al.*, 2008). Mechanistically, it would appear that *V. parahaemolyticus* has similar response systems for dealing with acid stress as does *V. cholerae,* as the amino acid decarboxylase, lysine decarboxylase is one of the main systems for acid tolerance. The *cadBA* operon of *V. parahaemolyticus* was shown to be highly homologous to the operon found in *V. cholerae*. To demonstrate the importance of the lysine decarboxylase system in *V. parahaemolyticus* experiments were performed examining the adapted acid tolerance response in a no-carbon minimal medium with or without lysine supplementation. When lysine was added to the minimal media, *V. parahaemolyticus* was able to successfully mount an acid tolerance response, and viable cells were able to be recovered from pH 4.0 up to 2 hours after acidification. However, when no lysine was supplied, there was a 4-log reduction in viable cells after 60 minutes compared with cells that were in the presence of exogenous lysine (Tanaka *et al.*, 2008). By 90 minutes, there were no viable cells in the media that had no lysine. Like in *V. cholerae,* transcript analysis revealed that the *cadBA* operon demonstrates a multiple pattern of transcription, with monocistronic *cadA* and *cadB* transcripts as well as polycistronic *cadBA* transcripts being detected. Additionally, transcription of the operon appears to be regulated by external pH, as all three transcriptional patterns are able to be detected in acid adapted cells, whereas only minimal levels of *cadA* transcript are able to be detected in non-acid adapted cells. Not surprisingly, when the *cadA* gene is deleted from the *V. parahaemolyticus* genome, the deletion mutant is unable to elicit an adapted acid tolerance response when pre-adapted in pH 5.5 and exhibited >3-log decrease in survival after 60 minutes in pH 4.0 compared with the isogenic wild-type strain. Likewise, supplementation with lysine had no effect on the adapted acid tolerance response of the *cadA* deletion mutant (Tanaka *et al.*, 2008).

There is a ToxRS system present in *V. parahaemolyticus,* showing 52 and 62% homology to the *V. cholerae* system. This system was first described by Lin *et al.* where it was implicated in the regulation of *tdh* (Lin *et al.*, 1993), but apart from this initial study, little is known about the ToxRS regulon in *V. parahaemolyticus.* No ToxT homologue is present in *V. parahaemolyticus.* There is however, an OmpU homologue (66% identity to *V. cholerae),* but no OmpT homologue. Additionally, a second ToxR homologue has been identified on chromosome 2 within the T3SS-2 island, which was subsequently named VtrA by Kodama and colleagues (Kodama *et al.*, 2010). Recent studies published concurrently with our examination of VtrA suggest that this regulator plays an indirect role in the regulation of *tdh* and T3SS-2 expression (Gotoh *et al.*, 2010; Kodama *et al.*, 2010). Given what is known of the ToxRS role in *V. cholerae* acid tolerance response, the role of this system in *V. parahaemolyticus* has also been recently investigated (Whitaker *et al.*, 2010). Transcription analysis has revealed that the *toxR* gene in *V. parahaemolyticus* is induced under several different environmental conditions, such as mild pH stress (pH 5.0) and low-salt conditions (1% NaCl) (Whitaker *et al.*, 2010). An increase in *toxR* transcript could indicate the importance of this regulator in responding to alterations in the extracellular environment. To confirm this hypothesis, a double deletion mutation of the *toxR* and *toxS* genes was constructed in *V. parahaemolyticus*. The *V. parahaemolyticus toxRS* deletion mutant was unable to mount an adapted acid tolerance response to either inorganic or organic acid stress (Whitaker *et al.*, 2010), which is a different phenotype from the *V. cholerae toxR* deletion mutant, which is only defective in mounting a response to organic acid stress (Merrell and Camilli, 1999; Whitaker *et al.*, 2010). Although at this time the exact mechanism for ToxRS mediated resistance to acid stress in *V. parahaemolyticus* is unknown, this organism, like *V. cholerae*, does possess OmpU and it is likely that this protein will have an effect on acid tolerance as well.

The role of salt in stress response

Perhaps owing to the notably halophilic nature of *V. parahaemolyticus*, it would appear that external salt concentrations can have a profound effect on the ability of this organism to withstand various environmental stress conditions. It is well known that LB broth supplemented with 3% NaCl is optimum laboratory growth conditions for this organism. When the organism is grown in the presence of low salt such as 1%, there is a slight decrease in growth rate and a noticeable decrease in final culture concentration (Whitaker *et al.*, 2010). This effect is magnified when the salt concentration is lowered to 0.5% NaCl. Interestingly, environmental salt concentrations also appear to play a role in the ability of *V. parahaemolyticus* to withstand other types of extracellular stresses. When *V. parahaemolyticus* is grown to mid-exponential phase in media containing 3% NaCl or media containing 1% NaCl and then transferred to media adjusted to pH 4.0 (HCl), there is a significant difference in the ability to withstand lethal acid stress. Cultures that were grown in the presence of 3% NaCl exhibited a 2- and 4-fold decrease in viability after 30 and 60 minutes in pH 4, respectively. However, when cells were pre-cultured in media containing 1% NaCl prior to acid challenge, no viable cells were recovered by the 30 minute time point. Interestingly, when cells that were pre-cultured in low-salt conditions were shifted to media containing 3% NaCl for 30 minutes prior to acidification, there was a significant increase in survival compared with the cells that were only cultured in media containing 1% NaCl (Whitaker *et al.*, 2010). Additionally, cells grown in media containing 1% NaCl are unable to mount a robust adapted acid tolerance response when pre-adapted in media with a pH of 5.5 before subsequently being transferred to lethal pH conditions (personal observation). The phenomenon of external salt concentrations influencing survival of *V. parahaemolyticus* is not restricted solely to acid tolerance. *V. parahaemolyticus* cells grown in media containing 3% NaCl exhibit higher growth rate and a significantly higher final growth yield when grown at 42°C compared with cells that are grown in media containing 1% NaCl. Similarly, when *V. parahaemolyticus* cells grown in the presence of 3% NaCl, they significantly higher survival rates in −20°C when compared with cells that were grown under low-salt conditions. Taken together, this demonstrates external salt concentrations play an important role in either strengthening or weakening this organism's ability to withstand additional environmental stress conditions (Whitaker *et al.*, 2010).

In an attempt to explain why cells that were grown in media supplemented with 3% NaCl were better equipped at surviving in lethal acid stress conditions, expression of the *cadA* gene was examined. When cells were grown in 3% and 1% NaCl (no pH stress) there were no statistically significant changes in *cadA* expression. However, when these cells were then subjected to 30 minutes in medium that was adjusted to pH 5, cells that were grown in 3% NaCl produced approximately three times as much *cadA* transcript as those that were only grow in the presence of 1% NaCl (Whitaker *et al.*, 2010).

Iron limitation and scavenging

Vibrioferrin is a unique siderophore that was initially described in *V. parahaemolyticus* (Amin *et al.*, 2009) and is structurally unique to the catechol siderophore produced by *V. cholerae* (Fig. 8.3A and C). While its role as a virulence factor is currently unknown, it has a relatively weak dissociation constant of 10^{18} M^{-} for iron. This indicates that this siderophore may play no role or at least a diminished role in the pathogenicity of this organism. Another unique property of this siderophore is that it is photosensitive when complexed with Fe^{3+}. In the presence of light, ferri-vibrioferrin undergoes photolysis, whereby the siderophore's carboxylate group is lost in the form of CO_2 and the bound Fe^{3+} is reduced to Fe^{2+} (Amin *et al.*, 2009). This process appears irreversible, as the resulting siderophore after photolysis loses its affinity for the Fe^{3+} ion. The Fe^{2+} that results from the photolysis of the siderophore is transiently oxidized to a ferric–hydroxo complex, which remains relatively soluble. ^{55}Fe experiments have demonstrated that this soluble, transient iron pool which is created after photolysis of ferri-vibrioferrin is more readily taken up by bacteria than the ferri-vibrioferrin itself. This siderophore is not exclusive to *V. parahaemolyticus* and has also been found in several species of *Marinobacter*. This is of special interest because vibrioferrin producing species of *Marinobacter* are characterized as being symbiotic with toxic, bloom-forming dinoflagellate species. Since free iron in marine waters has been previously implicated as a causing factor of algal blooms, it has been hypothesized that these dinoflagellates rely on the transiently soluble iron pools generated by the photo-breakdown of the vibrioferrin produced by their symbiotic bacterial counterparts (Amin *et al.*, 2009).

Vibrio vulnificus

Vibrio vulnificus was first isolated in 1964 by the Center for Disease Control and was later classified as a distinct species in 1976. Like *V. parahaemolyticus*, *V. vulnificus* is moderate halophile, though not as osmotolerant as *V. parahaemolyticus*, and resides within marine and estuarine environments throughout the world, including the United States. This organism also tends to prefer environmental conditions that are similar to those of *V. parahaemolyticus*. Infections by this organism can occur throughout the year, though there is typically a spike in *V. vulnificus* cases during the summer months, where water temperatures are commonly >22°C. At this time, *V. vulnificus* is commonly isolated from the water column and can also be isolated in high numbers from oysters and other filter-feeding shellfish. *Vibrio vulnificus* infections are often associated with the consumption of raw or improperly cooked shellfish, especially oysters. However, unlike *V. cholerae* and *V. parahaemolyticus*, the two primary syndromes of *V. vulnificus* infection are wound infection (45%) and primary septicaemia (43%). Gastrointestinal illness associated with *V. vulnificus* is much less common (5%). Mortality associated with *V. vulnificus* infection is very high (>50%), making *V. vulnificus* one of the most virulent of the *Vibrio* pathogens and also the leading cause of death in the United States associated with the consumption of seafood (Gulig *et al.*, 2005; Horseman and Surani, 2011).

Much less is known about the pathogenesis of *V. vulnificus* than compared with *V. cholerae* and even *V. parahaemolyticus*. Thus far, it would appear that not all strains of this bacterium are pathogenic towards humans, and as such, there have been several attempts to separate *V. vulnificus* isolates into groups based on whether they are clinical or environmental isolates using biochemical, serological, or genomic methods. Thus far, *V. vulnificus* strains have been split into three biotypes and two major phylogenetic lineages, lineage I (predominantly biotype 1 isolates), lineage II (biotype 1 and 2 isolates). A third divergent lineage also exists consisting of biotype 3 isolates (Blake *et al.*, 1979; Tison *et al.*, 1982; Dalsgaard *et al.*, 1996; Bisharat *et al.*, 1999; Cohen *et al.*, 2007). Lineage I almost entirely comprises strains that cause disease in humans, whereas lineage II comprises mainly either environmental isolates or fish pathogens. Lineage III comprises human pathogens, but strains constituting this group appear to be isolated to Israel (Cohen *et al.*, 2007).

Acid tolerance

Much like *V. cholerae* and *V. parahaemolyticus*, *V. vulnificus* is capable of mounting an adapted acid tolerance response when pre-adapted to mild pH stress. Furthermore, *V. vulnificus* also contains the lysine decarboxylase system (*cadBA*) and this appears to play a main role in low pH survival (Rhee *et al.*, 2002). Again, like with the other vibrios, this system is regulated by CadC (Rhee *et al.*, 2005). However, there is evidence that other factors may play a role in regulation of *cadBA*, and vice versa. Under conditions of low-pH stress, Kim *et al.* demonstrated that MnSOD was induced in cells that were shifted to media that was adjusted to pH 5 (Kim *et al.*, 2005). Furthermore, induction of MnSOD (*sodA*) required the regulator SoxR. When either the gene encoding MnSOD or SoxR was deleted, *V. vulnificus* cells demonstrated a defect in the ability to survive in low-pH conditions when compared with the isogenic mutant (Kim *et al.*, 2005). From these data, it would appear that the shift to low pH causes the accumulation of superoxide in the cytoplasm of bacterial cells. This results in the induction of superoxide dismutases, which remove cytosolic superoxides, and then allow the cell to resume the induction of other acid shock systems.

Conversely, lysine decarboxylase also appears to be important to *V. vulnificus* in dealing with superoxide stress. While it is well known that the *cadA* gene is expressed under low-pH conditions, Kim *et al.* observed that the *cadBA* operon was induced in the presence of methyl viologen (superoxide stress conditions), even in neutral culture conditions (Kim *et al.*, 2006). In the presence of methyl viologen, expression of *cadBA* increased in both wild-type cells and in a *cadC* deletion mutant, indicating that this response is independent of CadC, the canonical *cadBA* regulator. Additionally, the superoxide stress regulator SoxR, was demonstrated to be able to bind to the promoter region of the *cadBA* operon, further indicating that this operon may be expressed in a CadC independent manner and that it is up-regulated during times of superoxide stress (Kim *et al.*, 2006). Apart from its role in acid stress survival, lysine decarboxylase also appears to function in times of superoxide stress in *V. vulnificus.* The *cadA* deletion mutant in *V. vulnificus* exhibited decreased survival rates in the presence of methyl viologen, which indicates that lysine decarboxylase plays a direct role in overcoming superoxide stress. This is perhaps not surprising given that polyamines, such as cadaverine, have been shown to have the ability to scavenge reactive oxygen species and thus exert a protective effect when present in cells during times of oxidative stress (Kim *et al.*, 2006).

Phase variation

Phase variation is a phenomenon that has been observed in a number of bacterial pathogens and has been observed and characterized in *V. vulnificus*. Phenotypically, phase variation leads to what is known as translucent colony morphology, as opposed to the typical opaque colonies this organism produces. This switch from opaque to translucent colony morphology is typically associated with a decrease in the amount of capsular polysaccharide produced by the cells in the colony (Wright *et al.*, 2001; Hilton *et al.*, 2006). Capsular polysaccharide is thus far considered to be the only virulence factor known to date that is absolutely required for full virulence of this organism. This is demonstrated by comparing the LD_{50} of opaque and translucent strains of *V. vulnificus* in the iron overloaded intraperitoneal mouse model. Opaque strains have been shown to have an LD_{50} as low as a single cell, whereas translucent strains have an LD_{50} as high as 6.5×10^6 cells (Smith and Siebeling, 2003). It is believed that CPS, or a lack there of, is important in interacting with and also surviving interactions with the host immune system, though more work needs to be done to fully understand the role of CPS in *V. vulnificus* pathogenesis. The shifting between opaque and translucent phenotypes appears to occur randomly at low frequencies. However, altering culture conditions, such as shifting growing cultures from 30°C to 37°C has been shown to increase the shift from the opaque to translucent colony phenotype (Hilton *et al.*, 2006). Interestingly, strains belonging to the clinical biotype seem to resist the shift to translucent colonies, regardless of culture conditions, which may owe to the fact that high CPS production may be a necessary survival within the host, and lessened CPS production may be important for survival in the marine environment.

Survival in the environment: viable but non-culturable

The viable but non-culturable (VBNC) phenomenon was first described in *E. coli* and *V. cholerae*, but has since been shown to exist in other *Vibrio* species as well as many other non-*Vibrio* species. Historically, there existed doubt whether *V. cholerae* actually existed in the natural environment as a free living organism, as it was notoriously difficult to isolate outside of human or animal host organism. Likewise, many *Vibrio* species such as *V. vulnificus* and *V. parahaemolyticus,* which are typically planktonic, become very difficult to isolate during the winter months of the year. While it has been hypothesized that they may overwinter in the intestines of marine animals, it has also been demonstrated that they can also enter into a state known as the viable but non-culturable state (VBNC) (Oliver, 2005). As the name implies, these organisms enter into a state where they are still viable organisms, but at the same time lose the ability to be cultured using standard laboratory media and conditions. Entry into the VBNC state appears to be governed by unfavourable environmental conditions, such as starvation, prolonged exposure to temperatures outside the preferred range of the organism, increased osmotic conditions, etc. VBNC cells are characterized as undergoing a noticeable dwarfing of size, as well as experiencing alterations of the cell membrane and peptidoglycan structures (Oliver, 2005). Essentially, these cells are entering a state of dormancy during conditions that do not favour growth. One explanation for why these cells lose the ability to be cultured after entering into the VBNC state is the loss of catalase activity that accompanies induction of VBNC. Typical bacterial growth plates contain a number of reactive oxygen species, which are generated by the cells own metabolic processes. Thus, as VBNC cells lose the ability to properly produce catalase, they also lose the ability to grow on plated media. However, if these cells are grown on plates that contain anti-reactive oxygen species added, they retain their culturability (Kong *et al.*, 2004). Interestingly, VBNC cells appear to retain virulence, or are at least resuscitated upon entry into the host. Indeed, Oliver and Bockain demonstrated the ability to kill mice when infected with 10^5 VBNC cells of *V. vulnificus* (Oliver and Bockian, 1995), and similar finding have also been demonstrated in *V. cholerae* (Colwell *et al.*, 1996). These findings could have a significant impact on *Vibrio* monitoring programmes that use traditional (i.e. plating methods) to enumerate the presence of these organisms in food sources in hopes of preventing disease outbreaks.

Other *Vibrio* species

Vibrio fischeri

Vibrio fischeri is perhaps best known for two things: bioluminescence and its symbiosis with the Hawaiian bobtail squid, *Euprymna scolopes* (Ruby, 1996). In contrast to the previously mentioned *Vibrio* species, *V. fischeri* is not pathogenic and is typically used as a model organism to study the evolution of symbiotic relationships between microorganisms and their eukaryotic hosts. *Vibrio fischeri* colonizes the light organ of *E. scolopes,* where it conveys bioluminescent properties to its host (Ruby, 1996). It is believed that the ability to bioluminesce enhances the ability of the squid to mimic the night sky when it comes out to feed at night, thus evading predators. While the presence of *V. fischeri* is highly beneficial to the squid, it must still overcome the immune system of the squid in order to colonize the light organ, which seeks to expel all microorganisms from the light organ. Studies have shown that mRNA for halide peroxidase is abundant in the squid light organ. Halide peroxidase converts hydrogen peroxide into toxic hypohalous acids, which may in turn exert an antimicrobial effect in the light organ (Visick and Ruby, 1998). *V. fischeri* has been shown to possess a catalase (KatA) that is localized to the periplasm of the cells. KatA has high homology to catalases found in a variety of other organisms and even bares high amino acid identity with found in higher organisms. Expression of the *katA* gene is increased after the transition into stationary phase as well as in the presence of hydrogen peroxide. This indicates that *katA* is regulated by OxyR, a conserved activator of catalases, as wells the alternate sigma factor RpoS (Visick and Ruby, 1998). When the *katA* gene was deleted in *V. fischeri,* the mutant strain showed greater sensitivity to the presence of hydrogen peroxide than the parental wild-type strain. More importantly, the *katA* deletion strain was defective in its ability to colonize juvenile *E. scolopes* during a co-infection assay (Visick and Ruby, 1998). This indicates that the ability of *V. fischeri* to overcome oxidative stress is important in its symbiotic relationship with *E. scolopes.*

Vibrio splendidus

While not pathogenic to humans, *V. splendidus* is capable of infecting the oyster species *Crassostrea gigas* as well a number of other molluscs and fish species. *Vibrio splendidus* is often the dominant culturable *Vibrio* species in marine environments and in host organisms in which it is present (Gay *et al.*, 2004; Le Roux *et al.*, 2007). Within last decade, this organism is becoming more recognized as a serious pathogen to marine organisms. It has been demonstrated that *V. splendidus* can cause necrosis in larval molluscs, also known as larval vibriosis (Gay *et al.*, 2004; Le Roux *et al.*, 2007). This disease can have severe economic impacts to hatcheries, as it can cause large losses in production. Very little is known about the pathogenesis of this organism, and it has been until very recently that factors effecting colonization and survival within the host have been examined for this organism. As with most *Vibrio* species, *V. splendidus* contains a homologue of the *V. cholerae* outer membrane porin OmpU (64% amino acid identity) (Duperthuy *et al.*, 2010). When the *ompU* gene was deleted in *V. splendidus*, the deletion strain showed a significantly altered growth in oyster plasma when compared with the isogenic wild-type strain. Not surprisingly, the *ompU* deletion mutant exhibited increased sensitivity to polymyxin B, human BPI, and the synthetic anti-microbial peptide protegrin (Duperthuy *et al.*, 2010). Like with what is seen in *V. cholerae*, propidium iodide staining did not reveal any outer membrane permeability/defects in the *ompU* deletion mutant compared with the wild-type strain. Additionally, the *V. splendidus* OmpU protein contains a C-terminal Y-D-F motif, indicating that it too may act as an important regulator of RpoE activity (Duperthuy *et al.*, 2010). Novel to what is seen in *V. cholerae*, the *V. splendidus* OmpU has been shown to play a role in fibronectin binding as well as adhesion and subsequent invasion of the C. gigas haemocyte (Duperthuy *et al.*, 2010, 2011). These studies have shown that OmpU is an important stress response protein needed to survive antimicrobial peptide stress as well as an important colonization factor necessary for the intracellular invasion of the host.

Vibrio anguillarum

V. anguillarum is another member of the genus best known as a marine and freshwater fish pathogen throughout the world. This species causes vibriosis, which comprises a highly fatal haemorrhagic septicaemia, primarily in salmonids and other select fish, such as eels. This pathogen was first described in the early 1900s as causing disease in European eels and was first identified in North America in the 1950s when it was isolated from salmon (Frans *et al.*, 2011). Mortality in fish infected with *V. anguillarum* is typically between 15% and 25% and outbreaks of this organism in fish farms can result in significant economic loss. The ability of this pathogen to successfully colonize its host and cause disease appears to be tied to its ability to overcome iron starvation. Only strains that have horizontally acquired a 65-kb virulence plasmid, pJM1, can cause disease in aquatic wildlife (Frans *et al.*, 2011). Encoded on this plasmid is the iron sequestering siderophore anguibactin. Structurally, this siderophore is similar to yersiniabactinin that it contains a thiazoline ring. However, unlike yersiniabactin, the thiazoline ring of anguibactin is attached to a catechol ring instead of a phenolic ring (Ratledge and Dover, 2000) (Fig. 8.3B). Also encoded on the pJM1 plasmid is the uptake system fatABCD, which encodes the genes necessary of the bacterium to sequester the ferric–anguibactin complex. The pMJ1 plasmid also encodes the majority of the regulator elements that control expression of the iron uptake systems, namely the positive regulators AngR and Taf, as well as the antisense RNAα, which acts as a negative regulator. Additionally, the chromosomally encoded Fur protein also acts as a negative regulator of these systems. Not all *V. anguillarum* strains harbour the siderophore baring plasmid, and these strains are unable to cause vibriosis in fish, indicating again, a need for iron acquisition for this species to efficiently infect and cause disease its host (Wolf and Crosa, 1986).

Conclusions and future trends

The genus *Vibrio* is of significant importance, both from a human health standpoint, as well as an economic one. *Vibrio cholerae* is perhaps the most studied and most severe of the human pathogens in terms of cases and deaths associated with this organism. However, some of the less studied members of this genus (*V. parahaemolyticus* and *V. vulnificus*) are emerging pathogens that deserve a great deal of focus as well. These organisms occur in a wide range of environments and encounter a large number of stressors both in the natural and host environments, and have thus developed survival mechanisms to cope with these conditions. Perhaps even more interestingly, is the fact that there exists cross-talk (in the case of the ToxRS system in *V. cholerae)* between stress response and virulence systems. Likewise, the pathogenic potential of certain *Vibrio* species, such as *V. anguillarum,* is tied to the organism's ability to horizontally acquire systems that counter act the stress conditions seen in the host. In some cases, such as with *V. parahaemolyticus*, environmental growth conditions can have profound effects on the organism's ability to survive additional stress conditions, and this phenomenon should not be overlooked when considering food storage or shellfish post-harvesting conditions. While much research has been focused on elucidating novel virulence mechanisms in these organism, survival and colonization mechanism are perhaps of equal importance as these are the key first steps in pathogenesis. Understanding both survival and virulence mechanisms of these species is key to gaining knowledge about their ecology, evolution and emergence as fish and human pathogens.

Acknowledgements

We thank members of the group for their enthusiasm and hard work; Megan R. Carpenter, Olivia Cosides, Sai Kalburge, Sarah Gaffney, Brandy Haines Menges, Nityananda Chowdhury, Jean Bernard Lubin, and Serge Ongagna-Thombi. Research on *Vibrio* stress response mechanisms in the Boyd group is supported by a National Science Foundation grant IOS-0918429 and funding from the United States Department of Agriculture, NRI CSREES grant 2008–01198.

References

Amin, S.A., Green, D.H., Kupper, F.C., and Carrano, C.J. (2009). Vibrioferrin, an unusual marine siderophore: iron binding, photochemistry, and biological implications. Inorg. Chem. *48*, 11451–11458.

Andrews, S.C., Robinson, A.K., and Rodriguez-Quinones, F. (2003). Bacterial iron homeostasis. FEMS Microbiol. Rev. *27*, 215–237.

Arnold, C.N., McElhanon, J., Lee, A., Leonhart, R., and Siegele, D.A. (2001). Global analysis of *Escherichia coli* gene expression during the acetate-induced acid tolerance response. J. Bacteriol. *183*, 2178–2186.

Bashyam, M.D., and Hasnain, S.E. (2004). The extracytoplasmic function sigma factors: role in bacterial pathogenesis. Infect. Genet. Evol. *4*, 301–308.

Begley, M., Gahan, C.G., and Hill, C. (2005). The interaction between bacteria and bile. FEMS Microbiol. Rev. *29*, 625–651.

Billington, D., Evans, C.E., Godfrey, P.P., and Coleman, R. (1979). Effects of bile salts on the plasma membranes of isolated hepatocytes [proceedings]. Biochem. Soc. Trans. *7*, 947.

Bisharat, N., Agmon, V., Finkelstein, R., Raz, R., Ben-Dror, G., Lerner, L., Soboh, S., Colodner, R., Cameron, D.N., Wykstra, D.L., *et al.* (1999). Clinical, epidemiological, and microbiological features of *Vibrio vulnificus* biogroup 3 causing outbreaks of wound infection and bacteraemia in Israel. Israel Vibrio Study Group. Lancet *354*, 1421–1424.

Blake, P.A., Merson, M.H., Weaver, R.E., Hollis, D.G., and Heublein, P.C. (1979). Disease caused by a marine *Vibrio*. Clinical characteristics and epidemiology. N. Engl. J. Med. *300*, 1–5.

Butler, A. (2003). Iron acquisition: straight up and on the rocks? Nat. Struct. Biol. *10*, 240–241.

Cai, S.J., and Inouye, M. (2002). EnvZ–OmpR interaction and osmoregulation in *Escherichia coli*. J. Biol. Chem. *277*, 24155–24161.

Canny, G., Levy, O., Furuta, G.T., Narravula-Alipati, S., Sisson, R.B., Serhan, C.N., and Colgan, S.P. (2002). Lipid mediator-induced expression of bactericidal/permeability-increasing protein (BPI) in human mucosal epithelia. Proc. Natl. Acad. Sci. U.S.A. *99*, 3902–3907.

Cash, R.A., Music, S.I., Libonati, J.P., Snyder, M.J., Wenzel, R.P., and Hornick, R.B. (1974). Response of man to infection with *Vibrio cholerae*. I. Clinical, serologic, and bacteriologic responses to a known inoculum. J. Infect. Dis. *129*, 45–52.

Chiang, M.L., Yu, R.C., and Chou, C.C. (2005). Fatty acid composition, cell morphology and responses to challenge by organic acid and sodium chloride of heat-shocked *Vibrio parahaemolyticus*. Int. J. Food Microbiol. *104*, 179–187.

Cohen, A.L., Oliver, J.D., DePaola, A., Feil, E.J., and Boyd, E.F. (2007). Emergence of a virulent clade of *Vibrio vulnificus* and correlation with the presence of a 33-kilobase genomic island. Appl. Environ. Microbiol. *73*, 5553–5565.

Colwell, R.R., Brayton, P., Herrington, D., Tall, B., Huq, A., and Levine, M.M. (1996). Viable but non-culturable *Vibrio cholerae*; O1 revert to a cultivable state in the human intestine. World J. Microbiol. Biotechnol. *12*, 28–31.

Colwell, R.R., Kaper, J., and Joseph, S.W. (1977). *Vibrio cholerae, Vibrio parahaemolyticus*, and other vibrios:

occurrence and distribution in Chesapeake Bay. Science *198*, 394–396.

Culham, D.E., Lu, A., Jishage, M., Krogfelt, K.A., Ishihama, A., and Wood, J.M. (2001). The osmotic stress response and virulence in pyelonephritis isolates of *Escherichia coli*: contributions of RpoS, ProP, ProU and other systems. Microbiology *147*, 1657–1670.

Dalsgaard, A., Frimodt-Moller, N., Bruun, B., Hoi, L., and Larsen, J.L. (1996). Clinical manifestations and molecular epidemiology of *Vibrio vulnificus* infections in Denmark. Eur. J. Clin. Microbiol. Infect. Dis. *15*, 227–232.

Daniels, N.A., MacKinnon, L., Bishop, R., Altekruse, S., Ray, B., Hammond, R.M., Thompson, S., Wilson, S., Bean, N.H., Griffin, P.M., *et al.* (2000). *Vibrio parahaemolyticus* infections in the United States, 1973–1998. J. Infect. Dis. *181*, 1661–1666.

DePaola, A., Hopkins, L.H., Peeler, J.T., Wentz, B., and McPhearson, R.M. (1990). Incidence of *Vibrio parahaemolyticus* in U.S. coastal waters and oysters. Appl. Environ. Microbiol. *56*, 2299–2302.

DiRita, V.J., and Mekalanos, J.J. (1991). Periplasmic interaction between two membrane regulatory proteins, ToxR and ToxS, results in signal transduction and transcriptional activation. Cell *64*, 29–37.

DiRita, V.J., Parsot, C., Jander, G., and Mekalanos, J.J. (1991). Regulatory cascade controls virulence in *Vibrio cholerae*. Proc. Natl. Acad. Sci. U.S.A. *88*, 5403–5407.

Duperthuy, M., Binesse, J., Le Roux, F., Romestand, B., Caro, A., Got, P., Givaudan, A., Mazel, D., Bachere, E., and Destoumieux-Garzon, D. (2010). The major outer membrane protein OmpU of *Vibrio splendidus* contributes to host antimicrobial peptide resistance and is required for virulence in the oyster *Crassostrea gigas*. Environ. Microbiol. *12*, 951–963.

Duperthuy, M., Schmitt, P., Garzon, E., Caro, A., Rosa, R.D., Le Roux, F., Lautredou-Audouy, N., Got, P., Romestand, B., de Lorgeril, J., *et al.* (2011). Use of OmpU porins for attachment and invasion of *Crassostrea gigas* immune cells by the oyster pathogen *Vibrio splendidus*. Proc. Natl. Acad. Sci. U.S.A. *108*, 2993–2998.

Dziejman, M., and Mekalanos, J.J. (1994). Analysis of membrane protein interaction: ToxR can dimerize the amino terminus of phage lambda repressor. Mol. Microbiol. *13*, 485–494.

Empadinhas, N., and da Costa, M.S. (2008). Osmoadaptation mechanisms in prokaryotes: distribution of compatible solutes. Int. Microbiol. *11*, 151–161.

Fang, F.C., Libby, S.J., Buchmeier, N.A., Loewen, P.C., Switala, J., Harwood, J., and Guiney, D.G. (1992). The alternative sigma factor *katF* (rpoS) regulates *Salmonella* virulence. Proc. Natl. Acad. Sci. U.S.A. *89*, 11978–11982.

Foster, J.W. (1999). When protons attack: microbial strategies of acid adaptation. Curr. Opin. Microbiol. *2*, 170–174.

Foster, J.W. (2004). *Escherichia coli* acid resistance: tales of an amateur acidophile. Nat. Rev. Microbiol. *2*, 898–907.

Frans, I., Michiels, C.W., Bossier, P., Willems, K.A., Lievens, B., and Rediers, H. (2011). *Vibrio anguillarum* as a fish pathogen: virulence factors, diagnosis and prevention. J. Fish Dis. *34*, 643–661.

Fujino, T., Okuno, Y., Nakada, D., Aoyama, A., Fukai, K., Mukai, T., and Uebo, T. (1953). On the bacteriological examination of Shirasu food poisoning. Med. J. Osaka Univ. *4*, 299–304.

Gay, M., Berthe, F.C., and Le Roux, F. (2004). Screening of *Vibrio* isolates to develop an experimental infection model in the Pacific oyster *Crassostrea gigas*. Dis. Aquat. Organ. *59*, 49–56.

Gotoh, K., Kodama, T., Hiyoshi, H., Izutsu, K., Park, K.S., Dryselius, R., Akeda, Y., Honda, T., and Iida, T. (2010). Bile acid-induced virulence gene expression of *Vibrio parahaemolyticus* reveals a novel therapeutic potential for bile acid sequestrants. PLoS One *5*, e13365.

Gulig, P.A., Bourdage, K.L., and Starks, A.M. (2005). Molecular pathogenesis of *Vibrio vulnificus*. J. Microbiol. *43 Spec. No*, 118–131.

Gupta, S., and Chowdhury, R. (1997). Bile affects production of virulence factors and motility of *Vibrio cholerae*. Infect. Immun. *65*, 1131–1134.

Heiskanen, P., Taira, S., and Rhen, M. (1994). Role of rpoS in the regulation of *Salmonella* plasmid virulence (spv) genes. FEMS Microbiol. Lett. *123*, 125–130.

Helmann, J.D., and Chamberlin, M.J. (1988). Structure and function of bacterial sigma factors. Annu. Rev. Biochem. *57*, 839–872.

Hilton, T., Rosche, T., Froelich, B., Smith, B., and Oliver, J. (2006). Capsular polysaccharide phase variation in *Vibrio vulnificus*. Appl. Environ. Microbiol. *72*, 6986–6993.

Hofmann, A.F., and Eckmann, L. (2006). How bile acids confer gut mucosal protection against bacteria. Proc. Natl. Acad. Sci. U.S.A. *103*, 4333–4334.

Holmgren, J. (1981). Actions of cholera toxin and the prevention and treatment of cholera. Nature *292*, 413–417.

Holmgren, J., Lonnroth, I., Mansson, J., and Svennerholm, L. (1975). Interaction of cholera toxin and membrane GM1 ganglioside of small intestine. Proc. Natl. Acad. Sci. U.S.A. *72*, 2520–2524.

Horseman, M.A., and Surani, S. (2011). A comprehensive review of *Vibrio vulnificus*: an important cause of severe sepsis and skin and soft-tissue infection. Int. J. Infect. Dis. *15*, e157–e166.

Hung, D.T., and Mekalanos, J.J. (2005). Bile acids induce cholera toxin expression in *Vibrio cholerae* in a ToxT-independent manner. Proc. Natl. Acad. Sci. U.S.A. *102*, 3028–3033.

Jermyn, W.S., and Boyd, E.F. (2002). Characterization of a novel *Vibrio* pathogenicity island (VPI-2) encoding neuraminidase (nanH) among toxigenic *Vibrio cholerae* isolates. Microbiology *148*, 3681–3693.

Jermyn, W.S., and Boyd, E.F. (2005). Molecular evolution of *Vibrio* pathogenicity island-2 (VPI-2): mosaic structure among *Vibrio cholerae* and *Vibrio mimicus* natural isolates. Microbiology *151*, 311–322.

Kaneko, T., and Colwell, R.R. (1973). Ecology of *Vibrio parahaemolyticus* in Chesapeake Bay. J. Bacteriol. *113*, 24–32.

Kaper, J.B., Remmers, E.F., Lockman H., and Colwell R.R. (1981). Distribution of *Vibrio parahaemolyticus* in Chesapeake Bay during the summer season. Estuaries *4*, 321–327.

Karaolis, D.K., Johnson, J.A., Bailey, C.C., Boedeker, E.C., Kaper, J.B., and Reeves, P.R. (1998). A *Vibrio cholerae* pathogenicity island associated with epidemic and pandemic strains. Proc. Natl. Acad. Sci. U.S.A. *95*, 3134–3139.

Kazmierczak, M.J., Wiedmann, M., and Boor, K.J. (2005). Alternative sigma factors and their roles in bacterial virulence. Microbiol. Mol. Biol. Rev. *69*, 527–543.

Kim, J.S., Sung, M.H., Kho, D.H., and Lee, J.K. (2005). Induction of manganese-containing superoxide dismutase is required for acid tolerance in *Vibrio vulnificus*. J. Bacteriol. *187*, 5984–5995.

Kim, J.S., Choi, S.H., and Lee, J.K. (2006). Lysine decarboxylase expression by *Vibrio vulnificus* is induced by SoxR in response to superoxide stress. J. Bacteriol. *188*, 8586–8592.

Koch, R. (1884). An address on *Cholera* and its *Bacillus*. Br. Med. J. *2*, 403–407.

Kodama, T., Gotoh, K., Hiyoshi, H., Morita, M., Izutsu, K., Akeda, Y., Park, K.S., Cantarelli, V.V., Dryselius, R., Iida, T. *et al.* (2010). Two regulators of *Vibrio parahaemolyticus* play important roles in enterotoxicity by controlling the expression of genes in the Vp-PAI region. PLoS One *5*, e8678.

Kong, I.S., Bates, T.C., Hulsmann, A., Hassan, H., Smith, B.E., and Oliver, J.D. (2004). Role of catalase and *oxyR* in the viable but nonculturable state of *Vibrio vulnificus*. FEMS Microbiol. Ecol. *50*, 133–142.

Krantz, G.E., Colwell, R.R., and Lovelace, E. (1969). *Vibrio parahaemolyticus* from the blue crab *Callinectes sapidus* in Chesapeake Bay. Science *164*, 1286–1287.

Krewulak, K.D., and Vogel, H.J. (2011). TonB or not TonB: is that the question? Biochem. Cell Biol. Biochimie Biol. Cell. *89*, 87–97.

Kuehl, C.J., and Crosa, J.H. (2010). The TonB energy transduction systems in *Vibrio* species. Future Microbiol. *5*, 1403–1412.

Lamark, T., Kaasen, I., Eshoo, M.W., Falkenberg, P., McDougall, J., and Strom, A.R. (1991). DNA sequence and analysis of the *bet* genes encoding the osmoregulatory choline-glycine betaine pathway of *Escherichia coli*. Mol. Microbiol. *5*, 1049–1064.

Lange, R., and Hengge-Aronis, R. (1991). Identification of a central regulator of stationary-phase gene expression in *Escherichia coli*. Mol. Microbiol. *5*, 49–59.

Le Roux, F., Binesse, J., Saulnier, D., and Mazel, D. (2007). Construction of a *Vibrio splendidus* mutant lacking the metalloprotease gene *vsm* by use of a novel counterselectable suicide vector. Appl. Environ. Microbiol. *73*, 777–784.

Mathur, J., Davis, B.M., and Waldor, M.K. (2007). Antimicrobial peptides activate the *Vibrio cholerae* sigmaE regulon through an OmpU-dependent signalling pathway. Mol. Microbiol. *63*, 848–858.

Mathur, J., and Waldor, M.K. (2004). The *Vibrio cholerae* ToxR-regulated porin OmpU confers resistance to antimicrobial peptides. Infect. Immun. *72*, 3577–3583.

Merrell, D.S., and Camilli, A. (1999). The *cadA* gene of *Vibrio cholerae* is induced during infection and plays a role in acid tolerance. Mol. Microbiol. *34*, 836–849.

Merrell, D.S., and Camilli, A. (2000). Regulation of *Vibrio cholerae* genes required for acid tolerance by a member of the 'ToxR-like' family of transcriptional regulators. J. Bacteriol. *182*, 5342–5350.

Merrell, D.S., and Camilli, A. (2002). Acid tolerance of gastrointestinal pathogens. Curr. Opin. Microbiol. *5*, 51–55.

Merrell, D.S., Bailey, C., Kaper, J.B., and Camilli, A. (2001). The ToxR-mediated organic acid tolerance response of *Vibrio cholerae* requires OmpU. J. Bacteriol. *183*, 2746–2754.

Merrell, D.S., Hava, D.L., and Camilli, A. (2002). Identification of novel factors involved in colonization and acid tolerance of *Vibrio cholerae*. Mol. Microbiol. *43*, 1471–1491.

Merrell, D.S., Tischler, A.D., Lee, S.H., and Camilli, A. (2000). *Vibrio cholerae* requires rpoS for efficient intestinal colonization. Infect. Immun. *68*, 6691–6696.

Mey, A.R., Wyckoff, E.E., Kanukurthy, V., Fisher, C.R., and Payne, S.M. (2005). Iron and fur regulation in *Vibrio cholerae* and the role of fur in virulence. Infect. Immun. *73*, 8167–8178.

Miller, V.L., and Mekalanos, J.J. (1984). Synthesis of cholera toxin is positively regulated at the transcriptional level by toxR. Proc. Natl. Acad. Sci. U.S.A. *81*, 3471–3475.

Miller, V.L., and Mekalanos, J.J. (1988). A novel suicide vector and its use in construction of insertion mutations: osmoregulation of outer membrane proteins and virulence determinants in *Vibrio cholerae* requires toxR. J. Bacteriol. *170*, 2575–2583.

Miller, V.L., Taylor, R.K., and Mekalanos, J.J. (1987). Cholera toxin transcriptional activator toxR is a transmembrane DNA binding protein. Cell *48*, 271–279.

Morris, J.G., Jr. and Black, R.E. (1985). Cholera and other vibrioses in the United States. N. Engl. J. Med. *312*, 343–350.

Nair, G.B., Ramamurthy, T., Bhattacharya, S.K., Dutta, B., Takeda, Y., and Sack, D.A. (2007). Global dissemination of *Vibrio parahaemolyticus* serotype O3:K6 and its serovariants. Clin. Microbiol. Rev. *20*, 39–48.

Naughton, L.M., Blumerman, S.L., Carlberg, M., and Boyd, E.F. (2009). Osmoadaptation among *Vibrio* species and unique genomic features and physiological responses of *Vibrio parahaemolyticus*. Appl. Environ. Microbiol. *75*, 2802–2810.

Nielsen, A.T., Dolganov, N.A., Otto, G., Miller, M.C., Wu, C.Y., and Schoolnik, G.K. (2006). RpoS controls the *Vibrio cholerae* mucosal escape response. PLoS Pathog. *2*, e109.

Okuda, J., Ishibashi, M., Hayakawa, E., Nishino, T., Takeda, Y., Mukhopadhyay, A.K., Garg, S., Bhattacharya, S.K., Nair, G.B., and Nishibuchi, M. (1997). Emergence of a unique O3:K6 clone of *Vibrio parahaemolyticus* in Calcutta, India, and isolation of strains from the same clonal group from Southeast Asian travelers arriving in Japan. J. Clin. Microbiol. *35*, 3150–3155.

Oliver, J.D. (2005). The viable but nonculturable state in bacteria. J. Microbiol. *43 Spec No*, 93–100.

Oliver, J.D., and Bockian, R. (1995). *In vivo* resuscitation, and virulence towards mice, of viable but nonculturable cells of *Vibrio vulnificus*. Appl. Environ. Microbiol. *61*, 2620–2623.

Ottemann, K.M., and Mekalanos, J.J. (1995). Analysis of *Vibrio cholerae* ToxR function by construction of novel fusion proteins. Mol. Microbiol. *15*, 719–731.

Pflughoeft, K.J., Kierek, K., and Watnick, P.I. (2003). Role of ectoine in *Vibrio cholerae* osmoadaptation. Appl. Environ. Microbiol. *69*, 5919–5927.

Provenzano, D., and Klose, K.E. (2000). Altered expression of the ToxR-regulated porins OmpU and OmpT diminishes *Vibrio cholerae* bile resistance, virulence factor expression, and intestinal colonization. Proc. Natl. Acad. Sci. U.S.A. *97*, 10220–10224.

Provenzano, D., Schuhmacher, D.A., Barker, J.L., and Klose, K.E. (2000). The virulence regulatory protein ToxR mediates enhanced bile resistance in *Vibrio cholerae* and other pathogenic *Vibrio* species. Infect. Immun. *68*, 1491–1497.

Provenzano, D., Lauriano, C.M., and Klose, K.E. (2001). Characterization of the role of the ToxR-modulated outer membrane porins OmpU and OmpT in *Vibrio cholerae* virulence. J. Bacteriol. *183*, 3652–3662.

Ratledge, C., and Dover, L.G. (2000). Iron metabolism in pathogenic bacteria. Ann. Rev. Microbiol. *54*, 881–941.

Reidl, J., and Klose, K.E. (2002). *Vibrio cholerae* and cholera: out of the water and into the host. FEMS Microbiol. Rev. *26*, 125–139.

Rhee, J.E., Rhee, J.H., Ryu, P.Y., and Choi, S.H. (2002). Identification of the *cadBA* operon from *Vibrio vulnificus* and its influence on survival to acid stress. FEMS Microbiol. Lett. *208*, 245–251.

Rhee, J.E., Kim, K.S., and Choi, S.H. (2005). CadC activates pH-dependent expression of the *Vibrio vulnificus cadBA* operon at a distance through direct binding to an upstream region. J. Bacteriol. *187*, 7870–7875.

Ruby, E.G. (1996). Lessons from a cooperative, bacterial–animal association: the *Vibrio fischeri–Euprymna scolopes* light organ symbiosis. Ann. Rev. Microbiol. *50*, 591–624.

Santos, H., and da Costa, M.S. (2002). Compatible solutes of organisms that live in hot saline environments. Environ. Microbiol. *4*, 501–509.

Seliger, S.S., Mey, A.R., Valle, A.M., and Payne, S.M. (2001). The two TonB systems of *Vibrio cholerae*: redundant and specific functions. Mol. Microbiol. *39*, 801–812.

Smith, A., Hooper, N.I., Shipulina, N., and Morgan, W T. (1996). Heme binding by a bacterial repressor protein, the gene product of the ferric uptake regulation (fur) gene of *Escherichia coli*. J. Protein Chem. *15*, 575–583.

Smith, A.B., and Siebeling, R.J. (2003). Identification of genetic loci required for capsular expression in *Vibrio vulnificus*. Infect. Immun. *71*, 1091–1097.

Snow, J. (1855). On the Mode of Communication of Cholera (John Churchill, London.)

Stock, A.M., Robinson, V.L., and Goudreau, P.N. (2000). Two-component signal transduction. Annu. Rev. Biochem. *69*, 183–215.

Su, Y.C., and Liu, C. (2007). *Vibrio parahaemolyticus*: a concern of seafood safety. Food Microbiol. *24*, 549–558.

Tanaka, Y., Kimura, B., Takahashi, H., Watanabe, T., Obata, H., Kai, A., Morozumi, S., and Fujii, T. (2008). Lysine decarboxylase of *Vibrio parahaemolyticus*: kinetics of transcription and role in acid resistance. J. Appl. Microbiol. *104*, 1283–1293.

Taylor, R.K., Miller, V.L., Furlong, D.B., and Mekalanos, J.J. (1987). Use of *phoA* gene fusions to identify a pilus colonization factor coordinately regulated with cholera toxin. Proc. Natl. Acad. Sci. U.S.A. *84*, 2833–2837.

Tison, D.L., Nishibuchi, M., Greenwood, J.D., and Seidler, R.J. (1982). *Vibrio vulnificus* biogroup 2: new biogroup pathogenic for eels. Appl. Environ. Microbiol. *44*, 640–646.

Visick, K.L., and Ruby, E.G. (1998). The periplasmic, group III catalase of *Vibrio fischeri* is required for normal symbiotic competence and is induced both by oxidative stress and by approach to stationary phase. J. Bacteriol. *180*, 2087–2092.

Waldor, M.K., and Mekalanos, J.J. (1996). Lysogenic conversion by a filamentous phage encoding cholera toxin. Science *272*, 1910–1914.

Walsh, N.P., Alba, B.M., Bose, B., Gross, C.A., and Sauer, R.T. (2003). OMP peptide signals initiate the envelope-stress response by activating DegS protease via relief of inhibition mediated by its PDZ domain. Cell *113*, 61–71.

Watnick, P.I., Eto, T., Takahashi, H., and Calderwood, S.B. (1997). Purification of *Vibrio cholerae* fur and estimation of its intracellular abundance by antibody sandwich enzyme-linked immunosorbent assay. J. Bacteriol. *179*, 243–247.

Weiss, J., Elsbach, P., Olsson, I., and Odeberg, H. (1978). Purification and characterization of a potent bactericidal and membrane active protein from the granules of human polymorphonuclear leukocytes. J. Biol. Chem. *253*, 2664–2672.

Whitaker, W.B., Parent, M.A., Naughton, L.M., Richards, G.P., Blumerman, S.L., and Boyd, E.F. (2010). Modulation of responses of *Vibrio parahaemolyticus* O3:K6 to pH and temperature stresses by growth at different salt concentrations. Appl. Environ. Microbiol. *76*, 4720–4729.

Wolf, M.K., and Crosa, J.H. (1986). Evidence for the role of a siderophore in promoting *Vibrio anguillarum* infections. J. Gen. Microbiol. *132*, 2949–2952.

Wong, H.C., Peng, P.Y., Han, J.M., Chang, C.Y., and Lan, S.L. (1998). Effect of mild acid treatment on the survival, enteropathogenicity, and protein production in *Vibrio parahaemolyticus*. Infect. Immun. *66*, 3066–3071.

Wood, J.M. (2011). Bacterial osmoregulation: a paradigm for the study of cellular homeostasis. Annu. Rev. Microbiol. *65*, 215–238.

Wright, A.C., Powell, J.L., Kaper, J.B., and Morris, J.G., Jr. (2001). Identification of a group 1-like capsular polysaccharide operon for *Vibrio vulnificus*. Infect. Immun. *69*, 6893–6901.

Wyckoff, E.E., Mey, A.R., and Payne, S.M. (2007). Iron acquisition in *Vibrio cholerae*. Biometals *20*, 405–416.

Wyckoff, E.E., Stoebner, J.A., Reed, K.E., and Payne, S.M. (1997). Cloning of a *Vibrio cholerae* vibriobactin gene cluster: identification of genes required for early steps in siderophore biosynthesis. J. Bacteriol. *179*, 7055–7062.

Yildiz, F.H., and Schoolnik, G.K. (1998). Role of rpoS in stress survival and virulence of *Vibrio cholerae*. J. Bacteriol. *180*, 773–784.

Ziegler, C., Bremer, E., and Kramer, R. (2010). The BCCT family of carriers: from physiology to crystal structure. Mol. Microbiol. *78*, 13–34.

Stress Responses in *Mycobacterium*

9

Richard W. Stokes

Abstract

There are many species of *Mycobacteria*, some of which are pathogens of man. *Mycobacterium tuberculosis*, the aetiological agent of tuberculosis, is a major pathogen of man with about one third of the world's population being infected. It resides within host macrophages where it can survive in a dormant state for the lifetime of the host with about 10% of all infections resulting in disease. This environment results in the bacteria being exposed to numerous stresses including nutrient deprivation, reduced oxygen availability, exposure to pH changes and exposure to the antimicrobial activities of the host's cell-mediated immune response. The bacterium responds with its own defence mechanisms that include the increased expression of stress proteins (also called heat shock proteins). This review describes the regulation and function of the major stress proteins within mycobacteria such as the GroEL, GroES and DnaK homologues along with hspX (α-crystallin) and others. The multiple copies of cpn60 (GroEL homologue) that are found within mycobacteria are discussed along with their putative roles as chaperonins but also as 'moonlighting' proteins with roles in immunomodulation and receptor/ligand interactions that facilitate the pathogenesis of *M. tuberculosis*.

Introduction

Most of the members of the genus *Mycobacterium* are saprophytic inhabitants of soil and water. Some species, however, infect ungulates, birds and other animals. A few mycobacterial species, including *M. leprae*, *M. ulcerans* and some members of the *Mycobacterium tuberculosis* complex (*M. tuberculosis*, *M. africanum* and *M. canettii*) are major pathogens of man, while others (e.g. the *M. avium–intracellulare* complex, *M. marinum* and *M. fortuitum*) are opportunistic pathogens. Probably the best known of these pathogenic species is *M. tuberculosis*, the causative agent of tuberculosis (TB). TB is the leading cause of infectious death in adults due to a single bacterial species and it is estimated that every year there are up to 10 million new cases of TB (Dye and Williams, 2010) and that TB is the cause of around 2 million deaths worldwide (Anonymous, 2009). Recent data indicate that there is a resurgence in the incidence of TB in industrialized nations. In addition, the emergence of multidrug resistant and extensively drug-resistant TB, such as the Beijing strain, has become a serious problem and threatens to make the disease incurable (Jain and Mondal, 2008; Mitnick *et al.*, 2008). To compound matters further, the emergence of AIDS has accelerated the spread of TB as a dual infection with *M. tuberculosis* and HIV represents a deadly combination (Getahun *et al.*, 2010). The WHO has estimated that between 2000 and 2020, nearly one billion people will be newly infected, 200 million will get sick, and 35 million will die from TB. Clearly, TB remains a devastating disease of mankind that, despite recent successes in treatment, continues to infect increasingly higher numbers of people (Dye and Williams, 2010).

M. tuberculosis is an intracellular pathogen possessing mechanisms for evading and subverting the arsenal of host immune responses. It is significant, however, that only 10% of people infected with *M. tuberculosis* develop clinical disease over their lifetime. For the other 90% of infected individuals the bacterium can survive in a dormant, yet viable state for decades. There is obviously a conflict between pathogen and host with the ongoing balance being tipped in the favour of disease in some cases, whereas usually the host can keep bacterial replication in check. Whether a disease state develops in an infected individual is greatly affected by a variety of environmental factors that the host may experience, such as nutritional status, alcohol or drug abuse and socioeconomic status (Lienhardt, 2001; van der Eijk *et al.*, 2007). In addition, genetic variation in host susceptibility (Forget *et al.*, 1981; Orme *et al.*, 1986; Doffinger *et al.*, 2006) and variation in the virulence of the infecting *M. tuberculosis* isolate (Collins and Smith, 1969; North and Izzo, 1993; Sassetti and Rubin, 2003) may also play a role in whether infection leads to active TB.

Pathogenesis of tuberculosis

M. tuberculosis is predominantly an intracellular pathogen, able to survive within components of the host's defence system. The bacteria are generally found within host macrophages (MΦs) which are part of the effector arm of cell-mediated immunity. These infected MΦs are usually found within a granuloma, an organized collection of MΦs, giant cells and lymphocytes that is the hosts attempt to wall off an infection that it is unable to eliminate. The bacteria within a granuloma may stay dormant for decades until some extrinsic factor (e.g. a decline in cell mediated immunity that accompanies ageing) leads to reactivation of bacterial replication, a breakdown of the granuloma's integrity and breakthrough into the airways of the lung, whereupon viable bacteria are expelled by coughing, completing the infectious cycle when inhaled by the next host. Unfortunately, the recent AIDS pandemic has compounded the infection rates and morbidity of TB, owing to the diminished CD4 T cell-mediated immunity of the AIDS patient (Getahun *et al.*, 2010).

Although there is not a complete understanding of the molecular details controlling the pathogenesis of TB, recent research has made inroads into understanding many aspects of the microbiology and immunology of this phenomenon. Of major relevance to this chapter is the understanding of the stresses encountered by *M. tuberculosis* on its entry into MΦs and its transcriptional response to this situation. Related to this is the improved understanding of the bacterial shift into so called 'dormancy' with, again, a description of the associated transcriptome. These topics have been recently reviewed (Stokes and Waddell, 2009), however a brief synopsis is warranted. The first stage of the interaction of bacterium with MΦ is attachment followed by uptake via phagocytosis. There does not appear to be any active bacterial mechanism of entry into the host cell. During uptake, *M. tuberculosis* may be exposed to MΦ killing mechanisms such as reactive oxygen intermediates (ROI) and will trigger the production of various chemokines (e.g. RANTES and Mip-1) and cytokines (e.g. TNFα and IL-1) by the MΦs (Schorey and Cooper, 2003). These immune effector molecules are required for recruitment and activation of leucocytes and are important mediators of innate and adaptive immunity. After being ingested, *M. tuberculosis* resides within a subcellular compartment called a phagosome where they can be exposed to several host defence mechanisms such as lysosomal acid hydrolases, p47 GTPases and antimicrobial peptides. The bacteria may also induce host cell apoptosis and autophagy, innate host killing mechanisms resulting in programmed cell death. Additionally, the intracellular bacteria must acquire nutrients. Avoidance of host cell killing mechanisms and acquisition of nutrients by *M. tuberculosis* is thought to be facilitated by the bacterium's ability to prevent the maturation of the phagosome past the early endosome stage, when pH remains relatively close to neutral and access to the extracellular compartment is maintained (Russell, 2001; Pieters, 2008; de Chastellier, 2009). Another adaptation *M. tuberculosis* must make is the move to a lower oxygen tension within the MΦ and the granuloma (see below) resulting in its shift from aerobic respiration to anoxic respiration with the utilization of nitrate and fumarate as electron acceptors.

Obviously in some cases the *M. tuberculosis* bacteria survives the initial move to the intracellular environment and begins to replicate. One outcome of this replication is that bacterial antigens are processed by the host resulting in the development of adaptive immunity resulting in the production of interferon-γ (IFNγ) by T cells, which, in turn, activates the infected MΦs. The activated MΦs are able to control bacterial replication by enhancing the killing mechanisms seen in resting MΦs (such as ROI and apoptosis) and by inducing new killing mechanisms mediated by reactive nitrogen intermediates (RNI) or phagosomal maturation to phagolysosomes. The exposure to IFNγ activated MΦs triggers the physiology of *M. tuberculosis* to change towards a so called dormant or chronic phase. In this state, *M. tuberculosis* switches from carbohydrates to lipids as a carbon source and to the glyoxalate shunt for energy production. The host responds to the persisting bacteria by forming a highly organized collection of immune cells that wall off the microbes in an anoxic environment where they are commonly found within foamy MΦs, surrounded by T cells and fibroblasts. This is called a granuloma (or a tubercle in clinical terminology) and may calcify over time or, alternatively, cavitate and break through to the airways, releasing bacteria to begin the cycle again.

During the pathogenesis of TB the *M. tuberculosis* bacilli encounter numerous stresses resulting from the move to an intracellular environment, exposure to MΦ killing mechanisms, exposure to the effector mechanisms of adaptive immunity and changes to oxygen and nutrient availability (Ehrt and Schnappinger, 2009). The pathogen responds to these stresses in several ways (Ehrt and Schnappinger, 2009; Stokes and Waddell, 2009) including the induction of stress proteins.

Mycobacterial stress proteins

Since the 1960s it has been known that stress at the cellular level results in a specific set of gene products being expressed. These are variously called stress proteins, heat shock proteins (Hsps) and chaperonins (Cpns), with the latter referring to a common role of these products in ensuring the correct assembly and interaction of macromolecules (Laskey *et al.*, 1978). Much of the early work was done using bacterial and plant Hsps, resulting in the description of the ubiquitous protein folding chaperone Hsp60/Hsp65/Cpn60/GroEL (Georgopoulos, 1971, 1977; Georgopoulos and Hohn, 1978; Hemmingsen *et al.*, 1988). In *E. coli*, GroEL maintains bacterial viability by facilitating the correct folding of many intracellular proteins through the formation of a large ATPase-dependent multimeric complex with GroES, a 10-kDa Hsp (Ellis, 2005). Another ubiquitous Hsp, DnaK (Hsp70), working with DnaJ (Hsp40), also promotes correct protein folding (Young *et al.*, 2004). Both GroEL and DnaK function through the recognition of exposed hydrophobic areas on misfolded proteins (Young *et al.*, 2004).

Following the early work with plants and *E.coli*, the identification of a stress response in mycobacteria was discovered. The pathogens, *M. tuberculosis* and *M. leprae*, had been known to induce an immune response in humans directed to at least 12 proteins and mice injected with mycobacterial lysates also strongly recognized six *M. tuberculosis* and six *M. leprae* antigens of which four immunodominant proteins were homologues of the Hsp60 and Hsp70 families, the *E. coli* GroES (a 10–12 kDa protein in mycobacteria) and an 18-kDa eukaryotic low molecular weight Hsp (Young *et al.*, 1988; Young, 1990). Identification of the mycobacterial heat shock response at both transcriptional and translational levels identified Hsps 40, 65, 70 and 90 in *M. bovis* BCG (an attenuated strain widely used as a vaccine against TB) in response to heat shock (Patel *et al.*, 1991). Later studies demonstrated that a 16-kDa member of the family of Hsp20 α-crystallin-like chaperones (Acr) were the most strongly induced by growth in microaerobic and anaerobic conditions associated with late stages of *M. tuberculosis* pathogenesis and bacterial dormancy (Cunningham and Spreadbury, 1998). It has also been shown that the 16-kDa Acr is a prominent antigen in TB patients (Wilkinson *et al.*, 1998). Another family of proteins expressed by mycobacteria (*M. tuberculosis*, *M. bovis* BCG and *M. smegmatis*) during oxygen deprivation have been identified as counterparts to the universal stress protein or UspA family found in other organisms (reviewed in O'Toole and Williams, 2003).

Hsp16 (α-crystallin-like chaperone/ Acr1/Acr2/hspX)

The Hsp16 of *M. tuberculosis* was shown to be induced following exposure to reduced oxygen (Cunningham and Spreadbury, 1998) nitric oxide donors (Garbe *et al.*, 1999) heat shock (Stewart *et al.*, 2002) and uptake by MΦs (Schnappinger *et al.*, 2003). An in depth study of the mycobacterial Acr chaperones identified homologues falling into three classes, *acr1*, *acr2* and *acr3* with each sharing 25–30% identity with the other two classes (Stewart *et al.*, 2005). However, within each class there was 60–75% identity between mycobacterial species, e.g. *M. tuberculosis acr1* shares 60% and 75% identity with the corresponding orthologue in *M. marinum* and *M. smegmatis* (Stewart *et al.*, 2005). Representation of each *acr* class is not found in every species of mycobacteria, e.g. *M. tuberculosis* has only *acr1* (*hspX, M. tuberculosis* genome accession number Rv2031c) and *acr2* (*M. tuberculosis* genome accession number Rv0251c) whereas *M. leprae* has just acr3 and *M. marinum* and *M. smegmatis* have a representative of each class (Stewart *et al.*, 2005). Expression of *M. tuberculosis acr1* is up-regulated by exposure to low oxygen or nitric oxide as part of the dormancy regulon (Voskuil *et al.*, 2003) whereas *acr2* expression is up-regulated by low oxygen and internalization by MΦs (Schnappinger *et al.*, 2003). *Acr1* is strongly induced under hypoxic conditions and is part of the *devR/devS* (*dosR/dosS*) two-component dormancy regulon (Sherman *et al.*, 2001; Stokes and Waddell, 2009). The gene is also strongly induced during the course of an *in vitro* MΦ infection (Monahan *et al.*, 2001). Within the mouse model of TB, *acr1* expression increases following production of IFNγ resulting from the induction of adaptive immunity at the end of the acute phase of the disease (around day 14 post infection) and then declines progressively through the chronic phase of the disease (approximately day 21 onwards) (Shi *et al.*, 2003). In contrast, expression of *acr2* remains elevated throughout the whole time course of the mouse infection at levels similar to that seen on entry of the bacterium into MΦs (Stewart *et al.*, 2005).

Mutation of *M. tuberculosis acr1* and *acr2* has identified significant differences between these two paralogues with regards TB pathogenesis. A Δ*acr1* mutant was shown to have impaired growth within macrophages, demonstrating the requirement for this gene for intracellular pathogenesis (Yuan *et al.*, 1998). However, later studies showed that the Δ*acr1* mutant grew more rapidly during the acute phase of an *in vivo* infection in mice, resulting in a log higher bacterial load in the lung than that seen in mice infected with wild-type *M. tuberculosis* (Hu *et al.*, 2006; Stewart *et al.*, 2006). An explanation for these seemingly contradictory results may reside within differences in the complexities and time scales of the *in vitro* and *in vivo* models. Mutation of *acr2* had no effect on the persistence of the mutant within IFNγ-activated human MΦs (Wilkinson *et al.*, 2005) or resting murine MΦs (Stewart *et al.*, 2005). However, while there was no difference in the bacterial load of mice infected with wild type or mutant, the Δ*acr2* infected mice were significantly healthier than their wild-type infected counterparts corresponding to a delay in granulomatous development and reflecting a reduced pathology (Stewart *et al.*, 2005) and implying a reduced virulence of the mutant. Unfortunately, whether this difference resulted in a prolonged survival of the Δ*acr2* infected mice was not determined.

The Acr family of chaperones in mycobacteria are obviously an important factor in the pathogenesis of, at least, *M. tuberculosis*. The fact that Acr expression correlates with a shift in the bacterial physiology towards a dormant/latent phenotype has been exploited to try and improve the diagnosis of TB so that active disease can be discriminated from latent disease and from the BCG vaccine (Geluk *et al.*, 2007). In addition, as Acr is a strong immunogen recognized by TB patients, its potential as part of a new vaccine candidate has also been investigated with success as either a constituent in an improved BCG vaccine overexpressing Acr (Shi *et al.*, 2010), or as part of a subunit protein vaccine candidate (Jeon *et al.*, 2011).

Hsp60 (Hsp65/Cpn60/GroEL) and Hsp10 (GroES)

As with many bacteria, the mycobacterial Hsp60 homologue has received considerable attention since it was first shown to be a member of the 'common antigen' family (Thole *et al.*, 1988a) and an immunodominant antigen in TB patients and animal infections (Young *et al.*, 1988, 1990). Members of the Hsp60 family (often called Hsp65 in mycobacteria) are major antigens in several pathogens and are strongly induced under stress. Strong antibody responses to Hsp60 (and Hsp70 and Hsp90) are to be found in bacterial, protozoan and helminth infections (Young, 1990). Early studies on the mycobacterial Hsp65 concentrated on mapping epitopes on the protein that were recognized by monoclonal antibodies. This identified numerous epitopes with varying degrees of species specificity and species cross-reactivity (Anderson *et al.*, 1988; Thole *et al.*, 1988b). Similar strategies were also used to identify T-cell epitopes, showing that the Hsp65 antigen was able to stimulate T-cell responses in human subjects, both with and without a concomitant mycobacterial infection (Lamb *et al.*, 1986; Thole *et al.*, 1988b). This raised a question about recognition of conserved epitopes in the Hsp60 family resulting in autoimmunity (Lamb *et al.*, 1989) which continues to be of interest (Dubaniewicz, 2010). It has also been shown that the mycobacterial GroES homologue, in addition to its partnership with GroEL as a chaperonin, is a strong immunogen (Young and Garbe, 1991; Mehra *et al.*, 1992; Young *et al.*, 1992).

The observations that Hsp60 is highly conserved through evolution and is also a strong immunogen appears, at first, surprising as it could lead to autoimmunity. Of particular note, experimental adjuvant arthritis in rats and mice (which resembles rheumatoid arthritis (RA) in humans) can be induced by injection of intact mycobacteria or mycobacterial components, e.g. complete Freund's adjuvant (McLean *et al.*, 1990), which is a mixture of mycobacterial compounds within a mineral oil vehicle (Cohen, 1991). T-cell cloning showed that T cells with specificity for a particular epitope of mycobacterial Hsp65 (180–186 epitope) could either induce the disease or, induce resistance to the disease, depending on the administration of the T cells (van Eden, 1991). Furthermore, other studies using either peptides containing this same Hsp65 epitope, recombinant Hsp65 or virally expressed Hsp65 found that immunization with these preparations led to suppression and/or protection from adjuvant arthritis (Billingham *et al.*, 1990; Yang *et al.*, 1990; Haque *et al.*, 1996; Lopez-Guerrero *et al.*, 1994). Further studies of the immune responses to Hsps in models of autoimmune disease in animals indicated that autoimmunity does not necessarily lead to autoimmune disease. Modulation of the autoimmune response can prevent autoimmune disease as has been shown for the adjuvant arthritis model in rats and insulin dependent diabetes mellitus (IDDM) in non-obese diabetic (NOD) mice (Feige and van Eden, 1996). During the course of adjuvant arthritis, self Hsp60 directed T-cell responses are involved in mediating remission from the acute inflammatory arthritis induced by *M. tuberculosis* Hsp65 (Moudgil, 1998). In addition, adjuvant arthritis can be transferred with a T-cell clone specific for the 180–186 *M. tuberculosis* epitope, whereas prior immunization with whole mycobacterial Hsp65, protects against nearly all forms of experimental arthritis. Again, this protective effect is mediated by the induction of a T-cell response against self-Hsp60 (Prakken *et al.*, 2003). Studies of autoimmunity in the IDDM/NOD mouse model found that the incidence of IDDM in NOD mice can be reduced by immunizing these mice with mycobacterial Hsp65 (Elias *et al.*, 1990), yet the same research group also found that adoptive transfer of CD4+ T-cell clones raised against mycobacterial Hsp65 accelerated the onset of IDDM (Elias *et al.*, 1991).

The exact mechanisms behind autoimmunity induced by bacterial Hsps are not fully understood, but it appears that frequent interactions with bacteria, especially pathogens, that contain proteins with a high similarity to self antigens can affect the host's ability to distinguish between self and non-self antigens (Moudgil and Sercarz, 1994). Considering that mycobacterial and mammalian Hsp60 homologues share 60% homology (Jindal *et al.*, 1989) and that mycobacterial infections are commonly chronic and may remain with the patient all their lives, it can be seen why mycobacterial Hsps are commonly implicated in the causes of autoimmune diseases. However, the 'self versus non-self' model may not be an appropriate way to evaluate the development of mammalian immunity. An alternative, the 'Danger' model, may better explain why autoimmunity to self Hsps develops (Gallucci and Matzinger, 2001).

In this model it is thought that the host mounts immune responses when it receives particular danger signals from damaged/infected host cells. These endogenous signals include the release, or surface presentation, of compounds from stressed or damaged cells (Gallucci and Matzinger, 2001). As mammalian Hsps are normally concentrated within the cytosol, their appearance on the cell surface and beyond may represent a potential signal of cellular damage (Gallucci and Matzinger, 2001). Indeed it has been found that mammalian Hsp60 is localized to the cell surface when cells are stressed (Belles *et al.*, 1999), and that necrotic cells efficiently release cytosolic Hsps while apoptotic cells do not (Basu *et al.*, 2000). Thus, while the host may retain a natural level of self Hsp recognition, this delicate balance may be influenced by ongoing exposure to certain bacterial Hsps.

The strong immunogenicity of mycobacterial Hsp65 has also been utilized in vaccination and immunomodulatory strategies for protection against tumours and TB. A 'bystander' effect for mycobacterial Hsp65 was identified when it was shown that *hsp65* transfected into a murine macrophage tumour cell line J774 resulted in these cells no longer being tumorigenic (Lukacs *et al.*, 1993). Additionally, when mice were immunized with the J774-Hsp65 cells and then challenged with the parent J774 cells they did not develop tumours, indicating that the presence of the mycobacterial Hsp65 enhanced recognition of the parent tumour cells (Lukacs *et al.*, 1993). A number of vaccine constructs utilizing Hsp65 as either naked DNA (Tacson *et al.*, 1996; Bonato *et al.*, 1998; Baek *et al.*, 2003; Franco *et al.*, 2008; Rosada *et al.*, 2008) or a protein fusion (Wang *et al.*, 2008) have been tested and found to be at least as protective as the 'standard' *M. bovis* BCG-attenuated whole bacteria vaccine.

Another aspect of mycobacterial Hsp effects on the host immune response is their induction of cytokine, reactive oxygen intermediate (ROI) and reactive nitrogen intermediate (RNI) secretion. Early studies showed that Hsp65 of *M. tuberculosis* induced the production of TNF, IL-6 and IL-8 by a human macrophage-like cell line, THP-1 (Friedland *et al.*, 1993) and TNF and IL-6 by murine peritoneal MΦs (Peetermans *et al.*, 1995). Interestingly, murine MΦs also produced RNI in response to Hsp65 which was TNF dependent and inhibited intracellular replication of the protozoan pathogen, *Toxoplasma gondii* (Peetermans *et al.*, 1995). In contrast, Hsp65 treatment of human monocyte-derived MΦs induced the pro-inflammatory cytokines TNF and IL-1 and an increase in the expression of the surface complement receptor 3, but did not result in increased induction of reactive oxygen intermediates or MHCII expression, indicating a lack of classical MΦ activation (Peetermans *et al.*, 1994). The induction of cytokines by *M. tuberculosis* Hsp65 will be further described below, under 'Moonlighting chaperones'.

In most cases, bacterial GroEL and GroES are arranged in a bicistronic groESL operon to allow for co-ordinated expression in keeping with their usual biological function, as chaperones, in maintaining the post-translational function and transport of other proteins. Most bacteria contain a single *groEL* gene but it has been shown that in approximately 30% of bacteria sequenced thus far, additional copies of groEL exist that are not arranged in operon form with *groES* genes. This has been shown for *M. leprae* (Rinke de Wit *et al.*, 1992), *M. tuberculosis* (Kong *et al.*, 1993), *M. bovis* (Wang *et al.*, 2011) and *M. avium* paratuberculosis (Goyal *et al.*, 2006) which all have two copies of the groEL homologue, whereas *M. smegmatis* has three copies (Rao and Lund 2010). Phylogenetic analysis suggests that these genes resulted from a single gene duplication event followed by varied rates of evolutionary change (Hughes 1993) except in *M. smegmatis*, where the third homologue appears to have been acquired by horizontal gene transfer (Rao and Lund 2010). The two paralogues of Hsp60 in pathogenic mycobacteria are designated Cpn60.1 (GroEL1, *M. tuberculosis* genome accession number Rv3417c) and Cpn60.2 (GroEL2, Hsp65, *M. tuberculosis* genome accession number Rv0440). The Cpn60.1 and Cpn60.2 proteins from *M. tuberculosis* only share 61% sequence identity (Kong *et al.*, 1993) which greatly differs from the 95% identity between Cpn60.2 of *M. tuberculosis* and *M. leprae* (Shinnick *et al.*, 1987). This difference seems to confer several divergent functions for these two proteins (Qamra *et al.*, 2005). Like GroEL in *E. coli*, Cpn60.2 shows hydrophobicity-based protein folding activity.

This function seems to result from the formation of a Cpn60.2 homodimer that is less ATP-dependent than is GroEL (Qamra *et al.*, 2004). Qamra and colleagues have shown that Cpn60.1 and Cpn60.2 behave as dimers *in vivo* and *in vitro* which is unusual as most known Hsp60s exist as tetradecamers (Qamra *et al.*, 2004). It appears that Cpn60.2 is acting as the GroEL equivalent so it is surprising to find that it is cpn60.1 that appears to be arranged in a putative operon with cpn10 (GroES, *M. tuberculosis* genome accession number Rv3418c) (Kong *et al.*, 1993). However, recent studies show that the apical domains of *M. tuberculosis* GroEL paralogues are conserved in their three-dimensional structure and are similar to *E. coli* GroEL. Thus, while Cpn60.2 likely functions as the general housekeeping chaperonin, Cpn60.1, like Cpn60.2, is probably a chaperonin as well (Sielaff *et al.*, 2010).

Of major interest is the observation that a knockout mutant can be obtained in both *M. bovis* BCG (Wang *et al.*, 2011) and *M. tuberculosis* (Hu *et al.*, 2008) cpn60.1 but not in cpn60.2, which has been shown to be an essential gene required for the survival of *M. tuberculosis* (Hu *et al.*, 2008). These results further supported the interpretation that Cpn60.2 was the GroEL equivalent acting with GroES as the main housekeeping chaperone. Deletion of *M. tuberculosis* Cpn60.1 did not result in a dramatic phenotype with the mutant growing in broth, in MΦs and in mice at rates equal to that of the wild-type parent (Hu *et al.*, 2008). However, the mutant failed to produce granulomatous inflammation in both mice and guinea pigs when compared to the wild-type which was associated with a concomitant reduction in cytokine induction by the mutant (Hu *et al.*, 2008). This suggested that Cpn60.1 is essential for the induction of normal granuloma formation in the infected host.

A Cpn60.1 mutant in BCG grew in broth at the same rate as wild-type but appeared to secrete more proteins into the culture media, of which a significant portion was due to increased Cpn60.2 production (Wang *et al.*, 2011). Components of the cell wall lipids were also altered in the mutant and it was more susceptible to hydrogen peroxide (Wang *et al.*, 2011). In mice, the mutant was slightly less persistent in the lungs and spleen of infected animals but maintained its ability to protect vaccinated mice against a challenge with *M. tuberculosis* (Wang *et al.*, 2011). Thus, in *M. bovis* BCG, Cpn60.1 is necessary for bacterial cell wall integrity, resistance to the ROI, H_2O_2, but is not essential for the vaccine potential of BCG. Further differences between Cpn60.1 and Cpn60.2 will be covered in the section 'Moonlighting chaperones'.

Hsp70 (DnaK) and Hsp40 (DnAJ)

Following its biochemical and antigenic characterization (Mehlert and Young, 1989), Hsp70 (*M. tuberculosis* genome accession number Rv0350) was shown to be immunodominant in TB patients and in infected mice (Young *et al.*, 1988, 1990; Patel *et al.*, 1991). Comparable to the mycobacterial Hsp65, it has been shown that mycobacterial Hsp70 acts as a chaperone but is also a strong inducer of host cytokine production. *M. tuberculosis* Hsp70 induced MΦ mRNA production for the cytokines IL-1, IL-6, TNFα and GMCSF (Retzlaff *et al.*, 1994). Again, like Hsp65, the Hsp70 of mycobacteria has been investigated as a possible vaccine or as an immunomodulator. A prime boost strategy using BCG and plasmids expressing Hsp70 or Hsp65 DNA was shown to give better protection against a *M. tuberculosis* challenge than did BCG alone (Ferraz *et al.*, 2004). Overexpression of Hsp70 in *M. tuberculosis* resulted in a reduced ability of the mutant to persist in the chronic stage of a mouse infection, presenting the possibility of inducing mycobacterial Hsp expression as a means to boost the host response to an infection (Stewart *et al.*, 2001).

The association of mycobacterial Hsp70 with autoimmune disease is also comparable to that seen with Hsp65. Hsp70 from BCG induced greater levels of IL-10, TGFβ and IFNγ production by rheumatoid arthritis patients peripheral blood mononuclear cells (PBMC) indicating an increased T-regulatory response in these patients compared to controls (Sheikhi *et al.*, 2008). In an experimental study of proteoglycan induced arthritis, it was shown that previous immunization or even post-treatment with mycobacterial Hsp70 inhibited the clinical and histological manifestation of arthritis (Wieten *et al.*, 2009). Use of IL-10 knockout mice revealed that IL-10 was critical in mediating the protective effect of Hsp70 (Wieten *et al.*, 2009). Furthermore, it has been shown that Hsp70 induces IL-10 production from synoviocytes from arthritis patients resulting in a decrease in TNFα from the synovial cells suggesting that the Hsp70 may modulate the cytokine response in arthritis through an anti-inflammatory role of the protein (Detanico *et al.*, 2004). Following on from these observations that Hsp70 can down-modulate a damaging inflammatory response, it has been shown that mycobacterial Hsp70 can also be used to inhibit allograft rejection in two separate murine models through the action of T-regulatory cells (Borges *et al.*, 2010).

The role of Hsp40/DnaJ chaperonin is to regulate the activity of Hsp70/DnaK proteins, generally in combination with a dedicated Hsp70 partner. There are two Hsp40 homologues in *M. tuberculosis*, DnaJ1 (*M. tuberculosis* genome accession number Rv0352) and DnaJ2 (*M. tuberculosis* genome accession number Rv2373c). DnaJ1 is part of the Hsp70 operon (Cole *et al.*, 1998), while DnaJ2 appears to lack an hsp70/DnaK partner (Stewart *et al.*, 2001, 2002). Using microarray profiling, it was shown that overexpression of DnaJ1 led to increased transcription of both *dnaK* and the *hsp60/groE* genes along with genes flanking the origin of chromosomal replication, suggesting a role for DnaJ1 in the control of the Hsp70 chaperones interaction with DnaA (Stewart *et al.*, 2004). Conversely, overexpression of DnaJ2 had no measurable effect on the transcription of other genes (Stewart *et al.*, 2004).

Hsp90

Apart from noting its expression in response to heat shock (Patel *et al.*, 1991), very little attention has been paid to Hsp90 in *M. tuberculosis*.

Moonlighting chaperones

As can be seen from the studies discussed above, it appears that mycobacterial chaperones have additional roles besides that of mediating intracellular protein folding. There is increasing evidence that mycobacterial Hsps are secreted signalling molecules, inducers of host immunity, surface located bacterial ligands and bacterial cell wall components. These additional functions, beyond their normal chaperonin role, have been referred to as 'moonlighting' functions by Henderson and his colleagues (Cehovin *et al.*, 2010; Henderson *et al.*, 2010) – a term I shall adopt for this review. It is worth initially mentioning some of the more unusual examples of bacterial Hsp 'moonlighting'.

The Hsp60 of the oral bacterium *Actinobacillus actinomycetemcomitans* stimulates osteoclast function resulting in the breakdown of periodontal bone (Kirby *et al.*, 1995; Henderson *et al.*, 2003). For reasons not yet understood, the Hsp60 proteins from both humans and some other bacteria (e.g. *E. coli*) also have this function (Reddi *et al.*, 1998; Meghji *et al.*, 2003). Amongst the mycobacteria, *M. tuberculosis* Cpn10 (Meghji *et al.*, 1997) also stimulates osteoclast mediated periodontal bone breakdown. *M. tuberculosis* Cpn60.2 does not have this activity, whereas *M. tuberculosis* Cpn60.1 actually inhibits osteoclast function (Winrow *et al.*, 2008). Perhaps the most interesting 'moonlighting' functions of bacterial Hsp60 are those involving insect symbiosis (Fares *et al.*, 2004). In one example, *Myrmeleontidae* larvae, commonly called antlions, immobilize their arthropod prey through the delivery of a neurotoxin upon biting. It has been determined that the actual source of this neurotoxin is an oral bacterium of the antlion called *Enterobacter aerogenes*. Furthermore, the toxin itself is the bacterial GroEL protein, and it has been shown that a single amino acid substitution in *E. coli* GroEL will result in producing the neurotoxin (Yoshida *et al.*, 2001).

Most 'moonlighting' functions of mycobacterial Hsps, however, can be related to the bacterial pathogenesis within the host. Prior to the elucidation of the protein folding roles of molecular chaperones, some of these proteins were already recognized as being potent antigens. In fact, several bacterial Hsp60 homologues had come to be termed 'common antigen' as they each invoked a robust immune response in mammalian hosts (Shinnick, 1987; Thole *et al.*, 1988a). Antibodies raised against one particular Hsp60 are often cross reactive for homologous Hsp60 proteins from other species (Gillis *et al.*, 1985) (see above). However, responses to pathogen derived Hsp70 are usually more restricted and are commonly species-specific (Zugel and Kaufmann, 1999). That an immune response initiated against a bacterial Hsp60 protein can result in cross reactivity with

Hsp60 homologues from other species (especially those of the mammalian host of mycobacterial infection) may potentially lead to autoimmunity (see above). Mycobacterial Hsp65 and Hsp70 were shown in early studies to be strong immunomodulators but it was not clear how this was mediated. As described above, Hsp65 and Hsp70 of *M. tuberculosis* were found to induce both pro- and anti-inflammatory cytokine production from MΦ.

In addition to inducing cytokine secretion, mycobacterial Hsp70 has also been found to have other 'moonlighting' functions. Overexpression of Hsp70 results in a loss of *M. tuberculosis* virulence during the chronic phase of a mouse infection (Stewart *et al.*, 2001), possibly through its ability to enhance antigen delivery to dendritic cells (DC) and stimulate DC via the CCR5 chemokine receptor resulting in the generation of effector immune responses (Floto *et al.*, 2006). However, other studies have found that carefully controlled LPS-free *M. tuberculosis* Hsp70 inhibits murine DC maturation *in vitro*, while LPS-contaminated TBhsp70 induces DC maturation (Detanico *et al.*, 2004; Motta *et al.*, 2007). Thus, the action of Hsp70 on DC is still not fully clear. One characteristic of Hsp70 activity that is commonly found is its propensity to stimulate immune cell production of IL-10 (Detanico *et al.*, 2004; Motta *et al.*, 2007; Wieten *et al.*, 2009) which then acts to modulate inflammation in arthritis models (Detanico *et al.*, 2004; Wieten *et al.*, 2009). Hsp70 also has been shown to be a major inducer of CC chemokines such as CCL3 (Mip1α), CCL4 (Mip1β) and CCL5 (RANTES) (Lehner *et al.*, 2000) through binding to MΦ and DC CD40 (a member of the TNF super family; Wang *et al.*, 2001), and CCR5 (Whittall *et al.*, 2006).

Comparisons of *M. tuberculosis* Cpn60.1 and Cpn60.2 revealed surprising differences between them in both protein folding activity (see above) and their extracellular signalling capability. While both Cpn60.1 and Cpn60.2 stimulate human MΦ to produce IL-1, IL-6, IL-8, IL-10, IL-12, TNFα and GM-CSF but not IL-4 or IFNγ, two orders less Cpn60.1 was required to get comparable levels of production (Lewthwaite *et al.*, 2001). Furthermore, Cpn60.1 but not Cpn60.2 signalling involved CD14 (Lewthwaite *et al.*, 2001), though both utilize the ERK/1 and MAPK signalling pathways to activate the MΦ (Lewthwaite *et al.*, 2007). A more recent study has extended these findings to show that, when whole blood leucocyte populations are stimulated with the chaperonins, only IL-1βand IL-6 and not IL-8, IL-10, IL-12 or IFNγ were detected. Furthermore, Cpn60.2 was a more potent stimulator than was Cpn60.1 and was the only one that induced TNFα production (Cehovin *et al.*, 2010). The contrasting results with those previously reported by this group (Lewthwaite *et al.*, 2001) indicated that the chaperonins interaction with mixed leucocytes was very different from that seen with pure MΦ. When direct binding to leucocytes was investigated, Cpn60.1 was found to bind to approximately 90% of circulating monocytes compared to <50% binding with Cpn60.2 (Cehovin *et al.*, 2010). While both proteins only showed a partial requirement of MyD88 for MΦ cytokine induction, both had a requirement for Toll-like receptor (TLR)-4 for MΦ activation with Cpn60.2 having an additional requirement for TLR2 (Cehovin *et al.*, 2010).

The partial requirement of TLR/MyD88 for cell signalling in response to Cpn60.1 and Cpn60.2 does not explain the identity of the receptor for Cpn60. Even the demonstration that antibodies recognizing CD14 could block cytokine induction by Cpn60.1 (Lewthwaite *et al.*, 2001) does not identify CD14 as the receptor for Cpn60.1. It is certainly possible that TLRs and CD14 are necessary for appropriate signalling to take place in response to Cpn60 but not that they are necessarily the receptors that bind Cpn60. The demonstration that CD14, TLR4, MD2 and other cell surface moieties form an intricate complex that mediates cell signalling in response to lipopolysaccharide (Triantafilou and Triantafilou, 2005) shows that this may well be the situation for other bacterial molecules, including mycobacterial glycolipids and proteins. The search for MΦ receptors for Cpn60 has indicated that Cpn60.1 may bind to a number of cell surface proteins (Henderson and Mesher, 2007). In contrast, it appears that Cpn60.2 binds strongly to the MΦ surface receptor, CD43 (sialophorin, leukosialin) and that this receptor/ligand interaction accounts for 30–40% of all binding of *M. tuberculosis* to MΦs (Hickey *et al.*, 2009, 2010). Hsp70 was also shown to bind to isolated CD43, albeit with lesser affinity than that of Cpn60.2, but it did not act as a ligand for binding of *M. tuberculosis* bacteria to MΦs (Hickey *et al.*, 2009, 2010). The Hsp65/Cpn60.2 homologue for *M. avium* has also been shown to bind to MΦ and mediate binding of the bacteria but the receptor was not identified (Rao *et al.*, 1994). Binding of Hsp70 and Cpn60.1 from *M. bovis* BCG to DC-SIGN has also been reported (Carroll *et al.*, 2010). The demonstration that *M. tuberculosis* Cpn60.2 acts as a ligand mediating interactions of the bacterium with its host cell, the MΦ, is not without precedence. Several other bacteria have been shown to utilize Hsps (especially the GroEL homologue) as surface ligands to mediate their interaction with host cells (Hickey *et al.*, 2009). Interestingly, while there is some level of cross-reactivity between bacterial GroEL homologues with respect to their ability to bind to MΦ receptors, mammalian homologues of GroEL show no cross-reactivity in this function (Hickey *et al.*, 2010).

The identification of MΦ receptors that mediate binding of intact mycobacteria via Hsps necessitates that the Hsps are accessible to the receptors and must therefore be located at the cell surface of the mycobacteria. This idea has met with some resistance due, at least in part, to the idea that Hsps act only as chaperonins and are therefore located intracellularly where they can function to mediate protein folding. With the demonstration that mycobacterial chaperonins have 'moonlighting' functions and the fact that mycobacteria have multiple copies of Cpn60, it becomes easier to accept that they may have functional roles that involve their location on the bacterial surface. Evidence that mycobacteria do contain Hsps within their outer cell wall has in fact been presented on several occasions using various methodologies such as electron microscopy (Esaguy and Aguas, 1997), antibody binding (Gillis *et al.*, 1985; Esaguy and Aguas, 1997; Hickey *et al.*, 2009) and proteomics (Stokes, unpublished observations; Wolfe *et al.*, 2010). In fact Hsp70 along with Cpn60.1, Cpn60.2, Hsp10 and Hsp16 are among the most prevalent of proteins within the outer cell wall layer of *M. tuberculosis* (Stokes, unpublished observations; Cunningham and Spreadbury, 1998; Hickey *et al.*, 2009; Wolfe *et al.*, 2010) and Cpn60.1 has been shown to be secreted by *M. tuberculosis* cultures (Cehovin *et al.*, 2010). Interestingly, at the same time point that Cpn60.1 first appears in culture filtrates (6 days), no Cpn60.2 can be found (Hickey *et al.*, 2009; Cehovin *et al.*, 2010) even though it is on the surface of the bacteria (Hickey *et al.*, 2009). This would imply that Cpn60.1 is actively secreted, perhaps to facilitate its actions on host cells. To date no mechanism for the active secretion of Cpn60.1 has been identified, nor for that matter a mechanism for how Cpn60.2 and the other Hsps access their outer cell wall location.

The finding that Cpn60.2 (and Hsp70) can interact with purified CD43 (Hickey *et al.*, 2009, 2010), does not necessarily mean that they interact with MΦ surface CD43 in isolation. It is possible that CD43 interacts with mycobacteria within the context of a group of MΦ surface molecules, akin to the finding by Triantafilou *et al.* that LPS can interact with a CD14-independent cluster of four different surface proteins (Triantafilou *et al.*, 2001). In this model, CD43 co-operates with other surface MΦ receptors to facilitate efficient bacterial binding and/or signal transduction via interaction with one or more bacterial surface molecules. In addition to Cpn60.2, numerous other mycobacterial cell wall constituents have been identified as ligands for MΦ binding and several MΦ receptors have been shown to be involved in this binding (El-Etr and Cirillo, 2001; Schafer *et al.*, 2009; Mishra *et al.*, 2011). The demonstration that soluble CD43 can overcome the deficiency of mycobacterial binding in $CD43^{-/-}$ MΦ suggests that, although *M. tuberculosis* can bind CD43 directly, it appears that it may also interact with other MΦ receptors (Fratazzi *et al.*, 2000). Alternatively, it may be that Cpn60.2 and CD43 do interact in isolation, but this interaction then facilitates subsequent ligand–receptor interactions to effectively take place. Much of the knowledge about CD43 suggests that it often plays the role of an intercellular binding modulator, allowing some receptor–ligand interactions to take place more readily, while limiting other interactions (Ostberg *et al.*, 1998). In this regard, although the interaction between *M. tuberculosis* Cpn60.2 and CD43 is relatively weak (Hickey *et al.*, 2010), it may provide an initial 'tethering' of bacillus and MΦ that allows for stronger subsequent bonds to occur, such as binding by the phagocytic CR3 receptor (Melo *et al.*, 2000; Rooyakkers and Stokes, 2005), or signalling via TLRs (Means *et al.*, 1999; Thoma-Uszynski *et al.*, 2001; Reiling *et al.*, 2008).

Conclusions and future trends

Like so many other species across all phyla, *M. tuberculosis* and other mycobacteria respond to stress such as temperature change, pH change, nutrient deprivation, hypoxia and exposure to antimicrobials by expression of a set of Hsps. These Hsps act as chaperonins, ensuring the correct folding and transport of proteins to maintain their function. However, mycobacteria unusually have multiple copies of one Hsp, the GroEL homologue Cpn60 that do not appear to necessarily act as chaperonins. In addition, many of the Hsps of mycobacteria, especially *M. tuberculosis,* have multiple 'moonlighting' functions that appear, at least in many cases, to be associated with the pathogenesis of the bacterium. These proteins have been identified to act as immunomodulators, bacterial ligands, intercellular signalling molecules and inducers of autoimmunity. Future studies will continue to characterize these multiple 'moonlighting' functions of the mycobacterial Hsps with regard to identifying regulation of these different functions, active sites within the Hsp for each function and their role in bacterial survival and replication within the host. Of particular interest will be the identification of the mechanism(s) by which the Hsps access the outer layers of the bacterial cell wall and the extracellular environment. For mycobacteria this is an especially difficult passage, as besides the usual plasma membrane and peptidoglycan layers of the cell wall, any proteins accessing the extracellular space have to traverse the strongly hydrophobic mycolate layer of the mycobacterial cell wall and then the outer hydrophilic capsule.

Web resources

TubercuList: http://tuberculist.epfl.ch/

The TubercuList knowledge base integrates genome details, protein information, drug and transcriptome data, mutant and operon annotation, bibliography, structural views and comparative genomics, in a structured manner required for the rational development of new diagnostic, therapeutic and prophylactic measures against tuberculosis.

Mycobacterium blast server: http://www.sanger.ac.uk/cgi-bin/blast/submitblast/mycobacterium

The research at the Wellcome Trust Sanger Institute builds understanding of gene function in health and disease as well as creating resources of lasting value to biomedical research.

TB Database: http://www.tbdb.org/

Tuberculosis (TB) is a public health challenge of paramount importance. Control of TB will require a multifaceted approach integrating efficient public health interventions with the discovery and use of new vaccines and drugs. TBDatabase (TBDB) makes available the tools and resources available at the Stanford Microarray Database and the Broad Institute.

WHO Stop TB: http://www.who.int/tb/en/

The WHO Stop TB Partnership Mission is to provide leadership in strategic and technical aspects of TB control worldwide, in order to reverse the epidemic and eventually eliminate TB.

The Tuberculosis Antimicrobial Acquisition and Coordinating Facility (TAACF): http://www.taacf.org/index.html

TAACF, managed by Southern Research Institute, assists commercial, academic and government laboratories worldwide in identifying new chemical classes of compounds for evaluation in *in vitro* and *in vivo* models.

b.e.i. resources: http://www.beiresources.org/TBVTRMResearchMaterials/tabid/1431/Default.aspx

BEI Resources are a leading source for high-quality cultures and reagents for microbiology and infectious diseases research.

Bacterial Microarray Group at St George's (BμG@S): http://bugs.sgul.ac.uk/

The BμG@S project forms the hub of a collaborative network of over 80 research groups and 250 individual scientists, based in the UK, Europe and worldwide, providing microarrays, expertise, practical training, data analysis and database resources on the basis of scientific collaboration.

References

Anderson, D.C., Barry, M.E., and Buchanan, T.M. (1988). Exact definition of species-specific and cross-reactive epitopes of the 65-kilodalton protein of *Mycobacterium leprae* using synthetic peptides. J. Immunol. *141*, 607–613.

Anonymous (2009). World Health Organization Report. http://wwwwhoint/tb/publications/global_report/2009/en/index.html.

Baek, K.M., Ko, S.Y., Lee, M., Lee, J.S., Kim, J.O., Ko, H.J., Lee, J.W., Lee, S.H., Cho, S.N., and Kang, C.Y. (2003). Comparative analysis of effects of cytokine gene adjuvants on DNA vaccination against *Mycobacterium tuberculosis* heat shock protein 65. Vaccine *21*, 3684–3689.

Basu, S., Binder, R.J., Suto, R., Anderson, K.M., and Srivastava, P.K. (2000). Necrotic but not apoptotic cell death releases heat shock proteins, which deliver a partial maturation signal to dendritic cells and activate the NF-kappa B pathway. Int. Immunol. *12*, 1539–1546.

Belles, C., Kuhl, A., Nosheny, R., and Carding, S.R. (1999). Plasma membrane expression of heat shock protein 60 *in vivo* in response to infection. Infect. Immun. *67*, 4191–4200.

Billingham, M.E., Carney, S., Butler, R., and Colston, M.J. (1990). A mycobacterial 65-kD heat shock protein induces antigen-specific suppression of adjuvant arthritis, but is not itself arthritogenic. J. Exp. Med. *171*, 339–344.

Bonato, V.L.D., Lima, V.M.F., Tascon, R.E., Lowrie, D.B., and Silva, C.L. (1998). Identification and characterization of protective T cells in hsp65 DNA-vaccinated and *Mycobacterium tuberculosis*-infected mice. Infect. Immun. *66*, 169–175.

Borges, T.J., Porto, B.N., Teixeira, C.A., Rodrigues, M., Machado, F.D., Ornaghi, A.P., de Souza, A.P., Maito, F., Pavanelli, W.R., Silva, J.S., *et al.* (2010). Prolonged survival of allografts induced by mycobacterial Hsp70 is dependent on CD4+CD25+ regulatory T cells. PLoS One *5*, e14264.

Carroll, M.V., Sim, R.B., Bigi, F., Jakel, A., Antrobus, R., and Mitchell, D.A. (2010). Identification of four novel DC-SIGN ligands on *Mycobacterium bovis* BCG. Protein Cell *1*, 859–870.

Cehovin, A., Coates, A.R., Hu, Y., Riffo-Vasquez, Y., Tormay, P., Botanch, C., Altare, F., and Henderson, B. (2010). Comparison of the moonlighting actions of the two highly homologous chaperonin 60 proteins of *Mycobacterium tuberculosis*. Infect. Immun. *78*, 3196–3206.

de Chastellier, C. (2009). The many niches and strategies used by pathogenic mycobacteria for survival within host macrophages. Immunobiology *214*, 526–542.

Cohen, I.R. (1991). Autoimmunity to chaperonins in the pathogenesis of arthritis and diabetes. Annu. Rev. Immunol. *9*, 567–589.

Cole, S.T., Brosch, R., Parkhill, J., Garnier, T., Churcher, C., Harris, D., Gordon, S.V., Eiglmeier, K., Gas, S., Barry, C.E., *et al.* (1998). Deciphering the biology of *Mycobacterium tuberculosis* from the complete genome sequence. Nature *393*, 537–544.

Collins, F.M., and Smith, M.M. (1969). A comparative study of the virulence of *Mycobacterium tuberculosis* measured in mice and guinea pigs. Am. Rev. Respir. Dis. *100*, 631–639.

Cunningham, A.F., and Spreadbury, C.L. (1998). Mycobacterial stationary phase induced by low oxygen tension – cell wall thickening and localization of the 16-kilodalton α-crystallin homolog. J. Bacteriol. *180*, 801–808.

Detanico, T., Rodrigues, L., Sabritto, A.C., Keisermann, M., Bauer, M.E., Zwickey, H., and Bonorino, C. (2004). Mycobacterial heat shock protein 70 induces interleukin-10 production: immunomodulation of synovial cell cytokine profile and dendritic cell maturation. Int. Clin. Exp. Immunol. *135*, 336–342.

Doffinger, R., Patel, S.Y., and Kumararatne, D.S. (2006). Host genetic factors and mycobacterial infections: lessons from single gene disorders affecting innate and adaptive immunity. Microbes Infect. *8*, 1141–1150.

Dubaniewicz, A. (2010). Mycobacterium tuberclosis heat shock proteins and autoimmunity in sarcoidosis. Autoimmun. Rev. *9*, 419–424.

Dye, C., and Williams, B.G. (2010). The population dynamics and control of tuberculosis. Science *328*, 856–861.

van Eden, W. (1991). Heat-shock proteins as immunogenic bacterial antigens with the potential to induce and regulate autoimmune arthritis. Immunol. Rev. *121*, 5–28.

Ehrt, S., and Schnappinger, D. (2009). Mycobacterial survival strategies in the phagosome: Defense against host stresses. Cell. Microbiol. *11*, 1170–1178.

van der Eijk, E.A., van de Vosse, E., Vandenbroucke, J.P., and van Dissel, J.T. (2007). Heredity versus environment in tuberculosis in twins: the 1950s United Kingdom Prophit Survey Simonds and Comstock revisited. Am. J. Respir. Crit. Care Med. *176*, 1281–1288.

El-Etr, S.H., and Cirillo, J.D. (2001). Entry mechanisms of mycobacteria. Frontiers in Bioscience *6*, D737–D747.

Elias, D., Markovits, D., Reshef, T., van der Zee, R., and Cohen, I.R. (1990). Induction and therapy of autoimmune diabetes in the non-obese diabetic (NOD/Lt) mouse by a 65-kDa heat shock protein. Proc. Natl. Acad. Sci. U.S.A. *87*, 1576–1580.

Elias, D., Reshef, T., Birk, O.S., van der Zee, R., Walker, M.D., and Cohen, I.R. (1991). Vaccination against autoimmune mouse diabetes with a T-cell epitope of the human 65-kDa heat shock protein. Proc. Natl. Acad. Sci. U.S.A. *88*, 3088–3091.

Ellis, R.J. (2005). Chaperomics: *in vivo* GroEL function defined. Curr. Biol. *15*, R661–663.

Esaguy, N., and Aguas, A.P. (1997). Subcellular localization of the 65-kDa heat shock protein in mycobacteria by immunoblotting and immunogold ultracytochemistry. J. Submicrosc. Cytol. Pathol. *29*, 85–90.

Fares, M.A., Moya, A., and Barrio, E. (2004). GroEL and the maintenance of bacterial endosymbiosis. Trends Genet. *20*, 413–416.

Feige, U., and van Eden, W. (1996). Infection, autoimmunity and autoimmune disease. EXS *77*, 359–373.

Ferraz, J.C., Stavropoulos, E., Yang, M., Coade, S., Espitia, C., Lowrie, D.B., Colston, M.J., and Tascon, R.E. (2004). A heterologous DNA priming-*Mycobacterium bovis* BCG boosting immunization strategy using mycobacterial Hsp70, Hsp65, and Apa antigens improves protection against tuberculosis in mice. Infect. Immun. *72*, 6945–6950.

Floto, R.A., MacAry, P.A., Boname, J.M., Mien, T.S., Kampmann, B., Hair, J.R., Huey, O.S., Houben, E.N., Pieters, J., Day, C., *et al.* (2006). Dendritic cell stimulation by mycobacterial Hsp70 is mediated through CCR5. Science *314*, 454–458.

Forget, A., Skamene, E., Gros, P., Miailhe, A.-C., and Turcotte, R. (1981). Differences in response among inbred mouse strains to infection with small doses of *Mycobacterium bovis* BCG. Infect. Immun. *32*, 42–47.

Franco, L.H., Wowk, P.F., Silva, C.L., Trombone, A.P., Coelho-Castelo, A.A., Oliver, C., Jamur, M.C., Moretto, E.L., and Bonato, V.L. (2008). A DNA vaccine against tuberculosis based on the 65 kDa heat-shock protein differentially activates human macrophages and dendritic cells. Genet. Vaccines Ther. *6*, 3.

Fratazzi, C., Manjunath, N., Arbeit, R.D., Carini, C., Gerken, T.A., Ardman, B., Remold-O'Donnell, E., and Remold, H.G. (2000). A macrophage invasion mechanism for mycobacteria implicating the extracellular domain of CD43. J. Exp. Med. *192*, 183–192.

Friedland, J.S., Shattock, R., Remick, D.G., and Griffin, G.E. (1993). Mycobacterial 65-kd heat shock protein induces release of proinflammatory cytokines from human monocytic cells. Clin. Exp. Immunol. *91*, 58–62.

Gallucci, S., and Matzinger, P. (2001). Danger signals: SOS to the immune system. Curr. Opin. Immunol. *13*, 114–119.

Garbe, T.R., Hibler, N.S., and Deretic, V. (1999). Response to reactive nitrogen intermediates in *Mycobacterium tuberculosis*: induction of the 16-kilodalton alpha-crystallin homolog by exposure to nitric oxide donors. Infect. Immun. *67*, 460–465.

Geluk, A., Lin, M.Y., van Meijgaarden, K.E., Leyten, E.M., Franken, K.L., Ottenhoff, T.H., and Klein, M.R. (2007). T-cell recognition of the HspX protein of *Mycobacterium tuberculosis* correlates with latent M. tuberculosis infection but not with *M. bovis* BCG vaccination. Infect. Immun. *75*, 2914–2921.

Georgopoulos, C.P. (1971). Bacterial mutants in which the gene N function of bacteriophage lambda is blocked have an altered RNA polymerase. Proc. Natl. Acad. Sci. U.S.A. *68*, 2977–2981.

Georgopoulos, C.P. (1977). A new bacterial gene (groPC) which affects lambda DNA replication. Mol. Gen. Genet. *151*, 35–39.

Georgopoulos, C.P., and Hohn, B. (1978). Identification of a host protein necessary for bacteriophage morphogenesis (the *groE* gene product). Proc. Natl. Acad. Sci. U.S.A. *75*, 131–135.

Getahun, H., Gunneberg, C., Granich, R., and Nunn, P. (2010). HIV infection-associated tuberculosis: the epidemiology and the response. Clin. Infect. Dis. *50*, S201–207.

Gillis, T.P., Miller, R.A., Young, D.B., Khanolkar, S.R., and Buchanan, T.M. (1985). Immunochemical characterization of a protein associated with *Mycobacterium leprae* cell wall. Infect. Immun. *49*, 371–377.

Goyal, K., Qamra, R., and Mande, S.C. (2006). Multiple gene duplication and rapid evolution in the groEL gene: functional implications. J. Mol. Evol. *63*, 781–787.

Haque, M.A., Yoshino, S., Inada, S., Nomaguchi, H., Tokunaga, O., and Kohashi, O. (1996). Suppression of adjuvant arthritis in rats by induction of oral tolerance to mycobacterial 65-kDa heat shock protein. Eur. J. Immunol. *26*, 2650–2656.

Hemmingsen, S.M., Woolford, C., van der Vies, S.M., Tilly, K., Dennis, D.T., Georgopoulos, C.P., Hendrix, R.W., and Ellis, R.J. (1988). Homologous plant and bacterial proteins chaperone oligomeric protein assembly. Nature *333*, 330–334.

Henderson, B., and Mesher, J. (2007). The search for the chaperonin 60 receptors. Methods *43*, 223–228.

Henderson, B., Nair, S.P., Ward, J.M., and Wilson, M. (2003). Molecular pathogenicity of the oral opportunistic pathogen *Actinobacillus actinomycetemcomitans*. Annu. Rev. Microbiol. *57*, 29–55.

Henderson, B., Lund, P.A., and Coates, A.R. (2010). Multiple moonlighting functions of mycobacterial molecular chaperones. Tuberculosis *90*, 119–124.

Hickey, T.B., Thorson, L.M., Speert, D.P., Daffe, M., and Stokes, R.W. (2009). *Mycobacterium tuberculosis* Cpn60.2 and DnaK are located on the bacterial surface, where Cpn60.2 facilitates efficient bacterial association with macrophages. Infect. Immun. *77*, 3389–3401.

Hickey, T.B., Ziltener, H.J., Speert, D.P., and Stokes, R.W. (2010). *Mycobacterium tuberculosis* employs Cpn60.2 as an adhesin that binds CD43 on the macrophage surface. Cell. Microbiol. *12*, 1634–1647.

Hu, Y., Movahedzadeh, F., Stoker, N.G., and Coates, A.R.M. (2006). Deletion of the α-crystallin-like *hspX* gene causes increased bacterial growth *in vivo*. Infect. Immun. *74*, 861–868.

Hu, Y., Henderson, B., Lund, P.A., Tormay, P., Ahmed, M.T., Gurcha, S.S., Besra, G.S., and Coates, A.R. (2008). A *Mycobacterium tuberculosis* mutant lacking the groEL homologue cpn60.1 is viable but fails to induce an inflammatory response in animal models of infection. Infect. Immun. *76*, 1535–1546.

Hughes, A.L. (1993). Contrasting evolutionary rates in the duplicate chaperonin genes of *Mycobacterium tuberculosis* and *M. leprae*. Mol. Biol. Evol. *10*, 1343–1359.

Jain, A., and Mondal, R. (2008). Extensively drug-resistant tuberculosis: current challenges and threats. FEMS Immunol. Med. Microbiol. *53*, 145–150.

Jeon, B.Y., Kim, S.C., Eum, S.Y., and Cho, S.N. (2011). The immunity and protective effects of antigen 85A and heat-shock protein X against progressive tuberculosis. Microbes Infect. *13*, 284–290.

Jindal, S., Dudani, A.K., Singh, B., Harley, C.B., and Gupta, R.S. (1989). Primary structure of a human mitochondrial protein homologous to the bacterial and plant chaperonins and to the 65-kilodalton mycobacterial antigen. Mol. Cell. Biol. *9*, 2279–2283.

Kirby, A.C., Meghji, S., Nair, S.P., White, P., Reddi, K., Nishihara, T., Nakashima, K., Willis, A.C., Sim, R., Wilson, M., *et al.* (1995). The potent bone-resorbing mediator of *Actinobacillus actinomycetemcomitans* is homologous to the molecular chaperone GroEL. J. Clin. Invest. *96*, 1185–1194.

Kong, T.H., Coates, A.R.M., Butcher, P.D., Hickman, C.J., and Shinnick, T.M. (1993). *Mycobacterium tuberculosis* expresses 2 chaperonin-60 homologs. Pro. Nat. Acad. Sci. U.S.A. *90*, 2608–2612.

Lamb, J.R., Ivanyi, J., Rees, A., Young, R., A., and Young, D.B. (1986). The identification of T cell epitopes in *Mycobacterium tuberculosis* using human T lymphocyte clones. Leprosy Rev. *57*, 131–137.

Lamb, J.R., Bal, V., Rothbard, J.B., Mehlert, A., Mendez-Samperio, P., and Young, D.B. (1989). The mycobacterial groel stress protein: a common target of T cell recognition in infection and autoimmunity. J. Autoimmun. *2*, 93–100.

Laskey, R.A., Honda, B.M., Mills, A.D., and Finch, J.T. (1978). Nucleosomes are assembled by an acidic protein which binds histones and transfers them to DNA. Nature *275*, 416–420.

Lehner, T., Bergmeier, L.A., Wang, Y., Tao, L., Sing, M., Spallek, R., and van der Zee, R. (2000). Heat shock proteins generate beta-chemokines which function as innate adjuvants enhancing adaptive immunity. Eur. J. Immunol. *30*, 594–603.

Lewthwaite, J.C., Coates, A.R.M., Tormay, P., Singh, M., Mascagni, P., Poole, S., Roberts, M., Sharp, L., and Henderson, B. (2001). *Mycobacterium tuberculosis* Ccn 60.1 is a more potent cytokine stimulator than chaperonin 60.2 (Hsp 65) and contains a CD14-çn. Infect. Immun. *69*, 7349–7355.

Lewthwaite, J.C., Clarkin, C.E., Coates, A.R., Poole, S., Lawrence, R.A., Wheeler-Jones, C.P., Pitsillides, A.A., Singh, M., and Henderson, B. (2007). Highly homologous *Mycobacterium tuberculosis* chaperonin 60 proteins with differential CD14 dependencies stimulate cytokine production by human monocytes through cooperative activation of p38 and ERK1/2 mitogen-activated protein kinases. Int. Immunopharmacol. *7*, 230–240.

Lienhardt, C. (2001). From exposure to disease: the role of environmental factors in susceptibility to and development of tuberculosis. Epidemiol. Rev. *23*, 288–301.

Lopez-Guerrero, J.A., Ortiz, M.A., Paez, E., Bernabeu, C., and Lopez-Bote, J.P. (1994). Therapeutic effect of recombinant vaccinia virus expressing the 60-kd heat-shock protein on adjuvant arthritis. Arthritis Rheum. *37*, 1462–1467.

Lukacs, K.V., Lowrie, D.B., Stokes, R.W., and Colston, M.J. (1993). Tumor cells transfected with a bacterial heat-shock gene lose tumorigenicity and induce protection against tumors. J. Exp. Med. *178*, 343–348.

McLean, I.L., Archer, J.R., Cawley, M.I., Pegley, F.S., Kidd, B.L., and Thompson, P.W. (1990). Specific antibody response to the mycobacterial 65 kDa stress protein in ankylosing spondylitis and rheumatoid arthritis. Br. J. Rheumatol. *29*, 426–429.

Means, T.K., Wang, S.Y., Lien, E., Yoshimura, A., Golenbock, D.T., and Fenton, M.J. (1999). Human toll-like receptors mediate cellular activation by *Mycobacterium tuberculosis*. J. Immunol. *163*, 3920–3927.

Meghji, S., White, P.A., Nair, S.P., Reddi, K., Heron, K., Henderson, B., Zaliani, A., Fossati, G., Mascagni, P., Hunt, J.F., *et al.* (1997). *Mycobacterium tuberculosis* chaperonin 10 stimulates bone resorption: a potential contributory factor in Pott's disease. J. Exp. Med. *186*, 1241–1246.

Meghji, S., Lillicrap, M., Maguire, M., Tabona, P., Gaston, J.S., Poole, S., and Henderson, B. (2003). Human chaperonin 60 (Hsp60) stimulates bone resorption: structure/function relationships. Bone *33*, 419–425.

Mehlert, A., and Young, D.B. (1989). Biochemical and antigenic characterization of the *Mycobacterium tuberculosis* 71kD antigen, a member of the 70kD heat-shock protein family. Mol. Microbiol. *3*, 125–130.

Mehra, V., Bloom, B.R., Bajardi, A.C., Grisso, C.L., Sieling, P.A., Alland, D., Convit, J., Fan, X.D., Hunter, S.W., Brennan, P.J., *et al.* (1992). A major T-cell antigen of *Mycobacterium-leprae* is a 10-kD heat-shock cognate protein. J. Exp. Med. *175*, 275–284.

Melo, M.D., Catchpole, I.R., Haggar, G., and Stokes, R.W. (2000). Utilization of CD11b knockout mice to characterize the role of complement receptor 3 (CR3, CD11b/CD18) in the growth of *Mycobacterium tuberculosis* in macrophages. Cell. Immunol. *205*, 13–23.

Mishra, A.K., Driessen, N.N., Appelmelk, B.J., and Besra, G.S. (2011). Lipoarabinomannan and related glycoconjugates: structure, biogenesis and role in *Mycobacterium tuberculosis* physiology and host–pathogen interaction. FEMS Microbiol. Rev. *35*, 1126–1157.

Mitnick, C.D., Appleton, S.C., and Shin, S.S. (2008). Epidemiology and treatment of multidrug resistant tuberculosis. Semin. Respir. Crit. Care Med. *29*, 499–524.

Monahan, I.M., Betts, J., Banerjee, D.K., and Butcher, P.D. (2001). Differential expression of mycobacterial proteins following phagocytosis by macrophages. Microbiology *147*, 459–471.

Motta, A., Schmitz, C., Rodrigues, L., Ribeiro, F., Teixeira, C., Detanico, T., Bonan, C., Zwickey, H., and Bonorino, C. (2007). *Mycobacterium tuberculosis* heat-shock protein 70 impairs maturation of dendritic cells from bone marrow precursors, induces interleukin-10 production and inhibits T-cell proliferation *in vitro*. Immunol. *121*, 462–472.

Moudgil, K.D. (1998). Diversification of response to hsp65 during the course of autoimmune arthritis is regulatory rather than pathogenic. Immunol. Rev. *164*, 175–184.

Moudgil, K.D., and Sercarz, E.E. (1994). The T cell repertoire against cryptic self determinants and its involvement in autoimmunity and cancer. Clin. Immunol. Immunopathol. *73*, 283–289.

North, R.J., and Izzo, A.A. (1993). Mycobacterial virulence – virulent strains of *Mycobacteria tuberculosis* have faster *in vivo* doubling times and are better equipped to resist growth-inhibiting functions of macrophages in the presence and absence of specific immunity. J. Exp. Med. *177*, 1723–1733.

O'Toole, R., and Williams, H.D. (2003). Universal stress proteins and *Mycobacterium tuberculosis*. Res. Microbiol. *154*, 387–392.

Orme, I.M., Stokes, R.W., and Collins, F.M. (1986). Genetic control of natural resistance to nontuberculous mycobacterial infections in mice. Infect. Immun. *54*, 56–62.

Ostberg, J.R., Barth, R.K., and Frelinger, J.G. (1998). The Roman god Janus: a paradigm for the function of CD43. Immunol. Today *19*, 546–550.

Patel, B.K.R., Banerjee, D.K., and Butcher, P.D. (1991). Characterization of the heat shock response in mycobacterium-bovis BCG. J. Bacteriol. *173*, 7982–7987.

Peetermans, W.E., Raats, C.J., Langermans, J.A., and van Furth, R. (1994). Mycobacterial heat-shock protein 65 induces proinflammatory cytokines but does not activate human mononuclear phagocytes. Scand. J. Immunol. *39*, 613–617.

Peetermans, W.E., Raats, C.J., van Furth, R., and Langermans, J.A. (1995). Mycobacterial 65-kilodalton heat shock protein induces tumor necrosis factor alpha and interleukin 6, reactive nitrogen intermediates, and toxoplasmastatic activity in murine peritoneal macrophages. Infect. Immun. *63*, 3454–3458.

Pieters, J. (2008). *Mycobacterium tuberculosis* and the macrophage: maintaining a balance. Cell Host Microbe *3*, 399–407.

Prakken, B.J., Roord, S., Ronaghy, A., Wauben, M., Albani, S., and van Eden, W. (2003). Heat shock protein 60 and adjuvant arthritis: a model for T cell regulation in human arthritis. Springer Semin. Immunopathol. *25*, 47–63.

Qamra, R., Srinivas, V., and Mande, S.C. (2004). *Mycobacterium tuberculosis* GroEL homologues unusually exist as lower oligomers and retain the ability to suppress aggregation of substrate proteins. J. Mol. Biol. *342*, 605–617.

Qamra, R., Mande, S.C., Coates, A.R., and Henderson, B. (2005). The unusual chaperonins of *Mycobacterium tuberculosis*. Tuberculosis (Edinb) *85*, 385–394.

Rao, S.P., Ogata, K., Morris, S.L., and Catanzaro, A. (1994). Identification of a 68 kD surface antigen of *Mycobacterium avium* that binds to human macrophages. J. Lab. Clin. Med. *123*, 526–535.

Rao, T., and Lund, P.A. (2010). Differential expression of the multiple chaperonins of *Mycobacterium smegmatis*. FEMS Microbiol. Lett. *310*, 24–31.

Reddi, K., Meghji, S., Nair, S.P., Arnett, T.R., Miller, A.D., Preuss, M., Wilson, M., Henderson, B., and Hill, P. (1998). The *Escherichia coli* chaperonin 60 (groEL) is a potent stimulator of osteoclast formation. J. Bone Miner. Res. *13*, 1260–1266.

Reiling, N., Ehlers, S., and Holscher, C. (2008). MyDths and un-TOLLed truths: sensor, instructive and effector immunity to tuberculosis. Immunol. Lett. *116*, 15–23.

Retzlaff, C., Yamamoto, Y., Hoffman, P.S., Friedman, H., and Klein, T.W. (1994). Bacterial heat shock proteins directly induce cytokine mRNA and interleukin-1 secretion in macrophage cultures. Infect. Immun. *62*, 5689–5693.

Rinke de Wit, T.F., Bekelie, S., Osland, A., Miko, T.L., Hermans, P.W., van Soolingen, D., Drijfhout, J.W., Schoningh, R., Janson, A.A., and Thole, J.E. (1992). Mycobacteria contain two groEL genes: the second *Mycobacterium leprae* groEL gene is arranged in an operon with groES. Mol. Microbiol. *6*, 1995–2007.

Rooyakkers, A.W., and Stokes, R.W. (2005). Absence of complement receptor 3 results in reduced binding and ingestion of *Mycobacterium tuberculosis* but has no significant effect on the induction of reactive oxygen and nitrogen intermediates or on the survival of the bacteria in resident and interferon-gamma activated macrophages. Microb. Pathog. *39*, 57–67.

Rosada, R.S., de la Torre, L.G., Frantz, F.G., Trombone, A.P., Zarate-Blades, C.R., Fonseca, D.M., Souza, P.R., Brandao, I.T., Masson, A.P., Soares, E.G. *et al.* (2008). Protection against tuberculosis by a single intranasal administration of DNA-hsp65 vaccine complexed with cationic liposomes. BMC Immunol. *9*, 38.

Russell, D.G. (2001). *Mycobacterium tuberculosis*: here today, and here tomorrow. Nat. Rev. Mol. Cell. Biol. *2*, 569–577.

Sassetti, C.M., and Rubin, E.J. (2003). Genetic requirements for mycobacterial survival during infection. Proc. Natl. Acad. Sci. U.S.A. *100*, 12989–12994.

Schafer, G., Jacobs, M., Wilkinson, R.J., and Brown, G.D. (2009). Non-opsonic recognition of *Mycobacterium tuberculosis* by phagocytes. J. Inn. Immun. *1*, 231–243.

Schnappinger, D., Ehrt, S., Voskuil, M.I., Liu, Y., Mangan, J.A., Monahan, I.M., Dolganov, G., Efron, B., Butcher, P.D., Nathan, C., *et al.* (2003). Transcriptional Adaptation of *Mycobacterium tuberculosis* within Macrophages: Insights into the Phagosomal Environment. J. Exp. Med. *198*, 693–704.

Schorey, J.S., and Cooper, A.M. (2003). Macrophage signalling upon mycobacterial infection: the MAP kinases lead the way. Cell Microbiol. *5*, 133–142.

Sheikhi, A., Nazarian, M., Khadem-Al-Melleh, A., Nasab, N.M., Esmaeilzadeh, A., Yahaghi, N., and Sheikhi, R. (2008). In-vitro effects of *Mycobacterium bovis* BCG-lysate and its derived heat shock proteins on cytokines secretion by blood mononuclear cells of rheumatoid arthritis patients in comparison with healthy controls. Int. Immunopharmacol. *8*, 887–892.

Sherman, D.R., Voskuil, M., Schnappinger, D., Liao, R.L., Harrell, M.I., and Schoolnik, G.K. (2001). Regulation of the *Mycobacterium tuberculosis* hypoxic response gene encoding alpha-crystallin. Proc. Natl. Acad. Sci. U.S.A. *98*, 7534–7539.

Shi, C., Chen, L., Chen, Z., Zhang, Y., Zhou, Z., Lu, J., Fu, R., Wang, C., Fang, Z., and Fan, X. (2010). Enhanced protection against tuberculosis by vaccination with recombinant BCG over-expressing HspX protein. Vaccine *28*, 5237–5244.

Shi, L., Jung, Y.J., Tyagi, S., Gennaro, M.L., and North, R.J. (2003). Expression of Th1-mediated immunity in mouse lungs induces a *Mycobacterium tuberculosis* transcription pattern characteristic of nonreplicating persistence. Proc. Natl. Acad. Sci. U.S.A. *100*, 241–246.

Shinnick, T.M. (1987). The 65-kilodalton antigen of *Mycobacterium tuberculosis*. J. Bacteriol. *169*, 1080–1088.

Shinnick, T.M., Sweetser, D., Thole, J., van Embden, J., and Young, R.A. (1987). The etiologic agents of leprosy and tuberculosis share an immunoreactive protein antigen with the vaccine strain *Mycobacterium bovis* BCG. Infect. Immun. *55*, 1932–1935.

Sielaff, B., Lee, K.S., and Tsai, F.T. (2010). Structural and functional conservation of *Mycobacterium tuberculosis* GroEL paralogs suggests that GroEL1 is a chaperonin. J. Mol. Biol. *405*, 831–839.

Stewart, G.G., Robertson, B.D., and Young, D.B. (2004). Analysis of the function of mycobacterial DnaJ proteins by overexpression and microarray profiling. Tuberculosis *84*, 180–187.

Stewart, G.R., Snewin, V.A., Walzl, G., Hussell, T., Tormay, P., O'Gaora, P., Goyal, M., Betts, J., Brown, I.N., and Young, D.B. (2001). Overexpression of heat-shock proteins reduces survival of *Mycobacterium tuberculosis* in the chronic phase of infection. Nat. Med. *7*, 732–737.

Stewart, G.R., Wernisch, L., Stabler, R., Mangan, J.A., Hinds, J., Laing, K.G., Young, D.B., and Butcher, P.D. (2002). Dissection of the heat-shock response in *Mycobacterium tuberculosis* using mutants and microarrays. Microbiology *148*, 3129–3138.

Stewart, G.R., Newton, S.M., Wilkinson, K.A., Humphreys, I.R., Murphy, H.N., Robertson, B.D., Wilkinson, R.J., and Young, D.B. (2005). The stress-responsive chaperone α-crystallin 2 is required for pathogenesis of *Mycobacterium tuberculosis*. Mol. Microbiol. *55*, 1127–1137.

Stewart, J.N., Rivera, H.N., Karls, R., Quinn, F.D., Roman, J., and Rivera-Marrero, C.A. (2006). Increased pathology in lungs of mice after infection with an alpha-crystallin mutant of *Mycobacterium tuberculosis*: changes in cathepsin proteases and certain cytokines. Microbiology *152*, 233–244.

Stokes, R.W., and Waddell, S.J. (2009). Adjusting to a new home: *Mycobacterium tuberculosis* gene expression in response to an intracellular lifestyle. Future Microbiol. *4*, 1317–1335.

Tascon, R.E., Colston, M.J., Ragno, S., Stavropoulos, E., Gregory, D., and Lowrie, D.B. (1996). Vaccination against tuberculosis by DNA injection. Nat. Med. *2*, 888–892.

Thole, J., Hindersson, P., de Bruyn, J., Cremers, F., van der Zee, J., de Cock, H., Tommassen, J., van Eden, W., and van Embden, J. (1988a). Antigenic relatedness of a strongly immunogenic 65 kDA mycobacterial protein antigen with a similarly sized ubiquitous bacterial common antigen. Microb. Pathog. *4*, 71–83.

Thole, J.E.R., Van Schooten, W.C.A., Keulen, W.J., Hermans, P.W.M., Janson, A.A.M., De Vries, R.R.P., Kolk, A.H.J., and Van Embden, J.D.A. (1988b). Use of recombinant antigens expressed in *Escherichia coli* K-12 to map B-cell epitopes on the immunodominant 65-kilodalton protein of *Mycobacterium bovis* BCG. Infect. Immun. *56*, 1633–1640.

Thoma-Uszynski, S., Stenger, S., Takeuchi, O., Ochoa, M.T., Engele, M., Sieling, P.A., Barnes, P.F., Rollinghoff, M., Bolcskei, P.L., Wagner, M., *et al.* (2001). Induction of direct antimicrobial activity through mammalian toll-like receptors. Science *291*, 1544–1547.

Triantafilou, M., and Triantafilou, K. (2005). The dynamics of LPS recognition: complex orchestration of multiple receptors. J. Endotoxin. Res. *11*, 5–11.

Triantafilou, K., Triantafilou, M., and Dedrick, R.L. (2001). A CD14-independent LPS receptor cluster. Nat. Immunol. *2*, 338–345.

Voskuil, M.I., Schnappinger, D., Visconti, K.C., Harrell, M.I., Dolganov, G.M., Sherman, D.R., and Schoolnik, G.K. (2003). Inhibition of respiration by nitric oxide induces a *Mycobacterium tuberculosis* dormancy program. J. Exp. Med. *198*, 705–713.

Wang, L.M., Bai, Y.L., Shi, C.H., Gao, H., Xue, Y., Jiang, H., and Xu, Z.K. (2008). Immunogenicity and protective efficacy of a DNA vaccine encoding the fusion protein of mycobacterium heat shock protein 65 (Hsp65) with human interleukin-2 against *Mycobacterium tuberculosis* in BALB/c mice. APMIS *116*, 1071–1081.

Wang, X.M., Lu, C., Soetaert, K., S'heeren, C., Peirs, P., Laneelle, M.A., Lefevre, P., Bifani, P., Content, J., Daffe, M., *et al.* (2011). Biochemical and immunological characterization of a cpn60.1 knockout mutant of *Mycobacterium bovis* BCG. Microbiology *157*, 1205–1219.

Wang, Y., Kelly, C.G., Karttunen, J.T., Whittall, T., Lehner, P.J., Duncan, L., MacAry, P., Younson, J.S., Singh, M., Oehlmann, W., *et al.* (2001). CD40 is a cellular receptor mediating mycobacterial heat shock protein 70 stimulation of CC-chemokines. Immunity *15*, 971–983.

Whittall, T., Wang, Y., Younson, J., Kelly, C., Bergmeier, L., Peters, B., Singh, M., and Lehner, T. (2006). Interaction between the CCR5 chemokine receptors and microbial HSP70. Eur. J. Immunol. *36*, 2304–2314.

Wieten, L., Berlo, S.E., Ten Brink, C.B., van Kooten, P.J., Singh, M., van der Zee, R., Glant, T.T., Broere, F., and van Eden, W. (2009). IL-10 is critically involved in mycobacterial HSP70 induced suppression of proteoglycan-induced arthritis. PLoS ONE *4*, e4186.

Wilkinson, K.A., Stewart, G.R., Newton, S.M., Vordermeier, H.M., Wain, J.R., Murphy, H.N., Horner, K., Young, D.B., and Wilkinson, R.J. (2005). Infection biology of a novel a-crystallin of *Mycobacterium tuberculosis*: Acr2. J. Immunol. *174*, 4237–4243.

Wilkinson, R.J., Wilkinson, K.A., De Smet, K.A., Haslov, K., Pasvol, G., Singh, M., Svarcova, I., and Ivanyi, J. (1998). Human T- and B-cell reactivity to the 16kDa alpha-crystallin protein of *Mycobacterium tuberculosis*. Scand. J. Immunol. *48*, 403–409.

Winrow, V.R., Mesher, J., Meghji, S., Morris, C.J., Maguire, M., Fox, S., Coates, A.R., Tormay, P., Blake, D.R., and Henderson, B. (2008). The two homologous chaperonin 60 proteins of *Mycobacterium tuberculosis* have distinct effects on monocyte differentiation into osteoclasts. Cell. Microbiol. *10*, 2091–2104.

Wolfe, L.M., Mahaffey, S.B., Kruh, N.A., and Dobos, K.M. (2010). Proteomic definition of the cell wall of *Mycobacterium tuberculosis*. J. Proteome Res. *9*, 5816–5826.

Yang, X.D., Gasser, J., and Feige, U. (1990). Prevention of adjuvant arthritis in rats by a nonapeptide from the 65-kD mycobacterial heat-shock protein. Clin. Exp. Immunol. *81*, 189–194.

Yoshida, N., Oeda, K., Watanabe, E., Mikami, T., Fukita, Y., Nishimura, K., Komai, K., and Matsuda, K. (2001). Protein function. Chaperonin turned insect toxin. Nature *411*, 44.

Young, D., Lathigra, R., Hendrix, R., Sweetser, D., and Young, R.A. (1988). Stress proteins are immune targets in leprosy and tuberculosis. Proc. Natl. Acad. Sci. U.S.A. *85*, 4267–4270.

Young, D.B. (1990). The immune response to mycobacterial heat shock proteins. Autoimmunity *7*, 237–244.

Young, D.B., and Garbe, T.R. (1991). Heat Shock Proteins and Antigens of *Mycobacterium tuberculosis*. Infect. Immun. *59*, 3086–3093.

Young, D.B., Kaufmann, S.H.E., Hermans, P.W.M., and Thole, J.E.R. (1992). Mycobacterial protein antigens – a compilation. Mol. Microbiol. *6*, 133–145.

Young, J.C., Agashe, V.R., Siegers, K., and Hartl, F.U. (2004). Pathways of chaperone-mediated protein folding in the cytosol. Nat. Rev. Mol. Cell. Biol. *5*, 781–791.

Yuan, Y., Crane, D.D., Simpson, R.M., Zhu, Y.Q., Hickey, M.J., Sherman, D.R., and Barry, C.E.r. (1998). The 16-kda alpha-crystallin (Acr) protein of *Mycobacterium tuberculosis* is required for growth in macrophages. Proc. Natl. Acad. Sci. U.S.A. *95*, 9578–9583.

Zugel, U., and Kaufmann, S.H. (1999). Role of heat shock proteins in protection from and pathogenesis of infectious diseases. Clin. Microbiol. Rev. *12*, 19–39.

Stress Response in Mycoplasmas

Melissa L. Madsen and F. Chris Minion

10

Abstract

Because mycoplasmas lack cell walls and have a limited genome capacity, there has been interest in their regulation of gene expression in these unique organisms. Their restriction to the host environment only adds to the intrigue. That environment, however, is not constant, changing with the host adaption to colonization and disease by the mycoplasma. This review focuses on the types of stresses the mycoplasma might encounter *in vivo* including heat shock, oxidative stress, osmolarity shifts, hormone exposure, and iron deprivation. Biofilm studies are included because of their use by other pathogens as a defence measure to resist killing in the host. The field of mycoplasmology is still in its infancy, particularly in regards to gene regulatory mechanisms. There have even been suggestions that mycoplasmas may not have the capacity to respond to changing environmental conditions. The studies reported here, however, show unequivocally that mycoplasmas do respond to their environment by altering transcription rates. How that is accomplished is still unknown except in one instance, heat shock. In summary, like all bacteria, mycoplasmas respond to their environment. That response may be limited, but it appears essential to their survival.

Introduction

Mycoplasmas are unique prokaryotes in that they lack cell walls, have small genomes and exist only in host environments. They have only a single limiting membrane, which is composed of phospholipids, cholesterol and proteins. Owing to a restricted genome size, their biosynthetic pathways are missing, such that the precursors needed for DNA, RNA, protein and membrane synthesis must be obtained from their host environment. For most species, growth in the laboratory is an unusual occurrence since they are transmitted from host to host by various mechanisms. They rarely survive outside the host for any length of time except under laboratory conditions so their 'normal' growth environment is one filled with stressful conditions. They must compete with the host for nutrients and survive the host immune responses. As disease progresses, the environment changes by altering the pH, increasing oxidative stresses and temperatures, restricting access to iron, and other insults from the host. How these simple organisms respond to these changes and the stresses imposed upon them by the host has not been well characterized.

A second important issue is that in addition to the loss of biosynthetic pathways is the relative lack of recognizable gene regulatory elements. There is only a single sigma factor in each of the sequenced genomes and few known transcription factors. In fact, for many years it was thought that mycoplasmas respond poorly at best to changing environmental conditions. Since mycoplasmas do not exist for the most part outside their hosts, it was believed the need for such adaptive responses was not great. However, the recent advances in gene sequencing and microarray technology have shown unequivocally that mycoplasmas do respond to stress conditions.

Most *in vitro* studies have focused on a single stressor and the pathways associated with the bacterial response to this individual stressor. These include studies on heat shock, oxidative stress, pH imbalances, catecholamine exposure, and iron depletion, to name a few. This attempt to focus on a single stressor to identify the mechanisms underlying the survival of the bacteria in an attempt to shed light on virulence often provides only a single snapshot of gene regulation. Assembly of these studies into an overarching model of pathogenesis has not yet been accomplished.

In contrast to *in vitro* studies, *in vivo* studies generate data encompassing the total effect of all host stresses on the mycoplasma. These studies are difficult to perform even with mycoplasma species from natural hosts amenable to laboratory manipulation because the mycoplasma has to be isolated in sufficient quantities and in sufficient purity to allow quantitation of mycoplasma RNA or protein while controlling for host contaminants. Techniques that have been used with more conventional bacteria may not be useful with the cell wall-less mycoplasmas. These studies can also show great variability depending on the individual host and its response to the pathogen. In some systems, outbred animals must be used, further increasing the variability. Thus, *in vitro* experiments can be much better controlled, and they generally focus on a single environment change while *in vivo* experiments more accurately model the disease itself. In light of this variability, it can be difficult to analyse these data and determine their importance without increasing the number of observations. Also the large number of hypothetical genes in each mycoplasma genome compounds the analysis and prevents determining what many of the results signify in regards to virulence and pathogenesis.

Heat shock

One significant host response encountered by mycoplasmas is an elevated host temperature or febrile state. To overcome this challenge, bacteria have acquired a set of genes know as heat shock proteins (Hsp). Hsps help to prevent damage to the bacteria by maintaining proper protein conformation and folding. The Hsps are also shown to be protective during exposure to many stressors, maintaining protein integrity and cell survival. In bacteria there are two major cytoplasmic chaperone systems, DnaK/DnaJ/GrpE and GroES/GroEL. The DnaK/DnaJ/GrpE system seems to be universal in the mycoplasmas, but the GroEs/GroEl system may be absent in a number of species.

The genome sequence of *Mycoplasma genitalium*, the first to be obtained, identified a number of heat shock genes throughout the genome (Fraser *et al.*, 1995). Of interest are three paralogues of DnaJ identified by Musatovova *et al.* (Musatovova *et al.*, 2006), DnaJ1, DnaJ2, and DnaJ3. In *M. genitalium*, DnaJ lies between genes associated with DNA replication and repair. DnaJ2 is associated with genes for RNA processing and degradation of proteins, peptides and glycopeptides. DnaJ3 is linked to the Hsp encoding the putative nucleotide exchange factor GrpE. The studies performed by Musatovova *et al.* exposed cells to 42°C for 15, 30 and 60 minutes. At the 30- and 60-minute time points, increased transcripts of *clpB* and *dnaK* were identified. At 60 minutes, *dnaJ3* and *grpE* showed significantly increased transcript levels. All four of these proteins are part of a stress-induced multichaperone system. Of note, the authors incubated the cells at 37°C post incubation at 42°C for 60 minutes and found that all transcript levels returned to control levels indicating that *M. genitalium* is actively responsive to heat shock. This was an important observation since so little is known about mycoplasma gene regulation.

In this study, the role of CIRCE sequence elements and the transcriptional repressor HrcA in the regulation of gene transcription to heat shock was first suggested in mycoplasmas. *lon*, *clpB* and *dnaK* showed the greatest response to heat shock and each has a CIRCE element sequence associated with it within their promoter regions. More recent studies by Chang *et al.* (Chang *et al.*, 2008) show unequivocal evidence that HrcA interacted with CIRCE DNA sequences in the promoter region of the *Mycoplasma hyopneumoniae dnaK* gene by gel electrophoretic shift assays. Since *M. hyopneumoniae* lacks GroEL and GroES, these authors suggested that a single mechanism of heat shock transcriptional regulation might be more economical for the minimalist mycoplasma. In congruence with this hypothesis was the surprising finding by Musatovova *et al.* in their study of *M. genitalium* (Musatovova *et al.*, 2006) of the lack of significant transcript changes to GroES/GroEL, often identified as key heat shock proteins in most bacteria. Like *M. hyopneumoniae*, many of the mycoplasmas do not possess GroES/GroEL, indicating that mycoplasmas may have other protein folding protection schemes in place.

The study by Musatovova *et al.* (2006) shows concordance with the results obtained by Weiner *et al.* in their heat shock study of *M. pneumoniae* (Weiner III *et al.*, 2003). The Weiner *et al.* study was important for several reasons; it was the first study using an array-based technology to study global gene regulation in mycoplasmas, and it established incubation conditions for optimal gene expression changes during heat shock in mycoplasmas. Since mycoplasmas live in a relatively constant environment within the host, this simple observation served to guide similar studies in *M. hyopneumoniae* (Madsen *et al.*, 2006). However, the Weiner study does not show a significant ($P < 0.01$) increase in transcript levels of the *dnaJ2* orthologue. This has led to the assertion by Musatovova *et al.* that CIRCE elements are not the only requirement for transcriptional regulation during heat shock (Musatovova *et al.*, 2006). An alternative explanation is that the CIRCE-like sequence found between the −10 sequence and the ATG start in *dnaJ2* is non-functional in *M. genitalium* (Musatovova *et al.*, 2006) and *M. pneumoniae* (Weiner III *et al.*, 2003).

Microarray analysis by Madsen *et al.* identified numerous up-regulated *M. hyopneumoniae* transcripts following exposure to heat shock. Among those with the most significant increase were *dnaK*, *parC*, *dnaJ*, and *clpB*. Proteins encoded by these genes function to assist in protein folding, in degrading improperly folded proteins, and in cell partitioning activities associated with DNA structure (Madsen *et al.*, 2006). As stated previously, *dnaK* and *dnaJ* have upstream CIRCE elements that bind HrcA to control transcription during heat shock (Hecker *et al.*, 1996).

First reported by Madsen *et al.* is the regulation of *ffh*, the 54-kDa subunit homologue of the eukaryotic signal recognition particle. It was speculated that Ffh might play a role in membrane stabilization or integrity. In contrast, Weiner *et al.* showed that *M. pneumoniae* did not have significant transcript changes in *ffh* after exposure to varying temperatures. The observation of transcriptional changes with *M. hyopneumoniae ffh* has been observed previously only with *Streptococcus mutans* during acid stress (Gutierrez *et al.*, 1999; Kremer *et al.*, 2001) and *Bifidobacterium longum* during growth phase and acid adaption (Jin *et al.*, 2011) suggesting that ffh may play yet undefined roles in adaptation to the host environment during stress in some organisms, especially pH shifts.

Down-regulation of transcription and translation machinery is common across heat shock studies. As the mycoplasma adapts to survive the increasingly harsh host environment, the need to conserve energy, minimize growth and maintain membrane integrity becomes the main objective.

Oxidative stress

Oxidative stress *in vivo* can result from numerous reactive molecules produced in the host to combat infection. Of note here are methionine sulfoxide, H_2O_2, hydroxyl radicals and organic hyperperoxides. Mycoplasmas overcome these stressors presumably through the regulation of various genes within their reduced genome that has adapted for survival in specific host niches.

Dhandayuthapani *et al.* studied the effects of an *msrA* (methoinine sulfoxide reductase A) mutation on the survivability and pathogenicity of *M. genitalium* (Dhandayuthapani *et al.*, 2001). MsrA reverses oxidation of methionine residues in proteins by oxidative compounds. The oxidized methiones have decreased biological activity leading to a proposed decrease in cell adherence (Moskovitz *et al.*, 2001). The *M. genitalium msrA* mutants have increased sensitivity to H_2O_2, and survival in a hamster infection model indicating a role for MsrA in response to oxidative stress and pathogenesis (Dhandayuthapani *et al.*, 2001).

In a follow up to this study, Sankaralingam *et al.* (2009) identified MG_454, organic hydroperoxide reductase (Ohr), as a reducer of reactive oxygen species in *M. genitalium*. The primary role of Ohr is to prevent toxicity from organic hydroperoxide; however, there was no significant transcript increase when exposed to H_2O_2, suggesting that the gene is constitutively expressed as previously shown by Jenkins *et al.* (2008) for *M. gallisepticum*. Whether Ohr is the only oxidative stress responder is unknown. Unless the mycoplasma Ohr has enhanced substrate specificity, there remain a number of additional factors to be detoxified, reactive oxygen species, superoxide and hydrogen peroxide. As suggested by Sankaralingam *et al.*, it is possible that these additional factors may be converted to other compounds, i.e. superoxide and hydrogen peroxide to hydroperoxides. If MsrA and MsrB play a role in detoxifying these compounds, it must be through reduction of methionine residues in protein intermediates. A peroxiredoxin has been identified and characterized from *M. hyopneumoniae* (Machado et al., 2009). Its role is thought to be in protection against H_2O_2 produced by the host during infection.

Like other mycoplasmas, *M. hyopneumoniae* must likewise survive the exposure to hydrogen peroxide, nitric oxide and superoxide anions during respiratory bursts produced by the immune cells recruited to the site of infection. In examining the response of *M. hyopneumoniae* to H_2O_2 exposure, Schafer *et al.* identified *atpB, glyS, spoU* and *mglA* as being significantly up-regulated (Schafer *et al.*, 2007). *spoU* and *glyS* have assigned functions within tRNA synthesis and regulation, indicating that oxidative stress may lead to an increase in protein synthesis. The up-regulation of *atpB* and *mglA* transcripts indicates ATP requirements increase during oxidative damage, and the increased levels of ATP would subsequently be available for repair of damaged cellular components. The down-regulation of a number of ribosomal assembly and processing genes were also noted by Schafer *et al.* These included *rpl23, rplM, rpl4* and *rluC*. Thus, it seems that in *M. hyopneumoniae*, oxidative damage results in lower ribosome synthesis, possibly as a way to conserve energy. Since both *gyrA* (DNA gyrase subunit A) and *ftsY* (cell division) were also down-regulated following H_2O_2 exposure, one must assume that both DNA replication and cell division were slowed as well. A mechanism to slow cellular processes to allow time for DNA repair to take place seems reasonable. With that hypothesis in mind, *pkfA*, the gene encoding 6-phosphofructokinase and the enzyme catalysing the committed step of glycolysis, was also down-regulated. Interestingly, no genes for DNA-repair enzymes were significantly regulated during oxidative damage suggesting that this response is not similar to the SOS response seen in other organisms but may be constitutive in nature.

Lipoproteins are often implicated as major virulence factors in bacteria. *Mycoplasma pneumoniae* cells exposed to H_2O_2 showed increased transcript levels in two lipoproteins (Hallamaa *et al.*, 2008). As a part of this study in cells exposed to pH 5.5 for 10 minutes, seven lipoprotein genes were down-regulated. This study again revealed hypothetical genes are playing prominent roles in stress responses by mycoplasmas.

Osmolarity shifts

Like all bacteria, maintaining cell osmolarity is a challenge for mycoplasmas. Few studies have been reported on studies of changing osmolarity and the effects on mycoplasmas. It is generally thought that the lack of a cell wall might render the organisms more susceptible to osmolarity differences. However, mycoplasmas are generally resistant to osmotic lysis despite their lack of cell walls (Razin, 1983). Sayed and Kenny in an early study with *Ureaplasma urealyticum* showed that increased osmotic pressure inhibited its growth as well (Sayed and Kenny, 1978). Brown *et al.* showed that *Mycoplasma felis* and *Mycoplasmas gateae* failed to grow under the hyperosmotic conditions present in urine (Brown *et al.*, 1991).

The effect of osmolarity on gene regulation has not been studied in mycoplasmas until recently. A microarray study conducted by Zhang and Baseman identified *M. genitalium* genes regulated under increasing concentrations of extracellular NaCl (Zhang and Baseman, 2011b). Fifteen of the 39 up-regulated genes coded for membrane and membrane-associated proteins. The significance of these up-regulated genes indicates that *M. genitalium* undergoes surface remodelling when exposed to osmotic stress. Down-regulated genes included those involved in metabolism-related functions, including glycolysis (*fba, pgi, pgk, gap, eno, gpml, ldh, pta, ackA, pdhA* and *pdhB*); ribosomal protein function proteins (*rpsP, rplL, rplS, rplK, rplA, rpsD* and *rpmF*), translation factors (*tuf* and *tsf*), and tRNA synthesis (*trmD*). In this study, MG_454, an osmotically inducible protein in other bacteria, was down-regulated in *M. genitalium*.

Also of interest is the differential regulation of MG_289 (Zhang and Baseman, 2011a). This protein is involved in the phosphonate ABC transporter. Of interest is the homology it shares with the *Mycoplasma hyorhinis* p37 protein (renamed Cypl) (Sippel *et al.*, 2009). Sippel *et al.* show by structure studies that Cypl is an extracytoplasmic thiamine-binding protein. The interest in p37 arises from the observation that it may play a role in carcinogenesis (Gong *et al.*, 2008). Whether p37 is also a thiamine-binding protein, and whether that activity is the source of the carcinogenic activity associated with p37 (Gong *et al.*, 2008) has not been established.

Hormone exposure

Studies on the effects of hormones on mycoplasmas have mostly focused on hormone effects with genital and respiratory tract infections through colonization or pathogenesis (Furr and Taylor-Robinson, 1993a,b; Taylor-Robinson and Jensen, 2011). Exposure of asthmatic children to corticosteroids did not affect the infection rate or load with *M. pneumoniae* (Sasaki *et al.*, 2009). A study by Thacker *et al.* (2006) showed enhanced resistance to infection by *M. hyopneumoniae* when pigs were pre-exposed to growth-hormone releasing hormone. Only recently have defined studies focused directly on mycoplasmas and the effect that hormone exposure has on gene expression.

Oneal *et al.* used microarrays to examine the overall transcriptional profile of *M. hyopneumoniae* upon exposure to noradrenaline (Oneal *et al.*, 2008). Unlike other bacteria, *M. hyopneumoniae* slowed its growth in response to the hormone (Anderson and Armstrong, 2006; Dowd, 2007; Freestone *et al.*, 2007; Nakano *et al.*, 2007; Methner *et al.*, 2008; Li *et al.*, 2009). One reason for the slowed growth in *M. hyopneumoniae* could be the down-regulation of *ftsZ*, a component in cell division. Transcription and translation factors including *dnaN*, the DNA polymerase β sliding clamp, showed increased transcription as did components of the DNA repair machinery, *mutsS*, *lig and gyrA*. Topoiosmerase I, *topA*, responded not only to noradrenaline, but was first identified as an oxidative stress response gene (Schafer *et al.*, 2007).

It is unknown how the mycoplasma senses the hormone and responds to the catecholamine. The simplest explanation is through a membrane-associated signal transduction process, but these have not been characterized in mycoplasmas. It is also possible that the hormone could traverse the membrane either through an active or passive process and interact with an intracellular molecule resulting in transcriptional changes.

Iron deprivation

For most living organisms, iron is an absolute requirement. One exception is the Lyme disease agent *Borrelia burgdorferi* (Posey and Gherardini, 2000). The ability of iron to form complexes with oxygen, sulfur and nitrogen emphasizes the processes it is engaged in, electron transport, oxygen activation and H_2O_2 reduction, amino acid and pyrimidine biosynthesis, the tricarboxylic acid cycle, DNA replication, and protection from superoxide radicals among others (Earhart, 1996). In fact, the importance of iron to life is best illustrated by the fact that iron sequestration by the host is one means to mount a defence against infection, and thus its importance to the invading pathogen. Iron deprivation inhibits pathogenesis and reduces virulence in bacteria. The host mucosal surfaces are rich in the iron-binding proteins transferrin and lactoferrin. To combat this, bacteria have evolved iron uptake systems to strip their host of sequestered iron. Despite the importance of iron in other pathogens, there have been no definitive studies demonstrating the absolute requirement for iron in mycoplasmas. The complexity of the media used to grow mycoplasmas may prevent some types of studies. Also, most pathways or enzymes that utilize iron are absent in mycoplasmas including electron transport systems, the TCA cycle, and the amino acid and pyrimidine biosynthetic pathways among others. Superoxide dismutase is often found to contain iron in many bacteria and often these organisms have enzymes with different cofactors. The peroxiredoxin identified in *M. hyopneumoniae* was shown to use iron as a cofactor in studies by Machado *et al.* (2009), but other metals (copper, zinc, manganese, nickel) were not tested for their ability to substitute for iron. Only a single superoxide dismutase has been characterized in mycoplasmas, a copper-zinc superoxide dismutase in *M. hyopneumoniae* (Chen *et al.*, 2000). No other mycoplasmas have been reported to contain this activity (Lynch and Cole, 1980; O'Brien *et al.*, 1981). Thus, it is not clear if iron is a requirement of these organisms.

There have been scattered reports of the interaction of iron or iron-containing host proteins with mycoplasmas. Previous studies have examined the interactions and binding abilities of *M. pneumoniae* and *M. genitalium* with lactoferrin and transferrin (Tryon and Baseman, 1987). These studies found that *M. pneumoniae* binds lactoferrin but not transferrin and *M. genitalium* binds neither protein. *M. capricolum* (Bauminger *et al.*, 1980) does have the ability to store iron, but further studies have not been conducted to determine how or if iron is acquired in this species.

To determine if iron availability has an effect on gene regulation in mycoplasmas, microarray studies by Madsen *et al.* with *M. hyopneumoniae* identified both transport (*mglA* and *potB*) and metabolic genes (*iolCB*, *iolC*, *glpQ* and *ackA*) up-regulated during iron deprivation. *mglA* is one of 13 ATP transporter homologues found in the genome. This has led the authors to speculate that MglA may be an iron uptake protein. It was also up-regulated during exposure to H_2O_2 (Schafer *et al.*, 2007) suggesting that it has an important role in stress responses. Down-regulation of the *oppB-D* operon gives support to the idea that metabolism is slowed during iron deprivation. Results of this study and the heat shock study identified three up-regulated (*mhp151*, *mhp152* and *mhp275*) and three down-regulated genes (*mhp030*, *mhp176* and *mhp337*) that overlapped between the two studies. These genes were considered to be a subset of stress response genes in *M. hyopneumoniae*.

Biofilms

Microorganisms are generally protected from stressful conditions by biofilms, allowing the organism to persist under conditions that ordinarily would result in elimination. They have been commonly found in a wide variety of infections and increase the resistance to host defences and antibiotics (Costerton *et al.*, 1999). Interest in the structure and function of biofilms has intensified over recent years, but not until recently have biofilms been observed and studied in mycoplasmas. McAuliffe *et al.* first reported biofilms in mycoplasma in 2006 (McAuliffe *et al.*, 2006) followed by a more definitive study of the biofilm of *Mycoplasma mycoides* subsp. *mycoides* SC in 2008 (McAuliffe *et al.*, 2008). The ability to produce biofilms was strain specific in this species (McAuliffe *et al.*, 2008). These studies were followed by reports by Simmons and Dybvig of biofilms with *Mycoplasma pulmonis* and showed that biofilm formation in this species was dependent on the length of a variable surface antigen (Simmons *et al.*, 2007; Simmons and Dybvig, 2007, 2009). Biofilms have also been observed with *M. hyopneumoniae* (Fig. 10.1). So it is likely that biofilm production is an approach used by mycoplasmas to circumvent the host immune response and protect it against antibiotics during active infections.

In vivo

The reduced genomes of mycoplasmas give researchers a sense for the essential genes required to combat stress and survive in the host environment (Fraser *et al.*, 1995; Mushegian and Koonin, 1996; Hutchison III *et al.*, 1999). *In vivo* studies attempt to provide a snapshot at a point in time of disease. These studies often do not provide an easily understood analysis of disease progression, but they do offer insight into the overall pathogenic process that would not be obvious from *in vitro* studies. Results from *in vivo* studies rarely point to a magic bullet for pathogenesis or disease elimination, however.

To examine gene transcription during infection is a difficult process because it rare to obtain sufficient quantities of organisms to yield suitable quantities of RNA for studies of this type, host RNA often contaminates the preparations, and the natural host may not be available or amenable for studies of this type. Only recently have transcriptome studies been performed with *in vivo* grown mycoplasmas.

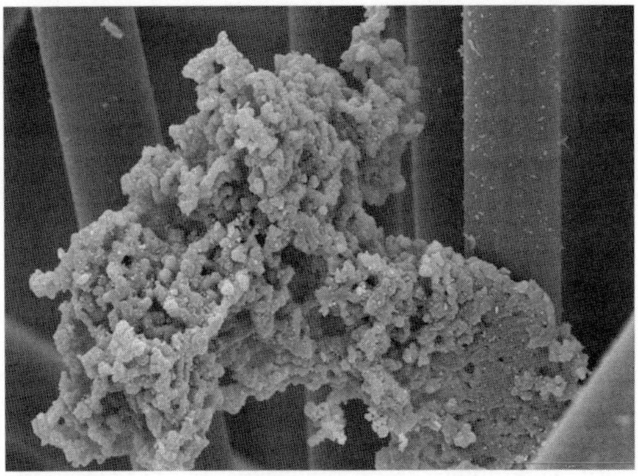

Figure 10.1 *M. hyopneumoniae* in biofilms on glass wool as observed by scanning electron microscopy (F.C. Minion *et al.*, unpublished).

A microarray study conducted by Madsen *et al.* (2008) examined the transcription response of *M. hyopneumoniae* 28 days post infection in swine. From these studies numerous genes were identified as significantly differentially regulated at this point in the infectious process. These included membrane transport genes *secA*, *abc*, *potA* and *ugpA*. Each of these genes appear to function in membrane translocation and transport of proteins across the cell membrane. During infection, mycoplasmas must respond rapidly to the changing host environment and the ability to quickly transport proteins across the cell membrane increases survivability by the mycoplasmas. This is the apparent role of PotA and the transport of spermidine across the cell membrane. Spermidine is able to bind to the ribosome and increase translation, while stabilizing DNA (Igarashi and Kashiwagi, 1999). In addition, host nitric oxide mediated immune functions are affected by spermidine. This may allow *M. hyopneumoniae* to respond to the recruitment of neutrophils and macrophages by the host. The up-regulation of *ugpA* (sn-glycerol-3-phosphate transport permease) indicates that *M. hyopneumoniae* may use glycerol as an alternative energy source or to maintain cell integrity in response to osmotic stress placed upon it by the host environment. It is noteworthy that Pilo *et al.* determined that *Mycoplasma mycoides* subsp. *mycoides* had enhanced virulence when glycerol was present in the media by enhancing hydrogen peroxide production (Pilo *et al.*, 2005).

The study by Madsen *et al.* (2008) in *M. hyopneumoniae* identified three lipoproteins differentially transcribed. Since lipoproteins can be regulated post-transcriptionally (Yogev *et al.*, 2002), these results are not surprising. This study did not identify specific virulence factors during infection. Most of the genes that showed differential expression were related to transcription and translation and may reflect the difference in the minimally available resources of the host environment compared to the nutrient rich environment of broth-grown cultures.

Conclusion and future trends

Each of the studies mentioned in this chapter provides insight into the regulation and mechanisms employed by mycoplasmas to combat the stressors encountered by the host environment. It is only the beginning, however, of understanding the complete mycoplasma response utilized to survive their respective host niches. It has been interesting to discover that organisms with a limited genome and confined to the host environment has the capacity to regulate its genes and respond to stress conditions. The mechanisms of this regulation, for the most part, are still elusive, and there is evidence that mycoplasmas do not use the conventional transcription start and stop signals as regulatory features (Gardner and Minion, 2010). In addition, the large percentage of unique and hypothetical genes in their genomes complicates our understanding of these processes. To overcome these limitations, future efforts should be focused on more intensive genetic and proteomic studies to draw a more complete picture of gene regulation and mycoplasmal survival in the harsh host environment. A more complete understanding of the changing host environment as adaption to the developing mycoplasma disease proceeds will also clearly be a research focus area in future studies.

Web resources

IOM: http://the-iom.org
The International Organization for Mycoplasmology is dedicated to the study of mycoplasmas.

MolliGen: http://cbi.labri.fr/outils/molligen
A site dedicated to genome sequence analysis of mycoplasma genomes.

NCBI: http://www.ncbi.nlm.nih.gov/Genomes
The National Center for Biotechnology Information provides tools for gene and protein analysis.

GOLD: http://genomesonline.org/cgi-bin/GOLD/bin/gold.cgi
The GOLD database is a listing of all known genome sequences.

GEO: http://www.ncbi.nlm.nih.gov/geo
The Gene Expression Omnibus is a public repository of MIAME-compliant microarray data submissions.

References

Anderson, M.T., and Armstrong, S.K. (2006). The *Bordetella bfe* system: growth and transcriptional response to siderophores, catechols, and neuroendocrine catecholamines. J. Bacteriol. *188*, 5731–5740.

Bauminger, E.R., Cohen, S.G., Labenski-Dekanter, F., Levy, A., Aster, S., Kessel, M., and Rottem, S. (1980). Iron storage in *Mycoplasma capricolum*. J. Bacteriol. *141*, 378–381.

Brown, M.B., Stoll, M., Maxwell, J., and Senior, D.F. (1991). Survival of feline mycoplasmas in urine. J. Clin. Microbiol. *29*, 1078–1080.

Chang, L.-J., Chen, W.-H., Minion, F.C., and Shiuan, D. (2008). Mycoplasmas regulate the expression of heat shock protein genes through CIRCE–HrcA interaction. Biochem. Biophys. Res. Commun. *367*, 213–218.

Chen, J.R., Weng, C.N., Ho, T.Y., Cheng, I.C., and Lai, S.S. (2000). Identification of the copper-zinc superoxide dismutase activity in *Mycoplasma hyopneumoniae*. Vet. Microbiol. *73*, 301–310.

Costerton, J.W., Stewart, P.S., and Greenberg, E.P. (1999). Bacterial biofilms: A common cause of persistent infections. Science *284*, 1318–1322.

Dhandayuthapani, S., Blaylock, M.W., Bebear, C.M., Rasmussen, W.G., and Baseman, J.B. (2001). Peptide methionine sulfoxide reductase (MsrA) is a virulence determinant in *Mycoplasma genitalium*. J. Bacteriol. *183*, 5645–5650.

Dowd, S.E. (2007). *Escherichia coli* O157:H7 gene expression in the presence of catecholamine norepinephrine. FEMS Microbiol. Lett. *273*, 214–223.

Earhart, C.F. (1996). Uptake and metabolism of iron and molybdenum. In *Escherichia coli* and *Salmonella typhimurium*: Cellular and Molecular Biology, Neidhardt, F.C., Curtiss, R.I., Ingraham, J.L., Lin, E.C.C., Lo, K.B., Magasanik, B., Reznikoff, W.S., Riley, M., Schaechter, M., and Umbarger, E.H., eds (ASM Press, Washington DC), pp. 1075–1090.

Fraser, C.M., Gocayne, J.D., White, O., Adams, M.D., Clayton, R.A., Fleischmann, R.D., Bult, C.J., Kerlavage, A.R., Sutton, G., Kelley, J.M., *et al.* (1995). The minimal gene complement of *Mycoplasma genitalium*. Science *270*, 397–403.

Freestone, P.P., Haigh, R.D., and Lyte, M. (2007). Specificity of catecholamine-induced growth in *Escherichia coli* O157:H7, *Salmonella enterica* and *Yersinia enterocolitica*. FEMS Microbiol. Lett. *269*, 221–228.

Furr, P.M., and Taylor-Robinson, D. (1993a). The contrasting effects of progesterone and oestrogen on the susceptibility of mice to genital infection with *Mycoplasma pulmonis*. J. Med. Microbiol. *38*, 160–165.

Furr, P.M., and Taylor-Robinson, D. (1993b). Factors influencing the ability of different mycoplasmas to colonize the genital tract of hormone-treated female mice. Int. J. Exp. Pathol. *74*, 97–101.

Gardner, S.W., and Minion, F.C. (2010). Detection and quantification of intergenic transcription in *Mycoplasma hyopneumoniae*. Microbiology *156*, 2305–2315.

Gong, M., Meng, L., Jiang, B., Zhang, J., Yang, H., Wu, J., and Shou, C. (2008). p37 from *Mycoplasma hyorhinis* promotes cancer cell invasiveness and metastasis through activation of MMP-2 and followed by phosphorylation of EGFR. Mol. Cancer Ther. *7*, 530–537.

Gutierrez, J.A., Crowley, P.J., Cvitkovitch, D.G., Brady, L.J., Hamilton, I.R., Hillman, J.D., and Bleiweis, A.S. (1999). *Streptococcus mutans ffh*, a gene encoding a homologue of the 54 kDa subunit of the signal recognition particle, is involved in resistance to acid stress. Microbiology *145*, 357–366.

Hallamaa, K.M., Tang, S.-L., Ficorilli, N., and Browning, G.F. (2008). Differential expression of lipoprotein genes in *Mycoplasma pneumoniae* after contact with human lung epithelial cells, and under oxidative and acidic stress. BMC Microbiol. *8*, 124.

Hecker, M., Schumann, W., and Volker, U. (1996). Heat-shock and general stress response in *Bacillus subtilis*. Mol. Microbiol. *19*, 417–428.

Hutchison III, C.A., Peterson, S.N., Gill, S.R., Cline, R.T., White, O., Fraser, C.M., Smith, H.O., and Venter, J.C. (1999). Global transposon mutagenesis and a minimal mycoplasma genome. Science *286*, 2165–2169.

Igarashi, K., and Kashiwagi, K. (1999). Polyamine transport in bacteria and yeast. Biochem. J. *344*, 633–642.

Jenkins, C., Samudrala, R., Geary, S.J., and Djordjevic, S.P. (2008). Structural and functional characterization of an organic hydroperoxide resistance protein from *Mycoplasma gallisepticum*. J. Bacteriol. *190*, 2206–2216.

Jin, J., Liu, S., Zhao, L., Ge, K., Mao, X., and Ren, F. (2011). Changes in *ffh*, *uvrA*, *groES* and *dnaK* mRNA abundance as a function of acid-adaptation and growth phase in *Bifidobacterium longum* BBMN68 isolated from healthy centenarians. Curr. Microbiol. *62*, 612–617.

Kremer, B.H., van der Kraan, M., Crowley, P.J., Hamilton, I.R., Brady, L.J., and Bleiweis, A.S. (2001). Characterization of the *sat* operon in *Streptococcus mutans*: evidence for a role of Ffh in acid tolerance. J. Bacteriol. *183*, 2543–2552.

Li, W., Lyte, M., Freestone, P.P., Ajmal, A., Colmer-Hamood, J.A., and Hamood, A.N. (2009). Norepinephrine represses the expression of *toxA* and the siderophore genes in *Pseudomonas aeruginosa*. FEMS Microbiol. Lett. *299*, 100–109.

Lynch, R.E., and Cole, B.C. (1980). *Mycoplasma pneumoniae*: a prokaryote which consumes oxygen and generates superoxide but which lacks superoxide dismutase. Biochem. Biophys. Res. Commun. *96*, 98–105.

Machado, C.X., Pinto, P.M., Zaha, A., and Ferreira, H.B. (2009). A peroxiredoxin from *Mycoplasma hyopneumoniae* with a possible role in H_2O_2 detoxification. Microbiology *155*, 3411–3419.

Madsen, M.L., Nettleton, D., Thacker, E.L., Edwards, R., and Minion, F.C. (2006). Transcriptional profiling of *Mycoplasma hyopneumoniae* during heat shock using microarrays. Infect. Immun. *74*, 160–166.

Madsen, M.L., Puttamreddy, S., Thacker, E.L., Carruthers, M.D., and Minion, F.C. (2008). Transcriptome changes in *Mycoplasma hyopneumoniae* during infection. Infect. Immun. *76*, 658–663.

McAuliffe, L., Ellis, R.J., Miles, K., Ayling, R.D., and Nicholas, R.A. (2006). Biofilm formation by mycoplasma species and its role in environmental persistence and survival. Microbiology *152*, 913–922.

McAuliffe, L., Ayling, R.D., Ellis, R.J., and Nicholas, R.A. (2008). Biofilm-grown *Mycoplasma mycoides* subsp. *mycoides* SC exhibit both phenotypic and genotypic variation compared with planktonic cells. Vet. Microbiol. *129*, 315–324.

Methner, U., Rabsch, W., Reissbrodt, R., and Williams, P.H. (2008). Effect of norepinephrine on colonisation and systemic spread of *Salmonella enterica* in infected animals: role of catecholate siderophore precursors and degradation products. Int. J. Med. Microbiol. *298*, 429–439.

Moskovitz, J., Bar-Noy, S., Williams, W.M., Requena, J., Berlett, B.S., and Stadtman, E.R. (2001). Methionine sulfoxide reductase (MsrA) is a regulator of antioxidant defense and lifespan in mammals. Proc. Natl. Acad. Sci. U.S.A. *98*, 12920–12925.

Musatovova, O., Dhandayuthapani, S., and Baseman, J.B. (2006). Transcriptional heat shock response in the smallest known self-replicating cell, *Mycoplasma genitalium*. J. Bacteriol. *188*, 2845–2855.

Mushegian, A.R., and Koonin, E.V. (1996). A minimal gene set for cellular life derived by comparison of complete bacterial genomes. Proc. Natl. Acad. Sci. U.S.A. *93*, 10268–10273.

Nakano, M., Takahashi, A., Sakai, Y., and Nakaya, Y. (2007). Modulation of pathogenicity with norepinephrine related to the type III secretion system of *Vibrio parahaemolyticus*. J. Infect. Dis. *195*, 1353–1360.

O'Brien, S.J., Simonson, J.M., Grabowski, M.W., and Barile, M.F. (1981). Analysis of multiple isoenzyme expression among twenty-two species of *Mycoplasma* and *Acholeplasma*. J. Bacteriol. *146*, 222–232.

Oneal, M.J., Schafer, E.R., Madsen, M.L., and Minion, F.C. (2008). Global transcriptional analysis of *Mycoplasma hyopneumoniae* following exposure to norepinephrine. Microbiology *154*, 2581–2588.

Pilo, P., Vilei, E.M., Peterhans, E., Bonvin-Klotz, L., Stoffel, M.H., Dobbelaere, D., and Frey, J. (2005). A metabolic enzyme as a primary virulence factor of *Mycoplasma mycoides* subsp. *mycoides* small colony. J. Bacteriol. *187*, 6824–6831.

Posey, J.E., and Gherardini, F.C. (2000). Lack of a role for iron in the Lyme disease pathogen. Science *288*, 1651–1653.

Razin, S. (1983). Cell lysis and isolation of membranes. In Methods in Mycoplasmology: Mycoplasma Characterization, Razin, S., and Tully, J.G., eds (Academic Press, New York), pp. 225–234.

Saikolappan, S., Sasindran, S.J., Yu, H.D., Baseman, J.B., and Dhandayuthapani, S. (2009). The *Mycoplasma genitalium* MG_454 gene product resists killing by organic hydroperoxides. J. Bacteriol. *191*, 6675–6682.

Sasaki, A., Ouchi, K., Makata, H., Hashimoto, K., Matsubara, T., and Furukawa, S. (2009). The effect of inhaled corticosteroids on *Chlamydophila pneumoniae* and *Mycoplasma pneumoniae* infection in children with bronchial asthma. J. Infect. Chemother. *15*, 99–103.

Sayed, I.A., and Kenny, G.E. (1978). Effect of ammonium ion concentration and osmotic pressure on growth of *Ureaplasma urealyticum* (T-strain mycoplasma). J. Bacteriol. *134*, 967–972.

Schafer, E.R., Oneal, M.J., Madsen, M.L., and Minion, F.C. (2007). Global transcriptional analysis of *Mycoplasma hyopneumoniae* following exposure to hydrogen peroxide. Microbiology *153*, 3785–3790.

Simmons, W.L., and Dybvig, K. (2007). Biofilms protect *Mycoplasma pulmonis* cells from lytic effects of complement and gramicidin. Infect. Immun. *75*, 3696–3699.

Simmons, W.L., and Dybvig, K. (2009). Mycoplasma biofilms *ex vivo* and *in vivo*. FEMS Microbiol. Lett. *295*, 77–81.

Simmons, W.L., Bolland, J.R., Daubenspeck, J.M., and Dybvig, K. (2007). A stochastic mechanism for biofilm formation by *Mycoplasma pulmonis*. J. Bacteriol. *189*, 1905–1913.

Sippel, K.H., Robbins, A.H., Reutzel, R., Boehlein, S.K., Namiki, K., Goodison, S., Agbandje-McKenna, M., Rosser, C.J., and McKenna, R. (2009). Structural insights into the extracytoplasmic thiamine-binding lipoprotein p37 of *Mycoplasma hyorhinis*. J. Bacteriol. *191*, 2585–2592.

Taylor-Robinson, D., and Jensen, J.S. (2011). *Mycoplasma genitalium*: from Chrysalis to multicolored butterfly. Clin. Microbiol. Rev. *24*, 498–514.

Thacker, E.L., Holtkamp, D.J., Khan, A.S., Brown, P.A., and Draghia-Akli, R. (2006). Plasmid-mediated growth hormone-releasing hormone efficacy in reducing disease associated with *Mycoplasma hyopneumoniae* and porcine reproductive and respiratory syndrome virus infection. J. Animal Sci. *84*, 733–742.

Tryon, V.V., and Baseman, J.B. (1987). The acquisition of human lactoferrin by *Mycoplasma pneumoniae*. Microb. Pathog. 3, 437–443.

Weiner III, J., Zimmerman, C.-U., Gohlmann, H.W.H., and Herrmann, R. (2003). Transcription profiles of the bacterium *Mycoplasma pneumoniae* grown at different temperatures. Nucl Acids Res. *31*, 6306–6320.

Yogev, D., Browning, G.F., and Wise, K.S. (2002). Genetic mechanisms of surface variation. In Molecular Biology and Pathogenicity of Mycoplasmas, Razin, S., and Herrmann, R., eds (Kluwer Academic/Plenum Publishers, New York), pp. 417–443.

Zhang, W., and Baseman, J.B. (2011a). Transcriptional regulation of MG_149, an osmoinducible lipoprotein gene from *Mycoplasma genitalium*. Mol. Microbiol. *81*, 327–339.

Zhang, W., and Baseman, J.B. (2011b). Transcriptional response of *Mycoplasma genitalium* to osmotic stress. Microbiology *157*, 548–556.

Stress Responses in Yeast

11

Eulàlia de Nadal and Francesc Posas

Abstract

Adaptation to environmental stress requires changes in many aspects of cellular physiology essential for cell survival, such as gene expression, translation, metabolism, morphogenesis or cell cycle progression. Accordingly, the ability of eukaryotic cells to survive and thrive within adverse environments depends on rapid and robust stress responses. Stress-activated protein kinases (SAPKs) pathways are key elements on intracellular stress-signalling networks to respond and adapt to extracellular changes. In this review, we describe the different mechanisms used by model yeasts *Saccharomyces cerevisiae* and *Schizosaccharomyces pombe*, as well as the pathogenic fungus *Candida albicans*, to sense and transduce stress signals to SAPKs in response to osmo, heat and oxidative stresses. Moreover, other signalling pathways related to stress are discussed. Although much remains to be learned, studies from yeast have served to understand how stress signalling molecules adjust precise and efficient adaptation strategies to maximize cell survival in response to extracellular stimuli.

Introduction

Yeast includes a wide range of very different single-celled eukaryotic fungi. What characterizes yeast as compared to the rest of the fungi is that they have at least one phase of their life cycle in a single cell. Moreover, depending on the species, they reproduce by budding or fission processes and some species alternate a unicellular phase (yeast) with a mycelium one (dimorphic yeast). Most of the niches that yeasts live in are not constant, but rather fluctuate frequently and suddenly. Among many others, changes in temperature, oxidative or osmolarity conditions are often occurring in the natural environment. These unexpected environmental changes lead to stressful situations for their inhabitants because affect directly the internal conditions and disrupt homeostasis and normal physiology. Consequently, yeasts have evolved sophisticated systems to sense and signal the properties of the environment and rapidly mount defence responses against environmental stress. Adaptive responses to stress in yeast cover almost any aspect of cell physiology from regulation of gene expression, cell cycle progression, metabolism, protein homeostasis, cytoskeleton organization and vesicular activities. Although many of these responses are common in different stress situations, some are specific to each particular type of stress.

Why is it worth study stress responses in yeast? First, yeasts are essential for industrial applications; the most important commercial yeasts are the baker's and brewer's yeasts, which have been significantly improved through the years by selection and genetic manipulation. Second, although single cells, yeasts are true eukaryotes easily manipulated in the laboratory, being excellent models for the study of fundamental cell processes of eukaryotic biology. Indeed, two species are commonly employed as models for biomedical research: the budding yeast *Saccharomyces cerevisiae* and the fission yeast *Schizosaccharomyces pombe* (reviewed in Forsburg, 2005). Last, a few yeast species are pathogenetic for animals and humans; for instance, the dimorphic fungus *Candida albicans* is an opportunistic pathogen causing human fungal infections, mainly in immunocompromised hosts. Thus, given their differences in virulence and habitat, the three yeast species represent biotechnologically and medically important eukaryotic microorganisms and they are frequently used in both fundamental and applied research. The main characteristics of the three yeast species are illustrated in Table 11.1. In this review, we describe how yeast cells (especially *S. cerevisiae*, *S. pombe* and *C. albicans*) detect, integrate and respond to extracellular changes.

Cellular stress in yeast

Because yeasts live in extremely heterogeneous environments, they need to deal with diverse range of stresses such as low availability of nutrients, pH and other relevant extracellular variations. In this chapter, implications for changes in external temperature, osmolarity or exposure to reactive oxygen species (ROS) are discussed.

Yeasts often encounter increases in external temperature. Heat stress causes protein unfolding and unspecific aggregation resulting in an imbalance of protein homeostasis, which damage essential cellular structures and interfere with vital functions such as defects of the cytoskeleton, loss of the correct organelle localization or a collapse of intracellular transport processes (Riezman, 2004; Richter *et al.*, 2010). These and other effects lead to an arrest of the cell cycle as well as a decrease of growth and proliferation, including the death of the cell depending on the duration and severity of the stress.

Osmostress is another frequent environmental change that yeasts have to cope with. Osmotic imbalances in external medium cause a rapid water outflow and cell shrinking. Water loss leads to an increased concentration of biomolecules and ions in the cell, which eventually results in an arrest of cellular activity and cytoskeleton reorganization, among others (Hohmann, 2002; Hohmann *et al.*, 2007). Oxidative stress is caused by high levels of ROS. ROS are partially reduced oxygen derivatives produced by cell metabolism (respiration) such as the superoxide anion, the hydroxyl free radical and hydrogen peroxide that are more reactive than oxygen itself. High concentrations of ROS cause damage in different cell macromolecules. For instance, proteins with sulphydryl groups are oxidized in response to stress, forming sulphonic derivatives in an irreversible process (Herrero *et al.*, 2008). Some heavy metals and various genotoxic agents, including UV irradiation and alkylating agents, also generate oxidative stress.

Adaptation responses to cellular stress need to be (1) very fast acting in order to allow maximal survival rates to the new extracellular conditions and (2) transient as their constitutive induction use to be detrimental for cell growth. The signal transduction mechanisms that are activated in response to stress are suited for such dynamic processes (see below).

Sensing and transducing stress signals in yeast

Stress-activated protein kinases (SAPKs) are members of the mitogen-activated protein kinase (MAPK) family and are key stress-signalling molecules in eukaryotic cells. MAPK modules are conserved kinase cascades that perform

Table 11.1 Characteristics of *S. cerevisiae*, *S. pombe* and *C. albicans*. *S. cerevisiae* was the first eukaryote to be sequenced (Goffeau *et al.*, 1996) whereas the complete genome sequence of *S. pombe* and *C. albicans* was published in 2002 (Wood *et al.*, 2002) and 2004 (Jones *et al.*, 2004), respectively. Comparison of genomes shows considerable differences between these species, including number of chromosomes, number of coding genes or presence of introns (based on Herrero *et al.*, 2003; Mitrovich *et al.*, 2007; Rhind *et al.*, 2011)

	Genome size	Number of chromosomes	Coding regions (approximately)	Presence of introns
S. cerevisiae (budding yeast)	13.4 Mb	16	6500	5% of genes
S. pombe (fission yeast)	13.8 Mb	3	5000	40% of genes
C. albicans (dimorphic yeast)	16 Mb	8	5900	6% of genes

central roles in intracellular signal transduction in eukaryotic cells. Hence, MAPK signalling occurs in response to almost any change in the external stimuli. Each MAPK module is composed of three sequentially activating kinases: a MAPK kinase kinase (MAPKKK), a MAPK kinase (MAPKK) and a MAPK, which transmit signals by sequential phosphorylation events in a hierarchical way (see Fig. 11.1). MAPKKKs activate MAPKKs by phosphorylation on serine and threonine residues within a conserved loop at the N-terminal lobe of the kinase domain. Then, MAPKKs phosphorylate MAPKs on a threonine (sometimes serine) and tyrosine residues located in the activation loop of the kinase catalytic domain. Both residues are adjacent to each other but separate by a single amino acid (Thr/Ser-X-Tyr) and dual phosphorylation is needed for activation of the MAPK. This phosphorylation increases the nuclear accumulation of the MAPK and it results in the phosphorylation of many substrates on serine/threonine followed by a proline, including transcription factors, other kinases, cell cycle regulators and membrane-associated proteins.

SAPKs are among the most evolutionarily conserved stress-signalling proteins in yeast (Nikolaou *et al.*, 2009). However, the mechanisms that have evolved to relay stress signals to the SAPK module have diverged significantly in different fungi. Besides SAPK pathways, other sensing and transducing systems in response to cellular stress exists (Smith *et al.*, 2010). In the following sections, we discuss some examples of stress signalling systems in budding, fission and dimorphic yeasts that are particularly well understood.

Stress signalling in *S. cerevisiae*

In budding yeast, specific pathways sense and transduce specific cellular stresses. The HOG (high osmolarity glycerol) pathway is the main mediator of cellular adaptation upon osmostress and it is the best-characterized osmoresponsive system in eukaryotes.

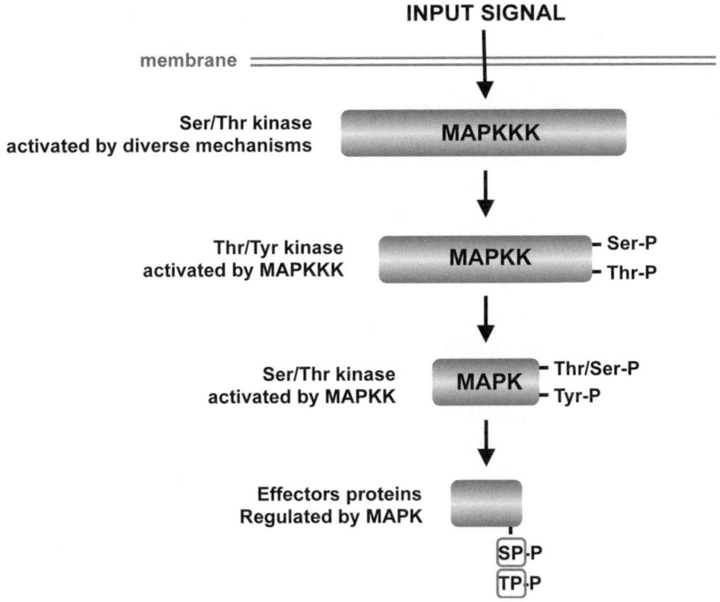

Figure 11.1 Schematic diagram of a MAPK core module. The core module of a MAPK cascade is composed of three consecutively activated tiers of kinases, a MAP kinase kinase (MAPKKK), a MAPK kinase (MAPKK) and a MAPK. In response to external stimuli, the MAPKKK phosphorylates and activates the MAPKK on serine and threonine. Then, the activated MAPKK phosphorylates the MAPK on a threonine (sometimes serine) and tyrosine residues. Activated MAPK phosphorylates its substrates on serine or threonine followed by a proline.

Osmostress signalling: the HOG SAPK pathway

The Hog1 MAPK is a prototype of the SAPK family, equivalent to the mammalian p38 kinase. In contrast to mammals or other yeast species where p38 related SAPKs are activated by a variety of stresses, in *S. cerevisiae* Hog1 SAPK is basically responding to osmostress and other signalling systems are important to sense and transduce oxidative and heat stresses.

In vivo replacement of components of the HOG pathway in *S. cerevisiae* by their mammalian counterparts demonstrated that there is a strong functional preservation of these MAPK pathways form yeast to mammals (Han *et al.*, 1994; Sheikh-Hamad and Gustin, 2004). The central core of the HOG pathway comprises three MAPKKKs, Ssk2, the functionally redundant Ssk22, and Ste11 that activate the Pbs2 MAPKK, which in turn phosphorylates the Hog1 MAPK.

This MAPK cascade is regulated by two independent upstream mechanisms that sense osmostress and relay stress signal to Hog1: the Sln1 two-component signalling pathway (Maeda *et al.*, 1994; Posas *et al.*, 1996) and a branch that contains the Sho1 transmembrane protein (Maeda *et al.*, 1995), which converge on the Pbs2 MAPKK (see Fig. 11.2A). The Sln1 branch involves a two-component osmosensor. Two-component sensors consist on a His-Asp phosphorelay system and are broadly used in prokaryotes to respond to environmental signals (Egger *et al.*, 1997). This sensing system is typically composed of a sensor protein with an extracellular input domain and a cytoplasmatic histidine kinase domain, and a response-regulator protein, which is a cytosolic protein containing a receiver and a DNA binding domains. Upon environmental stimulus, the sensor protein is able to autophosphorylate a histidine residue within its kinase domain. This phosphate group is then transferred to an aspartic acid residue in the receiver domain of the response-regulator molecule, resulting on its activation.

However, in eukaryotic cells, some two-component sensors have a more complex multistep phosphorelay system, involving three components. Specifically, the Sln1 pathway requires the primary osmosensor Sln1, which contains a histidine kinase and a receiver domain, the phosphorelay protein Ypd1 and the Ssk1 response-regulator (Maeda *et al.*, 1994; Posas *et al.*, 1996; Saito, 2001). At basal conditions, yeast Sln1 is autophosphorylated. This phosphate is then sequentially transferred to Sln1 (aspartic acid), then to Ypd1 (histidine) and finally to Ssk1 (aspartic acid) in a multistep phosphorelay mechanism. Phosphorylated Ssk1 represses the activity of the redundant Ssk2/22 MAPKKK and thus inhibits signalling via the HOG cascade (Posas *et al.*, 1996). In contrast, high osmolarity inactivates Sln1 and halts phosphorelay through Ypd1, leading to a rapid dephosphorylation of Ssk1 (Kaserer *et al.*, 2009). Unphosphorylated Ssk1 induces Ssk2 and Ssk22 to autophosphorylate and activate, triggering activation of Hog1 (Posas and Saito, 1998; Horie *et al.*, 2008). Although the physical mechanisms of osmosensing remain still to be elucidated, Sln1 might sense changes in membrane tension and/or turgor caused by high osmolarity (Reiser *et al.*, 2003).

The Sho1 branch contains an alternative sensor mechanism that leads to Pbs2 activation through the Ste11 MAPKKK. This branch is regulated by two mucin-like transmembrane sensors, Msb2 and Hrk1, which interact with Sho1 in response to osmostress (de Nadal *et al.*, 2007; Tatebayashi *et al.*, 2007). Mucins may monitor movements between the cell wall and the plasma membrane. Many proteins have been implicated in the transmission of signal though this branch, however some specific details remain still unknown. Upon high osmolarity, Msb2 and Hkr1, which are highly glycosylated, lead to activation of the PAK-like kinase Ste20 by inducing its association with the membrane-bound small G protein Cdc42 (Lamson *et al.*, 2002). Activated Ste20 and/or Cla1 phosphorylates and activates Ste11, which in turn phosphorylates Pbs2, signalling the osmostress (Drogen *et al.*, 2000).

How Ste11 and Pbs2 are recruited to the membrane to be activated is a key question. Here, the transmembrane Sho1 protein appears to play a role as membrane-localized scaffold protein since it anchors Pbs2 to the membrane upon stress (Maeda *et al.*, 1995; Reiser *et al.*, 2000; Tatebayashi *et al.*, 2006). Indeed, Pbs2 not only serves as MAPKK but also as scaffold protein (Posas and

Saito, 1997; Raitt *et al.*, 2000; Seet and Pawson, 2004; Zarrinpar *et al.*, 2004). Membrane localization of Ste11 is mediated by the Ste50 scaffold protein (Posas *et al.*, 1998; Wu *et al.*, 1999) and its association with the membrane anchor protein Opy2 as well as Cdc42 and Sho1 (Truckses *et al.*, 2006; Wu *et al.*, 2006; Tatebayashi *et al.*, 2007; Yamamoto *et al.*, 2010).

It is worth noting that there are other yeast MAPK pathways in addition to Hog1 with some overlapping signalling components. For instance, the Ste11 MAPKKK is common for three different MAPK cascades. However, different strategies such as scaffold proteins or kinetic insulation are utilized for the prevention of cross-talk between any MAPK modules (Patterson *et al.*, 2010; Saito, 2010).

Each sensor branch alone can activate Hog1 in response to osmostress, although the Sln1 branch is more prominent in pathway control (O'Rourke and Herskowitz, 2004), and display higher sensitivity that allow the cell to respond over a wide range of osmolarity changes (Maeda *et al.*, 1995; Macia *et al.*, 2009). Whereas the Sln1 branch is

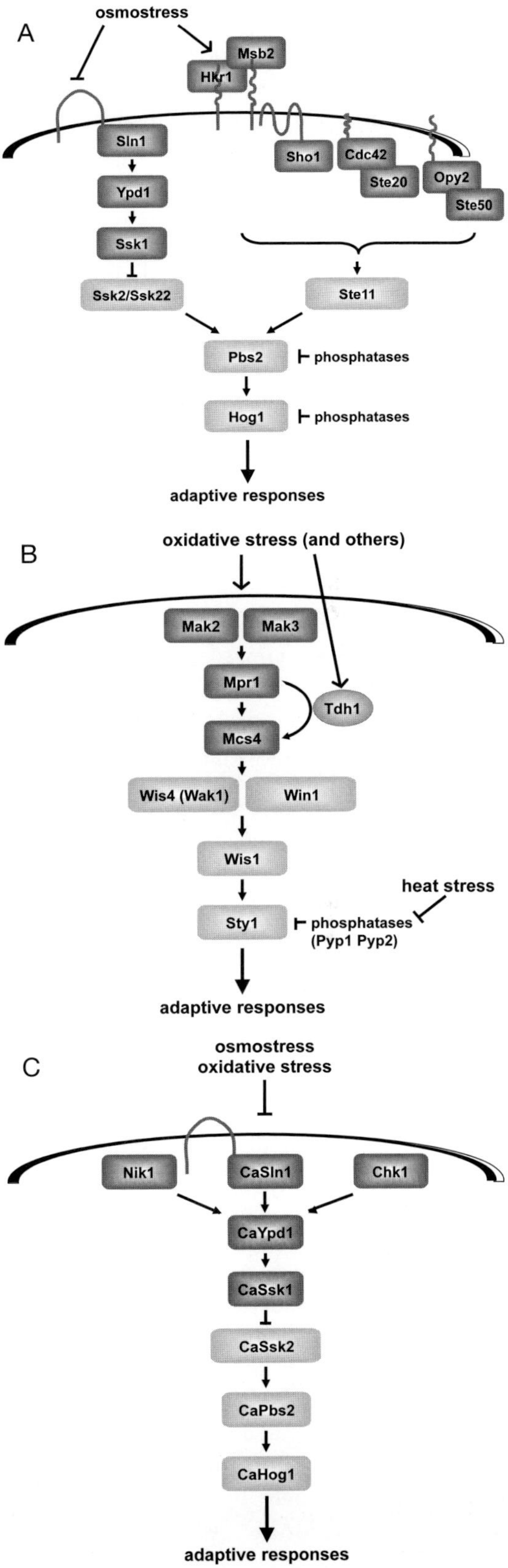

Figure 11.2 Sensing and transducing stress signals in yeast. (A) Schematic diagram of the *S. cerevisiae* HOG pathway. In response to osmostress, two independent upstream osmosensing mechanisms, the Sln1 and Sho1 branches, lead to the activation of specific MAPKKKs (Ssk2/22 and Ste11) that converge on the common Pbs2 MAPKK. Activated Pbs2 phosphorylates the Hog1 MAPK, which induces a set of adaptive responses. (B) Schematic diagram of the *S. pombe* Sty1 pathway. Mak2 and Mak3 sense oxidative stress though the Mpr1 phosphotransfer protein and the Msc4 response-regulator. Upon stress, Msc4 activates the Wis4 (Wak1) and Win1 MAPKKKs, which phosphorylate the Wis1 MAPKK. Wis1 in turn activates Sty1. Of note, Tdh1 is required for oxidative stress signal through the two-component sensor. Upon heat stress, the interaction between Sty1 and the phosphatases that inactivate the MAPK is inhibited. C) Schematic diagram of the *C. albicans* CaHog1 pathway. Three distinct histidine kinases (CaSln1, Chs1 and Nik1/Cos1) may sense stress and transmit the signal through CaYpd1 phosphorelay protein and CaSsk1 response-regulator. Although, *C. albicans* have all the components of the osmosensing Sho1 branch, such system may play no role in CaHog1 regulation.

capable of fast signal integration to repeated stimuli, the Sho1 branch does not, suggesting that the signalling properties of the branches are different (Hersen *et al.*, 2008). Whatever branch is activated, the osmostress signal flows to Pbs2, which in turn phosphorylates and activates Hog1, which elicits the programme for cell adaptation required for survival under osmostress (Hohmann, 2002; de Nadal *et al.*, 2002; O'Rourke *et al.*, 2002; Westfall *et al.*, 2004; Chen and Thorner, 2007).

Although HOG is the main pathway to sense osmostress, other signalling proteins have been implicated in the osmostress response in yeast. This is the case for the Sch9 kinase (Pascual-Ahuir and Proft, 2007) and protein kinase A (PKA) signalling pathway (Martinez-Pastor *et al.*, 1996; Schmitt and McEntee, 1996), which regulates osmostress activated gene expression (see below).

Regulation of the HOG SAPK pathway

To maximize cell survival, signalling mechanisms need to be highly reversible and fast acting, being activated within seconds of exposure to stimuli. The use of phosphorylation cascades is a mechanism that has been harnessed to achieve this, as these pathways are much faster than systems relying exclusively on gene expression. The response to stress should be rapid but also transient. Indeed, constitutive induction of the HOG pathway is detrimental for cell growth (Maeda *et al.*, 1993; Yaakov *et al.*, 2003). Sustained activation of Hog1 leads to an increase on ROS that leads to apoptosis (Vendrell *et al.*, 2011). Hog1 becomes activated rapidly upon osmostress but its level of phosphorylation drops again within a period of time that depends on the degree of stress (Klipp *et al.*, 2005). This fast but transient nature of the response is reflected not only at level of Hog1 phosphorylation but also by transient nuclear localization of the kinase (Ferrigno *et al.*, 1998; Reiser *et al.*, 1999; Mettetal *et al.*, 2008; Macia *et al.*, 2009) and transient transcriptional response upon stress (see below). Thus, signalling through the MAPK pathway is controlled not only by activity of kinases upstream of the MAPK (when cells start to adapt, there is a concomitant inactivation of the sensors), but also by protein phosphatases and various feedback systems to adjust the dynamic rage of the response.

The type 2C Serine/Threonine protein phosphatases (Ptc1, Ptc2 and Ptc3) (Jacoby *et al.*, 1997; Wurgler-Murphy *et al.*, 1997; Mattison and Ota, 2000) and the protein tyrosine phosphatases (Ptp2 and Ptp3) (Maeda *et al.*, 1993, 1994; Warmka *et al.*, 2001; Mapes and Ota, 2004) decrease Hog1 and Pbs2 activities (reviewed in Saito and Tatebayashi, 2004). It has been also described that Ssk1 stability depends on its phosphorylation state (Sato *et al.*, 2003). In addition, phosphorylation of Sho1 or Ste11 upon Hog1 activation seems to be important to decrease signalling through this branch (Sato *et al.*, 2003; Hao *et al.*, 2007, 2008; Molina *et al.*, 2010).

Besides down-regulation of the inducers, temporal restriction of adaptive responses is achieved by several negative feedback regulatory loops, which modulate the HOG pathway. Mathematical modelling and single-cell analyses have shown that inactivation of the pathway by a mechanism independent of transcription might be important to modulate acute responses; whereas transcription-dependent mechanisms might be important for proper adaptation to future stimuli (Klipp *et al.*, 2005; Mettetal *et al.*, 2008; Hohmann, 2009; Muzzey *et al.*, 2009). Moreover, protein complex formation and scaffolding are important contributors to the robustness of the HOG pathway (Krantz *et al.*, 2009). On the other hand, the HOG pathway is not just an ON and OFF signalling system, only activated in response to stress, but rather it is a system that it is constantly ON and counteracted by an internal negative feedback. It is described that there is intrinsic HOG signalling activity that is counteracted by a mechanism that requires Hog1 activity and depends solely on the Sln1 branch. This basic signalling may allow higher efficiency in terms of faster response and fine-tuning of signalling thresholds upon stress (Macia *et al.*, 2009).

Thus, the properties of the HOG signal transduction have been addressed in a high degree of detail, although not all of them are completely understood (reviewed in Hohmann, 2009).

Heat and oxidative stress signalling

The HOG cascade of *S. cerevisiae* was initially described as an osmosensing pathway but some studies showed that this SAPK might also be activated in response to other stimuli, including oxidative (Bilsland *et al.*, 2004) and heat (Winkler *et al.*, 2002) stresses. However, much less is known regarding how theses stresses are sensed by Hog1. The Sho1, but not the Sln1, activation branch is required for heat stress (Winkler *et al.*, 2002). In contrast, upstream signalling events relaying oxidative stress are still unclear (Bilsland *et al.*, 2004). Moreover, the kinetics of Hog1 activation upon these stresses are considerably slower and the contribution in the adaptive responses is minor compared to osmostress.

Other signalling pathways and/or sensor systems in addition to the HOG pathway are more relevant to trigger heat and oxidative stress responses. Measuring temperature can be reached either directly through primary thermosensory structures such as DNA, RNA, proteins and lipids (Klinkert and Narberhaus, 2009) or indirectly through sensing a general accumulation of denatured proteins (Nollen and Morimoto, 2002; Franzmann *et al.*, 2008). Moreover, the cell integrity pathway responds to heat shock by activating the MAPK Mpk1/Slt2 that regulates the MADS-box Rlm1 transcription factor (Jung and Levin, 1999; Jung *et al.*, 2002; Nadeau and Landry, 2007). However, regulation of gene expression by this signalling pathway is restricted to a group of genes that complements the activation of the Hsf1 transcription factor, the key player in the regulation of heat stress gene expression (see below).

On the other hand, Yap1 is a leucine-zipper AP-1 transcriptional factor that acts as the main regulator in response to oxidative stress in budding yeast (Lee *et al.*, 1999). Upon exposure to oxidants Yap1 changes its subcellular localization. Under non-stress conditions, Yap1 localizes in the cytoplasm through regulation of its nuclear export by the exporter receptor Crm1 (Yan *et al.*, 1998; Kuge *et al.*, 2001). Stress disrupts Yap1–Crm1 complex interaction, promoting Yap1 nuclear accumulation (Coleman *et al.*, 1999; Delaunay *et al.*, 2000, 2002; Wood *et al.*, 2004). On the other hand, the Gpx3 peroxidase functions as the sensor upon oxidative stress and transduces the signal to Yap1 (Delaunay *et al.*, 2002). Thus, Yap1 functions as a two-component system where Gpx3 senses and transduces the signal to Yap1, which then leads the adaptive responses (Toledano *et al.*, 2004). A third player, Ybp1, is also crucial in this pathway, although its molecular function remains still unknown (Veal *et al.*, 2003; Okazaki *et al.*, 2005). Oxidative stress is also sensed by the protein kinase C (PKC)-dependent MAPK pathway (Alic *et al.*, 2003; Vilella *et al.*, 2005) and Skn7 (Brown *et al.*, 1993, 1994). Skn7 is a response-regulated protein with a receiver and DNA-binding domains, which functions as a transcription factor and it appears to be important for oxidative stress response (Krems *et al.*, 1996; Morgan *et al.*, 1997).

Stress signalling in *S. pombe*

The homologous Hog1 in fission yeast is known as the Sty1/Spc1 MAPK (Millar *et al.*, 1995; Shiozaki and Russell, 1995). In contrast to the budding yeast HOG pathway, the Sty1 MAPK is robustly activated by several stresses, including osmo, heat and oxidative stress (Degols *et al.*, 1996; Wilkinson *et al.*, 1996; Degols and Russell, 1997; Buck *et al.*, 2001).

The multistep phosphorelay system responsible for MAPK activation is conserved among yeasts (Santos and Shiozaki, 2001) and plants (Urao *et al.*, 2001), but not in animals. Indeed, there is a high structural similarity between both, *S. cerevisiae* and *S. pombe*, sensors (see Fig. 11.2B). However, the homologous phosphorelay system in fission yeast seems to be dedicated for peroxide stress sensing rather than osmostress signalling. Moreover, while *S. cerevisiae* has only one sensor kinase (Sln1), *S. pombe* has three *SLN1* orthologues, named *mak1*$^+$, *mak2*$^+$ and *mak3*$^+$ (Buck *et al.*, 2001; Quinn *et al.*, 2011). On the other hand, the fission yeast equivalents of Ypd1 and Ssk1 are *mpr1*$^+$ (Aoyama *et al.*, 2000; Nguyen *et al.*, 2000) and *mcs4*$^+$ (Cottarel, 1997; Shieh *et al.*, 1997) respectively.

Much less is known regarding the precise mechanism underlying two-component signalling of stress signals to Sty1. Several studies showed that Mak2 and Mak3, but not Mak1, sense oxidative stress though the Mpr1 phosphotransfer protein and the Msc4 response-regulator (Aoyama *et al.*, 2000; Nguyen *et al.*, 2000; Buck *et al.*, 2001; Quinn *et al.*, 2002). Of note, Mak kinases have additional functional domains besides the histidine kinase domain that are not found in *S. cerevisiae* Sln1. These are the PAS/PAC and a GAF domains, which could confer the ability to sense oxidative stress stimuli (Taylor and Zhulin, 1999; Hurley, 2003; Quinn *et al.*, 2011).

The Mak1 sensor kinase is not required for signalling to the Sty1 pathway, but may regulate a second response regulator, Prr1 (Buck *et al.*, 2001), orthologue of Skn7 in budding yeast (Ohmiya *et al.*, 1999), which is essential for oxidative stress-induced gene expression. On the other hand, it is described that the glycolytic enzyme GAPDH (Tdh1) is required for oxidative stress signal through the two-component sensor (Morigasaki *et al.*, 2008).

Upon stress, Msc4 binds the Wis4 (also known as Wak1) and Win1 MAPKKKs, which are redundant functionally, and initiate the activation of the Sty1 cascade (Shieh *et al.*, 1997; Buck *et al.*, 2001). The MAPKKKs phosphorylate the Wis1 MAPKK (Shiozaki and Russell, 1996; Samejima *et al.*, 1997, 1998; Shieh *et al.*, 1997, 1998; Shiozaki *et al.*, 1997, 1998), which in turn activates Sty1 (Millar *et al.*, 1995; Shiozaki and Russell, 1995; Degols *et al.*, 1996; Kato *et al.*, 1996), translocating from the cytoplasm to the nucleus (Gaits *et al.*, 1998).

Much less is known about the relay of other stress stimuli to the Sty1 MAPK. In contrast to peroxide stress, it is not established which components of the pathway are required to sense and transduce heat or osmotic stresses. Upon heat stress, the interaction between Sty1 and Pyp1, one of the tyrosine phosphatase that inactivates the MAPK, is inhibited (Shiozaki, 1998; Nguyen and Shiozaki, 1999). Subsequently, the action of the type 2C serine/threonine phosphatases Ptc1 and Ptc3 dephosphorylates Sty1 (Nguyen and Shiozaki, 1999). On the other hand, it is still unknown how osmostress signals are relayed to the Sty1 SAPK.

Besides the Sty1, *S. pombe* triggers the Pap1 signalling pathway depending on the severity of the oxidative stress (reviewed in Ikner and Shiozaki, 2005; Vivancos *et al.*, 2006). Pap1, the orthologue of budding yeast Yap1, is an AP-like transcription factor, which cellular localization is regulated by stress, concentrating into the nucleus when cells are exposed to oxidative stress (Toone *et al.*, 1998; Kudo *et al.*, 1999). Whereas the Sty1 cascade is activated by higher levels of peroxide hydrogen, Pap1 responds to moderate increases in oxidative stress. Indeed, Pap1 activation upon high stress is quickly shut down by a negative feedback mechanism (Quinn *et al.*, 2002; Vivancos *et al.*, 2004). While Gpx3 is the peroxide hydrogen-sensor of Yap1 in *S. cerevisiae*, the thioredoxin peroxidase Tpx1 is the redox sensor of the Pap1 pathway by forming a transient Pap1–Tpx1 complex (Delaunay *et al.*, 2002; Vivancos *et al.*, 2005). This sensor is inactivated temporally at high doses of peroxide hydrogen and, in consequence, it can only transfer the redox signal to Pap1 under mild stress (Bozonet *et al.*, 2005; Vivancos *et al.*, 2005).

The Skn7 homologous Prr1 is another component of the fission yeast response against oxidative stress (Ohmiya *et al.*, 1999), although its specific function still remains to be elucidated.

Stress signalling in *C. albicans*

The *C. albicans* SAPK gene, *CaHOG1*, was isolated through genetic complementation of the budding yeast *hog1* mutant (San Jose *et al.*, 1996). As *S. pombe*, this pathway is activated by oxidative stress (Alonso-Monge *et al.*, 2003) (see Fig. 11.2C). Indeed, to successfully colonize the host body, *C. albicans* needs to sense and respond to ROS (see below). In addition, it responds to osmostress, but not to heat shock (Alonso-Monge *et al.*, 2003; Smith *et al.*, 2004).

Two-component sensors have also been identified in *C. albicans* (reviewed in Alonso-Mongue *et al.*, 2006; Kruppa and Calderone, 2006). There are three distinct histidine kinases that act as sensors of environmental stimuli: CaSln1, Chs1 and Nik1/Cos1 (Alex *et al.*, 1998; Nagahashi *et al.*, 1998; Calera and Calderone, 1999a,b), a single phosphorelay protein (CaYpd1) and a response-regulator (CaSsk1) (Calera and Calderone, 1999b; Calera *et al.*, 2000; Singh *et al.*, 2004). Although this pathway is sensing oxidative stress (Chauhan *et al.*, 2003; Menon *et al.*, 2006), it still remains unclear which of the three described histidine kinases is involved in this process (Buck *et al.*, 2001; Li *et al.*, 2004; Kruppa and Calderone, 2006). On the other hand, it seems that the CaSln1-CaYpd1-CaSsk1 system relays osmostress signals to a SAPK module (Nagahashi *et al.*, 1998; Roman *et al.*, 2005).

In some yeast species, such as *S. pombe*, the Sho1 branch is absent. In *C. albicans*, all the components of the Sho1 branch are identified (Krantz *et al.*, 2006), although does not seem to activate CaHog1 in response to osmostress (Roman *et al.*, 2005; Cheetham *et al.*, 2007; Roman *et al.*, 2009). Indeed, genetic studies suggest that additional pathway(s) transmits osmostress signals to the SAPK in parallel with the Sln1 branch (Roman *et al.*, 2005). Of note, it has been identified a second response-regulator, CaSkn7, which has also a role in protection against oxidants (Singh *et al.*, 2004).

Stress adaptation responses

Although stress signalling systems are not essential for growth in absence of stress, activation of these pathways are crucial for cell survival in stress conditions. Indeed, mutant strains lacking components of the MAPK cascades show a stress sensitive phenotype in *S. cerevisiae*, *S. pombe* and *C. albicans* (Brewster *et al.*, 1993; Degols *et al.*, 1996; Quinn *et al.*, 2002; Alonso-Monge *et al.*, 2003; Arana *et al.*, 2005).

Which are the yeast functions required for successful adaptation to these environmental stresses? Gene disruption in yeast permits to determine the consequence of loose of gene function. Because yeast genome sequence is known (see Table 11.1), it is possible to systematically inactive each gene and analyse the mutant phenotype in response to specific stress. Using this genome-wide approach, the fitness of each mutant in *S. cerevisiae* was scored under osmotic (Giaever *et al.*, 2002; Zapater *et al.*, 2007), oxidative (Weiss *et al.*, 2004) and heat stresses (Ruiz-Roig *et al.*, 2010). Although it has been described some overlap of genes required for stress survival (Auesukaree *et al.*, 2009), there is a dedicated adaptive response to each specific stress.

Stress-essential genes include almost all general cell features, from metabolic adjustment (carbon and energy metabolism) to mRNA synthesis, cell type differentiation, cellular transport and cytoskeleton organization, which are altered in response to stress. Thus, activation of intracellular signalling pathways in response to stress leads to the modulation of almost any aspect of cell physiology. In the next sections, we will describe two main adaptive responses upon cellular stress: control of the cell cycle progression and mRNA biogenesis.

Cell cycle regulation

The cell cycle process, by which one cell grows and divides into two cells, is organized in four coordinated phases: (1) G1 phase, in which the cell grows and make the decision to commit cell division, (2) S phase (synthesis), the period for DNA replication, (3) G_2 phase, when the cell prepares to the next phase and (4) M phase (mitosis), in which the chromosomes separate and the cell divides in two. To progress adequately through cell cycle, eukaryotic organisms have cell cycle checkpoints, which are activated and delay cell cycle progression when potential lethal alterations are detected to ensure that cellular events occur in the correct order and timing.

In yeast, progression through the cell cycle is driven by a highly regulated and complex network of proteins, where phase-specific cyclin subunits bind to mainly a single cyclin-dependent kinase (CDK), Cdc28, which transcription is constant during cell cycle. Specificity of each cyclin is assured by a variety of mechanisms: They are (1) sensitive to cell cycle-regulated inhibitors, (2) differentially restricted to subcellular locations, (3) timely controlled by specific transcriptional activators and (4) degraded by phase-specific events (Bloom and Cross, 2007).

Cell cycle regulation by stress in *S. cerevisiae*

Modulation of the cell cycle progression is essential for adaptation to stress. How SAPKs control cell cycle progression in response to stress has been studied in recent years in *S. cerevisiae*. The HOG pathway has been involved in the control of cell cycle progression at different cell cycle phases such as G1/S, S and G2/M (Reiser *et al.*, 2006; Clotet and Posas, 2007; Yaakov *et al.*, 2009).

Hog1 activity induces a transient arrest in G1 via a dual mechanism that involves the direct phosphorylation of the cell cycle inhibitor Sic1 (Escote *et al.*, 2004; Zapater *et al.*, 2005; Migdal *et al.*, 2008) (Fig. 11.3A). Sic1 is a potent Clb-Cdc28 inhibitor playing a role in the timing and robustness of regulation of DNA replication by holding under inhibition Clb5,6–Cdc28 complexes and setting a threshold for Clb-Cdc28 activation (Cross *et al.*, 2007). The levels of Sic1 are maximal at early G1, whereas at late G1 Cln-Cdc28 phosphorylates Sic1 to target it for ubiquitination by the Cdc4 ubiquitin ligase and the posterior degradation by the proteosome (Nash *et al.*, 2001). Once Sic1 is eliminated, Clb5,6-Cdc28 activity is liberated from inhibition, and free to initiated DNA replication (Schwob *et al.*, 1994). Under osmostress conditions, Hog1 stabilizes Sic1 by interfering its binding to Cdc4 and thus reduces its degradation arresting cells in G1 phase. This Hog1-mediated G1 arrest certainly plays an important role in osmoadaptation, as cells lacking Sic1 or that carry a mutant form of Sic1 that is not able to be phosphorylated, are unable to grow in high osmolarity conditions (Escote *et al.*, 2004).

In addition to Sic1 stabilization, Hog1 regulates G1 phase progression by down-regulating *CLN1*, *CLN2* and *CLB5* expression. Cln1,2 cyclins mediate the expression of a specific set of genes that encompasses the subsequent S-phase cyclins and are targeted by the Swe1 cell cycle box binding factor (SBF), whereas the B-type Clb5,6 initiate S-phase in a timely fashion and are regulated by Mlu cell cycle box binding factor (MBF) (Nasmyth and Dirick, 1991; Koch *et al.*, 1993; Breeden, 1996). The molecular mechanism through which Hog1 down-regulates the mRNAs of these SBF and MBF regulated cyclins is still to be elucidated (Escote *et al.*, 2004; Adrover *et al.*, 2011). Thus, Hog1 activation modulates cell cycle progression at G_1 by two mechanisms, the direct phosphorylation of Sic1 and the delay of the onset of Cln1,2 and Clb5. Recently, mathematical modelling with quantitative *in vivo* experiments, defined the temporal role and the direct contribution of the individual components of the G1–S network downstream of Hog1. These analyses showed that there is a different temporal and qualitative contribution of Sic1, Cln1,2 and Clb5 regulation to the delay upon osmostress (Adrover *et al.*, 2011).

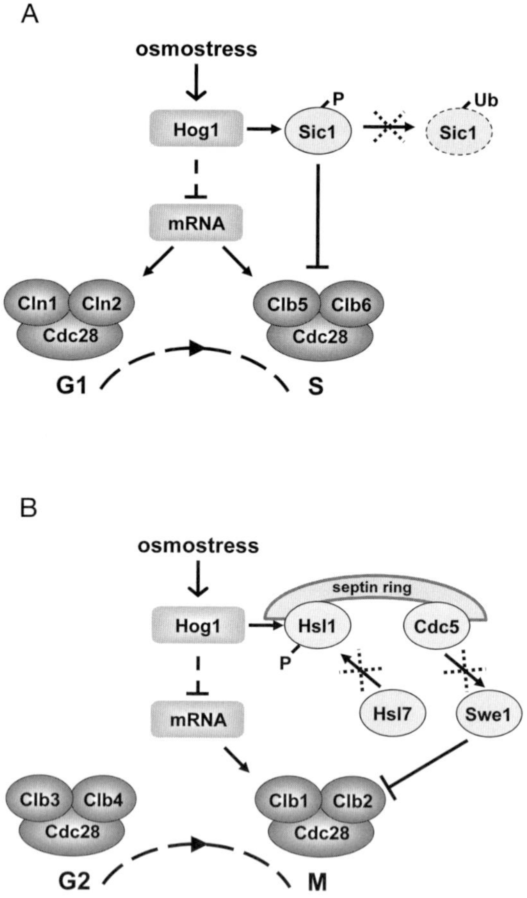

Figure 11.3 Cell cycle regulation by stress in budding yeast. (A) Hog1-mediated arrest at G1 phase. Upon stress, Hog1 induces a transient arrest in G1 via a dual mechanism. First, Hog1 phosphorylates the cell cycle inhibitor Sic1 and stabilizes it by interfering its binding to the proteosome complex. Second, Hog1 down-regulates both G1 cyclins (*CLN1* and *CLN2*) and the S-phase *CLB5*. (B) Hog-mediated arrest at G2 phase. In G2, Swe1 is tethered to the septin ring by Hsl1-Hsl7. Then, it is phosphorylated by Cdc5 and subsequently degraded, relieving Clb2-Cdc28 inhibition. Upon stress, the direct phosphorylation of Hsl1 by Hog1 leads to the delocalization of Hsl7 and, therefore, Swe1 cannot be recruited to be phosphorylated by Cdc5 and targeted for ubiquitination and degradation. Hog1 also down-regulates transcription of *CLB1* and *CLB2*, contributing in the inhibition of mitosis onset.

Control of S phase by Hog1 is exerted

independently on the S-phase checkpoint (Yaakov *et al.*, 2009). Moreover, in G2, activation of Hog1 leads to the prolonged half-life of the cell cycle inhibitor Swe1 and the down-regulation of the G2 cyclin *CLB2* transcript levels (Alexander *et al.*, 2001; Clotet *et al.*, 2006) (Fig. 11.3B). Entry into mitosis is controlled by the activity of the Clb2–Cdc28 complex, which is held by Swe1. Swe1 degradation depends both on Clb-Cdc28 activity and a conserved upstream kinase, Hsl1, in association with the adaptor protein Hsl7 (McMillan *et al.*, 1999; Tyers and Jorgensen, 2000; Asano *et al.*, 2005). In G2, Swe1 tethering to the septin ring by Hsl1–Hsl7, allows the polo-like Kinase (Cdc5) to phosphorylate it, which labels Swe1 for degradation, relieving Clb2-Cdc28 inhibition. Thus, this mechanism links proper assembly of septin filaments with efficient passage through G2 into mitosis.

Upon stress, Swe1 is stabilized by a Hog1-mediated phosphorylation on Hsl1. This phosphorylation leads to the delocalization of Hsl7 from the neck that results in the lack of recruitment of Swe1 nearby Cdc5 and therefore it is not phosphorylated to be targeted for ubiquitination and degradation (Clotet *et al.*, 2006). Of note, *swe1* mutant cells neither accumulate Swe1 nor arrest in G2 nor become osmosensitive. Similarly to G1, the mechanism by which Hog1 controls *CLB2* mRNA levels remains unclear.

It is worth noting that the combined deletion of *SIC1* and *SWE1* results in a synergistic osmosensitivity, pinpointing the relevance of proper cell cycle progression under osmostress at different stages of cell cycle (Clotet *et al.*, 2006). Thus, SAPKs modulate the timing of cell cycle progression by different molecular strategies being able to act immediately in different phases of the cycle.

Cell cycle regulation by stress in *S. pombe*

In fission yeast, the cell cycle progression is also controlled by a single CDK (Cdc2) associated to distinct cyclins. The role of SAPKs in cell cycle control was first proposed in *S. pombe* for the Sty1 MAPK pathway, linking the control of cell cycle to environmental responses. Indeed, cells lacking Sty1 show an elongated phenotype, even in the absence of stress, and *sty1* mutants have a G2 delay that is greatly exacerbated by growth in high osmolarity (Shiozaki and Russell, 1995).

It is well known that in fission yeast, stress induces a temporal arrest at the G2/M transition, which is not dependent on other described cell cycle checkpoints (Kawasaki *et al.*, 2006). G2/M transition is regulated by inhibiting the Cdc25 phosphatase that is the key factor governing the initiation of mitosis. Cdc25 reverses the inhibition of the Cdc2–mitotic complex owing to phosphorylation by the tyrosine kinases Wee1 and Mik1 (Russell and Nurse, 1986; Gautier *et al.*, 1991; Millar *et al.*, 1991).

In response to osmostress, the Sty1 MAPK controls Cdc25 activity by targeting Srk1 (Sty1-regulated kinase) (Lopez-Aviles *et al.*, 2005, 2008), which is related to the mammalian calmodulin-dependent and MAPKAP kinases and to the budding yeast Rck1 and Rck2 kinases. It is shown that Srk1 controls mitotic entry by directly phosphorylating and inhibiting Cdc25 during a normal cell cycle (Lopez-Aviles *et al.*, 2005) and in response to osmostress (Lopez-Aviles *et al.*, 2008). Srk1 is present in a complex with Sty1 and, upon stress, it is directly phosphorylated by the MAPK (Smith *et al.*, 2002). Phosphorylated Srk1 dissociates from Sty1 and phosphorylates Cdc25. Then, phosphorylation of Cdc25 by Srk1 causes Cdc25 to bind to Rad24, a 14-3-3 protein family member, and this helps both to retain Cdc25 in the cytoplasm and protects it from degradation. Thus, activated Srk1 leads to the inactivation of mitosis entry by translocating Cdc25 from the nucleus to the cytoplasm (see Fig. 11.4). Moreover, Srk1 released from its interaction to Sty1 is targeted to degradation by the proteosome. Of note, Srk1 over expression causes cell cycle arrest in late G2 phase, whereas cells lacking Srk1 enter mitosis prematurely and have an osmosensitive phenotype (Lopez-Aviles *et al.*, 2005, 2008).

Cell cycle regulation by stress in *C. albicans*

Whereas the transcriptional control of cell cycle in *C. albicans* is similar to the transcriptional network of *S. cerevisiae* (Cote *et al.*, 2009), there are no studies linking cell cycle regulation with the CaHOG pathway.

Although no *ScSIC1* homologous exists in *C.*

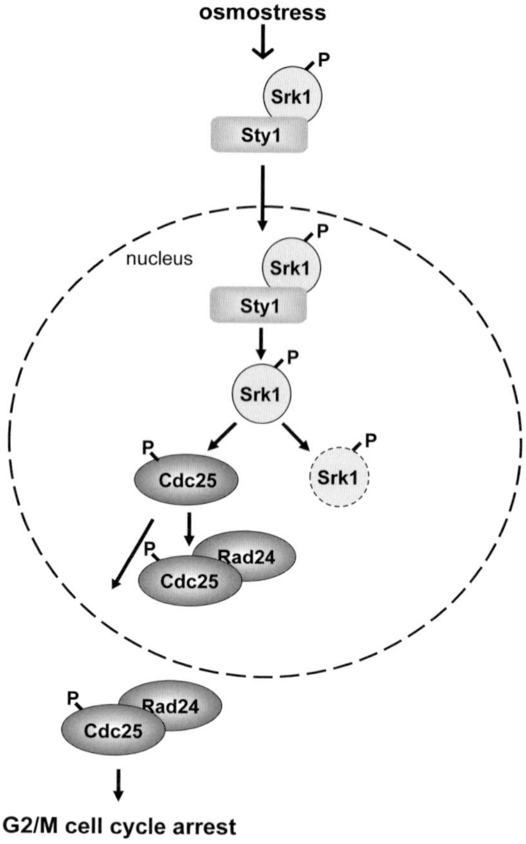

Figure 11.4 Cell cycle regulation by stress in fission yeast. Upon stress, Sty1 phosphorylates Srk1, inducing dissociation of both proteins and activation of Srk1. Activated Srk1 phosphorylates Cdc25, causing its binding to Rad24. This interaction helps to retain Cdc25 in the cytoplasm and protects it from degradation. Thus, Srk1 inhibits cell cycle progression by promoting Cdc25 cytoplasmatic localization. Then, Srk1 becomes unstable and it is degraded.

albicans, an open reading frame (named Sol1) is functional related to Sic1 (Donovan *et al.*, 1994; Atir-Lande *et al.*, 2005). Sol1 is a substrate for Cdc4-SCF degradation and it has the potential MAPK site conserved, although there are some data suggesting functional differences between the two homologous genes (Atir-Lande *et al.*, 2005). In *C. albicans* genome, there are Swe1 and Hsl1, but no Hsl7, *S. cerevisiae* homologues and no Hsl1 activation mechanism has been identified. Moreover, unlike *S. cerevisiae*, it has been suggested that the G2 delay may not happen in *Cahsl1* mutant under hyphal growth conditions (Umeyama *et al.*, 2005).

Control of mRNA biogenesis

After the initial minutes of exposure to stress, there is a major change in the transcriptional pattern of the cell. Thus, regulation of transcription represents a key step in the response of cells to changing environmental conditions (Lopez-Maury *et al.*, 2008). Different mutations in components of the transcriptional machinery render cells sensitive under osmostress in *S. cerevisiae* (de Nadal *et al.*, 2004; Zapater *et al.*, 2007; Mas *et al.*, 2009) as well as under oxidative stress in *S. pombe* (Sanso *et al.*, 2011). The relevance of such transcription changes is not clear. Gene expression seems to be important for maximizing cell survival on exposure to subsequent stresses and for cross-protection against unrelated stresses (Berry and Gasch, 2008) and depends on the strength and type of stress to which cells are subjected to (Westfall *et al.*, 2008). However, because of the lag-time of producing proteins *de novo* from stress-responsive genes, during the initial minutes of stress exposure, alternative, fast acting responses are required. Among others, these responses include the down-regulation of translation and the regulation of ion channels and glycerol transporters (Proft and Struhl, 2004; Hohmann *et al.*, 2007).

The adaptation response to stress is activated transiently and clearly affects the dynamics of gene activation. Analyses at the single cell level have revealed that the gene expression output at low stress of the HOG pathway has a 'bimodal' expression behaviour, seen as the presence of two distinct cell subpopulations, with non-responsive cells and cells that fully express an osmoresponsive gene in response to identical Hog1 activation. One of the determinants of this bimodal expression behaviour seems to be chromatin remodelling as well as the retention time and concentration of Hog1 in the nucleus (Pelet *et al.*, 2011). Remarkably, bimodal gene expression may be a general feature of yeast stress-induced genes since it is observed in other stresses such as oxidative or heat stresses (Pelet *et al.*, 2011).

The genomic landscape of stress-responsive genes: comparative genomics in *S. cerevisiae*, *S. pombe* and *C. albicans*

Analysis of the kinetic transcriptiome responses of yeast cells to a large battery of environmental changes show that regulation of gene expression exhibits exquisite selectivity, relatively fast response kinetics and controlled reversibility, following a characteristic 'pulse'-like pattern that represents an adaptation phase (Yosef and Regev, 2011). In general, the duration and amplitude of the transient changes in transcript levels vary with the magnitude on the severity of the stress, reflecting the dynamics of the signal transduction processes.

In addition to the specialized response to environmental stress, there is a functional module, the environmental stress response (ESR; refered as CESR in *S. pombe*), which contains genes whose expression is modulated by virtually every stress condition tested. Whereas this common response has been conserved between *S. cerevisiae* and *S. pombe*, it is not clear that exists in *C. albicans* (Gasch, 2007; Lelandais and Devaux, 2010), although some observations suggest a functional link between the dimorphic fungus orthologues of ESR genes (Enjalbert *et al.*, 2003; Gasch *et al.*, 2004; Smith *et al.*, 2004).

The ESR is a protective response to guard critical feature of cellular physiology in response to cellular stress. In budding yeast, the ESR contains 300 genes approximately (up to 600 when small changes are considered) whose expression is induced and 600 whose expression is reduced in response to diverse stresses, such as osmotic, oxidative or heat stress (Gasch *et al.*, 2000; Posas *et al.*, 2000; Causton *et al.*, 2001; Yale and Bohnert, 2001; Miller *et al.*, 2011). The genes induced in the ESR participate in a wide variety of functions, which mainly involve stress defence including carbohydrate metabolism, defence against ROS, protein metabolism, intracellular signalling and genes implicated in DNA-damage defence and repair. Notably, many of the ESR genes from fission yeast are direct orthologues of those induced and repressed in *S. cerevisiae* (Chen *et al.*, 2003; Quintales *et al.*, 2010). In contrast, nearly all of the genes repressed in both species are directly involved in protein synthesis and growth-related processes, which are highly expressed in actively growing genes but they should be repressed to conserve mass and energy during the stress period.

ESR initiation is negatively regulated by protein kinase A (PKA) and TOR signalling as well as for feedback mechanisms provided by inducing expression of ESR regulators such as stress-regulated transcription factors (Gorner *et al.*, 1998, 2002; Smith *et al.*, 1998; Beck and Hall, 1999; Garreau *et al.*, 2000; Gasch *et al.*, 2000; Chen *et al.*, 2003). On the other hand, most of genes induced in the ESR are regulated by the general transcription factors Msn2 and Msn4 in the budding yeast (Estruch and Carlson, 1993; Martinez-Pastor *et al.*, 1996; Schmitt and McEntee, 1996), whereas the main activator of the fission ESR is Atf1 (Shiozaki and Russell, 1996; Wilkinson *et al.*, 1996; Gaits *et al.*, 1998; Gaits and Russell, 1999; Lawrence *et al.*, 2007) (see below).

In addition to the ESR, there are a number of genes specific for each type of stress. Those are part of a specific adaptive programme that contains specialized genes required for yeast to respond to particular stresses such as osmostress, heat stress or increases on oxidants concentration (Alejandro-Osorio *et al.*, 2009; Bruno *et al.*, 2010; Martinez-Montanes *et al.*, 2010; Richter *et al.*, 2010).

Control of transcription in response to stress in *S. cerevisiae*

The fact that multiple genes are synchronously activated in response to stress is a clear indication that strong coordination is required to implement the proper gene expression regulation (de Nadal *et al.*, 2011).

In budding yeast, the reprogramming of gene expression in response to osmostress is coordinated by the HOG signalling pathway through different specific transcription factors, such as Smp1, Sko1 and Hot1 as well as the mentioned Msn2/Msn4 (de Nadal and Posas, 2010; Martinez-Montanes *et al.*, 2010). Indeed, there is a differential contribution of each of these transcription factors when working in cooperation, owing to different association dynamics to promoters (Capaldi *et al.*, 2008; Ni *et al.*, 2009). Hog1 regulates these transcription factors by modulating either their nuclear localization, recruitment to specific stress-responsive genes or activity (see Fig. 11.5). For instance, Hog1 phosphorylates and activates Sko1 and Smp1 (Proft *et al.*, 2001; Alepuz *et al.*, 2003; de Nadal *et al.*, 2003). Moreover, although Hog1 mediates phosphorylation of Hot1, this phosphorylation is not needed to induce transcription (Alepuz *et al.*, 2003). Instead, Hog1 and Hot1 associate interdependently at osmoresponsive genes (Alepuz *et al.*, 2001; Pascual-Ahuir *et al.*, 2006; Pokholok *et al.*, 2006; Proft *et al.*, 2006). The Msn2 and Msn4 transcription factors also serve to anchor Hog1 to Msn2–4 dependent genes. Thus, Hog1 cross-links with the osmostress target promoters via physical interaction with specific transcription factors (Alepuz *et al.*, 2001; Pascual-Ahuir *et al.*, 2006; Pokholok *et al.*, 2006). Once bound to osmostress dependent promoters, Hog1 is involved in the recruitment of basic components of the transcriptional machinery (e.g. Mediator and SAGA), RNA Pol II, Rpd3 histone deacetylase and Ubp3 ubiquitin protease (Proft and Struhl, 2002; Alepuz *et al.*, 2003; de Nadal *et al.*, 2004; Zapater *et al.*, 2007; Sole *et al.*, 2011).

In response to oxidative or heat stress, the ESR is also activated but is not regulated by the Hog1 MAPK. In addition to the Msn2 and Msn4 general transcription factors that are controlled by different signalling pathways, specific responses against oxidative stress mainly involve two additional transcription factors, Yap1 and Skn7. Although both activators participate cooperatively (Lee *et al.*, 1999), Yap1 is the specific regulator of oxidative stress defence and it appears to serve as oxidative stress sensor that directly regulates transcription (Ikner and Shiozaki, 2005; Herrero *et al.*, 2008). On the other hand, the heat shock transcription factor Hsf1 is the specific transcription factor responsible for gene expression upon heat shock (Ahn *et al.*, 2001; Pirkkala *et al.*, 2001; Hashikawa and Sakurai, 2004; Yamamoto *et al.*, 2005; Hashikawa *et al.*, 2007), although it also activates expression of some target genes upon oxidative stress (Yamamoto *et al.*, 2007).

Gene expression profiling analyses integrated

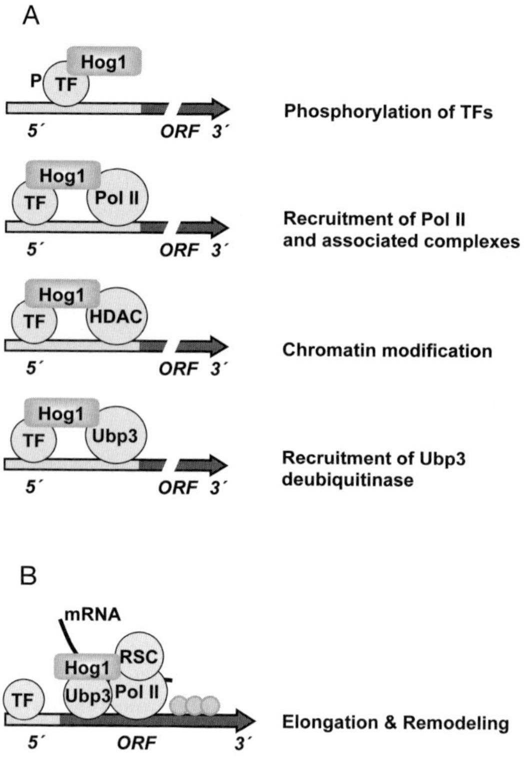

Figure 11.5 Control of transcription upon stress by the Hog1 MAPK. (A) In response to osmostress, activated Hog1 concentrates into the nucleus and binds to the osmo-dependent promoters through specific transcription factors. Once recruited, Hog1 modulates initiation of transcription by (a) direct phosphorylation of specific osmostress transcription factors (TF), (b) recruiting RNA Pol II and coactivators to the osmoresponsive promoters, (c) recruiting chromatin modifying activities such as the Rpd3 histone deacetylase complex (HDAC), and (d) recruiting the Ubp3 ubiquitin protease. (B) In addition to transcription initiation, Hog1 acts as a stress specific transcription elongation factor and it targets and controls Ubp3 ubiquitin protease activity to coding regions to increase mRNA production. Moreover, Hog1 displaces nucleosomes by targeting chromatin remodelling factors such as the RSC complex.

with genome wide chromatin binding of transcription factors and the analyses of mRNA stabilities have provided a clear overview of the landscape of mRNAs that are regulated upon various stresses, including osmotic, oxidative and heat stresses in budding yeast (Hahn *et al.*, 2004; Zanton and Pugh, 2004, 2006; Eastmond and Nelson, 2006; Capaldi *et al.*, 2008; Ni *et al.*, 2009; Molin *et al.*,

2009; Romero-Santacreu *et al.*, 2009; Miller *et al.*, 2011; Venters *et al.*, 2011).

Regulation of transcription in response to stress not only occurs at the initiation step of the cycle but also at the level of elongation. Indeed, in budding yeast, Hog1 is found throughout the transcribed regions of yeast osmo-inducible genes (Pascual-Ahuir *et al.*, 2006; Pokholok *et al.*, 2006; Proft *et al.*, 2006). Hog1 is recruited to these coding regions independently of the osmostress-specific transcription factors that are bound at the promoter but dependently on the 3′UTR regions (see Fig. 11.5). Once recruited, Hog1 is essential for an increased association of RNA Pol II in coding regions, suggesting that it directly affects the process of elongation (Proft *et al.*, 2006). In addition, Hog1 targets and controls Ubp3 ubiquitin protease activity to coding regions to increase mRNA production (Sole *et al.*, 2011).

Control of transcription in response to stress in *S. pombe*

As in budding yeast, the *S. pombe* Sty1 MAPK activates a complex transcriptional programme in response to stress through several transcription factors such as Atf1, Pap1 or Pcr1. It has been shown that Sty1 is also recruited to stress-dependent promoters upon osmotic, oxidative and heat stress (Lawrence *et al.*, 2007; Reiter *et al.*, 2008; Eshaghi *et al.*, 2010). Activated Sty1 accumulates into the nucleus and regulates ESR genes though the b-ZIP transcription factor Atf1, which is the main responsible for the ESR response (Shiozaki and Russell, 1996; Wilkinson *et al.*, 1996; Gaits *et al.*, 1998; Gaits and Russell, 1999; Lawrence *et al.*, 2007). Although Atf1 is the Sko1 orthologue, this regulator is not only activated by osmostress, but also under diverse stress conditions (Chen *et al.*, 2003) and its function seems to be related to the transcription factors Msn2 and Msn4. It has been long believed that Atf1 phosphorylation by Sty1 is essential for gene activation, although some data suggest that the main role of Sty1-dependent phosphorylation is to stabilize Atf1, rather than to activate it (Lawrence *et al.*, 2007). In addition, Sty1 is also required for *atf1* mRNA stability upon oxidative stress (Rodriguez-Gabriel *et al.*, 2003; Day and Veal, 2010). Atf1 activates transcription of some target genes by forming a heterodimer with the bZIP transcription factor Pcr1 (Wahls and Smith, 1994; Kanoh *et al.*, 1996; Watanabe and Yamamoto, 1996), which is essential for some but not all stress responses (Sanso *et al.*, 2008).

In addition to the ESR, subsets of Sty1-regulated genes are only induced in response to certain specific stress. Like Yap1, Pap1 relocalizes form the cytoplasm to the nucleus in response to oxidative stress and induces transcription of anti-oxidants genes (Toone *et al.*, 1998). Pap1 and the Sty1-Atf1 pathway share some target genes and cooperate in the transcriptional regulation upon oxidative stress (Nakagawa *et al.*, 2000). However, while Pap1 is found to be a redox sensor to low levels of hydrogen peroxide and could function independent of Sty1, Atf1 is a major player in response to high levels of oxidative stress (Quinn *et al.*, 2002; Vivancos *et al.*, 2004, 2006).

Control of transcription in response to stress in *C. albicans*

In *C. albicans*, some downstream factors of the HOG pathway have been identified (Drobna *et al.*, 2008). For instance, the Sko1 transcription factor is phosphorylated by CaHog1 in response to osmostress (Rauceo *et al.*, 2008) and it is an important mediator of the oxidative stress response (Alonso-Monge *et al.*, 2010). On the other hand, the *CAP1* gene in *C. albicans* encodes a transcription factor homologous to the *S. cerevisiae* Yap1 and the *S. pombe* Pap1 proteins that it is also required for the cellular defence against oxidative stress (Alarco and Raymond, 1999; Zhang *et al.*, 2000). Oxidative stress signalling to Cap1 is independent of the CaHog1 pathway, and like Yap1 and Pap1, Cap1 is also regulated through changes in cellular localization (Zhang *et al.*, 2000; Alonso-Monge *et al.*, 2003).

Chromatin remodelling and modification in response to stress

The packaging of DNA into nucleosomes affects all phases of the transcription cycle. The nucleosome is not a simple static unit, but it has dynamic properties that are tightly regulated by several protein complexes. The mechanisms by which chromatin is highly dynamic include histone variant incorporation, chromatin remodelling, histone eviction and histone covalent modification (Cairns, 2009;

Jiang and Pugh, 2009). In recent years, there have been huge advances in chromatin dynamics and its influence in the process of transcription.

In budding yeast, there is a dramatic change in the nucleosome organization of stress-responsive promoters and coding regions upon osmostress that depends on Hog1 and the RSC chromatin remodelling complex. Hog1 interacts with several subunits of RSC and targets them to coding regions of osmoresponsive genes, allowing for nucleosome rearrangements upon stress (Mas *et al.*, 2009). In addition, other chromatin remodelling complexes such as SWI/SNF are recruited to stress-dependent target promoters by Hog1 in a manner that correlates well with transcriptional induction (Proft and Struhl, 2002). On the other hand, there are some chromatin remodelling complexes such as INO80, which are responsible for nucleosome repositioning, limiting the extent of transcription induction (Klopf *et al.*, 2009). Thus, a dynamic balance among different chromatin remodelling complexes is required for proper expression of stress genes.

In response to heat stress, the nucleosomes of budding yeast are also evicted or displaced from heat stress-dependent promoters and coding regions (Erkine and Gross, 2003; Erkina and Erkine, 2006; Zanton and Pugh, 2006; Venters *et al.*, 2011). In this case, chromatin remodelling complexes such SWI/SNF, ISW1 and RSC play partially overlapping function (Gross *et al.*, 1993; Zhao *et al.*, 2005; Uffenbeck and Krebs, 2006; Erkina *et al.*, 2008, 2010; Shivaswamy and Iyer, 2008).

Complementing the role of nucleosome remodellers, recruitment of histone-modifying enzymes provides another means to modulate transcription. Genetic analyses in budding yeast have shown that multiple histone residues, when mutated, prevent modifications and render cells sensitive to different cellular stresses (Dai *et al.*, 2008; Huang *et al.*, 2009), indicating that several chromatin modifying complexes might be important for transcription initiation upon stress.

Some covalent histone modifications have been coupled to yeast stress-dependent changes in gene expression. For instance, the Rpd3 histone deacetylase complex is recruited by Hog1 to specific stress-activated promoters upon osmostress (de Nadal *et al.*, 2004). Other stresses such as heat shock and oxidative stress also require Rpd3 for full activation of gene expression (Alejandro-Osorio *et al.*, 2009; Ruiz-Roig *et al.*, 2010). On the other hand, in *S. pombe*, upon oxidative stress, Sty1 and Atf1 trigger the binding of the histone acetyltransferase SAGA complex and RNA Pol II to stress promoter. Deficient acetylation of histones in cells lacking the SAGA complex affects eviction of nucleosomes positioned downstream of promoters and RNA Pol II progression along genes (Johnsson *et al.*, 2009; Sanso *et al.*, 2011).

mRNA stability, export and control of translation in response to stress

Nascent RNA transcripts undergo splicing, nuclear export, stabilization and translation. Coordination of RNA processing creates several layers of regulation of gene expression in eukaryotic cells and the roles of SAPKs here are now starting to be elucidated. Comparison between proteomic and transcriptomic data in stress conditions sometimes shows weak correlation illustrating the importance of translational and post-translational modifications (Szopinska and Morsomme, 2010). In response to stress, transcript stability in *S. cerevisiae* makes substantial contributions to the difference in the steady-state mRNA levels (Molin *et al.*, 2009; Romero-Santacreu *et al.*, 2009; Miller *et al.*, 2011). Indeed, osmostress-responsive mRNAs are selectively stabilized or degraded at different rates, although the determinants for such selective stabilization are still undefined (Miller *et al.*, 2011). Changes in mRNA stability during stress also occur in response to other stresses such as heat shock and oxidative stress (Molina-Navarro *et al.*, 2008; Castells-Roca *et al.*, 2011).

Regulation of translation in response to stress is particularly valuable for an organism because translating of existing transcripts generates proteins more rapidly than with *de novo* transcriptional induction. Analyses of the global impact on translated mRNAs in osmostress show that many genes are translated more efficiently upon stress, including those that do not display changes in mRNA levels. Interestingly, the role of Hog1 in this translational regulation was even stronger than for the transcriptional level, highlighting the importance of translational control for fine tuning of the adaptive responses (Warringer *et al.*, 2010).

As other cellular stresses, osmotic shock causes transient inhibition of translation initiation and the HOG pathway seems not to be involved in the recovery of translation during adaptation (Uesono and Toh, 2002). It has been also reported that Sty1 binds to translation factors and its mutation results in defects in the recovery of translation after stress (Asp *et al.*, 2008). Moreover, Rck2, a yeast serine/threonine protein kinase homologous to Srk1 in fission yeast, is a Hog1 target. Phosphorylated Rck2 directly modulates the translation elongation factor EF-2 and this is required for translational efficiency in response to osmostress (Bilsland-Marchesan *et al.*, 2000; Teige *et al.*, 2001). Therefore, SAPKs play an important role in protein production from mRNA biogenesis to translation.

In addition to transcript synthesis and degradation, the reversible storage of transcripts provides an economical and rapid mechanism to regulate protein levels in response to stress. During stress responses, specific transcripts can associate to P-bodies or stress granules, re-entering in translation when environmental conditions recover (Lopez-Maury *et al.*, 2008). For instance, stress granules have been described in budding yeast during severe heat shock (Parker and Sheth, 2007; Buchan and Parker, 2009).

CaHOG MAPK pathway as regulator of fungal virulence

C. albicans is mainly an opportunistic pathogen that can grow as commensal in the gastrointestinal and vaginal tracts of humans, causing superficial or mucosal to life-threatening systemic infections. However, as a result of a defect in the host defences, *C. albicans* is able to disseminate through the bloodstream and reach target organs such as kidneys, brain or liver causing diseases collectively referred to as candidiasis (Berman and Sudbery, 2002; Diez-Orejas and Fernandez-Arenas, 2008). In the human host, cellular conditions such as pH, oxidant or osmolyte concentrations can differ substantially, having an essential role in the pathogenesis of fungal infections. Consequently, *C. albicans* have developed different survival strategies to colonize and produce disease in the host, which involve sensing and responding to these environmental signals. Indeed, one of the main factors that influence virulence in *C. albicans* is its ability to adapt to different environmental conditions mediated by MAPK pathways (Roman *et al.*, 2007).

CaHog1 plays a key role in the pathogenesis of candidiasis, copping with host immune defences. Indeed, a MAPK mutant appears to be less virulent in a mouse model of fungal infection (Alonso-Monge *et al.*, 1999). In order to defend against colonization of *C. albicans*, the host immune system, mainly macrophages, uses a method of killing through the production of ROS (Vazquez-Torres and Balish, 1997) and, consequently, *C. albicans* needs to sense and respond to ROS. The CaHOG pathway is activated in response to oxidative stress and it is involved in regulating ROS production and ATP demand by mitochondria (Alonso-Monge *et al.*, 2009). Moreover, Ca*hog1* mutants are more sensitive to oxidants (Alonso-Monge *et al.*, 2003) and more susceptible to phagocytic cells (Arana *et al.*, 2007). Moreover, mutants in the two-component response regulator Ssk1 are more easily killed by human neutrophils (Du *et al.*, 2005). Thus, the virulence defect of the MAPK mutants might be due to an enhanced susceptibility to phagocyte oxidative-mediated killing mechanisms. Of note, other *C. albicans* MAPK cascades such as the cell integrity Mkc1 pathway, which are activated by oxidative stress, are involved in virulence and influence the progression of the disease (Roman *et al.*, 2007).

Conclusions

Despite similarities, *S. cerevisiae*, *S. pombe* and *C. albicans* significantly diverge and each one has specific stress adaptation responses. We have described how yeast cells sense adverse environment conditions, transduce this information and which are the cellular responses triggered by stress, highlighting the differences among the three species.

Proper adaptation to stress is critical for cell survival in harsh environments, and cells have developed sensing and signal transduction mechanisms that permit the appropriate adaptive responses. SAPKs are the main pathways to respond and adapt to extracellular changes. Although the SAPK modules are highly conserved signalling molecules, the upstream sensors and signal transducers are less conserved among *S. cerevisiae, S. pombe* and *C. albicans*. Such differences may reflect an adaptation to particular environmental niches occupied by these fungi.

Activation of SAPKs results in the generation of a set of adaptive responses that leads to the modulation of several aspects of cell physiology essential for cell survival. The major adaptive responses consist in the modulation of cell cycle progression and the massive reorganization of gene expression in the cell. The ability of a single MAPK to regulate different stages of cell cycle might be necessary because cells need to respond to stress at any stage of their cell cycle. Therefore there ought to be mechanisms that mediate adaptation before progressing into more sensitive phases of the cell cycle. Hog1 controls G1, S and G2 transitions by complementary mechanisms that involve the stabilization of cell cycle inhibitors and the down-regulation of cyclin expression. In G1, this dual mechanism has distinct temporal roles to accommodate the appropriate signal to the basic cell cycle core mechanism. In contrast, the reason why the regulation of G2 by Hog1 is also structured with a dual mechanism is still not completely understood. Regulation of cell cycle by the Sty1 MAPK in fission yeast has also been deciphered. Sty1 mainly controls cell cycle in G2/M transition.

In response to stress there is a major change in the transcriptional pattern of the cell. The coordinated expression of many stress-responsive genes and the exquisite fine tuning of their expression levels is accomplished by the regulation of several steps in mRNA biogenesis and mRNA fate. Genome-wide studies have shown that signal transduction pathways control a variety of down-stream elements that permit a rapid change in the transcriptional landscape of a cell within minutes of exposure to stress. In budding yeast, the association of Hog1 to chromatin at specific osmoresponsive genes is a key step to orchestrate the appropriate recruitment of transcription factors to control initiation, elongation and mRNA stability. Although this broad scheme is common to many stresses, the underlying mechanistic details of how such regulation is achieved are highly dependent on any particular stress. As in budding yeast, the Sty1 MAPK has a pivotal role in regulation of transcription upon stress and its activation also modulates initiation, elongation and mRNA stability. In addition to MAPKs, other signalling pathways and specific transcription factors display the ability to sense and respond to specific stresses and participate in the regulation of gene expression.

Yeasts have been an interesting model organism to study stress responses for the evolutionary conservation of stress adaptive mechanisms among yeast and higher eukaryotes, its genetic amenability and the availability of genome-wide approaches. Although some aspects of the stress responses may be specific, the conservation of several of the regulatory mechanisms between yeast as a model organisms and higher eukaryotes are clear indications that data from model organisms is improving our understanding of stress responses in higher organisms. On the other hand, yeasts are organisms with biotechnological applications widely used for beer, wine, and bread production, as well as are also employed to produce industrial quantities of several metabolic products, such as vitamins or hormones, by heterologous gene expression. These biotechnological processes subject yeast cells to non-physiological and stressful conditions and, for that reason, the knowledge of adaptation processes is crucial for improving industrial processes. Last, some yeast species are relevant in medicine. The stress response is an important factor of virulence in the case of pathogenic fungi and thus the knowledge of these responses might be essential to limit the impact of those pathogenic fungi.

Future trends

Over the past two decades, many insights on SAPK pathways have emerged from basic research in yeast. However, several issues remain to be addressed in the future. How osmostress signals are funnelled into the Sty1, CaHog1 and mammalian SAPKs, as well as which are the accompanying pathways that serve to respond to osmostress are key questions to be answered. The budding yeast has provided unique evidence for a new role for mucins in osmostress sensing. However, a role for mammalian mucins in sensing osmostress has not yet been characterized.

SAPKs regulate cell cycle progression upon stress; however additional strategies to modulate cell cycle might exist such as the control of other cell cycle phases as replication or exit from mitosis. Last, further research is needed to investigate the possible impact of other stresses in cell cycle control, especially in pathogenic yeast species like *C. albicans*.

Quantitative analyses and combination of modelling and experimentation will provide new insight into the MAPK signalling as well as its biological roles in response to stress. Furthermore, single cell analyses will gain in importance over the next few years as they should provide us with a comprehensive view of basic characteristics such as the presence of thresholds and noise regulation on signalling pathways. Similarly, in recent years powerful genetic tools in *C. albicans* have emerged, which will allow focus in the CaHOG pathway and its relationship with oxidative stress response in human pathogenic fungi. The role of MAPK pathways within the human host is largely unexplored and their *in vivo* functions during infection of the human host are not yet identified. Presumably, such studies should contribute to identify novel therapeutic antifungal targets.

Acknowledgements

We thank all members of the Cell Signaling Unit. We apologize to colleagues whose work could not be cited owing to size limitation. The laboratory of F. Posas and E. de Nadal is being supported by Fundación Marcelino Botín (FMB), Ministerio de Ciencia y Innovación; BFU program (BFU2008-00530 and BFU2011-26722 to E.N. and BIO2009-07762 to F.P.), Consolider Ingenio 2010 programme (grant CSD2007-0015) and FP7 (*UNICELLSYS*) framework programmes. F.P. is recipient of an ICREA Acadèmia (Generalitat de Catalunya).

Web resources

SGD: http://yeastgenome.org/

The *Saccharomyces* Genome Database is a scientific database of the molecular biology and genetics of the yeast *S. cerevisiae*.

MIPS: http://mips.helmholtz-muenchen.de/genre/proj/yeast/

The MIPS Comprehensive Yeast Genome Database (CYGD) aims to present information on the molecular structure and functional network of *S. cerevisiae*.

GeneDB: http://old.genedb.org/genedb/pombe/

This database contains all *S. pombe* known and predicted protein coding genes, pseudogenes, transposons, tRNAs, rRNAs, snRNAs, snoRNAs and other known and predicted non-coding RNAs.

CGD: http://www.candidagenome.org/

The Candida Genome Database is a resource for genomic sequence data and gene and protein information for *C. albicans*.

General topics: http://wiki.yeastgenome.org/index.php/General_Topics

This website is focused on describing the yeast model organisms: *S. cerevisiae*, *S. pombe* and *C. albicans*, including educational resources.

External links: http://wiki.yeastgenome.org/index.php/External_Links

This website provides several links to yeast research data as nucleic acid, genome and protein, expression, localization and phenotype data.

References

Adrover, M.A., Zi, Z., Duch, A., Schaber, J., Gonzalez-Novo, A., Jimenez, J., Nadal-Ribelles, M., Clotet, J., Klipp, E., and Posas, F. (2011). Multi-component control of the G1–S network by the SAPK Hog1 upon osmostress. Sci. Signal. *4*, 63.

Ahn, S.G., Liu, P.C., Klyachko, K., Morimoto, R.I., and Thiele, D.J. (2001). The loop domain of heat shock transcription factor 1 dictates DNA-binding specificity and responses to heat stress. Genes Dev. *15*, 2134–2145.

Alarco, A.M., and Raymond, M. (1999). The bZip transcription factor Cap1p is involved in multidrug resistance and oxidative stress response in *Candida albicans*. J. Bacteriol. *181*, 700–708.

Alejandro-Osorio, A.L., Huebert, D.J., Porcaro, D.T., Sonntag, M.E., Nillasithanukroh, S., Will, J.L., and Gasch, A.P. (2009). The histone deacetylase Rpd3p is required for transient changes in genomic expression in response to stress. Gen. Biol. *10*, R57.

Alepuz, P.M., Jovanovic, A., Reiser, V., and Ammerer, G. (2001). Stress-induced map kinase Hog1 is part of transcription activation complexes. Mol. Cell *7*, 767–777.

Alepuz, P.M., de Nadal, E., Zapater, M., Ammerer, G., and Posas, F. (2003). Osmostress-induced transcription by Hot1 depends on a Hog1-mediated recruitment of the RNA Pol II. EMBO J. *22*, 2433–2442.

Alex, L.A., Korch, C., Selitrennikoff, C.P., and Simon, M.I. (1998). COS1, a two-component histidine kinase that is involved in hyphal development in the opportunistic pathogen *Candida albicans*. Proc. Natl. Acad. Sci. U.S.A. *95*, 7069–7073.

Alexander, M.R., Tyers, M., Perret, M., Craig, B.M., Fang, K.S., and Gustin, M.C. (2001). Regulation of cell cycle progression by Swe1p and Hog1p following hypertonic stress. Mol. Biol. Cell *12*, 53–62.

Alic, N., Higgins, V.J., Pichova, A., Breitenbach, M., and Dawes, I.W. (2003). Lipid hydroperoxides activate the mitogen-activated protein kinase Mpk1p in *Saccharomyces cerevisiae*. J. Biol. Chem. *278*, 41849–41855.

Alonso-Monge, R., Navarro-Garcia, F., Molero, G., Diez-Orejas, R., Gustin, M., Pla, J., Sanchez, M., and Nombela, C. (1999). Role of the mitogen-activated protein kinase Hog1p in morphogenesis and virulence of *Candida albicans*. J. Bacteriol. *181*, 3058–3068.

Alonso-Monge, R., Navarro-Garcia, F., Roman, E., Negredo, A.I., Eisman, B., Nombela, C., and Pla, J. (2003). The Hog1 mitogen-activated protein kinase is essential in the oxidative stress response and chlamydospore formation in *Candida albicans*. Eukaryot. Cell *2*, 351–361.

Alonso-Monge, R., Carvaihlo, S., Nombela, C., Rial, E., and Pla, J. (2009). The Hog1 MAP kinase controls respiratory metabolism in the fungal pathogen *Candida albicans*. Microbiology *155*, 413–423.

Alonso-Monge, R., Roman, E., Arana, D.M., Prieto, D., Urrialde, V., Nombela, C., and Pla, J. (2010). The Sko1 protein represses the yeast-to-hypha transition and regulates the oxidative stress response in *Candida albicans*. Fungal Genet. Biol. *47*, 587–601.

Aoyama, K., Mitsubayashi, Y., Aiba, H., and Mizuno, T. (2000). Spy1, a histidine-containing phosphotransfer signaling protein, regulates the fission yeast cell cycle through the Mcs4 response regulator. J. Bacteriol. *182*, 4868–4874.

Arana, D.M., Nombela, C., Alonso-Monge, R., and Pla, J. (2005). The Pbs2 MAP kinase kinase is essential for the oxidative-stress response in the fungal pathogen *Candida albicans*. Microbiology *151*, 1033–1049.

Arana, D.M., Alonso-Monge, R., Du,C., Calderone, R., and Pla, J. (2007). Differential susceptibility of mitogen-activated protein kinase pathway mutants to oxidative-mediated killing by phagocytes in the fungal pathogen *Candida albicans*. Cell. Microbiol. *9*, 1647–1659.

Asano, S., Park, J.E., Sakchaisri, K., Yu,L.R., Song, S., Supavilai, P., Veenstra, T.D., and Lee, K.S. (2005). Concerted mechanism of Swe1/Wee1 regulation by multiple kinases in budding yeast. EMBO J. *24*, 2194–2204.

Asp, E., Nilsson, D., and Sunnerhagen, P. (2008). Fission yeast mitogen-activated protein kinase Sty1 interacts with translation factors. Eukaryot. Cell *7*, 328–338.

Atir-Lande, A., Gildor, T., and Kornitzer, D. (2005). Role for the SCFCDC4 ubiquitin ligase in *Candida albicans* morphogenesis. Mol. Biol. Cell *16*, 2772–2785.

Auesukaree, C., Damnernsawad, A., Kruatrachue, M., Pokethitiyook, P., Boonchird, C., Kaneko, Y., and Harashima, S. (2009). Genome-wide identification of genes involved in tolerance to various environmental stresses in *Saccharomyces cerevisiae*. J. Appl. Genet. *50*, 301–310.

Beck, T., and Hall, M.N. (1999). The TOR signalling pathway controls nuclear localization of nutrient-regulated transcription factors. Nature *402*, 689–692.

Berman, J., and Sudbery, P.E. (2002). *Candida albicans*: a molecular revolution built on lessons from budding yeast. Nat. Rev. Genet. *3*, 918–930.

Berry, D.B., and Gasch, A.P. (2008). Stress-activated genomic expression changes serve a preparative role for impending stress in yeast. Mol. Biol. Cell *19*, 4580–4587.

Bilsland, E., Molin, C., Swaminathan, S., Ramne, A., and Sunnerhagen, P. (2004). Rck1 and Rck2 MAPKAP kinases and the HOG pathway are required for oxidative stress resistance. Mol. Microbiol. *53*, 1743–1756.

Bilsland-Marchesan, E., Arino, J., Saito, H., Sunnerhagen, P., and Posas, F. (2000). Rck2 kinase is a substrate for the osmotic stress-activated mitogen-activated protein kinase Hog1. Mol. Cell. Biol. *20*, 3887–3895.

Bloom, J., and Cross, F.R. (2007). Multiple levels of cyclin specificity in cell-cycle control. Nat. Rev. Mol. Cell. Biol. *8*, 149–160.

Bozonet, S.M., Findlay, V.J., Day, A.M., Cameron, J., Veal, E.A., and Morgan, B.A. (2005). Oxidation of a eukaryotic 2-Cys peroxiredoxin is a molecular switch controlling the transcriptional response to increasing levels of hydrogen peroxide. J. Biol. Chem. *280*, 23319–23327.

Breeden, L. (1996). Start-specific transcription in yeast. Curr. Top. Microbiol. Immunol. *208*, 95–127.

Brewster, J.L., de Valoir, T., Dwyer, N.D., Winter, E., and Gustin, M.C. (1993). An osmosensing signal transduction pathway in yeast. Science *259*, 1760–1763.

Brown, J.L., North, S., and Bussey, H. (1993). SKN7, a yeast multicopy suppressor of a mutation affecting cell wall beta-glucan assembly, encodes a product with domains homologous to prokaryotic two-component regulators and to heat shock transcription factors. J. Bacteriol. *175*, 6908–6915.

Brown, J.L., Bussey, H., and Stewart, R.C. (1994). Yeast Skn7p functions in a eukaryotic two-component regulatory pathway. EMBO J. *13*, 5186–5194.

Bruno, V.M., Wang, Z., Marjani, S.L., Euskirchen, G.M., Martin, J., Sherlock, G., and Snyder, M. (2010). Comprehensive annotation of the transcriptome of the human fungal pathogen *Candida albicans* using RNA-seq. Genome Res. *20*, 1451–1458.

Buchan, J.R., and Parker, R. (2009). Eukaryotic stress granules: the ins and outs of translation. Mol. Cell *36*, 932–941.

Buck, V., Quinn, J., Soto, P.T., Martin, H., Saldanha, J., Makino, K., Morgan, B.A., and Millar, J.B. (2001). Peroxide sensors for the fission yeast stress-activated mitogen-activated protein kinase pathway. Mol. Biol. Cell *12*, 407–419.

Cairns, B.R. (2009). The logic of chromatin architecture and remodelling at promoters. Nature. *461*, 193–198.

Calera, J.A., and Calderone, R. (1999a). Histidine kinase, two-component signal transduction proteins of *Candida albicans* and the pathogenesis of candidosis. Mycoses *42*, 49–53.

Calera, J.A., and Calderone, R.A. (1999b). Identification of a putative response regulator two-component phosphorelay gene (CaSSK1) from *Candida albicans*. Yeast *15*, 1243–1254.

Calera, J.A., Herman, D., and Calderone, R. (2000). Identification of YPD1, a gene of *Candida albicans* which encodes a two-component phosphohistidine intermediate protein. Yeast *16*, 1053–1059.

Capaldi, A.P., Kaplan, T., Liu, Y., Habib, N., Regev, A., Friedman, N., and O'Shea, E.K. (2008). Structure and function of a transcriptional network activated by the MAPK Hog1. Nat. Genet. *40*, 1300–1306.

Castells-Roca, L., Garcia-Martinez, J., Moreno, J., Herrero, E., Belli, G., and Perez-Ortin, J.E. (2011). Heat shock response in yeast involves changes in both transcription rates and mRNA stabilities. PLoS. One. *6*, e17272.

Causton, H.C., Ren, B., Koh, S.S., Harbison, C.T., Kanin, E., Jennings, E.G., Lee, T.I., True, H.L., Lander, E.S., and Young, R.A. (2001). Remodeling of yeast genome expression in response to environmental changes. Mol. Biol. Cell *12*, 323–337.

Chauhan, N., Inglis, D., Roman, E., Pla, J., Li,D., Calera, J.A., and Calderone, R. (2003). *Candida albicans* response regulator gene SSK1 regulates a subset of genes whose functions are associated with cell wall biosynthesis and adaptation to oxidative stress. Eukaryot. Cell *2*, 1018–1024.

Cheetham, J., Smith, D.A., da Silva, D.A., Doris, K.S., Patterson, M.J., Bruce, C.R., and Quinn, J. (2007). A single MAPKKK regulates the Hog1 MAPK pathway in the pathogenic fungus *Candida albicans*. Mol. Biol. Cell *18*, 4603–4614.

Chen, D., Toone, W.M., Mata, J., Lyne, R., Burns, G., Kivinen, K., Brazma, A., Jones, N., and Bahler, J. (2003). Global transcriptional responses of fission yeast to environmental stress. Mol. Biol. Cell *14*, 214–229.

Chen, R.E., and Thorner, J. (2007). Function and regulation in MAPK signaling pathways: lessons learned from the yeast *Saccharomyces cerevisiae*. Biochim. Biophys. Acta. *1773*, 1311–1340.

Clotet, J., and Posas, F. (2007). Control of cell cycle in response to osmostress: lessons from yeast. Methods Enzymol. *428*, 63–76.

Clotet, J., Escote, X., Adrover, M.A., Yaakov, G., Gari, E., Aldea, M., de Nadal, E., and Posas, F. (2006). Phosphorylation of Hsl1 by Hog1 leads to a G2 arrest essential for cell survival at high osmolarity. EMBO J. *25*, 2338–2346.

Coleman, S.T., Epping, E.A., Steggerda, S.M., and Moye-Rowley, W.S. (1999). Yap1p activates gene transcription in an oxidant-specific fashion. Mol. Cell. Biol. *19*, 8302–8313.

Cote, P., Hogues, H., and Whiteway, M. (2009). Transcriptional analysis of the *Candida albicans* cell cycle. Mol. Biol. Cell *20*, 3363–3373.

Cottarel, G. (1997). Mcs4, a two-component system response regulator homologue, regulates the *Schizosaccharomyces pombe* cell cycle control. Genetics *147*, 1043–1051.

Cross, F.R., Schroeder, L., and Bean, J.M. (2007). Phosphorylation of the Sic1 inhibitor of B-type cyclins in *Saccharomyces cerevisiae* is not essential but contributes to cell cycle robustness. Genetics *176*, 1541–1555.

Dai, J., Hyland, E.M., Yuan, D.S., Huang, H., Bader, J.S., and Boeke, J.D. (2008). Probing nucleosome function: a highly versatile library of synthetic histone H3 and H4 mutants. Cell *134*, 1066–1078.

Day, A.M., and Veal, E.A. (2010). Hydrogen peroxide-sensitive cysteines in the Sty1 MAPK regulate the transcriptional response to oxidative stress. J. Biol. Chem. *285*, 7505–7516.

Degols, G., and Russell, P. (1997). Discrete roles of the Spc1 kinase and the Atf1 transcription factor in the UV response of *Schizosaccharomyces pombe*. Mol. Cell. Biol. *17*, 3356–3363.

Degols, G., Shiozaki, K., and Russell, P. (1996). Activation and regulation of the Spc1 stress-activated protein kinase in *Schizosaccharomyces pombe*. Mol. Cell. Biol. *16*, 2870–2877.

Delaunay, A., Isnard, A.D., and Toledano, M.B. (2000). H_2O_2 sensing through oxidation of the Yap1 transcription factor. EMBO J. *19*, 5157–5166.

Delaunay, A., Pflieger, D., Barrault, M.B., Vinh, J., and Toledano, M.B. (2002). A thiol peroxidase is an H2O2 receptor and redox-transducer in gene activation. Cell *111*, 471–481.

Diez-Orejas, R., and Fernandez-Arenas, E. (2008). *Candida albicans*–macrophage interactions: genomic and proteomic insights. Future. Microbiol. *3*, 661–681.

Donovan, J.D., Toyn, J.H., Johnson, A.L., and Johnston, L.H. (1994). P40SDB25, a putative CDK inhibitor, has a role in the M/G1 transition in *Saccharomyces cerevisiae*. Genes Dev. *8*, 1640–1653.

Drobna, E., Bialkova, A., and Subik, J. (2008). Transcriptional regulators of seven yeast species: comparative genome analysis. Rev. Folia Microbiol. (Praha) *53*, 275–287.

Drogen, F., O'Rourke, S.M., Stucke, V.M., Jaquenoud, M., Neiman, A.M., and Peter, M. (2000). Phosphorylation of the MEKK Ste11p by the PAK-like kinase Ste20p is required for MAP kinase signaling *in vivo.* Curr. Biol. *10,* 630–639.

Du, C., Calderone, R., Richert, J., and Li,D. (2005). Deletion of the SSK1 response regulator gene in *Candida albicans* contributes to enhanced killing by human polymorphonuclear neutrophils. Infect. Immun. *73,* 865–871.

Eastmond, D.L., and Nelson, H.C. (2006). Genome-wide analysis reveals new roles for the activation domains of the *Saccharomyces cerevisiae* heat shock transcription factor (Hsf1) during the transient heat shock response. J. Biol. Chem. *281,* 32909–32921.

Egger, L.A., Park, H., and Inouye, M. (1997). Signal transduction via the histidyl-aspartyl phosphorelay. Genes Cells *2,* 167–184.

Enjalbert, B., Nantel, A., and Whiteway, M. (2003). Stress-induced gene expression in *Candida albicans*: absence of a general stress response. Mol. Biol. Cell *14,* 1460–1467.

Erkina, T.Y., and Erkine, A.M. (2006). Displacement of histones at promoters of *Saccharomyces cerevisiae* heat shock genes is differentially associated with histone H3 acetylation. Mol. Cell. Biol. *26,* 7587–7600.

Erkina, T.Y., Tschetter, P.A., and Erkine, A.M. (2008). Different requirements of the SWI/SNF complex for robust nucleosome displacement at promoters of heat shock factor and Msn2- and Msn4-regulated heat shock genes. Mol. Cell. Biol. *28,* 1207–1217.

Erkina, T.Y., Zou, Y., Freeling, S., Vorobyev, V.I., and Erkine, A.M. (2010). Functional interplay between chromatin remodeling complexes RSC, SWI/SNF and ISWI in regulation of yeast heat shock genes. Nucleic Acids Res. *38,* 1441–1449.

Erkine, A.M., and Gross, D.S. (2003). Dynamic chromatin alterations triggered by natural and synthetic activation domains. J. Biol. Chem. *278,* 7755–7764.

Escote, X., Zapater, M., Clotet, J., and Posas, F. (2004). Hog1 mediates cell-cycle arrest in G1 phase by the dual targeting of Sic1. Nat. Cell Biol. *6,* 997–1002.

Eshaghi, M., Lee, J.H., Zhu, L., Poon, S.Y., Li,J., Cho, K.H., Chu, Z., Karuturi, R.K., and Liu, J. (2010). Genomic binding profiling of the fission yeast stress-activated MAPK Sty1 and the bZIP transcriptional activator Atf1 in response to H2O2. PLoS. One. *5,* e11620.

Estruch, F., and Carlson, M. (1993). Two homologous zinc finger genes identified by multicopy suppression in a SNF1 protein kinase mutant of *Saccharomyces cerevisiae.* Mol. Cell. Biol. *13,* 3872–3881.

Ferrigno, P., Posas, F., Koepp, D., Saito, H., and Silver, P.A. (1998). Regulated nucleo/cytoplasmic exchange of HOG1 MAPK requires the importin beta homologs NMD5 and XPO1. EMBO J. *17,* 5606–5614.

Forsburg, S.L. (2005). The yeasts *Saccharomyces cerevisiae* and *Schizosaccharomyces pombe*: models for cell biology research. Gravit. Space Biol. Bull. *18,* 3–9.

Franzmann, T.M., Menhorn, P., Walter, S., and Buchner, J. (2008). Activation of the chaperone Hsp26 is controlled by the rearrangement of its thermosensor domain. Mol. Cell *29,* 207–216.

Gaits, F., and Russell, P. (1999). Active nucleocytoplasmic shuttling required for function and regulation of stress-activated kinase Spc1/StyI in fission yeast. Mol. Biol. Cell *10,* 1395–1407.

Gaits, F., Degols, G., Shiozaki, K., and Russell, P. (1998). Phosphorylation and association with the transcription factor Atf1 regulate localization of Spc1/Sty1 stress-activated kinase in fission yeast. Genes Dev. *12,* 1464–1473.

Garreau, H., Hasan, R.N., Renault, G., Estruch, F., Boy-Marcotte, E., and Jacquet, M. (2000). Hyperphosphorylation of Msn2p and Msn4p in response to heat shock and the diauxic shift is inhibited by cAMP in *Saccharomyces cerevisiae.* Microbiology *146,* 2113–2120.

Gasch, A.P. (2007). Comparative genomics of the environmental stress response in ascomycete fungi. Yeast. *24,* 961–976.

Gasch, A.P., Spellman, P.T., Kao, C.M., Carmel-Harel, O., Eisen, M.B., Storz, G., Botstein, D., and Brown, P.O. (2000). Genomic expression programs in the response of yeast cells to environmental changes. Mol. Biol. Cell *11,* 4241–4257.

Gasch, A.P., Moses, A.M., Chiang, D.Y., Fraser, H.B., Berardini, M., and Eisen, M.B. (2004). Conservation and evolution of cis-regulatory systems in ascomycete fungi. PLoS. Biol. *2,* e398.

Gautier, J., Solomon, M.J., Booher, R.N., Bazan, J.F., and Kirschner, M.W. (1991). cdc25 is a specific tyrosine phosphatase that directly activates p34cdc2. Cell *67,* 197–211.

Giaever, G., Chu, A.M., Ni, L., Connelly, C., Riles, L., Veronneau, S., Dow, S., Lucau-Danila, A., Anderson, K., Andre, *et al.* (2002). Functional profiling of the *Saccharomyces cerevisiae* genome. Nature *418,* 387–391.

Goffeau, A., Barrell, B.G., Bussey, H., Davis, R.W., Dujon, B., Feldmann, H., Galibert, F., Hoheisel, J.D., Jacq, C., Johnston, M., *et al.* (1996). Life with 6000 genes. Science *274,* 563–567.

Gorner, W., Durchschlag, E., Martinez-Pastor, M.T., Estruch, F., Ammerer, G., Hamilton, B., Ruis, H., and Schuller, C. (1998). Nuclear localization of the C2H2 zinc finger protein Msn2p is regulated by stress and protein kinase A activity. Genes Dev. *12,* 586–597.

Gorner, W., Durchschlag, E., Wolf, J., Brown, E.L., Ammerer, G., Ruis, H., and Schuller, C. (2002). Acute glucose starvation activates the nuclear localization signal of a stress-specific yeast transcription factor. EMBO J. *21,* 135–144.

Gross, D.S., Adams, C.C., Lee, S., and Stentz, B. (1993). A critical role for heat shock transcription factor in establishing a nucleosome-free region over the TATA-initiation site of the yeast HSP82 heat shock gene. EMBO J. *12,* 3931–3945.

Hahn, J.S., Hu,Z., Thiele, D.J., and Iyer, V.R. (2004). Genome-wide analysis of the biology of stress responses through heat shock transcription factor. Mol. Cell. Biol. *24,* 5249–5256.

Han, J., Lee, J.D., Bibbs, L., and Ulevitch, R.J. (1994). A MAP kinase targeted by endotoxin and hyperosmolarity in mammalian cells. Science *265*, 808–811.

Hao, N., Behar, M., Parnell, S.C., Torres, M.P., Borchers, C.H., Elston, T.C., and Dohlman, H.G. (2007). A systems-biology analysis of feedback inhibition in the Sho1 osmotic-stress-response pathway. Curr. Biol. *17*, 659–667.

Hao, N., Zeng, Y., Elston, T.C., and Dohlman, H.G. (2008). Control of MAPK specificity by feedback phosphorylation of shared adaptor protein Ste50. J. Biol. Chem. *283*, 33798–33802.

Hashikawa, N., and Sakurai, H. (2004). Phosphorylation of the yeast heat shock transcription factor is implicated in gene-specific activation dependent on the architecture of the heat shock element. Mol. Cell. Biol. *24*, 3648–3659.

Hashikawa, N., Yamamoto, N., and Sakurai, H. (2007). Different mechanisms are involved in the transcriptional activation by yeast heat shock transcription factor through two different types of heat shock elements. J. Biol. Chem. *282*, 10333–10340.

Herrero, E., de la Torre, M.A., and Valentin, E. (2003). Comparative genomics of yeast species: new insights into their biology. Int. Microbiol. *6*, 183–190.

Herrero, E., Ros, J., Belli, G., and Cabiscol, E. (2008). Redox control and oxidative stress in yeast cells. Biochim. Biophys. Acta *1780*, 1217–1235.

Hersen, P., McClean, M.N., Mahadevan, L., and Ramanathan, S. (2008). Signal processing by the HOG MAP kinase pathway. Proc. Natl. Acad. Sci. U.S.A. *105*, 7165–7170.

Hohmann, S. (2002). Osmotic stress signaling and osmoadaptation in yeasts. Microbiol. Mol. Biol. Rev. *66*, 300–372.

Hohmann, S. (2009). Control of high osmolarity signalling in the yeast *Saccharomyces cerevisiae*. FEBS Lett. *583*, 4025–4029.

Hohmann, S., Krantz, M., and Nordlander, B. (2007). Yeast osmoregulation. Methods Enzymol. *428*, 29–45.

Horie, T., Tatebayashi, K., Yamada, R., and Saito, H. (2008). Phosphorylated Ssk1 prevents unphosphorylated Ssk1 from activating the Ssk2 mitogen-activated protein kinase kinase kinase in the yeast high-osmolarity glycerol osmoregulatory pathway. Mol. Cell. Biol. *28*, 5172–5183.

Huang, H., Maertens, A.M., Hyland, E.M., Dai, J., Norris, A., Boeke, J.D., and Bader, J.S. (2009). HistoneHits: a database for histone mutations and their phenotypes. Genome Res. *19*, 674–681.

Hurley, J.H. (2003). GAF domains: cyclic nucleotides come full circle. Sci. STKE. *2003*, E1.

Ikner, A., and Shiozaki, K. (2005). Yeast signaling pathways in the oxidative stress response. Mutat. Res. *569*, 13–27.

Jacoby, T., Flanagan, H., Faykin, A., Seto, A.G., Mattison, C., and Ota, I. (1997). Two protein-tyrosine phosphatases inactivate the osmotic stress response pathway in yeast by targeting the mitogen-activated protein kinase, Hog1. J Biol. Chem. 272, 17749–17755.

Jiang, C., and Pugh, B.F. (2009). Nucleosome positioning and gene regulation: advances through genomics. Nat. Rev. Genet. *10*, 161–172.

Johnsson, A., Durand-Dubief, M., Xue-Franzen, Y., Ronnerblad, M., Ekwall, K., and Wright, A. (2009). HAT-HDAC interplay modulates global histone H3K14 acetylation in gene-coding regions during stress. EMBO Rep. *10*, 1009–1014.

Jones, T., Federspiel, N.A., Chibana, H., Dungan, J., Kalman, S., Magee, B.B., Newport, G., Thorstenson, Y.R., Agabian, N., Magee, P.T., *et al.* (2004). The diploid genome sequence of *Candida albicans*. Proc. Natl. Acad. Sci. U.S.A. *101*, 7329–7334.

Jung, U.S., and Levin, D.E. (1999). Genome-wide analysis of gene expression regulated by the yeast cell wall integrity signalling pathway. Mol. Microbiol. *34*, 1049–1057.

Jung, U.S., Sobering, A.K., Romeo, M.J., and Levin, D.E. (2002). Regulation of the yeast Rlm1 transcription factor by the Mpk1 cell wall integrity MAP kinase. Mol. Microbiol. *46*, 781–789.

Kanoh, J., Watanabe, Y., Ohsugi, M., Iino, Y., and Yamamoto, M. (1996). *Schizosaccharomyces pombe* gad7+ encodes a phosphoprotein with a bZIP domain, which is required for proper G1 arrest and gene expression under nitrogen starvation. Genes Cells *1*, 391–408.

Kaserer, A.O., Andi, B., Cook, P.F., and West, A.H. (2009). Effects of osmolytes on the SLN1-YPD1-SSK1 phosphorelay system from *Saccharomyces cerevisiae*. Biochemistry *48*, 8044–8050.

Kato, T., Okazaki, K., Murakami, H., Stettler, S., Fantes, P.A., and Okayama, H. (1996). Stress signal, mediated by a Hogl-like MAP kinase, controls sexual development in fission yeast. FEBS Lett. *378*, 207–212.

Kawasaki, Y., Nagao, K., Nakamura, T., and Yanagida, M. (2006). Fission yeast MAP kinase is required for the increased securin–separase interaction that rescues separase mutants under stresses. Cell Cycle *5*, 1831–1839.

Klinkert, B., and Narberhaus, F. (2009). Microbial thermosensors. Cell Mol. Life Sci. *66*, 2661–2676.

Klipp, E., Nordlander, B., Kruger, R., Gennemark, P., and Hohmann, S. (2005). Integrative model of the response of yeast to osmotic shock. Nat. Biotechnol. *23*, 975–982.

Klopf, E., Paskova, L., Sole, C., Mas, G., Petryshyn, A., Posas, F., Wintersberger, U., Ammerer, G., and Schuller, C. (2009). Cooperation between the INO80 complex and histone chaperones determines adaptation of stress gene transcription in the yeast *Saccharomyces cerevisiae*. Mol. Cell. Biol. *29*, 4994–5007.

Koch, C., Moll, T., Neuberg, M., Ahorn, H., and Nasmyth, K. (1993). A role for the transcription factors Mbp1 and Swi4 in progression from G1 to S phase. Science *261*, 1551–1557.

Krantz, M., Becit, E., and Hohmann, S. (2006). Comparative analysis of HOG pathway proteins to generate hypotheses for functional analysis. Curr. Genet. *49*, 152–165.

Krantz, M., Ahmadpour, D., Ottosson, L.G., Warringer, J., Waltermann, C., Nordlander, B., Klipp, E., Blomberg, A., Hohmann, S., and Kitano, H. (2009). Robustness and fragility in the yeast high osmolarity glycerol (HOG) signal-transduction pathway. Mol. Syst. Biol. *5*, 281.

Krems, B., Charizanis, C., and Entian, K.D. (1996). The response regulator-like protein Pos9/Skn7 of *Saccharomyces cerevisiae* is involved in oxidative stress resistance. Curr. Genet. *29*, 327–334.

Kruppa, M., and Calderone, R. (2006). Two-component signal transduction in human fungal pathogens. FEMS Yeast Res. *6*, 149–159.

Kudo, N., Taoka, H., Toda, T., Yoshida, M., and Horinouchi, S. (1999). A novel nuclear export signal sensitive to oxidative stress in the fission yeast transcription factor Pap1. J. Biol. Chem. *274*, 15151–15158.

Kuge, S., Arita, M., Murayama, A., Maeta, K., Izawa, S., Inoue, Y., and Nomoto, A. (2001). Regulation of the yeast Yap1p nuclear export signal is mediated by redox signal-induced reversible disulfide bond formation. Mol. Cell. Biol. *21*, 6139–6150.

Lamson, R.E., Winters, M.J., and Pryciak, P.M. (2002). Cdc42 regulation of kinase activity and signaling by the yeast p21-activated kinase Ste20. Mol. Cell. Biol. *22*, 2939–2951.

Lawrence, C.L., Maekawa, H., Worthington, J.L., Reiter, W., Wilkinson, C.R., and Jones, N. (2007). Regulation of *Schizosaccharomyces pombe* Atf1 protein levels by Sty1-mediated phosphorylation and heterodimerization with Pcr1. J. Biol. Chem. *282*, 5160–5170.

Lee, J., Godon, C., Lagniel, G., Spector, D., Garin, J., Labarre, J., and Toledano, M.B. (1999). Yap1 and Skn7 control two specialized oxidative stress response regulons in yeast. J. Biol. Chem. *274*, 16040–16046.

Lelandais, G., and Devaux, F. (2010). Comparative functional genomics of stress responses in yeasts. OMICS *14*, 501–515.

Li, D., Gurkovska, V., Sheridan, M., Calderone, R., and Chauhan, N. (2004). Studies on the regulation of the two-component histidine kinase gene CHK1 in *Candida albicans* using the heterologous *lacZ* reporter gene. Microbiology *150*, 3305–3313.

Lopez-Aviles, S., Grande, M., Gonzalez, M., Helgesen, A.L., Alemany, V., Sanchez-Piris, M., Bachs, O., Millar, J.B., and Aligue, R. (2005). Inactivation of the Cdc25 phosphatase by the stress-activated Srk1 kinase in fission yeast. Mol. Cell *17*, 49–59.

Lopez-Aviles, S., Lambea, E., Moldon, A., Grande, M., Fajardo, A., Rodriguez-Gabriel, M.A., Hidalgo, E., and Aligue, R. (2008). Activation of Srk1 by the mitogen-activated protein kinase Sty1/Spc1 precedes its dissociation from the kinase and signals its degradation. Mol. Biol. Cell *19*, 1670–1679.

Lopez-Maury, L., Marguerat, S., and Bahler, J. (2008). Tuning gene expression to changing environments: from rapid responses to evolutionary adaptation. Nat. Rev. Genet. *9*, 583–593.

Macia, J., Regot, S., Peeters, T., Conde, N., Sole, R., and Posas, F. (2009). Dynamic signaling in the Hog1 MAPK pathway relies on high basal signal transduction. Sci. Signal. *2*, ra13.

McMillan, J.N., Longtine, M.S., Sia, R.A., Theesfeld, C.L., Bardes, E.S., Pringle, J.R., and Lew, D.J. (1999). The morphogenesis checkpoint in *Saccharomyces cerevisiae*: cell cycle control of Swe1p degradation by Hsl1p and Hsl7p. Mol. Cell. Biol. *19*, 6929–6939.

Maeda, T., Tsai, A.Y., and Saito, H. (1993). Mutations in a protein tyrosine phosphatase gene (PTP2) and a protein serine/threonine phosphatase gene (PTC1) cause a synthetic growth defect in *Saccharomyces cerevisiae*. Mol. Cell. Biol. *13*, 5408–5417.

Maeda, T., Wurgler-Murphy, S.M., and Saito, H. (1994). A two-component system that regulates an osmosensing MAP kinase cascade in yeast. Nature *369*, 242–245.

Maeda, T., Takekawa, M., and Saito, H. (1995). Activation of yeast PBS2 MAPKK by MAPKKKs or by binding of an SH3-containing osmosensor. Science *269*, 554–558.

Mapes, J., and Ota, I.M. (2004). Nbp2 targets the Ptc1-type 2C Ser/Thr phosphatase to the HOG MAPK pathway. EMBO J. *23*, 302–311.

Martinez-Montanes, F., Pascual-Ahuir, A., and Proft, M. (2010). Toward a genomic view of the gene expression program regulated by osmostress in yeast. OMICS *14*, 619–627.

Martinez-Pastor, M.T., Marchler, G., Schuller, C., Marchler-Bauer, A., Ruis, H., and Estruch, F. (1996). The *Saccharomyces cerevisiae* zinc finger proteins Msn2p and Msn4p are required for transcriptional induction through the stress response element (STRE). EMBO J. *15*, 2227–2235.

Mas, G., de Nadal, E., Dechant, R., Rodriguez de la Concepcion M.L., Logie, C., Jimeno-Gonzalez, S., Chavez, S., Ammerer, G., and Posas, F. (2009). Recruitment of a chromatin remodelling complex by the Hog1 MAP kinase to stress genes. EMBO J. *28*, 326–336.

Mattison, C.P., and Ota, I.M. (2000). Two protein tyrosine phosphatases, Ptp2 and Ptp3, modulate the subcellular localization of the Hog1 MAP kinase in yeast. Genes Dev. *14*, 1229–1235.

Menon, V., Li,D., Chauhan, N., Rajnarayanan, R., Dubrovska, A., West, A.H., and Calderone, R. (2006). Functional studies of the Ssk1p response regulator protein of *Candida albicans* as determined by phenotypic analysis of receiver domain point mutants. Mol. Microbiol. *62*, 997–1013.

Mettetal, J.T., Muzzey, D., Gomez-Uribe, C., and van Oudenaarden A. (2008). The frequency dependence of osmo-adaptation in *Saccharomyces cerevisiae*. Science *319*, 482–484.

Migdal, I., Ilina, Y., Tamas, M.J., and Wysocki, R. (2008). Mitogen-activated protein kinase Hog1 mediates adaptation to G1 checkpoint arrest during arsenite and hyperosmotic stress. Eukaryot. Cell 7, 1309–1317.

Millar, J.B., McGowan, C.H., Lenaers, G., Jones, R., and Russell, P. (1991). p80cdc25 mitotic inducer is the tyrosine phosphatase that activates p34cdc2 kinase in fission yeast. EMBO J. *10*, 4301–4309.

Millar, J.B., Buck, V., and Wilkinson, M.G. (1995). Pyp1 and Pyp2 PTPases dephosphorylate an osmosensing MAP kinase controlling cell size at division in fission yeast. Genes Dev. *9*, 2117–2130.

Miller, C., Schwalb, B., Maier, K., Schulz, D., Dumcke, S., Zacher, B., Mayer, A., Sydow, J., Marcinowski, L., Dolken, L., *et al.* (2011). Dynamic transcriptome analysis measures rates of mRNA synthesis and decay in yeast. Mol. Syst. Biol. *7*, 458.

Mitrovich, Q.M., Tuch, B.B., Guthrie, C., and Johnson, A.D. (2007). Computational and experimental approaches double the number of known introns in the pathogenic yeast *Candida albicans*. Genome Res. *17*, 492–502.

Molin, C., Jauhiainen, A., Warringer, J., Nerman, O., and Sunnerhagen, P. (2009). mRNA stability changes precede changes in steady-state mRNA amounts during hyperosmotic stress. RNA. *15*, 600–614.

Molina, M., Cid, V.J., and Martin, H. (2010). Fine regulation of *Saccharomyces cerevisiae* MAPK pathways by post-translational modifications. Yeast *27*, 503–511.

Molina-Navarro, M.M., Castells-Roca, L., Belli, G., Garcia-Martinez, J., Marin-Navarro, J., Moreno, J., Perez-Ortin, J.E., and Herrero, E. (2008). Comprehensive transcriptional analysis of the oxidative response in yeast. J. Biol. Chem. *283*, 17908–17918.

Morgan, B.A., Banks, G.R., Toone, W.M., Raitt, D., Kuge, S., and Johnston, L.H. (1997). The Skn7 response regulator controls gene expression in the oxidative stress response of the budding yeast *Saccharomyces cerevisiae*. EMBO J. *16*, 1035–1044.

Morigasaki, S., Shimada, K., Ikner, A., Yanagida, M., and Shiozaki, K. (2008). Glycolytic enzyme GAPDH promotes peroxide stress signaling through multistep phosphorelay to a MAPK cascade. Mol. Cell *30*, 108–113.

Muzzey, D., Gomez-Uribe, C.A., Mettetal, J.T., and van Oudenaarden A. (2009). A systems-level analysis of perfect adaptation in yeast osmoregulation. Cell *138*, 160–171.

de Nadal, E., and Posas, F. (2010). Multilayered control of gene expression by stress-activated protein kinases. EMBO J. *29*, 4–13.

de Nadal, E., Alepuz, P.M., and Posas, F. (2002). Dealing with osmostress through MAP kinase activation. EMBO Rep. *3*, 735–740.

de Nadal, E., Casadome, L., and Posas, F. (2003). Targeting the MEF2-like transcription factor Smp1 by the stress-activated Hog1 mitogen-activated protein kinase. Mol. Cell. Biol. *23*, 229–237.

de Nadal, E., Zapater, M., Alepuz, P.M., Sumoy, L., Mas, G., and Posas, F. (2004). The MAPK Hog1 recruits Rpd3 histone deacetylase to activate osmoresponsive genes. Nature *427*, 370–374.

de Nadal, E., Real, F.X., and Posas, F. (2007). Mucins, osmosensors in eukaryotic cells? Trends Cell. Biol. *17*, 571–574.

de Nadal, E., Ammerer, G., and Posas, F. (2011). Controlling gene expression in response to stress. Nat. Rev. Genet. *12*, 833–845.

Nadeau, S.I., and Landry, J. (2007). Mechanisms of activation and regulation of the heat shock-sensitive signaling pathways. Adv. Exp. Med. Biol. *594*, 100–113.

Nagahashi, S., Mio, T., Ono, N., Yamada-Okabe, T., Arisawa, M., Bussey, H., and Yamada-Okabe, H. (1998). Isolation of CaSLN1 and CaNIK1, the genes for osmosensing histidine kinase homologues, from the pathogenic fungus *Candida albicans*. Microbiology *144*, 425–432.

Nakagawa, C.W., Yamada, K., and Mutoh, N. (2000). Role of Atf1 and Pap1 in the induction of the catalase gene of fission yeast *Schizosaccharomyces pombe*. J. Biochem. *127*, 233–238.

Nash, P., Tang, X., Orlicky, S., Chen, Q., Gertler, F.B., Mendenhall, M.D., Sicheri, F., Pawson, T., and Tyers, M. (2001). Multisite phosphorylation of a CDK inhibitor sets a threshold for the onset of DNA replication. Nature *414*, 514–521.

Nasmyth, K., and Dirick, L. (1991). The role of SWI4 and SWI6 in the activity of G1 cyclins in yeast. Cell *66*, 995–1013.

Nguyen, A.N., and Shiozaki, K. (1999). Heat-shock-induced activation of stress MAP kinase is regulated by threonine- and tyrosine-specific phosphatases. Genes Dev. *13*, 1653–1663.

Nguyen, A.N., Lee, A., Place, W., and Shiozaki, K. (2000). Multistep phosphorelay proteins transmit oxidative stress signals to the fission yeast stress-activated protein kinase. Mol. Biol. Cell *11*, 1169–1181.

Ni, L., Bruce, C., Hart, C., Leigh-Bell, J., Gelperin, D., Umansky, L., Gerstein, M.B., and Snyder, M. (2009). Dynamic and complex transcription factor binding during an inducible response in yeast. Genes Dev. *23*, 1351–1363.

Nikolaou, E., Agrafioti, I., Stumpf, M., Quinn, J., Stansfield, I., and Brown, A.J. (2009). Phylogenetic diversity of stress signalling pathways in fungi. BMC. Evol. Biol. *9*, 44.

Nollen, E.A., and Morimoto, R.I. (2002). Chaperoning signaling pathways: molecular chaperones as stress-sensing 'heat shock' proteins. J. Cell Sci. *115*, 2809–2816.

O'Rourke, S.M., and Herskowitz, I. (2004). Unique and redundant roles for HOG MAPK pathway components as revealed by whole-genome expression analysis. Mol. Biol. Cell *15*, 532–542.

O'Rourke, S.M., Herskowitz, I., and O'Shea, E.K. (2002). Yeast go the whole HOG for the hyperosmotic response. Trends Genet. *18*, 405–412.

Ohmiya, R., Kato, C., Yamada, H., Aiba, H., and Mizuno, T. (1999). A fission yeast gene (prr1(+)) that encodes a response regulator implicated in oxidative stress response. J. Biochem. *125*, 1061–1066.

Okazaki, S., Naganuma, A., and Kuge, S. (2005). Peroxiredoxin-mediated redox regulation of the nuclear localization of Yap1, a transcription factor in budding yeast. Antioxid. Redox. Signal. *7*, 327–334.

Parker, R., and Sheth, U. (2007). P bodies and the control of mRNA translation and degradation. Mol. Cell *25*, 635–646.

Pascual-Ahuir, A., Struhl, K., and Proft, M. (2006). Genome-wide location analysis of the stress-activated MAP kinase Hog1 in yeast. Methods *40*, 272–278.

Pascual-Ahuir, A., and Proft, M. (2007). The Sch9 kinase is a chromatin-associated transcriptional activator of osmostress-responsive genes. EMBO J. *26*, 3098–3108.

Patterson, J.C., Klimenko, E.S., and Thorner, J. (2010). Single-cell analysis reveals that insulation maintains signaling specificity between two yeast MAPK pathways with common components. Sci. Signal. *3*, ra75.

Pelet, S., Rudolf, F., Nadal-Ribelles, M., de Nadal E., Posas, F., and Peter, M. (2011). Transient activation of the HOG MAPK pathway regulates bimodal gene expression. Science *332*, 732–735.

Pirkkala, L., Nykanen, P., and Sistonen, L. (2001). Roles of the heat shock transcription factors in regulation of the heat shock response and beyond. FASEB J. *15*, 1118–1131.

Pokholok, D.K., Zeitlinger, J., Hannett, N.M., Reynolds, D.B., and Young, R.A. (2006). Activated signal transduction kinases frequently occupy target genes. Science *313*, 533–536.

Posas, F., and Saito, H. (1997). Osmotic activation of the HOG MAPK pathway via Ste11p MAPKKK: scaffold role of Pbs2p MAPKK. Science *276*, 1702–1705.

Posas, F., and Saito, H. (1998). Activation of the yeast SSK2 MAP kinase kinase kinase by the SSK1 two-component response regulator. EMBO J. *17*, 1385–1394.

Posas, F., Wurgler-Murphy, S.M., Maeda, T., Witten, E.A., Thai, T.C., and Saito, H. (1996). Yeast HOG1 MAP kinase cascade is regulated by a multistep phosphorelay mechanism in the SLN1-YPD1-SSK1 'two-component' osmosensor. Cell *86*, 865–875.

Posas, F., Witten, E.A., and Saito, H. (1998). Requirement of STE50 for osmostress-induced activation of the STE11 mitogen-activated protein kinase kinase kinase in the high-osmolarity glycerol response pathway. Mol. Cell. Biol. *18*, 5788–5796.

Posas, F., Chambers, J.R., Heyman, J.A., Hoeffler, J.P., de Nadal, E., and Arino, J. (2000). The transcriptional response of yeast to saline stress. J Biol. Chem. *275*, 17249–17255.

Proft, M., and Struhl, K. (2002). Hog1 kinase converts the Sko1-Cyc8-Tup1 repressor complex into an activator that recruits SAGA and SWI/SNF in response to osmotic stress. Mol. Cell *9*, 1307–1317.

Proft, M., and Struhl, K. (2004). MAP kinase-mediated stress relief that precedes and regulates the timing of transcriptional induction. Cell *118*, 351–361.

Proft, M., Pascual-Ahuir, A., de, N.E., Arino, J., Serrano, R., and Posas, F. (2001). Regulation of the Sko1 transcriptional repressor by the Hog1 MAP kinase in response to osmotic stress. EMBO J. *20*, 1123–1133.

Proft, M., Mas, G., de, N.E., Vendrell, A., Noriega, N., Struhl, K., and Posas, F. (2006). The stress-activated Hog1 kinase is a selective transcriptional elongation factor for genes responding to osmotic stress. Mol. Cell *23*, 241–250.

Quinn, J., Findlay, V.J., Dawson, K., Millar, J.B., Jones, N., Morgan, B.A., and Toone, W.M. (2002). Distinct regulatory proteins control the graded transcriptional response to increasing H(2)O(2) levels in fission yeast *Schizosaccharomyces pombe*. Mol. Biol. Cell *13*, 805–816.

Quinn, J., Malakasi, P., Smith, D.A., Cheetham, J., Buck, V., Millar, J.B., and Morgan, B.A. (2011). Two-component mediated peroxide sensing and signal transduction in fission yeast. Antioxid. Redox. Signal. *15*, 153–165.

Quintales, L., Sanchez, M., and Antequera, F. (2010). Analysis of DNA strand-specific differential expression with high density tiling microarrays. BMC. Bioinformatics. *11*, 136.

Raitt, D.C., Posas, F., and Saito, H. (2000). Yeast Cdc42 GTPase and Ste20 PAK-like kinase regulate Sho1-dependent activation of the Hog1 MAPK pathway. EMBO J. *19*, 4623–4631.

Rauceo, J.M., Blankenship, J.R., Fanning, S., Hamaker, J.J., Deneault, J.S., Smith, F.J., Nantel, A., and Mitchell, A.P. (2008). Regulation of the *Candida albicans* cell wall damage response by transcription factor Sko1 and PAS kinase Psk1. Mol. Biol. Cell *19*, 2741–2751.

Reiser, V., Ruis, H., and Ammerer, G. (1999). Kinase activity-dependent nuclear export opposes stress-induced nuclear accumulation and retention of Hog1 mitogen-activated protein kinase in the budding yeast *Saccharomyces cerevisiae*. Mol. Biol. Cell *10*, 1147–1161.

Reiser, V., Salah, S.M., and Ammerer, G. (2000). Polarized localization of yeast Pbs2 depends on osmostress, the membrane protein Sho1 and Cdc42. Nat. Cell Biol. *2*, 620–627.

Reiser, V., Raitt, D.C., and Saito, H. (2003). Yeast osmosensor Sln1 and plant cytokinin receptor Cre1 respond to changes in turgor pressure. J. Cell Biol. *161*, 1035–1040.

Reiser, V., D'Aquino, K.E., Ee, L.S., and Amon, A. (2006). The stress-activated mitogen-activated protein kinase signaling cascade promotes exit from mitosis. Mol. Biol. Cell *17*, 3136–3146.

Reiter, W., Watt, S., Dawson, K., Lawrence, C.L., Bahler, J., Jones, N., and Wilkinson, C.R. (2008). Fission yeast MAP kinase Sty1 is recruited to stress-induced genes. J. Biol. Chem. *283*, 9945–9956.

Rhind, N., Chen, Z., Yassour, M., Thompson, D.A., Haas, B.J., Habib, N., Wapinski, I., Roy, S., Lin, M.F., Heiman, D.I., *et al.* (2011). Comparative functional genomics of the fission yeasts. Science *332*, 930–936.

Richter, K., Haslbeck, M., and Buchner, J. (2010). The heat shock response: life on the verge of death. Mol. Cell *40*, 253–266.

Riezman, H. (2004). Why do cells require heat shock proteins to survive heat stress? Cell Cycle. *3*, 61–63.

Rodriguez-Gabriel, M.A., Burns, G., McDonald, W.H., Martin, V., Yates, J.R., III, Bahler, J., and Russell, P. (2003). RNA-binding protein Csx1 mediates global control of gene expression in response to oxidative stress. EMBO J. *22*, 6256–6266.

Roman, E., Nombela, C., and Pla, J. (2005). The Sho1 adaptor protein links oxidative stress to morphogenesis and cell wall biosynthesis in the fungal pathogen *Candida albicans*. Mol. Cell. Biol. *25*, 10611–10627.

Roman, E., Arana, D.M., Nombela, C., Alonso-Monge, R., and Pla, J. (2007). MAP kinase pathways as regulators of fungal virulence. Trends Microbiol. *15*, 181–190.

Roman, E., Cottier, F., Ernst, J.F., and Pla, J. (2009). Msb2 signaling mucin controls activation of Cek1 mitogen-activated protein kinase in *Candida albicans*. Eukaryot. Cell *8*, 1235–1249.

Romero-Santacreu, L., Moreno, J., Perez-Ortin, J.E., and Alepuz, P. (2009). Specific and global regulation of mRNA stability during osmotic stress in *Saccharomyces cerevisiae*. RNA. *15*, 1110–1120.

Ruiz-Roig, C., Vieitez, C., Posas, F., and de Nadal E. (2010). The Rpd3L HDAC complex is essential for the heat stress response in yeast. Mol. Microbiol. *76*, 1049–1062.

Russell, P., and Nurse, P. (1986). cdc25+ functions as an inducer in the mitotic control of fission yeast. Cell *45*, 145–153.

Saito, H. (2001). Histidine phosphorylation and two-component signaling in eukaryotic cells. Chem. Rev. *101*, 2497–2509.

Saito, H. (2010). Regulation of cross-talk in yeast MAPK signaling pathways. Curr. Opin. Microbiol. *13*, 677–683.

Saito, H., and Tatebayashi, K. (2004). Regulation of the osmoregulatory HOG MAPK cascade in yeast. J. Biochem. (Tokyo) *136*, 267–272.

Samejima, I., Mackie, S., and Fantes, P.A. (1997). Multiple modes of activation of the stress-responsive MAP kinase pathway in fission yeast. EMBO J. *16*, 6162–6170.

Samejima, I., Mackie, S., Warbrick, E., Weisman, R., and Fantes, P.A. (1998). The fission yeast mitotic regulator win1+ encodes an MAP kinase kinase kinase that phosphorylates and activates Wis1 MAP kinase kinase in response to high osmolarity. Mol. Biol. Cell *9*, 2325–2335.

San Jose, C., Monge, R.A., Perez-Diaz, R., Pla, J., and Nombela, C. (1996). The mitogen-activated protein kinase homolog HOG1 gene controls glycerol accumulation in the pathogenic fungus *Candida albicans*. J. Bacteriol. *178*, 5850–5852.

Sanso, M., Gogol, M., Ayte, J., Seidel, C., and Hidalgo, E. (2008). Transcription factors Pcr1 and Atf1 have distinct roles in stress- and Sty1-dependent gene regulation. Eukaryot. Cell *7*, 826–835.

Sanso, M., Vargas-Perez, I., Quintales, L., Antequera, F., Ayte, J., and Hidalgo, E. (2011). Gcn5 facilitates Pol II progression, rather than recruitment to nucleosome-depleted stress promoters, in *Schizosaccharomyces pombe*. Nucleic Acids Res. *39*, 6369–6379.

Santos, J.L., and Shiozaki, K. (2001). Fungal histidine kinases. Sci. STKE. *2001*, re1.

Sato, N., Kawahara, H., Toh-e A., and Maeda, T. (2003). Phosphorelay-regulated degradation of the yeast Ssk1p response regulator by the ubiquitin-proteasome system. Mol. Cell. Biol. *23*, 6662–6671.

Schmitt, A.P., and McEntee, K. (1996). Msn2p, a zinc finger DNA-binding protein, is the transcriptional activator of the multistress response in *Saccharomyces cerevisiae*. Proc. Natl. Acad. Sci. U.S.A. *93*, 5777–5782.

Schwob, E., Bohm, T., Mendenhall, M.D., and Nasmyth, K. (1994). The B-type cyclin kinase inhibitor p40SIC1 controls the G1 to S transition in S. cerevisiae. Cell *79*, 233–244.

Seet, B.T., and Pawson, T. (2004). MAPK signaling: Sho business. Curr. Biol. *14*, R708–R710.

Sheikh-Hamad, D., and Gustin, M.C. (2004). MAP kinases and the adaptive response to hypertonicity: functional preservation from yeast to mammals. Am. J. Physiol. Renal. Physiol. *287*, F1102–F1110.

Shieh, J.C., Wilkinson, M.G., Buck, V., Morgan, B.A., Makino, K., and Millar, J.B. (1997). The Mcs4 response regulator coordinately controls the stress-activated Wak1-Wis1-Sty1 MAP kinase pathway and fission yeast cell cycle. Genes Dev. *11*, 1008–1022.

Shieh, J.C., Wilkinson, M.G., and Millar, J.B. (1998). The Win1 mitotic regulator is a component of the fission yeast stress-activated Sty1 MAPK pathway. Mol. Biol. Cell *9*, 311–322.

Shiozaki, K. (1998). Negative regulation of the stress-activated protein kinase pathway by protein phosphatases. Tanpakushitsu Kakusan Koso *43*, 1055–1061.

Shiozaki, K., and Russell, P. (1995). Cell-cycle control linked to extracellular environment by MAP kinase pathway in fission yeast. Nature *378*, 739–743.

Shiozaki, K., and Russell, P. (1996). Conjugation, meiosis, and the osmotic stress response are regulated by Spc1 kinase through Atf1 transcription factor in fission yeast. Genes Dev. *10*, 2276–2288.

Shiozaki, K., Shiozaki, M., and Russell, P. (1997). Mcs4 mitotic catastrophe suppressor regulates the fission yeast cell cycle through the Wik1-Wis1-Spc1 kinase cascade. Mol. Biol. Cell *8*, 409–419.

Shiozaki, K., Shiozaki, M., and Russell, P. (1998). Heat stress activates fission yeast Spc1/StyI MAPK by a MEKK-independent mechanism. Mol. Biol. Cell *9*, 1339–1349.

Shivaswamy, S., and Iyer, V.R. (2008). Stress-dependent dynamics of global chromatin remodeling in yeast: dual role for SWI/SNF in the heat shock stress response. Mol. Cell. Biol. *28*, 2221–2234.

Singh, P., Chauhan, N., Ghosh, A., Dixon, F., and Calderone, R. (2004). SKN7 of *Candida albicans*: mutant construction and phenotype analysis. Infect. Immun. *72*, 2390–2394.

Smith, A., Ward, M.P., and Garrett, S. (1998). Yeast PKA represses Msn2p/Msn4p-dependent gene expression to regulate growth, stress response and glycogen accumulation. EMBO J. *17*, 3556–3564.

Smith, D.A., Morgan, B.A., and Quinn, J. (2010). Stress signalling to fungal stress-activated protein kinase pathways. FEMS Microbiol. Lett. *306*, 1–8.

Smith, D.A., Toone, W.M., Chen, D., Bahler, J., Jones, N., Morgan, B.A., and Quinn, J. (2002). The Srk1 protein kinase is a target for the Sty1 stress-activated MAPK in fission yeast. J. Biol. Chem. *277*, 33411–33421.

Smith, D.A., Nicholls, S., Morgan, B.A., Brown, A.J., and Quinn, J. (2004). A conserved stress-activated protein kinase regulates a core stress response in the human pathogen *Candida albicans*. Mol. Biol. Cell *15*, 4179–4190.

Sole, C., Nadal-Ribelles, M., Kraft, C., Peter, M., Posas, F., and de Nadal E. (2011). Control of Ubp3 ubiquitin protease activity by the Hog1 SAPK modulates transcription upon osmostress. EMBO J. *30*, 3274–3284.

Szopinska, A., and Morsomme, P. (2010). Quantitative proteomic approaches and their application in the study of yeast stress responses. OMICS. *14*, 639–649.

Tatebayashi, K., Yamamoto, K., Tanaka, K., Tomida, T., Maruoka, T., Kasukawa, E., and Saito, H. (2006). Adaptor functions of Cdc42, Ste50, and Sho1 in the yeast osmoregulatory HOG MAPK pathway. EMBO J. *25*, 3033–3044.

Tatebayashi, K., Tanaka, K., Yang, H.Y., Yamamoto, K., Matsushita, Y., Tomida, T., Imai, M., and Saito, H. (2007). Transmembrane mucins Hkr1 and Msb2 are putative osmosensors in the SHO1 branch of yeast HOG pathway. EMBO J. *26*, 3521–3533.

Taylor, B.L., and Zhulin, I.B. (1999). PAS domains: internal sensors of oxygen, redox potential, and light. Microbiol. Mol. Biol. Rev. *63*, 479–506.

Teige, M., Scheikl, E., Reiser, V., Ruis, H., and Ammerer, G. (2001). Rck2, a member of the calmodulin-protein kinase family, links protein synthesis to high osmolarity MAP kinase signaling in budding yeast. Proc. Natl. Acad. Sci. U.S.A. *98*, 5625–5630.

Toledano, M.B., Delaunay, A., Monceau, L., and Tacnet, F. (2004). Microbial H2O2 sensors as archetypical redox signaling modules. Trends Biochem. Sci. *29*, 351–357.

Toone, W.M., Kuge, S., Samuels, M., Morgan, B.A., Toda, T., and Jones, N. (1998). Regulation of the fission yeast transcription factor Pap1 by oxidative stress: requirement for the nuclear export factor Crm1 (Exportin) and the stress-activated MAP kinase Sty1/Spc1. Genes Dev. *12*, 1453–1463.

Truckses, D.M., Bloomekatz, J.E., and Thorner, J. (2006). The RA domain of Ste50 adaptor protein is required for delivery of Ste11 to the plasma membrane in the filamentous growth signaling pathway of the yeast *Saccharomyces cerevisiae*. Mol. Cell. Biol. *26*, 912–928.

Tyers, M., and Jorgensen, P. (2000). Proteolysis and the cell cycle: with this RING I do thee destroy. Curr. Opin. Genet. Dev. *10*, 54–64.

Uesono, Y., and Toh, E. (2002). Transient inhibition of translation initiation by osmotic stress. J Biol. Chem. *277*, 13848–13855.

Uffenbeck, S.R., and Krebs, J.E. (2006). The role of chromatin structure in regulating stress-induced transcription in *Saccharomyces cerevisiae*. Biochem. Cell Biol. *84*, 477–489.

Umeyama, T., Kaneko, A., Nagai, Y., Hanaoka, N., Tanabe, K., Takano, Y., Niimi, M., and Uehara, Y. (2005). *Candida albicans* protein kinase CaHsl1p regulates cell elongation and virulence. Mol. Microbiol. *55*, 381–395.

Urao, T., Yamaguchi-Shinozaki, K., and Shinozaki, K. (2001). Plant histidine kinases: an emerging picture of two-component signal transduction in hormone and environmental responses. Sci. STKE. *2001*, re18.

Vazquez-Torres, A., and Balish, E. (1997). Macrophages in resistance to candidiasis. Microbiol. Mol. Biol. Rev. *61*, 170–192.

Veal, E.A., Ross, S.J., Malakasi, P., Peacock, E., and Morgan, B.A. (2003). Ybp1 is required for the hydrogen peroxide-induced oxidation of the Yap1 transcription factor. J. Biol. Chem. *278*, 30896–30904.

Vendrell, A., Martínez-Pastor, M., González-Novo, A., Pascual-Ahuir, A., Sinclair, D.A., Proft, M., and Posas, F. (2011). Sir2 histone deacetylase prevents programmed cell death caused by sustained activation of the Hog1 SAPK. EMBO Reports. *12*, 1062–1068.

Venters, B.J., Wachi, S., Mavrich, T.N., Andersen, B.E., Jena, P., Sinnamon, A.J., Jain, P., Rolleri, N.S., Jiang, C., Hemeryck-Walsh, C., *et al.* (2011). A comprehensive genomic binding map of gene and chromatin regulatory proteins in Saccharomyces. Mol. Cell *41*, 480–492.

Vilella, F., Herrero, E., Torres, J., and de la Torre-Ruiz M.A. (2005). Pkc1 and the upstream elements of the cell integrity pathway in *Saccharomyces cerevisiae*, Rom2 and Mtl1, are required for cellular responses to oxidative stress. J. Biol. Chem. *280*, 9149–9159.

Vivancos, A.P., Castillo, E.A., Jones, N., Ayte, J., and Hidalgo, E. (2004). Activation of the redox sensor Pap1 by hydrogen peroxide requires modulation of the intracellular oxidant concentration. Mol. Microbiol. *52*, 1427–1435.

Vivancos, A.P., Castillo, E.A., Biteau, B., Nicot, C., Ayte, J., Toledano, M.B., and Hidalgo, E. (2005). A cysteine-sulfinic acid in peroxiredoxin regulates H2O2-sensing by the antioxidant Pap1 pathway. Proc. Natl. Acad. Sci. U.S.A. *102*, 8875–8880.

Vivancos, A.P., Jara, M., Zuin, A., Sanso, M., and Hidalgo, E. (2006). Oxidative stress in *Schizosaccharomyces pombe*: different H2O2 levels, different response pathways. Mol. Genet. Genomics *276*, 495–502.

Wahls, W.P., and Smith, G.R. (1994). A heteromeric protein that binds to a meiotic homologous recombination hot spot: correlation of binding and hot spot activity. Genes Dev. *8*, 1693–1702.

Warmka, J., Hanneman, J., Lee, J., Amin, D., and Ota, I. (2001). Ptc1, a type 2C Ser/Thr phosphatase, inactivates the HOG pathway by dephosphorylating the mitogen-activated protein kinase Hog1. Mol. Cell. Biol. *21*, 51–60.

Warringer, J., Hult, M., Regot, S., Posas, F., and Sunnerhagen, P. (2010). The HOG pathway dictates the short-term translational response after hyperosmotic shock. Mol. Biol. Cell *21*, 3080–3092.

Watanabe, Y., and Yamamoto, M. (1996). *Schizosaccharomyces pombe* pcr1+ encodes a CREB/ATF protein involved in regulation of gene expression for sexual development. Mol. Cell. Biol. *16*, 704–711.

Weiss, A., Delproposto, J., and Giroux, C.N. (2004). High-throughput phenotypic profiling of gene–environment interactions by quantitative growth curve analysis in *Saccharomyces cerevisiae*. Anal. Biochem. *327*, 23–34.

Westfall, P.J., Ballon, D.R., and Thorner, J. (2004). When the stress of your environment makes you go HOG wild. Science *306*, 1511–1512.

Westfall, P.J., Patterson, J.C., Chen, R.E., and Thorner, J. (2008). Stress resistance and signal fidelity independent of nuclear MAPK function. Proc. Natl. Acad. Sci. U.S.A. *105*, 12212–12217.

Wilkinson, M.G., Samuels, M., Takeda, T., Toone, W.M., Shieh, J.C., Toda, T., Millar, J.B., and Jones, N. (1996). The Atf1 transcription factor is a target for the Sty1 stress-activated MAP kinase pathway in fission yeast. Genes Dev. *10*, 2289–2301.

Winkler, A., Arkind, C., Mattison, C.P., Burkholder, A., Knoche, K., and Ota, I. (2002). Heat stress activates the yeast high-osmolarity glycerol mitogen-activated protein kinase pathway, and protein tyrosine phosphatases are essential under heat stress. Eukaryot. Cell *1*, 163–173.

Wood, M.J., Storz, G., and Tjandra, N. (2004). Structural basis for redox regulation of Yap1 transcription factor localization. Nature *430*, 917–921.

Wood, V., Gwilliam, R., Rajandream, M.A., Lyne, M., Lyne, R., Stewart, A., Sgouros, J., Peat, N., Hayles, J., Baker, S., *et al.* (2002). The genome sequence of *Schizosaccharomyces pombe*. Nature *415*, 871–880.

Wu, C., Leberer, E., Thomas, D.Y., and Whiteway, M. (1999). Functional characterization of the interaction of Ste50p with Ste11p MAPKKK in *Saccharomyces cerevisiae*. Mol. Biol. Cell *10*, 2425–2440.

Wu, C., Jansen, G., Zhang, J., Thomas, D.Y., and Whiteway, M. (2006). Adaptor protein Ste50p links the Ste11p MEKK to the HOG pathway through plasma membrane association. Genes Dev. *20*, 734–746.

Wurgler-Murphy, S.M., Maeda, T., Witten, E.A., and Saito, H. (1997). Regulation of the *Saccharomyces cerevisiae* HOG1 mitogen-activated protein kinase by the PTP2 and PTP3 protein tyrosine phosphatases. Mol. Cell. Biol. *17*, 1289–1297.

Yaakov, G., Bell, M., Hohmann, S., and Engelberg, D. (2003). Combination of two activating mutations in one HOG1 gene forms hyperactive enzymes that induce growth arrest. Mol. Cell. Biol. *23*, 4826–4840.

Yaakov, G., Duch, A., Garcia-Rubio, M., Clotet, J., Jimenez, J., Aguilera, A., and Posas, F. (2009). The stress-activated protein kinase Hog1 mediates S phase delay in response to osmostress. Mol. Biol. Cell *20*, 3572–3582.

Yale, J., and Bohnert, H.J. (2001). Transcript expression in *Saccharomyces cerevisiae* at high salinity. J Biol. Chem. *276*, 15996–16007.

Yamamoto, A., Mizukami, Y., and Sakurai, H. (2005). Identification of a novel class of target genes and a novel type of binding sequence of heat shock transcription factor in *Saccharomyces cerevisiae*. J. Biol. Chem. *280*, 11911–11919.

Yamamoto, A., Ueda, J., Yamamoto, N., Hashikawa, N., and Sakurai, H. (2007). Role of heat shock transcription factor in *Saccharomyces cerevisiae* oxidative stress response. Eukaryot. Cell *6*, 1373–1379.

Yamamoto, K., Tatebayashi, K., Tanaka, K., and Saito, H. (2010). Dynamic control of yeast MAP kinase network by induced association and dissociation between the Ste50 scaffold and the Opy2 membrane anchor. Mol. Cell *40*, 87–98.

Yan, C., Lee, L.H., and Davis, L.I. (1998). Crm1p mediates regulated nuclear export of a yeast AP-1-like transcription factor. EMBO J. *17*, 7416–7429.

Yosef, N., and Regev, A. (2011). Impulse control: temporal dynamics in gene transcription. Cell *144*, 886–896.

Zanton, S.J., and Pugh, B.F. (2004). Changes in genomewide occupancy of core transcriptional regulators during heat stress. Proc. Natl. Acad. Sci. U.S.A. *101*, 16843–16848.

Zanton, S.J., and Pugh, B.F. (2006). Full and partial genome-wide assembly and disassembly of the yeast transcription machinery in response to heat shock. Genes Dev. *20*, 2250–2265.

Zapater, M., Clotet, J., Escote, X., and Posas, F. (2005). Control of cell cycle progression by the stress-activated Hog1 MAPK. Cell Cycle *4*, 6–7.

Zapater, M., Sohrmann, M., Peter, M., Posas, F., and de Nadal E. (2007). Selective requirement for SAGA in Hog1-mediated gene expression depending on the severity of the external osmostress conditions. Mol. Cell. Biol. *27*, 3900–3910.

Zarrinpar, A., Bhattacharyya, R.P., Nittler, M.P., and Lim, W.A. (2004). Sho1 and Pbs2 act as coscaffolds linking components in the yeast high osmolarity MAP kinase pathway. Mol. Cell *14*, 825–832.

Zhang, X., De, M.M., Coleman, S.T., Sanglard, D., and Moye-Rowley, W.S. (2000). Analysis of the oxidative stress regulation of the *Candida albicans* transcription factor, Cap1p. Mol. Microbiol. *36*, 618–629.

Zhao, J., Herrera-Diaz, J., and Gross, D.S. (2005). Domain-wide displacement of histones by activated heat shock factor occurs independently of Swi/Snf and is not correlated with RNA polymerase II density. Mol. Cell. Biol. *25*, 8985–8999.

12 Stress Response in the Human Malaria Parasite *Plasmodium falciparum*

Sylke Müller and Christian Doerig

Abstract

The life cycle of malaria parasites comprises a complex succession of developmental stages occurring in two different hosts, the human patient and the mosquito vector. In both hosts, the parasite encounters hostile environments and must deal with stresses such as immune responses, sharp temperature shifts and exposure to drugs; partly because of large-scale haemoglobin degradation in the infected erythrocyte and resulting haem release, oxidative stress is another challenge that the parasite must face. In contrast to other eukaryotes where stress response is largely mediated through a well-defined and robust transcriptional response, it appears that malaria parasites opted for a different strategy. In line with the largely fixed transcriptional programme that characterizes the progression of the organisms through their life cycle stages, the transcriptional response to several stresses (such as drug treatments) consists primarily of low-amplitude, genome-wide changes of transcript abundance. However, recent findings suggest that specific transcriptome adaptations, that affect selected aspects of the parasites' physiology, also occur. Overall, the absence in the parasite's kinome of classical stress response mediators such as SAPKs/JNKs, together with the relative scarcity in transcription factors, suggest a low level of flexibility of the parasite in implementing classical eukaryotic stress response pathways. Post-transcriptional mechanisms are expected to play crucial roles in stress response in *Plasmodium*, as exemplified by the demonstrated involvement of an eIF2α kinases in response to starvation stress.

Introduction

Five species of intracellular protozoan parasites of the genus *Plasmodium* infect humans (*P. falciparum, P. vivax, P. ovale, P. malariae* and *P. knowlesi*) and cause malaria, one of the most devastating tropical diseases, responsible for more than 1 million deaths annually (Snow *et al.*, 2005; Hay *et al.*, 2009). *P. falciparum* causes most of the fatalities and the victims are mainly children under the age of 5 and pregnant women in sub-Saharan Africa. Despite the availability of several efficient chemotherapies, it is imperative to identify new ways for disease intervention because the parasites develop resistance against these existing drugs and a protective and curative vaccine is still elusive (Hay *et al.*, 2009; van Eijk *et al.*, 2011).

Human malaria parasites undergo a complex life cycle involving two hosts, the *Anopheles* mosquito and the human (http://www.cdc.gov/malaria/about/biology/index.html). Infection of the human host begins with the exo-erythrocytic schizogonic cycle taking place in the liver after the parasite enters the human body through the infective bite of a female mosquito. The infective sporozoites invade hepatocytes where they multiply into thousands of merozoites that are released into the bloodstream. In the second part, the merozoites infect erythrocytes and multiply within 48 or 72 hours into 8 to 32 daughter merozoites (depending on the human *Plasmodium* species) that are released to infect fresh erythrocytes. This periodic release of merozoites causes the typical fever bouts observed in malaria patients. The intra-erythrocytic part of the life cycle causes the pathology of the disease and can be lethal. The third part of parasite development is the sexual cycle, which is initiated by the generation of male and female gametocytes in the human host, and which is triggered by cues that are not yet fully understood. Unlike asexual parasites, the gametocytes are infective to the *Anopheles* vector and therefore essential for the transmission of the parasites from human to vector. Once taken up by the mosquito, the gametocytes develop into the male and female gametes and, upon fertilization, the zygote matures into a motile form called the ookinete. The ookinete migrates through the midgut wall into the basal lamina of the mosquito's midgut, where it forms the oocyst, the site of multiple sporogonic mitotic divisions and development of sporozoites. Sporozoites migrate into the salivary gland where they become infective and are primed to be injected into a new mammalian host.

Clearly, through this complex life cycle the parasites encounter considerable environmental variation and are exposed to stresses and challenges that must be met to allow development and survival. The genomes of several species of malaria parasites have been sequenced over the past decade (Carlton *et al.*, 2002; Gardner *et al.*, 2002), allowing comparative genomics to identify species-specific adaptations (Bozdech *et al.*, 2008; Mackinnon *et al.*, 2009). These studies also offer an excellent basis for the identification of general cross-species characteristics that may expose features necessary for the fundamental requirements of *Plasmodium* to survive the changes they are exposed to during their development in the human and mosquito hosts.

The genome of *P. falciparum* encodes about 5300 genes, and the parasites have lost a number of metabolic abilities. For instance, they rely on amino acid supply from their host because they lack the ability to generate most amino acids *de novo* (Gardner *et al.*, 2002). On the other hand they have expanded families of proteins to allow effective interactions with their hosts, such as large families of proteins exported to the surface of the host erythrocyte (e.g. the rifins, or the PfEMP1 proteins encoded by the *var* genes), of gametocytes (e.g. the PfCCP family), and they possess highly specialized organelles geared towards the complex process of host cell invasion (Gardner *et al.*, 2002).

Model unicellular eukaryotes like yeast also have to adapt to fluctuations in the types and availability of nutrients as well as in temperature, pH and osmolarity of their environment. Drastic changes in the internal milieu must occur to allow survival under such fluctuations in environmental conditions. In general, yeast responds to environmental shifts with massive and rapid genome-wide changes in gene expression, the programme of which is specific to the growth conditions or stress applied (although overlaps between the responses to individual stresses were observed) (Jelinsky and Samson, 1999; Gasch and Werner-Washburne, 2002; Berry and Gasch, 2008) (see Chapter 11). The definition of transcriptomic response networks provided new insights into the way these organisms cope with various stresses, as well as information about mechanisms of adaptation to environmental and nutritional fluctuations, and about the mode of action of drugs and the mechanisms of resistance (Berry and Gasch, 2008; Teixeira *et al.*, 2011). In addition it was shown that for instance yeast and *Mycobacterium tuberculosis* (Wilson *et al.*, 1999; Bammert and Fostel, 2000) respond to exposure to antimetabolites (inhibitors and drugs) by up- or down-regulating target-related transcripts, suggesting that genome wide transcriptome analyses provide an excellent tool to reveal mechanisms of drug action and drug resistance.

Post-genomic approaches such as global transcriptome analyses initially indicated that, in contrast to the largely transcriptional stress response in yeast and other eukaryotes, malaria parasites follow the rigid implementation of a largely fixed transcriptional programme, at least during their intra-erythrocytic cycle (Bozdech *et al.*, 2003a,b). In agreement with this, it was found that externally applied stresses only resulted in low amplitude global transcriptome changes (Bozdech *et al.*, 2003a,b).

This chapter will summarize the current knowledge about the responses of the malaria parasites *Plasmodium falciparum* to a variety of stresses focusing, on drug treatments, changes in temperature and elevation of oxidative stress. It will also discuss the mechanisms that are known to affect gene expression profiles in the parasites to date, although relevant knowledge is still sparse and a clear understanding of how exactly malaria parasites cope with the environmental variations they encounter during their complex life cycle remains elusive.

Response to drugs

Chloroquine

Chloroquine, a quinoline compound, was used as the mainstay drug to treat malaria for almost four decades (Jensen and Mehlhorn, 2009). The increasing resistance against chloroquine had an enormous impact on public health in endemic countries. The drug acts on the development of the intra-erythrocytic parasites and it is widely accepted that the primary mode of action of chloroquine is the prevention of the biomineralization of toxic haem released upon the digestion of host cell haemoglobin by the parasite to inert haemozoin (Wellems and Plowe, 2001). The advent of genome-wide gene expression approaches lent hope that new insights into chloroquine action and the development of parasite resistance to the drug could be gained by comparative transcriptomics. However, the information obtained through transcriptome analyses of responses to chloroquine treatment provided rather disappointing results (Gunasekera *et al.*, 2003, 2007). Exposing *P. falciparum* to chloroquine did not result in the expected high-amplitude genome-wide transcriptomic response; instead, it triggered a broad, genome-wide but low-amplitude response with only a small number of transcripts significantly altered. These included the up-regulation of genes encoding antioxidant and heat shock proteins, while genes encoding mitochondrial proteins were down-regulated (Gunasekera *et al.*, 2003, 2007). These data suggested that chloroquine treatment increases the antioxidative capacity and availability of chaperones, possibly due to the fact that chloroquine treatment ultimately results in increased generation of reactive oxygen species (ROS), while mitochondrial function may be negatively affected by drug action. This was corroborated by a focused approach investigating, in chloroquine-sensitive and -resistant *P. falciparum* lines, the transcription levels of genes encoding antioxidant proteins as well as two ABC transporters known to be involved in drug expulsion. This study revealed that the amplitude of transcriptional changes was higher in the sensitive than in the resistant line, suggesting that the drug action is less pronounced in the chloroquine-resistant line, presumably because it is removed from the site of action by the mutated chloroquine resistant transporter (Sanchez *et al.*, 2010). The most pronounced up-regulation was detected for the genes encoding superoxide dismutase 1 (PF08_0071; gene identifier in the interactive database PlasmoDB, www.plasmodb.org), glucose-6-phosphate dehydrogenase-6-phosphogluconolactonase (PF14_0511) and glutathione peroxidase (PFL0595c), a protein utilizing thioredoxin as reducing partner rather than glutathione (Nogueira *et al.*, 2010; Sztajer *et al.*, 2001). In addition the gene encoding the multidrug resistance transporter MDR2 (PF14_0455) was up-regulated in the chloroquine-sensitive parasite line in response to drug treatment. The data support the hypothesis that a consequence of chloroquine treatment is the increase of oxidative stress in parasitized erythrocytes, which in turn triggers the up-regulation of the transcription of genes encoding antioxidant proteins. This was further substantiated by a redox proteomics approach that showed a significant increase in protein carbonylation, an indicator of oxidative stress, after exposure to chloroquine (Bautista *et al.*, 2008). Higher levels of carbonylated proteins were detected in mature stages of the parasites than ring stage parasites, in agreement with the current hypothesis that chloroquine inhibits the biomineralization of toxic haem, whose accumulation during parasite maturation results in increasing oxidative stress (Bautista *et al.*, 2008). Since *Plasmodium* are most metabolically active and digest most of the host cell haemoglobin from the early trophozoite stage onwards, it can be inferred that the effect of chloroquine on protein oxidation may be most pronounced during this time of parasite development (Bautista *et al.*, 2008). Accordingly, molecular chaperones, peptidases and glycolytic enzymes appeared to be most susceptible to oxidative damage. This is in agreement with reports that the same classes of proteins are also affected by oxidative damage in other organisms (Cabiscol *et al.*, 2000a,b; Reverter-Branchat *et al.*, 2004). Overall these data allow the interpretation that the response to chloroquine may be of a broad and generalized nature rather than defined and specific to this particular drug treatment; however it cannot be excluded that some of the transcriptional changes observed are specific to treatment with chloroquine. A more recent study reported that the exposure to chloroquine generates highly reproducible transcriptional responses that are dose-dependent and it was proposed that the low amplitude and the broad and genome-wide response may be accompanied by specific transcriptome adaptations in response to exposure to the drug that affect specific aspects of the parasites' physiology (Hu *et al.*, 2010).

One aspect of physiological adaptation that has been analysed in response to chloroquine treatment is the differentiation from schizogonic blood stage parasite to gametocyte. A known factor that initiates development of gametocytes *in vitro* is the exposure of parasites to stresses (Buckling *et al.*, 1997, 1999; Reece *et al.*, 2010). This has been interpreted as a diversion of metabolic activity from replication to reproduction and thus to lead to an increased generation of gametocytes in response to stresses to maximize transmission between the human and *Anopheles* vector (Taylor and Read, 1997; Read and Taylor, 2001). However, most of the studies performed applied high drug pressure for short times, which is not actually what the parasite experiences *in vivo*: chloroquine has a relatively long half-life and therefore remains present in the treated individual at low levels for a relatively long period (Reece *et al.*, 2010). Modelling the responses that were obtained when chloroquine-sensitive and chloroquine-resistant *P. falciparum* were exposed to sublethal concentrations of chloroquine for up to 11 days, indicated that chloroquine-sensitive parasites adopt reproductive restraint and therefore have a lower tendency to generate gametocytes when exposed to low doses of drug for extended periods of time. Parasites resistant to the drug significantly increased their gametocyte conversion rates in response to the drug (Reece *et al.*, 2010). This finding is in contrast to earlier studies, proposing that exposure to antimalarial drugs drives parasites into investing into reproduction rather than replication (Buckling *et al.*, 1999). Instead, it suggests that drug-sensitive parasites invest in replication and prioritize on 'survival', while resistant parasites increase their gametocyte conversion rate and thus their transmission potential to guarantee the spread of their genotype. These data suggest a degree of plasticity of the parasites in response to drug stresses that may not have been obvious from the transcriptome and proteome analyses, but that are highly relevant when designing treatment strategies and regimens for implementation in the field.

Overall, the studies on the transcriptome responses of *Plasmodium* to chloroquine show that under the same experimental conditions the data are highly reproducible while variation in experimental set-up leads to variable outcomes. Therefore, it appears imperative that some standardization is applied when responses to stresses are investigated against *Plasmodium*. However, a general response to chloroquine treatment appears to be the up-regulated ability of parasites to respond to oxidative stress, presumably because ROS production increases owing to treatment with the drug.

Antifolates

The targets of antifolates are dihydropteroate synthase (DHPS) and dihydrofolate reductase (DHFR). They have been used as chemotherapeutics against malaria for the past 50 years, but resistance has rendered them unsuitable in many malaria endemic areas (Muller and Hyde, 2010). The principal compounds are pyrimethamine, proguanil (which requires metabolism *in vivo* to form the active component cycloguanil) and the sulfonamides (sulfadoxine, sulfalene and the sulfone dapsone). The targets of these drugs are well defined with pyrimethamine and cycloguanil targeting DHFR and the sulfonamides acting against DHPS (Hyde, 2007). The thymidylate synthase (TS) inhibitors fluoroorotate and fluorouracil as well as the DHFR inhibitors pyrimethamine and WR99210 *in vitro* were used to assess whether they affect the transcription or translation of the gene encoding the bifunctional *Plasmodium* DHFR-TS. Transcript levels were unaffected, while the enzymatic activities of both proteins were significantly increased in treated parasites. The latter was attributed to newly synthesized protein (Nirmalan *et al.*, 2004b). This surprising result was suggested to be due to the interaction of *dhfr-ts* mRNA with DHFR-TS protein to form a complex that is responsible for the translational regulation of this particular protein (Nirmalan *et al.*, 2004b). In humans it has been shown that the *dhfr* mRNA is complexed with DHFR protein and this process controls translation of the protein without affecting its mRNA levels (Chu *et al.*, 1993; Keyomarsi *et al.*, 1993; Ercikan-Abali *et al.*, 1997). A similar mechanism of post-translational regulation has been suggested for serine hydroxymethyltransferase in *P. falciparum*, supporting the idea that such post-transcriptional regulatory mechanisms are operating in the parasites (Pang *et al.*, 2009). These data are corroborated by Ganesan and colleagues, who used lethal doses of WR99210 to inhibit *Plasmodium* DHFR activity, which did not lead to a specific up-regulation of the expression of the target pathway and consequently a protection against drug action. Analysing the genome-wide response to exposure with the DHFR inhibitor revealed a genome-wide but low-amplitude, marginal transcriptional response. The genes that showed transcriptional changes were termed antifolate death-related genes and were presumed to mediate changes that are typical for metabolic changes of a dying parasite cell. In fact, the study led to the notion that the malaria parasites are not able to mount a specific response to specific metabolic perturbations, which should make them particularly susceptible to inhibition of their vital metabolic activities with lethal antimetabolites (Ganesan *et al.*, 2008).

Treatment of parasites with pyrimethamine and subsequent analyses of their proteome suggested significant changes in protein expression levels in response to the antifolate drug that extend beyond the immediate target of the drug (Nirmalan *et al.*, 2004a). It cannot be excluded that changes in expression levels of selected proteins such as enolase, lactate dehydrogenase, phosphoethanolamine N-methyltransferase and heat shock protein 72 may represent an early death response of the parasites, although the authors of this study suggested that the changes observed were attributable mainly to interference with parasite metabolism as a consequence of drug exposure rather than non-specific death related responses (Nirmalan *et al.*, 2004a).

Artemisinin and derivatives

Artemisinin, a sesquiterpene lactone endoperoxide, and its derivatives in combination with other antimalarials (artemisinin combination therapy, ACT), are the current gold standard of treatment recommended by the World Health Organization against malaria caused by infection with *P. falciparum* (Nosten and White, 2007; Ward *et al.*, 2007; Eastman and Fidock, 2009). The success using this therapy has revived the goal of malaria eradication (Bosman and Mendis, 2007), but in recent years worrying observation of decreased sensitivity and treatment failures to ACT have been reported (Dondorp *et al.*, 2009, 2010; Mok *et al.*, 2011). The spread of ACT resistance to Asia and Africa could have disastrous consequences. It is therefore imperative to understand the underlying mechanisms that lead to delayed clearance of *Plasmodium* after ACT treatment to guide new treatment strategies and regimens and to map resistance to prevent its spread (Dondorp *et al.*, 2010; Mok *et al.*, 2011). ACT resistance is not associated to alterations in expression levels or mutations in known drug resistance markers (Dondorp *et al.*, 2009; Witkowski *et al.*, 2010) and surprisingly the *in vitro* susceptibility to the drug of resistant parasite isolates is not changed (Dondorp *et al.*, 2009). This suggests more complex mechanisms that provide the parasites with ways to escape ACT *in vivo*. A recent study by Mok and colleagues provides evidence that the ACT resistant parasites remodel their metabolic programme to avoid drug action (Mok *et al.*, 2011). It was shown that basic metabolic activities appear largely reduced during the first part of the intra-erythrocytic cycle of the resistant isolates, resulting in slower growth and maturation during this part of their development. In addition, it was found that during schizont stage, protein metabolism ranging from haemoglobin digestion to protein biosynthesis is up-regulated. The slowed progression through the initial stages of the intra-erythrocytic cycle is in agreement with the finding that ring stage parasites are less susceptible to artemisinin and thus this adaptation provides a protection against the action of the drug. This hypothesis is supported by a study where resistance to artemisinin was induced *in vitro* demonstrating that *Plasmodium* parasites employ a quiescence mechanism, which allows them to escape the action of the drug (Witkowski *et al.*, 2010). The increased capacity of protein metabolism during late stage development provides the parasites with the ability to replace oxidized or alkylated proteins, which are a consequence of drug exposure, and promotes survival. Obviously this change in transcriptional pattern has to be regulated; indeed, the transcriptional levels of transcriptional regulators, RNA binding proteins, chromatin remodelling factors, signalling factors and cell cycle regulators were affected in the artemisinin-resistant isolates.

These results suggest that tolerance against ACT involves complex mechanisms that differ considerably from other drug resistance mechanisms, such as the point mutations in specific proteins underlying resistance to pyrimethamine/sulfadoxine or chloroquine (Lu *et al.*, 2010; Muller and Hyde, 2010; Sanchez *et al.*, 2010). It appears that, owing to differential expression of regulatory proteins, the expression profile of the parasites is skewed to allow them to survive under ACT pressure. There may be an overall genetic predisposition for this transcriptional response, but firm evidence for this proposal is still lacking (Mok *et al.*, 2011).

An *in vitro* study analysing the transcriptomic response to artesunate, an artemisinin derivative, was performed to obtain insights into the mode of action of this family of drugs and to provide new information about networks that are specifically affected by the artemisinin derivatives. Short exposure with a high concentration of artesunate led to a significant but low-amplitude broad perturbation of the parasite transcriptome (Natalang *et al.*, 2008). Although not a single pathway or a single putative target was identified by this approach, dynamic changes of gene expression of components of intracellular trafficking and general organization of the parasite were detected. This included changes in expression in mRNA levels of chaperones, transporters and proteins involved in the remodelling of the erythrocytic space beyond the parasite membrane. Similar to the transcriptome study using field isolates resistant to artemisinin, the down-regulation of basic metabolic activities, differences in the transcription of genes involved in the maintenance of the intracellular redox homeostasis as well as an alteration of the expression of genes encoding proteins involved in protein turnover were observed.

Neither in field isolates nor in artemisinin-resistant parasites generated *in vitro* were mutations in *pfatp6*, a gene encoding sarco-endoplasmatic reticulum calcium ATPase6, which has been shown to be a target of artemisinin (Eckstein-Ludwig *et al.*, 2003; Krishna *et al.*, 2010), nor mutations of *pfmdr1*, a gene encoding a multidrug resistance transporter (Duraisingh and Cowman, 2005), associated with the ACT treatment failure (Dondorp *et al.*, 2009; Witkowski *et al.*, 2010). Taken together, the transcriptional responses to artemisinin and its derivatives may reflect specific ways the parasite employs to cope with the stresses that it experiences when exposed to this family of drugs. The studies described above therefore form the basis for new testable hypotheses that should improve the understanding of the mode of artemisinin action and lead to new insights into the mechanisms underlying drug action and resistance. This is imperative to prevent development and spread of drug resistance against this important group of antimalarials.

Response to temperature changes

Plasmodium has to deal with alterations of temperature during the progression through their developmental cycle from mosquito to mammalian host and vice versa. Furthermore, once in the human host, they have to cope with regular episodes of increased host body temperatures due to the cyclical fever induced by schizont rupture during the intra-erythrocytic cycle (Oakley *et al.*, 2007; Pesce *et al.*, 2008). The intra-erythrocytic cycle differs between human malaria species. *P. falciparum*, *P. ovale* and *P. vivax* require 48 hours to complete their cycle ('tertian fever') while *P. malariae* develops within 72 hours ('quartan fever'). The fever is induced by the release of end- and by-products of parasite metabolism upon synchronous schizont rupture and the subsequent immunological responses of the host. Clearly these events require the ability to respond adequately to variation in temperature to allow survival and multiplication as well as reproduction of the parasites. Elevated temperatures can lead to protein structural damage, may cause cell cycle arrest and have effects on the fluidity of membranes (Oakley *et al.*, 2011). In response to temperature fluctuations, induction of heat shock proteins to prevent protein denaturation and aggregation as well as to promote correct folding is a widespread mechanism in eukaryotes (Sorger, 1991), including *Plasmodium* (Joshi *et al.*, 1992; Biswas and Sharma, 1994; Banumathy *et al.*, 2003; Pavithra *et al.*, 2007; Pallavi *et al.*, 2010). Oakley and colleagues analysed the effect of febrile temperature on the global transcriptome of cultures of *P. falciparum*. The expression of 336 genes encoding proteins were consistently altered in parasites exposed to heat stress for 2 hours and a large number of these were genes encoding for hypothetical proteins without apparent orthologues in other organisms. Unsurprisingly, the induction of heat shock proteins was a prominent response to elevated temperatures; interestingly, expression of a family of DnaJ domain proteins that show an apicomplexan lineage-specific expansion was also up-regulated (Oakley *et al.*, 2007, 2011). Components of the protein degradation system were down-regulated, resulting in a significant decrease in protein ubiquination (Oakley *et al.*, 2007). Elevated temperatures appear to also negatively affect the expression of genes encoding proteins involved in DNA replication, transcriptional regulators and ribosomal proteins, suggesting that progression through the intra-erythrocytic cycle is hampered by this stress (Oakley *et al.*, 2007). This phenomenon had been described earlier (Kwiatkowski and Greenwood, 1989) and it was shown that prolonged exposure of *Plasmodium* to febrile temperature is lethal to the parasites. Exposure of early intra-erythrocytic stages of *Plasmodium* to temperature stress revealed however, that their growth was not impaired during the first half of the intra-erythrocytic cycle, but that the transition from trophozoites to schizonts was greatly affected. This led to the hypothesis that the second half of *Plasmodium* intra-erythrocytic development is selectively disrupted by heat stress. Ultimately, this selects parasites in the early stages of their intra-erythrocytic development and results in the synchronisation of the cultured parasites in response to elevated temperatures. Once the parasites were synchronised through the selective pressure of temperature elevation, they grew normally under repeated elevation of temperature, suggesting an adaptive response to febrile temperatures, which is a requirement for survival of *Plasmodium in vivo,* given that they are regularly exposed to increased body temperatures during the fever episodes of their human hosts (Kwiatkowski, 1989). These data were corroborated by a later study, which showed that repeated exposure to febrile temperatures led to tolerance to heat shock in the parasites and promoted *Plasmodium* growth *in vitro* (Pavithra *et al.*, 2004). The study further suggested that Hsp90 function is important for the transformation from ring to trophozoite stage and the response to heat shock. Hsp90, like many other Hsps, functions in complex with co-chaperones and other effectors. Hsp70-1, protein phosphatase 5 and α-tubulin as well as FKBP35 have been identified as co-effectors of Hsp90 (Pavithra *et al.*, 2004, 2007; Kumar *et al.*, 2005, 2007; Pallavi *et al.*, 2010). A recent systems biology study has given further insights into potential chaperone interaction networks and suggested additional interaction partners of the Hsp90 complex apart from those experimentally identified. This bioinformatics approach supports the notion that the Hsp90 complex has a critical role in stress response and that it interacts with a number of t-RNA synthetases as well as protein factors involved in chromatin remodelling. Taken together, these data suggest that Hsp90 is a crucial factor in regulating gene expression and protein translation, equipping the parasites with an effective response mechanism that allows an adaptive response to febrile temperatures and possibly other stresses (Pavithra *et al.*, 2007).

Response to oxidative stress

Plasmodium are particularly vulnerable to oxidative stress throughout their life cycle in the mammalian and mosquito hosts (Müller *et al.*,

2003; Becker *et al.*, 2004; Müller, 2004). The parasite is furnished with functional antioxidant and redox systems to prevent oxidative stress and to guarantee survival (Becker *et al.*, 2004; Bozdech and Ginsburg, 2004; Müller, 2004; Kehr *et al.* 2011) and they even commandeer host cell antioxidants to aid their own antioxidant defence (Fairfield *et al.*, 1983; Koncarevic *et al.*, 2009). It has been shown that not only the cytosol and the mitochondrion, but also the plastid-like organelle, the apicopast, as well as the nucleus, contain appropriate mechanisms to deal with oxidative challenges (Müller, 2004; Sienkiewicz *et al.*, 2004; McMillan *et al.*, 2005; Boucher *et al.*, 2006; Kehr *et al.*, 2011; Richard *et al.*, 2011), and that post-transcriptional mechanisms regulate the distribution of the antioxidant proteins to their respective locations (Kehr *et al.*, 2011). How this is achieved and whether exposure to environmental stresses affects the distribution of the antioxidant proteins in the parasites remains to be investigated. A focused study exposing *P. falciparum* to superoxide anions revealed that in addition to the up-regulation of the transcription of antioxidants, the parasites also mounted a concomitant heat shock response. These results, albeit on a much smaller scale than those obtained from genome wide transcriptome analyses, suggests that the exposure to oxidative stress leads to a general stress response also observed as a result of exposure to drug exposure (see above) (Akide-Ndunge *et al.*, 2009). A recent study analysing the response to hyperoxia on intra-erythrocytic stages of *Plasmodium* revealed that the exposure to increased oxygen *in vitro* results in a delay of the parasite's cell cycle by about 4 hours. Genome-wide transcriptomics revealed that hyperoxia led to significant changes in the transcription levels of 219 genes, of which 114 were up-regulated and 105 were down-regulated. From these data a network of the metabolic changes during hyperoxia was constructed and revealed the down-regulation of glycolytic activities and amino acid metabolism, while transcript levels of DNA repair proteins and heat shock proteins were increased. Expression of genes encoding proteins of the antioxidant machinery was only marginally altered, suggesting that the parasites possess sufficiently high levels of these proteins to cope with hyperoxia. However, analyses of the expression of mRNA or protein levels of individual genes encoding antioxidant proteins in *Plasmodium* showed the up-regulation of cytosolic 2-Cys peroxiredoxin (Akerman and Müller, 2003), superoxide dismutases (Sienkiewicz *et al.*, 2004), γ-glutamylcysteine synthetase (Müller, 2003), isocitrate dehydrogenase (Wrenger and Müller, 2003), mitochondrial lipoamide dehydrogenase (McMillan *et al.*, 2005)and pyridoxal phosphate synthase (Wrenger *et al.*, 2006), to name only a few. On the other hand, glutathione reductase, thioredoxin reductase as well as 1-Cys peroxiredoxin showed no up-regulation under oxidative stress conditions (Akerman and Müller, 2003; Akide-Ndunge *et al.*, 2009). These data strongly suggest that on either the transcriptome or proteome levels specific responses to oxidative stress are mounted by the parasites although the precise mechanisms as to how this is orchestrated are still unknown.

In other organisms, exposure to oxidative stress leads to an adaptive response that leaves the organism equipped with an increased level of antioxidants ready to cope with a subsequent challenge after the initial insult (Moradas-Ferreira and Costa, 2000; Birringer, 2011). Whether a similar mechanism is operating in *Plasmodium* needs further investigation. Interfering with the antioxidant and redox networks of the parasites has a severe impact on their viability in the insect vector, presumably owing to increased metabolic activities that require a constant and high level antioxidant network. Deletion of the gene encoding γ-glutamylcysteine synthetase, glutathione reductase and thioredoxin reductase arrested sexual development of *P. berghei* (Vega-Rodriguez *et al.*, 2009; Buchholz *et al.*, 2010; Pastrana-Mena *et al.*, 2010). Surprisingly, these pathways could not be disrupted in the human malaria parasite during their intra-erythrocytic development, suggesting that the antioxidant requirements of rodent and human malaria species may be distinct, possibly because the intra-erythorcytic cycle of *P. falciparum* is twice as long as that of *P. berghei* and therefore requires a more robust antioxidant machinery to guarantee their survival during intra-erythrocytic development (Krnajski *et al.*, 2002; Patzewitz and Müller, unpublished results). However, the precise mechanisms that regulate and govern the antioxidant capacities and responses to oxidative challenges are not well understood and need further extensive studies to dissect their precise functions.

Mechanisms governing stress response

Mechanisms involved in regulation of gene expression

Compared with other eukaryotes, the parasite possesses a scarcity of specific transcription factors despite the fact that they undergo multiple morphological and metabolic changes during their complex life cycle (Coulson *et al.*, 2004; Callabaut *et al.*, 2005; De Silva *et al.*, 2008). Recent studies, however, have revealed the presence of a family of transcription factors important for regulation of gene expression during the complex parasite life cycle (Balaji *et al.*, 2005; De Silva *et al.*, 2008; Painter *et al.*, 2011). Overall there is little knowledge about differential gene expression and how it is altered in response to external stresses in *Plasmodium*, and this is an area of parasite biology that still needs extensive investigation.

In line with the relatively low abundance of classical transcription factors, several studies pointed to the importance of epigenetic mechanisms to govern transcription, and of post-transcriptional mechanisms in the control of gene expression, such as mRNA stability and translational repression (Mair *et al.*, 2006; Deitsch *et al.*, 2007; Hakimi and Deitsch, 2007; Cui and Miao, 2010; Salcedo-Amaya *et al.*, 2010; Painter *et al.*, 2011; Wong *et al.*, 2011).

Chromatin remodelling is one of the best characterized mechanisms governing gene expression in malaria parasites (Lopez-Rubio *et al.*, 2009; Merrick *et al.*, 2010). Other studies have provided evidence that post-translational stability of mRNA is regulated in a stage dependent manner (Shock *et al.*, 2007; Sims *et al.*, 2009). Another mechanism, which is crucial for completion of the sexual cycle of the parasite in the mosquito vector is translational repression (a mechanism that is widespread in eukaryotes): a number of transcripts accumulate in female gametocytes in an inactive form, whose translation is initiated following ingestion of the gametocyte by the mosquito vector. The transcripts regulated in this way encode proteins that participate in the formation of the zygote and subsequent development stages, such as p25 and p28. A major player in this mechanism is the DOZI protein ('development of zygote inhibited'), which is involved in the storage of quiescent mRNAs in the gametocyte, and whose inactivation by reverse genetics results in the premature degradation of 370 mRNA species (Mair *et al.*, 2006); this does not affect fertilization but causes arrest of ookinete development soon after zygote formation (Mair *et al.*, 2011).

Signalling pathways controlling adaptive response to stress

The post-transcriptional mechanisms outlined in the previous section are implemented in the regular progression of the *Plasmodium* life cycle; whether they are also employed in stress response *sensu stricto* remains to be investigated. There is, however, documented evidence that modulating translation is indeed employed in response to starvation stress. This response is mediated by PfeIK1, one of the three members of the 'eukaryotic initiation factor 2α kinase' family present in the *P. falciparum* kinome (Ward *et al.*, 2004). In yeast and metazoans, part of the stress response is mediated through phosphorylation of eukaryotic translation initiation factor 2α (eIF2α) by protein kinases of this family, which results in (i) overall shut-down of translation, and (ii) selective translation of mRNAs encoding stress-response proteins. The *P. falciparum* orthologue of the eIF2α translation factor is phosphorylated in response to isoleucine depletion, which constitutes a serious nutritional stress as the parasite is unable to obtain sufficient isoleucine from haemoglobin digestion and cannot generate the amino acid *de novo* (Liu *et al.*, 2006). This phosphorylation event is abolished in *pfeik1*$^-$ null mutant parasites, which demonstrates that this response is mediated solely by this kinase (Fennell *et al.*, 2009). The signalling pathway leading from isoleucine depletion to activation of the PfeIK1 is, however, not understood.

A second member of this kinase family, PfeIK2, is expressed predominantly during sexual development; like PfeIK1, this enzyme is dispensable for asexual proliferation and *pfeik2*⁻ null mutants still produce sporozoites in the mosquito. Work in *P. berghei*, however, showed that salivary gland sporozoites lacking PbeIK2 display premature translation of liver stage-specific genes, impairing infection of the vertebrate host (Zhang *et al.*, 2010).

The third eIF2α kinase in the *P. falciparum* kinome, PfPK4, appears to be essential for asexual growth (Solyakov *et al.*, under revision), but whether or not this is associated with a deficit in stress response is not known. Interestingly, the related apicomplexan parasite *T. gondii* has been shown to differentiate from tachyzoites to bradyzoites on exposure to a number of cellular stresses, concomitant with an increase in phosphorylation of TgeIF2α, indicating a possible role for this mechanism in differentiation (Sullivan *et al.*, 2004).

The presence of a stress-responsive gene expression control mechanism based on translational repression (as mediated by PfeIK1, see above) is consistent with the evidence supporting a significant role for post-transcriptional regulation of stress responses in *Plasmodium* (Bozdech *et al.*, 2003a,b; Coulson *et al.*, 2004). However, even though *Plasmodium* transcription factors are less numerous than expected in relation to the parasite's genome size, it is reasonable to propose that at least some of those that are present may be controlled through phosphorylation events potentially activated by stresses the parasite encounters. Indeed, a recent phosphoproteomic analysis (Solyakov *et al.*, under revision) identified a number of phosphorylation sites on several transcription factors, including members of the newly described plant-like ApiAP2 family (De Silva *et al.*, 2008). It will be fascinating to elucidate whether *Plasmodium* transcription factors undergo stress-dependent phosphorylation, and if so, to identify the protein kinases responsible.

Possible candidates in the *P. falciaprum* kinome potentially involved in phosphorylation of transcription factors are the mitogen-activated protein kinases (MAPKs) (Ward *et al.*, 2004). Stress activated protein kinases (SAPKs, which include JNKs and p38 kinases) are subfamilies of the MAPK family that are present in most eukaryotic cells, where they mediate the adaptive response to a variety of stress conditions (Engelberg, 2004). As members of the MAPK family, the SAPKs are part of three-tier MAPKKK-MAPKK-MAPK kinase cascades, and their substrates include a range of transcription factors that control the expression of genes encoding stress-response proteins (Robinson and Cobb, 1997; Davis *et al.*, 2000; Palmer and Nebreda, 2000; Engelberg, 2004). However, both bioinformatics and experimental approaches demonstrated that classical three-component MAPK signalling pathways are not operating in malaria parasites (Ward *et al.*, 2004; Dorin *et al.*, 2005). Although the parasite kinome includes two MAPK homologues, none of these are members of the SAPK subfamily (Dorin *et al.*, 1999, 2005; Ward *et al.*, 2004). One of these, Pfmap-2, is essential for completion of the asexual cycle in erythrocyte in *P. falciparum*. This is not the case in the murine malaria parasite *P. berghei*, where it is dispensable during asexual development, but is required for male gametogenesis (Tewari *et al.*, 2005). The other MAPK, Pfmap-1, can be deleted in both *P. falciparum* and *P. berghei* without causing a developmental block phenotype (Dorin-Semblat *et al.*, 2007). Interestingly, Pfmap-2 is up-regulated in *pfmap-1*⁻ null parasites, suggesting that compensatory mechanisms are probably operating. It would be of interest to determine whether either or both of these parasite MAPKs function in stress response pathways, in particular with respect to the activation of downstream transcription factors.

In addition to the MAPKs, the plasmodial kinome possesses homologues of protein kinases known to regulate transcription in other systems, such as casein kinase 1 (PfCK1) or casein kinase 2 (PfCK2) (Holland *et al.*, 2009; Doerig *et al.*, 2010). To our knowledge, an implication of these enzymes in stress response has not been documented. The plasmodial kinome also includes a number of enzymes with no orthologues in other organisms, which might conceivably function in stress-responsive transcriptional control; targeted experimental approaches are needed to explore this possibility. A recent study identified a phospho-signalling pathway of the host erythrocyte as crucial for parasite survival (Sicard *et al.*, 2011). In line with the observation that the parasite utilizes host antioxidants to bolster their own antioxidant defence (Koncarevic *et al.*, 2009), a reliance of the parasite on host erythrocyte signalling elements to sense and transmit stress signals is highly likely and certainly worthwhile exploring.

Conclusions and future trends

The malaria parasite encounters numerous stresses throughout its complex life cycle and appears to be able to cope with these effortlessly. Numerous studies (some of them described above) have been performed aiming to identify the systems the parasites employ to guarantee their progression through their life cycle. However, we are only beginning to understand the mechanisms the parasites employ to controlling and regulating the response to the environmental fluctuations they encounter throughout their developmental cycle in both human and mosquito hosts.

As outlined above, understanding the regulation of gene expression in *Plasmodium* is in its infancy and for a long time it was generally accepted that the parasites undergo a programme of gene expression from which they do not divert even under elevated exogenous stresses (Bozdech *et al.*, 2003a; Hu *et al.*, 2010). However, even though the response to drug and temperature stresses only elicit low amplitude transcriptional responses, in some cases it is time and dose dependent as well as drug-dependent, suggesting that it is worthwhile re-visiting the idea that *Plasmodium* is able to specifically respond to the stresses it encounters (Hu *et al.*, 2010). Another finding that suggests a regulated response to exogenous stresses is that the parasites are able to adapt to elevated temperatures presumably through changes in their transcriptional profile (Kwiatkowski and Greenwood, 1989; Oakley *et al.*, 2007, 2011). Clearly this is a very important adaptation given that *Plasmodium* encounter fluctuations in their host's body temperature periodically during the fever episodes of the human host and during transmission from human to mosquito host and *vice versa*.

Some of the genome wide transcriptome analyses performed after exposure of parasites to exogenous stresses (e.g. drugs or hyperoxia) revealed a general stress response that often showed overlapping characteristics (Gunasekera *et al.*, 2003, 2007; Hu *et al.*, 2010). Scrutiny and standardization of experimental design resulted in the revision of the view that *Plasmodium* is unable to mount a response to exogenous and endogenous stresses and the discovery of the large family of AP2-like transcription factors has given new credibility to the notion that *Plasmodium* are capable to mount specific responses to exogenous stimuli. It is expected that the AP2-like transcription together with translational repression and epigenetic as well as post-transcriptional factors are of crucial importance for these distinct responses and will provide new insights into parasite gene expression mechanisms and post-transcriptional and post-translational capabilities. This has led to a renewed interest in the analyses of responses to external and internal signals that may regulate parasite gene expression and provide them with a much larger flexibility than thought after the initial transcriptome and proteome analyses (Le Roch *et al.*, 2004; De Silva *et al.*, 2008; Lindner *et al.*, 2010; Painter *et al.*, 2011). Apart from regulated gene expression, post-translational modifications through signalling cascades employing protein kinases and phosphatases will have an important part to play when it comes to adjustments and adaptations of parasite metabolic functions in response to stresses (Wilkes and Doerig, 2008; Doerig *et al.*, 2010), but their precise roles still have to be elucidated. Response to amino acid starvation is governed by eIF2α kinases at specific times of parasite development, suggesting that they possess a variety of mechanisms to deal with this potentially lethal stress by down-regulating protein translation. The precise signalling cascade governing this response still needs to be dissected. The lack of a classical MAPK-dependent stress response signalling pathway has also supported the hypothesis that a classical stress response does not exist in *Plasmodium*. However, this hypothesis has been re-analysed and protein kinases potentially involved in stress response in the parasites are currently under investigation.

Overall, little is known about the precise mechanisms that govern stress response in the malaria parasites and it is expected that the next decade will reveal the systems that allow the parasites to successfully survive and respond to stresses they constantly encounter. Some of these mechanisms will be crucial for parasite survival and therefore may be exploitable for future development of intervention strategies against these deadly parasites.

Web resources

CDC Malaria: http://www.cdc.gov/MALARIA/

This web site provides general information about malaria including facts about the disease, information for travellers, about diagnostics and treatment. It also informs about up-to-date policies with respect to malaria control programmes.

Malaria Research and Reference Reagent Resource Center: http://www.mr4.org/

The Malaria Research and Reference Reagent Resource Center (MR4) has provided a central resource for malaria reagents, protocols and technical support for the international research community since 1998. Supported by the National Institute of Allergy and Infectious Diseases (NIAID) and now part of the NIAID BEI collections, the MR4 acquires, authenticates, preserves and distributes parasites, mosquito vectors and associated biological and molecular reagents.

NLM Malaria Resources*:* www.nlm.nih.gov/mimcom/NLM_malaria_resources.html

This web resource provides data and information relevant to malaria genetics and genomics. These resources include organism specific sequence BLAST databases (*Plasmodium falciparum* only, all *Plasmodium*), genome maps, linkage markers, and information about genetic studies. Links are provided for other malaria web sites and genetic data on related apicomplexan parasites.

PlasmoDB database: http://www.plasmodb.org

The PlasmoDB database is a functional genomic database for malaria parasites. It is now part of EupathDB (EupathDB.org) which provides a unified entry point for the Eukaryotic Pathogen Bioinformatics Resource Centre. The database is an invaluable research tool and platform that supports genomic, microarray, proteomic, EST, SAGE-tag, SNP and metabolic pathway data sets for *Plasmodium*.

WHO Malaria: http://www.who.int/topics/malaria/en/

This web resource provides background facts and features about malaria. It also contains important information about prevention, control and elimination programmes and policy setting processes that have been reviewed and re-designed this year.

References

Akerman, S.E., and Müller, S. (2003). 2-Cys peroxiredoxin PfTrx-Px1 is involved in the antioxidant defence of *Plasmodium falciparum*. Mol. Biochem. Parasitol. *130*, 75–81.

Akide-Ndunge, O.B., Tambini, E., Giribaldi, G., McMillan, P.J., Müller, S., Arese, P., and Turrini, F. (2009). Co-ordinated stage-dependent enhancement of *Plasmodium falciparum* antioxidant enzymes and heat shock protein expression in parasites growing in oxidatively stressed or G6PD-deficient red blood cells. Malar. J. *8*, 113.

Balaji, S., Babu, M.M., Iyer, L.M., and Aravind, L. (2005). Discovery of the principal specific transcription factors of Apicomplexa and their implication for the evolution of the AP2-integrase DNA binding domains. Nucleic Acids Res. *33*, 3994–4006.

Bammert, G.F., and Fostel, J.M. (2000). Genome-wide expression patterns in *Saccharomyces cerevisiae*: comparison of drug treatments and genetic alterations affecting biosynthesis of ergosterol. Antimicrob. Agents Chemother. *44*, 1255–1265.

Banumathy, G., Singh, V., Pavithra, S.R., and Tatu, U. (2003). Heat shock protein 90 function is essential for *Plasmodium falciparum* growth in human erythrocytes. J. Biol. Chem. *278*, 18336–18345.

Bautista, J.M., Radfar, A., and Diez, A. (2008). Chloroquine mediates specific proteome oxidative damage across the erythrocytic cycle of resistant *Plasmodium falciparum*. Free Rad. Biol. Med. *44*, 2034–2042.

Becker, K., Tilley, L., Vennerstrom, J.L., Roberts, D., Rogerson, S., and Ginsburg, H. (2004). Oxidative stress in malaria parasite-infected erythrocytes: host–parasite interactions. Int. J. Parasitol. *34*, 163–189.

Berry, D.B., and Gasch, A.P. (2008). Stress-activated genomic expression changes serve a preparative role for impending stress in yeast. Mol. Biol. Cell *19*, 4580–4587.

Birringer, M. (2011). Hormetics: dietary triggers of an adaptive stress response. Pharm Res. *28*, 2680–2994.

Biswas, S., and Sharma, Y.D. (1994). Enhanced expression of *Plasmodium falciparum* heat shock protein PFHSP70-I at higher temperatures and parasite survival. FEMS Microbiol. Lett. *124*, 425–429.

Bosman, A., and Mendis, K.N. (2007). A major transition in malaria treatment: the adoption and deployment of artemisinin-based combination therapies. Am. J. Trop. Med. Hyg. 77, 193–197.

Boucher, I.W., McMillan, P.J., Gabrielsen, M., Akerman, S.E., Brannigan, J.A., Schnick, C., Brzozowski, A.M., Wilkinson, A.J., and Müller, S. (2006). Structural and biochemical characterization of a mitochondrial peroxiredoxin from *Plasmodium falciparum*. Mol. Microbiol. *61*, 948–959.

Bozdech, Z., and Ginsburg, H. (2004). Antioxidant defense in *Plasmodium falciparum*--data mining of the transcriptome. Malar. J. *3*, 23.

Bozdech, Z., Llinas, M., Pulliam, B.L., Wong, E.D., Zhu, J., and DeRisi, J.L. (2003a). The transcriptome of the intraerythrocytic developmental cycle of *Plasmodium falciparum*. PLoS Biol. *1*, E5.

Bozdech, Z., Zhu, J., Joachimiak, M.P., Cohen, F.E., Pulliam, B., and DeRisi, J.L. (2003b). Expression profiling of the schizont and trophozoite stages of *Plasmodium falciparum* with a long-oligonucleotide microarray. Genome Biol. *4*, R9.

Bozdech, Z., Mok, S., Hu, G., Imwong, M., Jaidee, A., Russell, B., Ginsburg, H., Nosten, F., Day, N.P., White, N.J., *et al.* (2008). The transcriptome of *Plasmodium vivax* reveals divergence and diversity of transcriptional regulation in malaria parasites. Proc. Natl. Acad. Sci. U.S.A. *105*, 16290–16295.

Buchholz, K., Putrianti, E.D., Rahlfs, S., Schirmer, R.H., Becker, K., and Matuschewski, K. (2010). Molecular genetics evidence for the *in vivo* roles of the two major NADPH-dependent disulfide reductases in the malaria parasite. J. Biol. Chem. *285*, 37388–37395.

Buckling, A., Ranford-Cartwright, L.C., Miles, A., and Read, A.F. (1999). Chloroquine increases *Plasmodium falciparum* gametocytogenesis *in vitro*. Parasitology *118*, 339–346.

Buckling, A.G., Taylor, L.H., Carlton, J.M., and Read, A.F. (1997). Adaptive changes in *Plasmodium* transmission strategies following chloroquine chemotherapy. Proc. Biol. Sci. *264*, 553–559.

Cabiscol, E., Piulats, E., Echave, P., Herrero, E., and Ros, J. (2000a). Oxidative stress promotes specific protein damage in *Saccharomyces cerevisiae*. J. Biol. Chem. *275*, 27393–27398.

Cabiscol, E., Tamarit, J., and Ros, J. (2000b). Oxidative stress in bacteria and protein damage by reactive oxygen species. Int. Microbiol. *3*, 3–8.

Carlton, J.M., Angiuoli, S.V., Suh, B.B., Kooij, T.W., Pertea, M., Silva, J.C., Ermolaeva, M.D., Allen, J.E., Selengut, J.D., Koo, H.L., *et al.* (2002). Genome sequence and comparative analysis of the model rodent malaria parasite *Plasmodium yoelii* yoelii. Nature *419*, 512–519.

Chu, E., Takimoto, C.H., Voeller, D., Grem, J.L., and Allegra, C.J. (1993). Specific binding of human dihydrofolate reductase protein to dihydrofolate reductase messenger RNA *in vitro*. Biochemistry *32*, 4756–4760.

Coulson, R.M., Hall, N., and Ouzounis, C.A. (2004). Comparative genomics of transcriptional control in the human malaria parasite *Plasmodium falciparum*. Genome Res. *14*, 1548–1554.

Cui, L., and Miao, J. (2010). Chromatin-mediated epigenetic regulation in the malaria parasite *Plasmodium falciparum*. Eukaryot. Cell *9*, 1138–1149.

Davis, P.J., Shih, A., Lin, H.Y., Martino, L.J., and Davis, F.B. (2000). Thyroxine promotes association of mitogen-activated protein kinase and nuclear thyroid hormone receptor (TR) and causes serine phosphorylation of TR. J. Biol. Chem. *275*, 38032–38039.

De Silva, E.K., Gehrke, A.R., Olszewski, K., Leon, I., Chahal, J.S., Bulyk, M.L., and Llinas, M. (2008). Specific DNA-binding by apicomplexan AP2 transcription factors. Proc. Natl. Acad. Sci. U.S.A. *105*, 8393–8398.

Deitsch, K., Duraisingh, M., Dzikowski, R., Gunasekera, A., Khan, S., Le Roch, K., Llinas, M., Mair, G., McGovern, V., Roos, D., *et al.* (2007). Mechanisms of gene regulation in Plasmodium. Am. J. Trop. Med. Hyg. 77, 201–208.

Doerig, C., Abdi, A., Bland, N., Eschenlauer, S., Dorin-Semblat, D., Fennell, C., Halbert, J., Holland, Z., Nivez, M.P., Semblat, J.P., *et al.* (2010). Malaria: targeting parasite and host cell kinomes. Biochim. Biophys. Acta *1804*, 604–612.

Dondorp, A.M., Nosten, F., Yi, P., Das, D., Phyo, A.P., Tarning, J., Lwin, K.M., Ariey, F., Hanpithakpong, W., Lee, S.J., *et al.* (2009). Artemisinin resistance in *Plasmodium falciparum* malaria. N. Engl. J. Med. *361*, 455–467.

Dondorp, A.M., Yeung, S., White, L., Nguon, C., Day, N.P., Socheat, D., and von Seidlein, L. (2010). Artemisinin resistance: current status and scenarios for containment. Nat. Rev. Microbiol. *8*, 272–280.

Dorin, D., Alano, P., Boccaccio, I., Ciceron, L., Doerig, C., Sulpice, R., and Parzy, D. (1999). An atypical mitogen-activated protein kinase (MAPK) homologue expressed in gametocytes of the human malaria parasite *Plasmodium falciparum*. Identification of a MAPK signature. J. Biol. Chem. *274*, 29912–29920.

Dorin, D., Semblat, J.P., Poullet, P., Alano, P., Goldring, J.P., Whittle, C., Patterson, S., Chakrabarti, D., and Doerig, C. (2005). PfPK7, an atypical MEK-related protein kinase, reflects the absence of classical three-component MAPK pathways in the human malaria parasite *Plasmodium falciparum*. Mol. Microbiol. *55*, 184–196.

Dorin-Semblat, D., Quashie, N., Halbert, J., Sicard, A., Doerig, C., Peat, E., and Ranford-Cartwright, L. (2007). Functional characterization of both MAP kinases of the human malaria parasite *Plasmodium falciparum* by reverse genetics. Mol. Microbiol. *65*, 1170–1180.

Duraisingh, M.T., and Cowman, A.F. (2005). Contribution of the pfmdr1 gene to antimalarial drug-resistance. Acta Trop. *94*, 181–190.

Eastman, R.T., and Fidock, D.A. (2009). Artemisinin-based combination therapies: a vital tool in efforts to eliminate malaria. Nat. Rev. Microbiol. 7, 864–874.

Eckstein-Ludwig, U., Webb, R.J., Van Goethem, I.D., East, J.M., Lee, A.G., Kimura, M., O'Neill, P.M., Bray,

P.G., Ward, S.A., and Krishna, S. (2003). Artemisinins target the SERCA of *Plasmodium falciparum*. Nature *424*, 957–961.

van Eijk, A.M., Hill, J., Alegana, V.A., Kirui, V., Gething, P.W., ter Kuile, F.O., and Snow, R.W. (2011). Coverage of malaria protection in pregnant women in sub-Saharan Africa: a synthesis and analysis of national survey data. Lancet Infect. Dis. *11*, 190–207.

Engelberg, D. (2004). Stress-activated protein kinases-tumor suppressors or tumor initiators? Semin. Cancer Biol. *14*, 271–282.

Ercikan-Abali, E.A., Banerjee, D., Waltham, M.C., Skacel, N., Scotto, K.W., and Bertino, J.R. (1997). Dihydrofolate reductase protein inhibits its own translation by binding to dihydrofolate reductase mRNA sequences within the coding region. Biochemistry *36*, 12317–12322.

Fairfield, A.S., Meshnick, S.R., and Eaton, J.W. (1983). Malaria parasites adopt host cell superoxide dismutase. Science *221*, 764–766.

Fennell, C., Babbitt, S., Russo, I., Wilkes, J., Ranford-Cartwright, L., Goldberg, D.E., and Doerig, C. (2009). PfeIK1, a eukaryotic initiation factor 2alpha kinase of the human malaria parasite *Plasmodium falciparum*, regulates stress-response to amino-acid starvation. Malar. J. *8*, 99.

Ganesan, K., Ponmee, N., Jiang, L., Fowble, J.W., White, J., Kamchonwongpaisan, S., Yuthavong, Y., Wilairat, P., and Rathod, P.K. (2008). A genetically hard-wired metabolic transcriptome in *Plasmodium falciparum* fails to mount protective responses to lethal antifolates. PLoS Pathog. *4*, e1000214.

Gardner, M.J., Hall, N., Fung, E., White, O., Berriman, M., Hyman, R.W., Carlton, J.M., Pain, A., Nelson, K.E., Bowman, S., *et al.* (2002). Genome sequence of the human malaria parasite *Plasmodium falciparum*. Nature *419*, 498–511.

Gasch, A.P., and Werner-Washburne, M. (2002). The genomics of yeast responses to environmental stress and starvation. Funct. Integr. Genomics 2, 181–192.

Gunasekera, A.M., Patankar, S., Schug, J., Eisen, G., and Wirth, D.F. (2003). Drug-induced alterations in gene expression of the asexual blood forms of *Plasmodium falciparum*. Mol. Microbiol. *50*, 1229–1239.

Gunasekera, A.M., Myrick, A., Le Roch, K., Winzeler, E., and Wirth, D.F. (2007). *Plasmodium falciparum*: genome wide perturbations in transcript profiles among mixed stage cultures after chloroquine treatment. Exp. Parasitol. *117*, 87–92.

Hakimi, M.A., and Deitsch, K.W. (2007). Epigenetics in Apicomplexa: control of gene expression during cell cycle progression, differentiation and antigenic variation. Curr. Opin. Microbiol. *10*, 357–362.

Hay, S.I., Guerra, C.A., Gething, P.W., Patil, A.P., Tatem, A.J., Noor, A.M., Kabaria, C.W., Manh, B.H., Elyazar, I.R., Brooker, S., *et al.* (2009). A world malaria map: *Plasmodium falciparum* endemicity in 2007. PLoS Med. *6*, e1000048.

Holland, Z., Prudent, R., Reiser, J.B., Cochet, C., and Doerig, C. (2009). Functional analysis of protein kinase CK2 of the human malaria parasite *Plasmodium falciparum*. Eukaryot. Cell *8*, 388–397.

Hu, G., Cabrera, A., Kono, M., Mok, S., Chaal, B.K., Haase, S., Engelberg, K., Cheemadan, S., Spielmann, T., Preiser, P.R., *et al.* (2010). Transcriptional profiling of growth perturbations of the human malaria parasite *Plasmodium falciparum*. Nat. Biotechnol. *28*, 91–98.

Hyde, J.E. (2007). Drug-resistant malaria – an insight. FEBS J. *274*, 4688–4698.

Jelinsky, S.A., and Samson, L.D. (1999). Global response of *Saccharomyces cerevisiae* to an alkylating agent. Proc. Natl. Acad. Sci. U.S.A. *96*, 1486–1491.

Jensen, M., and Mehlhorn, H. (2009). Seventy-five years of Resochin in the fight against malaria. Parasitol. Res. *105*, 609–627.

Joshi, B., Biswas, S., and Sharma, Y.D. (1992). Effect of heat-shock on *Plasmodium falciparum* viability, growth and expression of the heat-shock protein 'PFHSP70-I' gene. FEBS Lett. *312*, 91–94.

Kehr, S., Sturm, N., Rahlfs, S., Przyborski, J.M., and Becker, K. (2011). Compartmentation of redox metabolism in malaria parasites. PLoS Pathog. *6*, e1001242.

Keyomarsi, K., Samet, J., Molnar, G., and Pardee, A.B. (1993). The thymidylate synthase inhibitor, ICI D1694, overcomes translational detainment of the enzyme. J. Biol. Chem. *268*, 15142–15149.

Koncarevic, S., Rohrbach, P., Deponte, M., Krohne, G., Prieto, J.H., Yates, J., 3rd, Rahlfs, S., and Becker, K. (2009). The malarial parasite *Plasmodium falciparum* imports the human protein peroxiredoxin 2 for peroxide detoxification. Proc. Natl. Acad. Sci. U.S.A. *106*, 13323–13328.

Krishna, S., Pulcini, S., Fatih, F., and Staines, H. (2010). Artemisinins and the biological basis for the PfATP6/SERCA hypothesis. Trends Parasitol. *26*, 517–523.

Krnajski, Z., Gilberger, T.W., Walter, R.D., Cowman, A.F., and Müller, S. (2002). Thioredoxin reductase is essential for the survival of *Plasmodium falciparum* erythrocytic stages. J. Biol. Chem. *277*, 25970–25975.

Kumar, R., Adams, B., Musiyenko, A., Shulyayeva, O., and Barik, S. (2005). The FK506-binding protein of the malaria parasite, *Plasmodium falciparum*, is a FK506-sensitive chaperone with FK506-independent calcineurin-inhibitory activity. Mol. Biochem. Parasitol. *141*, 163–173.

Kumar, R., Pavithra, S.R., and Tatu, U. (2007). Three-dimensional structure of heat shock protein 90 from *Plasmodium falciparum*: molecular modelling approach to rational drug design against malaria. J. Biosci. *32*, 531–536.

Kwiatkowski, D., and Greenwood, B.M. (1989). Why is malaria fever periodic? A hypothesis. Parasitol. Today *5*, 264–266.

Le Roch, K.G., Johnson, J.R., Florens, L., Zhou, Y., Santrosyan, A., Grainger, M., Yan, S.F., Williamson, K.C., Holder, A.A., Carucci, D.J., *et al.* (2004). Global analysis of transcript and protein levels across the *Plasmodium falciparum* life cycle. Genome Res. *14*, 2308–2318.

Lindner, S.E., De Silva, E.K., Keck, J.L., and Llinas, M. (2010). Structural determinants of DNA binding by a

P. falciparum ApiAP2 transcriptional regulator. J. Mol. Biol. *395*, 558–567.

Liu, J., Istvan, E.S., Gluzman, I.Y., Gross, J., and Goldberg, D.E. (2006). *Plasmodium falciparum* ensures its amino acid supply with multiple acquisition pathways and redundant proteolytic enzyme systems. Proc. Natl. Acad. Sci. U.S.A. *103*, 8840–8845.

Lopez-Rubio, J.J., Mancio-Silva, L., and Scherf, A. (2009). Genome-wide analysis of heterochromatin associates clonally variant gene regulation with perinuclear repressive centers in malaria parasites. Cell Host Microbe *5*, 179–190.

Lu, F., Lim, C.S., Nam, D.H., Kim, K., Lin, K., Kim, T.S., Lee, H.W., Chen, J.H., Wang, Y., Sattabongkot, J., *et al.* (2010). Mutations in the antifolate-resistance-associated genes dihydrofolate reductase and dihydropteroate synthase in *Plasmodium vivax* isolates from malaria-endemic countries. Am. J. Trop. Med. Hyg. *83*, 474–479.

Mackinnon, M.J., Li, J., Mok, S., Kortok, M.M., Marsh, K., Preiser, P.R., and Bozdech, Z. (2009). Comparative transcriptional and genomic analysis of *Plasmodium falciparum* field isolates. PLoS Pathog. *5*, e1000644.

Mair, G.R., Braks, J.A., Garver, L.S., Wiegant, J.C., Hall, N., Dirks, R.W., Khan, S.M., Dimopoulos, G., Janse, C.J., and Waters, A.P. (2006). Regulation of sexual development of Plasmodium by translational repression. Science *313*, 667–669.

Mair, G.R., Lasonder, E., Garver, L.S., Franke-Fayard, B.M., Carret, C.K., Wiegant, J.C., Dirks, R.W., Dimopoulos, G., Janse, C.J., and Waters, A.P. (2010). Universal features of post-transcriptional gene regulation are critical for Plasmodium zygote development. PLoS Pathog. *6*, e1000767.

McMillan, P.J., Stimmler, L.M., Foth, B.J., McFadden, G.I., and Muller, S. (2005). The human malaria parasite *Plasmodium falciparum* possesses two distinct dihydrolipoamide dehydrogenases. Mol. Microbiol. *55*, 27–38.

Merrick, C.J., Dzikowski, R., Imamura, H., Chuang, J., Deitsch, K., and Duraisingh, M.T. (2010). The effect of *Plasmodium falciparum* Sir2a histone deacetylase on clonal and longitudinal variation in expression of the var family of virulence genes. Int. J. Parasitol. *40*, 35–43.

Mok, S., Imwong, M., Mackinnon, M.J., Sim, J., Ramadoss, R., Yi, P., Mayxay, M., Chotivanich, K., Liong, K.Y., Russell, B., *et al.* (2011). Artemisinin resistance in *Plasmodium falciparum* is associated with an altered temporal pattern of transcription. BMC Genomics *12*, 391.

Moradas-Ferreira, P., and Costa, V. (2000). Adaptive response of the yeast *Saccharomyces cerevisiae* to reactive oxygen species: defences, damage and death. Redox Rep. *5*, 277–285.

Muller, I.B., and Hyde, J.E. (2010). Antimalarial drugs: modes of action and mechanisms of parasite resistance. Future Microbiol. *5*, 1857–1873.

Müller, S. (2003). Thioredoxin reductase and glutathione synthesis in *Plasmodium falciparum*. Redox Rep. *8*, 251–255.

Müller, S. (2004). Redox and antioxidant systems of the malaria parasite *Plasmodium falciparum*. Mol. Microbiol. *53*, 1291–1305.

Müller, S., Liebau, E., Walter, R.D., and Krauth-Siegel, R.L. (2003). Thiol-based redox metabolism of protozoan parasites. Trends Parasitol. *19*, 320–328.

Natalang, O., Bischoff, E., Deplaine, G., Proux, C., Dillies, M.A., Sismeiro, O., Guigon, G., Bonnefoy, S., Patarapotikul, J., Mercereau-Puijalon, O., *et al.* (2008). Dynamic RNA profiling in *Plasmodium falciparum* synchronized blood stages exposed to lethal doses of artesunate. BMC Genomics *9*, 388.

Nirmalan, N., Sims, P.F., and Hyde, J.E. (2004a). Quantitative proteomics of the human malaria parasite *Plasmodium falciparum* and its application to studies of development and inhibition. Mol. Microbiol. *52*, 1187–1199.

Nirmalan, N., Sims, P.F., and Hyde, J.E. (2004b). Translational up-regulation of antifolate drug targets in the human malaria parasite *Plasmodium falciparum* upon challenge with inhibitors. Mol. Biochem. Parasitol. *136*, 63–70.

Nogueira, F., Diez, A., Radfar, A., Perez-Benavente, S., do Rosario, V.E., Puyet, A., and Bautista, J.M. (2010). Early transcriptional response to chloroquine of the *Plasmodium falciparum* antioxidant defence in sensitive and resistant clones. Acta Trop. *114*, 109–115.

Nosten, F., and White, N.J. (2007). Artemisinin-based combination treatment of falciparum malaria. Am. J. Trop. Med. Hyg. *77*, 181–192.

Oakley, M.S., Kumar, S., Anantharaman, V., Zheng, H., Mahajan, B., Haynes, J.D., Moch, J.K., Fairhurst, R., McCutchan, T.F., and Aravind, L. (2007). Molecular factors and biochemical pathways induced by febrile temperature in intraerythrocytic *Plasmodium falciparum* parasites. Infect. Immun. *75*, 2012–2025.

Oakley, M.S., Gerald, N., McCutchan, T.F., Aravind, L., and Kumar, S. (2011). Clinical and molecular aspects of malaria fever. Trends Parasitol. *27*, 442–449.

Painter, H.J., Campbell, T.L., and Llinas, M. (2011). The Apicomplexan AP2 family: integral factors regulating Plasmodium development. Mol. Biochem. Parasitol. *176*, 1–7.

Pallavi, R., Roy, N., Nageshan, R.K., Talukdar, P., Pavithra, S.R., Reddy, R., Venketesh, S., Kumar, R., Gupta, A.K., Singh, R.K., *et al.* (2010). Heat shock protein 90 as a drug target against protozoan infections: biochemical characterization of HSP90 from *Plasmodium falciparum* and *Trypanosoma evansi* and evaluation of its inhibitor as a candidate drug. J. Biol. Chem. *285*, 37964–37975.

Palmer, A., and Nebreda, A.R. (2000). The activation of MAP kinase and p34cdc2/cyclin B during the meiotic maturation of Xenopus oocytes. Prog. Cell Cycle Res. *4*, 131–143.

Pang, C.K., Hunter, J.H., Gujjar, R., Podutoori, R., Bowman, J., Mudeppa, D.G., and Rathod, P.K. (2009). Catalytic and ligand-binding characteristics of *Plasmodium falciparum* serine hydroxymethyltransferase. Mol. Biochem. Parasitol. *168*, 74–83.

Pastrana-Mena, R., Dinglasan, R.R., Franke-Fayard, B., Vega-Rodriguez, J., Fuentes-Caraballo, M., Baerga-Ortiz, A., Coppens, I., Jacobs-Lorena, M., Janse, C.J., and Serrano, A.E. (2010). Glutathione reductase-null malaria parasites have normal blood stage growth but arrest during development in the mosquito. J. Biol. Chem. *285*, 27045–27056.

Pavithra, S.R., Banumathy, G., Joy, O., Singh, V., and Tatu, U. (2004). Recurrent fever promotes *Plasmodium falciparum* development in human erythrocytes. J. Biol. Chem. *279*, 46692–46699.

Pavithra, S.R., Kumar, R., and Tatu, U. (2007). Systems analysis of chaperone networks in the malarial parasite *Plasmodium falciparum*. PLoS Comput. Biol. *3*, 1701–1715.

Pesce, E.R., Acharya, P., Tatu, U., Nicoll, W.S., Shonhai, A., Hoppe, H.C., and Blatch, G.L. (2008). The *Plasmodium falciparum* heat shock protein 40, Pfj4, associates with heat shock protein 70 and shows similar heat induction and localisation patterns. Int. J. Biochem. Cell. Biol. *40*, 2914–2926.

Read, A.F., and Taylor, L.H. (2001). The ecology of genetically diverse infections. Science *292*, 1099–1102.

Reece, S.E., Ali, E., Schneider, P., and Babiker, H.A. (2010). Stress, drugs and the evolution of reproductive restraint in malaria parasites. Proc. Biol. Sci. *277*, 3123–3129.

Reverter-Branchat, G., Cabiscol, E., Tamarit, J., and Ros, J. (2004). Oxidative damage to specific proteins in replicative and chronological-aged *Saccharomyces cerevisiae*: common targets and prevention by calorie restriction. J. Biol. Chem. *279*, 31983–31989.

Richard, D., Bartfai, R., Volz, J., Ralph, S.A., Muller, S., Stunnenberg, H.G., and Cowman, A.F. (2011). A genome-wide chromatin-associated nuclear peroxiredoxin from the malaria parasite *Plasmodium falciparum*. J. Biol. Chem. *286*, 11746–11755.

Robinson, M.J., and Cobb, M.H. (1997). Mitogen-activated protein kinase pathways. Curr. Opin. Cell Biol. *9*, 180–186.

Salcedo-Amaya, A.M., Hoeijmakers, W.A., Bartfai, R., and Stunnenberg, H.G. (2010). Malaria: could its unusual epigenome be the weak spot? Int. J. Biochem. Cell. Biol. *42*, 781–784.

Sanchez, C.P., Dave, A., Stein, W.D., and Lanzer, M. (2010). Transporters as mediators of drug resistance in *Plasmodium falciparum*. Int. J. Parasitol. *40*, 1109–1118.

Shock, J.L., Fischer, K.F., and DeRisi, J.L. (2007). Whole-genome analysis of mRNA decay in *Plasmodium falciparum* reveals a global lengthening of mRNA half-life during the intra-erythrocytic development cycle. Genome Biol. *8*, R134.

Sicard, A., Semblat, J.P., Doerig, C., Hamelin, R., Moniatte, M., Dorin-Semblat, D., Spicer, J.A., Srivastava, A., Retzlaff, S., Heussler, V., *et al.* (2011). Activation of a PAK-MEK signalling pathway in malaria parasite-infected erythrocytes. Cell. Microbiol. *13*, 836–845.

Sienkiewicz, N., Daher, W., Dive, D., Wrenger, C., Viscogliosi, E., Wintjens, R., Jouin, H., Capron, M., Muller, S., and Khalife, J. (2004). Identification of a mitochondrial superoxide dismutase with an unusual targeting sequence in *Plasmodium falciparum*. Mol. Biochem. Parasitol. *137*, 121–132.

Sims, J.S., Militello, K.T., Sims, P.A., Patel, V.P., Kasper, J.M., and Wirth, D.F. (2009). Patterns of gene-specific and total transcriptional activity during the *Plasmodium falciparum* intraerythrocytic developmental cycle. Eukaryot. Cell *8*, 327–338.

Snow, R.W., Guerra, C.A., Noor, A.M., Myint, H.Y., and Hay, S.I. (2005). The global distribution of clinical episodes of *Plasmodium falciparum* malaria. Nature *434*, 214–217.

Sorger, P.K. (1991). Heat shock factor and the heat shock response. Cell *65*, 363–366.

Sullivan, W.J., Jr., Narasimhan, J., Bhatti, M.M., and Wek, R.C. (2004). Parasite-specific eIF2 (eukaryotic initiation factor-2) kinase required for stress-induced translation control. Biochem. J. *380*, 523–531.

Sztajer, H., Gamain, B., Aumann, K.D., Slomianny, C., Becker, K., Brigelius-Flohe, R., and Flohe, L. (2001). The putative glutathione peroxidase gene of *Plasmodium falciparum* codes for a thioredoxin peroxidase. J. Biol. Chem. *276*, 7397–7403.

Taylor, L.H., and Read, A.F. (1997). Why so few transmission stages? Reproductive restraint by malaria parasites. Parasitol. Today *13*, 135–140.

Teixeira, M.C., Mira, N.P., and Sa-Correia, I. (2011). A genome-wide perspective on the response and tolerance to food-relevant stresses in *Saccharomyces cerevisiae*. Curr. Opin. Biotechnol. *22*, 150–156.

Tewari, R., Dorin, D., Moon, R., Doerig, C., and Billker, O. (2005). An atypical mitogen-activated protein kinase controls cytokinesis and flagellar motility during male gamete formation in a malaria parasite. Mol. Microbiol. *58*, 1253–1263.

Vega-Rodriguez, J., Franke-Fayard, B., Dinglasan, R.R., Janse, C.J., Pastrana-Mena, R., Waters, A.P., Coppens, I., Rodriguez-Orengo, J.F., Srinivasan, P., Jacobs-Lorena, M., *et al.* (2009). The glutathione biosynthetic pathway of Plasmodium is essential for mosquito transmission. PLoS Pathog. *5*, e1000302.

Ward, P., Equinet, L., Packer, J., and Doerig, C. (2004). Protein kinases of the human malaria parasite *Plasmodium falciparum*: the kinome of a divergent eukaryote. BMC Genomics *5*, 79.

Ward, S.A., Sevene, E.J., Hastings, I.M., Nosten, F., and McGready, R. (2007). Antimalarial drugs and pregnancy: safety, pharmacokinetics, and pharmacovigilance. Lancet Infect. Dis. *7*, 136–144.

Wellems, T.E., and Plowe, C.V. (2001). Chloroquine-resistant malaria. J. Infect. Dis. *184*, 770–776.

Wilkes, J.M., and Doerig, C. (2008). The protein-phosphatome of the human malaria parasite *Plasmodium falciparum*. BMC Genomics *9*, 412.

Wilson, M., DeRisi, J., Kristensen, H.H., Imboden, P., Rane, S., Brown, P.O., and Schoolnik, G.K. (1999). Exploring drug-induced alterations in gene expression in *Mycobacterium tuberculosis* by microarray hybridization. Proc. Natl. Acad. Sci. U.S.A. *96*, 12833–12838.

Witkowski, B., Lelievre, J., Barragan, M.J., Laurent, V., Su, X.Z., Berry, A., and Benoit-Vical, F. (2010). Increased tolerance to artemisinin in *Plasmodium falciparum* is mediated by a quiescence mechanism. Antimicrob. Agents Chemother. *54*, 1872–1877.

Wong, E.H., Hasenkamp, S., and Horrocks, P. (2011). Analysis of the molecular mechanisms governing the stage-specific expression of a prototypical housekeeping gene during intraerythrocytic development of *P. falciparum*. J. Mol. Biol. *408*, 205–221.

Wrenger, C., and Müller, S. (2003). Isocitrate dehydrogenase of *Plasmodium falciparum*. Eur. J. Biochem. *270*, 1775–1783.

Wrenger, C., Eschbach, M.L., Muller, I.B., Laun, N.P., Begley, T.P., and Walter, R.D. (2006). Vitamin B1 *de novo* synthesis in the human malaria parasite *Plasmodium falciparum* depends on external provision of 4-amino-5-hydroxymethyl-2-methylpyrimidine. Biol. Chem. *387*, 41–51.

Zhang, M., Fennell, C., Ranford-Cartwright, L., Sakthivel, R., Gueirard, P., Meister, S., Caspi, A., Doerig, C., Nussenzweig, R.S., Tuteja, R., *et al.* (2010). The Plasmodium eukaryotic initiation factor-2alpha kinase IK2 controls the latency of sporozoites in the mosquito salivary glands. J. Exp. Med. *207*, 1465–1474.

Toxoplasma gondii: Without Stress There is No Life 13

Maria J. Figueras, Sergio O. Angel, Verónica M. Cóceres and Maria L. Alomar

Abstract

Toxoplasma gondii is an important pathogen of human and domestic animals. It has a complex life cycle which includes the transition from one host to another, being only exposed to the environment during one stage, as highly resistant oocysts. Interestingly, in the intermediate host (non-feline mammalians and birds) the parasite presents an asexual cycle with two stages that can interconvert without its passage in the definite host (felines). The asexual cycle is very important in the establishment of the infection and on its pathogenesis and it could be driven by different kind of stressors. Therefore, the response to environmental and host stresses is essential to their viability and successful progression through their life cycle. The heat shock proteins are key molecules not only in the resistance to different stressors, but they are also involved in the optimal differentiation as well as in other biological processes in *T. gondii*. This chapter summarizes the findings on different aspects of *T. gondii* stress responses and the implication of these processes in the biology and pathogenesis of this parasite.

Introduction

Toxoplasma gondii is a coccidian protozoan parasite that belongs to the phylum Apicomplexa. This phylum includes *Plasmodium* spp., *Babesia* spp., *Eimeria* spp., *Cryptosporidium* spp., among others important animal and human pathogens. *T. gondii* is a unique and fascinating organism. Contrary to other genera, the *Toxoplasma* genus has only one species: *Toxoplasma gondii*. *Toxoplasma* can be carried by many warm-blooded animals (birds or mammals including humans) and is able to infect most of the nucleated cells. During the invasion process, the host-cell remains passive while this non-flagellated parasite actively penetrates into the cell, being the motility machinery that generates a characteristic movement named gliding, fundamental in this process (Baum *et al.*, 2008). Once inside the host cell the parasite replicates within a parasitophorous vacuole (PV) through a mechanism called endodiogeny characterized by two daughter cells generated within one mother cell (Gubbels *et al.*, 2008).

It is estimated that toxoplasmosis is present as a chronic asymptomatic form into 5 hundred million to 1 billion of people worldwide (Tenter *et al.*, 2000). Although *T. gondii* infection is usually asymptomatic in most individuals, it is of great medical significance for pregnant women and immunocompromised patients. During pregnancy, primary *T. gondii* infection may result in congenital infection, which rarely may lead to abortion or severe neonatal consequences, being the most common consequence that infants are clinically asymptomatic at birth, developing clinical symptoms, predominantly eye and/or central nervous system injuries (Wilson *et al.*, 1980; Frenkel, 1988). Furthermore, toxoplasmosis may lead to severe damage in immunocompromised patients as those with AIDS, neoplastic diseases or transplanted organs (Frenkel, 1988). Chronic infection has been correlated with risks of schizophrenia (Yolken *et al.*, 2009). In fact, a recent study reported that patients with high levels of *T. gondii*-specific IgG antibodies have a significantly higher risk of developing schizophrenia spectrum disorders (Pedersen *et al.*, 2011).

In veterinary medicine, *T. gondii* infection is of economic importance owing to abortion and neonatal loss in sheep and goats or as source of transmission to humans (Dubey, 1990; Dubey and Thulliez, 1993). In addition, *Toxoplasma* is listed as a Category B pathogen in the NIAID list of organisms of interest for biodefense research.

The life cycle of *T. gondii* is complex alternating between definitive and intermediate hosts (Fig. 13.1). The definitive hosts are members of cat family (*Felidae*) which were detected for the first time by Frenkel *et al.* (1970). The gametogenesis and mating that culminates in the generation of oocysts that are shed in the cat's faeces occur within the intestinal epithelium of felines, where they sporulate to become infectious (Fig. 13.1). These oocysts are highly infectious and extremely stable in the environment. The ingestion of oocyst by an intermediate host, including human, leads to the asexual cycle, which is characterized by two stages, the rapidly growing tachyzoites, and the latent bradyzoite tissue cysts. Tachyzoites are responsible for acute illness and congenital birth defects. However the infection is asymptomatic in immunocompetent individuals. The presence of interferon-gamma (IFNγ) and T-cell mediated immunity is the responsible to eliminate the tachyzoite (Khan *et al.*, 1999; Lykens *et al.*, 2010). Nevertheless, some tachyzoites may be able to survive and then differentiate into the more slowly dividing bradyzoite form. This form can remain as latent cyst (especially in neural and muscular tissue) for many years and still be capable of converting into the destructive tachyzoite form if host immunity wanes. These two developmental stages are essential for disease propagation and causation. Recently, a genetic polymorphism analysis showed that the high success of this parasite could be owing to the fact that it emerged from at least three clonal lineages within the last 10,000 years after a single genetic cross that resulted from the concurrent acquisition of direct oral infectivity (Su *et al.*, 2003). This key adaptation circumvented sexual recombination, simultaneously promoting transmission through successive hosts (Fig. 13.1).

Stress changes the life

In general, it has been recognized that heat shock stress is playing significant roles in host–parasite relationship. It is especially important in protozoan parasites such as *Leishmania* spp., *Plasmodium* spp. and *Trypanosoma* spp., which present adaptation to changes in temperature associated with transmission from poikilothermic insect vectors to homeothermic mammalian hosts. In the case of *Toxoplasma*, it is remarkable that oocysts released from cat faeces are not infective until sporulation, process that require an external environmental exposure of 24–48 hours (Dubey *et al.*, 1970). Interestingly, sporulation is affected by temperatures below 4°C or higher than 37°C (Frenkel *et al.*, 1970). Unfortunately, up to now, there are only few studies on parasite oocysts;

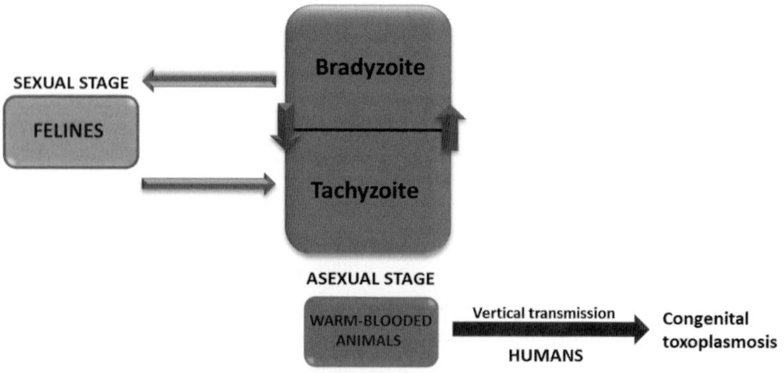

Figure 13.1 Schematic representation of the replication cycles in *T. gondii*. Felines are the definitive hosts where sexual replication takes place. Warm-blooded animals, including mammals and some birds, are the intermediate host where asexual replication occurs. Human congenital toxoplasmosis can be developed when vertical transmission occurs.

consequently, those factors associated to cold or heat stress during sporulation and environment aggression have not been identified yet.

As mentioned above, the success of *Toxoplasma* lies on its capacity of infection via oral consumption of tissue infected with cysts, which also implies that the asexual cycle can occur independently from the sexual cycle (Su *et al.*, 2003). In addition, the conversion between tachyzoites and bradyzoites (in both ways) plays a pivotal role in establishing the chronic infection and it is also responsible for disease reactivation. Therefore, the tachyzoite–bradyzoite interconversion that occurs during the asexual cycle is central to the pathogenesis and longevity of infection. It is thought that during infection the stage conversion is the parasite's response to environmental conditions such as inflammatory response from the host to the tachyzoite stage (Montoya and Liesenfeld, 2004) (Fig. 13.2). According to this vision, bradyzoites can reconvert to replicative tachyzoites in severely immunocompromised hosts (Gross *et al.*, 1996). Nevertheless, it is common to detect spontaneous bradyzoite to tachyzoite conversion within tissue (Weiss and Kim, 2000). To further analyse the mechanisms and factors underlying the tachyzoite to bradyzoite conversion, several cell culture differentiation systems were developed (Ferreira da Silva *et al.*, 2008). Infection of murine astrocytes with tachyzoites combined with the intermittent additions of IFNγ to the media produced cyst development (Jones *et al.*, 1986). Interestingly, the presence of IFNγ was not necessary for cyst formation, but it allowed cysts to remain for prolonged periods without rupturing. Several *in vitro* studies show that immunological factors are not

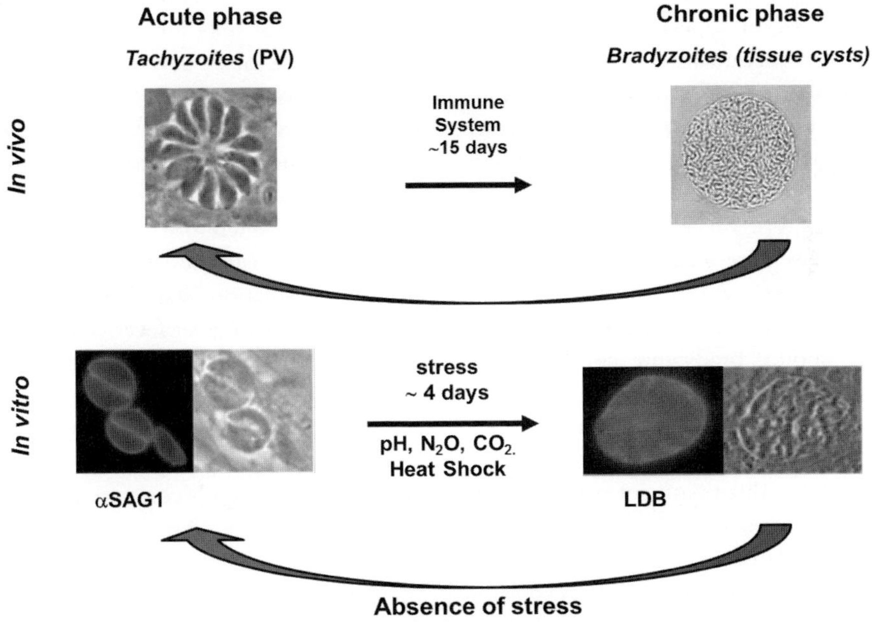

Figure 13.2 Asexual cycle of *T. gondii* has two stages: the highly replicative tachyzoite and the slowly replicative bradyzoite. *In vivo* the intracellular tachyzoite forms parasitophorous vacuoles (PV) whereas the bradyzoite stage forms intracellular cysts. The PVs and the cysts are different structures that can be easily distinguished by optical microscopy. *In vitro* parasites can be cultivated under standard tachyzoites conditions. However, when these tachyzoites are submitted to stress conditions they differentiate to bradyzoites. At the same time if the stress conditions are stopped, the bradyzoites turned over and become tachyzoites again. In this case, the structures are not easily to determine by bright field microscopy, but these two stages express different proteins, which allow identifying the stages through indirect immunofluorescence assays. As it is shown in the figure, SAG1 is a specific marker for tachyzoite membrane and *Dolichos biflorous* lectin is a specific dye that intercalates within the cyst wall.

necessarily required for cyst formation since high rates of spontaneous conversion from tachyzoites to bradyzoites were mainly dependent of the host cell type. As an example, Luder *et al.* (1999) observed the spontaneous appearance of bradyzoites in cells (neurons, astrocytes, and microglia) extracted from the central nervous system of rats at the fourth day post infection. More recently, Ferreira da Silva *et al.* (2009) showed spontaneous cyst formation of virulent (and apparently non-cystogenic strain) RH when skeletal muscle cells were infected, in a more successful way than fibroblast or another cell types. Noteworthy, the incubation of host-cells with a trisubstituted pyrrole 4-[2-(4-fluorophenyl)-5-(1-methylpiperidine-4-yl)-1H-pyrrol-3-yl] pyridine, designated as compound 1, reduced tachyzoite cell cycle and favoured bradyzoite generation, based on host cell transcription profile changes that includes autoantigen-1 (Radke *et al.*, 2006). These results imply that important questions concerning the host cell type and its relevance in *Toxoplasma* differentiation are still unanswered.

Despite the capacity of *T. gondii* to generate spontaneously cyst *in vitro*, especially in nervous or muscle-associated cell, one important finding was the induction of high rates of cyst formation in infected human foreskin cell (HFF) monolayer cultures when they were incubated under different stress conditions (Soete *et al.*, 1994). A change in pH and temperature, such as incubations at pH 8.1 and 43°C induce bradyzoite conversion of virulent RH strain in Vero and HFF cells. Moreover, the stage differentiation can be also obtained by treating tachyzoites with sodium arsenite, a stress inducer drug. This treatment produces the expression of bradyzoite genes but was inefficient to induce cyst formation, suggesting that parasite differentiation is a multistep process. The exposure of tachyzoites to nutrient factors by deprivation of the amino acid arginine or interfering with the pyrimidine pathway as well as temporary axenic cultivation triggers the differentiation to bradyzoites (Ferreira da Silva *et al.*, 2008). Interestingly, bradyzoite development can also be induced by exposure of extracellular tachyzoites to alkaline pH for only one hour prior to infection of the host cells and cultivation at physiological pH (Weiss *et al.*, 1998). This result suggests that commitment of *T. gondii* to differentiate to the bradyzoite stage can occur rapidly and as a direct response of the parasite to stressful conditions, without intracellular environment-dependent signal. Many of the genes altered during alkalyne stress in *Toxoplasma* have been previously described as being part of the common environmental response (CER) or environmental stress response (ESR) while others appear specific to the alkaline stress response (Naguleswaran *et al.*, 2010). CER/ESR is characterized by increased expression of genes involved in a wide array of cellular processes including metabolism detoxification, protein folding and degradation, metabolite transport, vacuolar and mitochondrial function, and intracellular signalling (Naguleswaran *et al.*, 2010).

Heat shock proteins make the difference

To date, there is no doubt that heat shock proteins (HSPs) are not only involved in stress response but they are also developmentally regulated, having important roles during life cycle. The family of HSP includes a large number of proteins constitutively expressed in high quantities. Nevertheless, their expression increases when the cell is subjected to stress conditions: elevated temperatures, osmotic shock, differences in the pH, UV irradiation, starvation conditions, viral infections, bleedings or fever (Gething and Sambrook, 1992). HSPs help the cell to respond to stress factors producing what is called the heat shock response or just stress response, which is characterized by the turn-on and turn-off of some genes. HSPs are highly conserved within the three main phylogenetic domains (Bacteria, Archea and Eukarya) which suggest an important role of them and other molecular chaperones. In protozoan systems a link between HSPs, development and intracellular survival has been established. For instance, *Leishmania donovani* HSP100 is expressed during axenic amastigote development (Krobitsch *et al.*, 1998), while *Leishmania chagasi* HSPs are associated with survival during oxidative stress and may play a role in promastigote to amastigote transition (Wilson *et al.*, 1994). In *L. donovani* it was proposed that HSP90 inactivation induces the differentiation from the promastigote to the amastigote stage (Wiesgigl and Clos, 2001). *P. falciparum* HSP90 is an essential protein during growth in human erythrocytes (Banumathy *et al.*, 2003).

In *T. gondii*, the inhibition of HSPs synthesis by quercetin coincided with a reduction in the number of alkaline-induced bradyzoites (Weiss *et al.*, 1998). It was also observed that the expression of *T. gondii* HSP70, HSP60, HSP90 and small heat shock (sHSP) proteins increased during bradyzoite formation (Weiss *et al.*, 1998; Toursel *et al.*, 2000; Echeverria *et al.*, 2005). The expression of HSP70 also increases in the conversion of bradyzoite to tachyzoite when working with cysts *in vivo* in a toxoplasmic reactivation murine model using anti-γIFN (Silva *et al.*, 1998).

A group of proteins that are still poorly studied but could have an important role during parasite differentiation are those belonging to small heat shock/alpha crystalline (sHSPs) family. *T. gondii* has five different sHSPs (HSP20, HSP21, HSP28, HSP29 and HSP30) which share high sequence homology in the α-crystalline domain and show increased expression upon heat stress conditions (de Miguel *et al.*, 2005). One of the *T. gondii* sHSPs, HSP28, is predominantly expressed in tachyzoites rather than in bradyzoites (de Miguel *et al.*, 2005), while another sHSP, HSP30 (also denominated bradyzoite antigen 1 or BAG1), is only expressed in bradyzoites (Bohne *et al.*, 1995; Parmley *et al.*, 1995). Interestingly, they also have different intracellular locations (Fig. 13.3). While HSP21, HSP29 and HSP30 are located in the cytosol (de Miguel *et al.*, 2005, 2009), HSP28 is located in the parasite mitochondrion (de Miguel *et al.*, 2005) and HSP20 is associated with the outer leaflet of the inner membrane complex (IMC) (de Miguel *et al.*, 2008), a continuous patchwork of flattened vesicles located beneath the plasma membrane associated to replication and gliding/invasion processes (de Melo and de Souza, 1997; Hu *et al.*, 2002). All five proteins form oligomers (12–24 subunits) and are potent chaperones *in vitro* (de Miguel *et al.*, 2009). However, the functional role of any *T. gondii* sHSP remains unknown except for HSP20 (see below). Surprisingly, the chaperon activity of HSP20, HSP21 and HSP28 were activated at temperatures ranging from 4°C to 25°C, below physiological conditions (de Miguel *et al.*, 2009). Asexual stages of *T. gondii* normally live in a range of temperature near 35°C. By contrast, *Toxoplasma* oocysts are subject of environmental stress, requiring for a correct sporulation a range of temperature within 5–37°C. Concerning the sexual cycle and according to the expression data at Toxodb (www.toxodb.org) HSP21 and HSP28 are expressed in unsporulated and sporulated oocysts, whereas HSP20 and HSP29 seem to be expressed only during parasite sporulation. In this scenario, an activation of some parasite sHSPs could be important to allow a proper sporulation and/or resist different environment conditions until the stress situation is overcome. Recent studies on cold stress demonstrated that several classical chaperones are involved in protein homeostasis during this kind of stress (Roobol *et al.*, 2009). Thus, the activation of the chaperone activity of *T. gondii* sHSPs under low, non-physiological temperatures suggests an important role for this protein family to cold stress tolerance in the sporozoite stage.

The expression of HSP30/BAG1 is restricted to the latent bradyzoite form of *T. gondii*, first appearing ~2–3 days after the initiation of bradyzoite differentiation (de Miguel *et al.*, 2005). The *bag1* null mutant was found to express several bradyzoite-specific markers with the same kinetics and frequency as the parental strain when parasites were treated under stress conditions *in vitro* (Bohne *et al.*, 1998). However, it could be observed a slight reduction in the efficiency to form cysts (Zhang *et al.*, 1999b). Infection of mice with HSP30/BAG1 null mutant revealed that tachyzoites were fully able to establish a chronic infection in C57BL/6 mice, producing brain cysts of a size, morphology and frequency indistinguishable from cysts formed by the parental control strain (Bohne *et al.*, 1998). These cysts were also able to establish an acute infection after oral administration, demonstrating that conversion of bradyzoites to tachyzoites is also unimpaired. In conclusion BAG1 is not essential for normal function of bradyzoite containing tissue cysts, at least in intermediate host stages. Since a slight effect was observed (Zhang *et al.*, 1999b), it could be hypothesized that BAG1 could be a secondary modulator of some bradyzoite-pathways. Another possibility could be that another cytosolic sHSP

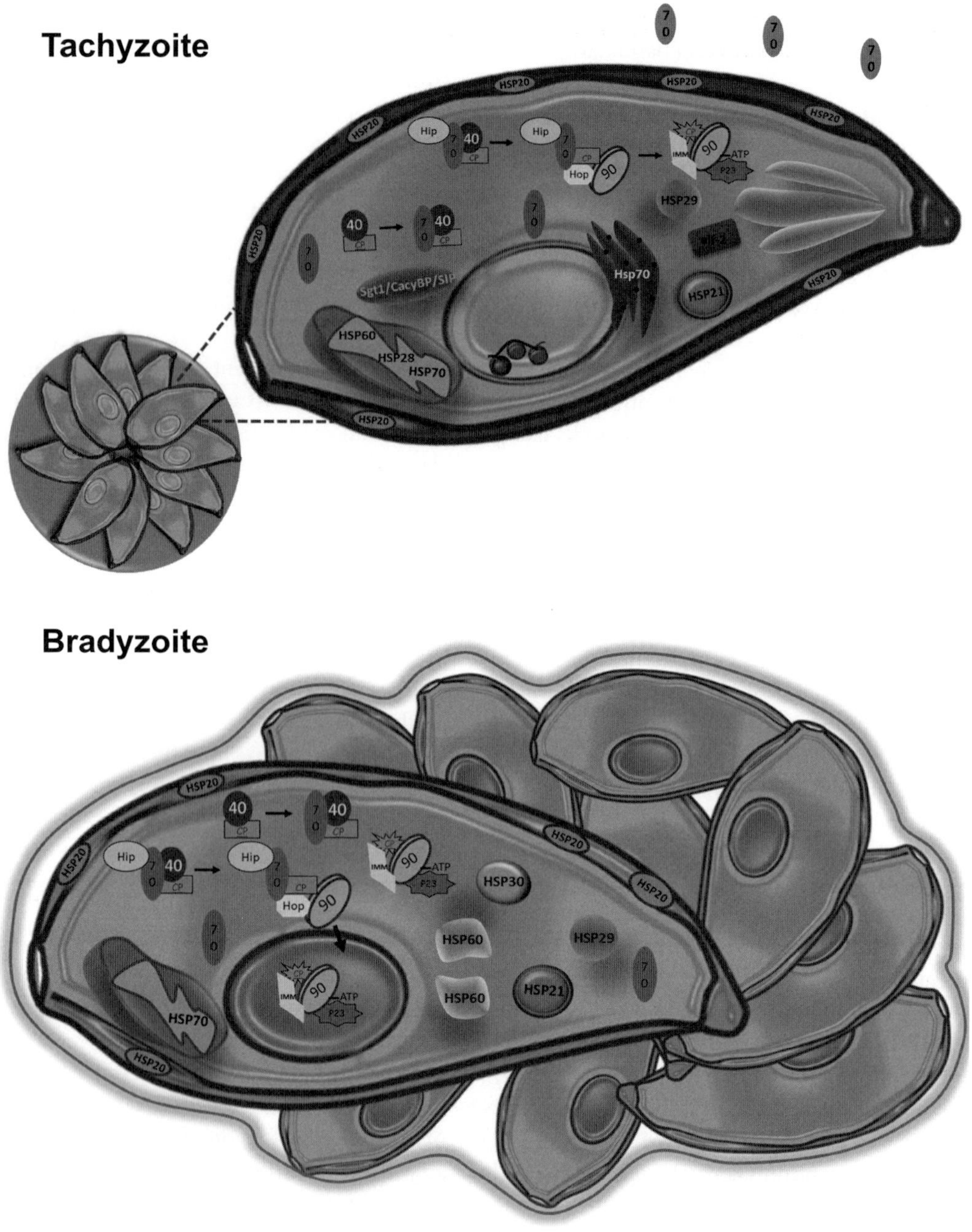

Figure 13.3 Schematic representation of the subcellular localization of heat shock proteins in tachyzoite and bradyzoite stages. Upper panel: On the left hand a graphical representation of a parasitophorous vacuole full with tachyzoites and on the right hand an amplification of a single tachyzoite are shown. Lower panel: A bradyzoite cyst is represented graphically beneath a single magnified bradyzoite For references: Hip: Hsp70/ Hsp90 Interacting protein, 40: HSP40, 70: HSP70, 90: HSP90, CP, client protein; Imm, immunophiline; eIF-2, elongation factor 2; Sgt1/CacyBP/SIP, calcyclin binding protein; HSP70: mitochondrial HSP70, HSP30: HSP30 or BAG1 (bradyzoite specific antigen 1).

(e.g. HSP21) chaperone could compensate the lack of BAG1/HSP30.

In addition to heat shock proteins the phosphorylation of eukaryotic initiation factor-2 (eIF2) also has a role during stress by regulating translation of some proteins as well as reduction of translation of other group of proteins (Gebauer and Hentze, 2004). Treatment of *T. gondii* under alkaline, arsenite or heat shock stress induced the phosphorylation of eIF2α, triggering the differentiation to bradyzoites and consequently the reduction of general protein synthesis in the quiescent bradyzoite (Sullivan *et al.*, 2004; Narasimhan *et al.*, 2008). Recently it was observed that extracellular tachyzoites could also support stressing condition where the TgIF2α phosphorylation is critical to maintain the parasite virulence when it is outside the host cell (Joyce *et al.*, 2010).

Giving a little of (hetero) complexity to parasite differentiation

One important aspect concerning HSPs is that their biological roles and expression are not restricted to stress response, but their expression can be constitutive or developmentally regulated. Among HSPs, eukaryotic HSP90 has a highly selective activity in stressed and unstressed cells, where this chaperone is responsible of the recovery of misfolded proteins, protein maturation, intracellular transport of proteins, and regulated activities of nuclear hormone receptors as well as other transcription factors, protein kinases involved in signal transduction and translation control (Pratt and Toft, 2003). The function of HSP90 is highly dependent on ATP and on its ATPase activity. This feature allowed the generation of drugs that specifically inhibit the binding of HSP90 to its substrates (e.g. geldanamycin or GA, a benzoquinone ansamycin antibiotic). GA binds to the HSP90 protein by interaction with its ATP binding pocket with much higher affinity that ATP itself (Stebbins *et al.*, 1997), altering the function or folding of proteins that bind with HSP90, a process that drives unbound proteins to the degradation pathway (Mimnaugh *et al.*, 1996). HSP90 was also studied during the development of a wide variety of organisms (Zhang *et al.*, 1999a; Graefe *et al.*, 2002; Segui-Simarro *et al.*, 2003). It was demonstrated that the HSP90 protein was essential in *Saccharomyces cerevisiae, Schizosaccharomyces pombe* and *Drosophila melanogaster* (Nathan *et al.*, 1997; Yue *et al.*, 1999; Graefe *et al.*, 2002). HSP90 proved to be important in controlling the cell cycle in *S. cerevisiae* (Nathan *et al.*, 1997). In *P. falciparum* the protein HSP90 is essential during the development inside the human erythrocytes (Banumathy *et al.*, 2003). In *Plasmodium yoelii,* it was observed a nuclear stage-specific localization of the HSP90 protein, which suggested that the protein might be transducing the necessary signals for the development of the parasite (Graefe *et al.*, 2002).

HSP90 is also regulated by other proteins, so called co-chaperones. Despite that almost all studies were done investigating the assembly of the glucocorticoid receptor (GR) to HSP90, it is thought that the same model explains the assembly of other client proteins (Pratt and Toft, 2003). Based on this, it could be defined five proteins involved in the HSP90-heterocomplex cycle: HSP90, HSP70, HOP (HSP70/HSP90 organization protein), p23 and HSP40. In addition, another co-chaperone, HSP70 interacting protein (HIP), was detected as an early marker of the HSP90-heterocomplex. By contrast, the p23 co-chaperone was detected as a marker of the mature heterocomplex (Freeman *et al.*, 2000; McLaughlin *et al.*, 2006).

As it was mentioned above, the expression of *T. gondii* HSP90 increases under stress treatment (Echeverria *et al.*, 2005). Since stressors produce parasite differentiation, this finding links HSP90 with a putative role in the *T. gondii* development. Nevertheless, treatment of *T. gondii* tachyzoites with GA blocks host cell invasion, parasite replication and tachyzoite to bradyzoite differentiation (Ahn *et al.*, 2003; Echeverria *et al.*, 2005). Furthermore, the subcellular localization also changes. HSP90 is only present in cytoplasm of tachyzoites and interestingly it occupies both cytoplasm and nucleus of bradyzoites as it was evident from immunofluorescence assays and confocal analysis (Echeverria *et al.*, 2005). The nuclear localization of HSP90 was observed *in vitro* in bradyzoites as well as obtained from brain cysts of infected mice (Echeverria *et al.*, 2005). Moreover, the treatment of these bradyzoites with GA delayed their conversion to tachyzoite. Once the bradyzoites became tachyzoites, the HSP90 localization changed and it is excluded from the nucleus being only present in the cytoplasm (Echeverria *et al.*, 2005). Thus, owing to its pleiotropic role, affecting invasion and differentiation, *T. gondii* HSP90 emerges as an interesting target for drug development.

In *T. gondii* all the co-chaperones of the HSP90-heterocomplex (HIP, HSP40, HOP and p23) are present (Fig. 13.3). They were also linked to the existence of at least two different complex (early: HIP, HSP90 and HSP70) and late (P23 and HSP90) (Echeverria *et al.*, 2010). HIP showed to be a cytosolic co-chaperone in both tachyzoites and bradyzoites stages. By contrast, p23 was detected in cytoplasm and nucleus of bradyzoites, while it was only present in the cytoplasm of tachyzoites, following the localization of HSP90. These data strongly suggest that HSP90-heterocomplex is functional in *T. gondii*, probably having a role during parasite differentiation.

The protein–protein interactome of the *Toxoplasma* p23 and HSP90 showed only highly expressed proteins involved in glycolysis, ATP production, protein synthesis, protein assistance (chaperones and foldases) and some kinases and phosphatases, but none of the putative HSP90-interactors seems to be nuclear proteins except for the histones H2AZ and H4 (Echeverria *et al.*, 2010). Glycolitic/ATP production enzymes are crucial during parasite development. In fact, the bradyzoite down-regulates the mitochondrion function and accumulates amylopectin granules (Denton *et al.*, 1996), indicating that in this stage the predominant pathway to obtain energy is glycolysis. According to this concept, pyruvate kinase and lactate dehydrogenase (LDH) enzymes increase their activity in the bradyzoite cells (Liwak and Ananvoranich, 2009). The importance of LDHs in bradyzoite conversion is evident since their overexpression enhances parasite differentiation under alkaline conditions (Liwak and Ananvoranich, 2009). In this context, p23/HSP90 could be assisting parasite glycolytic enzymes during bradyzoite conversion to ensure enough ATP generation in the new parasite scenario. Similarly, p23/HSP90 could be assisting the translational machinery to generate new proteins during early bradyzoite transformation (Echeverria *et al.*, 2010). The predicted pleiotropic role of parasite HSP90 and in consequence p23, is evidenced by the intracellular localization of its targets from the nucleus and apicoplast to the secretory organelles dense granules, ropthries and micronemes and by the wide range of annotated and predicted function of their interactors (Echeverria *et al.*, 2010).

Oxidative phosphorylation or glycolisis: a matter of stage

T. gondii possess a single tubular mitochondrion (Seeber *et al.*, 1998; Melo *et al.*, 2000) that contributes to the Krebs's cycle (tricarboxylic acid [TCA] cycle), and oxidative respiration (Vercesi *et al.*, 1998; Fleige *et al.*, 2008). As a unique cell, *T. gondii* mitochondrion lacks the pyruvate dehydrogenase complex (Fleige *et al.*, 2007), an important enzyme in the carbohydrate metabolism, and expresses type II NADH dehydrogenases (NDH2s), a non-proton pumping enzyme, instead of a canonical complex I at the inner mitochondrial membrane (Lin *et al.*, 2008). As mentioned above, the parasite mitochondrion is down-regulated in the bradyzoite stage. However, in the highly replicative tachyzoite the oxidative phosphorylation is essential for sufficient ATP generation (Bohne *et al.*, 1994; Tomavo and Boothroyd, 1995). Therefore, it is clear that *T. gondii* adapts its metabolism during transition from tachyzoites to long-term persistent bradyzoites (Bohne *et al.*, 1994; Weiss and Kim, 2000).

Mitochondria of higher eukaryotes contain nuclear-encoded heat shock proteins such as HSP10, HSP60 and organellar HSP70 which function selectively as mitochondrial chaperones in the post-translational assembly of multimeric proteins encoded by both nuclear and mitochondrial genes (Hemmingsen *et al.*, 1988; Cheng *et al.*, 1989; Georgopoulos and Welch, 1993; Bukau and Horwich, 1998). All of these chaperones are present in *T. gondii* (www.toxodb.org). HSP60 is an essential chaperone for the folding and assembly of mitochondrial proteins. In *T. gondii*, although *hsp60* gene could be detected in both tachyzoites and bradyzoites, it is overexpressed at the transcriptional level in the *in vivo* cysts (Toursel *et al.*, 2000). Noteworthy, the HSP60 protein has the classical mitochondrial localization in tachyzoites, but it is located within two unknown vesicles in the bradyzoite (Toursel *et al.*, 2000). The role of the others mitochondrial chaperones like HSP10, organellar HSP70 as well as organellar HSP90 and HSP28 still remained unstudied.

The stressing life of living outside the cell

There is new evidence suggesting that cell physiology and behaviour of extracellular parasites could be genuinely different to intracellular ones. Moreover, based on gene expression profiles it was recently suggested that the normal asexual cycle of *T. gondii* consists of three distinct stages: extracellular tachyzoites, intracellular tachyzoites and bradyzoites (Lescault *et al.*, 2010). Noteworthy, parasite glycolytic enzymes relocate from the cytosol to the pellicle when parasites egress from the host cells. Moreover, these glycolytic enzymes remain associated to the pellicle during extracellular period relocating to the cytosol when the parasite completes the host cell invasion process (Pomel *et al.*, 2008). In this context, the extracellular *T. gondii* is capable of maintaining a constant ATP level without the acquisition of external carbon sources during its first hour (Lin *et al.*, 2011). This ability is dependent on glycolysis even in the absence of glucose. Collectively, these findings suggest that extracellular *T. gondii* can utilize their own internal carbon sources to power the ATP demanding gliding motility and thus possess a high degree of independence from environmental conditions to establish new host cell infections (Lin *et al.*, 2011).

Members of the sHSP family can be found associated to nuclei, cytoskeleton, membranes, chloroplast, mitochondria and endoplasmic reticulum (Sun and MacRae, 2005; Nakamoto and Vigh, 2007). Several studies have demonstrated that sHSPs can participate in cytoskeleton protection, modulation of the apoptotic process and basic chaperoning activity (Liang and MacRae, 1997; Haslbeck *et al.*, 1999; Bruey *et al.*, 2000). *Toxoplasma* HSP20 is located at the IMC (de Miguel *et al.*, 2009), a complex of flattened vesicles beneath the plasma membrane (Fig. 13.3), that plays a central role during intracellular parasite replication and motility. During the extracellular state of the parasite, the IMC anchors the machinery that drives gliding motility and invasion of the host cell (Soldati and Meissner, 2004; Striepen *et al.*, 2007). The presence of HSP20 at the IMC suggests a putative role in apicomplexan motility and invasion. Preliminary results on *T. gondii* HSP20 null mutants support this hypothesis (Cóceres, unpublished results). HSP20 was also described in *Babesia* spp. (Precigout *et al.*, 1993). Similar to *Toxoplasma* HSP20, the *Babesia* counterpart is also located at the pellicle of the parasite, beneath the plasma membrane. The expression of *Babesia* HSP20 increases both at the level of specific mRNA and protein, when the parasite was subjected to nutrient stress caused by serum deprivation (Montero *et al.*, 2008). This result suggests that *Babesia* HSP20 could probably be of importance for parasite survival in the case of abnormal growth conditions. In addition, the incubation of *B. divergens* extracellular tachyzoites with monoclonal antibodies against *Babesia* HSP20 dramatically inhibits parasite intra-erythrocytic growth *in vitro* (Precigout *et al.*, 1993). More recently, Montagna *et al.* (2012) reported that *P. falciparum* HSP20-deficient sporozoites show an aberrant speed and trajectories. The mechanism by which HSP20 alters gliding/invasion process during extracellular life of Apicomplexan parasites is a topic that deserves particular attention and should be investigated in the next future.

The possibility that *T. gondii* HSP20 also plays a crucial role during sporozoite (another extracellular stage) remains to be studied as wells. As mentioned above, *Toxoplasma* HSP20 expression is low in the unsporulated oocyst increasing its expression under sporulation as indicated at Toxodb page (www.toxodb.org). Since the parasite sporulation requires a range of 4–37°C (Frenkel *et al.*, 1970), and HSP20 is *in vitro* activated at low temperatures (de Miguel *et al.*, 2009), it could be speculated that HSP20 could be having an important role under cold stress during this sporozoite stage. Alternatively, it could be preparing the sporozoite machinery to the stress shift when oral ingestion occurs and the parasite must invade the host cell.

Toxoplasma HSP70 to the attack

Similar to HSP90, members of HSP70 family have shown to be important chaperones in different cellular processes: folding and assembly of newly synthesized or misfolded proteins, membrane translocation of organellar and secretory proteins, and control of regulatory proteins (Bukau and Horwich, 1998). Some roles or expression patterns of *Toxoplasma* HSP70 were already described above, mainly as a chaperone associated to HSP90-heterocomplex. Nevertheless, *T. gondii* HSP70 shows novel and interesting roles associated with the course of immune response elicited by the host during parasite dissemination. The early events in the immune response are very important in the control of the potential pathological effects of *T. gondii* during infection. The entry of the parasite in the host stimulates a strong T-cell response with the production of interleukin 12 (IL-12) and T-cells that produce IFNγ. It was also observed that IFNγ and tumour necrosis factor-alpha (TNFα) act synergistically to mediate killing of tachyzoites by macrophages, enhancing the production of free radicals and nitric oxide (NO) (Miller *et al.*, 1999). The NO antiparasitic activity was also mentioned as a factor that triggers bradyzoite conversion *in vitro* by inhibition of mitochondrial respiration (Bohne *et al.*, 1994). In addition, IFNγ, TNFα and reactive nitrogen intermediates derived from NO are particularly important inducers of HSP70 production (Kaufmann, 1990). At the same time, parasite HSP70 down regulates host protective immune responses against *T. gondii* infection by inhibiting NO release by peritoneal macrophages (Mun *et al.*, 2000), and polarizing from Th1 to Th2 immune responses in *T. gondii*–infected mice (Ahmed *et al.*, 2004). Parasite HSP70 also has B-cell but not T-cell mitogenic properties (Aosai *et al.*, 2002) and induces maturation of bone marrow-derived dendritic cells priming them to activate the adaptive immune response (Aosai *et al.*, 2006). Lymphocyte B and dendritic cell stimulation occurs through Toll-like receptor 4 (TLR4). Macrophages also have TLR4 on its surface and they showed to be susceptible to the interaction with parasite's HSP70. This interaction is responsible of the suppression of NO release and mediates IFN-β activation that suppresses NF-κB-inducing kinase (Mun *et al.*, 2005). Simultaneously, *T. gondii* HSP70 induces NO release mediated via the TLR2/MyD88 signal transduction pathway. Collectively, these data suggest a role of this chaperone in the fine regulation of the immune response.

Noteworthy, the sequence analysis of HSP70 of virulent and avirulent *T. gondii* strains reveals differences in the seven residues repeat units (GGMPGGM) present at the end of the gene (Miller *et al.*, 1999). Similar repeat units (mainly of GGMP) are found in HSP70s of other protozoa without any specific function still described. It is possible that this repeat adopts a secondary structure consisting of a series of turns, resulting in differences at the structural level and protein stability, between HSP70s of virulent and avirulent strains. In agreement with this suggestion, parasite HSP70 presents differences at the expression level between both types of *T. gondii* strains (Miller *et al.*, 1999).

HSP40: a large family

The HSP40 family (or J family) comprises a large number of chaperones that could be divided in different types (I to IV) depending of their domain/motif organization (Cheetham and Caplan, 1998; Qiu *et al.*, 2006; Botha *et al.*, 2007). All of them present a DNAJ domain (J-domain), a 70 amino acid in length region with a conserved HPD tripeptide (Types I to III), that is the signature motif of this protein family (Qiu *et al.*, 2006). Members of HSP40 family participate in protein folding, suppression of amyloidal plaque formation, endocytosis, and protein translocation across membranes, signal transduction, DNA replication, protein degradation and prion propagation (Cyr *et al.*, 1994; Qiu *et al.*, 2006). Every HSP40 member is able to bind non-native proteins and target them to the peptide binding site of HSP70, modulating at the same time the HSP70 ATP hydrolytic cycle through the J-domain (Liberek *et al.*, 1991; Wickner *et al.*, 1991; Cyr *et al.*, 1992).

HSP40s of Type I are represented by human Hdj-2 and yeast Ydj-1 (Cheetham and Caplan, 1998), which were linked to HSP90-hetercomplex pathway (see above). Types II are represented by human HdjI and yeast Sis1 (Sha *et al.*, 2000; Lee *et al.*, 2002; Fan *et al.*, 2004). Hdj-2 is associated with translating ribosomes, facilitating the assembly of translation initiation complexes (Zhong and Arndt, 1993; Horton *et al.*, 2001). Sis1 is essential in yeast and in protozoan parasites (Luke *et al.*, 1991; Salmon *et al.*, 2001; Maier *et al.*, 2008). Type III HSP40s are represented by proteins such as yeast Swa2, zuotin and sec63 that have specialized structures which target them to bind specific proteins, nucleic acids or intracellular membranes (Cyr *et al.*, 1994; Lee *et al.*, 2002; Fan *et al.*, 2004). Recently, a novel J-protein was described in *P. falciparum* that was annotated as a type IV HSP40, in which the J-domain lacks the 'HPD' motif (Botha *et al.*, 2007). Type IV Hsp40s seem to be important for the erythrocyte remodelling caused by the parasite since some of them are predicted to be exported or have been located in the host cell.

Searching in the genome database of *T. gondii* (www.toxodb.org), 38 sequences containing DNAJ domains were retrieved and grouped according their structural features (Figueras *et al.* unpublished data). Three type I sequences were identified and one of them was definded as a putative *T. gondii* Ydj-like protein, most likely to be participating in the HSP90-heterocomplex (Echeverria *et al.*, 2010). In addition it could be detected putative Sis1-like and Sec63-like HSP40s (Figueras *et al.* unpublished results). Putative candidates for type III and IV were also found. Clearly, the *Toxoplasma* J family is a vast and fertile area of research to investigate conserved but also parasite-specific HSP40s. It is expected that a variety of significant findings that shed light on *Toxoplasma* cell biology will arise of such studies.

Co-chaperones and cell cycle

Toxoplasma tachyzoites and yeast appear grossly similar in the global transcriptional dynamics of their cell cycles while having many differences in the morphogenetic processes and the crucial molecular components involved (Behnke *et al.*, 2010). In the budding yeast the kinetochore is structurally analogous to multiple microtubule attachment sites found in higher eukaryotes (Steensgaard *et al.*, 2004). Kinetochores assemble on centromeric DNA to mediate the interaction of chromosomes with the mitotic spindle (Cleveland *et al.*, 2003). Mammalian kinetochores exist during interphase as pre-kinetochores, poorly characterized structures containing a subset of the mitotic kinetochore's markers (Pluta *et al.*, 1995). Sgt1 is a co-chaperone, critical for the mammalian kinetochore. This protein lends credit to the hypothesis that the kinetochore assembly pathway is conserved from yeast to human (Steensgaard *et al.*, 2004). Budding yeast Sgt1p binds another protein named S-phase kinase-associated (Skp1) protein to assemble the Cbf3p complex, the core component of the kinetochore in this organism (Kitagawa *et al.*, 1999).

The Sgt1 protein is composed of three major domains. The C-terminal region of Sgt1, called SGS, is unique for the Sgt1 isoforms and for its homologue, calcyclin binding protein (CacyBP/SIP) (Nowotny *et al.*, 2003). The N-terminal part of the Sgt1 protein contains the TPR domain which was described in proteins other than Sgt1 as a module interacting with Skp1p (Catlett and Kaplan, 2006; Zhang *et al.*, 2008). The central part of the Sgt1 molecule contains the CS domain which was shown to bind with HSP90 (Lee *et al.*, 2004) while the SGS domain of yeast Sgt1 was shown to bind to adenylyl cyclase (Dubacq *et al.*, 2002). Similar as Sgt1, CacyBP/SIP also consists of three domains. The flexible linker located between the N-terminal and the CS domain of CacyBP/SIP is responsible for interaction with Siah-1, a component of ubiquitin ligase (Matsuzawa and Reed, 2001). As it was observed, CacyBP/SIP also binds to the Skp1 protein (Filipek *et al.*, 2008). Sgt1 is also able to stabilize animal and plant nucleotide binding-leucine rich repeat (NB-LRR) proteins, which are conserved immune sensors to recognize pathogens (Shirasu, 2009).

T. gondii showed a putative CacyBP/SIP (gene ID TGME49_105820) and a putative Sgt1 (gene ID TGME49_031590). Interestingly, these two co-chaperones have only two domains (CS and SGS) (Alomar, unpublished results). By searching the database it could be also retrieved Skp1 (gene ID TGME49_007680).These data seem to indicate that *T. gondii* also share the kinetochore assembly machinery with other eukaryotes. It is intriguing which of the two putative candidates is the responsible of assemble the complex participating in the cell cycle progression. This topic is also of high interest for future studies.

Conclusions and future trends

T. gondii is an obligated intracellular protozoan parasite that is subjected to different stressors that include heat changes, immune response pressure, and extracellular environment. When the conditions are adverse during its life cycle, it needs to invade successfully a host cell and differentiates into latent forms. In this context, heat shock proteins have shown to be key molecules in this process.

Despite of the numerous studies that were presented in this chapter, the role and/or impact of HSPs and stress response in the biology of *T. gondii*, is still little understood. For example, the requirements of sporozoites present in oocyst to resist environmental stress and the molecules responsible of maintaining the invasion machinery ready to invade a new animal host are important field of studies. Since this stage can only be reproduced by infection of cats, it is not easy to assess or study. However findings in this area will be of great contribution to go deeper inside the cell biology of *Toxoplasma gondii* and may lead to an answer to how the parasite manages to go through this stage of its life cycle.

Here we further described the roles of chaperones like HSP90, HSP70 and HSP20 in tachyzoite to bradyzoite interconversion and parasite invasion, respectively. However, we still do not know which mechanism these chaperones use to play their roles. The discovery of key interactors will be important to define the biological role of these proteins.

Toxoplasma presents several HSPs that are still poorly studied but have high potential to lead to the comprehension in detail of different aspects of cell cycle progression. For example, we do not know the importance of the different mitochondrial HSPs in the protection of this organelle to different stressors or their role in mitochondrion homeostasis in the different parasite stages. On the other hand, there are some chaperones that their study has not being tackled yet (e.g. CacyBP, SGT1 and Sis1-like chaperones) that could be essential to parasite viability or replication.

In conclusion, in this chapter we presented the current knowledge of different stress response pathways and HSPs of *T. gondii* (Fig. 13.3), most of them representing interesting fields of study not only to know in detail the cell biology of this parasite, but also to gain information for future novel therapeutic approaches.

Acknowledgements

S. O. Angel (Researcher), M. J. Figueras (Doctoral Fellow), L. Alomar (Doctoral Fellow) and V. Coceres (Postdoctoral Fellow) are members of National Research Council of Argentina (CONICET). This work was supported by: ANPCyT grant BID1728 OC-ARPICT 05-34415 (to S.O.A), UNSAM SA08/006 (to S.O.A) and National Institutes of Health-National Institute of Allergy and Infectious Diseases (NIH-NIAID) grant AI083162 (to S.O.A).

Web resources

CDC: http://www.cdc.gov/parasites/toxoplasmosis/index.html
The Centers for Disease Control and Prevention is one of the major operating components of the Department of Health and Human Services in USA.

Clinic for Obstetrics: http://www.trojovsky.net/toxo/
This page is based upon Dr Michael Hyde's experiences at the neonatal ward of the clinic for obstetrics at the University Hospital of Graz, Austria.

The *Toxoplasma* Blog: http://toxoplasmaparasite.blogspot.com/
The Antitoxo page posts the latest information regarding the protozoan parasite *T. gondii*, as well as job opportunities and meetings relevant to this pathogen.

Toxodb: http://toxodb.org/toxo/

The *Toxoplasma* genome resource is a genome and functional genomic database for the protozoan parasite *T. gondii*. Sequence information is integrated with various other genomic-scale data, including community annotation, ESTs, gene expression and proteomics data.

References

Ahmed, A.K., Mun, H.S., Aosai, F., Piao, L.X., Fang, H., Norose, K., and Yano, A. (2004). Roles of *Toxoplasma gondii*-derived heat shock protein 70 in host defense against *T. gondii* infection. Microbiol. Immunol. *48*, 911–915.

Ahn, H.J., Kim, S., and Nam, H.W. (2003). Molecular cloning of the 82-kDa heat shock protein (HSP90) of *Toxoplasma gondii* associated with the entry into and growth in host cells. Biochem. Biophys. Res. Commun. *311*, 654–659.

Aosai, F., Chen, M., Kang, H.K., Mun, H.S., Norose, K., Piao, L.X., Kobayashi, M., Takeuchi, O., Akira, S., and Yano, A. (2002). *Toxoplasma gondii*-derived heat shock protein HSP70 functions as a B cell mitogen. Cell Stress Chaperones *7*, 357–364.

Aosai, F., Rodriguez Pena, M.S., Mun, H.S., Fang, H., Mitsunaga, T., Norose, K., Kang, H.K., Bae, Y.S., and Yano, A. (2006). *Toxoplasma gondii*-derived heat shock protein 70 stimulates maturation of murine bone marrow-derived dendritic cells via Toll-like receptor 4. Cell Stress Chaperones *11*, 13–22.

Banumathy, G., Singh, V., Pavithra, S.R., and Tatu, U. (2003). Heat shock protein 90 function is essential for *Plasmodium falciparum* growth in human erythrocytes. J. Biol. Chem. *278*, 18336–18345.

Baum, J., Gilberger, T.W., Frischknecht, F., and Meissner, M. (2008). Host-cell invasion by malaria parasites: insights from *Plasmodium* and *Toxoplasma*. Trends Parasitol. *24*, 557–563.

Behnke, M.S., Wootton, J.C., Lehmann, M.M., Radke, J.B., Lucas, O., Nawas, J., Sibley, L.D., and White, M.W. (2010). Coordinated progression through two subtranscriptomes underlies the tachyzoite cycle of *Toxoplasma gondii*. PLoS One *5*, e12354.

Bohne, W., Heesemann, J., and Gross, U. (1994). Reduced replication of *Toxoplasma gondii* is necessary for induction of bradyzoite-specific antigens: a possible role for nitric oxide in triggering stage conversion. Infect. Immun. *62*, 1761–1767.

Bohne, W., Gross, U., Ferguson, D.J., and Heesemann, J. (1995). Cloning and characterization of a bradyzoite-specifically expressed gene (hsp30/bag1) of *Toxoplasma gondii*, related to genes encoding small heat-shock proteins of plants. Mol. Microbiol. *16*, 1221–1230.

Bohne, W., Hunter, C.A., White, M.W., Ferguson, D.J., Gross, U., and Roos, D.S. (1998). Targeted disruption of the bradyzoite-specific gene BAG1 does not prevent tissue cyst formation in *Toxoplasma gondii*. Mol. Biochem. Parasitol. *92*, 291–301.

Botha, M., Pesce, E.R., and Blatch, G.L. (2007). The Hsp40 proteins of *Plasmodium falciparum* and other apicomplexa: regulating chaperone power in the parasite and the host. Int. J. Biochem. Cell. Biol. *39*, 1781–1803.

Bruey, J.M., Ducasse, C., Bonniaud, P., Ravagnan, L., Susin, S.A., Diaz-Latoud, C., Gurbuxani, S., Arrigo, A.P., Kroemer, G., Solary, E., *et al.* (2000). Hsp27 negatively regulates cell death by interacting with cytochrome c. Nat. Cell Biol. *2*, 645–652.

Bukau, B., and Horwich, A.L. (1998). The Hsp70 and Hsp60 chaperone machines. Cell *92*, 351–366.

Catlett, M.G., and Kaplan, K.B. (2006). Sgt1p is a unique co-chaperone that acts as a client adaptor to link Hsp90 to Skp1p. J. Biol. Chem. *281*, 33739–33748.

Cleveland, D.W., Mao, Y., and Sullivan, K.F. (2003). Centromeres and kinetochores: from epigenetics to mitotic checkpoint signaling. Cell *112*, 407–421.

Cyr, D.M., Lu, X., and Douglas, M.G. (1992). Regulation of Hsp70 function by a eukaryotic DnaJ homolog. J. Biol. Chem. *267*, 20927–20931.

Cyr, D.M., Langer, T., and Douglas, M.G. (1994). DnaJ-like proteins: molecular chaperones and specific regulators of Hsp70. Trends Biochem. Sci. *19*, 176–181.

Cheetham, M.E., and Caplan, A.J. (1998). Structure, function and evolution of DnaJ: conservation and adaptation of chaperone function. Cell Stress Chaperones *3*, 28–36.

Cheng, M.Y., Hartl, F.U., Martin, J., Pollock, R.A., Kalousek, F., Neupert, W., Hallberg, E.M., Hallberg, R.L., and Horwich, A.L. (1989). Mitochondrial heat-shock protein hsp60 is essential for assembly of proteins imported into yeast mitochondria. Nature *337*, 620–625.

Denton, H., Brown, S.M., Roberts, C.W., Alexander, J., McDonald, V., Thong, K.W., and Coombs, G.H. (1996). Comparison of the phosphofructokinase and pyruvate kinase activities of Cryptosporidium parvum, Eimeria tenella and *Toxoplasma gondii*. Mol. Biochem. Parasitol. *76*, 23–29.

Dubacq, C., Guerois, R., Courbeyrette, R., Kitagawa, K., and Mann, C. (2002). Sgt1p contributes to cyclic AMP pathway activity and physically interacts with the adenylyl cyclase Cyr1p/Cdc35p in budding yeast. Eukaryot. Cell *1*, 568–582.

Dubey, J.P. (1990). Status of toxoplasmosis in sheep and goats in the United States. J. Am. Vet. Med. Assoc. *196*, 259–262.

Dubey, J.P., and Thulliez, P. (1993). Persistence of tissue cysts in edible tissues of cattle fed *Toxoplasma gondii* oocysts. Am. J. Vet. Res. *54*, 270–273.

Dubey, J.P., Miller, N.L., and Frenkel, J.K. (1970). The *Toxoplasma gondii* oocyst from cat feces. J. Exp. Med. *132*, 636–662.

Echeverria, P.C., Matrajt, M., Harb, O.S., Zappia, M.P., Costas, M.A., Roos, D.S., Dubremetz, J.F., and Angel, S.O. (2005). *Toxoplasma gondii* Hsp90 is a potential drug target whose expression and subcellular localization are developmentally regulated. J. Mol. Biol. *350*, 723–734.

Echeverria, P.C., Figueras, M.J., Vogler, M., Kriehuber, T., de Miguel, N., Deng, B., Dalmasso, M.C., Matthews, D.E., Matrajt, M., Haslbeck, M., *et al.* (2010). The Hsp90 co-chaperone p23 of *Toxoplasma gondii*: Identification, functional analysis and dynamic interactome determination. Mol. Biochem. Parasitol. *172*, 129–140.

Fan, C.Y., Lee, S., Ren, H.Y., and Cyr, D.M. (2004). Exchangeable chaperone modules contribute to specification of type I and type II Hsp40 cellular function. Mol. Biol. Cell *15*, 761–773.

Ferreira-da-Silva Mda, F., Rodrigues, R.M., Andrade, E.F., Carvalho, L., Gross, U., Luder, C.G., and Barbosa, H.S. (2009). Spontaneous stage differentiation of mouse-virulent *Toxoplasma gondii* RH parasites in skeletal muscle cells: an ultrastructural evaluation. Mem. Inst. Oswaldo Cruz *104*, 196–200.

Ferreira da Silva Mda, F., Barbosa, H.S., Gross, U., and Luder, C.G. (2008). Stress-related and spontaneous stage differentiation of *Toxoplasma gondii*. Mol. Biosyst. *4*, 824–834.

Filipek, A., Michowski, W., and Kuznicki, J. (2008). Involvement of S100A6 (calcyclin) and its binding partners in intracellular signaling pathways. Adv. Enzyme Regul. *48*, 225–239.

Fleige, T., Fischer, K., Ferguson, D.J., Gross, U., and Bohne, W. (2007). Carbohydrate metabolism in the *Toxoplasma gondii* apicoplast: localization of three glycolytic isoenzymes, the single pyruvate dehydrogenase complex, and a plastid phosphate translocator. Eukaryot. Cell *6*, 984–996.

Fleige, T., Pfaff, N., Gross, U., and Bohne, W. (2008). Localisation of gluconeogenesis and tricarboxylic acid (TCA)-cycle enzymes and first functional analysis of the TCA cycle in *Toxoplasma gondii*. Int. J. Parasitol. *38*, 1121–1132.

Freeman, B.C., Felts, S.J., Toft, D.O., and Yamamoto, K.R. (2000). The p23 molecular chaperones act at a late step in intracellular receptor action to differentially affect ligand efficacies. Genes Dev. *14*, 422–434.

Frenkel, J.K. (1988). Pathophysiology of toxoplasmosis. Parasitol. Today *4*, 273–278.

Frenkel, J.K., Dubey, J.P., and Miller, N.L. (1970). *Toxoplasma gondii* in cats: fecal stages identified as coccidian oocysts. Science *167*, 893–896.

Gebauer, F., and Hentze, M.W. (2004). Molecular mechanisms of translational control. Nat. Rev. Mol. Cell. Biol. *5*, 827–835.

Georgopoulos, C., and Welch, W.J. (1993). Role of the major heat shock proteins as molecular chaperones. Annu. Rev. Cell Biol. *9*, 601–634.

Gething, M.J., and Sambrook, J. (1992). Protein folding in the cell. Nature *355*, 33–45.

Graefe, S.E., Wiesgigl, M., Gaworski, I., Macdonald, A., and Clos, J. (2002). Inhibition of HSP90 in *Trypanosoma cruzi* induces a stress response but no stage differentiation. Eukaryot. Cell *1*, 936–943.

Gross, U., Bohne, W., Soete, M., and Dubremetz, J.F. (1996). Developmental differentiation between tachyzoites and bradyzoites of *Toxoplasma gondii*. Parasitol. Today *12*, 30–33.

Gubbels, M.J., White, M., and Szatanek, T. (2008). The cell cycle and *Toxoplasma gondii* cell division: tightly knit or loosely stitched? Int. J. Parasitol. *38*, 1343–1358.

Haslbeck, M., Walke, S., Stromer, T., Ehrnsperger, M., White, H.E., Chen, S., Saibil, H.R., and Buchner, J. (1999). Hsp26: a temperature-regulated chaperone. EMBO J. *18*, 6744–6751.

Hemmingsen, S.M., Woolford, C., van der Vies, S.M., Tilly, K., Dennis, D.T., Georgopoulos, C.P., Hendrix, R.W., and Ellis, R.J. (1988). Homologous plant and bacterial proteins chaperone oligomeric protein assembly. Nature *333*, 330–334.

Horton, L.E., James, P., Craig, E.A., and Hensold, J.O. (2001). The yeast hsp70 homologue Ssa is required for translation and interacts with Sis1 and Pab1 on translating ribosomes. J. Biol. Chem. *276*, 14426–14433.

Hu, K., Mann, T., Striepen, B., Beckers, C.J., Roos, D.S., and Murray, J.M. (2002). Daughter cell assembly in the protozoan parasite *Toxoplasma gondii*. Mol. Biol. Cell *13*, 593–606.

Jones, T.C., Bienz, K.A., and Erb, P. (1986). *In vitro* cultivation of *Toxoplasma gondii* cysts in astrocytes in the presence of gamma interferon. Infect. Immun. *51*, 147–156.

Joyce, B.R., Queener, S.F., Wek, R.C., and Sullivan, W.J., Jr. (2010). Phosphorylation of eukaryotic initiation factor-2{alpha} promotes the extracellular survival of obligate intracellular parasite *Toxoplasma gondii*. Proc. Natl. Acad. Sci. U.S.A. *107*, 17200–17205.

Kaufmann, S.H. (1990). Heat shock proteins and the immune response. Immunol. Today *11*, 129–136.

Khan, I.A., Green, W.R., Kasper, L.H., Green, K.A., and Schwartzman, J.D. (1999). Immune CD8(+) T cells prevent reactivation of *Toxoplasma gondii* infection in the immunocompromised host. Infect. Immun. *67*, 5869–5876.

Kitagawa, K., Skowyra, D., Elledge, S.J., Harper, J.W., and Hieter, P. (1999). SGT1 encodes an essential component of the yeast kinetochore assembly pathway and a novel subunit of the SCF ubiquitin ligase complex. Mol. Cell *4*, 21–33.

Krobitsch, S., Brandau, S., Hoyer, C., Schmetz, C., Hubel, A., and Clos, J. (1998). *Leishmania donovani* heat shock protein 100. Characterization and function in amastigote stage differentiation. J. Biol. Chem. *273*, 6488–6494.

Lee, S., Fan, C.Y., Younger, J.M., Ren, H., and Cyr, D.M. (2002). Identification of essential residues in the type II Hsp40 Sis1 that function in polypeptide binding. J. Biol. Chem. *277*, 21675–21682.

Lee, Y.T., Jacob, J., Michowski, W., Nowotny, M., Kuznicki, J., and Chazin, W.J. (2004). Human Sgt1 binds HSP90 through the CHORD-Sgt1 domain and not the tetratricopeptide repeat domain. J. Biol. Chem. *279*, 16511–16517.

Lescault, P.J., Thompson, A.B., Patil, V., Lirussi, D., Burton, A., Margarit, J., Bond, J., and Matrajt, M. (2010). Genomic data reveal *Toxoplasma gondii* differentiation mutants are also impaired with respect to switching into a novel extracellular tachyzoite state. PLoS One 5, e14463.

Liang, P., and MacRae, T.H. (1997). Molecular chaperones and the cytoskeleton. J. Cell Sci. *110*, 1431–1440.

Liberek, K., Marszalek, J., Ang, D., Georgopoulos, C., and Zylicz, M. (1991). *Escherichia coli* DnaJ and GrpE heat shock proteins jointly stimulate ATPase activity of DnaK. Proc. Natl. Acad. Sci. U.S.A. *88*, 2874–2878.

Lin, S.S., Kerscher, S., Saleh, A., Brandt, U., Gross, U., and Bohne, W. (2008). The *Toxoplasma gondii* type-II NADH dehydrogenase TgNDH2-I is inhibited by 1-hydroxy-2-alkyl-4(1H)quinolones. Biochim. Biophys. Acta *1777*, 1455–1462.

Lin, S.S., Blume, M., von Ahsen, N., Gross, U., and Bohne, W. (2011). Extracellular *Toxoplasma gondii* tachyzoites do not require carbon source uptake for ATP maintenance, gliding motility and invasion in the first hour of their extracellular life. Int. J. Parasitol. *41*, 835–841.

Liwak, U., and Ananvoranich, S. (2009). *Toxoplasma gondii*: over-expression of lactate dehydrogenase enhances differentiation under alkaline conditions. Exp. Parasitol. *122*, 155–161.

Luder, C.G., Giraldo-Velasquez, M., Sendtner, M., and Gross, U. (1999). *Toxoplasma gondii* in primary rat CNS cells: differential contribution of neurons, astrocytes, and microglial cells for the intracerebral development and stage differentiation. Exp. Parasitol. *93*, 23–32.

Luke, M.M., Sutton, A., and Arndt, K.T. (1991). Characterization of SIS1, a *Saccharomyces cerevisiae* homologue of bacterial dnaJ proteins. J. Cell Biol. *114*, 623–638.

Lykens, J.E., Terrell, C.E., Zoller, E.E., Divanovic, S., Trompette, A., Karp, C.L., Aliberti, J., Flick, M.J., and Jordan, M.B. (2010). Mice with a selective impairment of IFN-gamma signaling in macrophage lineage cells demonstrate the critical role of IFN-gamma-activated macrophages for the control of protozoan parasitic infections *in vivo*. J. Immunol. *184*, 877–885.

McLaughlin, S.H., Sobott, F., Yao, Z.P., Zhang, W., Nielsen, P.R., Grossmann, J.G., Laue, E.D., Robinson, C.V., and Jackson, S.E. (2006). The co-chaperone p23 arrests the Hsp90 ATPase cycle to trap client proteins. J. Mol. Biol. *356*, 746–758.

Maier, A.G., Rug, M., O'Neill, M.T., Brown, M., Chakravorty, S., Szestak, T., Chesson, J., Wu, Y., Hughes, K., Coppel, R.L., *et al.* (2008). Exported proteins required for virulence and rigidity of *Plasmodium falciparum*-infected human erythrocytes. Cell *134*, 48–61.

Matsuzawa, S.I., and Reed, J.C. (2001). Siah-1, SIP, and Ebi collaborate in a novel pathway for beta-catenin degradation linked to p53 responses. Mol. Cell *7*, 915–926.

Melo, E.J., Attias, M., and De Souza, W. (2000). The single mitochondrion of tachyzoites of *Toxoplasma gondii*. J. Struct. Biol. *130*, 27–33.

de Melo, E.J., and de Souza, W. (1997). A cytochemistry study of the inner membrane complex of the pellicle of tachyzoites of *Toxoplasma gondii*. Parasitol. Res. *83*, 252–256.

de Miguel, N., Echeverria, P.C., and Angel, S.O. (2005). Differential subcellular localization of members of the *Toxoplasma gondii* small heat shock protein family. Eukaryot. Cell *4*, 1990–1997.

de Miguel, N., Lebrun, M., Heaslip, A., Hu, K., Beckers, C.J., Matrajt, M., Dubremetz, J.F., and Angel, S.O. (2008). *Toxoplasma gondii* Hsp20 is a stripe-arranged chaperone-like protein associated with the outer leaflet of the inner membrane complex. Biol. Cell *100*, 479–489.

de Miguel, N., Braun, N., Bepperling, A., Kriehuber, T., Kastenmuller, A., Buchner, J., Angel, S.O., and Haslbeck, M. (2009). Structural and functional diversity in the family of small heat shock proteins from the parasite *Toxoplasma gondii*. Biochim. Biophys. Acta *1793*, 1738–1748.

Miller, C.M., Smith, N.C., and Johnson, A.M. (1999). Cytokines, nitric oxide, heat shock proteins and virulence in Toxoplasma. Parasitol. Today *15*, 418–422.

Mimnaugh, E.G., Chavany, C., and Neckers, L. (1996). Polyubiquitination and proteasomal degradation of the p185c-erbB-2 receptor protein-tyrosine kinase induced by geldanamycin. J. Biol. Chem. *271*, 22796–22801.

Montagna, G.N., Matuschewski, K., and Buscaglia, C.A. (2012). Plasmodium sporozoite motility: an update. Front. Biosci. *17*, 726–744.

Montero, E., Rodriguez, M., Gonzalez, L.M., and Lobo, C.A. (2008). Babesia divergens: identification and characterization of BdHSP-20, a small heat shock protein. Exp. Parasitol. *119*, 238–245.

Montoya, J.G., and Liesenfeld, O. (2004). Toxoplasmosis. Lancet *363*, 1965–1976.

Mun, H.S., Aosai, F., Norose, K., Chen, M., Hata, H., Tagawa, Y.I., Iwakura, Y., Byun, D.S., and Yano, A. (2000). *Toxoplasma gondii* Hsp70 as a danger signal in *Toxoplasma gondii*-infected mice. Cell Stress Chaperones *5*, 328–335.

Mun, H.S., Aosai, F., Norose, K., Piao, L.X., Fang, H., Akira, S., and Yano, A. (2005). Toll-like receptor 4 mediates tolerance in macrophages stimulated with *Toxoplasma gondii*-derived heat shock protein 70. Infect. Immun. *73*, 4634–4642.

Naguleswaran, A., Elias, E.V., McClintick, J., Edenberg, H.J., and Sullivan, W.J., Jr. (2010). *Toxoplasma gondii* lysine acetyltransferase GCN5-A functions in the cellular response to alkaline stress and expression of cyst genes. PLoS Pathog. *6*, e1001232.

Nakamoto, H., and Vigh, L. (2007). The small heat shock proteins and their clients. Cell. Mol. Life Sci. *64*, 294–306.

Narasimhan, J., Joyce, B.R., Naguleswaran, A., Smith, A.T., Livingston, M.R., Dixon, S.E., Coppens, I., Wek, R.C., and Sullivan, W.J., Jr. (2008). Translation regulation by eukaryotic initiation factor-2 kinases in the development of latent cysts in *Toxoplasma gondii*. J. Biol. Chem. *283*, 16591–16601.

Nathan, D.F., Vos, M.H., and Lindquist, S. (1997). *In vivo* functions of the *Saccharomyces cerevisiae* Hsp90 chaperone. Proc. Natl. Acad. Sci. U.S.A. *94*, 12949–12956.

Nowotny, M., Spiechowicz, M., Jastrzebska, B., Filipek, A., Kitagawa, K., and Kuznicki, J. (2003). Calcium-regulated interaction of Sgt1 with S100A6 (calcyclin) and other S100 proteins. J. Biol. Chem. *278*, 26923–26928.

Parmley, S.F., Weiss, L.M., and Yang, S. (1995). Cloning of a bradyzoite-specific gene of *Toxoplasma gondii* encoding a cytoplasmic antigen. Mol. Biochem. Parasitol. *73*, 253–257.

Pedersen, M.G., Stevens, H., Pedersen, C.B., Norgaard-Pedersen, B., and Mortensen, P.B. (2011). Toxoplasma infection and later development of schizophrenia in mothers. Am. J. Psych. *168*, 814–821.

Pluta, A.F., Mackay, A.M., Ainsztein, A.M., Goldberg, I.G., and Earnshaw, W.C. (1995). The centromere: hub of chromosomal activities. Science *270*, 1591–1594.

Pomel, S., Luk, F.C., and Beckers, C.J. (2008). Host cell egress and invasion induce marked relocations of glycolytic enzymes in *Toxoplasma gondii* tachyzoites. PLoS Pathog. *4*, e1000188.

Pratt, W.B., and Toft, D.O. (2003). Regulation of signaling protein function and trafficking by the hsp90/hsp70-based chaperone machinery. Exp. Biol. Med. (Maywood) *228*, 111–133.

Precigout, E., Valentin, A., Carcy, B., Gorenflot, A., Nakamura, K., Aikawa, M., and Schrevel, J. (1993). Babesia divergens: characterization of a 17-kDa merozoite membrane protein. Exp. Parasitol. *77*, 425–434.

Qiu, X.B., Shao, Y.M., Miao, S., and Wang, L. (2006). The diversity of the DnaJ/Hsp40 family, the crucial partners for Hsp70 chaperones. Cell. Mol. Life Sci. *63*, 2560–2570.

Radke, J.R., Donald, R.G., Eibs, A., Jerome, M.E., Behnke, M.S., Liberator, P., and White, M.W. (2006). Changes in the expression of human cell division autoantigen-1 influence *Toxoplasma gondii* growth and development. PLoS Pathog. *2*, e105.

Roobol, A., Carden, M.J., Newsam, R.J., and Smales, C.M. (2009). Biochemical insights into the mechanisms central to the response of mammalian cells to cold stress and subsequent rewarming. FEBS J. *276*, 286–302.

Salmon, D., Montero-Lomeli, M., and Goldenberg, S. (2001). A DnaJ-like protein homologous to the yeast co-chaperone Sis1 (TcJ6p) is involved in initiation of translation in *Trypanosoma cruzi*. J. Biol. Chem. *276*, 43970–43979.

Seeber, F., Ferguson, D.J., and Gross, U. (1998). *Toxoplasma gondii*: a paraformaldehyde-insensitive diaphorase activity acts as a specific histochemical marker for the single mitochondrion. Exp. Parasitol. *89*, 137–139.

Segui-Simarro, J.M., Testillano, P.S., and Risueno, M.C. (2003). Hsp70 and Hsp90 change their expression and subcellular localization after microspore embryogenesis induction in *Brassica napus* L. J. Struct. Biol. *142*, 379–391.

Sha, B., Lee, S., and Cyr, D.M. (2000). The crystal structure of the peptide-binding fragment from the yeast Hsp40 protein Sis1. Structure *8*, 799–807.

Shirasu, K. (2009). The HSP90-SGT1 chaperone complex for NLR immune sensors. Annu. Rev. Plant Biol. *60*, 139–164.

Silva, N.M., Gazzinelli, R.T., Silva, D.A., Ferro, E.A., Kasper, L.H., and Mineo, J.R. (1998). Expression of *Toxoplasma gondii*-specific heat shock protein 70 during *in vivo* conversion of bradyzoites to tachyzoites. Infect. Immun. *66*, 3959–3963.

Soete, M., Camus, D., and Dubremetz, J.F. (1994). Experimental induction of bradyzoite-specific antigen expression and cyst formation by the RH strain of *Toxoplasma gondii in vitro*. Exp. Parasitol. *78*, 361–370.

Soldati, D., and Meissner, M. (2004). Toxoplasma as a novel system for motility. Curr. Opin. Cell Biol. *16*, 32–40.

Stebbins, C.E., Russo, A.A., Schneider, C., Rosen, N., Hartl, F.U., and Pavletich, N.P. (1997). Crystal structure of an Hsp90–geldanamycin complex: targeting of a protein chaperone by an antitumor agent. Cell *89*, 239–250.

Steensgaard, P., Garre, M., Muradore, I., Transidico, P., Nigg, E.A., Kitagawa, K., Earnshaw, W.C., Faretta, M., and Musacchio, A. (2004). Sgt1 is required for human kinetochore assembly. EMBO Rep. *5*, 626–631.

Striepen, B., Jordan, C.N., Reiff, S., and van Dooren, G.G. (2007). Building the perfect parasite: cell division in apicomplexa. PLoS Pathog. *3*, e78.

Su, C., Evans, D., Cole, R.H., Kissinger, J.C., Ajioka, J.W., and Sibley, L.D. (2003). Recent expansion of Toxoplasma through enhanced oral transmission. Science *299*, 414–416.

Sullivan, W.J., Jr., Narasimhan, J., Bhatti, M.M., and Wek, R.C. (2004). Parasite-specific eIF2 (eukaryotic initiation factor-2) kinase required for stress-induced translation control. Biochem. J. *380*, 523–531.

Sun, Y., and MacRae, T.H. (2005). Small heat shock proteins: molecular structure and chaperone function. Cell. Mol. Life Sci. *62*, 2460–2476.

Tenter, A.M., Heckeroth, A.R., and Weiss, L.M. (2000). *Toxoplasma gondii*: from animals to humans. Int. J. Parasitol. *30*, 1217–1258.

Tomavo, S., and Boothroyd, J.C. (1995). Interconnection between organellar functions, development and drug resistance in the protozoan parasite, *Toxoplasma gondii*. Int. J. Parasitol. *25*, 1293–1299.

Toursel, C., Dzierszinski, F., Bernigaud, A., Mortuaire, M., and Tomavo, S. (2000). Molecular cloning, organellar targeting and developmental expression of mitochondrial chaperone HSP60 in *Toxoplasma gondii*. Mol. Biochem. Parasitol. *111*, 319–332.

Vercesi, A.E., Rodrigues, C.O., Uyemura, S.A., Zhong, L., and Moreno, S.N. (1998). Respiration and oxidative phosphorylation in the apicomplexan parasite *Toxoplasma gondii*. J. Biol. Chem. *273*, 31040–31047.

Weiss, L.M., and Kim, K. (2000). The development and biology of bradyzoites of *Toxoplasma gondii*. Front Biosci. *5*, D391–405.

Weiss, L.M., Ma, Y.F., Takvorian, P.M., Tanowitz, H.B., and Wittner, M. (1998). Bradyzoite development in *Toxoplasma gondii* and the hsp70 stress response. Infect. Immun. *66*, 3295–3302.

Wickner, S., Hoskins, J., and McKenney, K. (1991). Function of DnaJ and DnaK as chaperones in origin-specific DNA binding by RepA. Nature *350*, 165–167.

Wiesgigl, M., and Clos, J. (2001). The heat shock protein 90 of *Leishmania donovani*. Med. Microbiol. Immunol. *190*, 27–31.

Wilson, C.B., Remington, J.S., Stagno, S., and Reynolds, D.W. (1980). Development of adverse sequelae in children born with subclinical congenital Toxoplasma infection. Pediatrics *66*, 767–774.

Wilson, M.E., Andersen, K.A., and Britigan, B.E. (1994). Response of *Leishmania chagasi* promastigotes to oxidant stress. Infect. Immun. *62*, 5133–5141.

Yolken, R.H., Dickerson, F.B., and Fuller Torrey, E. (2009). Toxoplasma and schizophrenia. Parasite Immunol. *31*, 706–715.

Yue, L., Karr, T.L., Nathan, D.F., Swift, H., Srinivasan, S., and Lindquist, S. (1999). Genetic analysis of viable Hsp90 alleles reveals a critical role in *Drosophila spermatogenesis*. Genetics *151*, 1065–1079.

Zhang, M., Boter, M., Li, K., Kadota, Y., Panaretou, B., Prodromou, C., Shirasu, K., and Pearl, L.H. (2008). Structural and functional coupling of Hsp90- and Sgt1-centred multi–protein complexes. EMBO J. *27*, 2789–2798.

Zhang, M., Hisaeda, H., Tsuboi, T., Torii, M., Sakai, T., Nakano, Y., Ishikawa, H., Maekawa, Y., Good, R.A., and Himeno, K. (1999a). Stage-specific expression of heat shock protein 90 in murine malaria parasite *Plasmodium yoelii*. Exp. Parasitol. *93*, 61–65.

Zhang, Y.W., Kim, K., Ma, Y.F., Wittner, M., Tanowitz, H.B., and Weiss, L.M. (1999b). Disruption of the *Toxoplasma gondii* bradyzoite-specific gene BAG1 decreases *in vivo* cyst formation. Mol. Microbiol. *31*, 691–701.

Zhong, T., and Arndt, K.T. (1993). The yeast SIS1 protein, a DnaJ homolog, is required for the initiation of translation. Cell 73, 1175–1186.

The Stressful Life of Pathogenic *Leishmania* Species

14

Jose M. Requena

Abstract

Leishmania parasites are unicellular protozoa descending from one of the oldest eukaryotic lineages. During its digenetic life cycle, *Leishmania* alternates between the alimentary tract of the sandfly vector as an extracellular promastigote and the acidic phagolysosomes of macrophage cells as an intracellular amastigote. Parasites must cope with varied and heterogeneous environments: changes in temperature, in pH, in nutrient and oxygen concentrations. Also, they must face the immune defences, such as complement factors, free radicals and other antimicrobial effectors. The focus of this chapter will be on our current knowledge of the different stress responses in *Leishmania*, ranging from description of the prototypical heat shock response to more specific responses found in this parasite. A comprehensive view on the implications of the stress response in parasite survival, in cytodifferentiation and in apoptotic processes will be presented. Future studies, which should be directed mainly to the uncovering of the stress sensors, signal transduction pathways and regulatory mechanisms leading to the induction of the appropriate stress response will be also discussed.

Introduction

Protists of the genus *Leishmania* belong to the order Trypanosomatida and form, together with the bodonids, the class Kinetoplastea (Moreira *et al.*, 2004). The name of this group derives from the presence within the single mitochondrion of a characteristic structure, the kinetoplast, where mitochondrial DNA, consisting of thousands of concatenated, circular DNA molecules, forms an enormous network termed 'kinetoplast DNA' (or kDNA). The trypanosomatids constitute an early branching group of unicellular eukaryotes including several other important parasitic pathogens of human and animals besides *Leishmania*, i.e. *Trypanosoma brucei* (causing sleeping sickness in humans and the Nagana disease in cattle; see Chapter 16) and *Trypanosoma cruzi* (causing Chagas' disease in humans; see Chapter 15).

Protozoan parasites of the genus *Leishmania* cause a severe disease, leishmaniasis, which threatens 350 million people worldwide (WHO, 2010). There are estimated to be over two million new cases of leishmaniasis each year, and mortality and morbidity from leishmaniasis worldwide show a worrying increasing trend (Desjeux, 2004; Dujardin *et al.*, 2008). Included within the term leishmaniasis is a diverse spectrum of disease manifestations that are caused by different species of *Leishmania*. The clinical manifestations of each disease are determined mainly by the species of *Leishmania* that infects the host, but also by the response of the host to the parasite (Murray *et al.*, 2005). The clinical syndromes of leishmaniasis range from self-healing cutaneous lesions through metastasizing mucocutaneous forms to often lethal visceral diseases. Several species produce cutaneous leishmaniasis (CL or Oriental sore), primarily *L. major, L. tropica* and *L. aethiopica* in the Old World, and *L. mexicana, L. amazonensis, L. braziliensis* and *L. guyanensis* in the New World. Mucosal dissemination (espundia) is a complication associated with the infection of New World species of the *Viannia* subgenus (mainly *L. braziliensis* and *L. guyanensis*); this clinical manifestation occurs in approximately 10% of infections, developing 1–5 years after CL has healed. Visceral leishmaniasis (VL), or kala-azar, is caused by *L. donovani* in the Indian subcontinent, Asia, and Africa, and by *L. infantum* in the Mediterranean region, southwest and central Asia, and South America (in the last region, it is called *L. chagasi*). In VL, parasites home to visceral organs and result in the pentad of syndromes comprising fever, weight loss, splenomegaly, hepatomegaly and anaemia. If left untreated, the disease has a high mortality rate (Chappuis *et al.*, 2007). In addition, it is now apparent that asymptomatic infection is quite common and represents a potential reservoir in its own right; thus, it is estimated that there are 30–100 subclinical infections for every overt case of VL (Guerin *et al.*, 2002). Risk factors for development of clinical disease include malnutrition and immunosuppression; in fact, a large number of *Leishmania*-HIV co-infections exist and most of them represent reactivation of prior subclinical infection (Alvar *et al.*, 2008). Nevertheless, despite the enormous number of patients worldwide, leishmaniases still remain as neglected diseases.

Current control measures rely on chemotherapy, vector control, and control of reservoir host populations. The basic treatment of leishmaniasis consists of the administration of pentavalent antimony (SbV) in the form of sodium stibogluconate (Pentostan®) or *N*-methylglucamine antimonate (Glucantime®). Although antimony-based drugs have been used since 1947, their mode of action still remains not fully understood. If antimony-based drugs were not effective, pentamidine (lomidine) and amphotericin B (fungizone) are administrated as second-line drugs. More recently, new therapies by liposomal amphotericin, oral miltefosine and paromomycin have been introduced. Nevertheless, these are not ideal drugs, and improved, less toxic and cheaper therapies are required (Alvar *et al.*, 2006). Furthermore, resistance to antimony and other drugs is now common in some parts of the world, including South America, Europe, The Middle East, and India (Croft *et al.*, 2006). Abundant clinical and experimental evidence indicates that leishmaniasis should be preventable by vaccination, but the development of a vaccine appears to be difficult and, at present, there is no vaccine against this parasitic disease. Current laboratory efforts are focused on the use of recombinant antigens, novel adjuvants, live-attenuated vaccines, and naked DNA with the goal of developing an effective and affordable vaccine against leishmaniasis (Requena *et al.*, 2004; Dunning, 2009). Owing to the lack of safe and effective treatments for this disease, leishmaniasis is classified by the World Health Organization into category I, as emerging and uncontrolled disease. Progress in controlling leishmaniasis requires improved appreciation of the biology of the parasite to allow novel treatment strategies to be designed.

Despite the spectrum of clinical pathologies associated with the different *Leishmania* species (20 *Leishmania* species are pathogenic for humans), these parasites possess similar morphologies and life cycles. *Leishmania* parasites are transmitted between vertebrate hosts by phlebotomine sand flies of the genera *Phlebotomus* (Old World) and *Lutzomyia* (New World). These parasitic protozoa exhibit two principal life cycle stages: an extracellular flagellated promastigote, which attaches to cells in the alimentary tract of the phlebotomine insect vector, and an obligatory intracellular amastigote that thrives inside phagolysosomes of vertebrate host macrophages (Fig. 14.1). These two life cycle stages exhibit dramatically different morphologies, promastigotes being spindle shaped, flagellated organisms that are highly motile, while amastigotes are oval shaped, non-motile forms that possess only a residual flagellum. These two developmental forms live in very distinct physiological environments. Promastigotes are extracellular, reside at the ambient temperature of the sand fly vector and at a pH that is close to neutral, where sugars are the main nutritional source at this stage. Amastigotes are intracellular, are exposed to the more elevated temperature of the vertebrate host and the acidic environment of the phagolysosome, and are thought to prevail in a relatively carbohydrate-poor environment (McConville *et al.*, 2007).

In the insect stage, promastigotes survive and proliferate in the alimentary canal of the vector, and differentiate into non-replicative, infective metacyclics prior to transmission to the mammalian host. Sandflies are pool feeders and when

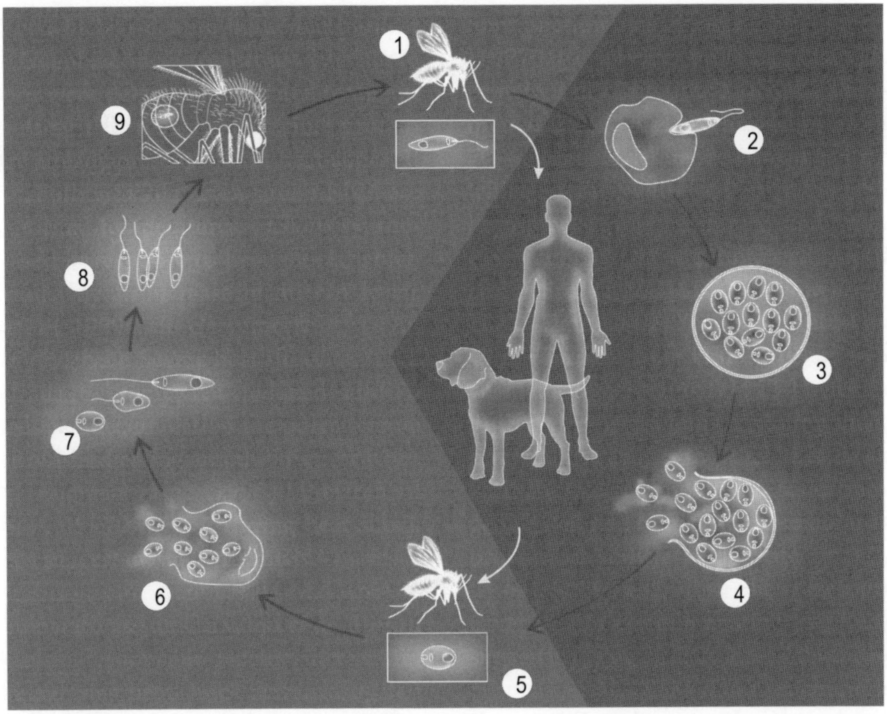

Figure 14.1 *Leishmania* life cycle. *Leishmania* parasites live as motile extracellular promastigotes in the midgut of their dipteran vector, the sandfly (1), but differentiate into non-replicative, infective metacyclics prior to transmission to the mammalian host (8–9). Once inoculated, the metacyclics are phagocytosed by macrophages (2), in which they transform into non-motile, intracellular, replicative amastigotes and become resident within one or more specialized parasitophorous vacuoles (phagolysosome). Amastigotes divide by binary fission within macrophage phagolysosomes (3) and eventually burst free from infected macrophages (4), spreading the disease within the mammalian host. Amastigotes can subsequently infect other macrophages or be ingested by a feeding sand fly (5). In the gut of the sand fly, amastigotes differentiate back into promastigotes (6–7), thereby completing the parasite life cycle.

the infective sandfly bites a mammal for its blood meal, it introduces metacyclic promastigotes into the skin where they are subsequently phagocytosed by macrophages, either directly or after transient infection of neutrophils (Dominguez and Toraño, 1999). It may seem surprising that *Leishmania* has chosen one of the most efficacious effector components of the immune system as host cell, i.e. the mononuclear phagocytes. Beyond doubt, this cell type provides an extreme environment to the predator, mainly if the pathogen has not designed egression strategies from the phagosome into the cytosol. This is the case for *Leishmania,* which remains inside the phagosome. Most intriguing is the finding that *Leishmania* containing-phagosome fuses with lysosomes soon after entry into macrophages. In the phagolysosomes of resident macrophages, surviving promastigotes transform and replicate as amastigotes, which infect additional macrophages either locally or in distant tissues after dissemination. Thus, in the macrophage phagolysosomes, the parasites confront an acidic and oligotrophic hostile environment where they differentiate to amastigotes (Chang *et al.*, 1990). The capacity of mononuclear phagocytes to actively engulf microbes was probably a critical step to pave the way for acquisition of an intracellular lifestyle, and this may be the reason for the predilection of *Leishmania* for these cells of the host defence. Moreover, macrophages are long-lived, and this distinguishes them from the neutrophils, which are also professional phagocytes but with a much shorter lifespan (Kaufmann, 2011).

In summary, *Leishmania* parasites must cope with varied and heterogeneous stressful situations during their life cycle (Fig. 14.1). Thus, in the insect vector, after blood meal excretion, the parasite experiences nutrient starvation and a local acidification through secretion of H^+; these environmental conditions induce parasite differentiation towards the highly motile virulent metacyclic form specialized for *Leishmania* transmission (Zakai *et al.*, 1998; Mazareb *et al.*, 1999; Sacks and Kamhawi, 2001). After inoculation into the mammalian host, the parasite is submitted successively to rapid changes in: (i) temperature, from ambient temperature in the sandfly vector to 37°C in the mammalian host; (ii) pH, after internalization into the macrophage, in the phagolysosome, the environmental pH decreases to very acidic values; (iii) nutrient and oxygen availability; and (iv) to immune attack by complement components, and other antimicrobial molecules produced by phagocytic cells (Alexander *et al.*, 1999). To cope with these hostile environmental conditions, *Leishmania* has evolved mechanisms of environmental sensing that are able to distinguish between the physiology of vector and host during the infectious cycle, and control the expression of *Leishmania* stage-specific factors. Furthermore, this parasite has evolved means for survival in the presence of hydrolytic enzymes and in the acid environment of the phagolysosomes. The aim of this chapter is to show our current knowledge of the stress response in *Leishmania*, and its involvement either as a defence mechanism to combat the harsh environmental conditions or as key trigger for the developmental differentiation of the parasite.

The heat shock response

For all living organisms, a moderate increase of just a few degrees above the optimum growth temperature represents a challenging problem for survival (Richter *et al.*, 2010). This temperature stress impacts many physiological processes in the cell, mainly affecting the function of proteins, as these molecules tend to denature in response to temperature increases. To facilitate protein folding, and prevent and reverse protein misfolding, cells use molecular chaperones. In response to heat stress, cells activate signalling pathways leading to the increased expression of the so called heat shock proteins (HSPs), whose main objective is alleviating the deleterious effect of heat increase on the cellular homeostasis and organization. Many of the HSPs are evolutionarily ancient proteins showing a universal distribution in all cell types and organisms. The classical HSPs, also named molecular chaperones, comprise five major families: HSP100, HSP90, HSP70, HSP60 and small heat shock proteins (sHSPs) (Lindquist and Craig, 1988). In general, molecular chaperones prevent the formation of non-specific protein aggregates and assist proteins in the acquisition of their native structures. More recently, a new role for molecular chaperones as mediators of intercellular signalling has been recognized, and the term 'chaperokine' has been coined to denote this function (Asea, 2006).

Temperature belongs to the many physical and chemical parameters that microorganisms need to monitor in their constantly changing environment. There are numerous ways to register a sudden change in temperature. Most of cells do not measure the ambient temperature but detect indirect consequences of a temperature shift, e.g. the accumulation of misfolded proteins after a heat shock. In fact, the heat shock response is triggered as a consequence of accumulation of unfolded proteins that may result of a variety of stresses, including heat, oxidative stress, heavy metals, ethanol and other toxic substances (Lindquist, 1986). Thus, the synthesis of HSPs has a dynamic character as the cells accommodate their levels to the amount of unfolded proteins (Richter *et al.*, 2010). Obviously, the function of molecular chaperones is also needed during normal cell growth and, therefore, the HSPs (or constitutively expressed variants) are present under normal physiological conditions too.

A particularly interesting case is found in digenetic organisms that along their life cycle alternate between poikilothermic and homeothermic hosts. In these organisms, like *Leishmania*, sudden temperature variations at the transmission steps are natural events of their life cycles. Nevertheless, incubation of *Leishmania* promastigotes at 37°C leads to the induction of a heat shock response, in which HSP83/90 (hereafter referred to as HSP90) and HSP70 are actively synthesized (Brandau *et al.*, 1995). Additionally, HSPs have been suggested to have a role in *Leishmania* virulence based on the observation of an increased infectivity of promastigotes after a heat shock treatment (Smejkal *et al.*, 1988). Thus, it has been suggested that HSPs induced on heat shock may pre-adapt promastigotes for stresses they encounter inside the macrophages. Furthermore, some HSPs seem to play relevant roles in the differentiation process (see below).

The HSP complement in *Leishmania* has been the subject of a comprehensive analysis (Folgueira and Requena, 2007). *Leishmania* contains representatives from all the evolutionarily conserved HSP families; Table 14.1 summarizes the composition and salient features of the main HSP families existing in this parasite. Among them, the HSP70 family is especially large, comprising nine subfamilies that localize in different cellular compartments (Louw *et al.*, 2010). Thus, HSP70, HSP70.c, HSP70.4 and HSP110 seem to be cytoplasmic proteins, GRP78 and GRP170 localize in the endoplasmic reticulum, MTP70 is mitochondrial, and HSP70.b might have a nuclear localization. It is probable that this large number of HSP70 variants is connected with the existence in *Leishmania*, and related trypanosomatids, of a numerous HSP40 family (e.g. 66 distinct HSP40s were predicted from the *L. major* genome database; Folgueira and Requena, 2007). HSP40s are partnership of HSP70, and they are responsible for conferring substrate specificity to HSP70 (Qiu *et al.*, 2006). Thus, in summary, *Leishmania* parasites possess an unusually large and diverse complement of HSP70s that may interact with an unusually large number of HSP40 partner proteins.

Some HSPs are present at high levels in *Leishmania* promastigotes; HSP70 and HSP90 make up 2.1% and 2.8%, respectively, of the total protein (Brandau *et al.*, 1995). In addition, the presence of low molecular proteolytic fragments of HSPs (mainly derived from either HSP70 or HSP90) is a consistent finding in proteome analyses of *Leishmania* (Mojtahedi *et al.*, 2008). Studies directed to investigate whether these molecular weight variants are the result of *in vivo* processing or protein degradation during sample preparation favour the existence of a controlled processing of these HSPs in the parasite (Drummelsmith *et al.*, 2003; Nugent *et al.*, 2004). On the other hand,

Table 14.1 Composition of the main families of heat shock proteins in *Leishmania*

Name	Members[a]	Remarks[b]
HSP100	HSP100 (LmjF29.1270) HSP78 (LmjF02.0710, LmjF27.2630)	HSP100 is amastigote specific
HSP90	HSP83/90[c] (LmjF33.0312-LmjF33.0365; 17 genes) GRP94 (LmjF29.0760) HSP75 (TRAP-1, LmjF33.2390)	Stage differentiation Drug resistance Anti-apoptotic *Leishmania*-secreted proteins
HSP70	GRP78 (BiP, LmjF28.1200) MTP70 (LmjF30.2460-LmjF30.2490, LmjF30.2550; five genes) HSP70 (LmjF28.2770, LmjF28.2780) HSP70.a (LmjF01.0640) HSP70.b (LmjF26.0900) HSP70.c (LmjF28.2820) HSP70.4 (LmjF26.1240) HSP110 (LmjF18.1370) GRP170 (LmjF35.4710)	Resistance to oxidative stress Drug resistance *Leishmania*-secreted proteins
HSP60	HSP60–1 (LmjF30.2820) HSP60–2 (LmjF32.1850) HSP60–3 (LmjF36.2020, LmjF36.2030)	*Leishmania*-secreted proteins
sHSPs	HSP20 (LmjF29.2450)	–

[a]Systematic names of the genes in the *L. major* GeneDB database.
[b]Special functions, others than molecular chaperone (see text for additional details).
[c]The *Leishmania* cytoplasmic member of the HSP90 family is often referred as HSP83.

recent studies in several *Leishmania* species have shown that HSPs are not restricted only to intracellular functions, but that some of them (HSP90, HSP70 and HSP60) are also abundantly secreted to the extracellular medium (Cuervo *et al.*, 2009; Silverman *et al.*, 2010), and their secretion increases after environmental temperature upshift to the mammalian host temperature (Hassani *et al.*, 2011). These findings may explain that some HSPs (like HSP70 and HSP90), during *Leishmania* infection, elicit strong antibody responses and behave as potent immunogens (Requena *et al.*, 2000). Furthermore, HSPs may act also as immunomodulators (chaperokines) for the host immune system (Rico *et al.*, 1999, 2002).

Regulation of heat shock response in *Leishmania*

In eukaryotes, the regulation of HSP synthesis is usually mediated by heat shock transcription factors (HSFs), which activate transcription from heat shock genes by interaction with cognate recognition sequences, the heat shock element or HSE (Santoro, 2000). In addition, for some organisms, regulation is also observed at the translational level (Semotok *et al.*, 2008). The stress response in *Leishmania*, in contrast to most organisms, is regulated exclusively at post-transcriptional levels (Brandau *et al.*, 1995; Quijada *et al.*, 1997). In *Leishmania*, there is no evidence for the existence of either HSE promoter elements or HSFs. Nevertheless, this is not so surprising taking into consideration that, indeed, the entire gene expression in *Leishmania* seems to be regulated mainly at post-transcriptional levels (Requena, 2011). Transcript stabilization and preferential translation of *HSP* mRNAs are the main mechanisms by which these parasites control HSP synthesis (Quijada *et al.*, 1997, 2000; Zilka *et al.*, 2001; Larreta *et al.*, 2004; Folgueira *et al.*, 2005; David *et al.*, 2010). Recently, phosphorylation of HSP70 (Thr498) and HSP90 (Thr233 and Ser526) has been demonstrated, postulating that regulation of the stress response in *Leishmania* may operate also post-translationally through phosphorylation/dephosphorylation (Morales *et al.*, 2010). In this regard, binding of *Leishmania* MAP kinases to HSP70 has been described (Morales *et al.*, 2007) and an interplay between chaperones and kinases might be regulating the stress response and the differentiation process in this parasite.

How changes in temperature and other environmental signals are sensed by *Leishmania* and converted to intracellular signals is poorly understood. In mammals, heat shock is one of the environmental factors causing the activation of the p38 MAP kinase cascade (Ono and Han, 2000). In *Leishmania*, during the promastigote-to-amastigote differentiation process, recent studies showed the existence of widespread changes in the phosphorylation state of many proteins, including phosphorylation of members of HSP families (HSP90, HSP70 and STI1) (Morales *et al.*, 2010). Furthermore, the *Leishmania* genomes encode for a large number of protein kinases and phosphatases that are probably involved in regulating parasite adaptive responses to the different conditions encountered in the insect and mammalian hosts (Parsons *et al.*, 2005). In particular, protein kinases, including *Leishmania* homologues of the mitogen-activated protein kinase (MAPK) and target of rapamycin (TOR) family of kinases, and a tyrosine-specific phosphatase have been shown to be essential for virulence (Agron *et al.*, 2005; Nascimento *et al.*, 2006; Wiese, 2007; Madeira da Silva and Beverley, 2010). However, the upstream signals that lead to the activation of these putative signalling pathways have yet to be defined.

Some studies have shown that the heat shock response and the differentiation of *Leishmania* promastigotes to axenic amastigotes are associated with a marked increase in intracellular Ca^{2+} levels (Sarkar and Bhaduri, 1995; Lu *et al.*, 1997). More recently, it was shown that extracellular Ca^{2+} influx is required for *Leishmania major* thermotolerance, and that genetic deletion of the regulatory calcineurin subunit leads to loss of thermotolerance, and inability to establish infection in susceptible BALB/c mice (Naderer *et al.*, 2011). In eukaryotes, changes in Ca^{2+} concentration are sensed by a number of proteins, including calmodulin and calcineurin. Calcineurin is a serine/threonine-specific protein phosphatase, activated by Ca^{2+} and calmodulin, that exists as a heterodimer composed of a catalytic and a regulatory subunit. When intracellular concentrations of Ca^{2+} rise, calmodulin binds to calcineurin and releases the catalytic subunit from the autoinhibitory, regulatory domain (Stie and Fox, 2008). Calcium influx and calcineurin signalling have been shown in a variety of eukaryotic organisms to be essential for the activation of cellular heat shock responses, including the increased expression of HSPs (Saidi *et al.*, 2009). In mammalian and yeast cells, calcineurin activation leads to the dephosphorylation of a number of transcription factors and activation of stress-responsive transcriptional programmes (Cyert, 2003). However, *Leishmania* lacks conventional transcription factors, and calcineurin signals must operate through different downstream events.

Atypical stress proteins

In addition to prototypical HSPs, a few *Leishmania* proteins have been found associated with the response to heat shock and stage differentiation, two imbricate processes in the life cycle of *Leishmania*. The *A2* gene was first identified in *L. donovani* by virtue of its expression, which is specific to the amastigote stage of the life cycle (Charest and Matlashewski, 1994). Expression of the amastigote-specific *A2* gene is induced in promastigotes when they are transferred from culture medium at 26°C and pH 7.4 to medium at 37°C and pH 4.5, conditions which mimic the macrophage phagolysosomal environment. Interestingly, mutants lacking the *HSP100* gene do not express A2 proteins, suggesting that HSP100 favours A2 expression (Krobitsch *et al.*, 1998). Remarkably, the A2 gene is a pseudogene in *L. major*, a species that causes dermotropic disease in humans, but the introduction of the *L. donovani A2* gene into *L. major* enhances the ability of this species to survive in visceral organs of BALB/c mice (Zhang *et al.*, 2003). Recently, it has been found that A2 co-localizes with chaperone BiP (GRP78; Table 14.1) in the endoplasmic reticulum, both in axenic amastigotes and in amastigotes inside macrophages, and also in heat shocked promastigotes (McCall and Matlashewski, 2010). Moreover, the interaction between A2 and BiP was found to be enhanced during heat shock. These observations argue that A2 plays an important role in *L. donovani* amastigote survival in internal organs and, perhaps, allowing the parasite to resist the typical fever processes associated with kala azar disease.

META1 gene was identified initially in *L. major* as a gene expressed predominantly in the infective metacyclic form of the parasite (Nourbakhsh *et al.*, 1996). Furthermore, overexpression of *META1* gene in *L. amazonensis* resulted in parasites with increased virulence (Uliana *et al.*, 1999). More recently, *META1* gene was found to be linked in the same gene cluster to *META2* gene, being both genes up-regulated in metacyclic promastigotes and conserved in several *Leishmania* species (Ramos *et al.*, 2004). Both proteins, META 1 and META 2, contain the META domain (PF03724), which is also present in several bacterial proteins of unknown function. META1 contains a single META domain, whereas META2 is composed of three copies of the META domain and a carboxyl-terminus calpain-like domain (Ramos *et al.*, 2004). *Leishmania* META2-null mutants are more sensitive to oxidative stress and heat shock, suggesting that this protein may confer resistance to heat and oxidative stresses upon entry of the parasite into the mammalian host (Ramos *et al.*, 2011).

Heat shock response and differentiation

Unsurprisingly, temperature variation is a major determinant of environmental sensing when endoparasites move from an invertebrate vector to a mammalian host, being heat stress an integral part of their life cycles. Thus, the differentiation of *Leishmania* parasites from the insect stage, the promastigote, towards the pathogenic mammalian stage, the amastigote, is triggered primarily by the rise in temperature encountered during the insect-to-mammal transmission (Clos and Krobitsch, 1999). For many *Leishmania* species, the process of differentiation may be mimicked *in vitro* by shifting cultured promastigotes (grown at 26°C, pH 7) to a lysosome-like environment (37°C and pH 5.5) (Zilberstein and Shapira, 1994; Saar *et al.*, 1998; Somanna *et al.*, 2002; Debrabant *et al.*, 2004; Barak *et al.*, 2005). Thus, the rise of temperature encountered during transmission can be viewed not as stress but rather as signal for cellular differentiation. On the other hand, there is a close correlation between the temperature tolerance of *Leishmania* species and their tropism in the mammalian host. Strictly cutaneotropic species like *L. major*, *L. mexicana* and *L. braziliensis* do not tolerate *in vitro* temperatures above 35°C, whereas viscerotropic species (*L. donovani* and *L. infantum*) tolerate temperatures exceeding 37°C, which is the temperature in the human abdominal cavity (Callahan *et al.*, 1996).

There exists substantial experimental evidence indicating that HSPs can themselves perform an important role in stage differentiation. An interesting observation was that ethanol and azetidine-2-carboxylic acid, two well-established modulators of the heat shock response resulting from increasing protein misfolding, induce promastigote to amastigote differentiation at acidic pH without raising the medium temperature (Barak *et al.*, 2005). This result suggests that HSPs, rather than the heat by itself, are ultimately responsible for induction of the differentiation process.

In *L. donovani*, inactivation of HSP90 with specific drugs such as geldanamycin or radicicol induces the differentiation from the promastigote to the amastigote form (Wiesgigl and Clos, 2001). Chemotherapeutic agents such as geldanamycin and radicicol bind to the ATP-binding pocket of HSP90, inhibiting specifically the function of this protein (Stebbins *et al.*, 1997). In *L. infantum*, fluctuations in both HSP90 mRNA abundance and HSP90 synthesis are observed during the promastigote-to-amastigote differentiation in axenic conditions (Larreta *et al.*, 2004). Thus, an increase in the steady-state level of *HSP90* transcripts occurs after 2 hours of incubation in the differentiation conditions (pH 5.5 and 37°C), but a drop in the *HSP90* RNA abundance is observed at 9 hours after the start of differentiation, decreasing further after 24 hours (Fig. 14.2). However, these changes in the steady-state levels of *HSP90* mRNAs are not due to a feedback response induced by a prolonged heat shock, since this down-regulation is not observed when promastigotes are continuously cultured at 37°C. Thus, the *HSP90* transcripts continuously accumulate under prolonged heat shock conditions (Larreta *et al.*, 2004). During the promastigote-to-amastigote differentiation in axenic conditions, the transient increase in the steady-state level of *HSP90* transcripts, followed by a decrease, is linked to paralleled variations in the *de novo* synthesis of HSP90. Thus, immunoprecipitation assays evidenced an 8-fold increase in HSP90 *de novo* synthesis after 3 hours of incubation in amastigote differentiation conditions relative to the synthesis in promastigotes cultures; thereafter, the level of HSP90 synthesis decreased to values even lower than those present in promastigotes. The parallelism between accumulation of *HSP90* transcripts and *de novo* HSP90 synthesis during the promastigote-to-amastigote differentiation suggests that both processes are controlled by the same regulatory mechanism. Another study, aimed at analysing the developmentally induced changes in the *L. donovani* proteome, identified HSP90 as a protein which showed an increase during heat shock (day 1) but resulted down-expressed in subsequent days of the amastigote differentiation course (Bente *et al.*, 2003). These findings reinforce the hypothesis that the stage differentiation in *Leishmania* is dependent on the HSP90 cellular homeostasis. This idea is based also on the proved involvement of HSP90 in the regulation of growth and development in different organisms (Nollen and Morimoto, 2002; Taipale *et al.*, 2010). Even in non-stressed cells, HSP90 is highly abundant and associates with a wide array of proteins (known as clients) that depend on its chaperoning function to acquire their active conformations. Although the best-studied clients of HSP90 are steroid hormone receptors and protein kinases, many others have been discovered, such as transcription factors, proteins involved in RNA processing, telomerase reverse transcriptase and chromatin proteins (Richter and Buchner, 2001; Taipale *et al.*, 2010). Taken together, current data suggest that removal of active HSP90, by either programmed down-regulation of its synthesis, or artificial inactivation by specific drugs (e.g. geldanamycin), would be a trigger for stage differentiation in *Leishmania*.

In addition, pharmacological inhibition of HSP90 leads to an increase in the synthesis of HSP100, HSP90 and HSP70 (Wiesgigl and Clos, 2001; our unpublished results), suggesting that HSP90 is also involved in the post-transcriptional control of the stress response in *Leishmania*.

HSP100 (ClpB) is the only HSP identified so far in *Leishmania* that shows a stage-specific expression. Thus, HSP100 is barely detectable in unstressed promastigotes but becomes abundant

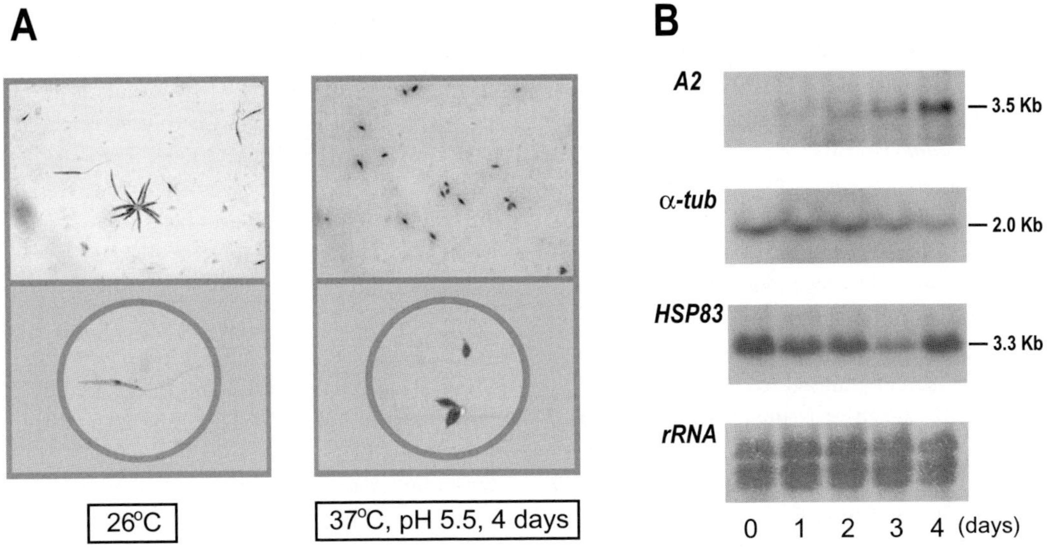

Figure 14.2 Promastigote to amastigote differentiation in axenic conditions. (A) The differentiation process can be mimicked in axenic culture by shifting promastigotes from an insect-like (26°C, pH 7) to an intralysosomal-like (37°C, pH 5.5) environment. (B) Analysis of *A2*, *α-tubulin*, *HSP90* and *rRNA* transcripts levels during the differentiation process (see text, and Larreta *et al.*, 2004, for additional details). These axenic amastigotes express A2 transcripts, a product of the A2 gene family that is detectable only in fully differentiated amastigote forms (Charest *et al.*, 1996).

both under heat stress and during axenic amastigote differentiation (Hubel *et al.*, 1995). Additionally, the presence of HSP100 appears to be required for synthesis of amastigote stage proteins, including the amastigote specific protein A2 (Krobitsch *et al.*, 1998). However, the inactivation of the *HSP100* gene does not lead to a block in the promastigote-to-amastigote differentiation, even though knock-out parasites have decreased virulence and diminished capacity to proliferate inside host macrophages (Hubel *et al.*, 1997). Remarkably, repeated cycles of mouse infection with this deletion mutant eventually led to the isolation of escape mutants with increased virulence (Reiling *et al.*, 2006). Characterization of these escape mutants has led to the identification of a 46-kDa protein (dubbed P46) of unknown function as responsible for this leaky phenotype, but the compensation mechanism is not known at present (Reiling *et al.*, 2010). Additionally, it was observed that P46 overexpression in wild-type *L. major* parasites boosts virulence, increases parasite multiplication into macrophages and widens the host range. Taken together, these data suggest that HSP100 is important for stage differentiation, but the parasite has the capacity to overcome an HSP100 deficiency by activating alternative pathways to progress through its life cycle. On the other hand, understanding the role and function of P46 will shed light on the compensation mechanism.

HSPs and resistance to drugs

The therapeutic mainstay against *Leishmania* is still based on the use of pentavalent antimonials, but resistance is increasing in several parts of the world (see above). Recently, by functional cloning experiments, two members of the HSP70 family have been found associated to antimony tolerance (Brochu *et al.*, 2004). It is likely that HSP70s are not directly involved in resistance to antimony but, clearly, increased levels of expression confer protection to the parasite against this metal. Proteomic analysis of differential protein expression associated with the resistance to methotrexate (an antifolate) in *L. major* revealed an increased expression of HSP70 and HSP90 (Drummelsmith *et al.*, 2004). More recently, overexpression of several members of the HSP70 family has been observed in antimony-resistant isolates of *Leishmania donovani* (Biyani *et al.*, 2011).

Antimonials appear to kill cells by a process with features of programmed cell death or apoptosis (Lee *et al.*, 2002). It has been suggested that other leishmanicide drugs like miltefosine (Paris *et al.*, 2004) and amphotericin B (Lee *et al.*, 2002) also kill *Leishmania* by apoptotic processes. A proteomics screen, aimed to compare Sb(V)-resistant and –sensitive *L. donovani* field isolates, identified two proteins, HSP90 and a small kinetoplastid calpain-related protein (SKCRP14.1) as responsible for modulating susceptibility to antimonials (Vergnes *et al.*, 2007). The HSP90 is overexpressed in the Sb(V)-resistant isolate, whereas protein SKCRP14.1 results in down-expression. Remarkably, the Sb(V)-resistant parasite exhibited cross-resistance to miltefosine and amphotericin B. Overexpression of HSP90 in the sensitive line conferred increased resistance to both Sb(III) and miltefosine. Interestingly, this resistance phenotype is associated with a decrease in drug-mediated apoptosis. Consistent with these findings, mammalian HSP90, the homologue to *Leishmania* HSP90, has been described as a negative regulator of the mitochondrial cytochrome c-dependent apoptosis pathway (Arya *et al.*, 2007). More recently, a direct link between HSP90 and apoptosis has been obtained analysing the effect of the HSP90 inhibitor geldanamycin (GA) during *L. donovani* promastigote-to-amastigote transformation. Promastigotes exhibited apoptotic morphological changes after geldanamycin treatment at a high temperature, and the effect was found to be dose and time dependent (Q. Li *et al.*, 2009).

Oxidative stress

In nearly all living organisms, maintenance of an intracellular reducing milieu is achieved by the presence of high concentrations of the sulfur-containing tripeptide glutathione (GSH). However, in *Leishmania* and other trypanosomatids, most of their GSH content is found in the form of N1, N8-bis(glytathionyl)spermidine, known as trypanothione, which is a conjugate of two GSH molecules with a spermidine molecule (Castro and Tomas, 2008). It is also the target for trivalent antimonial drugs which form conjugates with trypanothione (Fairlamb, 2003). Trypanothione protects the parasite against oxidative stress through peroxidase systems (see below), against reactive aldehydes through the glyoxalase system (Vickers *et al.*, 2004a), and against toxic xenobiotics through the trypanothione S-transferase (Vickers and Fairlamb, 2004). All these processes depend upon trypanothione being maintained in its dithiol, reduced form by trypanothione reductase (TR). In consequence, this NADPH-dependent flavoenzyme is central to the redox defence systems of trypanosomatids. In fact, attempts to obtain TR null mutants in *L. donovani* failed, suggesting that *TR* is an essential gene (Dumas *et al.*, 1997). The relevance of this protein for *Leishmania* can be deduced from the observation that parasites possessing only one *TR* allele experience a dramatic decrease in their capacity to survive within human macrophages.

A phylogenetic analysis comprising of the TR sequences of several trypanosomatids, together with representatives of the disulfide oxidoreductase family, showed that trypanosomatid TRs are most closely related to the glutathione reductases of plants, with which they forms a single and robust clade (Opperdoes and Szikora, 2006). It is remarkable that *Trypanosoma* and *Leishmania* contain many 'plant-like' genes encoding homologues of proteins found in either chloroplasts or the cytosol of plants and algae. Thus, other antioxidant enzymes, such as superoxide dismutase (SOD) and tryparedoxin peroxidase (TXNPx) are also more similar to plant than to animal homologue enzymes. These findings have led to the hypothesis that the ancestral trypanosomatid acquired those genes by horizontal transfer from a cyanobacterium or from an algal endosymbiont (Hannaert *et al.*, 2003).

Although the trypanothione system appears to be the principal defence against oxidant damage in *Leishmania*, in the insect stages ovothiol A (N^1-methyl-4-mercaptohistidine; OSH) is present in significant amounts. In fact, OSH can exceed trypanothione content particularly in late logarithmic and stationary phases of growth in the promastigote forms (Ariyanayagam and Fairlamb, 2001). Nevertheless, although OSH can act as a non-enzymatic scavenger of hydrogen peroxide, it is less efficient than trypanothione. In addition, given the absence of ovothiol reductase activity (Ariyanayagam and Fairlamb, 2001), any thiol-redox reactions resulting in the formation of OSSO will depend on the trypanothione/trypanothione reductase system. Hence, the physiological relevance and functional role of OSH in trypanosomatids remain unknown.

Throughout its life cycle, *Leishmania* parasites are exposed to a variety of reactive oxygen species (ROS) and reactive nitrogen species (RNS) that they must detoxify to survive. Antioxidant defence seems to be particularly important for amastigotes, although evidence shows that nitric oxide (NO) is important to limit parasite development also in the invertebrate host (Rivero, 2006). Upon being taken up by macrophages, parasites are exposed not only to heat shock, but also to various macrophage defence mechanisms. Two important macrophage-derived oxidants have been identified as critical in controlling *Leishmania* infection. During the first stages of infection, ROS, such as hydrogen peroxide (H_2O_2), OH^- and O_2^-, have been found to be produced as part of the respiratory burst of human and murine macrophages in response to phagocytosis. Superoxide production is catalysed by the NADPH oxidase, a haem-containing cytochrome family that includes cytosolic and membrane-bound members. Once assembled, the oxidase transfers an electron from NADPH to molecular oxygen, producing superoxide ($O_2^{\cdot-}$) anions, which spontaneously combine with water to form efficient oxidants (such as the hydroxyl radical OH^- and H_2O_2). A second anti-leishmanial oxidant produced by macrophages is NO (Liew and O'Donnell, 1993). Unlike O_2^-, which is generated during phagocytosis of the parasite, NO^- is generated after macrophage activation by IFN-γ and TNF-α and is most relevant to killing established intracellular amastigotes. Macrophages express cytokine-inducible nitric oxide synthase (iNOS), which catalyses the synthesis of high concentrations of NO from L-arginine and molecular oxygen. Thus, the presence of inhibitors for iNOS, such as N^G-monomethyl-L-arginine (L-NMMA), leads to an increase in amastigote survival and replication in murine macrophages (Iniesta *et al.*, 2001). Nevertheless, even though there is strong evidence that NO plays an important role in murine leishmaniasis, it remains controversial whether NO plays a similar role in the anti-leishmanial responses of human macrophages (Gantt *et al.*, 2001). The combination of NO with superoxide anions produces peroxynitrites ($ONOO^-$), which are very toxic substances. Furthermore, several of the clinically used drugs, such as the antimonials, may act through the induction of oxidative stress. In order to overcome those toxic effectors, intracellular parasites have evolved mechanisms to detoxify or counteract the reactive oxygen species. In addition, the promastigote's surface glycolipid lipophosphoglycan (LPG) protects the parasite by scavenging toxic oxygen products, and inhibits macrophage responses (Descoteaux and Turco, 2002). Therefore, resistance of the parasite to oxidant toxicity and inhibition of the macrophage respiratory burst seem to be relevant mechanisms by which the parasite survives phagocytosis by macrophages.

It has long been known that trypanosomatids lack catalase and classical selenium-containing glutathione peroxidases, two major hydroperoxide-eliminating enzymes, commonly present in eukaryotes. In agreement with this no catalase gene was found in any of the trypanosomatid genome databases (Opperdoes and Szikora, 2006). Instead, in these organisms, neutralization of hydroperoxide was found to depend on trypanothione metabolism (Opperdoes and Szikora, 2006; Castro and Tomas, 2008). Thus, hydroperoxide reduction is achieved by the concerted action of two enzymes, a tryparedoxin and a tryparedoxin peroxidase (TPX), belonging to the peroxiredoxin family. In addition, *Leishmania* possesses also peroxiredoxins (PRXs), a ubiquitous family of antioxidant enzymes with hydroperoxide, alkyl peroxide and peroxynitrite reducing activity (Levick *et al.*, 1998; Barr and Gedamu, 2001). Fig. 14.3 summarizes the main components of the protection arsenal of *Leishmania* against oxidant stress. Interestingly, the trypanothione reductase system seems to be regulated by the action of a subtilisin protease (encoded by gene LmjF13.1040). Thus, the predominant

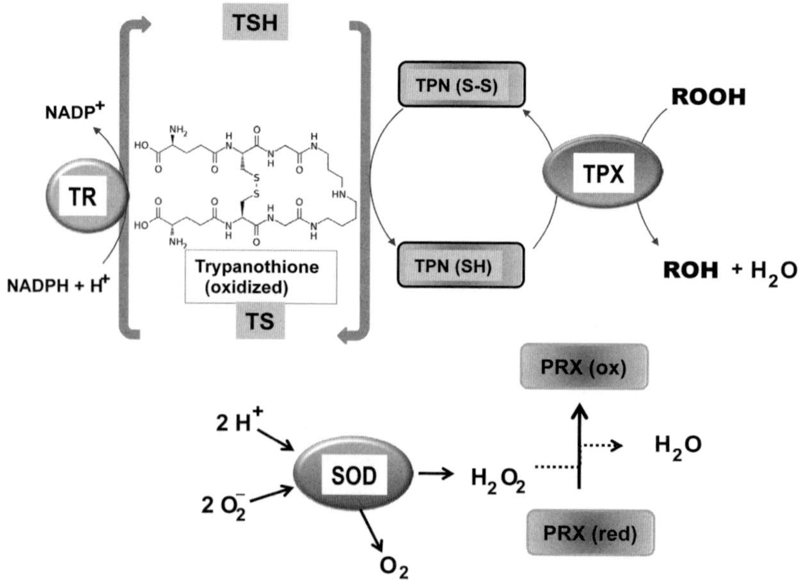

Figure 14.3 Thiol-redox system and peroxide metabolism in *Leishmania*. PRX, peroxiredoxin; SOD, superoxide dismutase; TPN, tryparedoxin; TPX, trypanothione peroxidase; TR, trypanothione reductase; TS and TSH, oxidized and reduced trypanothione, respectively. PRX is able to detoxify organic and inorganic peroxides.

tryparedoxin peroxidases were decreased in *L. donovani* knockout parasites, whereas higher molecular weight isoforms accumulate, indicating altered processing. Furthermore, knockout parasites showed increased sensitivity to hydroperoxide (Swenerton *et al.*, 2010).

Peroxiredoxins (PRXs) are found in a great variety of organisms, where they fulfil distinct functions, such as detoxification, signalling or differentiation (Hofmann *et al.*, 2002). In *Leishmania*, cytosolic as well as mitochondrial PRXs have been described (Castro *et al.*, 2002; Flohe *et al.*, 2002; Barr and Gedamu, 2003). PRXs localized in the cytosol appear to be adequate to protect the parasite against oxidative attack from the outside; PRXs localized in the mitochondrion seem to be of particular interest to protect this organelle from the oxidants produced by its own energy metabolism. Previous studies showed that the mitochondrial peroxiredoxin of *L. infantum* (LimTXNPx) and cytoplasmic peroxiredoxins from *L. infantum* (LicTXNPx), *L. chagasi* (LcPxn1, LcPxn2), and *L. donovani* (LdH6TXNPx) can all detoxify reactive oxygen species (ROS), with a preference for H_2O_2 (Castro *et al.*, 2002; Flohe *et al.*, 2002; Barr and Gedamu, 2003). It was further shown that overexpression of LcPxn1 in *L. chagasi* enhanced survival when exposed to either ROS or RNS and also enhanced survival within U937 macrophage cells (Barr and Gedamu, 2003). More recently, it was shown that mitochondrial peroxiredoxin (LdmPrx) protects *L. donovani* from H_2O_2-induced apoptosis, pointing to a vital role for this PRX in the survival of the parasite (Harder *et al.*, 2006). Mitochondria are important checkpoints for the control of apoptotic cell death (see below for further details). In addition, they are an important site for the production of ROS; in the electron transport chain, part of the consumed oxygen is released in the mitochondria as H_2O_2, O_2^-, and OH^-. Nevertheless, low concentrations of ROS are usually present in the cells, where they act in physiological processes. However, on the other hand, overproduction of these toxic ROS may induce damage to proteins, lipids and DNA and eventually can lead to cell death. In particular, mitochondrial H_2O_2 is a well-known inductor of apoptosis in mammalian cells (Kroemer and Reed, 2000). In consequence, mitochondrial PRXs, as important regulators of H_2O_2 levels, may result essential to prevent apoptosis.

During mitochondrial respiration, another

important ROS, the superoxide anion ($O_2^{\cdot-}$) is formed as a by-product. Respiratory chain complexes I and III are the main source of superoxide in mammals (St-Pierre *et al.*, 2002). The superoxide is then converted to H_2O_2 by spontaneous dismutation or by action of superoxide dismutases (SODs). High levels of superoxide dismutase (SOD) are present in *Leishmania* promastigotes, and two types of *SOD* genes, *SOD-A* and *SOD-B*, have been identified in *L. chagasi* (Paramchuk *et al.*, 1997; Plewes *et al.*, 2003). Overexpression of both types of genes conferred increased resistance to sodium nitroprusside (an NO donor) or paraquat, an O_2^- generator. In contrast, SOD-deficient parasites, generated by expressing the corresponding antisense RNA, have enhanced sensitivity to menadione (an O_2^--generating compound) and hydrogen peroxide in axenic cultures, as well as a reduced survival in mouse macrophages (Ghosh *et al.*, 2003). In addition, parasites containing a single allele for the *SOD-B* exhibited an important reduction in growth in the presence of paraquat and a significant decrease in survival within human macrophages (Plewes *et al.*, 2003).

Recently, an unusual plant-like ascorbate peroxidase (APX) from *L. major* (LmAPX) has been characterized, and it has been postulated that this protein may play an important role in detoxification of H_2O_2, generated by endogenous processes and, perhaps also, as a result of external influences such as the oxidative burst of infected host macrophages or during drug metabolism by *Leishmania* (Adak and Datta, 2005). Overexpression of LmAPX in *L. major* protects promastigotes against oxidative stress-induced apoptosis (Dolai *et al.*, 2009). Conversely, LmAPX-knockout parasites have higher levels of intracellular H_2O_2, and an increase of apoptotic cells are observed in the cultures. However, these knockout promastigotes are more virulent and induce larger cutaneous lesions than parental parasites (Pal *et al.*, 2010). This paradoxical observation may be explained on the basis of a remarkable association between apoptosis and virulence occurring during *Leishmania* transmission from insect vectors to mammalian hosts.

The leishmanial pteridine reductase, involved in pteridine salvage as well as in antifolate resistance, has been shown to protect intracellular amastigotes against reactive oxygen and nitrogen intermediates (Moreira *et al.*, 2009). Also, the *L. major* elongation factor 1B (eEF1b) complex presents trypanothione S-transferase and peroxidase activity, suggesting that, in addition to its role in protein synthesis, the *Leishmania* eEF1b may help to protect the parasite from peroxidation (Vickers *et al.*, 2004b). A comparative protein profile carried out on several phenotypic variants of *L. guyanensis* identified eEF1b and tryparedoxin peroxidase (TPX) as factors associated with metastasis (Walker *et al.*, 2006). Parasites of the *Viannia* subgenus are major causative agents of mucocutanous leishmaniasis (MCL), a disease characterized by parasite dissemination (metastasis) from the original cutaneous lesions to form debilitating secondary lesions in the nasopharyngeal mucosa. The association of TPX and eEF1b with metastasis in *Leishmania* strengthens the link between parasite virulence and antioxidant defence.

There are several reports involving HSPs as mediators of resistance to oxidative stress in *Leishmania*. Thus, exposure of promastigotes to heat shock causes them to become more resistant to H_2O_2. Conversely, incubation in sublethal concentrations of H_2O_2 or menadione causes a stress response in promastigotes (Wilson *et al.*, 1994). This oxidant-induced response was associated with an increase in the amount of HSP70. Subsequently, it was reported that overexpression of HSP70 bestows a remarkable resistance to oxidative stress in *L. chagasi* promastigotes, suggesting a relevant role of this protein in the antioxidant defence mechanisms (Miller *et al.*, 2000). Interestingly, exposure of *L. chagasi* promastigotes to sublethal concentrations of menadione significantly increased their virulence for BALB/c mice (Wilson *et al.*, 1994). This is reminiscent of the fact that the virulence of *Leishmania* promastigotes was found to increase after heat exposure (see above). On the other hand, HSP70 is also up-regulated in *Leishmania* promastigotes treated with sublethal doses of nitric oxide donors (Dea-Ayuela *et al.*, 2009).

A recent study reinforced further the interrelationship between heat shock and oxidative stresses in *Leishmania* (Alzate *et al.*, 2007). Parasites subjected to heat shock treatment at 38°C experience a rapid increase in superoxide levels. The reason seems to be an increase in the respiratory activity of the mitochondrion during heat stress, determined by the rate of O_2 consumption and the hyperpolarization of mitonchondrial membranes. In fact, inhibition of the respiratory chain with thenoyltrifluoroacetone (TTFA), antimycin A and cyanide leads to a decrease in superoxide intracellular concentration under heat-shock conditions (Alzate *et al.*, 2007).

Finally, a remarkable association between antioxidant response and resistance to antimonials, which is the mainstay therapy for leishmaniasis, has been put forward. Thus, a considerable body of evidence places trypanothione and other thiols at the centre of both clinical and laboratory-generated resistance mechanisms (Mukhopadhyay *et al.*, 1996; Arana *et al.*, 1998). Resistance to antimonials and arsenicals in *Leishmania* has been found to be linked to elevated levels of the enzymes involved in the synthesis of glutathione and polyamines, the precursor molecules of trypanothione (Haimeur *et al.*, 1999). Additionally, two antioxidant enzymes, trypanothione S-transferase and tryparedoxin peroxidase, may also have relevant roles in resistance to antimonials (Wyllie *et al.*, 2008).

Osmotic stress

Changes in the external chemical composition and osmotic strength are major physiological stress factors which *Leishmania* parasites encounter in the digestive tract of the sandfly (Fig. 14.1). It has been found that *Leishmania* promastigotes contain high concentrations of osmotically active amino acids, predominantly in the form of alanine. When challenged with hypotonic stress, the parasites swiftly release part of their amino acid content in order to reduce the increased cell volume (Vieira *et al.*, 1997). However, the underlying regulatory mechanisms remain largely unknown. Also little is known about volume regulation in the clinically relevant amastigote, the parasite form that multiplies within the parasitophorous vacuoles of mammalian host macrophages. A study documented the existence of coordinated changes in organic and inorganic osmolytes, and the involvement of the acidocalcidosomes in maintaining ionic homeostasis in the response of *Leishmania* amastigotes to hypo-osmotic stress (Lefurgey *et al.*, 2005).

Stress, apoptosis and virulence

Apoptosis, or programmed cell death (PCD), is a gene-regulated process relevant for the normal development and homeostasis of multicellular organisms. However, there is now increasing experimental evidence that PCD is also operative in unicellular organisms such as *Leishmania* (Smirlis *et al.*, 2010). Different stimuli have been described as inductors of apoptosis in *Leishmania*: heat shock (Moreira *et al.*, 1996; Alzate *et al.*, 2006), reactive oxygen species (Das *et al.*, 2001), drugs (Paris *et al.*, 2004; Q. Li *et al.*, 2009), serum deprivation (Zangger *et al.*, 2002), nitric oxide (Zangger *et al.*, 2002), and antimicrobial peptides (Luque-Ortega *et al.*, 2010). *Leishmania* also possesses autophagic cell death that is activated during starvation-induced stress and differentiation (Besteiro *et al.*, 2007). Nevertheless, the autophagy seems to be mainly a catabolic process and a survival mechanism whereby eukaryotic cells can degrade and recycle proteins and organelles.

Similar to metazoan apoptotic cell death, in *Leishmania*, many apoptogenic stresses are associated with a dysfunction of the unique mitochondrion of these organisms (Lee *et al.*, 2002). Increased levels of ROS, alteration of calcium homeostasis and mitochondrial dysfunction are interconnected effectors associated with apoptosis in *Leishmania* and related trypanosomatids (Smirlis *et al.*, 2010). Thus, for example, in a recent work, the endonuclease G (EndoG) homologue in *L. infantum* has been characterized. As occurs in mammalian cells, this mitochondrial endonuclease translocates to the nucleus upon oxidative and/or drug induced apoptosis (Rico *et al.*, 2009). Interestingly, overexpression in *Leishmania* of this endonuclease promoted apoptotic cell death under oxidant or differentiation-induced stresses, indicating that this protein may be an essential effector of apoptosis in this parasite.

In addition, evidence exists that proteins associated with proliferation, cell cycle control, chromosomal topology, and cytoskeleton are linked to apoptosis in *Leishmania* and related trypanosomatids (reviewed in Smirlis *et al.*, 2010). Furthermore, a link between endoplasmic reticulum stress and apopotosis has been experimentally evidenced in *Leishmania* (Dolai *et al.*, 2011). The endoplasmic reticulum (ER) is a multifunctional cellular structure involved in the folding and post-translational modification of proteins. The ER also plays important roles in Ca^{2+} signalling regulation, vesicle trafficking, drug metabolism, and lipid biogenesis. In addition, the ER can initiate apoptosis when accumulation of unfolded proteins and/or inhibition of the ER-Golgi transport occur (Rutkowski and Kaufman, 2004). Tunicamycin, a naturally occurring antibiotic, is an inductor of ER stress by inhibiting N-linked glycosylation of proteins. Consequently, tunicamycin was found to induce apoptosis in different cell types (Tabas and Ron, 2011). Dolai and co-workers (2011) showed that ER stress induced by tunicamycin in *Leishmania* leads to a rapid accumulation of ROS and an increase of the intracellular Ca^{2+} level, which causes mitochondrial depolarization and ultimately apoptosis. Furthermore, ER stress-induced ROS-dependent release of cytochrome c and EndoG from mitochondria to cytosol and subsequent translocation of EndoG to nucleus were observed by these authors. Recently, Goldshmidt *et al.* (2010) reported that persistent ER stress induces the spliced leader RNA silencing pathway (SLS), leading to apoptotic death in *T. brucei*. *Leishmania* and other trypanosomatids are known for their non-conventional gene expression mechanisms (reviewed in Clayton and Shapira, 2007; Requena, 2011). Among those, one may cite *trans*-splicing, which is required for the maturation of all mRNAs; in this process, a small common exon (named miniexon or spliced leader -SL), derived from a small RNA (the SL RNA), is joined to the 5′-end of all transcripts providing each with the 5′ cap structure for translation. The hallmark of SLS is shutoff of SL RNA transcription, leading to major reduction of mRNAs and, consequently, to a severe inhibition of protein synthesis (Lustig *et al.*, 2007). To date, the existence of such an aggressive mechanism in *Leishmania* has not been reported.

The relevance of apoptosis in the biology of *Leishmania,* and in general for unicellular organisms, is under discussion. However, recent studies have shown that apoptosis may be a parasite strategy to avoid an excessive inflammatory response during mammalian host infection. In the insect vector, the transformation of the procyclic promastigote into the non-replicating metacyclic form is accompanied by an improved ability to infect and survive in the vertebrate host, evading the immune response. This differentiation is also observed when parasites are cultured axenically: procyclic promastigotes are the predominant forms during the logarithmic phase of growth, but they transform to a highly infectious metacyclic promastigotes when they reach stationary phase (Sacks and Perkins, 1984). Moreover, it has been shown that promastigote cultures at the stationary phase contain up to 43% apoptotic cells, and when the latter are removed by sorting, the remaining non-apoptotic population is incapable of establishing and maintaining an infection (van Zandbergen *et al.*, 2006; Wanderley *et al.*, 2009). Thus, the presence of apoptotic parasites in the inoculum, or during natural infection, may be a *Leishmania* strategy for immune evasion that is based on the silent phagocytosis of apoptotic cells by mammalian macrophages (Kurosaka *et al.*, 2003). Since apoptotic cells do not represent danger, their uptake does not result in the activation of antimicrobial effector functions of phagocytes. Thus, the apoptotic parasites would promote inhibition of macrophage activation before invasion by viable *Leishmania* promastigotes. Furthermore, propagation of the parasite in the mammalian host is mediated by amastigotes that express phosphatidylserine (PS) on their membrane as a form of apoptotic mimicry (Wanderley *et al.*, 2006).

Stress response and extrachromosomal DNA amplification in *Leishmania*

Leishmania species exhibit considerable chromosomal plasticity in response to stress, and they are able to amplify both circular and linear extrachromosomal elements in response to selective pressures (reviewed in Beverley, 1991; Borst and Ouellette, 1995; Segovia and Ortiz, 1997). Thus, DNA amplification can be induced in this parasite by selection with a variety of drugs including methotrexate, oxyanions (antimony, arsenite), mycophenolic acid, tunicamycin and vinblastine, but amplicons are present even in unselected natural isolates (Tripp *et al.*, 1992; Segovia and Ortiz, 1997; Genest *et al.*, 2005). Most of DNA amplifications consist of extrachromosomal circular DNAs, ranging in size from 30 to greater than 200 kb, but linear amplifications have also been described. In addition, amplicons derived from equivalent chromosomal regions are present in different *Leishmania* species and isolates (Segovia and Ortiz, 1997). Current data indicate that these rearrangements would be generated by intramolecular homologous recombination involving repetitive DNA sequences (Ouellette *et al.*, 1991; Olmo *et al.*, 1995; Grondin *et al.*, 1996). Recently, it was described the existence of a family of repeated sequences, scattered in the different *Leishmania* chromosomes, whose peculiar distribution might be related with their involvement in the generation of extrachromosomal amplicons (Requena *et al.*, 2008; Smith *et al.*, 2009). These sequence elements, termed SIDER (for short interspersed degenerated retroposons), are related to retroposons existing in other trypanosomatids, but extinct along *Leishmania* evolution (Bringaud *et al.*, 2007). Interestingly, SIDER sequences, and particularly members of the subfamily SIDER2, were found in several *Leishmania* DNA amplicons (Requena *et al.*, 2008). Thus, it may be postulated that dissemination of SIDER sequences in the *Leishmania* genome has been evolutionarily favoured by their role in increasing the genomic plasticity.

Gene amplification in *Leishmania* seems to be an escape mechanism that *Leishmania* has evolved. Thus, when parasites are subjected to selective stresses, appropriate genomic DNA regions, containing particular genes, are amplified as extrachromosomal structures. This mechanism is used by the parasite to overexpress proteins responsible for drug resistance or able to tackle other stressful environmental conditions. Remarkably, the parasite also uses this mechanism to overcome artificial hurdles like knocking of essential genes made in the laboratory (Feng *et al.*, 2009; Fulwiler *et al.*, 2011). Relevant to this phenomenon is the fact that no specific sequences seem to be required for replication (and transcription) in *Leishmania*, as plasmids with a selectable marker, but devoid of *Leishmania* sequences, remain episomal under selective pressure (Papadopoulou *et al.*, 1994). Thus, the non-stringent sequence requirement for both expression and replication in *Leishmania* is conspicuous and might explain why gene amplification occurs so frequently in this parasite. In summary, the plasticity of the *Leishmania* genome should be regarded as a powerful mechanism evolved to face diverse stressful situations that the parasite may find along its complex life cycle. However, this mechanism adds difficulty to the objective of combating leishmaniasis through drug design for which the parasite cannot generate resistance.

Conclusions and future trends

During its digenetic life cycle, *Leishmania* alternates between the alimentary tract of the sandfly vector as an extracellular promastigote and the acidic phagolysosomes of macrophage cells as an intracellular amastigote. In these different environments, the parasites must adapt to variable conditions of temperature, pH and nutrient availability. Differentiation of the parasite is a prerequisite for its intracellular survival. The molecular bases of this transformation remain poorly understood although it is well established that several environmental factors such as pH and temperature trigger this cytodifferentiation process. Understanding the molecular mechanisms used by *Leishmania* parasites to survive under these harsh conditions is fundamental to our understanding of the disease, and ultimately for its control.

One fundamental but poorly understood aspect of parasite physiology concerns how *Leishmania* pathogenic microorganisms adapt to changes in their environments and what molecular mechanisms they employ to mediate these adaptations. In particular, the precise regulatory mechanism of the heat shock response is not understood. It will also be important to understand the function of HSP networks not only in the context of heat shock and other stresses, but also, and maybe more importantly, in the context of the differentiation processes. Despite the relevance of heat-induced stage differentiation for pathogenesis, mechanisms underlying the parasite heat-shock response and its role in the development and survival of the amastigote stage remain poorly understood. The absence of classical promoter-dependent mechanisms of gene regulation, and the constitutive expression of most of the HSPs in *Leishmania* raise important unresolved questions as to how the heat shock response is regulated and what are the stage-specific functions of *Leishmania* HSPs. Obviously, environmental sensing along the *Leishmania* life cycle should be of paramount relevance, but little is known about the pathways that transduce extracellular signals into stage-specific gene expression. Across most eukaryotic groups, highly conserved extracellular-regulated/mitogen-activated protein kinases (MAPKs) are implicated in environmental sensing and modulating gene expression in response to nutritional and physiological changes. They are well-known mediators of signal transduction, regulating important processes such as proliferation, differentiation, cell shape, stress response and apoptosis. In *Leishmania*, 17 MAPKs and MAPK-like kinases have been identified (Parsons *et al.*, 2005); however, our understanding of the molecular processes regulated by MAKPs in *Leishmania* is at the beginnings (Morales *et al.*, 2007).

Another important stress response is the unfolded protein response (UPR), which includes a set of signalling pathways that protect the cell from the accumulation of misfolded proteins in the endoplasmic reticulum (ER). In metazoans, activation of the UPR results in the transcriptional up-regulation of components of folding, trafficking and degradation machineries in the ER. In addition, UPR includes phosphorylation of eIF2α, leading to attenuation of protein synthesis (Rutkowski and Kaufman, 2004; Bernales *et al.*, 2006). At present, little is known about the mechanisms of UPR in *Leishmania*. In a recent article (Gosline *et al.*, 2011), it has been found that, as expected, UPR is not mediated by transcriptional control, evidencing instead the existence of a translational control of this pathway in *L. donovani*. The authors further observed that *L. donovani* is more sensitive to UPR inducing agents than host macrophages, suggesting that the UPR stress response might provide a new avenue for therapeutic treatment of parasitic infections.

The functions of HSPs, mainly those of HSP70 and HSP90 families, are not just limited to helping newly synthesized proteins to fold but also they are critical molecular chaperones in signal transduction events. In addition, genetic studies have evidenced that HSPs and their cochaperones are also involved in developing pathological processes like cancer, neurodegenerative disorders, and many other diseases. As such, they have gathered significant attention as potential, emerging drug targets (Evans *et al.*, 2010). It should be noted that a wide variety of chemical scaffolds have been identified by their affinity to HSP70 and HSP90 proteins (Shonhai *et al.*, 2011). And, more importantly, it is likely that the use of HSP90-specific drugs in the treatment of cancer may soon become a reality (Y. Li *et al.*, 2009). Given the relevance that HSPs also have for pathogens causing infectious diseases (as shown in the different chapters of this book), the effect of HSP inhibitors, specific for the microbial proteins, to control infections is being explored. For example, HSP90 is being studied as a potential drug target to combat protozoan diseases like toxoplasmosis (Echeverria *et al.*, 2005), trypanosomiasis and malaria (Pallavi *et al.*, 2010). Thus, considering the limitations in efficacy of current drugs used against leishmaniasis, it should be interesting to determine if the function of *Leishmania* HSPs can be pharmacologically manipulated in a specific manner.

Many signalling pathways in other organisms culminate in the activation of transcription factors; such factors have not been identified in trypanosomatids. Because the major mechanisms for gene regulation in these parasites appear to be post-transcriptional, it may be hypothesized that signalling pathways regulating gene expression might target molecules regulating RNA processing or turnover. Thus, greater emphasis may be placed on studying post-transcriptional regulatory mechanisms and *trans*-acting factors involved in the regulation of mRNA stability and translation, and also those responsible for protein turnover and post-translational modification of proteins. A more detailed knowledge of these pivotal regulatory processes in *Leishmania* might also present a new strategy for the control of the devastating epidemics caused by this pathogen, the leishmaniases.

Acknowledgements

Dr Kevin Portune is acknowledged for critically reading the manuscript. Also I would like to acknowledge Dr Ana Margarita Montalvo for the design of Fig. 14.1. The research done in the author's laboratory is supported by grants from the Ministerio de Ciencia y Tecnología (BFU2009-08986) and the Fondo de Investigaciones Sanitarias (ISCIII-RETIC RD06/0021/0008-FEDER).

Web resources

EuPathDomains: http://www.atgc-montpellier.fr/EuPathDomains/

An extended database of protein domains belonging to ten major eukaryotic human pathogens (three *Plasmodium* species, three *Leishmania* species, *Toxoplasma gondii*, *Giardia lamblia*, *Crystosporidium parvum*, and *Trypanosoma brucei*) (Ghouila *et al.*, 2011).

GEMINI (the Genome and Metabolome Integrated Initiative): www.leishrisk.net/gemini

A multicentre collaborative study documenting several *Leishmania donovani* genomes.

GeneDB: http://www.genedb.org

The GeneDB project provides reliable access to the latest sequence data and annotation/curation for more than 40 microorganisms, among them several *Leishmania* species.

International Leishmania Network (ILN): http://leishnet.net/site/

This site, devoted to *Leishmania*, contains valuable information about various topics including taxonomy, epidemiology, cryobanks and available hybridomas.

LeishCyc (database of *Leishmania* metabolic pathways): http://leishcyc.bio21.unimelb.edu.au/

Database of metabolic pathways for *Leishmania*.

TriTrypDB: http://tritrypdb.org/

The Kinetoplastid Genomics Resource provides tools and resources for searching genome information available for *Leishmania* and related trypanosomatids.

References

Adak, S., and Datta, A.K. (2005). *Leishmania major* encodes an unusual peroxidase that is a close homologue of plant ascorbate peroxidase: a novel role of the transmembrane domain. Biochem. J. *390*, 465–474.

Agron, P.G., Reed, S.L., and Engel, J.N. (2005). An essential, putative MEK kinase of *Leishmania major*. Mol. Biochem. Parasitol. *142*, 121–125.

Alexander, J., Satoskar, A.R., and Russell, D.G. (1999). *Leishmania* species: models of intracellular parasitism. J. Cell Sci. *112*, 2993–3002.

Alvar, J., Croft, S., and Olliaro, P. (2006). Chemotherapy in the treatment and control of leishmaniasis. Adv. Parasitol. *61*, 223–274.

Alvar, J., Aparicio, P., Aseffa, A., Den Boer, M., Cañavate, C., Dedet, J.P., Gradoni, L., Ter Horst, R., Lopez-Velez, R., and Moreno, J. (2008). The relationship between leishmaniasis and AIDS: the second 10 years. Clin. Microbiol. Rev. *21*, 334–359.

Alzate, J.F., Alvarez-Barrientos, A., Gonzalez, V.M., and Jimenez-Ruiz, A. (2006). Heat-induced programmed cell death in *Leishmania infantum* is reverted by Bcl-X(L) expression. Apoptosis *11*, 161–171.

Alzate, J.F., Arias, A.A., Moreno-Mateos, D., Alvarez-Barrientos, A., and Jimenez-Ruiz, A. (2007). Mitochondrial superoxide mediates heat-induced apoptotic-like death in *Leishmania infantum*. Mol. Biochem. Parasitol. *152*, 192–202.

Arana, F.E., Perez-Victoria, J.M., Repetto, Y., Morello, A., Castanys, S., and Gamarro, F. (1998). Involvement of thiol metabolism in resistance to glucantime in *Leishmania tropica*. Biochem. Pharmacol. *56*, 1201–1208.

Ariyanayagam, M.R., and Fairlamb, A.H. (2001). Ovothiol and trypanothione as antioxidants in trypanosomatids. Mol. Biochem. Parasitol. *115*, 189–198.

Arya, R., Mallik, M., and Lakhotia, S.C. (2007). Heat shock genes – integrating cell survival and death. J. Biosci. *32*, 595–610.

Asea, A. (2006). Initiation of the immune response by extracellular Hsp72: chaperokine activity of Hsp72. Curr. Immunol. Rev. *2*, 209–215.

Barak, E., Amin-Spector, S., Gerliak, E., Goyard, S., Holland, N., and Zilberstein, D. (2005). Differentiation of *Leishmania donovani* in host-free system: analysis of signal perception and response. Mol. Biochem. Parasitol. *141*, 99–108.

Barr, S.D., and Gedamu, L. (2001). Cloning and characterization of three differentially expressed peroxidoxin genes from *Leishmania chagasi*. Evidence for an enzymatic detoxification of hydroxyl radicals. J. Biol. Chem. *276*, 34279–34287.

Barr, S.D., and Gedamu, L. (2003). Role of peroxidoxins in *Leishmania chagasi* Survival. Evidence of an enzymatic defense against nitrosative stress. J. Biol. Chem. *278*, 10816–10823.

Bente, M., Harder, S., Wiesgigl, M., Heukeshoven, J., Gelhaus, C., Krause, E., Clos, J., and Bruchhaus, I. (2003). Developmentally induced changes of the proteome in the protozoan parasite *Leishmania donovani*. Proteomics *3*, 1811–1829.

Bernales, S., Papa, F.R., and Walter, P. (2006). Intracellular signaling by the unfolded protein response. Annu. Rev. Cell Dev. Biol. *22*, 487–508.

Besteiro, S., Williams, R.A.M., Coombs, G.H., and Mottram, J.C. (2007). Protein turnover and differentiation in *Leishmania*. Int. J. Parasitol. *37*, 1063–1075.

Beverley, S.M. (1991). Gene amplification in *Leishmania*. Annu. Rev. Microbiol. *45*, 417–444.

Biyani, N., Singh, A.K., Mandal, S., Chawla, B., and Madhubala, R. (2011). Differential expression of proteins in antimony-susceptible and -resistant isolates of *Leishmania donovani*. Mol. Biochem. Parasitol. *179*, 91–99.

Borst, P., and Ouellette, M. (1995). New mechanisms of drug resistance in parasitic protozoa. Annu. Rev. Microbiol. *49*, 427–460.

Brandau, S., Dresel, A., and Clos, J. (1995). High constitutive levels of heat-shock proteins in human-pathogenic parasites of the genus *Leishmania*. Biochem. J. *310*, 225–232.

Bringaud, F., Muller, M., Cerqueira, G.C., Smith, M., Rochette, A., El-Sayed, N.M.A., Papadopoulou, B., and Ghedin, E. (2007). Members of a large retroposon family are determinants of post-transcriptional gene expression in *Leishmania*. PLoS Pathog. *3*, 1291–1307.

Brochu, C., Haimeur, A., and Ouellette, M. (2004). The heat shock protein HSP70 and heat shock cognate protein HSC70 contribute to antimony tolerance in the protozoan parasite *Leishmania*. Cell Stress Chaperones *9*, 294–303.

Callahan, H.L., Portal, I.F., Bensinger, S.J., and Grogl, M. (1996). *Leishmania* spp: temperature sensitivity of promastigotes *in vitro* as a model for tropism *in vivo*. Exp. Parasitol. *84*, 400–409.

Castro, H., and Tomas, A.M. (2008). Peroxidases of trypanosomatids. Antioxid. Redox Signal. *10*, 1593–1606.

Castro, H., Sousa, C., Santos, M., Cordeiro-da-Silva, A., Floh, L., and Toms, A.M. (2002). Complementary antioxidant defense by cytoplasmic and mitochondrial peroxiredoxins in *Leishmania infantum*. Free Radic. Biol. Med. 33, 1552–1562.

Chang, K.P., Chaudhuri, G., and Fong, D. (1990). Molecular determinants of *Leishmania virulence*. Annu. Rev. Microbiol. *44*, 499–529.

Chappuis, F., Sundar, S., Hailu, A., Ghalib, H., Rijal, S., Peeling, R.W., Alvar, J., and Boelaert, M. (2007). Visceral leishmaniasis: what are the needs for diagnosis, treatment and control? Nat. Rev. Microbiol. *5*, 873–882.

Charest, H., and Matlashewski, G. (1994). Developmental gene expression in *Leishmania donovani*: differential cloning and analysis of an amastigote-stage-specific gene. Mol. Cell. Biol. *14*, 2975–2984.

Charest, H., Zhang, W.W., and Matlashewski, G. (1996). The developmental expression of *Leishmania donovani* A2 amastigote- specific genes is post-transcriptionally mediated and involves elements located in the 3′-untranslated region. J. Biol. Chem. *271*, 17081–17090.

Clayton, C., and Shapira, M. (2007). Post-transcriptional regulation of gene expression in trypanosomes and leishmanias. Mol. Biochem. Parasitol. *156*, 93–101.

Clos, J., and Krobitsch, S. (1999). Heat shock as a regular feature of the life cycle of *Leishmania* parasites. Am. Zool. *39*, 848–856.

Croft, S.L., Sundar, S., and Fairlamb, A.H. (2006). Drug resistance in leishmaniasis. Clin. Microbiol. Rev. *19*, 111–126.

Cuervo, P., De Jesus, J.B., Saboia-Vahia, L., Mendonça-Lima, L., Domont, G.B., and Cupolillo, E. (2009). Proteomic characterization of the released/secreted proteins of *Leishmania (Viannia) braziliensis* promastigotes. J. Proteomics *73*, 79–92.

Cyert, M.S. (2003). Calcineurin signaling in *Saccharomyces cerevisiae*: how yeast go crazy in response to stress. Biochem. Biophys. Res. Commun. *311*, 1143–1150.

Das, M., Mukherjee, S.B., and Shaha, C. (2001). Hydrogen peroxide induces apoptosis-like death in *Leishmania donovani* promastigotes. J. Cell Sci. *114*, 2461–2469.

David, M., Gabdank, I., Ben-David, M., Zilka, A., Orr, I., Barash, D., and Shapira, M. (2010). Preferential translation of Hsp83 in *Leishmania* requires a thermosensitive polypyrimidine-rich element in the 3′ UTR and involves scanning of the 5′ UTR. RNA *16*, 364–374.

Dea-Ayuela, M.A., Ordonez-Gutierrez, L., and Bolas-Fernandez, F. (2009). Changes in the proteome and infectivity of *Leishmania infantum* induced by *in vitro* exposure to a nitric oxide donor. Int. J. Med. Microbiol. *299*, 221–232.

Debrabant, A., Joshi, M.B., Pimenta, P.F.P., and Dwyer, D.M. (2004). Generation of *Leishmania donovani* axenic amastigotes: their growth and biological characteristics. Int. J. Parasitol. *34*, 205–217.

Descoteaux, A., and Turco, S.J. (2002). Functional aspects of the *Leishmania donovani* lipophosphoglycan during macrophage infection. Microbes Infect. *4*, 975–981.

Desjeux, P. (2004). Leishmaniasis: current situation and new perspectives. Comp. Immunol. Microbiol. Infect. Dis. *27*, 305–318.

Dolai, S., Yadav, R.K., Pal, S., and Adak, S. (2009). Overexpression of mitochondrial *Leishmania major* ascorbate peroxidase enhances tolerance to oxidative stress-induced programmed cell death and protein damage. Eukaryot. Cell 8, 1721–1731.

Dolai, S., Pal, S., Yadav, R.K., and Adak, S. (2011). Endoplasmic reticulum stress-induced apoptosis in *Leishmania* through Ca2+-dependent and caspase-independent mechanism. J. Biol. Chem. *286*, 13638–13646.

Dominguez, M., and Toraño, A. (1999). Immune adherence-mediated opsonophagocytosis: the mechanism of *Leishmania* infection. J. Exp. Med. *189*, 25–35.

Drummelsmith, J., Brochu, V., Girard, I., Messier, N., and Ouellette, M. (2003). Proteome mapping of the protozoan parasite *Leishmania* and application to the study of drug targets and resistance mechanisms. Mol. Cell Proteomics *2*, 146–155.

Drummelsmith, J., Girard, I., Trudel, N., and Ouellette, M. (2004). Differential protein expression analysis of *Leishmania major* reveals novel roles for methionine adenosyltransferase and S-adenosylmethionine in methotrexate resistance. J. Biol. Chem. *279*, 33273–33280.

Dujardin, J.C., Campino, L., Cañavate, C., Dedet, J.P., Gradoni, L., Soteriadou, K., Mazeris, A., Ozbel, Y., and Boelaert, M. (2008). Spread of vector-borne diseases and neglect of Leishmaniasis. Eur. Emerg. Infect. Dis. *14*, 1013–1018.

Dumas, C., Ouellette, M., Tovar, J., Cunningham, M.L., Fairlamb, A.H., Tamar, S., Olivier, M., and Papadopoulou, B. (1997). Disruption of the trypanothione reductase gene of *Leishmania* decreases its ability to survive oxidative stress in macrophages. EMBO J. *16*, 2590–2598.

Dunning, N. (2009). *Leishmania* vaccines: from leishmanization to the era of DNA technology. Biosci. Horizons *2*, 73–82.

Echeverria, P.C., Matrajt, M., Harb, O.S., Zappia, M.P., Costas, M.A., Roos, D.S., Dubremetz, J.F., and Angel, S.O. (2005). *Toxoplasma gondii* Hsp90 is a potential drug target whose expression and subcellular localization are developmentally regulated. J. Mol. Biol. *350*, 723–734.

Evans, C.G., Chang, L., and Gestwicki, J.E. (2010). Heat shock protein 70 (hsp70) as an emerging drug target. J. Med. Chem. *53*, 4585–4602.

Fairlamb, A.H. (2003). Chemotherapy of human African trypanosomiasis: current and future prospects. Trends Parasitol. *19*, 488–494.

Feng, X., Rodriguez-Contreras, D., Buffalo, C., Bouwer, H.G.A., Kruvand, E., Beverley, S.M., and Landfear, S.M. (2009). Amplification of an alternate transporter gene suppresses the avirulent phenotype of glucose transporter null mutants in *Leishmania mexicana*. Mol. Microbiol. *71*, 369–381.

Flohe, L., Budde, H., Bruns, K., Castro, H., Clos, J., Hofmann, B., Kansal-Kalavar, S., Krumme, D., Menge, U., Plank-Schumacher, K., *et al.* (2002). Tryparedoxin peroxidase of *Leishmania donovani*: molecular cloning, heterologous expression, specificity, and catalytic mechanism. Arch. Biochem. Biophys. *397*, 324–335.

Folgueira, C., and Requena, J.M. (2007). A postgenomic view of the heat shock proteins in kinetoplastids. FEMS Microbiol. Rev. *31*, 359–377.

Folgueira, C., Quijada, L., Soto, M., Abanades, D.R., Alonso, C., and Requena, J.M. (2005). The translational efficiencies of the two *Leishmania infantum* HSP70 mRNAs, differing in their 3′-untranslated regions, are affected by shifts in the temperature of growth through different mechanisms. J. Biol. Chem. *280*, 35172–35183.

Fulwiler, A.L., Boitz, J.M., Yates, P.A., Carter, N.S., and Ullman, B. (2011). Characterization of amplicons that suppress the conditional lethal growth phenotype of a *Leishmania donovani* mutant lacking normal purine salvage mechanisms. Mol. Biochem. Parasitol. *175*, 76–82.

Gantt, K.R., Goldman, T.L., McCormick, M.L., Miller, M.A., Jeronimo, S.M., Nascimento, E.T., Britigan, B.E., and Wilson, M.E. (2001). Oxidative responses of human and murine macrophages during phagocytosis of *Leishmania chagasi*. J. Immunol. *167*, 893–901.

Genest, P.A., ter Riet, B., Dumas, C., Papadopoulou, B., van Luenen, H.G.A.M., and Borst, P. (2005). Formation of linear inverted repeat amplicons following targeting of an essential gene in *Leishmania*. Nucleic Acids Res. *33*, 1699–1709.

Ghosh, S., Goswami, S., and Adhya, S. (2003). Role of superoxide dismutase in survival of *Leishmania* within the macrophage. Biochem. J. *369*, 447–452.

Ghouila, A., Terrapon, N., Gascuel, O., Guerfali, F.Z., Laouini, D., Marechal, E., and Brehelin, L. (2011). EuPathDomains: the divergent domain database for eukaryotic pathogens. Infect. Genet. Evol. *11*, 698–707.

Goldshmidt, H., Matas, D., Kabi, A., Carmi, S., Hope, R., and Michaeli, S. (2010). Persistent ER stress induces the spliced leader RNA silencing pathway (SLS), leading to programmed cell death in *Trypanosoma brucei*. PLoS Pathog. *6*, e1000731.

Gosline, S.J.S., Nascimento, M., McCall, L.I., Zilberstein, D., Thomas, D.Y., Matlashewski, G., and Hallett, M. (2011). Intracellular eukaryotic parasites have a distinct unfolded protein response. PLoS One *6*, e19118.

Grondin, K., Roy, G., and Ouellette, M. (1996). Formation of extrachromosomal circular amplicons with direct or inverted duplications in drug-resistant *Leishmania tarentolae*. Mol. Cell. Biol. *16*, 3587–3595.

Guerin, P.J., Olliaro, P., Sundar, S., Boelaert, M., Croft, S.L., Desjeux, P., Wasunna, M.K., and Bryceson, A.D. (2002). Visceral leishmaniasis: current status of control, diagnosis, and treatment, and a proposed research and development agenda. Lancet Infect. Dis. *2*, 494–501.

Haimeur, A., Guimond, C., Pilote, S., Mukhopadhyay, R., Rosen, B.P., Poulin, R., and Ouellette, M. (1999). Elevated levels of polyamines and trypanothione resulting from overexpression of the ornithine decarboxylase gene in arsenite- resistant *Leishmania*. Mol. Microbiol. *34*, 726–735.

Hannaert, V., Saavedra, E., Duffieux, F., Szikora, J.P., Rigden, D.J., Michels, P.A.M., and Opperdoes, F.R. (2003). Plant-like traits associated with metabolism of *Trypanosoma* parasites. Proc. Natl. Acad. Sci. U.S.A. *100*, 1067–1071.

Harder, S., Bente, M., Isermann, K., and Bruchhaus, I. (2006). Expression of a mitochondrial peroxiredoxin prevents programmed cell death in *Leishmania donovani*. Eukaryot. Cell *5*, 861–870.

Hassani, K., Antoniak, E., Jardim, A., and Olivier, M. (2011). Temperature-induced protein secretion by *Leishmania mexicana* modulates macrophage signalling and function. PLoS One *6*, e18724.

Hofmann, B., Hecht, H.J., and Flohe, L. (2002). Peroxiredoxins. Biol. Chem. *383*, 347–364.

Hubel, A., Brandau, S., Dresel, A., and Clos, J. (1995). A member of the ClpB family of stress proteins is expressed during heat shock in *Leishmania* spp. Mol. Biochem. Parasitol. *70*, 107–118.

Hubel, A., Krobitsch, S., Horauf, A., and Clos, J. (1997). *Leishmania major* Hsp100 is required chiefly in the mammalian stage of the parasite. Mol. Cell. Biol. *17*, 5987–5995.

Iniesta, V., Gomez-Nieto, L.C., and Corraliza, I. (2001). The inhibition of arginase by N(omega)-hydroxy-l-arginine controls the growth of *Leishmania* inside macrophages. J. Exp. Med. *193*, 777–784.

Kaufmann, S.H.E. (2011). Intracellular pathogens: living in an extreme environment. Immunol. Rev. *240*, 5–10.

Krobitsch, S., Brandau, S., Hoyer, C., Schmetz, C., Hubel, A., and Clos, J. (1998). *Leishmania donovani* heat shock protein 100. Characterization and function in amastigote stage differentiation. J. Biol. Chem. *273*, 6488–6494.

Kroemer, G., and Reed, J.C. (2000). Mitochondrial control of cell death. Nat. Med. *6*, 513–519.

Kurosaka, K., Takahashi, M., Watanabe, N., and Kobayashi, Y. (2003). Silent cleanup of very early apoptotic cells by macrophages. J. Immunol. *171*, 4672–4679.

Larreta, R., Soto, M., Quijada, L., Folgueira, C., Abanades, D.R., Alonso, C., and Requena, J.M. (2004). The expression of HSP83 genes in *Leishmania infantum* is affected by temperature and by stage-differentiation and is regulated at the levels of mRNA stability and translation. BMC Mol. Biol. *5*, 3.

Lee, N., Bertholet, S., Debrabant, A., Muller, J., Duncan, R., and Nakhasi, H.L. (2002). Programmed cell death in the unicellular protozoan parasite *Leishmania*. Cell Death Diff. *9*, 53–64.

Lefurgey, A., Gannon, M., Blum, J., and Ingram, P. (2005). *Leishmania donovani* amastigotes mobilize organic and inorganic osmolytes during regulatory volume decrease. J. Eukaryot. Microbiol. *52*, 277–289.

Levick, M.P., Tetaud, E., Fairlamb, A.H., and Blackwell, J.M. (1998). Identification and characterisation of a functional peroxidoxin from *Leishmania major*. Mol. Biochem. Parasitol. *96*, 125–137.

Li, Q., Zhou, Y., Yao, C., Ma, X., Wang, L., Xu, W., Wang, Z., and Qiao, Z. (2009). Apoptosis caused by Hsp90 inhibitor geldanamycin in *Leishmania donovani* during promastigote-to-amastigote transformation stage. Parasitol. Res. *105*, 1539–1548.

Li, Y., Zhang, T., Schwartz, S.J., and Sun, D. (2009). New developments in Hsp90 inhibitors as anti-cancer therapeutics: mechanisms, clinical perspective and more potential. Drug Resist. Updat. *12*, 17–27.

Liew, F.Y., and O'Donnell, C.A. (1993). Immunology of leishmaniasis. Adv. Parasitol. *32*, 161–259.

Lindquist, S. (1986). The heat-shock response. Annu. Rev. Biochem. *55*, 1151–1191.

Lindquist, S., and Craig, E.A. (1988). The heat-shock proteins. Annu. Rev. Genet. *22*, 631–677.

Louw, C.A., Ludewig, M.H., Mayer, J., and Blatch, G.L. (2010). The Hsp70 chaperones of the Tritryps are characterized by unusual features and novel members. Parasitol. Int. *59*, 497–505.

Lu, H.G., Zhong, L., Chang, K.P., and Docampo, R. (1997). Intracellular Ca2+ pool content and signaling and expression of a calcium pump are linked to virulence in *Leishmania mexicana amazonesis* amastigotes. J. Biol. Chem. *272*, 9464–9473.

Luque-Ortega, J.R., Cruz, L.J., Albericio, F., and Rivas, L. (2010). The antitumoral depsipeptide ib-01212 kills *Leishmania* through an apoptosis-like process involving intracellular targets. Mol. Pharmaceutics *7*, 1608–1617.

Lustig, Y., Sheiner, L., Vagima, Y., Goldshmidt, H., Das, A., Bellofatto, V., and Michaeli, S. (2007). Spliced-leader RNA silencing: a novel stress-induced mechanism in *Trypanosoma brucei*. EMBO Rep. *8*, 408–413.

McCall, L.I., and Matlashewski, G. (2010). Localization and induction of the A2 virulence factor in *Leishmania*: evidence that A2 is a stress response protein. Mol. Microbiol. *77*, 518–530.

McConville, M.J., de Souza, D., Saunders, E., Likic, V.A., and Naderer, T. (2007). Living in a phagolysosome; metabolism of *Leishmania* amastigotes. Trends Parasitol. *23*, 368–375.

Madeira da Silva, L., and Beverley, S.M. (2010). Expansion of the target of rapamycin (TOR) kinase family and function in *Leishmania* shows that TOR3 is required for acidocalcisome biogenesis and animal infectivity. Proc. Natl. Acad. Sci. U.S.A. *107*, 11965–11970.

Mazareb, S., Fu, Z.Y., and Zilberstein, D. (1999). Developmental regulation of proline transport in *Leishmania donovani*. Exp. Parasitol. *91*, 341–348.

Miller, M.A., McGowan, S.E., Gantt, K.R., Champion, M., Novick, S.L., Andersen, K.A., Bacchi, C.J., Yarlett, N., Britigan, B.E., and Wilson, M.E. (2000). Inducible resistance to oxidant stress in the protozoan *Leishmania chagasi*. J. Biol. Chem. *275*, 33883–33889.

Mojtahedi, Z., Clos, J., and Kamali-Sarvestani, E. (2008). *Leishmania major*: identification of developmentally regulated proteins in procyclic and metacyclic promastigotes. Exp. Parasitol. *119*, 422–429.

Morales, M.A., Renaud, O., Faigle, W., Shorte, S.L., and Spath, G.F. (2007). Over-expression of *Leishmania major* MAP kinases reveals stage-specific induction of phosphotransferase activity. Int. J. Parasitol. *37*, 1187–1199.

Morales, M.A., Watanabe, R., Dacher, M., Chafey, P., Osorio y Fortea, J., Scott, D.A., Beverley, S.M., Ommen, G., Clos, J., Hem, S., *et al.* (2010). Phosphoproteome dynamics reveal heat-shock protein complexes specific to the *Leishmania donovani* infectious stage. Proc. Natl. Acad. Sci. U.S.A. *107*, 8381–8386.

Moreira, D., Lopez-Garcia, P., and Vickerman, K. (2004). An updated view of kinetoplastid phylogeny using environmental sequences and a closer outgroup: proposal for a new classification of the class Kinetoplastea. Int. J. Syst. Evol. Microbiol. *54*, 1861–1875.

Moreira, M.E., Del Portillo, H.A., Milder, R.V., Balanco, J.M., and Barcinski, M.A. (1996). Heat shock induction of apoptosis in promastigotes of the unicellular organism *Leishmania (Leishmania) amazonensis*. J. Cell. Physiol. *167*, 305–313.

Moreira, W., Leblanc, E., and Ouellette, M. (2009). The role of reduced pterins in resistance to reactive oxygen and nitrogen intermediates in the protozoan parasite *Leishmania*. Free Radic. Biol. Med. *46*, 367–375.

Mukhopadhyay, R., Dey, S., Xu, N., Gage, D., Lightbody, J., Ouellette, M., and Rosen, B.P. (1996). Trypanothione overproduction and resistance to antimonials and arsenicals in *Leishmania*. Proc. Natl. Acad. Sci. U.S.A. *93*, 10383–10387.

Murray, H.W., Berman, J.D., Davies, C.R., and Saravia, N.G. (2005). Advances in leishmaniasis. Lancet *366*, 1561–1577.

Naderer, T., Dandash, O., and McConville, M.J. (2011). Calcineurin is required for *Leishmania major* stress response pathways and for virulence in the mammalian host. Mol. Microbiol. *80*, 471–480.

Nascimento, M., Zhang, W.W., Ghosh, A., Houston, D.R., Berghuis, A.M., Olivier, M., and Matlashewski, G. (2006). Identification and characterization of a protein-tyrosine phosphatase in *Leishmania*: Involvement in virulence. J. Biol. Chem. *281*, 36257–36268.

Nollen, E.A., and Morimoto, R.I. (2002). Chaperoning signaling pathways: molecular chaperones as stress-sensing 'heat shock' proteins. J. Cell Sci. *115*, 2809–2816.

Nourbakhsh, F., Uliana, S.R., and Smith, D.F. (1996). Characterisation and expression of a stage-regulated gene of *Leishmania major*. Mol. Biochem. Parasitol. *76*, 201–213.

Nugent, P.G., Karsani, S.A., Wait, R., Tempero, J., and Smith, D.F. (2004). Proteomic analysis of *Leishmania mexicana* differentiation. Mol. Biochem. Parasitol. *136*, 51–62.

Olmo, A., Arrebola, R., Bernier, V., Gonzalez-Pacanowska, D., and Ruiz-Perez, L.M. (1995). Co-existence of circular and multiple linear amplicons in methotrexate- resistant *Leishmania*. Nucleic Acids Res. *23*, 2856–2864.

Ono, K., and Han, J. (2000). The p38 signal transduction pathway: activation and function. Cell Signal. *12*, 1–13.

Opperdoes, F.R., and Szikora, J.P. (2006). In silico prediction of the glycosomal enzymes of *Leishmania major* and trypanosomes. Mol. Biochem. Parasitol. *147*, 193–206.

Ouellette, M., Hettema, E., Wust, D., Fase-Fowler, F., and Borst, P. (1991). Direct and inverted DNA repeats associated with P-glycoprotein gene amplification in drug resistant *Leishmania*. EMBO J. *10*, 1009–1016.

Pal, S., Dolai, S., Yadav, R.K., and Adak, S. (2010). Ascorbate peroxidase from *Leishmania major* controls the virulence of infective stage of promastigotes by regulating oxidative stress. PLoS One *5*, e11271.

Pallavi, R., Roy, N., Nageshan, R.K., Talukdar, P., Pavithra, S.R., Reddy, R., Venketesh, S., Kumar, R., Gupta, A.K., Singh, R.K., *et al.* (2010). Heat shock protein 90 as a drug target against protozoan infections: biochemical characterization of HSP90 from *Plasmodium falciparum* and *Trypanosoma evansi* and evaluation of its inhibitor as a candidate drug. J. Biol. Chem. *285*, 37964–37975.

Papadopoulou, B., Roy, G., and Ouellette, M. (1994). Autonomous replication of bacterial DNA plasmid oligomers in *Leishmania*. Mol. Biochem. Parasitol. *65*, 39–49.

Paramchuk, W.J., Ismail, S.O., Bhatia, A., and Gedamu, L. (1997). Cloning, characterization and overexpression of two iron superoxide dismutase cDNAs from *Leishmania chagasi*: role in pathogenesis. Mol. Biochem. Parasitol. *90*, 203–221.

Paris, C., Loiseau, P.M., Bories, C., and Breard, J. (2004). Miltefosine induces apoptosis-like death in *Leishmania donovani* promastigotes. Antimicrob. Agents Chemother. *48*, 852–859.

Parsons, M., Worthey, E.A., Ward, P.N., and Mottram, J.C. (2005). Comparative analysis of the kinomes of three pathogenic trypanosomatids: *Leishmania major*, *Trypanosoma brucei* and *Trypanosoma cruzi*. BMC Genomics *6*, 127.

Plewes, K.A., Barr, S.D., and Gedamu, L. (2003). Iron superoxide dismutases targeted to the glycosomes of *Leishmania chagasi* are important for survival. Infect. Immun. *71*, 5910–5920.

Qiu, X.B., Shao, Y.M., Miao, S., and Wang, L. (2006). The diversity of the DnaJ/Hsp40 family, the crucial partners for Hsp70 chaperones. Cell. Mol. Life Sci. *63*, 2560–2570.

Quijada, L., Soto, M., Alonso, C., and Requena, J.M. (1997). Analysis of post-transcriptional regulation operating on transcription products of the tandemly linked *Leishmania infantum* hsp70 genes. J. Biol. Chem. *272*, 4493–4499.

Quijada, L., Soto, M., Alonso, C., and Requena, J.M. (2000). Identification of a putative regulatory element in the 3′-untranslated region that controls expression of HSP70 in *Leishmania infantum*. Mol. Biochem. Parasitol. *110*, 79–91.

Ramos, C.S., Franco, F.A.L., Smith, D.F., and Uliana, S.R.B. (2004). Characterisation of a new *Leishmania* META gene and genomic analysis of the META cluster. FEMS Microbiol. Lett. *238*, 213–219.

Ramos, C.S., Yokoyama-Yasunaka, J.K.U., Guerra-Giraldez, C., Price, H.P., Mortara, R.A., Smith, D.F., and Uliana, S.R.B. (2011). *Leishmania amazonensis* META2 protein confers protection against heat shock and oxidative stress. Exp. Parasitol. *127*, 228–237.

Reiling, L., Jacobs, T., Kroemer, M., Gaworski, I., Graefe, S., and Clos, J. (2006). Spontaneous recovery of pathogenicity by *Leishmania major* hsp100-/- alters the immune response in mice. Infect. Immun. *74*, 6027–6036.

Reiling, L., Chrobak, M., Schmetz, C., and Clos, J. (2010). Overexpression of a single *Leishmania major* gene enhances parasite infectivity *in vivo* and *in vitro*. Mol. Microbiol. 76, 1175–1190.

Requena, J.M. (2011). Lights and shadows on gene organization and regulation of gene expression in *Leishmania*. Front. Biosci. *17*, 2069–2085.

Requena, J.M., Alonso, C., and Soto, M. (2000). Evolutionarily conserved proteins as prominent immunogens during *Leishmania* infections. Parasitol. Today *16*, 246–250.

Requena, J.M., Iborra, S., Carrion, J., Alonso, C., and Soto, M. (2004). Recent advances in vaccines for leishmaniasis. Exp. Opin. Biol. Ther. *4*, 1505–1517.

Requena, J.M., Folgueira, C., Lopez, M.C., and Thomas, M.C. (2008). The SIDER2 elements, interspersed repeated sequences that populate the *Leishmania* genomes, constitute subfamilies showing chromosomal proximity relationship. BMC Genomics *9*, 263.

Rico, A.I., Angel, S.O., Alonso, C., and Requena, J.M. (1999). Immunostimulatory properties of the *Leishmania infantum* heat shock proteins HSP70 and HSP83. Mol. Immunol. *36*, 1131–1139.

Rico, A.I., Girones, N., Fresno, M., Alonso, C., and Requena, J.M. (2002). The heat shock proteins, Hsp70 and Hsp83, of *Leishmania infantum* are mitogens for mouse B cells. Cell Stress Chaperones *7*, 339–346.

Rico, E., Alzate, J.F., Arias, A.A., Moreno, D., Clos, J., Gago, F., Moreno, I., Dominguez, M., and Jimenez-Ruiz, A. (2009). *Leishmania infantum* expresses a mitochondrial nuclease homologous to EndoG that migrates to the nucleus in response to an apoptotic stimulus. Mol. Biochem. Parasitol. *163*, 28–38.

Richter, K., and Buchner, J. (2001). Hsp90: chaperoning signal transduction. J. Cell. Physiol. *188*, 281–290.

Richter, K., Haslbeck, M., and Buchner, J. (2010). The heat shock response: life on the verge of death. Mol. Cell *40*, 253–266.

Rivero, A. (2006). Nitric oxide: an antiparasitic molecule of invertebrates. Trends Parasitol. *22*, 219–225.

Rutkowski, D.T., and Kaufman, R.J. (2004). A trip to the ER: coping with stress. Trends Cell Biol. *14*, 20–28.

Saar, Y., Ransford, A., Waldman, E., Mazareb, S., Amin-Spector, S., Plumblee, J., Turco, S.J., and Zilberstein, D. (1998). Characterization of developmentally regulated activities in axenic amastigotes of *Leishmania donovani*. Mol. Biochem. Parasitol. *95*, 9–20.

Sacks, D., and Kamhawi, S. (2001). Molecular aspects of parasite-vector and vector–host interactions in leishmaniasis. Annu. Rev. Microbiol. *55*, 453–483.

Sacks, D.L., and Perkins, P.V. (1984). Identification of an infective stage of *Leishmania* promastigotes. Science *223*, 1417–1419.

Saidi, Y., Finka, A., Muriset, M., Bromberg, Z., Weiss, Y.G., Maathuis, F.J., and Goloubinoff, P. (2009). The heat shock response in moss plants is regulated by specific calcium-permeable channels in the plasma membrane. Plant Cell *21*, 2829–2843.

Santoro, M.G. (2000). Heat shock factors and the control of the stress response. Biochem. Pharmacol. *59*, 55–63.

Sarkar, D., and Bhaduri, A. (1995). Temperature-induced rapid increase in cytoplasmic free Ca2+ in pathogenic *Leishmania donovani* promastigotes. FEBS Lett. *375*, 83–86.

Segovia, M., and Ortiz, G. (1997). LD1 amplifications in *Leishmania*. Parasitol. Today *13*, 342–348.

Semotok, J.L., Luo, H., Cooperstock, R.L., Karaiskakis, A., Vari, H.K., Smibert, C.A., and Lipshitz, H.D. (2008). *Drosophila* maternal Hsp83 mRNA destabilization is directed by multiple SMAUG recognition elements in the open reading frame. Mol. Cell. Biol. *28*, 6757–6772.

Shonhai, A., Maier, A.G., Przyborski, J.M., and Blatch, G.L. (2011). Intracellular protozoan parasites of humans: the role of molecular chaperones in development and pathogenesis. Protein Pept. Lett. *18*, 143–157.

Silverman, J.M., Clos, J., de'Oliveira, C.C., Shirvani, O., Fang, Y., Wang, C., Foster, L.J., and Reiner, N.E. (2010). An exosome-based secretion pathway is responsible for protein export from *Leishmania* and communication with macrophages. J. Cell Sci. *123*, 842–852.

Smejkal, R.M., Wolff, R., and Olenick, J.G. (1988). *Leishmania braziliensis panamensis*: increased infectivity resulting from heat shock. Exp. Parasitol. *65*, 1–9.

Smirlis, D., Duszenko, M., Jimenez-Ruiz, A., Scoulica, E., Bastien, P., Fasel, N., and Soteriadou, K. (2010). Targeting essential pathways in trypanosomatids gives insights into protozoan mechanisms of cell death. Parasit. Vectors *3*, 107.

Smith, M., Bringaud, F., and Papadopoulou, B. (2009). Organization and evolution of two SIDER retroposon subfamilies and their impact on the *Leishmania* genome. BMC Genomics *10*, 240.

Somanna, A., Mundodi, V., and Gedamu, L. (2002). In vitro cultivation and characterization of *Leishmania chagasi* amastigote-like forms. Acta Trop. *83*, 37–42.

St-Pierre, J., Buckingham, J.A., Roebuck, S.J., and Brand, M.D. (2002). Topology of superoxide production from different sites in the mitochondrial electron transport chain. J. Biol. Chem. *277*, 44784–44790.

Stebbins, C.E., Russo, A.A., Schneider, C., Rosen, N., Hartl, F.U., and Pavletich, N.P. (1997). Crystal structure of an Hsp90–geldanamycin complex: targeting of a protein chaperone by an antitumor agent. Cell *89*, 239–250.

Stie, J., and Fox, D. (2008). Calcineurin regulation in fungi and beyond. Eukaryot. Cell *7*, 177–186.

Swenerton, R.K., Knudsen, G.M., Sajid, M., Kelly, B.L., and McKerrow, J.H. (2010). *Leishmania* subtilisin is a maturase for the trypanothione reductase system and contributes to disease pathology. J. Biol. Chem. *285*, 31120–31129.

Tabas, I., and Ron, D. (2011). Integrating the mechanisms of apoptosis induced by endoplasmic reticulum stress. Nat. Cell Biol. *13*, 184–190.

Taipale, M., Jarosz, D.F., and Lindquist, S. (2010). HSP90 at the hub of protein homeostasis: emerging mechanistic insights. Nat. Rev. Mol. Cell. Biol. *11*, 515–528.

Tripp, C.A., Wisdom, W.A., Myler, P.J., and Stuart, K.D. (1992). A multicopy, extrachromosomal DNA in *Leishmania infantum* contains two inverted repeats of the 27.5-kilobase LD1 sequence and encodes numerous transcripts. Mol. Biochem. Parasitol. *55*, 39–50.

Uliana, S.R., Goyal, N., Freymuller, E., and Smith, D.F. (1999). *Leishmania*: overexpression and comparative structural analysis of the stage-regulated meta 1 gene. Exp. Parasitol. *92*, 183–191.

Vergnes, B., Gourbal, B., Girard, I., Sundar, S., Drummelsmith, J., and Ouellette, M. (2007). A proteomics screen implicates HSP83 and a small kinetoplastid calpain-related protein in drug resistance in *Leishmania donovani* clinical field isolates by modulating drug-induced programmed cell death. Mol. Cell Proteomics *6*, 88–101.

Vickers, T.J., and Fairlamb, A.H. (2004). Trypanothione S-transferase activity in a trypanosomatid ribosomal elongation factor 1B. J. Biol. Chem. *279*, 27246–27256.

Vickers, T.J., Greig, N., and Fairlamb, A.H. (2004a). A trypanothione-dependent glyoxalase I with a prokaryotic ancestry in *Leishmania major*. Proc. Natl. Acad. Sci. U.S.A. *101*, 13186–13191.

Vickers, T.J., Wyllie, S., and Fairlamb, A.H. (2004b). *Leishmania major* elongation factor 1B complex has trypanothione S-transferase and peroxidase activity. J. Biol. Chem. *279*, 49003–49009.

Vieira, L.L., Lafuente, E., Blum, J., and Cabantchik, Z.I. (1997). Modulation of the swelling-activated amino acid channel of *Leishmania major* promastigotes by protein kinases. Mol. Biochem. Parasitol. *90*, 449–461.

Walker, J., Acestor, N., Gongora, R., Quadroni, M., Segura, I., Fasel, N., and Saravia, N.G. (2006). Comparative protein profiling identifies elongation factor-1beta and tryparedoxin peroxidase as factors associated with metastasis in *Leishmania guyanensis*. Mol. Biochem. Parasitol. *145*, 254–264.

Wanderley, J.L.M., Moreira, M.E.C., Benjamin, A., Bonomo, A.C., and Barcinski, M.A. (2006). Mimicry of apoptotic cells by exposing phosphatidylserine participates in the establishment of amastigotes of *Leishmania (L) amazonensis* in mammalian hosts. J. Immunol. *176*, 1834–1839.

Wanderley, J.L., Pinto da Silva, L.H., Deolindo, P., Soong, L., Borges, V.M., Prates, D.B., de Souza, A.P.A., Barral, A., Balanco, J.M.F., do Nascimento, M.T.C., *et al.* (2009). Cooperation between apoptotic and viable metacyclics enhances the pathogenesis of Leishmaniasis. PLoS One *4*, e5733.

WHO (2010). WHO Technical Report Series 949. Control of the leishmaniases. WHO Press, Geneva, Switzerland. Available at http://whqlibdoc.who.int/trs/WHO_TRS_949_eng.pdf.

Wiese, M. (2007). *Leishmania* MAP kinases--familiar proteins in an unusual context. Int. J. Parasitol. *37*, 1053–1062.

Wiesgigl, M., and Clos, J. (2001). Heat shock protein 90 homeostasis controls stage differentiation in *Leishmania donovani*. Mol. Biol. Cell *12*, 3307–3316.

Wilson, M.E., Andersen, K.A., and Britigan, B.E. (1994). Response of *Leishmania chagasi* promastigotes to oxidant stress. Infect. Immun. *62*, 5133–5141.

Wyllie, S., Vickers, T.J., and Fairlamb, A.H. (2008). Roles of trypanothione S-transferase and tryparedoxin peroxidase in resistance to antimonials. Antimicrob. Agents Chemother. *52*, 1359–1365.

Zakai, H.A., Chance, M.L., and Bates, P.A. (1998). In vitro stimulation of metacyclogenesis in *Leishmania braziliensis, L. donovani, L. major* and *L. mexicana*. Parasitology *116*, 305–309.

van Zandbergen, G., Bollinger, A., Wenzel, A., Kamhawi, S., Voll, R., Klinger, M., Muller, A., Holscher, C., Herrmann, M., Sacks, D., *et al.* (2006). *Leishmania* disease development depends on the presence of apoptotic promastigotes in the virulent inoculum. Proc. Natl. Acad. Sci. U.S.A. *103*, 13837–13842.

Zangger, H., Mottram, J.C., and Fasel, N. (2002). Cell death in *Leishmania* induced by stress and differentiation: programmed cell death or necrosis? Cell Death Diff. *9*, 1126–1139.

Zhang, W.W., Mendez, S., Ghosh, A., Myler, P., Ivens, A., Clos, J., Sacks, D.L., and Matlashewski, G. (2003). Comparison of the A2 gene locus in *Leishmania donovani* and *Leishmania major* and its control over cutaneous infection. J. Biol. Chem. *278*, 35508–35515.

Zilberstein, D., and Shapira, M. (1994). The role of pH and temperature in the development of *Leishmania* parasites. Annu. Rev. Microbiol. *48*, 449–470.

Zilka, A., Garlapati, S., Dahan, E., Yaolsky, V., and Shapira, M. (2001). Developmental regulation of heat shock protein 83 in *Leishmania*. J. Biol. Chem. *276*, 47922–47929.

The Stress Response of *Trypanosoma cruzi* 15

Turán P. Ürményi, Deivid C. Rodrigues, Rosane Silva and Edson Rondinelli

Abstract

Trypanosoma cruzi, the causal agent of Chagas' disease, is a flagellated protozoan parasite with a complex life cycle that involves infecting an insect and a mammalian host. Several environmental stresses occur during its life cycle, such as heat, reactive oxygen species, and osmolarity changes, and the parasite has evolved a variety of stress responses to cope with these challenges. The stress responses range from synthesis of several proteins and small molecules to modulation of the activity of organelles, and they are essential for the parasite's viability and survival in both hosts. Here we review the components and operation of *T. cruzi*'s stress response with emphasis on its relevance to the parasite's biology and to Chagas' disease transmission, pathogenesis and treatment.

Introduction

Cellular and molecular biology

Trypanosoma cruzi, the causal agent of Chagas' disease, is a flagellated protozoan parasite of the family Trypanosomatidae, order Kinetoplastida, that exhibits a complex life cycle characterized by several developmental stages and the ability to infect an insect and a mammal host. The parasite is an evolutionarily ancient, genetically highly diverse, organism that shows several peculiar ultrastructural features (reviewed by Simpson *et al.*, 2006; de Souza, 2009; Sturm and Campbell, 2010). The cytoskeleton is composed of subpellicular microtubules that are organized below the plasma membrane, forming a 'corset' which confers resistance to mechanical stress. A single ramified mitochondrion is present, extending throughout the cell body and containing a network of catenated circular DNA molecules. This network is called kinetoplast and comprises the mitochondrial genome. The glycolytic enzymes are compartmentalized in a specialized peroxisome called glycosome, and are relevant to the parasite's defences against oxidative stress. Two other organelles are important for the stress response, in this case against osmotic and pH stress. An acidic vacuole called acidocalcisome stores divalent cations and phosphorous compounds, and functions in association with a contractile vacuole. These organelles are sometimes called stress organelles. Endocytosis and exocytosis occur exclusively in the flagellar pocket, an invagination of the plasma membrane next to the flagellum which is devoid of subpellicular microtubules. In association with the flagellar pocket, the cytostome, a specialized opening and invagination of the plasma membrane, is used for endocytosis in some cell forms. When the parasite is cultivated in rich media, the macromolecules internalized by endocytosis are stored in a membrane-bound organelle located in the posterior region of the cell. This organelle is called reservosome and is important during nutritional stress owing to its nutrient storage capacity.

The genome of *T. cruzi* contains more than 12,000 protein-coding genes and is composed of a large percentage of repetitive sequences (El-Sayed *et al.*, 2005; http://tritrypdb.org/). These repetitive sequences include surface protein gene families such as mucins, mucin-associated surface proteins (MASPs), trans-sialidases and gp63s, which are important for host cell invasion and immune evasion. The parasite's diploid genome is estimated to comprise 41 pairs of chromosomes (Weatherly *et al.*, 2009), and genome size varies from 44 Mb to 55 Mb, depending on the strain (Franzén *et al.*, 2011). This variability is mainly due to different copy numbers of the large gene families. The parasite also shows peculiar features in gene structure and expression. Protein-coding genes are arranged head-to-tail with usually small intergenic regions in large, directional gene clusters that may span a considerable portion of the chromosome and are separated by strand-switch regions. These regions are called convergent or divergent depending on whether transcription is oriented towards or away from them, respectively. Divergent strand-switch regions have specific epigenetic marks such as histone acetylation and methylation in a pattern that suggests the presence of bidirectional promoters (Respuela *et al.*, 2008). Except in a few cases, gene promoters have been very difficult to identify and characterize in the Trypanosomatidae family. *T. cruzi*'s gene clusters are transcribed as multicistronic RNAs that are subsequently processed into mature mRNAs by *trans*-splicing and polyadenylation. In *trans*-splicing, a capped, short spliced leader RNA is added to the 5′ untranslated region (UTR) of all coding sequences via two transesterification reactions (reviewed by Martínez-Calvillo *et al.*, 2010). A single intron-containing gene, coding for poly(A) polymerase, was found in the genome, and the intron is removed by canonical *cis*-splicing (Mair *et al.*, 2000). The presence of a single intron-containing gene in an otherwise intronless genome is unexpected. The pervasive generation of individual mRNAs by post-transcriptional processing from polycistronic transcripts implies that gene regulation is exerted by post-transcriptional mechanisms such as modulation of mRNA stability and translation (Clayton and Shapira, 2007). In addition, most mitochondrial mRNA precursors undergo RNA editing, a series of sometimes extensive uridine insertions and deletions specified by small RNAs called guide RNAs that result in functional mRNAs (Stuart *et al.*, 2005). Cell form-specific gene expression has also been investigated (Atwood *et al.*, 2005; Cuervo *et al.*, 2010). Proteome analysis of the four developmental stages showed that 30% of detected proteins are present in all stages, and that there is a shift from carbohydrate- to lipid-based energy metabolism in some of the cell form transitions. In addition, metacyclogenesis is accompanied by a reduction in ribosomal proteins, consistent with trypomastigotes being unable to divide, and by an increase in antioxidant enzymes, adequate for the intracellular environment of phagocytic cells. However, despite the wealth of biological information described above, sequence similarity-based annotation of the genome shows that at least half of the protein-coding genes identified are of unknown function, highlighting low little we understand this organism at the molecular level.

Life cycle

T. cruzi has a complex life cycle, in which several cell forms with different morphological and functional characteristics participate (reviewed by de Souza, 2002). Four main cell forms are recognized. Epimastigotes, found in the insect vector, are long, flagellated cells, have the kinetoplast positioned anterior to the nucleus, and are able to divide. Trypomastigotes also are elongated and flagellated, but are unable to divide, and have the kinetoplast positioned posterior to the nucleus. Metacyclic trypomastigotes are present in the insect vector, and bloodstream trypomastigotes are found in the mammalian host. Amastigotes have a spherical cell body, a very small flagellum, are able to divide and are present in the mammalian host. In the parasite's life cycle, an insect (usually from the *Rhodnius*, *Triatoma* or *Panstrongylus* genera) takes a blood meal from an infected mammalian host. Bloodstream trypomastigotes present in the blood reach the insect's digestive tract and then differentiate into epimastigotes. Epimastigotes bind to epithelial cells, multiply by binary fission and thereby colonize the insect's midgut. After reaching the hind gut these cells differentiate into metacyclic trypomastigotes, the infective form for the mammalian host, in a process called metacyclogenesis. During the next blood meal the parasites are deposited along with faeces near the bite wound. The metacyclic trypomastigotes enter the bloodstream either through the wound or intact mucosa, invading nearby cells. Once inside the cell the parasite differentiates into amastigotes, which start multiplying by binary fission. After nine rounds of division, amastigotes differentiate into bloodstream trypomastigotes, which are released after cell rupture. These trypomastigotes may reach the blood and/or invade other cells, maintaining the infection. The circulating parasites can then infect another insect host, closing the cycle. There are two types of life cycles with respect to the mammalian host, the sylvatic cycle and the domestic cycle. In the sylvatic cycle, widespread in nature, the parasite infects small wild mammals, usually without causing harm. On the other hand, in the domestic cycle, humans are the occasional mammal host and Chagas' disease may develop. *T. cruzi* isolates from the sylvatic and the domestic cycles vary considerably in biochemical and genetic characteristics, indicating that the parasite possesses a great deal of intraspecific heterogeneity (reviewed by Miles *et al.*, 2009; Sturm and Campbell, 2010). The variability of isolated strains affects several aspects of the parasite and the disease, including virulence, tissue tropism, and severity of the clinical manifestations (Luquetti *et al.*, 1986). Strain variability also affects the parasite's stress response (discussed below). Phylogenetic analysis of the strains shows that the parasite has evolved into several lineages with distinct characteristics. A recent revision of strain classification and nomenclature has been proposed, in which a division in discrete typing units numbered I to VI was recommended (Zingales *et al.*, 2009).

The ability of *T. cruzi* to colonize the insect vector's digestive tract depends on the cell interactions between epimastigotes and epithelial cells. Cell interactions also allow trypomastigotes to invade different cell types, which is crucial for initiating and maintaining infection in the mammalian host. Accordingly, the parasite presents a surface coat that provides cell adhesion and protection from digestive proteases in the insect and immune evasion in the mammalian host (for reviews see Andrade and Andrews, 2005; Buscaglia *et al.*, 2006). The surface coat is mainly composed of glycosylphosphatidylinositol-anchored glycoproteins called mucins. Also present in the coat are trans-sialidases which catalyse the transfer of sialic acid residues from host glycoconjugates to parasite mucins. During invasion of the mammalian host cell, the parasite first attaches to the cell surface by receptor-mediated recognition and is then internalized in a parasitophorous vacuole. The vacuole's content is acidified by lysosome fusion, and signal transduction pathways lead to Ca^{2+} mobilization from acidocalcisomes. The parasite eventually escapes the vacuole, replicating in the cytosol. The intracellular stage is considered part of the immune evasion strategy of *T. cruzi*, which is capable of invading many cell types, including macrophages, smooth and striated muscle cells and fibroblasts. However, cell invasion also exposes the parasite to the hazards of the intracellular environment, in particular the oxidative burst of phagocytic cells, eliciting a cellular response that will be discussed below.

Chagas' disease

T. cruzi is the causal agent of Chagas' disease (American trypanosomiasis), a debilitating disease affecting millions of people in Latin and South America (Chagas, 1909; Rassi *et al.*, 2010). Over the last several hundred years, the disease switched from accidental to endemic infection owing to deforestation with occupation of cleared areas and adaptation of the insect vector to human dwellings. Besides vector-borne transmission, the disease can also be transmitted by blood transfusion and vertically from mother to infant, with occasional cases of oral ingestion of contaminated food and laboratory infection. The number of affected people has been reduced in endemic countries from 18 million to 8 million in the last 20 years owing to vector control programs and blood bank screening, although the number of new cases is increasing in non-endemic countries owing to immigration (Coura and Viñas, 2010). Interestingly, there are variations in the clinical manifestation of the disease with geographical differences in morbidity and mortality. These differences are believed to reflect the heterogeneity of the parasite as evidenced from biological, biochemical and genetic characterization of isolates (Luquetti *et al.*, 1986; Miles *et al.*, 2009).

After infection, circulating trypomastigotes are taken up by macrophages and carried to the liver, spleen, lymphatic ganglia and skeletal and heart muscles. The disease has an acute and a chronic phase (reviewed by Rassi *et al.*, 2010). The acute phase is oligosymptomatic or asymptomatic, lasts 4–8 weeks and resolves spontaneously in

90% of the cases. In some cases, however, death may occur owing to severe myocarditis and/or meningoencephalitis. In 60–70% of the cases the patient does not develop clinically apparent disease, which is called the indeterminate form. The remainder of cases (30–40%) develops the chronic form of the disease, which usually manifests itself 10 to 30 years after the initial infection. Persistence of *T. cruzi* in the host is necessary for chronic disease, and organ and tissue damage is caused directly by the parasite and by the host immunoinflammatory response. There are several forms of chronic disease, the cardiac and digestive forms being the most common. In the cardiac form, low intensity but incessant myocarditis eventually causes impaired contractile function and dilatation, leading to ventricular arrhythmias (with frequent sudden death) and dilated cardiomyopathy (with heart failure). In the digestive form, destruction of intramural autonomic ganglia of the oesophagus and colon causes peristalsis disfunction and dilatation of hollow viscera, which leads to megaesophagus and megacolon. Treatment of choice in the acute phase consists in the use of the antitrypanosomal drugs benznidazole and nifurtimox (reviewed by Wilkinson and Kelly, 2009; Urbina, 2010). These drugs, however, show relatively low efficacy (specially in the chronic phase) but significant side effects, and several new classes of drugs are being developed based on a better understanding of the parasite's biology. These include inhibitors of ergosterol biosynthesis, cruzipain, farnesyl-pyrophosphate synthase, hexokinase, and trypanothione synthesis. No vaccine against the parasite is available.

The stress response

Variations in environmental conditions must be dealt with by the organism to maintain the structural and functional integrity of cellular components. At the cellular level, two kinds of responses can be discerned: the cellular stress response (CSR) and the cellular homeostasis response (CHR) (reviewed by Kültz, 2005). CSR is transient and deals with macromolecular integrity and redox potential. It is not specific to the type of stress but to the type of damaged macromolecule, and it leads to stress tolerance. Damage to protein, DNA and membrane lipid is monitored by the cell, and the response involves (a) cell cycle control; (b) protein chaperoning, repair or degradation; (c) DNA and chromatin stabilization and repair; and (d) modifications of cellular metabolism. The basic CSR is very evolutionarily conserved, but several species-specific aspects and components are present that reflect the particular circumstances of the organism. Typical CSR inducers are heat and oxidative stress. On the other hand, CHR is stressor-specific and deals with loss of homeostasis relative to some environmental variable. It leads to adaptation to the new conditions and is in general permanent until the conditions change. Osmotic stress is an example of a CHR elicitor. Here we review the current knowledge on *T. cruzi*'s response to the various environmental stresses with emphasis on experimentally characterized proteins and genes. Both CSR and CHR are called stress response in what follows. It should be noted that frequently one kind of stress elicits multiple cellular responses and the same response may be induced by different kinds of stress, so the division below is somewhat artificial in this respect.

Environmental stresses associated with life cycle

During its life cycle, *T. cruzi* has to cope with several environmental changes. Among them are shifts in temperature, low availability of nutrients, changes in osmolarity, and, when inside a mammalian host cell, the presence of highly reactive oxidative species (ROS). The stress response in all these situations is similar to that of other protozoan parasites (reviewed by Vonlaufen *et al.*, 2008), and is varied and sophisticated. *T. cruzi* may respond by increasing the synthesis of proteins such as heat shock proteins (HSPs) or DNA repair enzymes when specific classes of macromolecules are damaged. To neutralize ROS, the response may consist of the synthesis of non-enzyme redox molecules such as trypanothione and of an increase in enzymatic activity of superoxide dismutases and peroxidases. Even subcellular structures may be part of the parasite's response. Nutritional stress involves cytoplasmic granules, while osmotic stress is dealt with by organelles such as acidocalcisomes and contractile vacuoles. Although harmful in general, environmental stresses may be crucial for *T. cruzi*'s development, since nutritional stress can induce metacyclogenesis (Contreras *et al.*, 1985). In any case, the parasite's stress response is essential for viability in various situations, such as the heat and osmotic stress responses during transition from the insect to the mammalian host, and the oxidative stress response when invading mammalian cells. The parasite's stress response is therefore important for disease transmission and pathogenesis. It is also relevant to treatment, since the drugs of choice for Chagas' disease, nifurtimox and benznidazole, are thought to interfere with the parasite's oxidative stress response (Wilkinson and Kelly, 2009). The types of stress that occur during *T. cruzi*'s life cycle and the corresponding cellular response components are summarized in Fig. 15.1. A more detailed list of the proteins and small molecules that are discussed in the following sections is shown in Table 15.1.

Stress sensor and signalling

The different stress responses must be triggered by stress sensors that activate signal transduction pathways. Although our understanding of the nature of these stress sensors has increased over the last few years, much remains to be investigated, in particular in eukaryotes. The primary sensors of each kind of stress or of each kind of macromolecular damage have proven elusive,

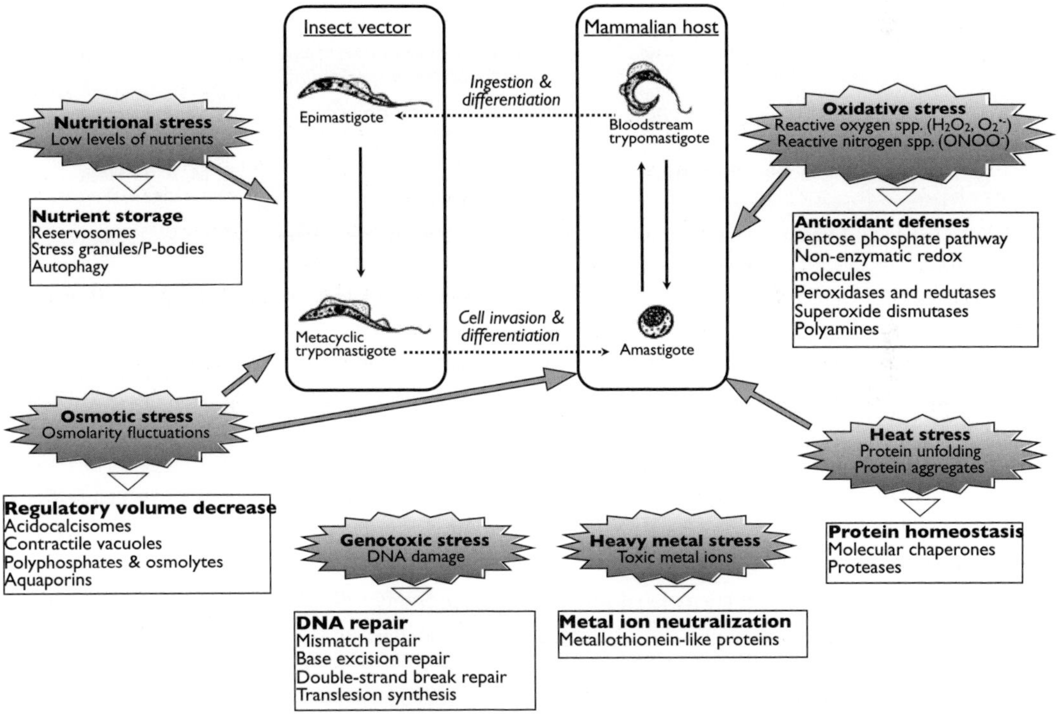

Figure 15.1 The stress response of *T. cruzi* to environmental stressors during its life cycle. The various types of stress that occur during the parasite's life cycle are summarized. Grey arrows indicate where in the cycle a given stress occurs or has been investigated in more detail. White arrowheads indicate the corresponding known cellular response. Nutritional stress takes place during (and in fact can induce) metacyclogenesis. Osmotic stress occurs in the insect's digestive tract owing to fluctuations in osmolarity, and during the transition from the insect hind gut to the mammalian interstitial fluid where there is a substantial drop in osmolarity. Heat stress also occurs mainly during the transition from insect vector to the mammalian host. The parasite undergoes oxidative stress after mammalian host cell invasion, particularly in macrophages, when it is exposed to the oxidative burst. Oxidative stress is frequently accompanied by genotoxic stress since reactive oxygen or nitrogen species may cause DNA damage. Both genotoxic and heavy metal stresses may occur in any cell form in either host.

Table 15.1 Proteins and small molecules of *T. cruzi*'s stress response

Molecule	Function	Role in stress response	Cell form[a]	Type of stress
AP endonuclease[b]	Base-excision repair	DNA repair	E	Genotoxic
TcRAD51	Double-strand break repair	DNA repair	E	Genotoxic
8-OxoG-DNA glycosylase[b]	Base-excision repair	DNA repair	E	Genotoxic, oxidative
DNA polymerase kappa[b]	Translesion synthesis	DNA repair	E	Genotoxic, oxidative
TcMSH2[c]	Mismatch repair	DNA repair	E	Genotoxic, oxidative
TcPARP	Poly(ADP-ribose) polymerase	DNA damage sensor	E	Genotoxic, oxidative
TcCALPx11	Calpain-like protein	Signal transduction[d]	E	Heat, pH
HSP10	Molecular chaperone	Restore protein homeostasis by assisted folding/refolding	E	Heat[e]
HSP104	Molecular chaperone	Restore protein homeostasis by assisted folding/refolding	E	Heat
HSP40s	HSP70-associated factor	Restore protein homeostasis by assisted folding/refolding	E, M	Heat
HSP60	Molecular chaperone	Restore protein homeostasis by assisted folding/refolding	E, M	Heat
HSP70s	Molecular chaperone	Restore protein homeostasis by assisted folding/refolding	E, M, A, T	Heat
HSP83	Molecular chaperone	Signal transduction[d] and cell cycle control	E, M, A, T	Heat
sHSP16	Molecular chaperone	Restore protein homeostasis by assisted folding/refolding	E	Heat
Metallothioneins	Metal ion binding	Prevent heavy metal toxicity	E	Heavy metal
TcMAPK2	Protein phosphatase	Signal transduction[d]	E, M, A, T	ND
Autophagins	Protease	Autophagy	E, M, A, T	Nutritional
Cleaved tRNAs	Unknown	Unclear	E	Nutritional
TcDHH1[f]	Helicase present in P-bodies/stress granules	mRNA storage	E	Nutritional
TcSTI-1	HSP83-associated factor	Signal transduction[c] and cell cycle control[c]	E	Nutritional
ENA-type ATPase[a]	Na^+/K^+-ATPase	Restore osmolarity	E, A, T	Osmotic
Exopolyphosphatase[f]	Polyphosphate metabolism	Restore osmolarity	E, A, T	Osmotic
Phosphatidylinositol 3-kinase	Lipid phosphatase	Signal transduction	E	Osmotic
Ascorbate peroxidase	Redox enzyme	Reduce reactive oxygen/nitrogen species	E, M, A	Oxidative
Fe–Superoxide dismutase	Redox enzyme	Reduce reactive oxygen/nitrogen species	M, A	Oxidative
Glutathione peroxidase	Redox enzyme	Reduce reactive oxygen/nitrogen species	T	Oxidative
Methionine sulfoxide reductase	Antioxidant enzyme	Revert oxidation of methionine residues	E, A	Oxidative
Polyamines	Antioxidant	Inhibit lipid peroxidation	E	Oxidative
TcG6PDH	Pentose phosphate pathway	Generation of reducing power	E, M, A, T	Oxidative

Molecule	Function	Role in stress response	Cell form[a]	Type of stress
Trypanothione	Redox molecule	Reduce reactive oxygen/ nitrogen species	E, M	Oxidative
Trypanothione reductase	Redox enzyme	Reduce reactive oxygen/ nitrogen species	E, M	Oxidative
Trypanothione synthetase	Redox enzyme	Reduce reactive oxygen/ nitrogen species	E, M	Oxidative
Tryparedoxin	Redox molecule	Reduce reactive oxygen/ nitrogen species	E, M, A, T	Oxidative
Tryparedoxin peroxidases	Redox enzyme	Reduce reactive oxygen/ nitrogen species	E, M, A, T	Oxidative
Arginine kinase	ATP generation	Unclear	E	Oxidative, nutritional, pH
Cytochrome P450 reductases[a]	Hydroxylation of toxic compounds	Neutralize xenobiotics	E, M, A, T	Toxic

[a]Cell form in which the protein or small molecule was shown to be part of the stress response. Not every cell form was investigated in each instance. E, epimastigotes; M, metacyclic trypomastigotes; A, amastigotes; T, tissue culture trypomastigotes.
[b]Induction was not detected but the protein is protective when the gene is overexpressed.
[c]Induction was not detected but single allele gene knockout leads to increased sensitivity.
[d]Presumed role; needs to be experimentally verified.
[e]Description has been mostly in the heat stress response, but any protein-denaturing agent may induce these proteins.
[f]Induction of the cytoplasmic granules was detected but not of the protein.
[g]Induction was not detected but the protein increases sensitivity when the gene is overexpressed.
ND, not determined. References are cited in the text.

and very difficult to distinguish from secondary sensors that are part of the signal transduction pathways. Much additional work is required to clarify the issue. In protozoan parasites in general, and in *T. cruzi* in particular, the necessary work has barely begun.

Environmental stresses activate signalling cascades that involve modulation of protein activity, changes in gene expression and even participation of whole organelles. These changes will ultimately result in a protective or adaptive cellular response. In eukaryotes, well-characterized stress signal transduction pathways include those mediated by eukaryotic initiation factor 2 (eIF2) kinases and mitogen-activated protein kinases (MAPKs), regulators of translation and proliferation, respectively. eIF2 kinases phosphorylate the translation initiation factor eIF2 and thereby arrest translation. Four types are described: general control non-inducible-2 (GCN2), dsRNA-dependent protein kinase (PKR), haemin-regulated inhibitor (HRI), and PKR-like endoplasmic reticulum-resident kinase (PEK/Perk) (reviewed by Hinnebusch, 2005). These eIF2 kinases are activated in response to nutrient starvation, viral infection, haemin availability and unfolded proteins in the endoplasmic reticulum, respectively. MAPKs are mediators of stress response that regulate proliferation, differentiation and apoptosis. These proteins are also divided into four types: extracellular signal-regulated kinase (ERK), c-Jun NH2-terminal kinase/stress activated protein kinase (JNK/SAPK), p38, and big mitogen-activated protein kinase-1/ERK5 (BMK-1/ERK5) (reviewed by Avruch, 2007). MAPK phosphorylation cascades have been described to be involved in both heat and osmotic stress responses. The participation of eIF2 kinases and MAPKs in several stress-initiated signal transduction pathways indicates that translation and proliferation arrests are frequently part of the stress response. In addition, it has recently become clear that the nucleolus is a multifunctional subnuclear compartment. Besides biogenesis of the ribosome and regulation of mitosis and cell cycle progression, its roles include stress sensor and stress response coordinator with the involvement of p53 and rDNA transcription, respectively (Boisvert *et al.*, 2007).

Signal transduction pathways involved in the stress response have been very poorly characterized in *T. cruzi*. eIF2 kinases have not been experimentally described, although four putative orthologue genes have been found through similarity searches in the parasite's genome draft sequence (Vonlaufen *et al.*, 2008). Among these putative orthologues, three of the four eIF2 kinase types appear to be represented. On the other hand, a member of the MAPK family has been recently described in this organism (Bao *et al.*, 2010). TcMAPK2 was identified as an ERK2 orthologue, and it is present in all cell forms but at lower levels in amastigotes. The subcellular localization of TcMAPK2 varies between cell forms. The protein is located mainly in the cytoplasm in epimastigotes, along the flagellum in trypomastigotes and in the plasma membrane of amastigotes. Levels of phosphorylated TcMAPK2 are higher in trypomastigotes, lower in epimastigotes and lower still in amastigotes. The functional significance of all these differences is unclear. Although no direct participation in the parasite's stress response has been described, overexpression of the TcMAPK2 gene in epimastigotes leads to cell death (Bao *et al.*, 2010). A member of another signal transduction pathway has been characterized in *T. cruzi* in the context of the stress response. Calpains are calcium-dependent cysteine proteases involved in signal transduction, differentiation and cytoskeleton remodelling (Sorimachi *et al.*, 2011). Among the calpains, a subfamily of calpain-like proteins comprise atypical calpains in which catalytic activity and/or EF-hand calcium-binding domains are missing. These proteins are also involved in signal transduction and are widespread in lower eukaryotes. A calpain-like protein was described in *T. cruzi* (Giese *et al.*, 2008). TcCALPx11, the *T. cruzi* orthologue, is present in differentiating epimastigotes but not in metacyclic trypomastigotes. Proteins levels increase in epimastigotes after heat, nutritional or pH stress, suggesting a general role for TcCALPx11 in the parasite's stress response. The authors speculate that it may be involved in stress response-related signal transduction pathways (Giese *et al.*, 2008). Additional work will be needed to clarify the signalling pathways of the stress response in this parasite.

Heat stress

Proteins need to be conformationally flexible to execute their cellular functions, and accordingly they are evolutionarily optimized to be only marginally stable at growth temperatures (reviewed by Richter *et al.*, 2010). Organisms are therefore sensitive to small temperature changes, which cause protein unfolding and, owing to the very high intracellular protein concentration, aggregation and in general loss of protein homeostasis. It is assumed that unfolded proteins trigger the stress response, and so any denaturing agent besides heat, such as oxidative species, heavy metals, ethanol or organic substances, also induces the stress response. Heat stress affects several cellular structures and processes. Among them are defects of the cytoskeleton, fragmentation of the endoplasmic reticulum and the Golgi apparatus, decreased number of lysosomes and mitochondria, changes in membrane permeability, and decreased translation. The stress response protects the cell from these adverse effects, and a previous heat shock results in the cell being much less affected by a subsequent heat shock, a phenomenon called thermotolerance. Heat stress-induced proteins are grouped in seven classes: (a) molecular chaperones, which help other proteins to return to their native conformation; (b) components of the proteolytic system, to degrade irreversibly denatured and/or aggregated proteins; (c) DNA- and RNA-modifying enzymes, to remove heat-induced modifications of nucleic acids; (d) metabolic enzymes, to reorganize and stabilize the energy supply of the cell; (e) regulatory proteins that initiate further stress response pathways or inhibit expression cascades; (f) cytoskeleton-sustaining proteins; and (g) transport proteins, detoxifying enzymes and membrane-modulating proteins that are needed to restore membrane stability and function (Richter *et al.*, 2010 and references therein). Most HSPs are molecular chaperones, which comprise the best studied and model group for the cellular response to heat stress. This class of proteins is divided in several conserved families according to molecular weight and broad function, among which the five major are: HSP104, HSP90, HSP70, chaperonins, and small HSPs (reviewed by Young *et al.*, 2004; Kriehuber *et al.*, 2010).

HSPs and the heat stress response of protozoan parasites have been the subject of several recent reviews (Folgueira and Requena, 2007; Vonlaufen *et al.*, 2008; Shonhai *et al.*, 2011). Digenetic parasites frequently are submitted to temperature changes during their life cycles, and HSPs are, as expected, part of their heat stress response. Although intracellular organisms and parasites frequently undergo genome reduction, the chaperone gene complement is sometimes expanded in protozoan parasites, highlighting the importance of these proteins in the biology of these organisms. HSP70 and HSP40 gene families are examples of these expansions (Folgueira and Requena, 2007; Shonhai *et al.*, 2011). HSPs have also been shown to be important for parasite intracellular survival, differentiation and virulence. In fact, HSPs are immunogenic and anti-HSP antibodies are present in Chagasic patients and in patients with other parasitic and bacterial infectious diseases.

T. cruzi 's response when exposed to elevated temperatures in culture has been extensively studied (Carvalho *et al.*, 1987; Alcina *et al.*, 1988; de Carvalho *et al.*, 1994; Rondinelli, 1994). Increased levels of HSP100, HSP90, HSP70 and HSP60 can be detected when the parasite is incubated at 37°C and 40°C. Different from higher eukaryotes, however, these proteins are already present in non-heat shocked parasites (Dragon *et al.*, 1987; Requena *et al.*, 1992; de Marval *et al.*, 1993). The typical general inhibition of protein synthesis during heat shock can be seen at 40°C but not at 37°C, and it can be prevented by incubating the parasite in serum (Carvalho *et al.*, 1987). The mechanism by which serum protects protein synthesis remains to be explained. *T. cruzi*'s heat stress response also confers thermotolerance, since pre-incubation at a high but permissible temperature allows the parasite to withstand a subsequent otherwise lethal temperature (J.F. Carvalho *et al.*, unpublished). In addition, a dose-dependent induction of HSP90 was seen after ethanol treatment, showing that a non-heat stress can also induce HSPs in this organism (Giambiagi-de-Marval *et al.*, 1993b).

Since *T. cruzi* naturally undergoes a heat shock when it transits from an insect to a mammalian host during its life cycle, it is assumed that the parasite synthesizes HSPs in preparation for infection. In fact, metacyclic trypomastigotes, the infective form, show increased levels of HSP70, HSP60 and HSP90 when compared with epimastigotes (Paba *et al.*, 2004), although it is still unclear whether HSP induction is part of the differentiation process or is just an adaptation to the new environment (Folgueira and Requena, 2007). However, elevated temperatures inhibited metacyclogenesis in infected insects but did not affect parasite survival (Garcia *et al.*, 2011). These results suggest that any preparedness to higher temperatures is restricted to the metacyclic trypomastigote. The possibility that elevated temperature by itself could trigger a differentiation event was also investigated. Consistent with the results in infected insects, heat treatment does not induce differentiation to the infective form in *T. cruzi* (Contreras *et al.*, 1985; Rondinelli *et al.*, 1988). However, incubation temperature can promote differentiation in an *in vitro* cycle of the parasite in axenic culture (Rondinelli *et al.*, 1988). In this cycle, the authors have shown that they are able to induce differentiation of *in vitro*-generated amastigote-like cells to epimastigotes or to trypomastigote-like cells depending on the temperature of incubation (29°C or 37°C) and culture medium type (LIT or serum). These results suggest that temperature alone can influence a differentiation programme in this parasite, although it is unclear whether the heat stimulus has such an effect in the natural life cycle. Recently, the effects of severe heat treatment on RNA-binding proteins have been investigated. In epimastigotes incubated at 40°C, TcSR62 and TcPTB2, an SR- and a polypyrimidine tract-binding protein, respectively, partially accumulate in the nucleolus in an active transport-dependent manner (Názer *et al.*, 2011). This accumulation is reversible upon return to normal temperatures, and suggests the participation of the nucleolus in the heat stress response of the parasite (Názer *et al.*, 2011).

The contributions of the main HSP families to *T. cruzi*'s heat stress response are reviewed in the following sections.

HSP70 family

The HSP70 family is the most conserved and studied of HSPs, and one of the most conserved families of proteins in general. There are induced and constitutive paralogues (the latter called HSC70, or HSP70 cognate protein), and family members localized in the cytoplasm, mitochondria and endoplasmic reticulum (where it is called Grp78). HSP70 proteins exert several different functions: (a) binding to nascent polypeptide chains to prevent premature folding; (b) disassembling of protein aggregates and assisting refolding of misfolded proteins; (c) directing proteolytic degradation of denatured or unstable proteins; (d) assisting membrane translocation of organelle and export proteins; and (e) modulating the activity of regulatory proteins. HSP70s contain an N-terminal ATP-binding domain and a C-terminal peptide-binding domain, and function by binding and releasing, in an ATP-dependent manner, an extended polypeptide substrate, in association with HSP40 as a co-factor (reviewed by Hartl and Hayer-Hartl, 2002; Young *et al.*, 2004). HSP70 may even act as a signal transducer by being released extracellularly and interacting with cells of the immune system with immunoregulatory effects, known as the chaperokine activity of the protein (Asea, 2003).

In *T. cruzi*, HSP70 genes of different subfamilies were initially characterized as present in ten tandemly repeated copies in the genome (Requena *et al.*, 1988; Engman *et al.*, 1989a,b, 1992; Requena *et al.*, 1989). Analysis of the draft sequence of the parasite's genome (El-Sayed *et al.*, 2005) showed that 11 HSP70 genes are present, with the HSP70.a, HSP70.b, HSP70.c, mitochondrial HSP70, HSP110, Grp170 and Grp78 subfamilies being represented together with some unusual members (Folgueira and Requena, 2007; Louw *et al.*, 2010; Shonhai *et al.*, 2011). There appears to be some stage-specific HSP70 members, although its biological significance is unclear. The cytosolic HSP70.4 is highly enriched in amastigotes but undetectable in trypomastigotes, while HSP70.a was found exclusively in trypomastigotes (Atwood *et al.*, 2005). Differences in activity and behaviour of HSP70 members were also described. Both cytosolic and mitochondrial HSP70s are potent ATPases, although only the mitochondrial protein possesses autophosphorylation activity (Edkins *et al.*, 2004; Olson *et al.*, 1994). In addition, cytosolic HSP70 migrates to the nucleus upon heat shock, while mitochondrial HSP70 remains associated with the kinetoplast (Martin *et al.*, 1993; Olson *et al.*, 1994; Klein *et al.*, 1995). The Grp78 member has also been characterized, and shown to be located in the endoplasmic reticulum (Tibbetts *et al.*, 1994). Grp78 is induced strongly by tunicamycin, a glycosylation inhibitor, but not by heat, a typical behaviour of this subfamily. Grp78 was also shown to bind and assist folding of the lysosomal protease cruzipain and delivery of this glycoprotein to calcireticulin (Labriola *et al.*, 2011).

Consistent with published reports that describe the immunogenicity of HSPs, the presence of anti-*T. cruzi*'s HSP70 antibodies was shown in Chagasic patients. Antibodies against the cytosolic, mitochondrial and endoplasmic reticulum HSP70 members have been detected. Although these results indicate the immunogenicity of the protein, it does not appear to play a role in the pathogenesis of the disease (Engman *et al.*, 1990; Levy Yeyati *et al.*, 1992; Olson *et al.*, 1994; Krautz *et al.*, 1998). Attempts have been made, however, to use the immune system's response to the parasite's HSP70 for diagnostic and vaccine-related purposes. Although anti-HSP antibodies are unable to distinguish Chagasic patients from those infected by other trypanosomatids, antibodies against a fragment of *T. cruzi*'s HSP70 can distinguish between healthy and infected people and between acute and chronic Chagasic patients (Flechas *et al.*, 2009). In addition, the parasite's HSP70 induces functional maturation of murine dendritic cells, and the authors suggest that *T. cruzi*'s HSP70 may be useful as a vehicle for dendritic cells-based immunoprophylaxis and therapy against infections (Planelles *et al.*, 2002).

As in other eukaryotes, the heat shock response in *T. cruzi* has been used as model for studies of gene regulation. In particular, HSP70 gene expression has been characterized in greater detail, and a few mechanisms of gene regulation have been described. In fact, it is the most detailed description in this organism of a stress response-related gene regulatory mechanism. At least eight HSP70 protein isoforms have been experimentally identified, of which five were inducible by heat treatment (Giambiagi-deMarval *et al.*, 1996). No new isoforms were detected at elevated temperatures, suggesting that the heat-inducible protein isoforms are already present in non-heat-shocked cells. In fact, both HSP70 protein and mRNA are present at normal temperatures. HSP70 synthesis is increased in response to heat shock, which is accompanied by an increase in mRNA levels (Engman *et al.*, 1989a; de Carvalho *et al.*, 1990; Requena *et al.*, 1992). Severe heat shock at 40°C, however, leads to somewhat reduced HSP70 mRNA levels and accumulation the polycistronic precursor RNA, indicating an inhibition of RNA processing (Engman *et al.*, 1995). Preferential translation of HSP70 mRNA during heat shock has also been suggested, since the mRNA increase in polysomes is greater than it is in total RNA at elevated temperatures (de Carvalho *et al.*, 1990). The mechanism responsible for the increase in mRNA levels during heat shock was investigated (Rodrigues *et al.*, 2010). HSP70 mRNA half-life increases from 1 hour at 29°C to 2 hours at 37°C, and this increase in stability is dependent on protein synthesis. Consistent with these results, degradation of the 3′-UTR of the mRNA occurs to a greater degree at 29°C in an *in vitro* assay, and a U-rich sequence appears to be the target of degradation. Interestingly, U-rich sequences can be found in the 3′-UTR of all heat-inducible mRNAs identified so far, which was first pointed out by Sullivan *et al.* (1994). In addition, a heat shock-responsive element is present in the HSP70 mRNA's UTRs, and both 5′- and 3′-UTRs act cooperatively to stabilize the mRNA during heat shock (Rodrigues *et al.*, 2010). Additional work is required to further characterize these RNA elements. Also, identification of the regulatory molecule, presumably a protein, that recognizes this heat shock-responsive element would advance our understanding of how an increase in temperature leads to increased synthesis of HSP70 by post-transcriptional regulatory mechanisms.

HSP40 family

HSP40 proteins, also called J proteins, function as cofactors of HSP70 and are responsible for much of the functional diversity of the HSP70 family by conferring substrate specificity. Several HSP40s may function with a single HSP70, and may target HSP70 to specific proteins. Like HSP70, the protein is also distributed in all cellular compartments. HSP40 is characterized by an N-terminal, 70-amino acid sequence known as the J-domain, which interacts with HSP70. In the HSP70s ATP-dependent protein-binding and release cycle, HSP40 binds to the unfolded peptide region and delivers it to HSP70 (reviewed by Hartl and Hayer-Hartl, 2002; Young *et al.*, 2004).

In *T. cruzi*, four HSP40 genes were initially characterized, TcJ1 to TcJ4 (Tibbetts *et al.*, 1998). TcJ2 and TcJ4 proteins contain prenyl modification motifs, suggesting that they are associated with the plasma membrane, and only the TcJ2 gene gives rise to a heat-inducible mRNA. TcJ2 protein was shown to stimulate the ATPase activity of HSP70 *in vitro*, and the corresponding gene is capable of complementing *ydj1* yeast mutant, suggesting that TcJ2 is the orthologue of YDJ1 (Edkins *et al.*, 2004). A mitochondrial HSP40 has also been characterized. The corresponding gene was shown to be developmentally regulated, since both mRNA and protein levels are higher in epimastigotes than in metacyclic trypomastigotes (Carreira *et al.*, 1998). The authors suggest that the protein may participate in the biosynthesis of the mitochondria. A member of the HSP40 family involved with translation initiation was also described (Salmon *et al.*, 2001). The protein was found to be associated with ribosomal subunits, 80S monosomes and smaller polysomes, and the corresponding gene is capable of complementing a yeast mutant deficient in the orthologue gene *Sis1*. Similarity searches in the parasite's genome sequence led to the identification of 67 HSP40 genes. This gene complement represents a greatly expanded family, but the significance of the expansion is unclear (Folgueira and Requena, 2007). Presumably, a larger number of HSP40 proteins implies a larger number of HSP70 protein substrates. It would be interesting to determine why a single celled organism requires an apparently more complex HSP70/HSP40 system.

HSP90 family

HSP90 is present at very high levels in the cell and it is strongly induced during heat stress. Unlike other HSPs, HSP90 is more selective regarding protein substrates and does not bind unfolded proteins. Most known substrates are involved in cell cycle control and signal transduction, and HSP90 functions in an ATP-dependent manner associated with HSC70, HSP40, HOP and p23. The mechanistic details of HSP90 function, however, are less well understood (reviewed by Young *et al.*, 2004; Richter *et al.*, 2010).

The HSP90 genes of *T. cruzi* are organized as six copies arranged in tandem, and give rise to an 83 kDa protein (Dragon *et al.*, 1987; Folgueira and Requena, 2007). Unlike mammalian HSP90, which binds ATP but has no ATPase activity, the parasite's HSP83 shows ATPase activity, is autophosphorylated and its ATPase activity is stimulated by peptides (Nadeau *et al.*, 1992). HSP83 can also functionally complement an hsp90 yeast mutant (Palmer *et al.*, 1995). Treatment of *T. cruzi* with the HSP90 inhibitor geldanamycin produces a dose-dependent increase in HSP100, HSP83 and HSP70 levels and proliferation arrest at G1, indicating the importance of HSP83 in the parasite's cell cycle control (Graefe *et al.*, 2002). The effects of geldanamycin on differentiation were also tested. Bloodstream trypomastigotes showed morphological changes in the presence of the drug, but no differentiation to epimastigotes occurred. The orthologue of HSP90-associated factor HOP, TcSTI-1, was also characterized and shown to interact with HSP83 (Schmidt *et al.*, 2011). TcSTI-1 does not appear to be readily induced by different stressors, as protein and mRNA levels remain unchanged under heat, nutritional or pH stress conditions in exponentially growing epimastigotes. However, TcSTI-1 is induced by nutritional stress in late-phase growth epimastigotes, indicating that growth phase affects the stress response in epimastigotes. Three orthologues of the endoplasmic reticulum and two of the mitochondrial members of the HSP90 family, Grp94 and TRAP-1/HSP75, respectively, have been found in the parasite's genome sequence (Folgueira and Requena, 2007; Shonhai *et al.*, 2011). However, these genes have not yet been characterized experimentally. It remains to be seen whether HSP83 is involved in signal transduction in the parasite.

Chaperonin family

The chaperonins form a large, barrel-shaped structure that encloses unfolded or misfolded proteins. In group I chaperonins, two heptameric HSP60 and one heptameric HSP10 rings comprise the structure, which provide an isolated environment where non-native proteins can achieve their final conformation without risk of aggregation. Group I chaperonins are found in bacteria, mitochondria and chloroplasts, and group II chaperonins are found in archea and the eukaryotic cytosol. Folding of the protein is assisted in an ATP-dependent binding and release cycle, where exposed hydrophobic regions bind to the internal surface of the barrel, trapping the partially folded protein. Once the protein reaches its native conformation, hydrophobic regions are no longer exposed and the protein is released to the cytoplasm (reviewed by Hartl and Hayer-Hartl, 2002; Young *et al.*, 2004).

In *T. cruzi*, both components of group I chaperonins have been characterized. HSP60 genes are organized in nine copies in tandem (Giambiagi-de Marval *et al.*, 1993a; Sullivan *et al.*, 1994), and the protein is located in the mitochondrial matrix, as expected (de Marval *et al.*, 1993; Sullivan *et al.*, 1994). Three isoforms have been identified, and only one is heat-induced (Sullivan *et al.*, 1994; Giambiagi-deMarval *et al.*, 1996). HSP60 mRNA levels, however, remain unchanged when epimastigotes are incubated at 37°C (Giambiagi-de-Marval *et al.*, 1993b), which suggests translational control. The other chaperonin, HSP10, is encoded by three tandemly repeated genes intercalated by genes of unknown function (Fernandes *et al.*, 2005). Interestingly, this arrangement is conserved in *Leishmania* spp. and *Trypanosoma brucei* (Folgueira and Requena, 2007). The HSP10 protein contains a mitochondrial targeting sequence and shows predicted structural features that are conserved in the family. In addition, HSP10 mRNA remains unaltered during heat shock, similarly to HSP60, although an alternative polyadenylation site is also used at 37°C and 40°C (Fernandes *et al.*, 2005). Since both HSP60 and HSP10 are part of the same molecular structure, and presumably are present in equimolar amounts, it would be interesting to investigate the regulatory mechanisms to determine whether both genes are coordinately regulated.

Small HSP family

Small HSPs (sHSPs) are found in all groups of organisms and comprise the least conserved family of molecular chaperones. sHSPs are heterogeneous in size and sequence and are characterized by low monomeric molecular mass and the presence of a 90 amino acid conserved region called the alpha-crystallin domain. sHSPs form large oligomers of often 24 subunits that bind unfolded proteins in an ATP-independent manner and prevent aggregation. It is believed that sHSPs associate with partially folded proteins during heat stress and store these proteins until another chaperone assists their refolding. sHSPs are one of the most strongly induced among the HSP families (reviewed by Richter *et al.*, 2010).

A single *T. cruzi* sHSP has recently been characterized (Pérez-Morales *et al.*, 2009). sHSP16 protein contains the characteristic alpha-crystallin domain, and the corresponding single-copy gene shows some sequence polymorphism among strains. Gene expression studies showed that mRNA levels are moderately increased after heat treatment. To test for chaperone activity, recombinant sHSP16 was incubated with malate dehydrogenase and shown to prevent aggregation of the enzyme *in vitro* (Pérez-Morales *et al.*, 2009). Further work is needed to determine the biological role of this protein in the different forms of the parasite.

HSP104 family

The HSP100 family of chaperone proteins, a member of the AAA+ superfamily of ATPases, comprises hexameric protein complexes that catalyse protein unfolding, disassembly, disaggregation and degradation. The group is subdivided in two classes based on domain organization. Class 1 proteins contain two nucleotide-binding domains (NBDs) and promote either disaggregation or proteolysis. Class 2 proteins have only one NBD and promote proteolysis. Bacterial ClpB and its eukaryotic orthologue, HSP104, are class 1 HSP100s that are characterized by the presence of a Walker-type NBD. These proteins that help organisms survive severe stress by being capable of disentangling even insoluble protein aggregates. ClpB and HSP104 are induced by heat shock, act in concert with HSP70 and are required for thermotolerance (reviewed by Doyle and Wickner, 2009; Richter *et al.*, 2010).

Orthologues of both HSP104 and of bacterial ClpB have been identified in the parasite's genome sequence, the latter presumably having a function in the mitochondrial protein metabolism (Folgueira and Requena, 2007). With the gene sequence available as a starting point, a member of the HSP104 family has been characterized in *T. cruzi* (Campos *et al.*, manuscript in preparation). The amino acid sequence contains the characteristic NBD domains, and the protein is induced at 37°C and 40°C. mRNA levels also increase during heat shock, similar to what has been described for HSP70. Additional work is needed to determine the subcellular localization of the protein, its contribution to thermotolerance in the parasite and the mechanisms of gene expression.

Oxidative stress

The unavoidable generation of free radicals by metabolism in an aerobic environment led to the emergence of scavenging systems the purpose of which is to minimize and repair oxidative damage. These systems include compounds such as ascorbate, glutathione, thioredoxin, and several antioxidant enzymes. In addition, ROS and reactive nitrogen species (RNS) may also act as second messengers in signal transduction pathways that include the stress response (Kültz, 2005).

During its life cycle, *T. cruzi* switches from an insect to a mammalian host and the infection pathway in the latter involves invasion by metacyclic trypomastigotes of several cell types such as macrophages, smooth and striated muscle cells and fibroblasts (Andrade and Andrews, 2005). Successful interaction with these initial cell types is crucial for dissemination of the parasite and the establishment of infection. A successful interaction depends on *T. cruzi*'s ability to evade and/or counteract the host's early innate immune response, which relies on complement-mediated defences and the oxidative burst generated by resident macrophages after phagocytosis of the parasite. Since metacyclic trypomastigotes are resistant to complement-mediated damage, the strength of the host's oxidative burst and the parasite's antioxidant network determine the outcome (recently reviewed by Piacenza *et al.*, 2009b). Therefore, *T. cruzi*'s antioxidant defence network strongly affects virulence (Piacenza *et al.*, 2009a). In addition, it has become increasingly clear that sustained, parasite-induced generation of reactive oxidative species in the heart of chronic Chagasic patients, together with an inadequate antioxidant response of the host, leads to long-term oxidative stress and eventually to oxidative damage of cellular components (Gupta *et al.*, 2009). In fact, chronic Chagasic patients have elevated levels of circulating nitrated proteins (Gupta *et al.*, 2009; Piacenza *et al.*, 2009b). However, additional studies are necessary to assess the relevance of oxidative stress to pathogenesis and clinical severity in Chagasic cardiopathy (Gupta *et al.*, 2009). Finally, it is believed that the mechanism of action of nifurtimox, one of the drugs currently in use for treatment of Chagas' disease, involves the production of ROS. *T. cruzi*'s antioxidant defences are so critical for successful infection that a search for inhibitors of the parasite's redox system is currently under way (Urbina, 2010).

After engulfing the metacyclic trypomastigote, non-stimulated macrophages induce ROS in the parasitophorous vacuole by activating a membrane-associated NADPH oxidase and producing superoxide anion and hydrogen peroxide for 90–120 minutes (Alvarez *et al.*, 2004). This initial oxidative burst appears insufficient, however, to eliminate the parasite (Alvarez *et al.*, 2004; Piacenza *et al.*, 2009b). When macrophages are stimulated by pro-inflammatory cytokines such as IL-12, INF-γ and TNF-α, an inducible nitric oxide synthase (iNOS) is synthesized during 24 hours, generating cytoplasmic nitric oxide that diffuses into the vacuole and reacts with superoxide anions yielding peroxynitrite ($ONOO^-$). This compound is a more powerful oxidant and is very damaging to the parasite. In fact, its survival depends on escaping from the vacuole to the cytoplasm in the first few hours of iNOS induction (Piacenza *et al.*, 2009b).

T. cruzi's antioxidant defence network is sophisticated and is composed of enzymes and non-enzymatic redox molecules. The redox molecules are trypanothione, glutathione, tryparedoxin (TXN; a thioredoxin homologue) and ascorbate, and enzymes consist of peroxidases and superoxide dismutases. These components are localized in the endoplasmic reticulum, the glycosome, the cytosol and the mitochondria, and ultimately utilize the reducing power of NADPH generated by the pentose phosphate pathway (PPP) (reviewed by Maugeri and Cazzulo, 2004; Piacenza *et al.*, 2009b). In the endoplasmic reticulum, hydrogen peroxide is metabolized by an ascorbate-dependent haemoperoxidase (TcAPX), while organic hydroperoxides are metabolized by glutathione-dependent peroxidase II (GPX-II). The oxidized components are reduced by trypanothione, which is in turn reduced by trypanothione reductase (TcTR). In the cytosol, hydrogen peroxide is metabolized by a cytosolic TXN peroxidase (TcCPX), while superoxide anions are metabolized by an iron-dependent superoxide dismutase (Fe–SOD). The resulting oxidized TXN is reduced by trypanothione, which is then reduced by TR. Trypanothione is synthesized in the cytosol by trypanothione synthetase (TcTS) from two glutathione molecules. In the mitochondria, superoxide anions produced by the respiratory chain or by the macrophages oxidative burst have two fates: to be metabolized by Fe–SOD to hydrogen peroxide, or to react with nitric oxide to form peroxynitrite. Both hydrogen peroxide and peroxynitrite are reduced and neutralized by a mitochondrial TXN peroxidase (TcMPX), and the catalytic detoxification of peroxynitrite is done via oxidation of fast reacting thiols (Trujillo *et al.*, 2004; Piacenza *et al.*, 2008). The resulting oxidized TXN is reduced by trypanothione, which is subsequently reduced by TR. Although the parasite's antioxidant defences appear formidable, macrophage-induced peroxynitrite is capable of controlling the parasite's proliferation by causing considerable oxidative damage and even lead to programmed cell death (Alvarez *et al.*, 2004; Piacenza *et al.*, 2007). In addition, the distinguishing aspects of the parasite's defences when compared to the mammalian antioxidant network may be exploited as potential drug targets for Chagas' disease treatment. Among the potential targets are (a) iron-containing superoxide dismutases, which are distinct from the mammalian copper-, zinc-, or manganese-containing superoxide dismutases; (b) the trypanothione-based antioxidant system (TcTR, TcTS and the TXN peroxidases) instead of mammalian selenium-containing glutathione peroxidase; and (c) TcAPX, used by *T. cruzi* instead of the mammalian catalase (Turrens, 2004; Bachega *et al.*, 2009).

Metacyclic trypomastigotes contain higher levels of TcTS, TXN, TcMPX, TcAPX and Fe–SOD when compared to epimastigotes, apparently in preparation for infection of the mammalian host (Atwood *et al.*, 2005; Parodi-Talice *et al.*, 2007). In addition, the activity of the PPP is increased during metacyclogenesis, ensuring a greater cellular source of reducing power to the infective form (Maugeri and Cazzulo, 2004; Mielniczki-Pereira *et al.*, 2007). The parasite's orthologue of glucose-6-phosphate dehydrogenase (TcG6PDH), the first enzyme of the oxidative branch of the PPP, is encoded by three genes with alternate start codons, the functional significance of which is unclear. TcG6PDH enzyme and mRNA contents are higher in metacyclic trypomastigotes than in other cell forms. Also, treatment of metacyclic trypomastigotes with hydrogen peroxide leads to higher enzyme levels and a 45-fold increase in enzyme activity, suggesting an important role for TcG6PDH in the parasite's stress response (Igoillo-Esteve and Cazzulo, 2006). On the other hand, TcCPX and TcMPX enzymes are present in similar amounts in all four life cycle stages. TcCPX is induced in a dose-dependent manner during hydrogen peroxide treatment of epimastigotes (Finzi *et al.*, 2004). In addition, overexpression of TcCPX and TcMPX genes leads to increased resistance to oxidative stress and more efficient infection but not invasion, indicating that both enzymes are virulence factors (Piñeyro *et al.*, 2008). In fact, several enzymes of the antioxidant network are determiners of *T. cruzi* virulence. Analysis of TcTS, TcAPX, TcCPX, TcMPX and TcTR enzyme levels in epimastigotes and metacyclic trypomastigotes from several strains shows that epimastigotes from virulent strains are more resistant to oxidative challenge. It also shows that TcTS, TcCPX and TcMPX enzyme levels increase during metacyclogenesis and are in general higher in virulent strains, measured in experimental mice infections (Piacenza *et al.*, 2009a). Comparison of two *T. cruzi* strains, Y and Tulahuen 2, shows that they differ in sensitivity to oxidative agents, with Tulahuen 2 being more resistant. The sensitivity correlates with the different TcCPX and TcG6PDH enzyme levels in the two strains. In addition, treatment with hydrogen peroxide increases TcG6PDH activity in the Tulahuen 2 strain (Mielniczki-Pereira *et al.*, 2007). The parasite's antioxidant network is also relevant to drug resistance. Comparison of benznidazole-sensitive and -resistant strains showed that resistance to the drug also correlates with higher levels of Fe–SOD and HSP70 mRNAs (Murta *et al.*, 2008). Recently, Díaz *et al.* (2011) have reported that *T. cruzi* isolates from acute Chagasic patients contain higher levels of TXN peroxidase and other enzymes associated with resistance to oxidative stress when compared to isolates from chronic Chagasic patients. The authors suggest that these proteins may represent virulence factors related to the development of the acute form of the disease and may have clinical relevance.

Complementing the antioxidant system described above, the parasite may have additional defence mechanisms. Polyamines such as putrescine, spermine, spermidine and cadaverine, which have been described to function as free radical scavengers in other eukaryotes, are present in tightly regulated concentrations in *T. cruzi* and are necessary for normal growth (Schwarcz de Tarlovsky *et al.*, 1993). Spermine and spermidine, in particular, inhibit lipid peroxidation, suggesting a protective role in oxidative damage in the parasite (Hernández *et al.*, 2006). Methionine sulfoxide reductase (TcMSR), an enzyme that reverses oxidation of methionine residues in proteins, has also been characterized in *T. cruzi*. TcMRSs interact with TXN, being present in epimastigotes and amastigotes but not in trypomastigotes, and overexpression of the corresponding genes in epimastigotes confers resistance to hydrogen peroxide treatment (Arias *et al.*, 2011). It has also been described that proline metabolism is important for the stress response. Inhibition with a proline analogue increases the sensitivity to oxidative stress, and addition of proline to the culture medium abolishes this effect (Magdaleno *et al.*, 2009). Finally, arginine kinase, an enzyme that catalyses the interconversion between phosphoarginine and ATP, appears to be induced by hydrogen peroxide and nifurtimox treatment. Overexpression of the corresponding gene leads to resistance to oxidative stress, although its role in the stress response remains unclear (Miranda *et al.*, 2006).

Nutritional stress

Nutritional stress in eukaryotes involves several cellular responses, which are activated depending of the severity of the stimulus. Mild to moderate nutritional stress leads to inhibition of protein synthesis, modulation of the energy metabolism and formation of two types of cytoplasmic foci called stress granules and P-bodies. Severe nutritional stress, or starvation, also leads to autophagy, whereby portions of the cytoplasm including whole organelles are directed for degradation and recycling. Other cellular responses of uncertain significance include the induction of cleaved tRNA molecules. Some of these responses to nutritional stress have been described in *T. cruzi* and are discussed below.

An important effect of nutritional stress in *T. cruzi* is the induction of metacyclogenesis by low nutrient levels. It is believed that the poor nutrient conditions of the insect vector's hind gut triggers metacyclogenesis in the parasite's life cycle. Accordingly, a chemically defined nutrient-poor medium mimicking the insect vector's urine induces epimastigotes to differentiate into metacyclic trypomastigotes (Contreras *et al.*, 1985). During differentiation, the parasite adheres to the substrate, and morphological changes in reservosomes, an epimastigote-specific, nutrient storage compartment, can be seen (Bonaldo *et al.*, 1988). Reservosomes gradually disappear during metacyclogenesis, and enriching the nutrient-poor medium inhibits the process (Figueiredo *et al.*, 2000). A proteomic analysis of nutritionally stressed epimastigotes showed the mobilization to polysomes of mRNAs coding for proteins related to DNA metabolism, membranes, motor activity and transcription and provided indirect evidence that nutritional stress leads to translation inhibition (Alves *et al.*, 2010).

Stress granules and P-bodies are frequently found in nutritionally stressed eukaryotes. Low levels of nutrients induce the phosphorylation of the translation initiation factor eIF2 by a GNC2-like eIF2 kinase that blocks translation and leads to stress granule formation, in which non-translating mRNAs and specific proteins are concentrated into cytoplasmic foci (Anderson and Kedersha, 2009). mRNAs in stress granules are believed to be dynamically sorted for storage, from which they return to translation after stress recovery or are degraded. mRNA degradation takes place in so called P-bodies (for processing bodies), cytoplasmic foci that are enriched in RNA-degrading enzymes and that can also respond to stressful conditions. The stored mRNA may be degraded by decapping followed by 5′ exonuclease digestion, or it may be delivered back to the translation machinery intact (Sheth and Parker, 2003; Teixeira *et al.*, 2005).

Evidence for the presence of P-bodies in *T. cruzi* was initially reported in 2007, when the parasite's orthologue of the DEAD box helicase DHH1, a marker of these cytoplasmic foci and known to interact with RNA degradation enzymes (Coller *et al.*, 2001), was cloned and characterized (Holetz *et al.*, 2007). The encoded protein is present in all cell forms in similar levels, and those levels remain unchanged during nutritional stress, suggesting that the TcDHH1 gene is not up-regulated during starvation. The TcDHH1 protein is associated with non-translating mRNA complexes and localizes to discrete cytoplasmic foci. In addition, similar to P-bodies in other eukaryotes, these foci are sensitive to the translation inhibitor cycloheximide but resistant to the translation inhibitor puromycin (Holetz *et al.*, 2007). TcDHH1-containing foci are present in all cell forms, but less defined in metacyclic trypomastigotes, and these foci increase in number in nutritionally stresses epimastigotes (Holetz *et al.*, 2010). Proteins directly or indirectly associated with TcDHH1 include translation factors, cytoskeleton proteins, HSPs and proteins related to metabolism, while mRNAs present in TcDHH1–containing complexes code for mucins, MASPs and amastin (Holetz *et al.*, 2010). However, although stress granules and P-bodies, characterized by the component proteins, are distinct structures in other eukaryotes, in *T. cruzi* mRNA-containing granules induced by carbon source starvation show mixed

P-bodies and stress granules characteristics (Cassola *et al.*, 2007). Inhibition of transcription by actinomycin D does not affect maintenance of the mRNA granules, suggesting that the mRNAs in the granule are stabilized. Finally, these mRNA granules have been detected in epimastigotes in the intestinal tract of the insect vector, where it is believed the parasite naturally undergoes nutritional stress (Cassola *et al.*, 2007).

Severe starvation leads to the bulk degradation of proteins and organelles in a process called autophagy, where portions of the cytoplasm are engulfed in double-membrane vesicles named autophagosomes and fused to lysosomes (Klionsky, 2005). Several autophagy-related genes have been characterized that control the process (Klionsky *et al.*, 2003). Orthologues of some of these autophagy-related genes have been found in the *T. cruzi*'s genome (Alvarez *et al.*, 2008). All the major genes of the Atg8 conjugation system are present, whereas the major components of the Atg12-Atg5 conjugation system are apparently lacking. TcAtg4 and TcAtg8 were found to substitute the *Saccharomyces cerevisiae* orthologues in functional assays, and TcAtg8 was found to be located in autophagosome-like vesicles. Using TcAtg8 as an autophagosome marker, the authors showed that autophagy occurred in nutritionally stressed but non-differentiating epimastigotes. It also occurred to a considerably larger degree during metacyclogenesis (Alvarez *et al.*, 2008). The authors suggest that the autophagy pathway could represent a novel target for drug development.

tRNAs have been increasingly recognized as not only adaptor molecules but also as regulators of protein synthesis, in which uncharged tRNAs modulate translation in response to nutritional stress (Li and Zhou, 2009). In addition, small RNAs derived from cleavage of tRNAs in the anticodon loop or left arm have been described. Although their biological significance is unknown, it has been proposed that in eukaryotes tRNA cleavage could down-regulate translation during nutritional stress by reducing tRNA availability or by direct translational repression (Lee and Collins, 2005; Li and Zhou, 2009). Small RNAs derived from tRNA cleavage have been detected in *T. cruzi* (Garcia-Silva *et al.*, 2010). The tRNA halves comprise about 25% of the small RNA population, and their abundance increased 2- to 4-fold during nutritional stress in epimastigotes. Most tRNA halves are derived from $tRNA^{Asp}$ and $tRNA^{Glu}$, and the 5′ and 3′ halves localize to different cytoplasmic granules (Garcia-Silva *et al.*, 2010). However, the functional significance of these molecules is still unclear, and further work is necessary to clarify the issue.

It has been proposed that arginine kinase may play a role in adaptation to stress, since the enzyme levels increase during growth in culture (Alonso *et al.*, 2001). No corresponding increase in mRNA levels was detected, suggesting a post-transcriptional control of gene expression. Although not induced by stress, overexpression of the arginine kinase gene improves the parasite's survival under nutrient-poor and high- or low-pH conditions (Pereira *et al.*, 2003). In addition, several aspects of the parasite's amino acid metabolism is crucial for energy and carbon sources and is involved in the cellular response to some stresses. For example, glutamate metabolism is involved in the resistance to heat, nutritional and oxidative stress, as evidenced by the effects of glutamate analogues that inhibit the corresponding metabolism (Magdaleno *et al.*, 2011). Glutamate also participates in the osmotic stress response (Rohloff *et al.*, 2003). Arginine is involved in energy metabolism during nutritional and pH stress (Pereira *et al.*, 2003), while proline is important in the oxidative stress response (Magdaleno *et al.*, 2009).

Genotoxic stress

Genomic integrity is ensured by several genome maintenance mechanisms that correct DNA lesions caused by both environmental and endogenous metabolic agents (Hoeijmakers, 2001; Branzei and Foiani, 2010; Huertas, 2010). These mechanisms involve a series of pathways. The direct repair pathway directly reverses base modifications through demethylation processes and through the activity of photolyases or dioxigenases, and the base excision repair (BER) pathway removes oxidized, methylated or misincorporated bases. The nucleotide excision repair (NER) pathway excises bulky, damaged nucleotides, while the mismatch repair (MMR) pathway removes misincorporated bases in the newly replicated DNA strand. The double-strand break (DSB) repair pathway joins the two ends of a DSB either by homologous recombination or by non-homologous end-joining (NHEJ). Lastly, the translesion synthesis (TLS) pathway activates error-prone DNA polymerases that can accommodate template strands with damaged bases and are therefore able to resume stalled DNA replication (Waters *et al.*, 2009).

The DNA repair machinery of *T. cruzi* has recently been reviewed (Passos-Silva *et al.*, 2010), and differences between the DNA maintenance mechanisms of this parasite and those of other eukaryotes were highlighted. The parasite apparently lacks NHEJ. In addition, the unexpected mitochondrial localization of some typically nuclear proteins was observed, which is probably due to kDNA demands. The mitochondrial localization of these proteins may also be important during higher metabolic activity. A bona fide photolyase orthologue is lacking, which may be complemented by transcription-coupled repair (TCR). The additional role of some proteins in response to oxidative stress was emphasized, and proposed to be critical in the intracellular developmental stage when confronted by host ROS and RNS attacks. Also, TCR may have a greater role in DNA repair in this organism owing to constitutive transcription of most or all of its genome (Passos-Silva *et al.*, 2010). Finally, it has been suggested that the large intraspecific genetic variability of the parasite may be important for adaptation to different hosts, and was achieved by varying DNA repair efficiencies (Machado *et al.*, 2006). Here we focus on the DNA repair pathways that are part of the genotoxic stress response induced by environmental agents.

DNA damage signalling

In eukaryotes, poly(ADP-ribose) polymerase (PARP) is a nuclear enzyme involved in processes of DNA repair (D'Amours *et al.*, 1999). PARP catalyses the transfer of ADP-ribose moieties from NAD+ to glutamate, aspartate or carboxy-terminal lysine residues of target nuclear proteins such as histones, generating branched polymers of up to 200 units and establishing a molecular link between DNA damage and chromatin structure remodelling. PARP is one of the first nuclear factors to recognize lesions in DNA, and poly(ADP-ribose) synthesized at the site of a DNA strand break could dissociate histones from DNA, thus granting the repair machinery access to damaged DNA (D'Amours *et al.*, 1999). Characterization of *T. cruzi*'s poly(ADP-ribose) polymerase (TcPARP) showed that it accumulates in the nucleus after treatment with the DNA damaging agents hydrogen peroxide and beta-lapachone (Fernández Villamil *et al.*, 2008). PARP, therefore, appears to also play a role as a DNA break sensor in *T. cruzi*. TcPARP translocates to the nucleus after DNA damage caused by UV-C light or hydrogen peroxide treatment, and induces DNA repair when damage is low or triggers cell death pathways when it is high (Larrea *et al.*, 2011). Poly(ADP-ribose) metabolism is involved in at least two DNA strand break signalling pathways, and different DNA-damaging agents induce distinct patterns of PARPylated proteins (Larrea *et al.*, 2011).

Histone variants or histone modifications are also very early marks of DNA damage. Histones are post-transcriptionally modified during the S phase of the cell cycle and during DNA repair (Herceg and Wang, 2005; Vidanes *et al.*, 2005). In trypanosomatids, histones H4 and H2A are preferentially acetylated while histones H3 and H2B are preferentially methylated (Alsford and Horn, 2004; Nardelli *et al.*, 2009 and references therein). In *T. cruzi*, acetylation at positions K4, K10 and K14 of histone H4 has a distinct nuclear distribution (Nardelli *et al.*, 2009). In addition, acetylation of histone H4 is relevant to the genotoxic response of this parasite, since DSBs induced by gamma irradiation lead within 24–48 h to decreased levels of H4-K4Ac and increased levels of H4-K10Ac and H4-K14Ac. The authors speculate that these changes in acetylated H4 levels may be related to nuclear reorganization after DNA damage (Nardelli *et al.*, 2009).

DNA repair pathways

T. cruzi appears to possess all the essential components of the BER repair pathway. In 1999, it was described that an apurinic/apyrimidinic (AP) endonuclease of the parasite, which is responsible for the cleavage of the phosphodiester bond at the 5′ side of an apurinic-apyrimidinic (AP) site, can confer resistance to DNA-alkylating and -oxidizing agents when the corresponding gene is inserted in DNA repair-deficient *Escherichia coli* (Pérez *et al.*, 1999). Also, the long interspersed repetitive element L1Tc found in the parasite's genome encodes an endonuclease which, when the corresponding gene is overexpressed, protects against DNA damage caused by the anthracycline daunorubicin and by gamma irradiation (Olivares *et al.*, 2003). The uracil-DNA glycosylase enzyme, which promotes the excision of uracil bases from DNA, was characterized in 2001, although the effects of DNA damaging agents was not investigated (Fárez-Vidal *et al.*, 2001). Two DNA polymerase beta enzymes, which have a role in filling short gaps in the BER pathway, have been characterized in the parasite and shown to be localized in the kinetoplast, where they may participate in correcting DNA lesions by oxidative damage in the mitochondria (Lopes *et al.*, 2008). Finally, 8-oxoG-DNA glycosylase, which removes the 7,8-dihydro-8-oxoguanine that mimics thymine functionally and is formed by oxidative damage, has been characterized in the parasite. Overexpression of the corresponding gene protects against hydrogen peroxide treatment, suggesting that the enzyme participates in the response to oxidative damage of the DNA (Passos-Silva *et al.*, 2010).

TcMSH2, a component of the MMR pathway, was identified in *T. cruzi* (Augusto-Pinto *et al.*, 2001), and TcMSH2 gene polymorphisms were observed among different strains (Augusto-Pinto *et al.*, 2003). Differences in TcMH2 enzymatic activity among strains were also observed, and presumably lead to variations in MMR efficiency (Campos *et al.*, 2011). These differences correlate with cell viability and detected DNA damage in the presence of the genotoxic agents cisplatin, *N*-methyl-*N*′-nitro-*N*-nitrosoguanine and hydrogen peroxide (Campos *et al.*, 2011). The authors also showed that single allele knockout of TcMSH2 leads to increased sensitivity to oxidative DNA damage induced by hydrogen peroxide, suggesting a role in the oxidative and genotoxic stress responses (Campos *et al.*, 2011).

Investigation of *T. cruzi*'s ability to repair DSBs by homologous recombination led to the finding that the parasite is highly resistant to gamma radiation (Regis-da-Silva *et al.*, 2006). In fact, *T. cruzi* appears to be one of the most radiation-resistant eukaryotic organisms described. Characterization of the parasite's RAD51 gene, orthologue of the bacterial RecA, and important in repair by homologous recombination, showed that gamma irradiation increases 4-fold the corresponding mRNA levels. In addition, epimastigotes overexpressing RAD51 show increased resistance to gamma irradiation and to treatment with the DSB-inducing drug zeocin (Regis-da-Silva *et al.*, 2006). These results indicate that RAD51 participates in the parasite's genotoxic stress response.

Finally, components of the TLS pathway were also characterized in this organism. DNA polymerase kappa localizes to the mitochondria, not the nucleus like other eukaryotes, and is probably involved in replicating damaged kinetoplast DNA (Rajão *et al.*, 2009). The polymerase efficiently bypasses 8-oxoguanine lesions in *in vitro* polymerization assays. Also, when the corresponding gene is overexpressed, the parasite shows greater resistance to hydrogen peroxide treatment and to gamma irradiation (Rajão *et al.*, 2009). In contrast, DNA polymerase eta, also of the TLS pathway, is a nuclear protein, is not induced by genotoxic agents such as UV light, gamma irradiation and hydrogen peroxide, and does not improve parasite survival in the presence of these agents when the corresponding gene is overexpressed (de Moura *et al.*, 2009). Considering the differences between *T. cruzi*'s DNA repair pathways and those of other eukaryotes, more work is needed to increase our understanding of the parasite's genotoxic stress response.

Osmotic stress

In vertebrate cells, several regulatory mechanisms operate to maintain cellular osmolarity. These regulatory mechanisms respond to changes in extracellular osmolarity and to changes resulting from transport of osmotically active molecules across the cell membrane and from appearance or disappearance of such molecules owing to metabolism (Lang *et al.*, 1998a,b). After a reduction of extracellular osmolarity, an initial swelling of the cell is rapidly corrected by what is called a regulatory volume decrease (RVD). The RVD is accomplished by an efflux of inorganic ions such as Na^+ and K^+ and organic osmolytes such as glycerophosphorylcholine, sorbitol, inositol, betaine and amino acids that is believed to occur through a non-specific, volume-sensitive, organic osmolyte anion channel (Lang *et al.*, 1998a,b; Fürst *et al.*, 2002). In protozoan parasites, adaptation to hyposmotic stress involves, in addition to efflux of ions and osmolytes, water release by a contractile vacuole complex that is linked to the acidocalcisome, a reservoir of osmotically active molecules (recently reviewed by Rohloff and Docampo, 2008).

In *T. cruzi*, the need for osmoregulation in physiological conditions arises when the parasite is in the insect vector's gut, where extreme fluctuations in osmolarity occur. It also occurs during transition from the insect vector to the mammalian host, where osmolarity is reduced from 600–700 mOsm in insect excreta to 300 mOsm in the mammalian interstitial fluid (Rohloff and Docampo, 2008). A model for RVD in *T. cruzi* has been proposed (Rohloff and Docampo, 2008). Hyposmotic stress leads to swelling of the parasite, which results in adenylyl cyclase activation, cyclic AMP increase and protein kinase activation. Protein kinase activation leads to microtubule-mediated fusion of acidocalcisomes with the contractile vacuole complex, with the resulting translocation of aquaporin to the latter. Ammonium ions produced in the cytoplasm are sequestered by and accumulate in the acidocalcisome, which activates the organelle's exopolyphosphatase. The resulting increased content of inorganic phosphate is released together with phosphate-chelated osmolytes, and the osmotic gradient sequesters water through aquaporins, which is then ejected from the cell through the flagellar pocket. The main components of the RVD and their relevance to the parasite's osmotic stress response are reviewed below.

The polyphosphate content within acidocalcisomes was shown to increase during differentiation from trypomastigotes to amastigotes and during the lag growth phase of epimastigotes (Ruiz *et al.*, 2001). In addition, enzymatic activities of polyphosphate modulation, such as polyphosphate kinase and exopolyphosphatase, were detected in acidocalcisomes (Ruiz *et al.*, 2001). However, only the exopolyphosphatase enzyme was characterized biochemically (Fang *et al.*, 2007). The authors showed that overexpression of the exopolyphosphatase gene leads to reduced polyphosphate content, increased swelling and a delay in cell volume recovery, suggesting the participation of the enzyme in the osmotic stress response. It was also shown that polyphosphate content sharply decreases after hyposmotic or alkaline stress and increases after hyperosmotic stress, suggesting an important role for acidocalcisomes in the parasite's response to environmental stresses (Ruiz *et al.*, 2001). Similar to vertebrate cells, a RVD was detected after hyposmotic stress in all of *T. cruzi*'s developmental stages, which is partially dependent on a strong efflux of neutral and anionic amino acids (Rohloff *et al.*, 2003). The parasite's RVD is also dependent on an increase in cyclic AMP, on swelling of acidocalcisomes, and displacement of aquaporins from acidocalcisomes to the contractile vacuole complex with fusion of these organelles, resulting in water movement (Rohloff *et al.*, 2004). An alkalization of acidocalcisomes by accumulation of ammonium ions produced in the cytosol, with concomitant acidification of the cytosol, was detected during hyposmotic stress (Rohloff and Docampo, 2006). The alkalization is believed to induce the reduction of polyphosphate content by stimulating the organelle's exopolyphosphatase, which has a high optimum pH (Rohloff and Docampo, 2006). It has also been described that phosphatidylinositol 3-kinase (PI3K) participates in osmoregulation in this parasite (Schoijet *et al.*, 2008). The authors showed that overexpression of the PI3K gene leads to enlarged contractile vacuoles and resistance to PI3K inhibitors, while in wild-type cells these inhibitors reduce the RVD response to hyposmotic stress (Schoijet *et al.*, 2008). Additional work is required to characterize this putative signal transduction pathway for activation of the osmotic stress response.

In addition to the RVD described above, transport proteins have been characterized that may participate in the parasite's stress response. P-type ATPases are ion pumps that carry out fundamental processes ranging from the generation of membrane potential to removal of toxic ions from cells (reviewed by Kühlbrandt, 2004). The regulation of ion concentrations is particularly relevant in *T. cruzi*, which must both invade and escape the mammalian host cell. Amastigotes and bloodstream trypomastigotes must survive in the very different Na^+ and K^+ concentrations of the intracellular and extracellular environments, and therefore regulation of ionic concentrations by Na^+/K^+-ATPases is important for adaptation of the parasite. A Na^+/K^+ ENA-type ATPase gene was characterized in *T. cruzi* and its mRNA was shown to be present in all cell forms but more abundantly in trypomastigotes (Iizumi *et al.*, 2006). The authors also showed that overexpression of this protein leads to increased resistance to high Na+ stress, or salt tolerance (Iizumi *et al.*, 2006). Characterization of the proteins involved in the regulation of ion concentrations is just beginning, and further investigations must be carried out to improve our understanding of the parasite's ability to respond to changes in intracellular and extracellular osmolarity.

Heavy metal stress

Metal ions are essential cofactors for a wide range of biochemical reactions, but in certain conditions may be extremely toxic to cells. In eukaryotes, the presence of toxic metal ions typically induces metallothioneins (MTs), a family of low-molecular-weight proteins with high sulfur content and a very high metal ion-binding capacity (Blindauer and Leszczyszyn, 2010). MTs bind both physiological metal ions, such as zinc, copper and selenium, and toxic metal ions, such as cadmium, mercury, silver and arsenic, through the thiol group of its cysteine residues, which represent approximately 30% of its residues. This family of proteins shows functional diversity, and some members are involved in the cellular response to heavy metal stress and oxidative stress. In this role, MTs typically participate in the detoxification of drugs and in scavenging of electrophilic compounds and free radicals.

There have been few reports of *T. cruzi*'s response to heavy metal stress. MT-like proteins have been detected and characterized biochemically, and variations in content observed between strains (Maya *et al.*, 2004). MT levels increase after $ZnCl_2$ addition, showing that the parasite responds to heavy metal ions. In addition, the oxidized state of MTs is reduced by nifurtimox and benznidazole (Maya *et al.*, 2004), highlighting the relevance of MT function to Chagas' disease current drug treatment. It has also been described that 2 mM arsenite induces the accumulation in the nucleus of the RNA-binding proteins TcUBP1, TcUBP2, TcRBP6 and the translation factor TceIF4E.1 (Cassola and Frasch, 2009). Accumulation takes 3–4 hours and is dependent on the RNA-binding capacity of these proteins, which is also important for their nuclear export and cytoplasmic localization. It is unclear, however, if this accumulation is part of the stress response or an indirect consequence of the heavy metal's effects in the cell. Much additional work is necessary to elucidate the parasite's response to heavy metal stress.

Toxic stress

Cellular responses to toxic compounds and xenobiotics usually consist of removing the compound from the cell or neutralizing it by chemical modification. Removal of the compound from the cell is accomplished through translocation across the plasma membrane by members of ATP-binding cassette superfamily of transporters that include multidrug resistance (MDR) 1 P-glycoproteins (PGPs) and multidrug resistance-associated proteins (MRPs) (recently reviewed by Doshi *et al.*, 2011). The drug resistance phenotype is typically accompanied by local amplification of PGP genes. Chemical modification of toxic compounds involves addition of chemically reactive functional groups, such as hydroxylation, which leads to detoxification by increasing solubility and by allowing further metabolism. Hydroxylation of toxic compounds is typically performed by cytochrome P450s, which are restored to their reduced state by cytochrome P450 reductases (CPRs) (Guengerich, 1994). Both MDR-like proteins and CPRs have been described in parasitic protozoa (Berger and Fairlamb, 1993; Jones and George, 2005). However, the molecular mechanisms by which *T. cruzi* removes or modifies toxic compounds are largely unknown. The few relevant proteins characterized in this organism are reviewed below.

Two MDR-like genes, TcPGP1 and TcPGP2, have been characterized in *T. cruzi* (Dallagiovanna *et al.*, 1996; Torres *et al.*, 1999). TcPGP2 mRNA has been detected only in the dividing cell forms epimastigotes and amastigotes (Dallagiovanna *et al.*, 1996). Surprisingly, drug resistance in *T. cruzi* is not associated with amplification or increased expression of PGP genes, as revealed by analyses of strains and clones with natural or induced resistance to benznidazole and nifurtimox (Murta *et al.*, 2001). CPRs, on the other hand, may be relevant to drug resistance in this parasite. TcCPR genes A, B and C have been characterized, and the corresponding gene products shown to be present in all developmental stages (Portal *et al.*, 2008). Higher levels of TcCPR-A were observed in trypomastigotes. In addition, overexpression of TcCPR-B leads to increased resistance to benznidazole and nifurtimox (Portal *et al.*, 2008). Similar to other stress situations, additional studies are needed to elucidate the parasite's response to toxic compounds and xenobiotics.

Conclusions and future trends

T. cruzi has a complex life cycle, in which cell forms with different morphology and functional characteristics interact with an insect and a mammalian host and in the process undergo several environmental variations. To successfully cope with these challenges, the parasite has evolved varied and sophisticated stress responses. Some of these responses are conserved among eukaryotes or even all cells, such as the induction of HSPs during heat stress or the modulation of DNA repair enzymes during genotoxic or oxidative stress. Others are similar to that of other eukaryotes but possess some distinguishing features, such as the antioxidant defence network during oxidative stress or the formation of stress granules/P-bodies during nutritional stress. Still others contain components that are restricted to a group of organisms, such as the role of acidocalcisomes and contractile vacuoles during osmotic stress in protozoan parasites. *T. cruzi*'s stress response is essential for the parasite's viability and survival in both hosts and therefore relevant to Chagas' disease transmission and pathogenesis. The stress response is also relevant to treatment, since the parasite uses the same set of defences against drugs. In this context, more knowledge of *T. cruzi*'s different stress responses would potentially be beneficial to the control of the disease, in particular of those which so far have been poorly characterized such as the toxic stress response. An improved understanding of *T. cruzi*'s stress response would also be of great value in the context of the parasite's basic biology. Studying how an organism responds to an environmental condition led in many cases to a better knowledge of signal transduction pathways with its component proteins and small molecules. A great deal of work is needed to clarify *T. cruzi*'s signal transduction pathways, since our current knowledge is rudimentary. Also, it would be interesting to understand how the parasite monitors macromolecular damage or the loss of homeostasis relative to some environmental variable. Investigation of the regulation of the stress response traditionally led to a better understanding of gene regulatory mechanisms, and progress has been made in this respect in *T. cruzi*. However, there are large gaps in our understanding of these mechanisms. The role of regulatory RNAs in the modulation of the stress response in eukaryotic organisms has recently been emphasized (Leung and Sharp, 2010), and microRNAs have been found in other trypanosomatids (Batista and Marques, 2011). Although these molecules have not been found in *T. cruzi*, which lacks the necessary protein machinery for RNA interference, a ssRNA-mediated pathway has been proposed to be present in this organism (Batista and Marques, 2011). Given that sequence-specific RNA-binding proteins are surprisingly difficult to identify in trypanosomatids, it is possible that RNA molecules may participate in gene regulation in *T. cruzi*. Future studies may shed light on these and other intriguing issues.

Web resources

TriTrypDB: http://tritrypdb.org/
A Bioinformatics Resource Center containing Genomics information and tools for the TriTryps: *T. cruzi*, *T. brucei* and *L. major*.

TcSNP: http://snps.tcruzi.org/
A resource database with information on genetic variation (polymorphisms, mutations) for different stocks, strains and isolates of *T. cruzi*.

EuPathDB: http://eupathdb.org/eupathdb/

A Bioinformatics Resource Center for accessing genomic-scale datasets associated with eukaryotic pathogens (*Babesia, Cryptosporidium, Encephalitozoon, Entamoeba, Enterocytozoon, Giardia, Leishmania, Neospora, Plasmodium, Theileria, Toxoplasma, Trichomonas* and *Trypanosoma*).

ExPASy: http://us.expasy.org/

A Bioinformatics Resource Portal providing access to databases and software tools in areas such as proteomics, genomics, phylogeny, systems biology, population genetics, and transcriptomics.

HSP Portal: http://www.heatshock.net/

A HSP research portal containing tools to access the heat shock literature, a database of HSP genes and proteins, heat shock labs, and other related information.

CDC: http://www.cdc.gov/parasites/chagas/

The Centers for Disease Control and Prevention provides information on diagnosis, treatment and epidemiology of Chagas' disease as well as on the biology of *T. cruzi*.

References

Alcina, A., Urzainqui, A., and Carrasco, L. (1988). The heat-shock response in *Trypanosoma cruzi*. Eur. J. Biochem. *172*, 121–127.

Alonso, G.D., Pereira, C.A., Remedi, M.S., Paveto, M.C., Cochella, L., Ivaldi, M.S., Gerez de Burgos, N.M., Torres, H.N., and Flawiá, M.M. (2001). Arginine kinase of the flagellate protozoa *Trypanosoma cruzi*. Regulation of its expression and catalytic activity. FEBS Lett. *498*, 22–25.

Alsford, S., and Horn, D. (2004). Trypanosomatid histones. Mol. Microbiol. *53*, 365–372.

Alvarez, M.N., Piacenza, L., Irigoín, F., Peluffo, G., and Radi, R. (2004). Macrophage-derived peroxynitrite diffusion and toxicity to *Trypanosoma cruzi*. Arch. Biochem. Biophys. *432*, 222–232.

Alvarez, V.E., Kosec, G., Sant'Anna, C., Turk, V., Cazzulo, J.J., and Turk, B. (2008). Autophagy is involved in nutritional stress response and differentiation in *Trypanosoma cruzi*. J. Biol. Chem. *283*, 3454–3464.

Alves, L.R., Avila, A.R., Correa, A., Holetz, F.B., Mansur, F.C.B., Manque, P.A., de Menezes, J.P.B., Buck, G.A., Krieger, M.A., and Goldenberg, S. (2010). Proteomic analysis reveals the dynamic association of proteins with translated mRNAs in *Trypanosoma cruzi*. Gene *452*, 72–78.

Anderson, P., and Kedersha, N. (2009). Stress granules. Curr. Biol. *19*, R397–8.

Andrade, L.O., and Andrews, N.W. (2005). The *Trypanosoma cruzi*-host-cell interplay: location, invasion, retention. Nat. Rev. Microbiol. *3*, 819–823.

Arias, D.G., Cabeza, M.S., Erben, E.D., Carranza, P.G., Lujan, H.D., Téllez Iñón, M.T., Iglesias, A.A., and Guerrero, S.A. (2011). Functional characterization of methionine sulfoxide reductase A from *Trypanosoma* spp. Free Radic. Biol. Med. *50*, 37–46.

Asea, A. (2003). Chaperokine-induced signal transduction pathways. Exerc. Immunol. Rev. *9*, 25–33.

Atwood, J.A., Weatherly, D.B., Minning, T.A., Bundy, B., Cavola, C., Opperdoes, F.R., Orlando, R., and Tarleton, R.L. (2005). The *Trypanosoma cruzi* proteome. Science *309*, 473–476.

Augusto-Pinto, L., Bartholomeu, D.C., Teixeira, S.M., Pena, S.D., and Machado, C.R. (2001). Molecular cloning and characterization of the DNA mismatch repair gene class 2 from the *Trypanosoma cruzi*. Gene *272*, 323–333.

Augusto-Pinto, L., Teixeira, S.M.R., Pena, S.D.J., and Machado, C.R. (2003). Single-nucleotide polymorphisms of the *Trypanosoma cruzi* MSH2 gene support the existence of three phylogenetic lineages presenting differences in mismatch-repair efficiency. Genetics *164*, 117–126.

Avruch, J. (2007). MAP kinase pathways: the first twenty years. Biochim. Biophys. Acta *1773*, 1150–1160.

Bachega, J.F.R., Navarro, M.V.A.S., Bleicher, L., Bortoleto-Bugs, R.K., Dive, D., Hoffmann, P., Viscogliosi, E., and Garratt, R.C. (2009). Systematic structural studies of iron superoxide dismutases from human parasites and a statistical coupling analysis of metal binding specificity. Proteins *77*, 26–37.

Bao, Y., Weiss, L.M., Ma, Y.F., Lisanti, M.P., Tanowitz, H.B., Das, B.C., Zheng, R., and Huang, H. (2010). Molecular cloning and characterization of mitogen-activated protein kinase 2 in *Trypanosoma cruzi*. Cell Cycle *9*, 2888–2896.

Batista, T.M., and Marques, J.T. (2011). RNAi pathways in parasitic protists and worms. J. Proteomics *74*, 1673–1682.

Berger, B.J., and Fairlamb, A.H. (1993). Cytochrome P450 in trypanosomatids. Biochem. Pharmacol. *46*, 149–157.

Blindauer, C.A., and Leszczyszyn, O.I. (2010). Metallothioneins: unparalleled diversity in structures and functions for metal ion homeostasis and more. Nat. Prod. Rep. *27*, 720–741.

Boisvert, F.-M., van Koningsbruggen, S., Navascués, J., and Lamond, A.I. (2007). The multifunctional nucleolus. Nat. Rev. Mol. Cell. Biol. *8*, 574–585.

Bonaldo, M.C., Souto-Padrón, T., de Souza, W., and Goldenberg, S. (1988). Cell-substrate adhesion during *Trypanosoma cruzi* differentiation. J. Cell Biol. *106*, 1349–1358.

Branzei, D., and Foiani, M. (2010). Maintaining genome stability at the replication fork. Nat. Rev. Mol. Cell. Biol. *11*, 208–219.

Buscaglia, C.A., Campo, V.A., Frasch, A.C.C., and Di Noia, J.M. (2006). *Trypanosoma cruzi* surface mucins: host-dependent coat diversity. Nat. Rev. Microbiol. *4*, 229–236.

Campos, P.C., Silva, V.G., Furtado, C., Machado-Silva, A., daRocha, W.D., Peloso, E.F., Gadelha, F.R., Medeiros, M.H.G., Lana, G. de C., Chen, Y., *et al.* (2011). *Trypanosoma cruzi* MSH2: Functional analyses on different parasite strains provide evidences for a role on the oxidative stress response. Mol. Biochem. Parasitol. *176*, 8–16.

Carreira, M.A., Tibbetts, R.S., Olson, C.L., Schuster, C., Renz, M., Engman, D.M., and Goldenberg, S. (1998). TcDJ1, a putative mitochondrial DnaJ protein in *Trypanosoma cruzi*. FEMS Microbiol. Lett. *166*, 141–146.

de Carvalho, E.F., de Castro, F.T., Rondinelli, E., Soares, C.M., and Carvalho, J.F. (1990). HSP 70 gene expression in *Trypanosoma cruzi* is regulated at different levels. J. Cell. Physiol. *143*, 439–444.

de Carvalho, E.F., de Castro, F.T., Rondinelli, E., and Carvalho, J.F. (1994). Physiological aspects of *Trypanosoma cruzi* gene regulation during heat-shock. Biol. Res. *27*, 225–231.

Carvalho, J.F., de Carvalho, E.F., Rondinelli, E., Silva, R., and de Castro, F.T. (1987). Protein biosynthesis changes in *Trypanosoma cruzi* induced by supra-optimal temperature. Exp. Cell Res. *168*, 338–346.

Cassola, A., and Frasch, A.C. (2009). An RNA recognition motif mediates the nucleocytoplasmic transport of a trypanosome RNA-binding protein. J.Biol. Chem. *284*, 35015–35028.

Cassola, A., De Gaudenzi, J.G., and Frasch, A.C. (2007). Recruitment of mRNAs to cytoplasmic ribonucleoprotein granules in trypanosomes. Mol. Microbiol. *65*, 655–670.

Chagas, C. (1909). Nova tripanozomiaze humana: estudos sobre a morfolojia e o ciclo evolutivo do Schizotrypanum cruzi *n*. gen., sp., ajente etiolojico de nova entidade morbida do homem. Mem. Inst. Oswaldo Cruz *1*, 159–218.

Clayton, C., and Shapira, M. (2007). Post-transcriptional regulation of gene expression in trypanosomes and leishmanias. Mol. Biochem. Parasitol. *156*, 93–101.

Coller, J.M., Tucker, M., Sheth, U., Valencia-Sanchez, M.A., and Parker, R. (2001). The DEAD box helicase, Dhh1p, functions in mRNA decapping and interacts with both the decapping and deadenylase complexes. RNA *7*, 1717–1727.

Contreras, V.T., Salles, J.M., Thomas, N., Morel, C.M., and Goldenberg, S. (1985). In vitro differentiation of *Trypanosoma cruzi* under chemically defined conditions. Mol. Biochem. Parasitol. *16*, 315–327.

Coura, J.R., and Viñas, P.A. (2010). Chagas disease: a new worldwide challenge. Nature *465*, S6–S7.

Cuervo, P., Domont, G.B., and De Jesus, J.B. (2010). Proteomics of trypanosomatids of human medical importance. J. Proteomics *73*, 845–867.

D'Amours, D., Desnoyers, S., D'Silva, I., and Poirier, G.G. (1999). Poly(ADP-ribosyl)ation reactions in the regulation of nuclear functions. Biochem. J. *342*, 249–268.

Dallagiovanna, B., Gamarro, F., and Castanys, S. (1996). Molecular characterization of a P-glycoprotein-related tcpgp2 gene in *Trypanosoma cruzi*. Mol. Biochem. Parasitol. *75*, 145–157.

Díaz, M.L., Solari, A., and González, C.I. (2011). Differential expression of *Trypanosoma cruzi* I associated with clinical forms of Chagas disease: Overexpression of oxidative stress proteins in acute patient isolate. J. Proteomics *74*, 1673–1682.

Doshi, R., Gutmann, D.A.P., Khoo, Y.S.K., Fagg, L.A., and van Veen, H.W. (2011). The choreography of multidrug export. Biochem. Soc. Trans *39*, 807–811.

Doyle, S.M., and Wickner, S. (2009). Hsp104 and ClpB: protein disaggregating machines. Trends Biochem. Sci. *34*, 40–48.

Dragon, E.A., Sias, S.R., Kato, E.A., and Gabe, J.D. (1987). The genome of *Trypanosoma cruzi* contains a constitutively expressed, tandemly arranged multicopy gene homologous to a major heat shock protein. Mol. Cell. Biol. *7*, 1271–1275.

Edkins, A.L., Ludewig, M.H., and Blatch, G.L. (2004). A *Trypanosoma cruzi* heat shock protein 40 is able to stimulate the adenosine triphosphate hydrolysis activity of heat shock protein 70 and can substitute for a yeast heat shock protein 40. Int. J. Biochem. Cell Biol. *36*, 1585–1598.

El-Sayed, N.M., Myler, P.J., Bartholomeu, D.C., Nilsson, D., Aggarwal, G., Tran, A.-N., Ghedin, E., Worthey, E.A., Delcher, A.L., Blandin, G., *et al.* (2005). The genome sequence of *Trypanosoma cruzi*, etiologic agent of Chagas disease. Science *309*, 409–415.

Engman, D.M., Kirchhoff, L.V., and Donelson, J.E. (1989a). Molecular cloning of mtp70, a mitochondrial member of the hsp70 family. Mol. Cell. Biol. *9*, 5163–5168.

Engman, D.M., Sias, S.R., Gabe, J.D., Donelson, J.E., and Dragon, E.A. (1989b). Comparison of HSP70 genes from two strains of *Trypanosoma cruzi*. Mol. Biochem. Parasitol. *37*, 285–287.

Engman, D.M., Dragon, E.A., and Donelson, J.E. (1990). Human humoral immunity to hsp70 during *Trypanosoma cruzi* infection. J. Immunol. *144*, 3987–3991.

Engman, D.M., Fehr, S.C., and Donelson, J.E. (1992). Specific functional domains of mitochondrial hsp70s suggested by sequence comparison of the trypanosome and yeast proteins. Mol. Biochem. Parasitol. *51*, 153–155.

Engman, D.M., Henkle-Dührsen, K., Kirchhoff, L.V., and Donelson, J.E. (1995). *Trypanosoma cruzi*: accumulation of polycistronic hsp70 RNAs during severe heat shock. Exp. Parasitol. *80*, 575–577.

Fang, J., Ruiz, F.A., Docampo, M., Luo, S., Rodrigues, J.C.F., Motta, L.S., Rohloff, P., and Docampo, R. (2007). Overexpression of a Zn2+-sensitive soluble exopolyphosphatase from *Trypanosoma cruzi* depletes polyphosphate and affects osmoregulation. J. Biol. Chem. *282*, 32501–32510.

Fárez-Vidal, M.E., Gallego, C., Ruiz-Pérez, L.M., and González-Pacanowska, D. (2001). Characterization of uracil-DNA glycosylase activity from *Trypanosoma cruzi* and its stimulation by AP endonuclease. Nucleic Acids Res. *29*, 1549–1555.

Fernandes, M., Silva, R., Rössle, S.C., Bisch, P.M., Rondinelli, E., and Ürményi, T.P. (2005). Gene characterization and predicted protein structure of the mitochondrial chaperonin HSP10 of *Trypanosoma cruzi*. Gene *349*, 135–142.

Fernández Villamil, S.H., Baltanás, R., Alonso, G.D., Vilchez Larrea, S.C., Torres, H.N., and Flawiá, M.M. (2008). TcPARP: A DNA damage-dependent poly(ADP-ribose) polymerase from *Trypanosoma cruzi*. Int. J. Parasitol. *38*, 277–287.

Figueiredo, R.C., Rosa, D.S., and Soares, M.J. (2000). Differentiation of *Trypanosoma cruzi* epimastigotes: metacyclogenesis and adhesion to substrate are triggered by nutritional stress. J. Parasitol. *86*, 1213–1218.

Finzi, J.K., Chiavegatto, C.W.M., Corat, K.F., Lopez, J.A., Cabrera, O.G., Mielniczki-Pereira, A.A., Colli, W., Alves, M.J.M., and Gadelha, F.R. (2004). *Trypanosoma cruzi* response to the oxidative stress generated by hydrogen peroxide. Mol. Biochem. Parasitol. *133*, 37–43.

Flechas, I.D., Cuellar, A., Cucunubá, Z.M., Rosas, F., Velasco, V., Steindel, M., Thomas, M.D.C., López, M.C., González, J.M., and Puerta, C.J. (2009). Characterising the KMP-11 and HSP-70 recombinant antigens' humoral immune response profile in chagasic patients. BMC Infect. Dis. *9*, 186.

Folgueira, C., and Requena, J.M. (2007). A postgenomic view of the heat shock proteins in kinetoplastids. FEMS Microbiol. Rev. *31*, 359–377.

Franzén, O., Ochaya, S., Sherwood, E., Lewis, M.D., Llewellyn, M.S., Miles, M.A., and Andersson, B. (2011). Shotgun Sequencing Analysis of *Trypanosoma cruzi* I Sylvio X10/1 and Comparison with *T. cruzi* VI CL Brener. PLoS Negl. Trop. Dis. *5*, e984.

Fürst, J., Gschwentner, M., Ritter, M., Bottà, G., Jakab, M., Mayer, M., Garavaglia, L., Bazzini, C., Rodighiero, S., Meyer, G., *et al.* (2002). Molecular and functional aspects of anionic channels activated during regulatory volume decrease in mammalian cells. Pflugers Arch. *444*, 1–25.

Garcia, S., Rodrigues, V., Garcia, N., and Mello, M. (2011). Prevalence and intensity of infection, metacyclogenesis and nuclear phenotypes in Panstrongylus megistus (Burmeister, 1835) after ingestion of *Trypanosoma cruzi* (Chagas, 1909) II and subjection to heat shock. Braz. J. Biol. *71*, 491–500.

Garcia-Silva, M.R., Frugier, M., Tosar, J.P., Correa-Dominguez, A., Ronalte-Alves, L., Parodi-Talice, A., Rovira, C., Robello, C., Goldenberg, S., and Cayota, A. (2010). A population of tRNA-derived small RNAs is actively produced in *Trypanosoma cruzi* and recruited to specific cytoplasmic granules. Mol. Biochem. Parasitol. *171*, 64–73.

Giambiagi-de Marval, M., Gottesdiener, K., Rondinelli, E., and van der Ploeg, L.H. (1993a). Predicted amino acid sequence and genomic organization of *Trypanosoma cruzi* hsp 60 genes. Mol. Biochem. Parasitol. *58*, 25–31.

Giambiagi-de-Marval, M., Lees, R.A., Monteiro, B.A.G., Carvalho, J.F.O., Gottesdiener, K., Castro, F.T. de, and Rondinelli, E. (1993b). The heat-shock response in *Trypanosoma cruzi* and *Crithidia fasciculata*. Ciênc. Cult. (São Paulo) *45*, 216–222.

Giambiagi-deMarval, M., Souto-Padrón, T., and Rondinelli, E. (1996). Characterization and cellular distribution of heat-shock proteins HSP70 and HSP60 in *Trypanosoma cruzi*. Exp. Parasitol. *83*, 335–345.

Giese, V., Dallagiovanna, B., Marchini, F.K., Pavoni, D.P., Krieger, M.A., and Goldenberg, S. (2008). *Trypanosoma cruzi*: a stage-specific calpain-like protein is induced after various kinds of stress. Mem. Inst. Oswaldo Cruz *103*, 598–601.

Graefe, S.E.B., Wiesgigl, M., Gaworski, I., Macdonald, A., and Clos, J. (2002). Inhibition of HSP90 in *Trypanosoma cruzi* induces a stress response but no stage differentiation. Eukaryot. Cell *1*, 936–943.

Guengerich, F.P. (1994). Catalytic selectivity of human cytochrome P450 enzymes: relevance to drug metabolism and toxicity. Toxicol. Lett. *70*, 133–138.

Gupta, S., Wen, J.-J., and Garg, N.J. (2009). Oxidative stress in Chagas disease. Interdis. Perspect. Infect. Dis. *2009*, 190354.

Hartl, F.U., and Hayer-Hartl, M. (2002). Molecular chaperones in the cytosol: from nascent chain to folded protein. Science *295*, 1852–1858.

Herceg, Z., and Wang, Z.-Q. (2005). Rendez-vous at mitosis: TRRAPed in the chromatin. Cell Cycle *4*, 383–387.

Hernández, S.M., Sánchez, M.S., and de Tarlovsky, M.N.S. (2006). Polyamines as a defense mechanism against lipoperoxidation in *Trypanosoma cruzi*. Acta Trop. *98*, 94–102.

Hinnebusch, A.G. (2005). Translational regulation of GCN4 and the general amino acid control of yeast. Annu. Rev. Microbiol. *59*, 407–450.

Hoeijmakers, J.H. (2001). Genome maintenance mechanisms for preventing cancer. Nature *411*, 366–374.

Holetz, F.B., Correa, A., Ávila, A.R., Nakamura, C.V., Krieger, M.A., and Goldenberg, S. (2007). Evidence of P-body-like structures in *Trypanosoma cruzi*. Biochem. Biophys. Res. Commun. *356*, 1062–1067.

Holetz, F.B., Alves, L.R., Probst, C.M., Dallagiovanna, B., Marchini, F.K., Manque, P., Buck, G., Krieger, M.A., Correa, A., and Goldenberg, S. (2010). Protein and mRNA content of TcDHH1-containing mRNPs in *Trypanosoma cruzi*. FEBS J. *277*, 3415–3426.

Huertas, P. (2010). DNA resection in eukaryotes: deciding how to fix the break. Nat. Struct. Mol. Biol. *17*, 11–16.

Igoillo-Esteve, M., and Cazzulo, J.J. (2006). The glucose-6-phosphate dehydrogenase from *Trypanosoma cruzi*: its role in the defense of the parasite against oxidative stress. Mol. Biochem. Parasitol. *149*, 170–181.

Iizumi, K., Mikami, Y., Hashimoto, M., Nara, T., Hara, Y., and Aoki, T. (2006). Molecular cloning and characterization of ouabain-insensitive Na(+)-ATPase in the parasitic protist, *Trypanosoma cruzi*. Biochim. Biophys. Acta *1758*, 738–746.

Jones, P.M., and George, A.M. (2005). Multidrug resistance in parasites: ABC transporters, P-glycoproteins and molecular modelling. Int. J. Parasitol. *35*, 555–566.

Klein, K.G., Olson, C.L., Donelson, J.E., and Engman, D.M. (1995). Molecular comparison of the mitochondrial and cytoplasmic hsp70 of *Trypanosoma cruzi*, *Trypanosoma brucei* and Leishmania major. J. Eukaryot. Microbiol. *42*, 473–476.

Klionsky, D.J. (2005). The molecular machinery of autophagy: unanswered questions. J. Cell. Sci. *118*, 7–18.

Klionsky, D.J., Cregg, J.M., Dunn, W.A., Emr, S.D., Sakai, Y., Sandoval, I.V., Sibirny, A., Subramani, S., Thumm, M., Veenhuis, M., *et al.* (2003). A unified nomenclature for yeast autophagy-related genes. Dev. Cell *5*, 539–545.

Krautz, G.M., Peterson, J.D., Godsel, L.M., Krettli, A.U., and Engman, D.M. (1998). Human antibody responses to *Trypanosoma cruzi* 70-kD heat-shock proteins. Am. J. Trop. Med. Hyg. *58*, 137–143.

Kriehuber, T., Rattei, T., Weinmaier, T., Bepperling, A., Haslbeck, M., and Buchner, J. (2010). Independent evolution of the core domain and its flanking sequences in small heat shock proteins. FASEB J. *24*, 3633–3642.

Kühlbrandt, W. (2004). Biology, structure and mechanism of P-type ATPases. Nat. Rev. Mol. Cell. Biol. *5*, 282–295.

Kültz, D. (2005). Molecular and evolutionary basis of the cellular stress response. Annu. Rev. Physiol. *67*, 225–257.

Labriola, C.A., Giraldo, A.M.V., Parodi, A.J., and Caramelo, J.J. (2011). Functional cooperation between BiP and calreticulin in the folding maturation of a glycoprotein in *Trypanosoma cruzi*. Mol. Biochem. Parasitol. *175*, 112–117.

Lang, F., Busch, G.L., Ritter, M., Völkl, H., Waldegger, S., Gulbins, E., and Häussinger, D. (1998a). Functional significance of cell volume regulatory mechanisms. Physiol. Rev. *78*, 247–306.

Lang, F., Busch, G.L., and Völkl, H. (1998b). The diversity of volume regulatory mechanisms. Cell. Physiol. Biochem. *8*, 1–45.

Larrea, S.C.V., Alonso, G.D., Schlesinger, M., Torres, H.N., Flawiá, M.M., and Villamil, S.H.F. (2011). Poly(ADP-ribose) polymerase plays a differential role in DNA damage-response and cell death pathways in *Trypanosoma cruzi*. Int. J. Parasitol. *41*, 405–416.

Lee, S.R., and Collins, K. (2005). Starvation–induced cleavage of the tRNA anticodon loop in Tetrahymena thermophila. J. Biol. Chem. *280*, 42744–42749.

Leung, A.K.L., and Sharp, P.A. (2010). MicroRNA functions in stress responses. Mol. Cell *40*, 205–215.

Levy Yeyati, P., Bonnefoy, S., Mirkin, G., Debrabant, A., Lafon, S., Panebra, A., Gonzalez-Cappa, E., Dedet, J.P., Hontebeyrie-Joskowicz, M., and Levin, M.J. (1992). The 70-kDa heat-shock protein is a major antigenic determinant in human *Trypanosoma cruzi*/*Leishmania braziliensis* braziliensis mixed infection. Immunol. Lett. *31*, 27–33.

Li, Y., and Zhou, H. (2009). tRNAs as regulators in gene expression. Sci. China, C, Life Sci. *52*, 245–252.

Lopes, D. de O., Schamber-Reis, B.L.F., Regis-da-Silva, C.G., Rajão, M.A., DaRocha, W.D., Macedo, A.M., Franco, G.R., Nardelli, S.C., Schenkman, S., Hoffmann, J.-S., *et al.* (2008). Biochemical studies with DNA polymerase beta and DNA polymerase beta-PAK of *Trypanosoma cruzi* suggest the involvement of these proteins in mitochondrial DNA maintenance. DNA Repair (Amst.) *7*, 1882–1892.

Louw, C.A., Ludewig, M.H., Mayer, J., and Blatch, G.L. (2010). The Hsp70 chaperones of the Tritryps are characterized by unusual features and novel members. Parasitol. Int. *59*, 497–505.

Luquetti, A.O., Miles, M.A., Rassi, A., de Rezende, J.M., de Souza, A.A., Póvoa, M.M., and Rodrigues, I. (1986). *Trypanosoma cruzi*: zymodemes associated with acute and chronic Chagas' disease in central Brazil. Trans. R. Soc. Trop. Med. Hyg. *80*, 462–470.

Machado, C.R., Augusto-Pinto, L., McCulloch, R., and Teixeira, S.M.R. (2006). DNA metabolism and genetic diversity in Trypanosomes. Mutat. Res. *612*, 40–57.

Magdaleno, A., Ahn, I.-Y., Paes, L.S., and Silber, A.M. (2009). Actions of a proline analogue, L-thiazolidine-4-carboxylic acid (T4C), on *Trypanosoma cruzi*. PLoS ONE *4*, e4534.

Magdaleno, A., Suárez Mantilla, B., Rocha, S.C., Pral, E.M.F., and Silber, A.M. (2011). The Involvement of Glutamate Metabolism in the Resistance to Thermal, Nutritional, and Oxidative Stress in *Trypanosoma cruzi*. Enzyme Res. *2011*, 486928.

Mair, G., Shi, H., Li, H., Djikeng, A., Aviles, H.O., Bishop, J.R., Falcone, F.H., Gavrilescu, C., Montgomery, J.L., Santori, M.I., *et al.* (2000). A new twist in trypanosome RNA metabolism: cis-splicing of pre-mRNA. RNA *6*, 163–169.

Martin, F., Requena, J.M., Martin, J., Alonso, C., and Lopez, M.C. (1993). Cytoplasmic-nuclear translocation of the Hsp70 protein during environmental stress in *Trypanosoma cruzi*. Biochem. Biophys. Res. Commun. *196*, 1155–1162.

Martínez-Calvillo, S., Vizuet-de-Rueda, J.C., Florencio-Martínez, L.E., Manning-Cela, R.G., and Figueroa-Angulo, E.E. (2010). Gene expression in trypanosomatid parasites. J. Biomed. Biotechnol. *2010*, 525241.

de Marval, M.G., Souto-Padrón, T., Gottesdiener, K., Silva, R., van der Ploeg, L.H., and Rondinelli, E. (1993). Heat shock proteins in *Trypanosoma cruzi*: identification and localization of HSP70 and HSP60 proteins and structure of HSP60 genes (brief report). Biol. Res. *26*, 313–314.

Maugeri, D.A., and Cazzulo, J.J. (2004). The pentose phosphate pathway in *Trypanosoma cruzi*. FEMS Microbiol. Lett. *234*, 117–123.

Maya, J.D., Rodríguez, A., Pino, L., Pabón, A., Ferreira, J., Pavani, M., Repetto, Y., and Morello, A. (2004). Effects ofbuthionine sulfoximine nifurtimox and benznidazole upon trypanothione and metallothionein proteins in *Trypanosoma cruzi*. Biol. Res. *37*, 61–69.

Mielniczki-Pereira, A.A., Chiavegatto, C.M., López, J.A., Colli, W., Alves, M.J.M., and Gadelha, F.R. (2007). *Trypanosoma cruzi* strains, Tulahuen 2 and Y, besides the difference in resistance to oxidative stress, display differential glucose-6-phosphate and 6-phosphogluconate dehydrogenases activities. Acta Trop. *101*, 54–60.

Miles, M.A., Llewellyn, M.S., Lewis, M.D., Yeo, M., Baleela, R., Fitzpatrick, S., Gaunt, M.W., and Mauricio, I.L. (2009). The molecular epidemiology and phylogeography of *Trypanosoma cruzi* and parallel research on *Leishmania*: looking back and to the future. Parasitology *136*, 1509–1528.

Miranda, M.R., Canepa, G.E., Bouvier, L.A., and Pereira, C.A. (2006). *Trypanosoma cruzi*: Oxidative stress induces arginine kinase expression. Exp. Parasitol. *114*, 341–344.

de Moura, M.B., Schamber-Reis, B.L.F., Passos-Silva, D.G., Rajão, M.A., Macedo, A.M., Franco, G.R., Pena, S.D.J., Teixeira, S.M.R., and Machado, C.R. (2009). Cloning and characterization of DNA polymerase eta from *Trypanosoma cruzi*: roles for translesion bypass of oxidative damage. Environ. Mol. Mutagen. *50*, 375–386.

Murta, S.M., Santos, dos, W.G., Anacleto, C., NIRDE, P., Moreira, E.S., and Romanha, A.J. (2001). Drug resistance in *Trypanosoma cruzi* is not associated with amplification or overexpression of P-glycoprotein (PGP) genes. Mol. Biochem. Parasitol. *117*, 223–228.

Murta, S.M.F., Nogueira, F.B., Santos, Dos, P.F., Campos, F.M.F., Volpe, C., Liarte, D.B., Nirdé, P., Probst, C.M., Krieger, M.A., Goldenberg, S., *et al.* (2008). Differential gene expression in *Trypanosoma cruzi* populations susceptible and resistant to benznidazole. Acta Trop. *107*, 59–65.

Nadeau, K., Sullivan, M.A., Bradley, M., Engman, D.M., and Walsh, C.T. (1992). 83-kilodalton heat shock proteins of trypanosomes are potent peptide-stimulated ATPases. Protein Sci. *1*, 970–979.

Nardelli, S.C., da Cunha, J.P.C., Motta, M.C.M., and Schenkman, S. (2009). Distinct acetylation of *Trypanosoma cruzi* histone H4 during cell cycle, parasite differentiation, and after DNA damage. Chromosoma *118*, 487–499.

Názer, E., Verdún, R.E., and Sanchez, D.O. (2011). Nucleolar localization of RNA binding proteins induced by actinomycin D and heat Shock in *Trypanosoma cruzi*. PLoS ONE *6*, e19920.

Olivares, M., Lopez, M.C., García-Pérez, J.L., Briones, P., Pulgar, M., and Thomas, M.C. (2003). The endonuclease NL1Tc encoded by the LINE L1Tc from *Trypanosoma cruzi* protects parasites from daunorubicin DNA damage. Biochim. Biophys. Acta *1626*, 25–32.

Olson, C.L., Nadeau, K.C., Sullivan, M.A., Winquist, A.G., Donelson, J.E., Walsh, C.T., and Engman, D.M. (1994). Molecular and biochemical comparison of the 70-kDa heat shock proteins of *Trypanosoma cruzi*. J. Biol. Chem. *269*, 3868–3874.

Paba, J., Santana, J.M., Teixeira, A.R.L., Fontes, W., Sousa, M.V., and Ricart, C.A.O. (2004). Proteomic analysis of the human pathogen *Trypanosoma cruzi*. Proteomics *4*, 1052–1059.

Palmer, G., Louvion, J.F., Tibbetts, R.S., Engman, D.M., and Picard, D. (1995). *Trypanosoma cruzi* heat-shock protein 90 can functionally complement yeast. Mol. Biochem. Parasitol. *70*, 199–202.

Parodi-Talice, A., Monteiro-Goes, V., Arrambide, N., Avila, A.R., Duran, R., Correa, A., Dallagiovanna, B., Cayota, A., Krieger, M., Goldenberg, S., *et al.* (2007). Proteomic analysis of metacyclic trypomastigotes undergoing *Trypanosoma cruzi* metacyclogenesis. J. Mass Spectrom. *42*, 1422–1432.

Passos-Silva, D.G., Rajão, M.A., Nascimento de Aguiar, P.H., Vieira-da-Rocha, J.P., Machado, C.R., and Furtado, C. (2010). Overview of DNA Repair in *Trypanosoma cruzi*, *Trypanosoma brucei*, and *Leishmania major*. J. Nucleic Acids *2010*, 1–14.

Pereira, C.A., Alonso, G.D., Ivaldi, S., Silber, A.M., Alves, M.J.M., Torres, H.N., and Flawiá, M.M. (2003). Arginine kinase overexpression improves *Trypanosoma cruzi* survival capability. FEBS Lett. *554*, 201–205.

Pérez, J., Gallego, C., Bernier-Villamor, V., Camacho, A., González-Pacanowska, D., and Ruiz-Pérez, L.M. (1999). Apurinic/apyrimidinic endonuclease genes from the trypanosomatidae leishmania major and *Trypanosoma cruzi* confer resistance to oxidizing agents in DNA repair-deficient *Escherichia coli*. Nucleic Acids Res. *27*, 771–777.

Pérez-Morales, D., Ostoa-Saloma, P., and Espinoza, B. (2009). *Trypanosoma cruzi* SHSP16: Characterization of an alpha-crystallin small heat shock protein. Exp. Parasitol. *123*, 182–189.

Piacenza, L., Irigoín, F., Alvarez, M.N., Peluffo, G., Taylor, M.C., Kelly, J.M., Wilkinson, S.R., and Radi, R. (2007). Mitochondrial superoxide radicals mediate programmed cell death in *Trypanosoma cruzi*: cytoprotective action of mitochondrial iron superoxide dismutase overexpression. Biochem. J. *403*, 323–334.

Piacenza, L., Peluffo, G., Alvarez, M.N., Kelly, J.M., Wilkinson, S.R., and Radi, R. (2008). Peroxiredoxins play a major role in protecting *Trypanosoma cruzi* against macrophage- and endogenously derived peroxynitrite. Biochem. J. *410*, 359–368.

Piacenza, L., Zago, M.P., Peluffo, G., Alvarez, M.N., Basombrio, M.A., and Radi, R. (2009a). Enzymes of the antioxidant network as novel determiners of *Trypanosoma cruzi* virulence. Int. J. Parasitol. *39*, 1455–1464.

Piacenza, L., Alvarez, M.N., Peluffo, G., and Radi, R. (2009b). Fighting the oxidative assault: the *Trypanosoma cruzi* journey to infection. Curr. Opin. Microbiol. *12*, 415–421.

Piñeyro, M.D., Parodi-Talice, A., Arcari, T., and Robello, C. (2008). Peroxiredoxins from *Trypanosoma cruzi*: virulence factors and drug targets for treatment of Chagas disease? Gene *408*, 45–50.

Planelles, L., Thomas, M., Pulgar, M., Marañón, C., Grabbe, S., and Lopez, M.C. (2002). *Trypanosoma cruzi* heat-shock protein-70 kDa, alone or fused to the parasite KMP11 antigen, induces functional maturation of murine dendritic cells. Immunol. Cell Biol. *80*, 241–247.

Portal, P., Villamil, S.F., Alonso, G.D., De Vas, M.G., Flawiá, M.M., Torres, H.N., and Paveto, C. (2008). Multiple NADPH-cytochrome P450 reductases from *Trypanosoma cruzi* suggested role on drug resistance. Mol. Biochem. Parasitol. *160*, 42–51.

Rajão, M.A., Passos-Silva, D.G., DaRocha, W.D., Franco, G.R., Macedo, A.M., Pena, S.D.J., Teixeira, S.M., and Machado, C.R. (2009). DNA polymerase kappa from *Trypanosoma cruzi* localizes to the mitochondria, bypasses 8-oxoguanine lesions and performs DNA synthesis in a recombination intermediate. Mol. Microbiol. *71*, 185–197.

Rassi, A., Rassi, A., and Marin-Neto, J.A. (2010). Chagas disease. Lancet *375*, 1388–1402.

Regis-da-Silva, C.G., Freitas, J.M., Passos-Silva, D.G., Furtado, C., Augusto-Pinto, L., Pereira, M.T., DaRocha, W.D., Franco, G.R., Macedo, A.M., Hoffmann, J.-S., *et al.* (2006). Characterization of the *Trypanosoma cruzi* Rad51 gene and its role in recombination events associated with the parasite resistance to ionizing radiation. Mol. Biochem. Parasitol. *149*, 191–200.

Requena, J.M., Lopez, M.C., Jimenez-Ruiz, A., la Torre, de, J.C., and Alonso, C. (1988). A head-to-tail tandem organization of hsp70 genes in *Trypanosoma cruzi*. Nucleic Acids Res. *16*, 1393–1406.

Requena, J.M., Lopez, M.C., Jimenez-Ruiz, A., Morales, G., and Alonso, C. (1989). Complete nucleotide sequence of the hsp70 gene of T.cruzi. Nucleic Acids Res. *17*, 797–797.

Requena, J.M., Jimenez-Ruiz, A., Soto, M., Assiego, R., Santarén, J.F., Lopez, M.C., Patarroyo, M.E., and Alonso, C. (1992). Regulation of hsp70 expression in *Trypanosoma cruzi* by temperature and growth phase. Mol. Biochem. Parasitol. *53*, 201–211.

Respuela, P., Ferella, M., Rada-Iglesias, A., and Åslund, L. (2008). Histone acetylation and methylation at sites initiating divergent polycistronic transcription in *Trypanosoma cruzi*. J. Biol. Chem. *283*, 15884–15892.

Richter, K., Haslbeck, M., and Buchner, J. (2010). The heat shock response: life on the verge of death. Mol. Cell *40*, 253–266.

Rodrigues, D.C., Silva, R., Rondinelli, E., and Ürményi, T.P. (2010). *Trypanosoma cruzi*: modulation of HSP70 mRNA stability by untranslated regions during heat shock. Exp. Parasitol. *126*, 245–253.

Rohloff, P., and Docampo, R. (2006). Ammonium production during hypo-osmotic stress leads to alkalinization of acidocalcisomes and cytosolic acidification in *Trypanosoma cruzi*. Mol. Biochem. Parasitol. *150*, 249–255.

Rohloff, P., and Docampo, R. (2008). A contractile vacuole complex is involved in osmoregulation in *Trypanosoma cruzi*. Exp. Parasitol. *118*, 17–24.

Rohloff, P., Rodrigues, C.O., and Docampo, R. (2003). Regulatory volume decrease in *Trypanosoma cruzi* involves amino acid efflux and changes in intracellular calcium. Mol. Biochem. Parasitol. *126*, 219–230.

Rohloff, P., Montalvetti, A., and Docampo, R. (2004). Acidocalcisomes and the contractile vacuole complex are involved in osmoregulation in *Trypanosoma cruzi*. J. Biol. Chem. *279*, 52270–52281.

Rondinelli, E. (1994). Conservation of heat-shock proteins in *Trypanosoma cruzi*. Parasitol. Today (Regul. Ed.) *10*, 172–176.

Rondinelli, E., Silva, R., Carvalho, J.F., de Almeida Soares, C.M., de Carvalho, E.F., and de Castro, F.T. (1988). *Trypanosoma cruzi*: an *in vitro* cycle of cell differentiation in axenic culture. Exp. Parasitol. *66*, 197–204.

Ruiz, F.A., Rodrigues, C.O., and Docampo, R. (2001). Rapid changes in polyphosphate content within acidocalcisomes in response to cell growth, differentiation, and environmental stress in *Trypanosoma cruzi*. J. Biol. Chem. *276*, 26114–26121.

Salmon, D., Montero-Lomeli, M., and Goldenberg, S. (2001). A DnaJ-like protein homologous to the yeast co-chaperone Sis1 (TcJ6p) is involved in initiation of translation in *Trypanosoma cruzi*. J. Biol. Chem. *276*, 43970–43979.

Schmidt, J.C., Soares, M.J., Goldenberg, S., Pavoni, D.P., and Krieger, M.A. (2011). Characterization of TcSTI-1, a homologue of stress-induced protein-1, in *Trypanosoma cruzi*. Mem. Inst. Oswaldo Cruz *106*, 70–77.

Schoijet, A.C., Miranda, K., Girard-Dias, W., de Souza, W., Flawiá, M.M., Torres, H.N., Docampo, R., and Alonso, G.D. (2008). A *Trypanosoma cruzi* phosphatidylinositol 3-kinase (TcVps34) is involved in osmoregulation and receptor-mediated endocytosis. J. Biol. Chem. *283*, 31541–31550.

Schwarcz de Tarlovsky, M.N., Hernandez, S.M., Bedoya, A.M., Lammel, E.M., and Isola, E.L. (1993). Polyamines in *Trypanosoma cruzi*. Biochem. Mol. Biol. Int. *30*, 547–558.

Sheth, U., and Parker, R. (2003). Decapping and decay of messenger RNA occur in cytoplasmic processing bodies. Science *300*, 805–808.

Shonhai, A., Maier, A.G., Przyborski, J.M., and Blatch, G.L. (2011). Intracellular protozoan parasites of humans: the role of molecular chaperones in development and pathogenesis. Protein Pept. Lett. *18*, 143–157.

Simpson, A.G.B., Stevens, J.R., and Lukes, J. (2006). The evolution and diversity of kinetoplastid flagellates. Trends Parasitol. *22*, 168–174.

Sorimachi, H., Hata, S., and Ono, Y. (2011). Impact of genetic insights into calpain biology. J. Biochem. *150*, 23–37.

de Souza, W. (2002). Basic cell biology of *Trypanosoma cruzi*. Curr. Pharm. Des. *8*, 269–285.

de Souza, W. (2009). Structural organization of *Trypanosoma cruzi*. Mem. Inst. Oswaldo Cruz *104*, 89–100.

Stuart, K.D., Schnaufer, A., Ernst, N.L., and Panigrahi, A.K. (2005). Complex management: RNA editing in trypanosomes. Trends Biochem. Sci. *30*, 97–105.

Sturm, N.R., and Campbell, D.A. (2010). Alternative lifestyles: the population structure of *Trypanosoma cruzi*. Acta Trop. *115*, 35–43.

Sullivan, M.A., Olson, C.L., Winquist, A.G., and Engman, D.M. (1994). Expression and localization of

Trypanosoma cruzi hsp60. Mol. Biochem. Parasitol. *68*, 197–208.

Teixeira, D., Sheth, U., Valencia-Sanchez, M.A., Brengues, M., and Parker, R. (2005). Processing bodies require RNA for assembly and contain nontranslating mRNAs. RNA *11*, 371–382.

Tibbetts, R.S., Kim, I.Y., Olson, C.L., Barthel, L.M., Sullivan, M.A., Winquist, A.G., Miller, S.D., and Engman, D.M. (1994). Molecular cloning and characterization of the 78-kilodalton glucose-regulated protein of *Trypanosoma cruzi*. Infect. Immun. *62*, 2499–2507.

Tibbetts, R.S., Jensen, J.L., Olson, C.L., Wang, F.D., and Engman, D.M. (1998). The DnaJ family of protein chaperones in *Trypanosoma cruzi*. Mol. Biochem. Parasitol. *91*, 319–326.

Torres, C., Barreiro, L., Dallagiovanna, B., Gamarro, F., and Castanys, S. (1999). Characterization of a new ATP-binding cassette transporter in *Trypanosoma cruzi* associated to a L1Tc retrotransposon. Biochim. Biophys. Acta *1489*, 428–432.

Trujillo, M., Budde, H., Piñeyro, M.D., Stehr, M., Robello, C., Flohé, L., and Radi, R. (2004). *Trypanosoma brucei* and *Trypanosoma cruzi* tryparedoxin peroxidases catalytically detoxify peroxynitrite via oxidation of fast reacting thiols. J. Biol. Chem. *279*, 34175–34182.

Turrens, J.F. (2004). Oxidative stress and antioxidant defenses: a target for the treatment of diseases caused by parasitic protozoa. Mol. Aspects Med. *25*, 211–220.

Urbina, J.A. (2010). Specific chemotherapy of Chagas disease: relevance, current limitations and new approaches. Acta Trop. *115*, 55–68.

Vidanes, G.M., Bonilla, C.Y., and Toczyski, D.P. (2005). Complicated tails: histone modifications and the DNA damage response. Cell *121*, 973–976.

Vonlaufen, N., Kanzok, S.M., Wek, R.C., and Sullivan, W.J. (2008). Stress response pathways in protozoan parasites. Cell. Microbiol. *10*, 2387–2399.

Waters, L.S., Minesinger, B.K., Wiltrout, M.E., D'Souza, S., Woodruff, R.V., and Walker, G.C. (2009). Eukaryotic translesion polymerases and their roles and regulation in DNA damage tolerance. Microbiol. Mol. Biol. Rev. *73*, 134–154.

Weatherly, D.B., Boehlke, C., and Tarleton, R.L. (2009). Chromosome level assembly of the hybrid *Trypanosoma cruzi* genome. BMC Genom. *10*, 255.

Wilkinson, S.R., and Kelly, J.M. (2009). Trypanocidal drugs: mechanisms, resistance and new targets. Exp. Rev. Mol. Med. *11*, e31.

Young, J.C., Agashe, V.R., Siegers, K., and Hartl, F.U. (2004). Pathways of chaperone-mediated protein folding in the cytosol. Nat. Rev. Mol. Cell. Biol. *5*, 781–791.

Zingales, B., Andrade, S.G., Briones, M.R.S., Campbell, D.A., Chiari, E., Fernandes, O., Guhl, F., Lages-Silva, E., Macedo, A.M., Machado, C.R., *et al.* (2009). A new consensus for *Trypanosoma cruzi* intraspecific nomenclature: second revision meeting recommends TcI to TcVI. Mem. Inst. Oswaldo Cruz *104*, 1051–1054.

Stress Response in the Infective Stage of *Trypanosoma brucei* 16

Marcelo A. Comini, Andrea Medeiros and Bruno Manta

Abstract

African trypanosomes (*Trypanosoma brucei* spp.) are unicellular eukaryotic organisms that undergo a complex life cycle shuttling between an invertebrate (vector) and a mammalian host. The parasites have evolved sophisticated and efficient mechanisms to cope with, and adapt to, different environmental conditions. Distinct physical (temperature, pH, osmotic pressure) and biological (endo- and exo-biotic molecules, antibodies, proteases, etc.) stimuli acting individually or in a concerted manner induce an adaptive response in the parasite. Depending on the nature and extent of the stress, the cellular response can be transient or long-term and associated with minor or major morphological and metabolic changes. In this chapter we compile the most significant molecular and biological aspects related to the mechanisms and components of the stress response of *T. brucei* to adapt and survive in the bloodstream of mammals.

Introduction

Trypanosomatids comprise a large group of flagellated unicellular protists with free-living and parasitic life styles. Several members of this family are widely known for being vertebrate parasites of biomedical and veterinary importance. They belong to the order Kinetoplastida, which together with the groups Diplonemida (deep-see organisms) and Euglenida (photosynthetizing organisms) form the phylum Euglenozoa (Simpson *et al.*, 2006). Kinetoplastids represent one of the most ancient eukaryotic lineages that diverged after the acquisition of the mitochondrion and share unique features such as the nuclear and mitochondrial gene expression mechanism, the energy and thiol redox metabolism and some cellular structures (i.e. kinetoplast: mitochondrial DNA, glycosome: peroxisome-like organelle). Phylogenetic considerations revealed the monophyletic ancestry of trypanosomes, which can be traced back to the bodonids, a group of biflagellate kinetoplasts that comprise free-living and parasitic (fish and snails) species (Simpson *et al.*, 2002). The most prevailing theory about the evolution of haemoflagellate trypanosomes indicates that these organisms evolved from a free-living bodonid that later became a gut parasite of blood-sucking insects, which further transmitted the trypanosomatid predecessor to a vertebrate host where adaptation and parasitism finally took place (Simpson *et al.*, 2002; Hamilton *et al.*, 2004). Further evolution of pathogenic trypanosomes appears to have developed independently. This is supported by an early split of the genus *Leishmania* from the trypanosome taxon and the identification of two well-defined clades within the genus *Trypanosoma*: the '*brucei*' clade and the '*cruzi*' clade. The *brucei* clade includes trypanosomes with an origin confined to Africa (Old World Trypanosomes), whereas the *cruzi* clade contains species that mostly evolved in the New World continents (South America and Oceania). In contrast to members from the *cruzi* clade, several lines of evidence indicate that African trypanosomes co-evolved with their mammalian hosts and insect vectors (Maslov *et al.*, 1996; Stevens *et al.*, 1998, 2001; Hamilton *et al.*, 2007). The natural infectious cycle for this group of parasites is initiated when infective trypanosomes are mechanically (bites of tse-tse fly or bats) or sexually (e.g. *T. equiperdum*) transmitted to or between mammals, respectively. In the tse-tse fly, procyclic parasites migrate from the gut to the salivarian glands where final differentiation to the infective and non-dividing metacyclic trypomastigote is achieved (Fig. 16.1). Hundreds to thousands metacyclic trypomastigotes are inoculated with the insect saliva into the host dermis or superficial blood vessels during a blood meal. The parasites initially replicate in the interstitial space at the site of infection and then migrate and invade the haematolymphatic system. In the vascular system, trypomastigotes readily transform into

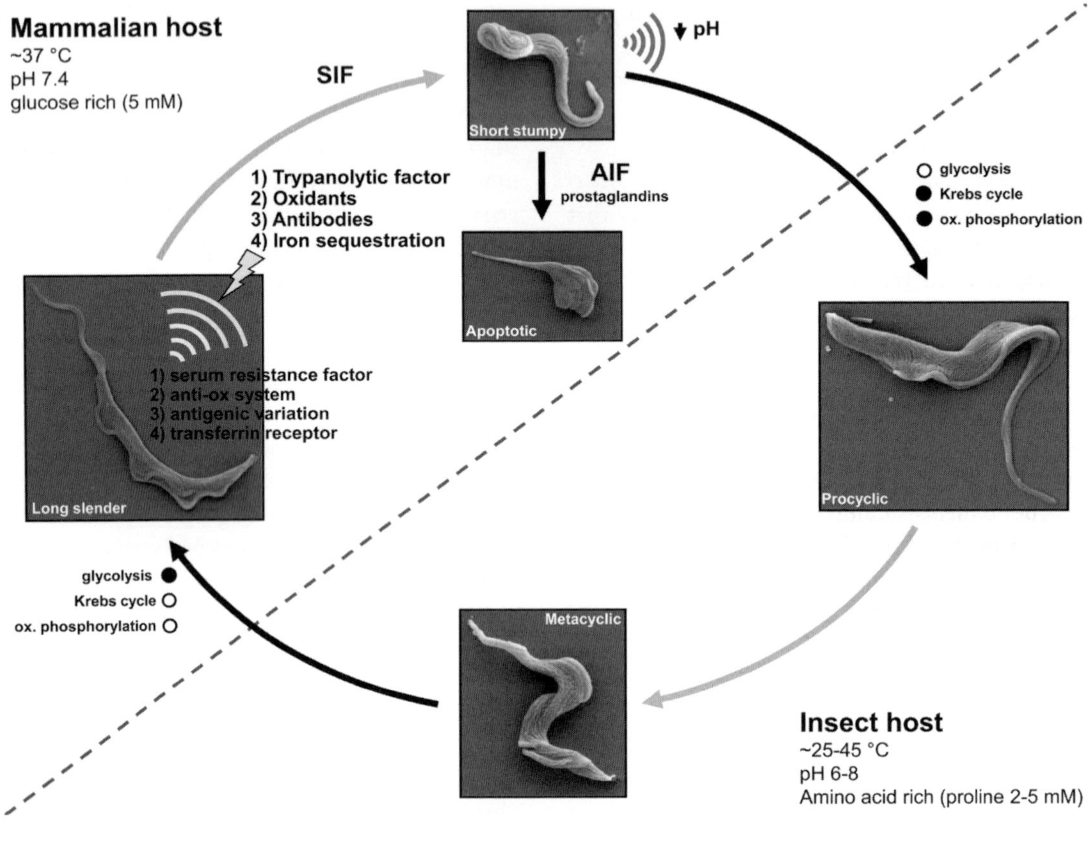

Figure 16.1 Life cycle of African trypanosomes and stress factors. During its life cycle, a trypanosome shuttles between an invertebrate host (tse-tse fly) and a mammalian host (i.e. humans, animals). In each host, the parasite develops from proliferative (long-slender and procyclic) to quiescent forms (short-stumpy and metacyclic; grey curved arrows) and vice versa upon transmission to a new host (black curved arrows). Differentiation to the non-dividing stages endows the cells with biochemical and morphological features that ensures a rapid and efficient adaptation to the different physiological environments from each host. For instance, changes in the energy metabolism (glycolysis vs. oxidative phosphorylation) are accompanied by ultrastructural remodelling of organelles (glycosomes vs. mitochondrion). Black and white dots denote active and inactive metabolic pathways, respectively. In the mammal bloodstream, *T. brucei* is confronted with components of the innate and acquired host immune response (lightning bolt) and has accordingly developed unique systems and mechanisms to counteract them (antigenic variation, antioxidant system, serum resistance protein, high affinity transferrin receptor, etc.). Transformation to the short-stumpy form is a cell density-dependent phenomena triggered by a yet unknown soluble factor (stumpy inducing factor, SIF). This stage displays resistant against proteolytic and pH stress, and a subset of it may undergo programmed cell death, probably to control parasite load inside the mammal host. Apoptosis may be triggered by the apoptosis inducing factor (AIF) and prostaglandins/oxidative stress. Surface electron microscopy images showing the representative life stages of *T. brucei brucei* (a kind gift of H. Lünsdorf, Helmholtz Centre for Infection Research, Braunschweig, Germany).

the replicative bloodstream form known as 'long slender' (hereon LS). Depending on the species, trypanosomes remain in the vascular system or colonize the interstitial space and different host fluids. LS parasites can undergo indefinite cycles of cell division in the bloodstream of the mammalian host but are not adapted to survive in the fly and must differentiate into the non-replicating 'short stumpy' (hereon SS) form to enable a perpetuation of the infection cycle. Finally, SS parasites ingested upon blood feeding will transform into the procyclic form in the insect gut to initiate a new transmission cycle (Fig. 16.1).

Different species and subspecies from the *brucei* clade are causative agents of highly disabling and often fatal diseases of humans and live stock. *T. b. rodhesiense* and *T. b. gambiense* cause human African trypanosomiasis (sleeping sickness), a fulminant disease that presents two clinical phases: an acute one, which affects the haematolymphatic system, and the chronic phase, which compromises the central nervous system (CNS). Thirty six million people are at risk of infection, and after continued control efforts the number of annual cases fell from 50,000–70,000 in 2005 to an estimated of 30,000 in 2009.[1] Trypanosomiasis caused by *T. b. rodhesiense* and *T. b. gambiense* differ in geographical distribution and disease development. The first is endemic of Eastern and Southern Africa, whereas the second is mostly confined to West and Central Africa. Rodhesian sleeping sickness develops rapidly as an acute disease that is characterized by a high parasite load, severe anaemia and thrombocytopenia accompanied by hypertrophy of the reticuloendothelial system and the liver. These parasites can traverse the blood–brain barrier, thereby inducing a semi-comatose state in a relatively short time after infection. In contrast, infection by *T. b. gambiense* presents a longer latency period with a slow, but still fatal, progress to a chronic state. There are no characteristic symptoms in the early phase of this disease (anaemia is rare and lymphoadenopathy is the major clinical manifestation) and invasion of the cerebrospinal fluid may occur three to 10 years after infection. The differential pathogenicity and clinical manifestations of both diseases have been formerly ascribed to mutual host–parasite adaptations that shaped pathogen virulence and host resistance. However, the hypothesis of a co-evolutionary virulence attenuation of *T. b. gambiense* has recently been challenged by studies on the phylogenetic relationships within the *T. brucei* taxon (Balmer *et al.*, 2011). The distinct virulence displayed by these two subspecies can be best explained by the expression of a single gene that confers resistance against host's innate immunity to *T. b. rodhesiense* (discussed in 'Stress response to innate and acquired host immunity'). Other gene products equally important for parasite virulence are currently being investigated (i.e. variable surface glycoprotein repertoire and proteases; discussed below). Domestic animals (e.g. cattle, sheep, pigs, goats, buffalo, camels and horses) are susceptible to infection by *T. brucei brucei*, *T. congolense* and *T. vivax* (Nagana disease), or *T. evansi* (Surra disease), all of which causes severe anaemia and/or immunosuppression. Both domestic and wild animals are parasite reservoirs from which trypanosomes are transmitted to humans. Apart from its impact on human nutrition, animal African trypanosomiasis results in total economic losses, in terms of agricultural Gross Domestic Product, of US$4.75 billion per year.[2] Members of the *brucei* clade, i.e. *T. evansi, T. vivax, T. equiperdum*, have also spread to Asia and Central and South America owing to cattle and horse importation and, thus, represent a serious threat even for the well-developed agricultural economies of these regions (Luckins and Dwinger, 2004).

As expected, the completion of a successful life cycle demands that *T. brucei* implements short- or long-lived adaptive responses to counteract the challenges arising from the different and sometimes hostile environments that endanger parasite viability. The evolved mechanisms of adaptation enabled one of the most ancient protozoan parasite lineages to still exist nowadays (Gadhela *et al.*, 2011). Some of the most relevant stress responses of African trypanosomes will be addressed in the next sections.

1 http://www.who.int/mediacentre/factsheets/fs259/en/

2 http://www.fao.org/ag/againfo/programmes/en/paat/disease.html

Control of gene expression in trypanosomes

As any other living organism, trypanosomes rely on the control of gene expression to adapt to stress, to differentiate, survive or die in response to environmental challenges. However, in contrast to most organisms, a major genetic feature of trypanosomes is the constitutive transcription of intronless genes as long polycistronic units, which are then converted into single mRNA via post-transcriptional processing: a single capped leader is added to the 5′ end of all pre-mRNAs (trans-splicing), while the 3′ end is polyadenylated (Vanhamme and Pays, 1995). To control the level of gene products, the parasite uses different post-transcriptional mechanisms that comprise degradation, stabilization and elongation of mRNA as well as initiation of translation (Clayton and Shapira, 2007). In fact, *T. brucei* has for long been recognized to have developed various strategies to cope with deleterious stresses (Muhich and Boothroyd, 1989). Although gene-specific expression has to be regulated individually, the fact that all mRNAs contain the same 5′ splice-leader sequence raises the possibility for a master regulation of transcription. The existence of such a mechanism has been recently demonstrated in *T. brucei*: down-regulation of a component of the protein translocation machinery triggered a marked depletion in the spliced-leader RNA with a subsequent rapid shut-off of RNA transcription and cell death (Lustig *et al.*, 2007). More importantly, a similar behaviour was induced by pH stress suggesting that this regulatory pathway may sense and respond to physiological stimuli (Lustig *et al.*, 2007; see 'Acidic, osmotic and proteolytic stress'). In recent years, epigenetic regulation (e.g. chromatin remodelling, nucleosome and histone structure modification) emerged as a novel phenomenon modulating the expression of certain subsets of genes that require high expression levels such as the surface glycoproteins (Figueiredo *et al.*, 2009; Rudenko *et al.*, 2010). The flanking 3′ untranslated sequences of mRNAs appear to play an important role in controlling the turnover of individual messengers through cis-acting signals that influence their processing. In this context, three independent studies revealed the existence of putative post-transcriptional regulons that contain both functionally unrelated and related genes whose expression is coordinatedly modulated in a stage-specific manner (Jenzen *et al.*, 2009; Kabani *et al.*, 2009; Queiroz *et al.*, 2009). Another study proposes that *T. brucei* has a limited transcriptional flexibility within each developmental stage, as evidenced by poor responsiveness to general stressors such as altered endocytosis, reductive stress without unfolded protein response, and lack of off-target phenomena upon genetic manipulation of gene expression, the response to iron deprivation being a marked exception (Koumandou *et al.*, 2008). These intrinsic transcriptional limitations have been attributed to the almost invariable conditions the trypanosomes face in each host, a fact that is particularly true for the bloodstream-stage (Koumandou *et al.*, 2008). This behaviour has important collateral implications: (i) the parasite relies exclusively on differentiation to acquire a new set of transcripts suitable for adaption; (ii) gene expression is overwhelmingly controlled at the post-transcriptional level, and accordingly (iii) changes in proteome reflect more accurately the response to environmental stimuli.

Temperature stress response

Temperature represents one of the major stresses *T. brucei* has to endure during development. In the poikilothermic arthropod the parasite must adapt to highly variable temperatures across the range 25–45°C (Edney and Barras, 1962). In the human bloodstream the temperature range is narrower and may vary between 37°C up to ~40°C during febrile episodes. While a shift in temperature is known to down-regulate genes associated with growth and to up-regulate certain transcripts such as heat shock proteins (HSP) believed to increase cell survival, the mechanisms involved in sensing and transducing the heat signal are barely understood. However, a recent study provided a clue how the overall rate of gene expression is regulated during heat shock in procyclic trypanosomes (Kramer *et al.*, 2008). Also molecular transducers of the heat signal have been reported. For instance, within the mitogen activated protein kinase (MAPK) family, MAPK4 has been shown to confer resistance to heat stress to procyclic parasites (Güttinger *et al.*, 2007), whereas ablation of the MKK1 gene impaired the growth at 39°C of pathogenic *T. brucei* (Jensen *et al.*, 2011).

The field of kinetoplastid's HSPs has recently been reviewed (Folgueira and Requena, 2007; Louw *et al.*, 2010). *T. brucei* genome survey identified several members of different HSP families, although only few of them have been biochemically or biologically characterized and will be briefly commented here. Members of the HSP60 family are chaperones that assist the mitochondrial import of proteins. In infective *T. brucei*, HSP60 localizes in the mitochondrion and close to the kinetoplast and its expression is significantly increased when LS parasites spontaneously differentiate to SS forms (Radawanska *et al.*, 2000a). Previously, Bringaud *et al.* (1995) reported a 2–4 folds higher HSP60 content in procyclic parasites than in the bloodstream stage, an expression profile directly linked to the distinct mitochondrial metabolic status (see the next section) and the chaperone function. Unexpectedly, HSP60 was identified as a highly immunogenic component of the flagellar pocket (Radwanska *et al.*, 2000b), the only site actively engaged in endo- and exocytic processes that involve the incorporation of extracellular molecules, the constant renewal of the glycoprotein surface or transferrin-receptor recycling (Mowatt and Clayton, 1987; Roditi *et al.*, 1987). As shown for the mammalian orthologues (Wells and Malkowsky, 2000), it is therefore tempting to speculate that HSP60 participates in endocytosis and intracellular trafficking probably by preserving protein conformation and supporting protein cargo to the cell surface. Moreover, a new role for *T. brucei* HSP60 in the generation of auto-antibodies and immunomodulation during infection has been proposed (Radawasnka *et al.*, 2000a).

Another class of highly conserved molecular chaperones comprises the HSP70, which are actively expressed in both life-stages of *T. brucei* (Jones *et al.*, 2006; Vertommen *et al.*, 2008). Thus far, proteomic approaches provided evidence for cytosolic and mitochondrial HSP70, HSP110 and the endoplasmic reticulum (ER) chaperones Grp78/BiP. The mRNA level of the cytosolic isoform of HSP70 was up-regulated in parasites cultured under conditions mimicking infection of the mammal host via a mechanism other than heat shock (van der Ploeg *et al.*, 1985). The 3′ UTR of HSP70 appears to dominate transcript stability against heat shock, stress that otherwise triggers massive RNA down-regulation (Lee, 1998). The role and partners of mitochondrial HSP70 await elucidation, since this organelle suffers important biochemical and structural changes during parasite life cycle. Members of the HSP40 family act as co-chaperones of HSP70 protein family. Infective *T. brucei* expresses an unusual large HSP40 that together with HSP70, but not by itself, supressed protein aggregation *in vitro* (Louw *et al.*, 2010). The biological relevance of HSP40 for *T. brucei* remains unknown. In agreement with the high demand for glycoprotein production (see section 6.1), BiP content was significantly increased in bloodstream trypanosomes compared to procyclic parasites (Bangs *et al.*, 1993). Besides its compartmentation in the endoplasmic reticulum, BiP was also detected in the junction complex between the body and the flagellum (*maculae adherentes*), pointing to a possible role in parasite motility (Bangs *et al.*, 1993). In contrast to the ER-stress response observed in other organisms, in *T. brucei* neither heat shock stress nor tunicamycin, an inhibitor of glycoprotein synthesis, induced an up-regulation of Grp78/BiP mRNA or protein levels (Izquierdo *et al.*, 2009). This and other experimental evidences led the authors to suggest that African trypanosomes possess an ER protein folding/quality control system that is constitutively active to allow a high-through-put synthesis and export of glycoproteins (Izquierdo *et al.*, 2009). HSP90 is a chaperone that participates in the functional maturation/activation of signalling proteins that control cell homeostasis, proliferation, differentiation, and cell death (Pearl and Prodromou, 2006). HSP90 appears pivotal for the maintenance of protein homeostasis in *T. brucei* as shown by genetic and chemical approaches (Jones *et al.*, 2008). Heat shock promotes HSP90 translocation from cytosol to nucleus along with a serine/threonine phosphatase protein (PP5), which may be involved in the protection of nuclear targets against temperature-dependent denaturation. Finally, HSP100 was reported to be strongly up-regulated upon heat shock stress in the procyclic stage (Güttinger *et al.*, 2007) but its role remains elusive.

Bloodstream trypanosomes are also subject to cold shock after ingested by the tse-tse fly. Engstler and Boshart (2004) show that a cold shock of $\Delta T > 15°C$ triggered the expression of procyclin, a membrane protein specific of the insect–stage, in bloodstream trypanosomes. The process is reversible and turns the stumpy forms, but not the replicative-stage, sensitive towards cis-aconitate/citrate-mediated differentiation to the procyclic stage. The authors proposed that differentiation commitment by lowering temperature involves three consecutive steps and components: sensing of temperature change, surface access of a putative receptor, and response to a chemical cue. The molecular details of this process are so far unknown.

Acidic, osmotic and proteolytic stress

From the physicochemical and biochemical point of view, the mammalian bloodstream represents a physiological 'paradise' for infective *T. brucei*. The blood provides a tightly buffered, relatively stable osmotic pressure and controlled environment when compared to the more adverse milieu of the tse-tse fly midgut (Wigglesworth, 1929; Cheeseman and Gooding, 1985). Therefore, proliferating bloodstream-stage trypanosomes are poorly concerned to confront with pH and proteolytic stresses and, hence, are extremely sensitive to mild acidic and trypsin treatment (Rolin *et al.*, 1996). However, it is during this bloodstream stage that major pre-adaptive changes occur to endow stumpy forms with the ability to survive the sudden change in extracellular environment during transmission to the insect. Nolan *et al.* (2000) demonstrated that the non-replicative forms of the parasite are more tolerant than LS forms to proteolytic and acidic challenges. Accordingly, several stress-induced signalling routes (i.e. adenylate cyclase and gycosylphosphoinositol-specific phospholypase C-mediated release of surface envelope) operating in the replicative stage were repressed in growth-arrested parasites. It was concluded that the intermediate stumpy stage possesses an efficient pH-regulating system that allows the parasite to maintain an intracellular pH around 7, similar to that detected in LS forms, even against large variations in the extracellular H^+ concentration (Nolan *et al.*, 2000).

Acidocalsisomes are small acidic organelles rich in calcium and polyphosphate that in trypanosomes are apparently involved in endo-lysosomal trafficking and osmoregulation via controlled polyphosphate release/uptake (Vercesi *et al.*, 1994; Docampo *et al.*, 2010). A recent study shows that interfering with the biogenesis of acidocalsisomes in bloodstream *T. brucei* leads to higher susceptibility towards osmotic stress and reduces virulence (Huang *et al.*, 2011). The basis for the impaired infectivity of the acidocalsisome-defective trypanosomes was in part ascribed to the lack of responsiveness to the drastic changes in extracellular osmolarity the parasites have to overcome when circulating through the renal medulla, where the osmolarity is estimated to be at least 3-fold higher than in the vascular system (1400 vs. 300 mOsm, respectively; Lang, 2007). Recently, a kinase from the TOR (target of rapamycin) family has been recognized as a key player in regulating the acidocalsisome and polyphosphate metabolism (de Jesus *et al.*, 2010). Interestingly, TOR signalling has also been implicated as a pivotal checkpoint for defining progression to autophagy depending on nutrient levels (Barquilla y Navarro, 2009; see 'Autophagy').

On the other hand, three aquaglyceroporins have been characterized in *T. brucei* (Uzcátegui *et al.*, 2004; Bassarak *et al.*, 2011). The proteins contribute to the transport of glycerol and other small uncharged molecules, and at least one of them (AQP2), of unknown subcellular localization, is critical for maintaining intracellular osmolarity (Bassarak *et al.*, 2011). It remains to be elucidated whether the expression and distribution of these transporters is affected by specific stimuli.

Nutritional stress response

As previously stated, the different developmental forms of the trypanosome are exposed to nutritionally contrasting environments that demand drastic changes in parasite's metabolism and morphology. Fine-tuning of energy metabolism, organelle reorganization and dedicated nutrient uptake are some of the mechanisms the parasite exploits to survive in the human host, as discussed below.

Energy metabolism and glycosomes

Long-slender trypanosomes dwell in a medium rich in glucose and other nutrients and have consequently reduced to a minimum their metabolic repertoire. Glucose catabolism is the major source of ATP (Seyfang and Duszenko, 1991), which takes place almost entirely in a specialized single-membrane organelle of peroxisomal origin but unique to Kinetoplastida, the glycosome (Hannaert *et al.*, 2003; Opperdoes, 2010). Rather than an increased metabolic efficiency, the selective advantage of glycolytic compartmentalization has been discussed as necessary to control the activity of hexokinase and phosphofructokinase that otherwise may lead to the accumulation of toxic hexose-phosphates (Kessler and Parsons, 2005; Michels *et al.*, 2006; Haanstra *et al.*, 2008). Glycosomes also house components of important metabolic pathways such as those involving gluconeogenesis, pentose phosphate shunt, purine salvage and pyrimidine biogenesis (Michels *et al.*, 2006). On the other hand, the mitochondrion of infective parasites lacks the typical cristae (inner fold membrane) and is metabolically repressed owing to the absence of the citric acid cycle and respiratory chain enzymes (Hannaert *et al.*, 2003). However, it still retains some essential functions such as the regeneration of NAD^+ to allow intra-glycosomal redox balance and constant glycolysis (reaction catalysed by a glygerol-3-phosphate oxidase, the plant-like trypanosomes alternative oxidase: TAO; Helfert *et al.*, 2001), the biogenesis of iron–sulfur clusters (Smíd *et al.*, 2006; Comini *et al.*, 2008) and the synthesis of a thymidine precursor (Roldán *et al.*, 2011). The opposite situation occurs in the procyclic parasite, where the mitochondrion is fully developed to accomplish amino acid (mainly proline) catabolism rather than complete glucose utilization, and glycosomal activity is heavily down-regulated (Besteiro *et al.*, 2005). In the lumen of tse-tse salivary glands, mature metacyclic trypomastigotes have acquired an unbranched mitochondrion and spherical glycosomes as its successor in the mammalian host (Vickerman, 1985). The signal and mechanism triggering these pre-adaptive changes in metacyclic *T. brucei* are unknown. Besides the global up-regulation of glycosomal enzymes occurring in the bloodstream stage compared to procyclic forms (Hart *et al.*, 1984; Parsons and Nielsen, 1990), the stable substrate concentration supply and the design for glucose metabolization point against any physiological need to respond to carbohydrate-based nutritional stress while being in the host. However, a recent finding opens the possibility for the existence of a differentiation signal in response to carbohydrate starvation that is not linked to energy production. *T. brucei* cannot incorporate galactose from the extracellular medium and depends on its synthesis from glucose in a reaction catalysed by UDP-glucose 4′ epimerase (*galE*), a glycosomal enzyme (Roper *et al.*, 2005). The sugar is then a building block of variant surface glycoproteins (VSGs, see below). Galactose deprivation using a *galE* conditional knockout mutant produced a significant increase in dihydrolipoamide dehydrogenase activity (typical marker of stumpy/intermediate and procyclic form; Tyler *et al.*, 1997), a stumpy-like morphology, signs of secretory/trafficking dysregulation and growth cessation (Urbaniak *et al.*, 2006). Although the phenotype may have developed as a consequence of metabolic starvation due to impairment of the endolysosomal system, the authors were tempted to speculate that galactose depletion could have triggered differentiation by the inability of the parasites to provide glycans for VSG modification. Further investigations are needed to determine whether certain type of physiological stresses may induce a shortage of galactose and to confirm the importance of its potential regulatory role in the parasite life cycle.

Autophagy

Autophagy is a degradative process that involves different cellular components and serves multiple purposes, such as (i) recycling of macromolecules during starvation, (ii) removal of damaged molecules and structures, (iii) metabolic remodelling for developmental differentiation or to withstand stress conditions. Hence, in contrast to programmed cell death (see 'Apoptosis'), autophagy is primarily meant to support cell survival under unfavourable conditions. However, if cellular homeostasis is not restored autophagy may result in cell death. In other eukaryotic organisms the autophagic process involves the participation of two protein complexes, the Atg complex that initiates autophagy and a second one containing the TORC1 kinase that acts as negative regulator (reviewed by Kiel, 2011). After the molecular events have begun, a double-layered membrane pre-autophagosomal structure, called phagophore, is formed. A series of cascade reactions involving lipids and proteins lead to the formation of a vesicle that fuses to the lysosome resulting in a single-membrane vacuole, the autophagosome. In contrast, microphagy does not require formation of the autophagosome and takes place by engulfment of the organelles by the lysosome. In both processes, permeases are in charge of transporting the products derived from hydrolytic digestion to the cytosol (Brennand *et al.*, 2011). The function and components of the machinery are starting to be unravelled for *T. brucei* (Herman *et al.*, 2006, 2008; Denninger *et al.*, 2008; Barquilla and Navarro, 2009; Koopmann *et al.*, 2009). Bearing in mind the structural and metabolic changes bloodstream trypanosomes must undergo to rapidly adapt to their hosts (Vickerman, 1985; Michels *et al.*, 2006), the glycosomes and mitochondrion are natural targets of autophagic processes. Following the identification of several members of the Atg complex in the different developmental phases of *T. brucei* (Denninger *et al.*, 2008), Herman *et al.* (2008) show very elegantly an stage-dependent turnover of glycosomes, where the degradation of 'old' organelles was synchronized to the biogenesis of new ones containing an enzyme repertoire adequate to the forthcoming environmental condition. Expectedly, 'microglycophagy' and the accompanying biochemical phenomena were accelerated during the transition from SS to procyclic form and involved post-transcriptional control of the protein effectors. Although starvation of procyclic trypanosomes by resuspension on a nutrient-free buffered solution rapidly induced glycophagy (Herman *et al.*, 2008), it remains elusive whether this phenomenon occurs under physiological nutrient depletion or stress. Thus far, three different molecules were reported to initiate autophagic-like processes in *T. brucei*: dihydroxyacetone (DHA; Uzcátegui *et al.*, 2007), the macrolide rapamycin (Denninger *et al.*, 2008) and neuropeptides (Campos-Salinas *et al.*, 2009; Delgado *et al.*, 2009). Among them, DHA appears as a physiological metabolite for which a regulatory role might be envisaged, since *T. brucei* contains a class I fructose-bisphosphate aldolase activity able to generate DHA and glyceraldehyde phosphate from fructose-1,6-diphosphate (Misset and Opperdoes, 1984).

As occurs in other eukaryotes, kinases from the TORC1 complex are apparently engaged as negative regulators of autophagy and prolonged survival of starved bloodstream-stage *T. brucei* (Barquilla y Navarro, 2009). Interestingly, TOR 1- and TOR 1-like knockdown induce quite distinct phenotypes characterized by autophagy and acidic vesicle enlargement, respectively. Regardless this difference, it is clear that TORC1-mediated signalling is a key checkpoint where a variety of physical and biochemical stimuli are integrated and cellular decisions are taken to counteract the perturbation.

Iron depletion

Iron is an essential element for all living organisms owing to its structural or functional role on several key metabolic enzymes. *T. brucei* confirms this rule and stands out by having developed exceptional mechanisms to guarantee metal supply from a host that un/intentionally will limit metal availability in response to infection. For example, iron deficiency (i.e. ferritin and lactoferrin up-regulation, and reduction of iron-transferrin saturation) is part of the innate immune response mounted during the acute phase of the infection to limit parasite proliferation (Chisi *et al.*, 2004). Also the chronic stage is accompanied by a notorious iron deprivation (e.g. anaemia) but, in this case, its origin is a chronic inflammatory disorder involving macrophage hyperactivation along with iron accumulation within the reticuloendothelial system (Stijelmans *et al.*, 2008).

In the bloodstream, trypanosomes acquire the metal via a high-affinity receptor-mediated endocytosis of iron-loaded transferrin (Grab *et al.*, 1992, 1993; Steverding *et al.*, 1995). The receptor (TfR) was identified as a heterodimeric complex of proteins encoded by the expression-site-associated-gene (ESAG) 6 and ESAG 7

(Salmon *et al.*, 1994; Steverding *et al.*, 1995). ESAG 6 and 7 are truncated forms of VSGs that lack the C-terminal domain, and, in the case of ESAG 7, the GPI anchor. Under normal growth conditions the TfR localizes at the flagellar pocket wherefrom bound-transferrin is internalized and proteolytically degraded in the endolysosome to release iron (Mussmann *et al.*, 2003). The relevance of the endocytic pathway as the solely source of iron for the parasite was demonstrated using antibodies anti-TfR that inhibited growth of *T. brucei in vitro* (Grab *et al.*, 1992). Expression of TfR is regulated by iron availability and post-transcriptional control that does not involve the iron responsive element/iron regulatory protein (IRE/IRP) system classical for eukaryote (Fast *et al.*, 1999). Iron starvation induced by iron chelators or species-specific transferrins leads to a 3- to 10-fold up-regulation in the expression of TfR with a concomitant redistribution of the receptor from the flagellar pocket to the entire parasite surface (Fast *et al.*, 1999; Mussmann *et al.*, 2004). This, together with the rapid recycling of TfR (Kabiri and Steverding, 2000) and gene-specific in/activation events (van Luenen *et al.*, 2004) allows trypanosomes to efficiently compete for limiting substrate and withstand iron deprivation until a new set of higher affinity TfR is expressed. For instance, sequence polymorphisms in ESAG 6 and 7 were proposed to determine the affinities of TfR for transferrins from different mammalian species (Bitter *et al.*, 1998) permitting for a rapid adaptation of the parasite to distinct hosts. It was also suggested that the TfR repertoire may allow the parasite to overcome anti-TfR antibody response by the host (Gerrit *et al.*, 2002). However, studies by Steverding (1998, 2003, 2006) disproved this hypothesis and supported the view that high affinity TfRs were evolutionary selected to enable the parasite to cope with the diversity of mammalian transferrins (Steverding, 2003, 2006). Monothiol glutaredoxins (1-C-Grxs) are ubiquitous iron–sulfur proteins involved in cellular iron homeostasis and iron–sulfur cluster biogenesis (Vilella *et al.*, 2004). Three different isoforms occur in infective *T. brucei*, which reach maximum intracellular content when culture enters in starvation phase, suggesting a role as iron or iron–sulfur stores until normal growth conditions (nutrient replenishment) are resumed (Comini *et al.*, 2008). Both iron-depletion and oxidative stress triggered a two-fold down- and up-regulation, respectively, of a mitochondrial isoform (1-C-Grx1), which was interpreted as a physiological response to suppress metal sequestration and promote cofactor biogenesis or repair, respectively. Strikingly, overexpression of the wild-type protein and an apo-mutant is detrimental for parasite survival in mice (Manta *et al.*, unpublished), suggesting an important function of 1-C-Grx1 in a tightly controlled iron-metabolism.

Stress response to innate and acquired host immunity

The extracellular lifestyle of African trypanosomes represents a major biological challenge for the parasite that will face the continuous attack of the mammalian host immune system. The host response to control and eliminate *T. brucei* from the bloodstream and tissue fluids does not differ from that employed against other infectious agents. The first line of defence involves non-specific mechanisms (innate immunity), such as humoral (e.g. acute phase proteins, cytokines) and cellular components (e.g. phagocytic cells). In a second stage, the adaptive immunity leads to the production and activation of more specific and persistent immune effectors (i.e. cytokines, phagocytic and regulatory cells, antibodies). The most peculiar strategies of *T. brucei* spp. to masterly circumvent the host's immune defence are summarized below.

Antigenic variation

African trypanosomes are widely known for their capacity to efficiently evade the humoral immune response of the host. The parasite body is coated with a dense layer of glycoproteins, the VSG,[3] which acts as a physical shield preventing components of the innate (i.e. complement system) or acquired (i.e. antibodies) immune system to access the pathogen plasma membrane and cause cell death. On the other hand, VSGs are highly

3 More than 15 million copies of VSG proteins cover the surface of the parasite and account for $2–$3 % of total protein from bloodstream form parasites (Vickerman, 1969; Cross, 1975).

immunogenic molecules that elicit the production of lytic antibodies via both T-cell-dependent and -independent B-cell responses. This makes VSGs a double-edged sword that forces the parasite to invest a huge amount of energy and genetic resources to accomplish a frequent replacement of VSG variants in order to escape immune recognition. The process is called antigenic variation and occurs stochastically at a frequency of 10^{-3} to 10^{-7} per cell and generation time depending on the strain (Pays *et al.*, 2004). Sequencing of the *T. brucei* genome revealed ~ 1500 VSG sequences representing almost 10% of the parasite genome but, strikingly, almost 93% of them are supposed to be pseudogenes (Berriman *et al.*, 2005). This finding revealed a new mechanism to maintain the diversity of the antigenic repertoire in *T. brucei* that depends on extensive recombinatory events to generate new arrays of chimeric molecules (Vanhamme and Pays, 1995; Pays *et al.*, 2004; Taylor and Rudenko, 2006). Hence, parasite survival inside the host is warranted by constant repopulation with cells carrying a new VSG variant. As a preadaptation to the future environment in the human host, VSG expression is initiated already in metacyclic trypanosomes in the salivary gland of the tse-tse fly. The 'metacyclic VSG' repertoire is indeed limited compared with that of the bloodstream form (Vanhamme and Pays, 1995). Nevertheless, as the metacyclic population is heterogeneous and each cell expresses a single VSG, the whole metacyclic VSG repertoire is already present in the trypanosomes that are ready to be transferred into the mammal host (Barry *et al.*, 1998). Expression of metacyclic VSG genes is believed to be triggered by specific factors present in the insect salivary gland (Lenardo *et al.*, 1986; Barry *et al.*, 1990; Graham *et al.*, 1990) and silenced when the stimuli cease and/or a single VSG is selected for expression in the bloodstream-stage. Active expression of VSG genes is restricted to 10–20 specific regions (telomers) of the genome known as 'expression sites' (ES). ESs are long polycistronic transcription units containing several ESAGs, where the VSG sequence is the most telomeric one. Only one ES is active at a time and, thus, only one VSG gene is expressed. To switch-on the expression of a VSG, trypanosomes use the following genetic mechanisms: (i) *in situ* activation of another VSG ES, while the old one is silenced, or (ii) DNA insertion of silent VSG sequences into an active ES by homologous recombination, gene conversion or telomere exchange events (Taylor and Rudenko, 2006). VSG-switching has been proposed to be consequence of a stress response triggered by decay in VSG transcript or protein levels (Borst and Ulbert, 2001). However, recent investigations disproved this model. The dissection of the mechanism controlling VSG-switching was addressed by means of VSG-targeted RNA interference that could induce clonal VSG variants *in vitro* and, thus, mimicked the effect of the immune system (Aitcheson *et al.*, 2005). The study showed that down-regulation of specific VSGs by RNAi does not trigger an immediate switch and statistic analysis corroborated the appearance of VSG variants being a spontaneous phenomenon occurring at rate of about 10^{-4} per cell division. Despite the strength of these results, the model employed a monomorphic strain uncapable to undergo the full life cycle and did not consider other host factors that may serve as co-signalling molecules for VSG-switching. Clearly, this question awaits further experimental analysis. Interestingly, related research aiming at the role of VSG in proliferation and pathogenicity revealed that VSG levels are sensed by trypanosomes to control cell-cycle progression. A blockage of VSG synthesis triggers a global and reversible stress response characterized by a severe inhibition of protein synthesis (Sheader *et al.*, 2005; Smith *et al.*, 2009). Since VSG is synthesized in the ER where it is modified rapidly by core N-linked glycosylation and GPI addition (Bangs *et al.*, 1997), ER stress was first thought to be the primary response. However, the typical markers of the unfolded protein response (i.e. up-regulation of BiP chaperone and apoptotic-related events; Berrnales *et al.*, 2006) are absent in parasites when VSG knockdown is induced. It was also possible to rule out the involvement of the spliced-leader RNA silencing response (SLS), which takes place upon induction of stress in the ER by pH stress, or by interrupting synthesis of the signal recognition particle (SRP) receptor that translocates secretory proteins across the ER (Lustig *et al.*, 2007). Although the signalling pathway participating in down-regulation of VSG remains elusive, the authors proposed that the reduced deposition of VSG on the cell surface is responsible for causing the global translation arrest (Smith *et al.*, 2009). Moreover, the morphological and biochemical changes induced by VSG RNAi resemble those occurring naturally during differentiation from LS to SS form in the bloodstream. Thus, it is worth considering this novel stress response as an important physiological check-point during VSG-switching that defines cell fate towards (i) cell division of LS form and substitution of the old VSG coat, when the levels of the new VSG are high enough to warrants a rapid reposition, or (ii) cell arrest and transformation to SS form, when the content of the new VSG variant does not surpass a minimal threshold.

Finally, the fact that down-regulation of actin, a major component of the vesicular trafficking pathway, induces a remarkable translation arrest (Smith *et al.*, 2009) opens up the possibility that other molecules involved in intracellular trafficking participate in the stress response.

Serum resistance factors

In the human host, infection with African trypanosomes induces a strong inflammatory response with high levels of interleukin (IL) 1, IL-6, interferon (IFN)-γ and tumour necrosis factor (TNF) α, which further stimulates the production and secretion by the liver of acute phase proteins, some of which confer natural resistance against certain subspecies of *T. brucei*. The so-called trypanolytic factor (TLF) of the human sera is a supramolecular assembly of non-specific components from the innate immune response that comprises apolipoprotein L1 (ApoL1, a component of the serum amyloid A), the haptoglobin related protein (Hpr, a subproduct of the haem-transport protein haptoglobin) and haemoglobin (Hb), all associated with high-density lipoprotein particles (HDL3) or to an IgM/apolipoprotein A-I complex (Pays *et al.*, 2006). The innate defence mechanism of anti-trypanosome lytic activity is also present in the serum of diurnal non-arboreal African primates like gorillas and baboons (Lugli *et al.*, 2004; Poelvoorde *et al.*, 2004), suggesting that the resistance factor appeared early in evolution driven by the frequent contact of the terrestrial primates with infected tse-tse flies (Pays and Vanhollebeke, 2009). Bloodstream parasites efficiently internalize the HDL particles via endocytosis mediated by specific surface receptors against the haptoglobin–Hb complex. The progressive acidification of the endosome throughout the endocytic pathway induces the dissociation of ApoL1 from the HDL fraction. Free ApoL1 integrates into the lysosomal membrane where it forms a pore that drives an uncontrolled flux of ions and water from the cytosol leading to osmotic swelling of the lysosome. The integrity of the parasite plasma membrane is disrupted probably as a consequence of the increased intracellular pressure and/or the release of lysosomal components (Pays and Vanhollebeke, 2009). Whereas *T. b. brucei* is susceptible to lysis by human serum, the subspecies *rhodesiense* and *gambiense* have developed resistance against this killing factor (Brun *et al.*, 1987). The entity conferring serum resistance (serum resistance associated, SRA) to *T. b. rhodesiense*, has been identified as a truncated VSG that lacks immunogenic regions (surface exposed loops) and is selectively expressed upon parasite exposure to human serum (De Greef and Hamers, 1994). Genetic analysis of SRA and related sequences supports the idea that the SRA gene originated only once in a *T. brucei* subset as a consequence of a deletion event in a precursor VSG gene (Campillo and Carrington, 2003). SRA is present as a single copy gene per haploid genome in a VSG expression locus, the so-called resistant expression site (RES), that lacks the majority of ESAGs co-transcribed with the VSG (Xhong *et al.*, 1998). Activation of the RES must occur rapidly upon differentiation of metacyclic trypomastigotes into LS forms and must be followed by the expression of VSG variants to endow the parasite with full protection against innate and humoral host immune responses. This is achieved by the homologous recombination-mediated insertion of VSG genes into the active RES in each cycle of antigenic variation (Xhong *et al.*, 1998; Pays *et al.*, 2006). In non-human hosts, expression of SRA is counter-selected, probably because many of the ESAGs that are not co-expressed with SRA are somehow indispensable (Van Meirvenne *et al.*, 1975; Xhong *et al.*, 1998). Once synthesized, SRA enters the secretory pathway becoming N-glylcosylated and then targeted to the cell membrane wherefrom it is rapidly endocytosed (De Greef and Hamers, 1994; Stephens and Hajduk, 2011). Within this compartment, SRA and ApoL1 bind through strong hydrophobic coiled-coil interactions via their corresponding N- and C-terminal helices (Campillo and Carrington, 2003). Formation of this heterocomplex inhibits the lytic process by preventing the integration of ApoL1 to the endo-lysosomal membrane (Vanhamme *et al.*, 2003) and/or by accelerating the proteolytic degradation of the lipoprotein in the lysosome (Stephens and Hajduk, 2011).

T. b. gambiense lacks the SRA gene but is able to circumvent lysis mediated by the TLF by a mechanism that is not yet fully understood but probably involves a reduced expression of and/or point mutations in the haptoglobin–Hb complex receptor that lower ligand affinity (Kieft *et al.*, 2010), altogether resulting in a decreased uptake of TLF by the parasite. Moreover, the search for putative components responsible for serum resistance in *T. b. gambiense*, led to the identification of a chromosomal region that undergoes rearrangement and deletion of a sequence encoding a putative AutI gene (Berberof *et al.*, 2001; Felu *et al.*, 2007), a protein involved in autophagy in yeast. Since autophagy typically involves the participation of lysosomes, the organelle where TLF degradation occurs, it remains to be demonstrated whether a defect in this pathway may additionally contribute to serum resistance in *T. b. gambiense*.

Oxidant defence

Other innate defence mechanism to restrain the proliferation and spread of infection by *T. brucei* includes the production of reactive oxygen (ROS: $^{\cdot}OH$, H_2O_2, $O_2^{\cdot-}$, ROOH, $^{\cdot}OR$, OCl^-) and nitrogen (RNS: $^{\cdot}NO$, $^{\cdot}NO_2$, $ONOO^-$) species by the host. The first step in this pathway consists in the recognition of parasite molecules, usually VSG epitopes and parasite DNA, by pattern recognition receptors located on antigen presenting cells (APC) and other phagocytes (Drennan *et al.*, 2005). APCs will activate the release of ROS and RNS by monocytic cells and the vascular endothelium. Production of oxidants occurs in a series of sequential spontaneous or enzyme- or metal-catalysed reactions. The plasma membrane-bound NADPH oxidase complex of activated phagocytic cells transfers one electron from NADPH to reduce molecular oxygen to superoxide radical, $O_2^{\cdot-}$. At low pH or catalysed by superoxide dismutase, $O_2^{\cdot-}$ dismutates to O_2 and the oxidant species hydrogen peroxide (H_2O_2). These cells are also equipped with inducible nitric oxide ($^{\cdot}NO$) synthase and myeloperoxidase that lead to the formation of stronger oxidants such as peroxynitrite ($ONOO^-$) and hypochlorous acid (HClO). Transition metals are capable of reacting with H_2O_2 to generate the highly reactive hydroxyl radical ($^{\cdot}OH$). Highly toxic hydroperoxy- and hydroxyl-fatty acids can be produced by activated lipoxygenases or free radical chain reactions. Cyclo-oxygenases and 5-lipoxygenase generate prostaglandins and leukotrienes, both signalling molecules that trigger and potentiate the inflammatory response during infection. If not neutralized by the parasite, ROS and RNS can cause a wide range of damage to macromolecules resulting in physiological dysfunction, loss of cellular integrity and cell death (reviewed in Irigoín *et al.*, 2008). Several studies documented the production of ROS and RNS in association with the inflammatory response during infection with *T. brucei* (Anosa and Kaneko, 1983; Grosskinsky *et al.*, 1983; Vray *et al.*, 1991; Igbokwe *et al.*, 1994; Mabbot *et al.*, 1995; Gobert *et al.*, 1998; Sternberg, 2004; Vincendeau and Bouteille, 2006; Bosschaerts *et al.*, 2008; Guilliams *et al.*, 2008). A critical role for NO-mediated trypanosome killing during acute infection has been recently discussed and linked to a striking parasite-elicited subversion of host arginase pathway (Gobert *et al.*, 2000; Duleu *et al.*, 2004).

Redox-active protein thiols or low molecular mass bio-compounds play a key role in the neutralization of these oxidants or acting as redox sensors to regulate diverse cellular functions. In trypanosomatids, the thiol redox homeostasis is maintained by the unique trypanothione/trypanothione reductase couple, which substitutes for the glutathione/glutathione reductase and the thioredoxin/thioredoxin reductase systems of the host (Irigoín *et al.*, 2008; Krauth-Siegel and Comini, 2008). Trypanothione is a bis(glutathionyl) conjugate of spermidine that represents the main low molecular mass thiol-redox metabolite of trypanosomes. As first demonstrated in *Crithidia fasciculata* (Nogoceke *et al.*, 1997), trypanothione fuels with its reducing equivalents a most complex peroxidase system and a plethora of other essential cellular functions (Schlecker *et al.*, 2007; Krauth-Siegel and Comini, 2008). The dithiol is synthesized by monoglutathionylspermidine synthetase and, or only, trypanothione synthetase, which catalyses the sequential addition of one or two glutathione molecules, respectively, to one spermidine moiety (Comini *et al.*, 2003; Oza *et al.*, 2003). The enzymes form a specific subfamily of ATP-dependent C-N ligases that contains an N-terminal domain with a papain-like cysteine-protease activity which can hydrolyse the synthesis products back to substrates. The synthetic activity, however, prevails, while the amidase activity appears to be dispensable *in vivo* (Wyllie *et al.*, 2009). Trypanothione can decompose H_2O_2 and $ONOO^-$ and trap $^{\cdot}NO$ and labile iron in a dinitrosyl–iron complex with higher efficiency than glutathione (Carnieri *et al.*, 1993; Thomson *et al.*, 2003; Bocedi *et al.*, 2010). However, removal of a substantial load of oxidants at rates that guarantee the intracellular redox homeostasis requires catalysis. This is mediated by the action of several peroxidases that are functionally and/or structurally related to the peroxiredoxin- (TXN-Px) and glutathione-peroxidase (GPx)-type (Schlecker *et al.*, 2007), which depends on the thioredoxin-paralogue oxidoreductase tryparedoxin (TXN; Nogoceke *et al.*, 1997; Lüdemann *et al.*, 1998). Both classes of peroxidases depend on TXN for obtaining the electrons that will convert the hydroperoxides, lipoperoxides and peroxynitrite into water and the corresponding redox inactive alcohol-derivatives or nitrite (Nogoceke *et al.*, 1997; Tetaud *et al.*, 2001; Budde *et al.*, 2003; Hillebrand *et al.*, 2003; Trujillo *et al.*, 2004; Schlecker *et al.*, 2005). While the tryparedoxin specificity of the peroxiredoxin-type peroxidases complies with the common (thio)redoxin specificity of 2-cysteine-peroxiredoxins, the identical specificity of trypanosomal GPxs may surprise. However, they belong to a large subfamily of GPxs that differ from the name-giving glutathione peroxidases in preferentially using redoxins as reducing substrate (Maiorino *et al.*, 2007; Toppo *et al.*, 2009). Supporting earlier biochemical evidences (Fulton and Spooner 1956; Boveris *et al.*, 1980; Penketh and Klein, 1986), the genome of *T. brucei* lacks the efficient hydroperoxide catalysts of mammals such as the selenocysteine-containing glutathione peroxidases and catalase (Berriman *et al.*, 2005). Indeed, the infective form of trypanosomes is admittedly less competent to deal with oxidative challenge than mammalian cells, which is ascribed to its limited antioxidant armamentarium (Meshnick *et al.*, 1977; Arrick *et al.*, 1981; Wilkinson *et al.*, 2003; Comini *et al.*, 2004). This biological deficiency can hardly be explained by the kinetic inefficiency of the trypanosomal anti-oxidant machinery but rather by the absence of back-up systems (Krauth-Siegel and Comini, 2008) and the lower capacity of the antioxidant pools when compared to the mammalian GSH pool. Bloodstream parasites have a notoriously higher intracellular concentration of different components of the hydroperoxide detoxifying cascade compared to procyclic forms (Schlecker *et al.*, 2005; Comini *et al.*, 2007), which is indicative of an adaptation to withstand the oxidant environment of the host. On the other hand, subjecting bloodstream parasites to conditions mimicking oxidative stress (e.g. H_2O_2 challenge, TryS- and TXN-down-regulation) yielded fast compensatory responses such as increase in low molecular mass thiols and of the redox-active enzymes TXN and TXN-Px (Comini *et al.*, 2007; Medeiros, Fiestas, Comini, unpublished). It remains to be elucidated whether these mechanisms operate under physiological conditions of infection. The enzymatic entities participating in oxidant defence are distributed in different subcellular compartments and their indispensability for survival of the infective stage of African trypanosomes under *in vitro* or *in vivo* conditions has been largely documented: trypanothione reductase (Krieger *et al.*, 2000), trypanothione synthetase (Comini *et al.*, 2004; Wyllie *et al.*, 2009), cytosolic TXN (Wilkinson *et al.*, 2005; Comini *et al.*, 2007), cytosolic TXN-Px (Wilkinson *et al.*, 2005), mitochondrial tryparedoxin-dependent GPx-like peroxidase (Schlecker *et al.*, 2005; Diechtierow and Krauth-Siegel, 2011) and glycosomal superoxide dismutase (Wilkinson *et al.*, 2006) proved to be essential. These studies confirmed the crucial dependence of the thiol-redox metabolism on trypanothione and highlighted the lack of redundancy in its antioxidant system. Thus, this antioxidant system evolved with a minimalistic trait that, nevertheless, ensures parasite survival by efficient neutralization of the host's oxidant attack.

Response to xenobiotics and heavy metals

Xenobiotics are defined as chemical compounds that stem from the extracellular medium and cannot be consumed but, eventually, can be metabolized and/or neutralized by a living organism. These substances may be artificial (e.g. drugs, chemicals) or naturally occurring compounds (e.g. toxins, hormones). As to the first class of compounds concerns, inside the cells, chemicals are subjected to a series of biotransformation reactions and elimination processes that involve (i) oxido-reductive and hydrolytic reactions (phase I); (ii) enzyme-mediated conjugation to different low molecular mass molecules (phase II); (iii) the active extrusion of the modified xenobiotic from the cytosol by membrane transporters (phase III). Operating as a whole, this system constitutes an efficient mechanism to detoxify and confer resistance against drugs.

Inside the mammal host, *T. brucei* is exposed to two major sources of xenobiotic stress, the serum trypanolytic factor, discussed in detail before, and some drugs used to treat African trypanosomiasis. For the last, experimental evidences point that the trivalent arsenic compound melarsoprol and the nitroheterocycles nifurtimox and fexinidazole undergo metabolization by the parasite (Berger *et al.*, 1993; Nok, 2003; Lepesheva *et al.*, 2008; Sokolova *et al.*, 2010; Hall *et al.*, 2011). Phase I reactions are typically catalysed by enzymes from the cytochrome P450 superfamily, aldo-keto reductasas, nitroreductases, esterases, phosphatases, sulfatases, amidases, among others. Cytochrome P450 activity has been reported earlier in *T. brucei* and other trypanosomatids (Berger and Fairlamb, 1993). Recent studies ruled out a potential role of two cytochrome P450 reductases from *T. brucei* in metabolization and biodefence against nifurtimox (Hall *et al.*, 2011) but confirmed that a member of this family, CYP51, is indispensable for ergosterol biosynthesis (Lepesheva *et al.*, 2004; Lepesheva *et al.*, 2008). This contrasts with the situation in *T. cruzi*, where overexpression of a CYP homologue conferred resistance against nifurtimox and benznidazole (Portal *et al.*, 2008), a nitroheterocycle used to treat Chagas disease. Trypanosomes are equipped with nitroreductases (NTR) of the type I, which are FMN-dependent and oxygen-insensitive enzymes, and type II, which use FAD or FMN as coenzyme and are oxygen sensitive enzymes (Wilkinson *et al.*, 2008). Reduction of the nitro-group by class I and II NTR is achieved via two- and one-electron transfer reactions, respectively, that generates the corresponding nitroso intermediate and an unstable nitroradical (Hall *et al.*, 2011). These intermediate species are highly reactive and toxic and must be neutralized, for instance by conjugation to thiol groups. Functional genetic studies carried out on the bloodstream form of *T. brucei* identified mitochondrial NTR I as the major determinant for parasite susceptibility or resistance against nitrofurans (Hall *et al.*, 2011). Thus, loss of a single NTR allele confers cross-resistance to a wide range of nitroheterocyclic drugs without affecting growth rate or the ability to establish an infection, whereas overexpression of an ectopic copy of the gene increased by ten-fold drug sensitivity (Wilkinson *et al.*, 2008). Although nitrofuran resistance has been generated in laboratory strains, it does not seems unreasonable to speculate that African trypanosomes may develop it in the field (i.e. by down-regulation or mutation of NTR I gene[4]). This potential scenario raises concerns about the future performance of nifurtimox and fexinidazole, two drugs that recently entered in phase III and II clinical trials (DNDi *et al.*, 2009; Priotto *et al.*, 2009).

Phase II and phase III reactions involve conjugation and drug–conjugate export systems. Conjugation to low molecular mass thiols such as cysteine or glutathione is a common physiological reaction, the last catalysed by glutathione-s-transferases. Glutathione-S-transferases have a central role in detoxification metabolism, and the overexpression of several isoenzymes with

4 *T. brucei* null mutants of NTR I are not viable (Wilkinson *et al.* 2008).

characteristic substrate specificities is a common mechanism of drug resistance in several organisms (Sau *et al.*, 2010). Trypanothione has been regarded as the primary target and the reason for the selectivity of aromatic arsenical derivatives against African trypanosomes. The stable thiol–drug conjugate has been proposed to cause an intracellular thiol-redox imbalance by depleting the level of reduced trypanothione (Fairlamb *et al.*, 1989). Conjugation of trypanothione to electrofilic compounds may occur spontaneously or mediated by a trypanothione-S-transferase activity associated to the eukaryotic elongation factor 1B complexes (eEF1B) from different trypanosomatids (Vickers *et al.*, 2004a,b). Genes encoding for a eEF1B homologue and a putative glutathione-S-transferase-related glutaredoxin are present in the genome of *T. brucei* (Berriman *et al.*, 2005) but none of them has been characterized thus far. With respect to the role of trypanothione in drug resistance, the content of this dithiol and glutathione was not significantly different between melarsen-resistant and –sensitive trypanosomes (Fairlamb *et al.*, 1992) and depletion of trypanothione reductase neither affected drug susceptibility (Krieger *et al.*, 2000). These results contrasted with the overwhelming evidences demonstrating the relevance of trypanothione and a multidrug resistant-like transporter for resistant to antimonials and arsenicals in *Leishmania* (reviewed in Perez-Victoria *et al.*, 2002; Mäser *et al.*, 2003) and stimulated new investigations. In this regard, overexpression of a gene encoding for an ABC-related transporter (TbMRPA) specialized in the export of the adduct trypanothione-arsenical derivative (Mäser and Kaminiski, 1998) resulted in a tenfold increase in *in vitro* EC_{50} for melarsoprol (Shahi *et al.*, 2002). Also overexpression of the rate-limiting enzymes of trypanothione synthesis, ornithine decarboxylase and γ-glutamyl-cysteine synthetase, reduced melarsoprol sensitivity but, strikingly, co-expression of TbMRPA did not have an accumulative effect on drug resistance (Shahi *et al.*, 2002). However, further work failed to detect MRPA overexpression in four melarsoprol-resistant field isolates and to reproduce drug resistance *in vivo* as achieved *in vitro* (Alibu *et al.*, 2006). Although the analysis of a higher number of field isolates is necessary to elaborate final conclusions, the current knowledge points against a preponderant role of trypanothione in melarsoprol resistance.

Unlike free-living trypanosomatids, which are exposed to environmental toxic heavy metals, parasitic trypanosomes are likely to have a limited contact with these poisoning ions inside the mammalian host under physiological conditions. The case of the organic trivalent arsenic drug melarsoprol has been discussed above. As a general strategy to minimize the toxicity of metal ions, cells rely on ATP-driven efflux pumps, metal ion/H^+ antiporters, redox-dependent metabolization systems and metal scavengers. A metal ion sequestering role may be envisaged for thiol groups present in low molecular mass biocompounds such as glutathione and trypanothione, and putative metallothionein-like proteins (Maya *et al.*, 2004). For instance, trypanothione has been mentioned in protection against Cd^{2+} and Hg^{2+} for the non-pathogenic trypanosomatid *C. fasciculata* (Fairlamb and Cerami, 1992) and the presence of Zn-inducible metallothioneins were reported in *T. cruzi* (Maya *et al.*, 2004). Free or porphyrin-bound Hg^{2+}, Cd^{2+} and Pb^{2+} were reported to be cytotoxic to bloodstream *T. brucei* but the potential neutralizing role of trypanothione and metallothioneins was not investigated (Nyarko *et al.*, 2002). The biological implications of the parasite thiol-based defence mechanism may be particularly important under host oxidative insult or drug treatment leading to metal (Cu, Zn) release from endogenous proteins or to the formation of radicals and electrophilic byproducts.

Apoptosis

To establish a sustained infection and increase the chances of transmission to a new host, parasites are almost obligated to avoid hyperparasitism, which would compromise host and parasite survival equally. In this regard, several findings support the hypothesis that programmed cell death (PCD) in protozoan parasites contributes to the regulation of population density, at least under certain conditions (Vasella *et al.*, 1997; Welburn *et al.*, 2006). PCD involves at least three possible death mechanisms[5] from which apoptosis (from the Greek 'falling off', like a leaf in autumn, Kerr *et al.*, 1972) is the better understood and deeply studied (see Elmore, 2007, for a review). Apoptosis is a genetically encoded and highly efficient programme initiated, in most cases, by an environmental signal that activates a cascade of energy-dependent processes committed to terminate with cell integrity in a sort of altruistic and well-orchestrated suicide (Duszenko *et al.*, 2006; Elmore, 2007). In principle, this death mechanism was assumed to be exclusive of complex eukaryotic organisms that demand a constant tissue renewal and reshaping (Welburn *et al.*, 2006). However, this concept, together with the belief that unicellular lineages are 'immortal' and only subject of accidental or predator-related death (Welburn *et al.*, 2006; Nedelcu *et al.*, 2010), has been challenged by the identification of PCD in diverse unicellular organisms (Cornillon *et al.*, 1994; reviewed in Lewis, 2000). Within the Kingdom Protista this phenomenon has been mostly studied in members of the family *Trypanosomatidae* and the genus *Plasmodium*. In these organisms, apoptosis has been seen as a means to control parasite number and modulate the host immune response during the establishment and progress of the infection (Welburn *et al.*, 2006; Lüder *et al.*, 2010; Nedelcu *et al.*, 2010).

As noted early (Ameisen *et al.*, 1995, 1996; Welburn *et al.*, 1996), the morphological and biochemical phenotype that characterizes unicellular eukaryotic PCD is almost identical to that described for classical apoptosis (reviewed in Jimenez-Ruiz *et al.*, 2010). However, there are still major gaps in knowledge regarding the molecular components responsible for triggering and controlling parasite PCD. For instance, the main players of metazoan apoptosis such as TNF- or Fas-receptors, caspases, proteins from the Bcl2/Bax family and several apoptotic nucleases like the caspase-activated DNAse (CAD, see below) are missing in the protozoan genomes (Kaczanoswski *et al.*, 2011). Instead, caspase-related sequences, the so called 'metacaspases' (EC 3.4.22.-), have been identified in several parasites, including *Plasmodium* spp., *Trypanosoma* and *Leishmania* (Jimenez-Ruiz *et al.*, 2010). The active site of metacaspases comprises a conserved histidine-cysteine catalytic dyad, the cysteine being the active site nucleophile. All metacaspases examined to date have a basic P1 specificity towards arginine or lysine, whereas caspases have an acidic P1 specificity towards aspartic acid (Vercammen *et al.*, 2007), explaining why they are not inhibited by classical caspase inhibitors. In the specific case of *T. brucei*, five genes encoding metacaspases are expressed (TbMCA1-TbMCA5, Helms *et al.*, 2006). At the protein level, TbMCA2 and TbMCA3 are highly similar differing only in a putative organelle targeting-signal at the N-terminus of TbMCA3. Importantly, TbMCA1 and TbMCA4 have amino acid substitutions in their conserved catalytic residues that probably make them enzymatically inactive (Mottram *et al.*, 2003). TbMCA2, 3 and 5 are actively expressed by bloodstream *T. brucei* (Helms *et al.*, 2006). RNA-interference disclosed that metacaspase activity is essential for this parasite stage but failed to ascribe the essentiality to any of these proteins owing to functional redundancy (Helms *et al.*, 2006). In addition, the anti-metacaspase activity of specific inhibitors against recombinant TbMCA2 does not correlate with the low anti-trypanosomal effect they exerted (Berg *et al.*, 2010). Clearly, the precise role of trypanosomal metacaspases remains to be elucidated (Deponte, 2008). Nucleosomal fragmentation of DNA is a classical hallmark of apoptosis in metazoans and is mediated by specific nucleases like the CAD or endonuclease G (EndoG). Nuclease activation and/or DNA fragmentation was detected in apoptotic-like cells of *T. brucei* (Ridgley *et al.*, 1999; Pearson *et al.*, 2000; Figarella *et al.*, 2006; Welburn *et al.*, 2006) but the enzymatic entity responsible for it remains yet undisclosed. Only a homologue of mitochondrial EndoG (Ganavaran *et al.*, 2008) has been unequivocally identified in *T. brucei* but the signal triggering its activation is unknown (Kaczanowski *et al.*, 2010). In addition, the mitochondrial flavoprotein apoptosis-inducing factor (AIF) that translocates to the nucleus during

5 Since the introduction of the 'programmed cell death' idea by Lockshin and colleagues in the sixties, several types of PCD has been described, the most important being apoptosis, paraptosis and autophagic cell death. In this section we focus on apoptosis whereas autophagy is discussed in the section 'Nutritional stress response'.

apoptosis in higher eukaryotes (Elmore, 2007) appears to be absent from trypanosome genomes (Berriman *et al.*, 2005; Kaczanowski *et al.*, 2010). The molecular events induced by expression of a pro-apoptotic protein (human Bax) were recently dissected for *T. brucei* (Esseiva *et al.*, 2004). They involve the release of cytochrome c, the depolarization of the mitochondrial membrane potential and mitochondrial fission. In contrast to mammalian systems, these events do not occur simultaneously and are reversible (Esseiva *et al.*, 2004). Other factors reported to trigger apoptosis in *T. brucei* are: prostaglandin- (Figarella *et al.*, 2005, 2006), flavonoid-induced oxidative stress (Mamani-Matsuda *et al.*, 2004), persistent ER stress (Goldshmidt *et al.*, 2010) and lectins (Welburn *et al.*, 1996). ROS-signal pathway was efficiently blocked by antioxidants (glutathione or N-acetylcysteine; Mamani-Matsuda *et al.*, 2004; Figarella *et al.*, 2005), an inhibitor of protein synthesis (cycloheximide; Figarella *et al.*, 2006) or overexpression of the mitochondrial plant-like TAO (Tsuda *et al.*, 2005, 2006), an enzyme that contributes to support the glycosomal redox balance and continuous glycolysis (Clarkson *et al.*, 1989). The biochemical basis behind TAO protection against apoptosis has not been deciphered yet but it might function as a sensor linking the cytosolic energetic metabolism with the mitochondrial redox status to control ROS production. ER stress was induced in procyclic cells genetically manipulated to accumulate unfolded protein in the ER lumen or chemically by lowering the medium pH (Lustig *et al.*, 2007). The effect of lectins was described for procyclic forms. Whether any of these signals operate *in vivo* remains unknown. However, given that parasite and host are capable to produce and secrete prostaglandins (Luder *et al.*, 2010), these molecules appear as suitable physiological candidates.

Apoptosis is considered an important process closely linked to differentiation and perpetuation of *T. brucei* in the human host (Jimenez-Ruiz *et al.*, 2010; Smirlis *et al.*, 2010). It is currently accepted that a parasite-derived, so far undiscovered, soluble signal, the stumpy induction factor (SIF; Black *et al.*, 1985; Vasella *et al.*, 1997), induces the transformation of LS into SS trypanosomes. Differentiation is cell density-dependent, indicating that a critical concentration of SIF must be achieved to initiate the process (Figarella *et al.*, 2005, 2006). Two different cell populations are distinguished within the SS parasites, one that is fly infective and a second one susceptible to undergo apoptosis (Seed and Wenck, 2003; Figarella *et al.*, 2005, 2006). The signal (pathogen and/or host catabolite) and interdependence of the differentiation and apoptotic processes observed in *T. brucei* resemble a quorum sensing mechanism, whose major goal is to ensure a persistent parasite infection and life cycle (MacGregor *et al.*, 2011). In the scenario of a human infection, non-dividing parasites will serve to the elimination of the major LS and SS variant antigen population, whereas apoptotic SS trypanosomes will offer a rapid, and probably reversible (Esseiva *et al.*, 2004), way to control the SS density. At this point it is important to remark that dying by apoptosis might have the additional advantage of preserving host homeostasis by diminishing the overwhelming host inflammatory response that takes place during massive parasite death by necrosis or from an uncontrolled pathogen number.

Conclusions and future trends

The past two decades have produced a significant growth in our knowledge and understanding of trypanosome's biology, mostly driven by the development of tools for the genetic manipulation of trypanosomes and the completion of the genome sequencing of three major representatives from each trypanosoma genus. Fundamental questions related to the genetic (transcriptome), metabolic (energy metabolism) and morphological (glycosomal and mitochondrial reshaping) changes that accompany each transitional stage of the parasite during its life cycle have been unravelled. However, we know considerably less about the precise molecular machinery and mechanisms that underlie these processes, especially those related to stress response and host–parasite relationship. Evolution has shaped unique features on African trypanosomes that pose still intriguing questions regarding the origin, basis and fate of the drastic genetic and biochemical changes they underwent. For instance, the proposed lack of transcriptional flexibility and the almost entire dependency on post-transcriptional regulation of gene expression raises the question to the biological need of wasting a substantial amount of energetic resources in synthesizing and then degrading large amounts of RNA. It is however tempting to speculate that the great repertoire of transcripts, readily available for translation, may confer the parasites an evolutionary advantage to quickly respond to the changing environmental conditions of their hosts. The recent finding of epigenetic phenomena occurring in highly expressed regions of the chromatin and the existence of stage-specific transcriptional regulons add new levels of complexity to the regulation of gene expression in trypanosomes, which merit further investigations. Various components of the systems sensing and transducing different stress signal also remain to be addressed. For instance, the study of new stress conditions to disclose novel pathways (in)activated by kinases and their relevance for trypanosome survival. Notably, the possible existence of host-derived signals promoting differentiation to non-replicative forms and apoptosis is a research area that deserves attention. At the clinical level, it should be noted that during peaks of parasitaemia the host suffers from hypoglycaemic shock, low oxygen tension and metabolic acidosis, factors that combined might trigger a stress response in the parasite. Whether apoptosis plays an altruistic role in the life cycle of *T. brucei* or selectively evolved because of a dual function of the molecular entities participating in PCD in other indispensable cellular processes, is open to debate and will require further investigations.

The mutual self-defence mechanisms developed by host (TLF) and parasite (antigenic variation and SRA) are an interesting example on how the host–parasite interaction shapes their evolution. The long-termed parasite-mediated subversion of the host immune system is another research area of interest since the identification of the molecules involved may disclose treatments to ameliorate disease symptoms and outcome. In relation to therapy, although the limited antioxidant system of the parasite confers efficient protection against host-derived oxidative stress, the distinctive molecular features of its components offers excellent opportunities for more efficacious and specific chemotherapeutic interventions.

In this chapter, we provided a review on the most relevant subjects related to stress response in the infective form of African trypanosomes, with the hope that substantial advance in this area will potentially lead to the development of novel therapeutic strategies to combat these fascinating but lethal microorganisms.

Acknowledgements

This work was supported by Agencia Nacional de Investigación e Innovación (Innova Uruguay, Agreement No. DCI–ALA/2007/19.040 between Uruguay and the European Commission). Professor Leopold Flohé is gratefully acknowledged for a number of helpful suggestions on an early draft of this chapter.

Web resources

International health and surveillance organizations

WHO-HAT: http://www.who.int/trypanosomiasis_african/en
Web resources of WHO related to HAT.

WHO-TDR: http://apps.who.int/tdr/svc/diseases/african-trypanosomiasis
TDR is a global programme of scientific collaboration that supports and coordinates global efforts to combat major diseases of the poor and disadvantaged.

CDC: http://www.cdc.gov/parasites/sleepingsickness
Information about biology, epidemiology and treatment of *T. brucei* in the Center for Disease Control and Prevention, USA.

DNDi: www.dndi.org/diseases/hat.html
Comprehensive information about HAT epidemiology and treatments.

AU-IBAR: http://www.au-ibar.org/
Web page from the InterAfrican Bureau for Animal Resources, institutional framework responsible for trypanosomiasis research and control activities in Africa.

Pathogen Safety Data Sheets: http://www.msdsonline.com/resources/msds-resources/free-safety-data-sheet-index/
Infectious disease index from the Public Health Agency of Canada providing information about the hazardous properties of human pathogens.

Genomic resources

WT-Sanger: http://www.sanger.ac.uk/resources/downloads/protozoa
This page provides access to genomes sequenced at the Wellcome Trust Sanger Institute, including *T. brucei* genome and other trypanosomatids.

TriTryp: http://tritrypdb.org/tritrypdb
The most complete web resource for genomic and gene-related information for trypanosomes.

GeneDB: http://www.genedb.org/Homepage
The GeneDB project is part of the Sanger Institute's Pathogen Genomics and provides access to sequence data and annotation for several pathogenic organisms.

Web resources for specialists

Cross's Lab: http://tryps.rockefeller.edu
A useful set of resources for scientists working on trypanosomes.

Tree of Life web project: http://tolweb.org/Trypanosoma/98034
This project provides information about biodiversity, the characteristics of different groups of organisms, and their evolutionary history.

Bovine trypanosomiasis: http://www.genomics.liv.ac.uk/tryps/links.php
Web-page of a consortium studying bovine trypanosomiasis (i.e. functional genomics, cattle genome sequencing and trypanosomiasis-resistant cattle).

SGPP: http://www.sgpp.org
Structural Genomics of Pathogenic Protozoa consortium is a pilot project aimed to high-throughput structure determination from drug-target candidates from three major trypanosomatids.

Public interest

Rockhopper TV: http://www.rockhopper.tv/programmes/150/
Documentary film about HAT in the Democratic Republic of Congo.

References

Aitcheson, N., Talbot, S., Shapiro, J., Hughes, K., Adkin, C., Butt, T., Sheader, K., and Rudenko, G. (2005). VSG switching in *Trypanosoma brucei*: antigenic variation analysed using RNAi in the absence of immune selection. Mol. Microbiol. *57*, 1608–1622.

Alibu, V.P., Richter, C., Voncken, F., Marti, G., Shahi, S., Renggli, C.K., Seebeck, T., Brun, R., and Clayton, C. (2006). The role of *Trypanosoma brucei* MRPA in melarsoprol susceptibility. Mol. Biochem. Parasitol. *146*, 38–44.

Ameisen, J.C. (1996). The origin of programmed cell death. Science *272*, 1278–1279.

Ameisen, J.C., Idziorek, T., Billaut-Mulot, O., Loyens, M., Tissier, J.P., Potentier, A., and Ouaissi, A. (1995). Apoptosis in a unicellular eukaryote (*Trypanosoma cruzi*): implications for the evolutionary origin and role of programmed cell death in the control of cell proliferation, differentiation and survival. Cell Death Differ. *2*, 285–300.

Anosa, V.O., and Kaneko, J.J. (1983). Pathogenesis of *Trypanosoma brucei* infection in deer mice (*Peromyscus maniculatus*): hematologic, erythrocyte biochemical, and iron metabolic aspects. Am. J. Vet. Res. *44*, 639–644.

Arrick, B.A., Griffith, O.W., and Cerami, A. (1981). Inhibition of glutathione synthesis as a chemotherapeutic strategy for trypanosomiasis. J. Exp. Med. *153*, 720–725.

Balmer, O., Beadell, J.S., Gibson, W., and Caccone, A. (2011). Phylogeography and taxonomy of *Trypanosoma brucei*. PLoS Negl. Trop. Dis. *5*, e961.

Bangs, J.D., Uyetake, L., Brickman, M.J., Balber, A.E., and Boothroyd, J.C. (1993). Molecular cloning and cellular localization of a BiP homologue in *Trypanosoma brucei*. Divergent ER retention signals in a lower eukaryote. J. Cell Sci. *105*, 1101–1113.

Bangs, J.D., Ransom, D.M., McDowell, M.A., and Brouch, E.M. (1997). Expression of bloodstream variant surface glycoproteins in procyclic stage *Trypanosoma brucei*: role of GPI anchors in secretion. EMBO J. *16*, 4285–4294.

Barquilla, A., and Navarro, M. (2009). Trypanosome TOR as a major regulador of cell growth and autophagy. Autophagy *5*, 256–258.

Barquilla, A., Crespo, J.L., Navarro, M. (2008). Rapamycin inhibits trypanosome cell growth by preventing TOR complex 2 formation. Proc. Natl. Acad. Sci. U.S.A. *105*, 14579–14584.

Barry, J.D., Graham, S.V., Matthews, K.R., Shiels, P.G., and Shonekan, O.A. (1990). Stage specific mechanisms for activation and expression of variant surface glycoprotein genes in *Trypanosoma brucei*. Biochem. Soc. Trans. *8*, 708–710.

Barry, J.D., Graham, S.V., Fotheringham, M., Graham, V.S., Kobryn, K., and Wymer, B. (1998). VSG gene control and infectivity strategy of metacyclic stage *Trypanosoma brucei*. Mol. Biochem. Parasitol. *91*, 93–105.

Bassarak, B., Uzcátegui, N.L., Schönfeld, C., and Duszenko, M. (2011). Functional characterization of three aquaglyceroporins from *Trypanosoma brucei* in osmoregulation and glycerol transport. Cell Physiol. Biochem. *27*, 411–420.

Berberof, M., Pérez-Morga, D., and Pays, E. (2001). A receptor-like flagellar pocket glycoprotein specific to *Trypanosoma brucei gambiense*. Mol. Biochem. Parasitol. *113*, 127–138.

Berg, M., Van der Veken, P., Joossens, J., Muthusamy, V., Breugelmans, M., Moss, C.X., Rudolf, J., Cos, P., Coombs, G.H., Maes, L., *et al.* (2010). Design and evaluation of *Trypanosoma brucei* metacaspase inhibitors. Bioorg. Med. Chem. Lett. *20*, 2001–2006.

Berger, B.J., and Fairlamb, A.H. (1993). Cytochrome P450 in trypanosomatids. Biochem. Pharmacol. *46*, 149–157.

Berger, B.J., Carter, N.S., and Fairlamb, A.H. (1993). Polyamine and pentamidine metabolism in African trypanosomes. Acta Tropica. *54*, 215–224.

Bernales, S., Papa, F.R., and Walter, P. (2006). Intracellular signaling by the unfolded protein response. Annu. Rev. Cell Dev. Biol. *22*, 487–508.

Berriman, M., Ghedin, E., Hertz-Fowler, C., Blandin, G., Lennard, N.J., Bartholomeu, D., Renauld, H.J., Caler, E., Hamlin, N., Haas, B., *et al.* (2005). The genome of the African trypanosome, *Trypanosoma brucei*. Science *309*, 416–422.

Besteiro, S., Barrett, M.P., Rivière, L., and Bringaud, F. (2005). Energy generation in insect stages of *Trypanosoma brucei*: Metabolism in flux. Trends Parasitol. *21*, 185–191.

Bitter, W., Gerrits, H., Kieft, R., and Borst, P. (1998). The role of transferrinreceptor variation in the host range of *Trypanosoma brucei*. Nature *391*, 499–502.

Black, S.C., Huang, J.Q., Rezaiefar, P., Radinovic, S., Eberhart, A., Nicholson, D.W., and Rodger, I.W. (1998). Co-localization of the cysteine protease caspase-3 with apoptotic myocytes after *in vivo* myocardial ischemia and reperfusion in the rat. J. Mol. Cell Cardiol. *30*, 733–742.

Bocedi, A., Dawood, K.F., Fabrini, R., Federici, G., Gradoni, L., Pedersen, J.Z., and Ricci, G. (2010). Trypanothione efficiently intercepts nitric oxide as a harmless iron complex in trypanosomatid parasites. FASEB J. *24*, 1035–1042.

Borst, P., and Ulbert, S. (2001). Control of VSG gene expression sites. Mol. Biochem. Parasitol. *114*, 17–27.

Bosschaerts, T., Guilliams, M., Noel, W., Hérin, M., Burk, R.F., Hill, K.E., Brys, L., Raes, G., Ghassabeh, G.H., De Baetselier, P., *et al.* (2008). Alternatively activated myeloid cells limit pathogenicity associated with African trypanosomiasis through the IL-10 inducible gene selenoprotein P. J. Immunol. *180*, 6168–6175.

Boveris, A., Sies, H., Martino, E.E., Docampo, R., Turrens, J.F., and Stoppani, A.O. (1980). Deficient metabolic utilization of hydrogen peroxide in *Trypanosoma cruzi*. Biochem. J. *188*, 643–648.

Brasseur, R., and Pays, E. (2003). Apolipoprotein L-I is the trypanosome lytic factor of human serum. Nature *422*, 83–87.

Brennand, A., Gualdrón-López, M., Coppens, I., Rigden, D.J., Ginger, M.L., and Michels, P.A. (2011). Autophagy in parasitic protists: unique features and drug targets. Mol. Biochem. Parasitol. *177*, 83–99.

Bringaud, F., Peyruchauda, S., Baltz, D., Girouda, C., Simpson, L., and Baltz, T. (1995). Molecular characterization of the mitochondrial heat shock protein 60 gene from *Trypanosoma brucei*. Mol. Biochem. Parasitol. *74*, 119–123.

Brun, R., and Jenni, L. (1987). Human serum resistance of metacyclic forms of *Trypanosoma brucei brucei, T. brucei rhodesiense* and *T. brucei gambiense*. Parasitol. Res. *73*, 218–223.

Budde, H., Flohé, L., Hecht, H.J., Hofmann, B., Stehr, M., Wissing, J., and Lünsdorf, H. (2003). Kinetics and redox-sensitive oligomerisation reveal negative subunit cooperativity in tryparedoxin peroxidase of *Trypanosoma brucei brucei*. Biol. Chem. *384*, 619–633.

Campillo, N., and Carrington, M. (2003). The origin of the serum resistance associated (SRA) gene and a model of the structure of the SRA polypeptide from *Trypanosoma brucei rhodesiense*. Mol. Biochem. Parasitol. *127*, 79–84.

Campos-Salinas, J., and Gonzalez-Rey, E. (2009). Autophagy and neuropeptides at the crossroad for parasites: to survive or to die? Autophagy *5*, 551–554.

Carnieri, E.G.S., Moreno, S.N.J., and Docampo, R. (1993). Trypanothione-dependent peroxide metabolism in *Trypanosoma cruzi* different stages. Mol. Biochem. Parasitol. *61*, 79–86.

Clarkson, A.B., Brienen, E.J., Pollakis, G., and Grady, R.W. (1989). Respiration of bloodstream forms of the parasite *Trypanosoma brucei brucei* is dependent on a plant-like alternative oxidase. J. Biol. Chem. *264*, 17770–17776.

Clayton, C., and Shapira, M. (2007). Post-transcriptional regulation of gene expression in trypanosomes and leishmanias. Mol. Biochem. Parasitol. *156*, 93–101.

Comini, M., Menge, U., and Flohé. L. (2003). Biosynthesis of trypanothione in *Trypanosoma brucei brucei*. Biol. Chem. *384*, 653–656.

Comini, M.A., Guerrero, S.A., Haile, S., Menge, U., Lünsdorf, H., and Flohé, L. (2004). Validation of *Trypanosoma brucei* trypanothione synthetase as drug target. Free Radic. Biol. Med. *36*, 1289–1302.

Comini, M.A., Krauth-Siegel, R.L., and Flohé, L. (2007). Depletion of the thioredoxin homologue tryparedoxin impairs antioxidative defence in African trypanosomes. Biochem. J. *402*, 43–49.

Comini, M.A., Rettig, J., Dirdjaja, N., Hanschmann, E.M., Berndt, C., and Krauth-Siegel, R.L. (2008). Monothiol glutaredoxin-1 is an essential iron–sulfur protein in the mitochondrion of African trypanosomes. J. Biol. Chem. *283*, 27785–27798.

Cornillon, S., Foa, C., Davoust, J., Buonavista, N., Gross, J.D., and Golstein, P. (1994). Programmed cell death in Dictyostelium. J. Cell Sci. *107*, 2691–2704.

Cheeseman, M.T., and Gooding, R.H. (1985). Proteolytic enzymes from tsetse flies, *Glossina mositans* and *Glossina palpalis* (Diptera: Glossinidae). Insect Biochem. *15*, 677–680.

Chisi, J.E., Misiri, H., Zverev, Y., Nkhoma, A., and Sternberg, J.M. (2004). Anaemia in human African trypanosomiasis caused by *Trypanosoma brucei rhodesiense*. East Afr. Med. J. *81*, 505–508.

Delgado, M., Anderson, P., Garcia-Salcedo, J.A., Caro, M., and Gonzalez-Rey, E. (2009). Neuropeptides kill African trypanosomes by targeting intracellular compartments and inducing autophagic-like cell death. Cell Death Differ. *16*, 406–416.

Denninger, V., Koopmann, R., Muhammad, K., Barth, T., and Bassarak, B., and Schonfeld, C. (2008). Kinetoplastids model organisms for simple autophagic pathways? Methods Enzymol. *451*, 373–408.

Deponte, M. (2008). Programmed cell death in protists. Biochim. Biophys. Acta *1783*, 1396–1405.

Diechtierow, M., and Krauth-Siegel, R.L. (2011). A tryparedoxin-dependent peroxidase protects African trypanosomes from membrane damage. Free Radic. Biol. Med. *51*, 856–868.

DNDi. 2009. Annual report 2008–2009. Delivering innovation and building arobust pipeline. Drugs for Neglected Diseases Initiative, Geneva, Switzerland.

Docampo, R., Ulrich, P., and Moreno, S.N. (2010). Evolution of acidocalcisomes and their role in polyphosphate storage and osmoregulation in eukaryotic microbes. Philos. Trans. R. Soc. Lond. B. Biol. Sci. *365*, 775–784.

Drennan, M.B., Stijlemans, B., Van den Abbeele, J., Quesniaux, V.J., Barkhuizen, M., Brombacher, F., De Baetselier, P., Ryffel, B., and Magez, S. (2005). The induction of a type 1 immune response following a *Trypanosoma brucei* infection is MyD88 dependent. J. Immunol. *175*, 2501–2509.

Duleu, S., Vincendeau, P., Courtois, P., Semballa, S., Lagroye, I., Daulouède, S., Boucher, J.L., Wilson, K.T., Veyret, B., and Gobert, A.P. (2004). Mouse strain susceptibility to trypanosome infection: an arginase-dependent effect. J. Immunol. *172*, 6298–6303.

Duszenko, M., Figarella, K., Macleod, E.T., and Welburn, S.C. (2006). Death of a trypanosome: a selfish altruism. Trends Parasitol. *22*, 536–542.

Edney, E.B., and Barrass, R. (1962). The body temperature of the tsetse fly, *Glossina morsitans* Westwood (Diptera, Muscidae). J. Insect. Physiol. *8*, 469–481.

Elmore, S. (2007). Apoptosis: a review of programmed cell death. Toxicol. Pathol. *35*, 495–516.

Engstler, M., and Boshart, M. (2004). Cold shock and regulation of surface protein trafficking convey sensitization to inducers of stage differentiation in *Trypanosoma brucei*. Genes Dev. *18*, 2798–2811.

Esseiva, A.C., Chanez, A.L., Bochud-Allemann, N., Martinou, J.C., Hemphill, A., and Schneider, A. (2004). Temporal dissection of Bax-induced events leading to fission of the single mitochondrion in *Trypanosoma brucei*. EMBO Rep. *5*, 268–273.

Fairlamb, A.H., and Cerami, A. (1992) Metabolism and functions of trypanothione in the Kinetoplastida. Annu. Rev. Microbiol. *46*, 695–729.

Fairlamb, A.H., Henderson, G.B., and Cerami, A. (1989). Trypanothione is the primary target for arsenical drugs against African trypanosomes. Proc. Natl. Acad. Sci. U.S.A. *86*, 2607–2611.

Fairlamb, A.H., Carter, N.S., Cunningham, M., and Smith, K. (1992). Characterisation of melarsen-resistant *Trypanosoma brucei brucei* with respect to cross-resistance to other drugs and trypanothione metabolism. Mol. Biochem. Parasitol. *53*, 213–222.

Fast, B., Kremp, K., Boshart, M., and Steverding, D. (1999). Iron-dependent regulation of transferrin receptor expression in *Trypanosoma brucei*. Biochem. J. *342*, 691–696.

Felu, C., Pasture, J., Pays, E., and Pérez-Morga, D. (2007). Diagnostic potential of a conserved genomic rearrangement in the *Trypanosoma brucei gambiense*-specific TGSGP locus. Am. J. Trop. Med. Hyg. *76*, 922–929.

Figarella, K., Rawer, M., Uzcategui, N.L., Kubata, B.K., Lauber, K., Madeo, F., Wesselborg, S., and Duszenko, M. (2005). Prostaglandin D2 induces programmed cell death in *Trypanosoma brucei* bloodstream form. Cell Death Differ. *12*, 335–346.

Figarella, K., Uzcategui, N.L., Beck, A., Schoenfeld, C., Kubata, B.K., Lang, F., and Duszenko, M. (2006). Prostaglandin-induced programmed cell death in *Trypanosoma brucei* involves oxidative stress. Cell Death Differ. *13*, 1802–1814.

Figueiredo, L.M., Cross, G.A., and Janzen, C.J. (2009). Epigenetic regulation in African trypanosomes: a new kid on the block. Nat. Rev. Microbiol. *7*, 504–513.

Folgueira, C., and Requena, J.M. (2007). A postgenomic view of the heat shock proteins in kinetoplastids. FEMS Microbiol. *31*, 359–377.

Fulton, J.D., and Spooner, D.F. (1956). Inhibition of the respiration of *Trypanosoma rhodesiense* by thiols. Biochem. J. *63*, 475–481.

Gadelha, C., Holden, J.M., Allison, H.C., and Field, M.C. (2011) Specializations in a successful parasite: What makes the bloodstream-form African trypanosome so deadly? Mol. Biochem. Parasitol. *179*, 51–58.

Gannavaram, S., Vedvyas, C., and Debrabant, A. (2008). Conservation of the pro-apoptotic nuclease activity of endonuclease G in unicellular trypanosomatid parasites. J. Cell Sci. *121*, 99–109.

Gerrits, H., Mussmann, R., Bitter, W., Kieft, R., and Borst, P. (2002). The physiological significance of transferrin receptor variations in *Trypanosoma brucei*. Mol. Biochem. Parasitol. *119*, 237–247.

Gobert, A.P., Semballa, S., Daulouede, S., Lesthelle, S., Taxile, M., Veyret, B., and Vincendeau, P. (1998). Murine macrophages use oxygen- and nitric oxide-dependent mechanisms to synthesize

S-nitroso-albumin and to kill extracellular trypanosomes. Infect. Immun. *66*, 4068–4072.

Gobert, A.P., Daulouede, S., Lepoivre, M., Boucher, J.L., Bouteille, B., Buguet, A., Cespuglio, R., Veyre, B., and Vincendeau, P. (2000). L-Arginine availability modulates local nitric oxide production and parasite killing in experimental trypanosomiasis. Infect. Immun. *68*, 4653–4657.

Goldshmidt, H., Matas, D., Kabi, A., Carmi, S., Hope, R., and Michaeli, S. (2010). Persistent ER stress induces the spliced leader RNA silencing pathway (SLS), leading to programmed cell death in *Trypanosoma brucei*. PLoS Pathog. *6*, e1000731.

Grab, D.J., Wells, C.W., Shaw, M.K., Webster, P., and Russo, D.C. (1992). Endocytosed transferrin in African trypanosomes is delivered to lysosomes and may not be recycled. Eur. J. Cell Biol. *59*, 398–404.

Grab, D.J., Shaw, M.K., Wells, C.W., Verjee, Y., Russo, D.C., Webster, P., Naessens, J., and Fish, W.R. (1993). The transferrin receptor in African trypanosomes: identification, partial characterization and subcellular localization. Eur. J. Cell Biol. *62*, 114–126.

Graham, S.V., Matthews, K.R., Shiels, P.G., and Barry, J.D. (1990). Distinct developmental stage-specific activation mechanisms of trypanosome VSG genes. Parasitology *101*, 361–367.

De Greef, C., and Hamers, R. (1994). The serum resistance-associated (SRA) gene of *Trypanosoma brucei rhodesiense* encodes a variant surface glycoprotein-like protein. Mol. Biochem. Parasitol. *68*, 277–284.

Grosskinsky, C.M., Ezekowitz, R.A., Berton, G., Gordon, S., and Askonas, B.A. (1983). Macrophage activation in murine African trypanosomiasis. Infect. Immun. *39*, 1080–1086.

Guilliams, M., Bosschaerts, T., Hérin, M., Hünig, T., Loi, P., Flamand, V., De Baetselier.,P., and Beschin, A. (2008). Experimental expansion of the regulatory T cell population increases resistance to African trypanosomiasis. J. Infect. Dis. *198*, 781–791.

Güttinger, A., Schwab, C., Morand, S., Roditi, I., and Vassella, E. (2007). A mitogen-activated protein kinase of *Trypanosoma brucei* confers resistance to temperature stress. Mol. Biochem. Parasitol. *153*, 203–206.

Haanstra, J.R., van Tuijl, A., Kessler, P., Reijnders, W., Michels, P.A., Westerhoff, H.V., Parsons, M., and Bakker, B.M. (2008). Compartmentation prevents a lethal turbo-explosion of glycolysis in trypanosomes. Proc. Natl. Acad. Sci. U.S.A. *105*, 17718–17723.

Hall, B.S., Bot, C., and Wilkinson, S.R. (2011). Nifurtimox Activation by Trypanosomal Type I Nitroreductases Generates Cytotoxic Nitrile Metabolites. J. Biol. Chem. *286*, 13088–13095.

Hamilton, P.B., Stevens, J.R., Gaunt, M.W., Gidley, J., and Gibson, W.C. (2004). Trypanosomes are monophyletic: evidence from genes for glyceraldehyde phosphate dehydrogenase and small subunit ribosomal RNA. Int. J. Parasitol. *34*, 1393–1404.

Hamilton, P.B., Gibson, W.C., and Stevens, J.R. (2007). Patterns of co-evolution between trypanosomes and their hosts deduced from ribosomal RNA and protein-coding gene phylogenies. Mol. Phylogenetics Evol. *44*, 15–25.

Hannaert, V., Bringaud, F., Opperdoes F.R., and Michels P.A.M. (2003). Evolution of energy metabolism and its compartmentation in Kinetoplastida. Kinetoplastid Biol. Dis. *2*, 11.

Hart, D.T., Misset, O., Edwards, S.W., and Opperdoes, F.R. (1984). A comparison of the glycosomes (microbodies) isolated from *Trypanosoma brucei* bloodstream form and cultured procyclic trypomastigotes. Mol. Biochem. Parasitol. *12*, 25–35.

Helfert, S., Estévez, A.M., Bakker, B., Michels, P., and Clayton, C. (2001). Roles of triosephosphate isomerase and aerobic metabolism in *Trypanosoma brucei*. Biochem. J. *357*, 117–125.

Helms, M.J., Ambit, A., Appleton, P., Tetley, L., Coombs, G.H., and Mottram, J.C. (2006). Bloodstream form *Trypanosoma brucei* depend upon multiple metacaspases associated with RAB11-positive endosomes. J. Cell Sci. *119*, 1105–1117.

Herman, M., Gillies, S., Michels, P.A.M., and Rigden, D.J. (2006). Autophagy and related processes in trypanosomatids. Insights from genomic and bioinformatics analyses. Autophagy *2*, 107–118.

Herman, M., Pérez-Morga, D., Schtickzelle, N., and Michels, P.A.M. (2008). Turnover of glycosomes during life-cycle differentiation of *Trypanosoma brucei*. Autophagy 4. *3*, 294–308.

Hillebrand, H., Schmidt, A., and Krauth-Siegel, R.L. (2003). A second class of peroxidases linked to the trypanothione metabolism. J. Biol. Chem. *278*, 6809–6815.

Holmgren, A., Johansson, C., Berndt, C., Lönn, M.E., Hudemann, C., and Lillig, C.H. (2005). Thiol redox control via thioredoxin and glutaredoxin systems. Biochem. Soc. Trans. *33*, 1375–1377.

Huang, G., Fang, J., Sant'anna, C., Li, Z., Wellems, D.L., Rohloff, P., and Docampo, R. (2011). Adaptor protein-3 (AP-3) complex mediates the biogenesis of acidocalcisomes and is essential for growth and virulence of *Trypanosoma brucei*. J. Biol. Chem. *286*, 36619–36630.

Igbokwe, I.O., Esievo, K.A., Saror, D.I., and Obagaiye, O.K. (1994). Increased susceptibility of erythrocytes to *in vitro* peroxidation in acute *Trypanosoma brucei* infection of mice. Vet. Parasitol. *55*, 279–286.

Igbokwe, I.O., Umar, I.A., Omage, J.J., Ibrahim, N.D., Kadima, K.B., Obagaiye, O.K., Saror, D.I., and Esievo, K.A. (1996). Effect of acute *Trypanosoma vivax* infection on cattle erythrocyte glutathione and susceptibility to *in vitro* peroxidation. Vet. Parasitol. *63*, 215–224.

Irigoín, F., Cibils, L., Comini, M.A., Wilkinson, S.R., Flohé, L., and Radi, R. (2008). Insights into the redox biology of *Trypanosoma cruzi*: Trypanothione metabolism and oxidant detoxification. Free Radic. Biol. Med. *45*, 733–742.

Izquierdo, L., Schulz, B.L., Rodrigues, J.A., Güther, M.L., Procter, J.B., Barton, G.J., Aebi, M., and Ferguson, M.A. (2009). Distinct donor and acceptor specificities of *Trypanosoma brucei* oligosaccharyltransferases. EMBO J. *28*, 2650–2661.

Jensen, B.C., Sivam, D., Kifer, C.T., Myler, P.J., and Parsons, M. (2009). Widespread variation in transcript abundance within and across developmental stages of *Trypanosoma brucei*. BMC Genomics *10*, 482.

Jensen, B.C., Kifer, C.T., and Parsons, M. (2011). *Trypanosoma brucei*: Two mitogen activated protein kinase kinases are dispensable for growth and virulence of the bloodstream form. Exp. Parasitol. *128*, 250–255.

de Jesus, T.C., Tonelli, R.R., Nardelli, S.C., da Silva Augusto, L., Motta, M.C., Girard-Dias, W., Miranda, K., Ulrich, P., Jimenez, V., Barquilla, A., *et al.* (2010). Target of rapamycin (TOR)-like 1 kinase is involved in the control of polyphosphate levels and acidocalcisome maintenance in *Trypanosoma brucei*. J. Biol. Chem. *285*, 24131–24140.

Jiménez-Ruiz, A., Alzate, J.F., Macleod, E.T., Lüder, C.G., Fasel, N., and Hurd, H. (2010). Apoptotic markers in protozoan parasites. Parasit. Vectors *3*, 104.

Jones, A., Faldas, A., Foucher, A., Hunt, E., Tait, A., Wastling, J.M., and Turner, C.M. (2006). Visualisation and analysis of proteomic data from the procyclic form of *Trypanosoma brucei*. Proteomics *6*, 259–267.

Jones, C., Anderson, S., Singha, U.K., and Chaudhuri, M. (2008). Protein phosphatase 5 is required for Hsp90 function during proteotoxic stresses in *Trypanosoma brucei*. Parasitol. Res. *102*, 835–844.

Kabani, S., Fenn, K., Ross, A., Ivens, A., Smith, T.K., Ghazal, P., and Matthews, K. (2009). Genome-wide expression profiling of *in vivo*-derived bloodstream parasite stages and dynamic analysis of mRNA alterations during synchronous differentiation in *Trypanosoma brucei*. BMC Genomics *10*, 427.

Kabiri, M., and Steverding, D. (2000). Studies on the recycling of the transferrin receptor in *Trypanosoma brucei* using an inducible gene expression system. Eur. J. Biochem. *267*, 3309–3314.

Kaczanowski, S., Sajid, M., and Reece, S.E. (2011). Evolution of apoptosis-like programmed cell death in unicellular protozoan parasites. Parasit. Vectors *4*, 44.

Kerr, J.F., Wyllie, A.H., and Currie, A.R. (1972). Apoptosis: a basic biological phenomenon with wide-ranging implications in tissue kinetics. Br. J. Cancer. *26*, 239–257.

Kessler, P.S., and Parsons, M. (2005). Probing the role of compartmentation of glycolysis in procyclic form *Trypanosoma brucei*: RNA interference studies of PEX14, hexokinase, and phosphofructokinase. J. Biol. Chem. *280*, 9030–9036.

Kieft, R., Capewell, P., Turner, C.M., Veitch, N.J., MacLeod, A., and Hajduk, S. (2010). Mechanism of *Trypanosoma brucei gambiense* (group 1) resistance to human trypanosome lytic factor. Proc. Natl. Acad. Sci. U.S.A. *107*, 16137–16141.

Kiel, J.A. (2010). Autophagy in unicellular eukaryotes. Philos. Trans. R. Soc. Lond. B. Biol. Sci. *365*, 819–830.

Koumandou, V.L., Natesan, S.K.A., Sergeenko, T., and Fiel, M.C. (2008). The trypanosome transcriptome is remodelled during differentiation but displays limited responsiveness within life stages BMC Genomics *9*, 298.

Kramer, S., Queiroz, R., Ellis, L., Webb, H., Hoheisel, J.D., Clayton, C., and Carrington, M. (2008). Heat shock causes a decrease in polysomes and the appearance of stress granules in trypanosomes independently of eIF2(alpha) phosphorylation at Thr169. J. Cell Sci. *121*, 3002–3014.

Krauth-Siegel, R.L., and Comini, M.A. (2008). Redox control in trypanosomatids, parasitic protozoa with trypanothione-based thiol metabolism. Biochim. Biophys. Acta. *1780*, 1236–1248.

Koopman, R., Muhammad, K., Perbandt, M., Betzel, C., and Duszenko, M. (2009). *Trypanosoma brucei* ATG8. Structural insights into autophagic-like mechanisms in protozoa. Autophagy *5*, 1085–1091.

Krieger, S., Schwarz, W., Ariyanayagam, M.R., Fairlamb, A.H., Krauth-Siegel, R.L., and Clayton, C. (2000). Trypanosomes lacking trypanothione reductase are avirulent and show increased sensitivity to oxidative stress. Mol. Microbiol. *35*, 542–552.

Lang, F. (2007). Mechanisms and significance of cell volume regulation. J. Am. Coll. Nutr. *26*, 613S-623S.

Lee, M.G. (1998). The 3′ untranslated region of the hsp 70 genes maintains the level of steady state mRNA in *Trypanosoma brucei* upon heat shock. Nucleic Acids Res. *26*, 4025–4033.

Lenardo, M.J., Esser, K.M., Moon, A.M., Van der Ploeg, L.H.T., and Donelson, E. (1986). Metacyclic variant surface glycoprotein genes of *Trypanosoma brucei* subsp. *rhodesiense* are activated *in situ*, and their expression is transcriptionally regulated. Mol. Cell. Biol. *6*, 1991–1997.

Lepesheva, G.I., Nes, W.D., Zhou, W., Hill, G.C., and Waterman, M.R. (2004). CYP51 from *Trypanosoma brucei* is obtusifoliol-specific. Biochemistry. *43*, 10789–10799.

Lepesheva, G.I., Hargrove, T.Y., Kleshchenko, Y., Nes, W.D., Villalta, F., and Waterman, M.R. (2008). CYP51: A major drug target in the cytochrome P450 superfamily. Lipids. *43*, 1117–1125.

Lewis, K. (2000). Programmed death in bacteria. Microbiol. Mol. Biol. Rev. *64*, 503–514.

Lillig, C.H., and Holmgren, A. (2007). Thioredoxin and related molecules – from biology to health and disease. Antioxid. Redox Signal *9*, 25–47.

Louw, C.A., Ludewig, M.H., Mayer, J., and Blatch, G.L. (2010a). The Hsp70 chaperones of the Tritryps are characterized by unusual features and novel members. Parasitol. Int. *59*, 497–505.

Louw, C.A., Ludewig, M.H., and Blatch, G.L. (2010b). Overproduction, purification and characterisation of Tbj1, a novel Type III Hsp40 from *Trypanosoma brucei*, the African sleeping sickness parasite. Protein Expr. Purif. *69*, 168–177.

Lüdemann. H., Dormeyer, M., Sticherling, C., Stallmann, D., Follmann, H., and Krauth-Siegel, R.L. (1998). *Trypanosoma brucei* tryparedoxin, a thioredoxin-like protein in African trypanosomes. FEBS Lett. *431*, 381–385.

Lüder, C.G., Campos-Salinas, J., Gonzalez-Rey, E., and van Zandbergen, G. (2010). Impact of protozoan cell death on parasite–host interactions and pathogenesis. Parasitol. Vectors *3*, 116.

van Luenen, H.G., Kieft, R., Mussmann, R., Engstler, M., ter Riet, B., and Borst, P. (2004). Trypanosomes change their transferrin receptor expression to allow effective uptake of host transferrin. Mol. Microbiol. *58*, 151–165.

Lugli, E.B., Pouliot, M., Portela, M., del P., Loomis, M.R., and Raper, J. (2004). Characterization of primate trypanosome lytic factors. Mol. Biochem. Parasitol. *138*, 9–20.

Luckins, A.G., and Dwinger, R.H. (2004). Non-tsetse-transmitted trypanosomiasis. In The Trypanosomiases, Maudlin, I., Holmes P.H., and Miles, M.A., eds (CA International Press, Wallingford, UK), pp. 269–281.

Lustig, Y., Sheiner, L., Vagima, Y., Goldshmidt, H., Das, A., Bellofatto, V., and Michaeli, S. (2007). Spliced-leader RNA silencing: a novel stress-induced mechanism in *Trypanosoma brucei*. EMBO Rep. *8*, 408–413.

Mabbott, N.A., Sutherland, I.A., and Sternberg, J.M. (1995). Suppressor macrophages in *Trypanosoma brucei* infection: nitric oxide is related to both suppressive activity and lifespan *in vivo*. Parasite Immunol. *17*, 143–150.

MacGregor, P., Savill, N.J., Hall, D., and Matthews, K.R. (2011). Transmission stages dominate trypanosome within-host dynamics during chronic infections. Cell Host Microbe *9*, 310–318.

Maiorino, M., Ursini, F., Bosello, V., Toppo, S., Tosatto, S.C., Mauri, P., Becker, K., Roveri, A., Bulato, C., Benazzi, L., *et al.* (2007). The thioredoxin specificity of Drosophila GPx: a paradigm for a peroxiredoxin-like mechanism of many glutathione peroxidases. J. Mol. Biol. *365*, 1033–1046.

Mamani-Matsuda, M., Rambert, J., Malvy, D., Lejoly Boisseau, H., Daulouède, S., Thiolat, D., Coves, S., Courtois, P., Vincendeau, P., and Mossalayi, M.D. (2004). Quercetin induces apoptosis of *Trypanosoma brucei gambiense* and decreases the proinflammatory response of human macrophages. Antimicrob. Agents. Chemother. *48*, 924–929.

Mäser, P., and Kaminsky, R. (1998). Identification of three ABC transporter genes in *Trypanosoma brucei* spp. Parasitol. Res. *84*, 106–111.

Mäser, P., Lüscher, A., and Kaminsky, R. (2003). Drug transport and drug resistance in African trypanosomes. Drug Resist. Updat. *6*, 281–290.

Maslov, D.A., Lukes, J., Jirku, M., and Simpson, L. (1996). Phylogeny of trypanosomes as inferred from the small and large subunit rRNAs: implications for the evolution of parasitism in the trypanosomatid protozoa. Mol. Biochem. Parasitol. *75*, 197–205.

Maya, J.D., Rodríguez, A., Pino. L., Pabón, A., Ferreira, J., Pavani, M., Repetto, Y., and Morello, A. (2004). Effects of buthionine sulfoximine, nifurtimox and benznidazole upon trypanothione and metallothionein proteins in *Trypanosoma cruzi*. Biol. Res. *37*, 61–69.

van Meirvenne, N., Magnus, E., and Janssens, P.G. (1975). Antigenic variation in syringe passaged populations of *Trypanosoma* (Trypanozoon) *brucei*. 1. Rationalization of the experimental approach. Ann. Soc. Belg. Med. Trop. *55*, 1–23.

Meshnick, S.R., Chang, K.P., and Cerami, A. (1977). Heme lysis of the bloodstream forms of *Trypanosoma brucei*. Biochem. Pharmacol. *26*, 1923–1928.

Michels, P.A.M., Bringaud, F., Herman, M., and Hannaert, V. (2006). Metabolic functions of glycosomes in trypanosomatids. Biochim. Biophys. Acta. *1763*, 1463–1477.

Misset, O., and Opperdoes, F.R. (1984). Simultaneous purification of hexokinase, class-I fructose-bisphosphate aldolase, triosephosphate isomerase and phosphoglycerate kinase from *Trypanosoma brucei*. Eur. J. Biochem. *144*, 475–483.

Mottram, J.C., Helms, M.J., Coombs, G.H., and Sajid, M. (2003). Clan CD cysteine peptidases of parasitic protozoa. Trends. Parasitol. *19*, 182–187.

Mowatt, M.R., and Clayton, C.E. (1987). Developmental regulation of a novel repetitive protein of *Trypanosoma brucei*. Mol. Cell. Biol. *7*, 2838–2844.

Muhich, M.L., and Boothroyd, J.C. (1989). Synthesis of trypanosome hsp70 mRNA is resistant to disruption of trans-splicing by heat shock. J. Biol. Chem. *264*, 7107–7110.

Mussmann, R., Janssen, H., Calafat, J., Engstler, M., Ansorge, I., Clayton, C., and Borst, P. (2003). The expression level determines the surface distribution of the transferrin receptor in *Trypanosoma brucei*. Mol. Microbiol. *47*, 23–35.

Mussmann, R., Engstler, M., Gerrits, H., Kieft, R., Toaldo, C.B., Onderwater, J., Koerten, H., van Luenen, H.G., and Borst, P. (2004). Factors affecting the level and localization of the transferrin receptor in *Trypanosoma brucei*. J. Biol. Chem. *279*, 40690–40698.

Nedelcu, A.M., Driscoll, W.W., Durand, P.M., Herron, M.D., and Rashidi, A. (2011). On the paradigm of altruistic suicide in the unicellular world. Evolution *65*, 3–20.

Nogoceke, E., Gommel, D.U., Kiess, M., Kalisz, H.M., and Flohé, L. (1997). A unique cascade of oxidoreductases catalyses trypanothione-mediated peroxide metabolism in *Crithidia fasciculata*. Biol. Chem. *378*, 827–836.

Nok, A.J. (2003). Arsenicals (melarsoprol), pentamidine and suramin in the treatmentof human African tripanosomiasis. Parasitol. Res. *90*, 71–79.

Nolan, D.P., Rolin, S., Rodriguez, J.R., Van Den Abbeele, J., and Pays, E. (2000). Slender and stumpy bloodstream forms of *Trypanosoma brucei* display a differential response to extracellular acidic and proteolytic stress. Eur. J. Biochem. *267*, 18–27.

Nyarko, E., Hara, T., Grab, D.J., Tabata, M., and Fukuma, T. (2002). Toxic effects of mercury(II), cadmium(II) and lead(II) porphyrins on *Trypanosoma brucei brucei* growth. Chem. Biol. Interact. *139*, 177–185.

Nyden, S.J. (1984). Changes in ascorbic acid metabolism of the rat during infection with *Trypanosoma hippicum*. Proc. Soc. Exp. Biol. Med. *69*, 206–210.

Opperdoes, F.R. (2010). The glycosome of trypanosomatids. Microbiol. Monographs. *17*, 285–298.

Ouellette, M., and Papadopoulou, B. (2009). Coordinated gene expression by post-transcriptional regulons in African trypanosomes. J. Biol. *8*, 100.

Oza, S.L., Ariyanayagam, M.R., Aitcheson, N., and Fairlamb, A.H. (2003). Properties of trypanothione synthetase from *Trypanosoma brucei*. Mol. Biochem. Parasitol. *131*, 25–33.

Parsons, M., and Nielsen, B. (1990). *Trypanosoma brucei*: two-dimensional gel analysis of the major glycosomal proteins during the life cycle. Exp. Parasitol. *70*, 276–285.

Pays, E., and Vanhollebeke, B. (2009). Human innate immunity against African trypanosomes. Curr. Opin. Immunol. *21*, 493–498.

Pays, E., Vanhamme, L., and Pérez-Morga, D. (2004). Antigenic variation in *Trypanosoma brucei*: facts, challenges and mysteries. Curr. Opin. Microbiol. *7*, 369–374.

Pays, E., Vanhollebeke, B., Vanhamme, L., Paturiaux-Hanocq, F., Nolan, D.P., and Pérez-Morga, D. (2006). The trypanolytic factor of human serum. Nat. Rev. Microbiol. *4*, 477–486.

Pearl, L.H., and Prodromou, C. (2006). Structure and mechanism of the Hsp90 molecular chaperone machinery. Annu. Rev. Biochem. *76*, 271–294.

Pearson, T.W., Beecroft, R.P., Welburn, S.C., Ruepp, S., Roditi,I., Hwa, K.Y., Englund, P.T., Wells, C.W., Murphy, N.B. (2000). The major cell surface glycoprotein procyclin is a receptor for induction of a novel form of cell death in African trypanosomes *in vitro*. Mol. Biochem. Parasitol. *111*, 333–349.

Penketh, P.G., and Klein, R.A. (1986). Hydrogen peroxide metabolism in *Trypanosoma brucei*. Mol. Biochem. Parasitol. *20*, 111–121.

Pérez-Victoria, J.M., Di Pietro, A., Barron, D., Ravelo, A.G., Castanys, S., and Gamarro, F. (2002). Multidrug resistance phenotype mediated by the P-glycoprotein-like transporter in *Leishmania*: a search for reversal agents. Curr. Drug Targets. *3*, 311–333.

van der Ploeg, L.H.T., Giannini, S.H., and Cantor, C.R. (1985). Heat shock genes: regulatory role for differentiation in parasitic protozoa. Science *228*, 1443–1445.

Poelvoorde, P., Vanhamme, L., Van Den Abbeele, J., Switzer, W.M., and Pays, E. (2004). Distribution of apolipoprotein L-I and trypanosome lytic activity among primate sera. Mol. Biochem. Parasitol. *134*, 155–157.

Portal, P., Villamil, S.F., Alonso, G.D., De Vas, M.G., Flawiá, M.M., Torres, H.N., and Paveto, C. (2008). Multiple NADPH-cytochrome P450 reductases from *Trypanosoma cruzi* suggested role on drug resistance. Mol. Biochem. Parasitol. *160*, 42–51.

Priotto, G., Kasparian, S., Mutombo, W., Ngouama, D., Ghorashian, S., Arnold, U., Ghabri, S., Baudin, E., Buard, V., Kazadi-Kyanza, S., *et al.* (2009). Nifurtimox–eflornithine combination therapy for second-stage African *Trypanosoma brucei gambiense* trypanosomiasis: a multicentre, randomised, phase III, non-inferiority trial. Lancet. *374*, 56–64.

Queiroz, R., Benz, C., Fellenberg, K., Hoheisel, J.D., and Clayton, C. (2009). Transcriptome analysis of differentiating trypanosomes reveals the existence of multiple post-transcriptional regulons. BMC Genomics *10*, 495.

Radwanska, M., Magez, S., Michel, A., Stijlemans, B., Geuskens, M., and Pays, E. (2000a). Comparative analysis of antibody responses against HSP60, invariant surface glycoprotein 70, and variant surface glycoprotein reveals a complex antigen-specific pattern of immunoglobulin isotype switching during infection by *Trypanosoma brucei*. Infect. Immun. *68*, 848–860.

Radwanska, M., Magez, S., Dumont, N., Pays, A., Nolan, D., and Pays, E. (2000b). Antibodies raised against the flagellar pocket fraction of *Trypanosoma brucei* preferentially recognize HSP60 in cDNA expression library. Parasite Immunol. *22*, 639–650.

Ridgley, E.L., Xiong, Z.H., and Ruben, L. (1999). Reactive oxygen species activate a Ca2+-dependent cell death pathway in the unicellular organism *Trypanosoma brucei brucei*. Biochem. J. *340*, 33–40.

Roditi, I., Carrington, M., and Turner, M. (1987). Expression of a polypeptide containing a dipeptide repeat is confined to the insect stage of *Trypanosoma brucei*. Nature *325*, 272–274.

Roldán, A., Comini, M.A., Crispo, M., and Krauth-Siegel, R.L. (2011). Lipoamide dehydrogenase is essential for both bloodstream and procyclic *Trypanosoma brucei*. Mol. Microbiol. *81*, 623–639.

Rolin, S., Hanocq-Quertier, J., Paturiaux-Hanocq, F., Nolan, D., Salmon. D., Webb, H., Carrington, M., Voorheis, P., and Pays, E. (1996). Simultaneous but independent activation of adenylate cyclase and glycosylphosphatidylinositol-phospholipase C under stress conditions in *Trypanosoma brucei*. J. Biol. Chem. *271*, 10844–10852.

Roper, J.R., Güther, M.L., Macrae, J.I., Prescott, A.R., Hallyburton, I., Acosta-Serrano, A., and Ferguson, M.A. (2005). The suppression of galactose metabolism in procylic form *Trypanosoma brucei* causes cessation of cell growth and alters procyclin glycoprotein structure and copy number. J. Biol. Chem. *280*, 19728–19736.

Rudenko, G. (2010). Epigenetics and transcriptional control in African trypanosomes. Essays Biochem. *48*, 201–219.

Salmon, D., Geuskens, M., Hanocq, F., Hanocq-Quertier, J., Nolan, D., Ruben, L., and Pays, E. (1994). A novel heterodimeric transferrin receptor encoded by a pair of VSG expression site-associated genes in *T. brucei*. Cell *78*, 75–86.

Salmon, D., Hanocq-Quertier, J., Paturiaux-Hanocq, F., Pays, A., Tebabi, P., Nolan, D.P., Michel, A., and Pays, E. (1997). Characterization of the ligand-binding site of the transferrin receptor in *Trypanosoma brucei* demonstrates a structural relationship with the N-terminal domain of the variant surface glycoprotein. EMBO J. *16*, 7272–7278.

Schlesinger, M.J., Ashbumer, M., and Tissieres, A. (1982). Heat Shock from Bacteria to Man (Cold Spring Harbor Laboratory, Cold Spring Harbor, NY).

Schlecker, T., Schmidt, A., Dirdjaja, N., Voncken, F., Clayton, C., and Krauth-Siegel, R.L. (2005). Substrate specificity, localization, and essential role of the glutathione peroxidase-type tryparedoxin peroxidases in *Trypanosoma brucei*. J. Biol. Chem. *280*, 14385–14394.

Schlecker, T., Comini, M.A., Melchers, J., Ruppert, T., and Krauth-Siegel, R.L. (2007). Catalytic mechanism of the glutathione peroxidase-type tryparedoxin peroxidase of *Trypanosoma brucei*. Biochem J. *405*, 445–454.

Seed, J.R., Wenck, M.A. (2003). Role of the long slender to short stumpy transition in the life cycle of the african trypanosomes. Kinetop. Biol. Dis. *2*, 1–8.

Seyfang, A., and Duszenko, M. (1991). Specificity of glucose transport in *Trypanosoma brucei*. Eur. J. Biochem. *202*, 191–196.

Shahi, S.K., Krauth-Siegel, R.L., and Clayton, C.E. (2002). Overexpression of the putative thiol conjugate transporter TbMRPA causes melarsoprol resistance in *Trypanosoma brucei*. Mol. Microbiol. *43*, 1129–1138.

Sheader, K., Vaughan, S., Minchin, J., Hughes, K., Gull, K., and Rudenko, G. (2005). Variant surface glycoprotein RNA interference triggers a precytokinesis cell cycle arrest in African trypanosomes. Proc. Natl. Acad. Sci. U.S.A. *102*, 8716–8721.

Simpson, A.G., Lukes, J., and Roger, A.J. (2002). The evolutionary history of kinetoplastids and their kinetoplasts. Mol. Biol. Evol. *19*, 2071–2083.

Simpson, A.G., Stevens, J.R., and Lukes, J. (2006). The evolution and diversity of kinetoplastid flagellates. Trends. Parasitol. *22*, 168–174.

Smíd, O., Horáková, E., Vilímová, V., Hrdy, I., Cammack, R., Horváth, A., Lukes, J., and Tachezy, J. (2006) Knock-downs of iron–sulfur cluster assembly proteins IscS and IscU down-regulate the active mitochondrion of procyclic *Trypanosoma brucei*. J. Biol. Chem. *281*, 28679–28686.

Smirlis, D., Duszenko, M., Ruiz, A.J., Scoulica, E., Bastien, P., Fasel, N., and Soteriadou, K. (2010). Targeting essential pathways in trypanosomatids gives insights into protozoan mechanisms of cell death. Parasit. Vectors *3*, 107.

Smith, T.K., Vasileva, N., Gluenz, E., Terry, S., Portman, N., Kramer, S., Carrington, M., Michaeli, S., Gull, K., and Rudenko, G. (2009). Blocking variant surface glycoprotein synthesis in *Trypanosoma brucei* triggers a general arrest in translation initiation. PLoS One *4*, e7532.

Sokolova, A.Y., Wyllie, S., Patterson, S., Oza, S.L., Read, K.D., and Fairlamb, A.H. (2010). Cross-resistance to nitro drugs and implications for treatment of human African trypanosomiasis. Antimicrob. Agents Chemother. *54*, 2893–2900.

Sternberg, J.M. (2004). Human African trypanosomiasis: clinical presentation and immune response. Parasite Immunol. *26*, 469–476.

Stephens, N.A., and Hajduk, S.L. (2011). Endosomal localization of the serum resistance-associated protein in african trypanosomes confers human infectivity. Eukaryot. Cell *10*, 1023–1033.

Stevens, J., Noyes, H., and Gibson, W. (1998). The evolution of trypanosomes infecting humans and primates. Mem. Inst. Oswaldo Cruz. *93*, 669–676.

Stevens, J.R., Noyes, H.A., Schofield, C.J., and Gibson, W. (2001). The molecular evolution of Trypanosomatidae. Adv. Parasitol. *48*, 1–56.

Steverding, D. (1998). Bloodstream forms of *Trypanosoma brucei* require only small amounts of iron for growth. Parasitol. Res. *84*, 59–62.

Steverding, D. (2003). The significance of transferrin receptor variation in *Trypanosoma brucei*. Trends Parasitol. *19*, 125–127.

Steverding, D. (2006). On the significance of host antibody response to the *Trypanosoma brucei* transferrin receptor during chronic infection. Microbes Infect. *8*, 2777–2782.

Steverding, D., Stierhof, Y.D., Fuchs, H., Tauber, R., and Overath, P. (1995). Transferrin-binding protein complex is the receptor for transferrin uptake in *Trypanosoma brucei*. J. Cell Biol. *131*, 1173–1182.

Stijlemans, B., Vankrunkelsven, A., Brys, L., Magez, S., and De Baetselier, P. (2008). Role of iron homeostasis in trypanosomiasis-associated anemia. Immunobiology *213*, 823–835.

Taylor, J.E., and Rudenko, G. (2006). Switching trypanosome coats: what's in the wardrobe? Trends Gen. *22*, 614–620.

Tetaud, E., Giroud, C., Prescott, A.R., Parkin, D.W., Baltz, D., Biteau, N., Baltz, T., and Fairlamb, A.H. (2001). Molecular characterisation of mitochondrial and cytosolic trypanothione-dependent tryparedoxin peroxidases in *Trypanosoma brucei*. Mol. Biochem. Parasitol. *116*, 171–183.

Thomson, L., Denicola, A., and Radi, R. (2003). The trypanothione-thiol system in *Trypanosoma cruzi* as a key antioxidant mechanism against peroxynitrite-mediated cytotoxicity. Arch. Biochem. Biophys. *412*, 55–64.

Toppo, S., Flohé, L., Ursini, F., Vanin, S., and Maiorino, M. (2009). Catalytic mechanisms and specificities of glutathione peroxidases: variations of a basic scheme. Biochim. Biophys. Acta. *1790*, 1486–1500.

Trujillo, M., Budde, H.-, Piñeyro, M.-D.-, Stehr, M., Robello, C., Flohé, L., and Radi R. (2004). *Trypanosoma brucei* and *Trypanosoma cruzi* tryparedoxin peroxidases catalytically detoxify peroxynitrite via oxidation of fast reacting thiols. J. Biol. Chem. *279*, 34175–34182.

Tsuda, A., Witola, W.H., Ohashi, K., and Onuma, M. (2005). Expression of alternative oxidase inhibits programmed cell death-like phenomenon in bloodstream form of *Trypanosoma brucei rhodesiense*. Parasitol. Int. *54*, 243–251.

Tsuda, A., Witola, W.H., Konnai, S., Ohashi, K., and Onuma, M. (2006). The effect of TAO expression on PCD-like phenomenon development and drug resistance in *Trypanosoma brucei*. Parasitol. Int. *55*, 135–142.

Tyler, K.M., Matthews, K.R., and Gull, K. (1997). The bloodstream differentiation-division of *Trypanosoma brucei* studied using mitochondrial markers. Proc. R. Soc. Lond. B. *264*, 1481–1490.

Urbaniak, M.D., Turnock, D.C., and Ferguson, M.A.J. (2006). Galactose Starvation in a Bloodstream Form

Trypanosoma brucei UDP-Glucose 4-Epimerase Conditional Null Mutant. Euk. Cell 5, 1906–1913.

Uzcátegui, N.L., Szallies, A., Pavlovic-Djuranovic, S., Palmada, M., Figarella, K., Boehmer, C., Lang, F., Beitz, E., and Duszenko, M. (2004). Cloning, heterologous expression, and characterization of three aquaglyceroporins from *Trypanosoma brucei*. J. Biol. Chem. *279*, 42669–42676.

Uzcátegui, N.L., Denninger, V., Merkel, P., Schoenfeld, C., Figarella, K., and Duszenko, M. (2007). Dihydroxyacetone induced autophagy in African trypanosomes. Autophagy *3*, 626–629.

Vanhamme, L., and Pays, E. (1995). Control of gene expression in trypanosomes. Microbiol. Revs. *59*, 223–240.

Vanhamme, L., Paturiaux-Hanocq, F., Poelvoorde, P., Nolan, D.P., Lins, L., Van Den Abbeele, J., Pays, A., Tebabi, P., Van Xong, H., Jacquet, A., *et al.* (2003). Apolipoprotein L-I is the trypanosome lytic factor of human serum. Nature *422*, 83–87.

Vanhamme, L., Renauld, H., Lecordier, L., Poelvoorde, P., Van Den Abbeele, J., and Pays, E. (2004). The *Trypanosoma brucei* reference strain TREU927/4 contains *T. brucei rhodesiense*-specific SRA sequences, but displays a distinct phenotype of relative resistance to human serum. Mol. Biochem. Parasitol. *135*, 39–47.

Vassella, E., Reuner, B., Yutzy, B., and Boshart, M. (1997). Differentiation of African trypanosomes is controlled by a density sensing mechanism which signals cell cycle arrest via the cAMP pathway. J. Cell Sci. *110*, 2661–2671.

Vercammen, D., Declercq, W., Vandenabeele, P., and Van Breusegem, F. (2007). Are metacaspases caspases? J. Cell Biol. *179*, 375–380.

Vercesi, A.E., Moreno, S.N., and Docampo, R. (1994). Ca2+/H+ exchange in acidic vacuoles of *Trypanosoma brucei*. Biochem. J. *304*, 227–233.

Vertommen, D., Van Roy, J., Szikora, J.P., Rider, M.H., Michels, P.A., and Opperdoes, F.R. (2008). Differential expression of glycosomal and mitochondrial proteins in the two major life-cycle stages of *Trypanosoma brucei*. Mol. Biochem. Parasitol. *158*, 189–201.

Vickerman, K. (1985). Developmental cycles and biology of pathogenir trypanosomes. Br. Med. Bull. *41*, 105–114.

Vickers, T.J., and Fairlamb, A.H. (2004). Trypanothione S-transferase activity in a trypanosomatid ribosomal elongation factor 1B. J. Biol. Chem. *279*, 27246–27256.

Vickers, T.J., Wyllie, S., and Fairlamb, A.H. (2004a). *Leishmania major* elongation factor 1B complex has trypanothione S-transferase and peroxidase activity. J. Biol. Chem. *279*, 49003–49009.

Vilella, F., Alves, R., Rodríguez-Manzaneque, M.T., Bellí, G., Swaminathan, S., Sunnerhagen, P., and Herrero, E. (2004b). Evolution and cellular function of monothiol glutaredoxins: involvement in iron–sulphur cluster assembly. Comp. Funct. Genomics *5*, 328–341.

Vincendeau, P., and Bouteille, B. (2006). Immunology and immunopathology of African trypanosomiasis. An. Acad. Bras. Cienc. *78*, 645–665.

Vray, B., De Baetselier, P., Ouaissi, A., and Carlier, Y. (1991). *Trypanosoma cruzi* but not *Trypanosoma brucei* fails to induce a chemiluminescent signal in a macrophage hybridoma cell line. Infect. Immun. *59*, 3303–3308.

Welburn, S.C., Dale, C., Ellis, D., Beecroft, R., and Pearson, T.W. (1996). Apoptosis in procyclic *Trypanosoma brucei rhodesiense in vitro.* Cell Death Differ. *3*, 229–236.

Welburn, S.C., Macleod, E., Figarella, K., and Duzensko, M. (2006). Programmed cell death in African trypanosomes. Parasitology *132*, S7–S18.

Wells, A.D., and Malkowsky, M. (2000). Heat shock proteins, tumor immunogenicity and antigen presentation: an integrated view. Immunol. Today *21*, 129–132.

Wigglesworth, V.B. (1929). Phthiriasis in the Primates: a Sidelight on Phylogeny. Proc. R. Soc. Med. *22*, 829.

Wilkinson, S.R., Horn, D., Prathalingam, S.R., and Kelly, J.M. (2003). RNA interference identifies two hydroperoxide metabolizing enzymes that are essential to the bloodstream form of the African trypanosome. J. Biol. Chem. *278*, 31640–31660.

Wilkinson, S.R., Prathalingam, S.R., Taylor, M.C., Ahmed, A., Horn, D., and Kelly, J.M. (2006). Functional characterisation of the iron superoxide dismutase gene repertoire in *Trypanosoma brucei*. Free Radic. Biol. Med. *40*, 198–209.

Wilkinson, S.R., Taylor, M.C., Horn, D., Kelly, J.M., and Cheeseman, I. (2008). A mechanism for cross-resistance to nifurtimoxand benznidazole in trypanosomes. Proc. Natl. Acad. Sci. U.S.A. *105*, 5022–5027.

Wyllie, S., Oza, S.L., Patterson, S., Spinks, D., Thompson, S., and Fairlamb, A.H. (2009). Dissecting the essentiality of the bifunctional trypanothione synthetase-amidase in *Trypanosoma brucei* using chemical and genetic methods. Mol. Microbiol. *74*, 529–540.

Xong, H.V., Vanhamme, L., Chamekh, M., Chimfwembe, C.E., Van Den Abbeele, J., Pays, A., Van Meirvenne, N., Hamers, R., De Baetselier, P., and Pays, E. (1998). A VSG expression site-associated gene confers resistance to human serum in *Trypanosoma rhodesiense*. Cell *95*, 839–846.

Stress Response in *Entamoeba histolytica*

17

Alfonso Olivos-García, Emma Saavedra, Erika Rubí Luis-García, Mario Nequiz and Ruy Pérez-Tamayo

Abstract

Several species belonging to the genus *Entamoeba* can colonize the mouth or the human gut; only *Entamoeba histolytica,* however, is pathogenic to the host, causing the disease amoebiasis. This illness is responsible for one hundred thousand human deaths per year worldwide, affecting mainly underdeveloped countries. Throughout its entire life cycle, or invasion of human tissues, the parasite is constantly subjected to stress conditions. Under *in vitro* culture, this microaerophilic parasite can tolerate up to 5% oxygen concentrations; however, during tissue invasion the parasite has to cope with the higher oxygen content found in well-perfused tissues (4–14%) and with reactive oxygen and nitrogen species (ROS and NOS, respectively) derived from both host and parasite. In almost all living cells, a low-dose, tightly regulated generation of ROS and NOS mediates several physiological functions such as growth, differentiation and metabolism; an excess of ROS and NOS, however, damages DNA, proteins and lipids, leading to cell death. In this chapter we review the latest findings regarding the physiological and pathological molecular functions of oxidative and nitrosative stresses in *E. histolytica* and discuss whether the molecules involved in the antioxidant system of the parasite can be appropriate drug targets for the treatment of amoebiasis.

Introduction

Amoebiasis is the human parasitic disease caused by the protozoan *Entamoeba histolytica.* This intestinal illness is responsible for 100,000 deaths per year worldwide, affecting mainly underdeveloped countries in areas with poor sanitary conditions (WHO, 1997). The most frequent affection is amoebic colitis, whereas the most severe complications are intestinal perforations and extra-intestinal abscesses (Brandt and Pérez-Tamayo, 1970).

The life cycle of *E. histolytica* starts when the human host ingests water or foods contaminated with parasite cysts. The amoebal cysts endure the acidic pH and nitric oxide found in the stomach (Raat *et al.,* 2009); once reaching the small intestine, some of them excyst releasing four trophozoites which are the vegetative and proliferating form of the parasite. In the large intestine the trophozoites multiply and some of them encyst and are further evacuated in the faeces, reinitiating the life cycle.

In contrast to the intestinal parasite *Giardia* spp. for which several encystment–excystment mechanisms are well known (Lauwaet *et al.,* 2007), for *E. histolytica* such mechanisms remain unknown because of the inability of *in vitro* generation of amoebal cysts. The development of an animal model to replicate the complete parasite life cycle might help to overcome this problem. On this regard, the spider monkey could be a good candidate since it is reported to be susceptible to amoebic intestinal and extra-intestinal infections in a fashion similar to that which occurs in humans (Amyx *et al.,* 1978; Márquez-Monter *et al.,* 1991; Verweij *et al.,* 2003); however, this animal model has not yet been used in amoebiasis research. On the other hand, the model of experimental amoebic liver abscess in hamsters, gerbils and mice (Tsutsumi and Shibayama, 2006) has contributed during several decades of research to unravel the mechanisms of the extra-intestinal amoebiasis.

E. histolytica is considered a microaerophilic organism because (a) it lives and multiplies in the hypoxic environment of the lumen of the large intestine; (b) glucose stimulates oxygen (O_2) consumption by amoebae (Weinbach and Diamond, 1974), (c) in *in vitro* culture, strict anaerobiosis affects the adhesive capacity of the parasite and (d) its growth is favoured by O_2 concentrations below 5% in the gas phase; above this level the parasite dies (Gillin and Diamond, 1981). On this regard, in the lumen of the intestine the O_2 concentration diffused from epithelia (8.5%) (Lund *et al.*, 1999) could be dangerous for anaerobic microorganisms, including the different amoebic species living there; however, facultative bacteria can contribute to maintain a highly hypoxic microenvironment with 0.1–2.3% O_2 (Ladas *et al.*, 2007) by consuming O_2 through aerobic respiration. In turn, proliferation of anaerobic bacteria also appears to limit the growth of facultative microorganisms (some of them pathogenic) (Jones *et al.*, 2007). On the other hand, some bacteria can inhibit the growth of other luminal microorganisms by delivering hydrogen peroxide (H_2O_2) and/or redox cycling drugs that produce intracellular superoxide radical ($O_2^{\cdot-}$) (Jang and Imlay, 2010; Gu and Imlay, 2011). Furthermore, nitrite and nitrate molecules are present in the large intestine and their reduction by some bacteria or amoeba can lead to nitric oxide ($NO^{\cdot}$) production. This radical can also be generated by the inducible nitric oxide synthase (iNOS) present in inflammatory or epithelial cells (Vermeiren *et al.*, 2009).

It is clear that all microorganisms living in the intestine along with *Entamoeba histolytica*, namely *Entamoeba hartmanni, Entamoeba coli, Iodamoeba bütschlii, Dientamoeba fragilis, Endolimax nana* and *Entamoeba dispar* had to evolve strategies to cope against moderate heat shock and low concentrations of oxidant stressors (O_2, $O_2^{\cdot-}$, H_2O_2, hydroxyl radical ($^{\cdot}OH$) as well as $NO^{\cdot}$ and its reactive species) which are produced by the host, by cohabitants or by endogenous metabolism of the amoebae. Among the enteric amoebae, only *E. histolytica* can invade the intestinal epithelia and other tissues such as liver, lung, brain or skin (Brandt and Pérez-Tamayo, 1970). Therefore, during tissue invasion this parasite has to contend with higher concentrations of the same oxidants present in the colonic environment.

It is well documented that some amoebic functions, such as adhesion, haemolytic and proteolytic activities, complement resistance, phagocytosis and cytotoxicity are involved in *E. histolytica* virulence; however, little attention has been given to amoebic resistance to oxidative and nitrosative stresses as additional factors to consider in its virulent capabilities (Olivos-García *et al.*, 2009). We propose that intracellular hypoxia generated by an efficient antioxidant response is also a primary condition for *E. histolytica* tissue invasion and pathogenicity. In this chapter we review the basic general oxidative and nitrosative redox metabolism and its similarities and differences with the *E. histolytica* antioxidant systems and heat shock response. Also, we discuss the implications of these processes during experimental *in vitro* stress conditions and amoebic liver abscess development.

Oxidative stress

Thiol and disulfide oxidation

Thiol molecules play a central role in the ROS-mediated signalling pathways as they link the intracellular changes in redox state with various biochemical processes. The frequent oxidation of thiol groups in the cells under mild oxidizing conditions involves the thiol/disulfide redox transition $2R\text{-}SH \leftrightarrow R\text{-}S\text{-}S\text{-}R + 2e^{-} + 2H^{+}$. The disulfide molecules generated are very stable in the biological milieu and usually do not proceed to further oxidation (Bindoli *et al.*, 2008). The cytoplasm does not contain many protein–disulfide bonds owing to the presence of the highly efficient glutathione system (GSH)/GSH reductase (Le Moan *et al.*, 2006). However, if cellular oxidation is strong, thiols can be further oxidized to sulfenic acid → sulfinic acid → sulfonic acid; from these species, the first is considered to be important for redox signalling because of its instability (Poole *et al.*, 2004). Sulfenic acids are formed (a) after the thiolate group reacts with H_2O_2, alkylhydroperoxides and peroxynitrite; (b) from the hydrolysis of S-nitrosothiols; and (c) after the reaction of thiols with thiol-sulfinates. The most significant reaction of sulfenic acid is with thiol-forming disulfides (for example glutathionylation) (Bindoli *et al.*, 2008). This reaction is important in enzymes involved in redox maintenance because disulfide formation prevents the overoxidation of cysteine (Cys) residues in proteins or low-molecular weight thiols to sulfinic and/or sulfonic acids. In addition, oxidation, reduction and isomerization of disulfide bonds in nascent polypeptides are functions held by enzymes from the protein disulfide isomerase (PDI) family. It has been found that of 11 proteins belonging to this family, *E. histolytica* highly express a PDI enzyme involved in protein folding (Ramos *et al.*, 2011).

Hydrogen peroxide is recognized as a signalling molecule able to react with a broad range of molecules containing thiol groups. Oxidation of thiol-containing proteins by H_2O_2 leads to sulfenic acids formation which can further react with a nearby thiol to form a disulfide that later can be reduced by thioredoxin (Trx); if the H_2O_2 concentration increases, stable sulfinic or sulfonic acids can be produced (Little and O'Brien, 1969). In contrast to the irreversible process of sulfonic acid formation in proteins, the sulfinic modification in the active site of peroxiredoxin can be reduced enzymatically by *Saccharomyces cerevisiae* sulphiredoxin or bacterial sestrins in an ATP and Mg^{2+}-dependent manner (Biteau *et al.*, 2003). Such enzymes are stress-responsive proteins highly conserved throughout the phylogenetic scale. On this regard, it has been observed that when *E. histolytica* is exposed to oxidants, a peroxidase protein increases (Akbar, 2004), although the participation of sulphiredoxin or sestrin-like enzyme activities in *E. histolytica* has not been reported.

The reactivity of thiols with H_2O_2 is as follows: low molecular weight thiols (GSH, Cys, cysteamine, penicillamine, *N*-acetylcysteine, dithiothreitol and captopril), proteins containing thiol groups with low pKa values, peroxidases, the bacterial sensor OxyR and PFOR (pyruvate:ferredoxin oxidoreductase) of *Desulfovibrio* bacteria (Bindoli *et al.*, 2008). The reactivity of thiols is determined mainly by two parameters, a low pK_a value of the reactive Cys and the presence of adjacent amino acids able to stabilize the transition state intermediate sulfenic acid (Tosatto *et al.*, 2008). For example, the essential thiol group of papain is about 30,000 times more reactive with chloroacetate than the thiol group of Cys (O'Leary and Solberg, 1976). Treatment of bacteria, yeast and mammal cells with H_2O_2 does not lead to a generalized protein thiol oxidation but rather increases the oxidation state of a specific group of oxidizable proteins as translation factors, stress chaperones and those involved in H_2O_2 detoxification and carbohydrate metabolism, indicating both high selectivity and a restricted contribution of thiol oxidation to the cellular toxicity of this oxidant (Le Moan *et al.*, 2006). A similar selectivity for protein oxidation has been observed in *E. histolytica* during *in vitro* oxidative stress in which thiol-dependent proteolytic activity increased (Nandi *et al.*, 2010). In contrast, inactivation of PFOR from the bacteria *Desulfovibrio* by oxidants, through disulfide bond formation, has been proposed as a protective mechanism during oxidative stress since its activity is promptly recovered once the cell stress ends (Pieulle *et al.*, 2011). Such mechanism could be involved in PFOR of *E. histolytica* since it shows a similar behaviour with oxidants (Ramos-Martínez *et al.*, 2009; Pineda *et al*, 2010). Sometimes disulfide oxidation leads to thiolsulfinate → thiolsulfonate → disulfide trioxide → disulfone → sulfonic acid (Bindoli *et al.*, 2008). Thiosulfinates are formed *in vivo* by three major routes: oxidative (from disulfides), reductive (from sulfinic acids) and enzymatically (from sulfoxides). Disulfides are also able to reduce oxidative stressors, such as H_2O_2, forming a thiosulfinate (RS(O)SR) or thiosulfonate (RS(O)2SR) and water, both having antimicrobial activity (Jacob, 2006). Also, a thiyl radical (which is part of the catalytic cycle of many enzymes) can be formed by oxidation of thiols in a reaction catalysed by transition metals or by the action of peroxidases or myoglobin in the presence of H_2O_2 (D'Aquino *et al.*, 1994). This radical can dimerize to form a disulfide, or can react with O_2 to form a thioperoxyl radical intermediate which in turn can bind to thiol groups leading to sulfenic acid and regenerating the thiyl radical. Alternatively, the latter together with a thiolate anion can form a strong reductant disulfide radical anion that by reacting with O_2 can form a superoxide radical (Wardman, 1998). This reaction might be an important superoxide source in *E. histolytica* subjected to microanaerobiosis since all the reactants might be present.

Also, methionine (Met) contained in proteins can be oxidized to methionine sulfoxides by ROS and reactive nitrogen species (RNS). These methionine sulfoxides can be reduced to Met by methioninesulfoxide reductase avoiding its irreversible oxidation into sulfone derivatives. This system may function as a repair system and as a ROS scavenger (Boschi-Muller *et al.*, 2008). Methioninesulfoxide reductases protecting organisms from oxidative damage are essential regulators for longevity in animal species and can play a role in virulence in some bacterial pathogens (Lee *et al.*, 2009). In *E. histolytica* such enzymes are yet to be described.

Redox regulation of the nicotinamide adenine dinucleotide pools

Proton-translocating transhydrogenases which reduce $NADP^+$ at the expense of NADH oxidation are membrane proteins utilizing the electrochemical proton gradient for NADPH generation. When ATP is reduced the same enzyme can catalyse the reverse reaction (Pedersen *et al.*, 2008). It is known that proton-pumping transhydrogenases constitute one of the main sources of cytosolic NADPH in some prokaryotes and mitochondrial NADPH in eukaryotes. The NADPH is used to prevent oxidative stress by maintaining a high GSH/GSSG ratio. This also indirectly involves other defence systems linked to GSH/NADPH like Trx, glutaredoxin and peroxiredoxin (Rydström, 2006). The NADH generated by glycolysis in *E. histolytica* can be the principal electron source for $NADP^+$ reduction (Fig. 17.1). This parasite has a dimeric pyridine nucleotide transhydrogenase (EhPTN) able to catalyse such transhydrogenation (Weston *et al.*, 2002). EhPTN is distributed in the membrane vesicles and vacuoles, including lysosomes, phagosomes and endosomes (Yosuf *et al.*, 2010). Under aerobic conditions the bifunctional aldehyde-alcohol dehydrogenase (ADH2) mainly responsible of reoxidizing the NADH produced in glycolysis is inhibited (Pineda *et al.*, 2010), then, the transhydrogenase may help to maintain the pool of NAD^+ required for glycolysis to proceed (Fig. 17.1). The NADPH produced can be used by the enzymatic O_2 reduction pathway (Fig. 17.1) to maintain an intracellular hypoxia, or used by the enzymes responsible to maintain the thiol redox system (cysteine/cystine) in acceptable levels for metabolic work. In microaerobic environments, such as that present in the human colon (O_2 0.1–2.3%) (Ladas *et al.*, 2007), O_2 is the principal terminal electron acceptor. Physiologically, the EhPTN might dissipate energy by using ATP. At the same time O_2 could dissipate electrons by their reduction to H_2O which requires NADPH. Recently, *E. histolytica* was described to contain two NAPH-dependent oxidoreductases (EhNO1/2) able to reduce $NADP^+$ to NADPH through the utilization of electrons from reduced ferredoxin (Jeelani *et al.*, 2010) (Fig. 17.1). This whole set of data may explain why low O_2 concentrations are beneficial for parasite growth and justify the designation of *E. histolytica* as a microaerophilic organism. On the other hand, amoebic protection against oxidative stress is a process requiring high levels of NADPH. The amoebic vesicular acidification observed during moderate oxidative challenge (A. Olivos-García, unpublished) along with ATP could stimulate NADPH production by improving the EhPTN activity (pH optimal 5.5) through its forward reaction. In contrast, ATP and reduced ferredoxin, that decrease under strong O_2 stress (Ramos-Martínez *et al.*, 2009), could favour the reverse reaction of EhPTN (NADH formation) and reduce NADPH formation by EhNO1/2. Then, NADPH deprivation may lead to O_2 and ROS accumulation and amoebic death. Therefore, EhPTN and EhNO1/2 might be considered as possible therapeutic targets.

Lipid peroxidation

During oxidation of iron–sulfur (Fe^{2+}–S) clusters present in some proteins of anaerobic, microaerophilic or air-tolerant microorganisms, Fe^{3+} is produced. Especially when O_2 is the oxidant agent, $O_2^{\cdot-}$ is also produced (Outten and Theil, 2009). Then free iron or that present in proteins or DNA can produce the dangerous $^{\cdot}OH$ by

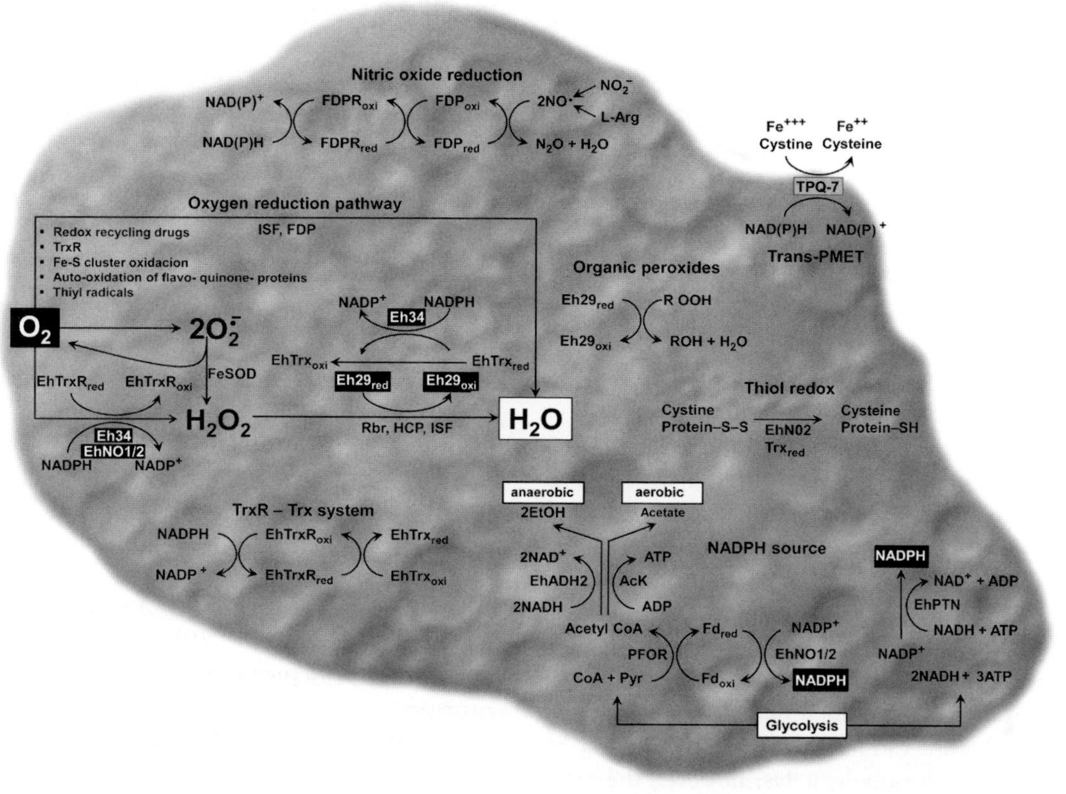

Figure 17.1 Antioxidant pathways in *E. histolytica*. The parasite protects itself from O_2 and ROS through a main oxygen reduction pathway assisted by other contributing systems. The NADH generated by glycolysis is transhydrogenated to produce NADPH which is further used as the reductant equivalent donor to reduce O_2 by the antioxidant machinery. Eh34, EhNO1/2 and EhTrxR generate H_2O_2 through divalent O_2 reduction. H_2O_2 is transformed to H_2O by Eh29 which is in turn reduced by Eh34 or Trx. Oxidized Trx is reduced by TrxR at expense of NADPH oxidation. The intracellular $O_2^{\cdot -}$ produced adventitiously is incorporated to this pathway by a FeSOD. Although other enzymes as ISF, FDP, Rbr and HCP with antioxidant capacity are present in the genome of *E. histolytica*, their specific function awaits to be revealed. Other functions attributed to Eh29 are detoxification of organic peroxides and together with EhNO and Trx they are proposed to maintain the intracellular redox state (cysteine/cystine). Also amoebae may diminish the redox potential of the surroundings through the trans-PMET. On the other hand, the FDP enzyme can detoxify the $NO^{\cdot}$ generated by the parasite or that diffused from the microenvironment. Abbreviations: Iron sulfur flavoprotein (ISF); flavo diiron protein (FDP); flavodiiron protein reductase (FDPR); rubrerythrin (Rbr); hybrid cluster protein (HCP); NADPH: flavin oxido reductase (Eh34); peroxiredoxin (Eh29); NADPH-dependent oxidoreductases (EhNO1/2); Fe-superoxide dismutase (FeSOD); transplasma membrane electron transport (Trans-PMET); thermoplasmaquinone-7 (TPQ-7); thioredoxin (EhTrx); thioredoxin reductase (EhTrxR); ferredoxin (Fd); pyruvate:ferredoxin oxidoreductase (PFOR); pyridine nucleotide transhydrogenase (EhPTN); bifunctional aldehyde – alcohol dehydrogenase (EhADH2); acetate kinase (AcK); reduced (red); oxidized (oxi).

the Fenton ($H_2O_2 + Fe^{2+} \rightarrow Fe^{3+} + OH^- + {}^{\cdot}OH$) and Haber–Weiss ($Fe^{3+} + O_2^{\cdot -} + H_2O_2 \rightarrow Fe^{3+} + {}^{\cdot}OH + OH^- + O_2$) reactions.

Alternatively, Fe^{3+} can be reduced by Cys or vitamin C (Wong *et al.*, 2010). All living cells avoid formation of ${}^{\cdot}OH$ radical because of its ability to oxidize proteins, lipids, carbohydrates and nucleic acids. Certain conditions might favour ${}^{\cdot}OH$ generation during amoebic oxidative stress, but its detection and contribution to amoebic damage during oxidative stress have not been achieved, owing to its high reactivity (10^9/M/s), which make it short lived in the site of generation.

It is well documented that polyunsaturated lipids present in biological membranes are targets of ${}^{\cdot}OH$. The reactive aldehydes generated can bind and damage proteins and DNA in most cells. In contrast, monounsaturated lipids found in most bacterial membranes are unreactive (Imlay, 2003). It has been observed that thiobarbituric acid reactive species (TBARS), an index of lipid peroxidation, did not increase in *E. histolytica* after a lethal *in vitro* oxidant stress (Ramos-Martínez *et al.*, 2009). Similarly to bacteria, *E. histolytica* could avoid membrane lipid oxidation because it contains mainly saturated lipids (Sawyer *et al.*, 1967).

Protein carbonylation

Protein carbonylation is present at physiological levels in living cells (1–4 nmol/mg protein) and increases with ageing or during pathological conditions related to oxidative stress (8 nmol/mg protein); therefore, it has been considered an index of cellular oxidant conditions (Møller *et al.*, 2011). Protein carbonylation can be generated by conjugation with aldehydes produced during peroxidation of polyunsaturated fatty acids or by reacting with carbonyl derivatives generated by reaction with carbohydrates; however, the most common way of formation is by reacting with the ·OH produced during the Fenton reaction (Maisonneuve *et al.*, 2009). Tyrosine and tryptophan residues are electron-rich sinks for ·OH formation (Imlay, 2003). Oxidation of proline or arginine side chains in proteins forms glutamate-semialdehyde, whereas lysine oxidation results in aminoadipate-semialdehyde compounds; these chemical transformations introduce carbonyl groups (RR'C=O) in the protein structure. The OH group in the threonine side chain can also be oxidized to form the carbonyl group. It has been suggested that protein carbonylation may play an important role in redox signalling because: (a) it not always leads to loss of protein function, (b) only a few, selected group of proteins are susceptible to carbonylation, and (c) the process can be reversible/irreversible (only metal-catalysed oxidation) (Wong *et al.*, 2010).

In the hypoxic environment present in the human colon (O_2 0.1–2.3%; Ladas *et al.*, 2007), *E. histolytica* can generate low levels of $O_2^{\cdot -}$ and H_2O_2 from O_2 reduction; however, the damage that these ROS may cause might be constantly repaired since low protein carbonylation levels have been detected under *in vitro* non-stressful conditions; however, this parameter increased during the oxidative stress caused by a challenge with a high O_2 concentration (Ramos-Martínez *et al.*, 2009). It has been argued that the only source of protein carbonylation in bacteria derives from the Fenton reaction (Møller *et al.*, 2011). A similar case for *E. histolytica* might occur because low H_2O_2 concentration (500 nM) and iron that are known to favour protein carbonylation (Wong *et al.*, 2010) can be produced from O_2 reduction and oxidation of the Fe–S cluster containing proteins. Protein carbonylation through lipid peroxidation products is unlikely to happen since, as mentioned above, amoeba contains saturated membrane lipids (Sawyer *et al.*, 1967). The role that protein carbonylation/decarbonylation process may play in amoebic physiology waits to be elucidated.

Irreversible degradation of oxidized proteins

Proteins that are irreversibly inactivated by formation of methionine sulfones, cysteine-sulfinic or -sulfonic acids and carbonyl derivatives cannot be repaired and are recognized and degraded by cellular proteolytic process. Chaperones can recognize hydrophobic surfaces exposed as a result of protein oxidation and partial denaturation, and assist in protein refolding. If protein recovery is unsuccessful, then chaperones may also assist in the unfolding of proteins preceding proteolysis (Costa *et al.*, 2007). The selective degradation of oxidized proteins is of critical importance to maintain cellular homeostasis and is performed by the proteasome system with the participation of lysosome/vacuole compartments.

Proteasomes

The proteasome consists of a cylindrical 20S core unit which is activated under oxidative stress and is capable of degrading oxidized proteins in an ATP- and ubiquitin-independent manner (Kästle and Grune, 2011). Its subunits are up-regulated in cells exposed to oxidant stressors and its activity increases during recovery after oxidative damage (Marques *et al.*, 2006). Degradation of carbonylated proteins does not require ubiquitination; protein multi-ubiquitination is not observed in

H_2O_2 treated cells (Inai and Nishikimi, 2002). Furthermore, the 26S proteasome is assembled by association of the 20S proteasome catalytic core with the 19S regulatory cap which confers the specificity ubiquitin- and ATP-dependent protein degrading activity. The full proteasome complex is responsible for the degradation of short lived regulatory proteins, as well as unfolded proteins tagged with hydrophobic polyubiquitin chains. In contrast with the 20S proteasome core activity, which is inhibited by ATP and induced under oxidative stress, the 26S proteasome activity is inhibited under oxidative stress (Marques *et al.*, 2006).

In mammalian cells the proteasome is located in both cytoplasm and nucleus whereas in yeast it has been found mainly in the nucleus. In *E. histolytica* both proteasome units have been identified showing only a cytosolic distribution (Sánchez *et al.*, 2002). By using lactacystin, the specific inhibitor of proteasome activity, it has been demonstrated that proteasome activities are essential for proliferation in *E. histolytica* trophozoites (Makioka *et al.*, 2002), whereas for *Entamoeba invadens* they are important only for encystation (Gonzalez *et al.*, 1999).

Protein degradation in lysosomes/vacuoles

In mammalian cells subjected to stress, the HSP70 family of chaperones recognize target signals in cytosolic oxidized proteins and mediate their binding, unfolding and translocation across the lysosomal membranes for proteolysis (Orenstein and Cuervo, 2010). In sorting of yeast proteins, the proteolytic activity of the vacuole and its internal acidification are core functions required for broad resistance to oxidative stress (Costa *et al.*, 2007). Also in *S. cerevisiae,* the genes associated with vacuolar proteolysis are up-regulated during recovery after oxidative damage including overexpression of the vacuolar protease PEP4 that prevents accumulation of carbonylated proteins in the cell (Marques *et al.*, 2006). In *E. histolytica,* vesicular acidification (A. Olivos-García, unpublished), HSP70 expression (Akbar *et al.*, 2004) and calpain activity (Nandi *et al.*, 2010) increase during oxidative stress, so oxidized protein degradation probably occurs. Further studies to elucidate such proteolytic mechanisms are required.

Oxidation of Fe–S clusters in proteins

Oxygen, $O_2^{\cdot -}$ and H_2O_2 are able to oxidize at different rates the Fe–S centres of some enzymes present in aerobic and anaerobic cells (Imlay, 2003). The most common $[4Fe–4S]^{2+}$ clusters in proteins are oxidized to the form $[3Fe–4S]^{1+}$ and then to $[2Fe–2S]^{2+}$ clusters. If oxidant conditions persist, the $[2Fe–2S]^{2+}$ can be further degraded yielding the apoenzyme form (Crack *et al.*, 2008). Since most of the Fe–S containing enzymes are involved in energy metabolism or antioxidant activity, their inactivation by oxidation can lead to cellular arrest (by reversible oxidation of the proteins) or even death (by irreversible oxidation). In contrast to mammalian cells, in anaerobic microorganisms the Fe–S protein cluster are exposed to the solvent in order to work efficiently at low redox potential (~−300 mV) but at the same time this property make them susceptible to oxidation in aerobic environment (Outten and Theil, 2009).

As many microaerophilic parasites do, *E. histolytica* obtains energy by fermentative glycolysis (Saavedra *et al.*, 2005). PFOR is involved in oxidative decarboxylation of pyruvate to produce acetyl-CoA which is further metabolized to ethanol by a bifunctional aldehyde-alcohol dehydrogenase (ADH2) (Reeves, 1984) (Fig. 17.1). PFOR contains three 4Fe–4S clusters whereas ADH2 contains an Fe^{2+} atom in the alcohol dehydrogenase domain which is essential to preserve the aldehyde dehydrogenase activity. It has been demonstrated that PFOR is inactivated by O_2 and H_2O_2 under *in vitro* incubation and that both enzymes are highly inactivated when the parasites are subjected to oxidant conditions (Ramos-Martínez *et al.*, 2009). Furthermore, PFOR and ADH2 inactivation leads to decreased ethanol production in a process that is reversible when oxidant conditions are removed (Pineda *et al.*, 2010).

Iron availability is essential for *de novo* synthesis and reparation of Fe–S clusters. The transplasma membrane electron transport (transPMET) system of *E. histolytica* is able to transfer electrons from cytosolic electron donors to non-permeable electron acceptors and (among other functions) it has been suggested to play an important role in iron acquisition (Bera *et al.*, 2006) (Fig. 17.1). Anaerobic or facultative cells have developed systems to repair oxidized Fe–S clusters; for example, *E. coli* has the iron–sulfur cluster (Isc) that is a constitutive system for assembly/repair of Fe–S clusters, but is inhibited by high oxidants concentrations. The function of the Fe–S containing enzymes is maintained at a metabolically acceptable level by the alternate sulfur mobilization (Suf) assembly/repair system, induced by the same oxidants; this system is more resistant to oxidative damage. Once the stressor diminishes, Suf is down-regulated and Isc becomes active because its expression remains during the oxidative stress. Assembly/repair of Fe–S clusters is essential for bacteria and eukaryote cell survival (Jang and Imlay, 2010). Microorganisms lacking the Isc system have instead the Suf or the nitrogen fixation (Nif) systems. *E. histolytica* only contains a Nif-like system (EhNifS/EhNifU) for assembly/repair of Fe–S clusters. This system is intracellularly distributed in areas where Fe–S cluster enzymes are present, i.e. the cytosol and mitosomes (Maralikova *et al.*, 2010). EhNifU is a tetrameric protein with a stable $[2Fe–2S]^{+2}$ per monomer that provides a scaffold onto which the nascent Fe–S cluster is assembled. The EhNifS is a monomeric protein with cysteine-desulfurase activity that provides a sulfur atom for cluster formation. The activity and susceptibility to aerobic conditions of the EhNif system has been demonstrated by ectopic expression in *E. coli* in which the isc and suf operons were deleted (Ali *et al.*, 2004). A similar Nif system has been described for *H. pylori* whose deletion is lethal for the cells (Olson *et al.*, 2000).

Another way to compensate the loss of function by oxidation of Fe–S enzymes is the synthesis of resistant isoenzymes. For example, when Fe–S centres of fumarase A and aconitase B of *E. coli* are oxidized, transcription of the isoenzymes fumarase C, which lacks Fe–S centres, and aconitase A, which has a Fe–S centre resistant to oxidation, is initiated (Imlay, 2008). A similar mechanism has not yet been described in *E. histolytica*.

The absence of an efficient oxidant-resistant assembly/repair system for Fe–S cluster proteins in anaerobic or microaerotolerant organisms (as might be the case for *E. histolytica*), makes them susceptible to O_2, O_2^- and H_2O_2. To circumvent this problem, most of them have developed mechanisms to maintain the O_2 and its reactive species in levels permissive for cell survival.

Oxygen detoxifying mechanisms in *E. histolytica*

Oxygen is able to cross cell membranes freely; however, some lipid components like cholesterol offer some resistance to its diffusion (Subczynski *et al.*, 2009; Al-Abdul-Wahid *et al.*, 2011). On the other hand, proliferation (Lujan and Diamond, 1997) and virulence (Meerovich and Ghadirian, 1978; Serrano-Luna *et al*, 2010) are stimulated by growing *E. histolytica* in cholesterol-enriched culture. Since amoebic virulence is correlated with its resistance to O_2 (Ramos-Martínez *et al.*, 2009), the cholesterol contribution to amoebic virulence through membrane resistance to O_2 diffusion waits to be examined.

Once inside the amoeba, O_2 is reduced to H_2O_2 by the NADPH:flavin oxidoreductase (Eh34) (Bruchhaus *et al.*, 1998) and then the product is transformed into H_2O by a peroxiredoxin (Eh29) (Fig. 17.1). The latter is two thousand times more efficient than the thiol-specific antioxidant protein (TSA) from *Saccharomyces* (Bruchhaus *et al.*, 1997a). The Eh29 peroxiredoxin differs from catalase in that it avoids O_2 recycling which is harmful to the organism. Both Eh34 and Eh29 enzymes are abundant in the amoebic cytoplasm (the latter is also associated with the plasmatic membrane) and their principal function is to avoid oxidation by detoxifying O_2 and its derivatives (Bruchhaus *et al.*, 1997b). Interestingly, the nucleus of mammalian cells is more reduced than the cytoplasm and the peroxiredoxin, GSH and thioredoxin 1 (Trx1) maintain its redox status and protect it from oxidation (Go and Jones, 2010). The peroxiredoxin Eh29 and the Trx system (Fig. 17.1) in *E. histolytica* could be candidates to carry out similar functions in the amoebic nucleus; the first has high efficiency whereas the GSH system is absent in the parasite. So far the Eh29 and Trx have not been located in the nucleus and it is not known whether they can translocate to this compartment during oxidative or nitrosative stress. If a nuclear localization is absent, other uncharacterized antioxidant enzymes including trypanothione could fulfil such functions.

An Fe-superoxide dismutase (FeSOD) has also been identified in *E. histolytica* as a dimeric enzyme responsible of $O_2^{\cdot-}$ detoxification (Tannich *et al.*, 1991). It has been suggested that its main function might be to cope with the $O_2^{\cdot-}$ generated by the host during the inflammatory response (Bruchhaus and Tannich, 1994). However, such hypothesis appears improbable since $O_2^{\cdot-}$ is a highly reactive charged molecule with a limited membrane diffusion capacity which rapidly reacts in the place of formation. In addition, similar FeSOD activity or content is present in some non-pathogenic *Entamoeba* species (Tannich *et al.*, 1991) that have no contact with inflammatory cells. That FeSOD is part of the main O_2 reduction pathway in the parasite (Fig. 17.1) appears improbable because O_2 is a byproduct of the reaction; furthermore, none of the enzymes that detoxify O_2 characterized in *E. histolytica* are able to generate $O_2^{\cdot-}$ as a deliberate stoichiometric product. In any case, once the FeSOD products O_2 and H_2O_2 are produced, they are detoxified by the Eh34 and Eh29 couple system that has a cytosolic distribution (Bruchhaus *et al.*, 1997b). It can be hypothesized that the possible functions of FeSOD could be to detoxify the $O_2^{\cdot-}$ generated inside the amoeba by: (a) bacterial redox recycling drugs (Imlay, 2008), (b) oxidation of Fe–S clusters (Crack *et al.*, 2008), (c) thioredoxin reductases (Leitsch *et al.*, 2007), (d) auto-oxidation of flavo- and quinone- containing proteins (Band and Cirrito, 1979; Imlay, 2002) and (e) thiyl radicals formed by the thiol oxidation (Wardman, 1998).

The thioredoxin/thioredoxin reductase system

The Trx and thioredoxin reductase (TrxR) system is present in all taxonomic groups, from *Archaea* to mammals. Trx is a small dithiol protein that participates as a reductant donor in different biological processes such as protection against oxidative stress, regulation of DNA synthesis, transcription, cellular growth, apoptosis and signalling *via* thiol redox control. Oxidized Trx is in turn reduced by TrxR at expense of NADPH oxidation (Fig. 17.1). The general oxido-reductase activity of Trx serves as: (a) an equivalent reductant donor to proteins or metabolites involved in the catalytic cycles of several enzymes, such as ribonucleotide reductase, thioredoxin peroxidases (peroxiredoxins), methionine sulfoxide reductases and sulfate reductases, and (b) it protects cytosolic proteins from aggregation or inactivation *via* oxidative formation of intra- or inter-molecular disulfides (Arnér and Holmgren, 2000).

The *E. histolytica* genome encodes four different genes of Trx *plus* two TrxR- related sequences which probably emerged by gene duplication. It has been demonstrated that EhTrxR together with EhTrx8 or EhTrx41 function as redox pairs (Fig. 17.1). In addition to reduce Trx, EhTrxR is able to catalyse the NADPH-dependent reduction of the artificial substrate 5,5′-dithiobis-2-nitrobenzoic acid (DTNB) and O_2 to H_2O_2, whereas EhTrx is able to reduce cystine, insulin, oxidized glutathione (GSSG), trypanothione and the Eh29 enzyme (Arias *et al.*, 2007, 2008). On the other hand, since *E. histolytica* lacks GSH, the amino acid Cys acts as the major intracellular thiol which can be synthesized *de novo* from sulfate and glycolytic intermediates (Nozaki *et al.*,1999). TrxR/Trx has been proposed to carry out cystine reduction since no cysteine reductase enzyme has been found in this parasite (Arias *et al.*, 2008). The TrxR protein content in *E. histolytica* is high, accounting for nearby 0.2% of the total cellular protein and it has been suggested to be in equimolar concentrations with Trx (Leitsch *et al.*, 2007). Mammalian TrxR/Trx proteins show variable intracellular distribution (plasmatic membrane, cytoplasmic, nuclear or extracellular, depending on the tissue) (Godoy *et al.*, 2011); hence, the association of EhTrxR/EhTrx41 within the underlying plasma membrane (Arias *et al.*, 2008) is intriguing since their functions are expected to be nuclear and cytoplasmic; therefore, it is probable that the others amoebic Trxs also have different intracellular distribution.

For anaerobic and Gram-positive anaerobic microorganisms such as *Bacillus subtilis, Staphylococcus aureus* or *Lactobacillus lactis* and probably for *E. histolytica*, Trx is the main control system of the intracellular redox status, since in these organisms the GSH/GSH reductase system is absent (Rocha *et al.*, 2007). The low redox potential of the Trxs present in *E. coli* and *E. histolytica* (−270 and −283 mV, respectively) turns them into the major intracellular dithiol reductants in these cells. Also, TrxR has been proposed as a cellular redox sensor because it can be inactivated transiently or permanently by oxidants such as H_2O_2 (Arnér and Holmgren, 2000). The Trx/TrxR system is essential for *E. histolytica* physiology; its inactivation by metronidazole through the formation of inactive EhTrxR/EhTrx adducts (among others) leads to toxicity (Leitsch *et al.*, 2007). In bacterial or mammalian cells, reduced Trx inhibits apoptosis whereas oxidized Trx caused by oxidative stress promotes it (Arnér and Holmgren, 2000). In a similar fashion, oxidized Trx might contribute to *E. histolytica* apoptosis observed during high O_2 exposure (Sen *et al.*, 2007).

The relevance of the Trx-dependent system was demonstrated in the opportunistic pathogen *Bacteroides fragilis* that (similar to *E. histolytica*) lives in the intestine and is able to invade tissues. The system was induced by oxidants like diamide, H_2O_2 and O_2, and in contrast to that of *E. coli* it is not controlled by the OxyR regulon. Moreover, a TrxR mutant of *B. fragilis* is more sensitive to O_2 and diamide and was unable to survive and induce abscess formation in the mouse peritoneal cavity that contains ~6% O_2; without TrxR this microorganism was unable to utilize the host reduced thiol compounds as sources of thiol reducing power (Rocha *et al.*, 2007). The induction and/or redistribution of the *E. histolytica* Trx system by oxidants as well as the role it may play during host invasion are yet to be examined.

E. histolytica in vitro exposure to oxygen, superoxide and hydrogen peroxide

Analysis of amoebae exposed to several oxidants under *in vitro* conditions has been reported. While both mRNA and Eh34 activity increased, Eh29 and FeSOD activity decreased gradually 3 to 9 hours after *E. histolytica* exposure to 0.8 mM H_2O_2 (Nandi *et al.*, 2010). In another work, gene transcription of FeSOD and Eh29 did no change after 1 hour exposure to 1 mM H_2O_2 (Vicente *et al.*, 2009a). Such discrepancies could be attributable to the sampling time after the stress since both experiments were made in the same culture media. Also, resistance to H_2O_2 and/or independent expression control could explain the increase of mRNA and higher activity of Eh34 after H_2O_2 challenge. On the other hand, exposure of *E. histolytica* to a high O_2 concentration during 3 hours leads to increased mRNAs of FeSOD and Eh29 and also higher protein content of the latter (Akbar *et al.*, 2004). In contrast when Eh29 expression was inhibited by antisense RNA, the cells became highly susceptible to O_2 exposure. In addition, apoptosis was observed with prolonged high O_2 challenge (after 6 hours) (Sen *et al.*, 2007). Also, accumulation of $O_2^{\cdot-}$ and H_2O_2 was observed during *E. histolytica* exposure to high O_2 concentrations (Ramos-Martínez *et al.*, 2009). Finally, the protein and activity of amoebic FeSOD was induced after 1.5 hours under an $O_2^{\cdot-}$-generating system or by iron chelation (Bruchhaus and Tannich, 1994). Taken together, these results suggest that *E. histolytica* resists short time periods of oxidative stress by both high basal activity levels and overexpression of enzymes belonging to the O_2 reduction pathway (Fig. 17.1). In contrast, under longer periods of oxidant stress the decrease in protein and/or activity of Eh29 and FeSOD leads to death by accumulation of H_2O_2 produced by overexpression of Eh34 and by disproportionation of $O_2^{\cdot-}$.

Energy metabolism failure could be also a consequence of longer time periods of amoebic O_2 and H_2O_2 stresses. In this context, PFOR inhibition in amoebae subjected to a high O_2 concentration leads to accumulation of glycolytic intermediaries and decrease of ATP and ethanol products of glycolysis (Ramos-Martínez *et al.*, 2009; Pineda *et al*, 2010). Under such conditions, it might be also possible that the flux of reductive power derived from glucose catabolism which is necessary to reduce the oxidants to maintain the redox homeostasis is impaired (Fig. 17.1).

Other important products of the oxidative stress that contribute to toxicity are cytosolic protons generated during NAD(P)H-dependent reduction of O_2 and ROS. In *E. histolytica* cytoplasmic acidification has been observed after 6 hours of a H_2O_2 stress (Nandi *et al.*, 2010). Cytosol amoebic acidification during moderate and/or short oxidative stress might be controlled by the activity of H^+-ATPases that increase vesicular and/or extracellular acidification. In agreement with this, vesicular acidification has been observed in *E. histolytica* after 1 hour of high O_2 exposure (A. Olivos-García, unpublished). Cytoplasmic acidification under long oxidative stress periods could be attributed at least in part to the decrease of ATP required for proton-pump activity.

Newly discovered oxido-reductase enzymes in *E. histolytica*

The NADPH-dependent oxidoreductases (EhNO1 and EhNO2) (Fig. 17.1) were initially annotated in the genome as small subunits of glutamate dehydrogenase in *E. histolytica* but recently it was demonstrated that instead they have oxidoreductase activity. They are involved in the redox homeostasis through L-cysteine and iron reduction (Jeelani *et al.*, 2010). Like Eh34, they might be involved in the O_2 reduction pathway since they are able to reduce O_2 to H_2O_2 and have cytosolic locations. The antioxidant contribution of EhNO1/2 during amoebic oxidative or nitrosative stresses has not yet been explored.

The hybrid-cluster proteins (HCPs) are able to reduce H_2O_2 to H_2O (Fig. 17.1). Although first isolated from anaerobic sulfate-reducing bacteria, the analysis of the genomic sequences reveals that genes encoding putative HCPs are present in a wide range of organisms from the Bacteria, Archaea and Eukarya domains (Almeida *et al.*, 2006; Andersson *et al.*, 2006). Whereas in *E. coli* the HCP genes are up-regulated by H_2O_2 exposure, in *E. histolytica* they did not change (Vicente el al., 2009a). The precise physiological function of these proteins in amoebae are still to be determined.

The flavodiiron protein (flavoprotein type A; FDP) is an enzyme present in microaerophilic organisms able to reduce O_2 to H_2O and/or $NO^{\cdot}$ to N_2O (Andersson *et al.*, 2006; Kurtz, 2007) (Fig. 17.1). From four FDP genes contained in the *E. histolytica* genome, only one was up-regulated during H_2O_2 stress. Also in basal conditions, FDPs have lower expression in non-virulent *E. dispar* than in *E. histolytica* (Vicente *et al.*, 2009a). Unlike most FDPs, the protozoan enzymes from the human pathogens *Giardia intestinalis* and *Trichomonas vaginalis* are selective towards O_2; they are considered to be metabolic relics and pathogenic factors (Vicente *et al.*, 2009b). Similar to the hybrid-cluster proteins, the EhFDPs may play new important roles in the *E. histolytica* O_2 reduction pathway since both enzymes safely scavenge O_2 without releasing ROS into the cell; the mechanism for the tetravalent O_2 reduction is under study.

The iron sulfur flavoproteins (Isf) are oxidoreductase enzymes widely distributed among strictly anaerobic prokaryotes and are generally present as multicopy genes. Isf from *Methanosarcina thermophila* is able to reduce O_2 and H_2O_2 to H_2O using ferredoxin as the electron donor (Cruz and Ferry, 2006). Although the Isf genes were up-regulated during H_2O_2 or NO stresses (Vicente *et al.*, 2009a) and kinetics properties are compatible with a role in combating oxidative stress, its inhibition by RNA antisense did no modify the *E. histolytica* susceptibility to H_2O_2 stress. Instead they showed a severe growth defect when incubated in Cys deprived medium (Husain *et al.*, 2011). The specific Isf function in *E. histolytica* remains undetermined.

Similar to Eh29, amoebal rubrerythrin (Rbr) can reduce H_2O_2 to H_2O avoiding O_2 recycling. The enzyme is confined to mitosomes which are cytoplasmic structures in which the Fe–S cluster assembly has been proposed to occur (Maralikova *et al.*, 2010). A protective function from H_2O_2 for cluster assembly was the suggested function for Rbr.

Nitrosative stress

When *E. histolytica* inhabits the human colon, reactive nitrogen species (RNS) are other strong oxidants harmful for the parasite; a concentration of 18 ppm of $NO^{\cdot}$ in the gas phase has been detected in the human colon (Kastner *et al.*, 1997). It is known that $NO^{\cdot}$ is produced by host epithelial cells in response to cytokines or bacterial products. Besides, it has been documented that $NO^{\cdot}$ can also be produced by microorganisms through bacterial nitric oxide synthase (bNOS), nitrate/nitrite reductases, or nitrite reduction under an acidic environment. Also, human faecal microbiota is able to produce $NO^{\cdot}$ from trace nitrate levels under gastrointestinal conditions by nitrate reduction to ammonium, but not by the L-arginine metabolic pathway (Vermeiren *et al.*, 2009). Despite the presence in human faeces of microorganisms capable of *in vitro* heterotrophic nitrification, the conditions in the lower gastrointestinal tract appear to strongly favour nitrate and nitrite reduction (Saul *et al.*, 1981). In humans, 0–20 nmol/g nitrite and 0–48 nmol/g nitrate enter the colon from the distal ileum or they might diffuse from circulation or be produced by inflammatory cells. Dietary factors and metabolic products from anaerobic or colonic epithelial cells could modify colonic $NO^{\cdot}$ concentration by altering the nitrite/nitrate levels (Radcliffe *et al.*, 1989).

In the colonic environment *E. histolytica* could produce $NO^{\cdot}$ from arginine (Hernández-Campos *et al.*, 2003), however, $NO^{\cdot}$ synthesis by nitrite reduction is an attractive possibility that has not been explored in amoebae. Also, it has been observed that *E. histolytica* supports higher $NO^{\cdot}$ concentrations than liver cells for short time periods (Pacheco-Yépez *et al.*, 2001). Such resistance might be developed by the constant exposure of the parasite to its own $NO^{\cdot}$ and/or that derived from the colonic environment. However, under prolonged time periods, $NO^{\cdot}$ shows amoebicidal activity through induction of apoptosis (Ramos *et al.*, 2007).

One of the most susceptible targets of $NO^{\cdot}$ are proteins containing Fe–S clusters. On this regard, *E. coli* contains the fumarate/nitrate reductase regulatory protein (FNR) that is an O_2 and $NO^{\cdot}$ sensor, showing reaction with $NO^{\cdot}$ as follow: $[4Fe\text{–}4S]^{2+} + 4NO^{\cdot} \rightarrow 2Fe(NO^{\cdot})_2 + 2Fe^{3+} + 4S^{2-}$ (Crack *et al.*, 2008). Therefore Fe–S oxidation by RNS might be important to signal nitrosative stress in the amoeba.

Nitric oxide is an endogenous diffusible transcellular messenger shown to participate in survival and death pathways and can alter protein function directly through post-translational modifications (nitration or s-nitrosylation) or indirectly through interactions with O_2, $O_2^{\cdot-}$, thiols and heavy metals (Uys *et al.*, 2011). S-nitrosylation (RSNO) is a thiol oxidation caused by $NO^{\cdot}$ derivatives. The proposed reaction in biological systems is $2NO^{\cdot} + O_2 \rightarrow 2NO_2^{\cdot}$ further $NO_2^{\cdot} + R\text{-}SH \rightarrow NO_2^{-} + R\text{-}S^{\cdot} + H^{+}$; further reaction of the thiyl radical with $NO^{\cdot}$ can then generate RSNO. This reaction is promoted in membranes or hydrophobic domains in proteins since $NO^{\cdot}$ and O_2 have favourably partition into lipophilic environments. An alternative mechanism of S-nitrosylation involves reaction of $NO^{\cdot}$ with the thiyl radical generated during the catalytic cycles of several enzymes like peroxidases or ribonucleotide reductases (Bindoli *et al.*, 2008). In contrast, denitrosylation is possible through enzymatic or non-enzymatic pathways; for example, denitrosylation of S-nitrosogluthathione is catalysed by class III alcohol dehydrogenase isoenzyme (Jensen *et al.*, 1998), whereas Trx catalyses the denitrosylation of a number of S-nitroisoproteins (Benhar *et al.*, 2008). Because of its reversibility, S-nitrosylation is considered to be a mechanism for dynamic, post-translational regulation of most or all major protein classes. In *E. histolytica* such regulation control could take place since $NO^{\cdot}$ is produced during normal amoebic culture (Hernández-Campos *et al.*, 2003) and proliferation is diminished by the addition of the specific $NO^{\cdot}$ scavenger PTIO (A. Olivos-García unpublished). Also, EhFDP is the only enzyme identified in the *E. histolytica* genome able to detoxify NO to N_2O (Fig. 17.1). The EhFDP gene is highly expressed in normal conditions and is up-regulated during NO stress (Vicente *et al.*, 2009a). It will be important to establish the physiological function of EhFDP.

On the other hand, as mentioned before, PDI is the most abundant chaperone in the endoplasmic reticulum (ER) of mammalian cells that facilitates protein folding via its oxido-reductase activity. Prolonged or acute nitrosative stresses interfere with PDI activity through formation of a complex PDI-SNO (S-nitrosylation) and PDI-SSG (S-glutathionylation) leading to accumulation of misfolded or unfolded proteins and activation of the unfolded protein response (UPR) (Uys *et al.*, 2011). The UPR function is to restore the cellular function by halting protein translation, activating the signalling pathways to increase chaperone synthesis and triggering the degradation of terminally misfolded proteins (Townsend, 2007). Recently, it has been observed that *E. histolytica* lacks a UPR after a $NO^{\cdot}$ challenge. Conversely, during an ER fission process its components are localized in the same compartments. ER fragmentation and/or eventual fusion may be one of the strategies of the parasite to rapidly recover ER function after NO· stress (Santi-Rocca *et al.*, 2012). In addition, and in parallel with mammalian or yeast cells, PDI of *E. histolytica* could function as a sensor for nitrosative and oxidative stress.

Heat shock stress

All cell proteins must fold into precise three-dimensional conformations to fulfil their biological functions. This fundamental process is procured by molecular chaperones to assist in folding, assembly, intracellular localization, secretion, regulation and degradation of different proteins by the ubiquitin–proteasome system. During several types of stress, chaperons increase and prevent protein misfolding/aggregation and at the same time promote degradation of oxidized proteins by the lysosome/vacuolar system.

Heat shock proteins (Hsp) were first described as gene products whose expression was induced by heat and other sources of stress (Muga and Moro 2008). They are ATP-dependent enzymes and are usually classified according to their molecular weights (Hsp40, Hsp60, Hsp70, Hsp90, Hsp100 and the so-called small Hsps). Chaperonins are large, double–ring complexes of approximately 800 kDa. Two groups of chaperonins exist, the Hsp60s present in bacteria, mitochondria and chloroplasts belong to group 1 and functionally they cooperate with Hsp10 which forms the lid of the folding cage. The chaperonins of group II are called thermosomes in *Archaea* and TRiC/CCT in eukaryotic cells; they have eight or nine membered rings. With some exceptions, members of all these classes are present throughout the three kingdoms of life (Vabulas *et al.*, 2010). Hsp70s generally function in cooperation with Hsp40 or J-proteins and with various nucleotide exchange factors (NEFs). As a common structural feature, all Hsp70 family members consist of an amino-terminal ATPase domain and a carboxy-terminal substrate binding domain (Muga and Moro, 2008). Hsp90 is another important chaperone system that functions downstream of Hsp70 during the final structural maturation and conformational regulation of a number of signalling molecules and transcription factors (Pearl and Prodromou, 2006; Zhao and Houry, 2007).

Several HSPs have been reported in *E. histolytica*: Hsp10, Hsp40, Hsp60, Hsp70 (two isoforms), Hsp90 (three isoforms), Hsp100 and Hsp101. All of them (transcripts or protein) are overexpressed by heat shock (Field *et al.*, 2000; Bernes *et al.*, 2005; Weber *et al.*, 2006) with exception of Hsp10 and Hsp60 (van der Giezen *et al.*, 2005). Their intracellular location might give clues about their functions. Mitosomes contain Hsp10, Hsp60 and Hsp70, and along with an ADP/ATP carrier they are suggested to assist in the function of the Nif-like system described above (Tovar *et al.*, 2007; Maralikova *et al.*, 2010). Also Hsp60 was demonstrated to play an important role in amoebic proliferation since its inhibition by RNA antisense led to a marked growth defect (Mi-Ichi *et al.*, 2011). Amoebic Hsp100 is located in the nucleus and it has been suggested to be regulated by epigenetic and classical pathways during differentiation and stress, respectively. The Hsp100 probably functions might be to restore and/or repair stress-induced nuclear alterations (Bernes *et al.*, 2005). Also, Hsp70 and proteasomes appear to be involved in *E. invadens* encystation; during this process it is likely that some amoebic Hsp are involved in degradation of cytoskeletal proteins as well (Field *et al.*, 2000).

A transcription profiling study during heat shock stress in *E. histolytica* has shown that a somewhat specific response is initiated through transcription factor, ubiquitin, proteasome 26S, proteases and HSP70 and HSP90 families, according to a classical heat shock response. Also, a significant down-regulation of genes involved in almost all physiological functions (including related virulence genes) was observed; such transcriptional changes might correspond with a general adapting response to the new environmental conditions (Weber *et al.*, 2006).

Other stressing conditions

Withdrawal of nutrients for 2 hours did not make *E. histolytica* trophozoites to encyst and did not increase the mRNA expression for heat shock proteins or chitinase (Field *et al.*, 2000). This intriguing result might be explained by the abundance of glucose equivalents in the form of glycogen that are depleted after 24 hours. In addition, when glucose is absent in the medium for 12 hours, the expression of Hsp70 does not increase and the parasite become slightly more sensitive to H_2O_2 (Tovy *et al.*, 2011). Moreover, during amoebic oxidative stress with H_2O_2, transcription of Hsp101 and Hsp70 increases, as well as that of Hsp40/DnaJ and an ubiquitin-conjugating enzyme (a marker of misfolded proteins for degradation). These enzymes can assist the protein folding since Hsp40 stimulates the ATPase activity of Hsp70 and Hsp101. A similar response is observed during $NO^{\cdot}$ stress (Vicente *et al.*, 2009a). Furthermore, contrary to what is expected, *E. histolytica* exposure to metronidazole leads to Hsp70 and Hsp101 protein decrease, with minimal variations at the transcriptome or proteome levels (Tazreiter *et al.*, 2008). It is possible that amoebic stress sensors (if they are present) do not recognize the stress caused by this hazardous chemical, contributing to its cytotoxicity by accumulation of misfolded proteins.

E. histolytica transcriptome has been analysed under different stress conditions such as heat shock (Weber *et al.*, 2006), cysteine deprivation (Husain *et al.*, 2011), during murine intestinal colonization (Gilchrist *et al.*, 2006), encystation (Ehrenkaufer *et al.*, 2007) and in the presence of H_2O_2, and $NO^{\cdot}$ (Vicente *et al.*, 2009a). Although results show different transcriptional responses, some similarities are present at different levels. The common responses reflect the overlapping regulatory mechanisms to these stresses. The amoebic proteomic and metabolomic analysis during the stress conditions will unravel the stress amoebic regulatory pathways at all their levels. This knowledge might help in the development of novel therapeutic strategies against amoebiasis.

E. histolytica stress during amoebic liver abscess

E. histolytica has to cope with many dangerous molecules derived from host or itself during amoebic liver abscess development. We here focus on those related to amoebic ROS and RNS stresses.

Pathogenicity is defined as the ability of a microorganism to produce disease in a susceptible host, whereas virulence is the degree of damage caused by the pathogen. *E. histolytica* virulence is frequently evaluated as the parasite ability to induce experimental amoebic liver abscess in hamsters (ALAH). Also for many years this model has been used to explore the molecular mechanisms of *E. histolytica* pathogenicity because of its similarity with that in humans. In early ALAH (before 12 hours) a massive dying population of parasites is observed (almost 70%); after this clearance time, the surviving amoebae proliferate until the animal dies (Rigothier *et al.*, 2002). Two important questions derive from these observations: (1) which is the amoebal phenotype enabling survival? and (2) Is there a selection of a specific O_2 resistant parasite subpopulation?

Resistance to oxidants could be the main characteristic of the surviving amoebae in experimental ALAH since non-virulent *E. histolytica* are unable to survive in the hamster liver despite sharing similar rates of several *in vitro* functions used as parameters of virulence excepting that of O_2 and H_2O_2 resistance (Olivos-García *et al.*, 2009; Ramos-Martínez *et al.*, 2009). Clonal selection possibility is unlikely since *in vitro* resistance to high O_2 exposure is similar between recovered parasites 2 hours after ALAH and non-injected control amoebae (A. Olivos-García unpublished). Also, the amoebic-resistant population (if that were the case) is unable to survive when few virulent parasites (less than 5×10^5) are injected into the hamster portal vein; little or no inflammatory reaction is observed (Ramos-Martínez *et al.*, 2009). An alternative explanation for parasite survival during early ALAH is next proposed.

Once the parasites are injected into the hamster portal vein they spread within the sinusoids of all lobules of the liver which contain variable O_2 concentrations toxic for the amoebae (6–13%) (Nauck *et al.*, 1981). The high Cys/cystine content in *E. histolytica* could help to resist this initial high O_2 challenge since it is the principal thiol redox pair for the parasite during *in vitro* oxygen resistance (before 12 hours) (Gillin and Diamond, 1981); Cys (0.1% in the culture medium) is also responsible for maintaining a low redox potential (−58 mV) necessary for parasite growth (Band and Cirrito, 1979). It is possible that (similar to the iron acquisition), the amoebic transPMET system could assist in cystine reduction for its incorporation to the cell (Fig. 17.1). Moreover, the parasite can uptake Cys from blood (which has a concentration of 125 μmol/l as cystine) (Johnson *et al.*, 2008), by phagocytosis of red blood cells (35 μmol/l) (Reid *et al.*, 2006) and from plasma proteins. In addition, since glucose catabolism provides the reductive power and the ATP for the amoebic O_2 detoxification system, blood glucose (3.6–5.8 mM) can stimulates the O_2 reduction as observed in *in vitro* conditions. Cystine and glucose have been shown to protect *E. histolytica* from *in vitro* O_2 challenge; the *in vivo* constant supply could compensate the low cystine blood levels. Also, the redox potential of the blood plasma Cys/cystine is calculated to be −80 mV (Jones, 2006) which is in favour of amoebic survival. In addition, *in vitro* oxygen resistance of *E. histolytica* is very similar when incubations are made in TYI-S- 33 media or de-complemented hamster serum but higher respect to phosphate buffer saline without Cys (A. Olivos-García unpublished).

To explain amoebic survival, local liver O_2 concentrations can be grouped as high, intermediate and low.

1 As observed during *in vitro* high O_2 challenge, the amoebic O_2 reduction pathway is unable to diminish the intracellular higher O_2 concentrations to a moderate hypoxic environment. This generates a multifactorial affection by indiscriminate and irreversible protein oxidation including that caused by $^{\cdot}OH$. Among others, the impairment of the glycolytic-fermentative pathway through the irreversible inhibition of enzymes sensitive to oxidant stress leads to ATP depletion, with consequent chaperon inability to repair oxidized proteins and DNA, cytosolic acidification, semi-permeability loss, etc.
2 In the intermediate aerobic environment, moderate O_2 concentrations and accumulation of $O_2^{\cdot-}$ and H_2O_2 may reversibly inhibit some antioxidant or glycolytic enzymes among others, leading to increased oxidant status and partial/reversible inhibition of the glycolytic pathway. These changes might generate increased cytosolic calcium, loss of membrane integrity, intracellular acidification, low ATP and NAD(P)H, even apoptosis. In such conditions, as in *in vitro* studies, the parasites could be easily disintegrated by complement (A. Olivos-García unpublished).
3 In a low-O_2 environment, the parasites are able to maintain an intracellular hypoxic environment enough to permit its metabolic flux for survival; moderate inhibition of the glycolytic pathway could be observed and this time chaperons are able to maintain protein and DNA in a low oxidized level. As it is observed in *in vitro* conditions, the cooperative association of the parasites could contribute to their survival by making the microenvironment more hypoxic. In this regard, amoebic viability improves *in vitro* with high amoebic density when they are incubated in PBS equilibrated with air that contains toxic levels of oxygen (~220 μM) (A. Olivos-García unpublished). Also, Cys plus ascorbic acid improves amoebic viability in Dulbecco's modified Eagle medium, Hanks' balanced salt solution or PBS (Wong *et al.*, 2011).

It is important to consider that in experimental ALAH, intrahepatic injection offers more methodological artifices than intra-portal amoebic administration; mechanical damage and local high amoebic concentration will favour parasite survival by both high oxygen reduction rate and strong inflammation.

In contrast to other amoebae inhabiting the human intestine, it is quite probable that the high efficiency of *E. histolytica* O_2 reduction systems enables it to support the initial episode of tissue invasiveness by maintaining intracellular hypoxia. In this context a higher content of Eh34 protein in *E. histolytica* than in *E. dispar* (A. Olivos-García unpublished) and more Eh29 in *E. histolytica* than in either *E. dispar* or *E. histolytica* Rahman have been observed (Choi *et al.*, 2005; Davis *et al.*, 2006). Also, in *in vitro* conditions, virulent *E. histolytica* is more resistant to O_2 than either non-virulent *E. histolytica* (Ramos-Martínez at al., 2009) or non-pathogenic *E. dispar* (A. Olivos-García unpublished). In addition, during the early steps of amoebic tissue invasion (intestine and/or liver) the presence of facultative bacteria could aid to: (a) generate intracellular amoebic hypoxia by consuming O_2, (b) stimulate strong inflammation that generates ischaemia, and (c) maintain a microaerobic environment. Such proposal derives from the following observations: (a) in *in vitro* conditions Gram-negative bacteria phagocytized by *E. histolytica* protect the parasite from H_2O_2 stress probably by scavenging oxidized molecules trough the antioxidant enzymatic and the electron transport systems (Bracha and Mirelman, 1984) and (b) ALA in hamsters increased 270% in the presence of *E. coli* (Anaya-Velázquez and Padilla-Vaca, 1992).

After approximately 5 hours, decreased local O_2 concentration in the liver is a necessary condition for subsequent amoebic survival. This proposal is supported by the following experimental observations: (1) during experimental ALAH focal ischaemia generated by inflammation is observed from 5 hours onwards and remains until the animal dies (Pérez-Tamayo *et al.*, 1992); (2) avoiding the ischaemia development in leukopenic animals, virulent amoebae are unable to survive in the hamster liver (Olivos-García *et al.*, 2004); (3) *E. histolytica* in the liver of mice is not able to induce an acute inflammatory response, and similar to what occurs in leukopenic hamsters, the parasite dies (A. Olivos-García unpublished). On the other hand, once the parasites resist low liver O_2 concentration (~after 5 hours of injection) they are surrounded by abundant polymorphonuclear cells (PMNs) that could produce H_2O_2 (only for a short time period) and $NO^{\cdot}$. In *in vitro* studies these oxidants along with HOCl generated by the myeloperoxidase from PMNs have shown to be the principal inflammatory cytotoxic compounds against *E. histolytica* (Ghadirian *et al.*, 1986; Lin *et al.*, 1994). However, since neither iNOS expression nor plasmatic $NO^{\cdot}$ levels increase during early ALAH and iNOS inhibitors do not increase amoebic survival (before 24 hours) (A. Olivos-García unpublished), $NO^{\cdot}$ should not be consider as an important stressor molecule for *E. histolytica* during early ALAH. The temporality of the amoebic oxidative stress during experimental ALAH is not known but it can be assumed to occur before 12 hours since after this time, an uninterrupted amoebic proliferation is established (Rigothier *et al.*, 2002), and the O_2 necessary for ROS production is diminished by the permanent ischaemia (Pérez-Tamayo *et al.*, 1992). Additionally, after 24 hours, the PMNs surrounding amoebae are being gradually replaced by macrophages that are unable to generate an oxidative burst (Denis and Chadee, 1988). On the other hand, in human ALA or after 24 hours in hamsters, a moderate heat shock response generated by the inflammatory process is present. Currently its relevance for amoebic survival it is not known since in *in vitro* studies only a moderate overall unspecific amoebic response has been observed after heat shock (Weber *et al.*, 2006).

An unexplained observation in ALA, either in humans or in hamsters, is the absence of parasites in the central areas of the necrotic foci (Brandt and Pérez-Tamayo, 1970). It could be attributed to the absence of glucose, Cys and the few O_2 that are necessary for amoebic survival. In order to have access to those and other molecules, the parasites move to the peripheral areas of the ischaemic foci in which they could incorporate nutrients by diffusion or phagocytosis of host cells. During progression of the tissue destruction caused mainly by the host, the necrotic central areas lacking amoebae increase. With inhibition of tissue destruction through anti-inflammatory (methyl prednisolone or cyclosporin A) and/or antioxidant (N-acetyl cysteine) therapy, in experimental ALAH abundant and well-preserved parasites remain in groups, probably receiving nutrients by diffusion and maintaining a local hypoxia by clustering instead of spreading to invade other regions of the organ (Olivos *et al.*, 2005; Olivos-García *et al.*, 2007). In addition, the activity of the amoebic transPMET system favoured by hypoxia could contribute to maintain a reduced microenvironment and promote acidity, providing protons to the surroundings. It could explain at least in part why human necrotic foci are acid (pH 5.2–6.7) (Ramachandran *et al.*, 1976). During the late stages of experimental ALAH, $NO^{\cdot}$ is another dangerous oxidant that is increased in the blood circulation (Pacheco-Yépez *et al.*, 2001) and iNOS mRNA in the liver tissues (Ramírez-Emiliano *et al.*, 2005). Since $NO^{\cdot}$ is toxic for the parasite under *in vitro* conditions; it may be asked why the parasite survives despite the high levels of NO˙? Two possible explanations can be suggested:

1 Amoebae survive because they are more resistant to NO˙ than liver cells, as it has been observed in *in vitro* experiments (Pacheco-Yépez *et al.*, 2001). But this explanation is unlikely since such resistance has been evaluated *in vitro* only during short time periods and at longer times amoebic apoptosis is induced (Ramos *et al.*, 2007). Also, NO˙ is involved in the *in vitro* amoebicidal activity of murine macrophages Lin *et al.*, 1994).
2 *E. histolytica* survives because decreased or no NO˙ production is present in the necrotic microenvironment. Such hypothesis is supported by the following observations: in the liver, macrophages and liver cells are able to produce NO˙ from L-arginine and O_2 as substrates by the iNOS. The concentrations of both substrates have not been determined in the abscess but it can be speculated to be low or absent since the lesions are hypoxic (ischaemia) and, as in some tumours, hypoxia may induce arginase synthesis by macrophages (Murdoch *et al.*, 2005). In addition, macrophages and liver cells surrounding the parasites in ALAH do not express iNOS whereas those of well-preserved areas do (A. Olivos-García unpublished).

From the above, the total high NO˙ levels detected in experimental ALAH could be produced only by macrophages and liver cells present in well oxygenated areas (not associated with necrosis).

A remarkable phenomenon is that few amoebae are observed outside the necrotic foci during 24–72 hours of experimental ALAH. Probably some suicidal parasites that leave ischaemic focus are killed by O_2 and complement. Also, after 72 hours, NO˙ from circulation (nitrate/nitrite) can diffuse to the necrotic surroundings and along with O_2 induce apoptosis, as recently observed for this time period (Villalba-Magdaleno *et al.*, 2011). On the other hand, but only on late stages of ALA either human (Brandt and Pérez-Tamayo, 1970) or hamster (A. Olivos-García unpublished), some well-preserved parasites are present in the sinusoidal space without inflammation and in contact with liver cells, without apparent cell damage. Although the high nitrate/nitrites and O_2 present in the blood could kill these parasites, they are alive. Increased amoebic resistance to such oxidants is an unlikely explanation since once recovered and cultured *in vitro* they show similar susceptibility levels than the original strain (A. Olivos-García unpublished). Instead it can be hypothesized that a local hypertensive phenomenon induced through NO˙ scavenging by $O_2^{\cdot-}$ could explain such survival. This proposal is supported by: (1) a high $O_2^{\cdot-}$ generation in sinusoids as determined *in vivo* by NBT precipitation, which is observed only on late stages of ALAH (A. Olivos-García unpublished); (2) in contrast to the overexpression of the mRNA for iNOS, that for the eNOS (the principal producer of endothelial NO˙) is not induced in well preserved liver areas of ALAH (Ramírez-Emiliano *et al.*, 2005) and (3) high NO˙ concentration in blood circulation (nitrate/nitrite) are detected on late stages of ALAH (Pacheco-Yépez *et al.*, 2001). As occurs in cirrhosis, the high sinusoidal $O_2^{\cdot-}$ concentration could cause a local hypertension by scavenging the low endothelial NO˙ generated by eNOS (García-Calderó *et al.*, 2011). Then, decreased blood flow can lead to hypoxia that promotes NO˙ generation through nitrite reduction by acidification (amoebic abscess pH 5.2–6.7) and by nitrite reductase activity of deoxyhaemoglobin and xantin oxidoreductase (Raat *et al.*, 2009). Also, acidification might contribute to hypoxia by decreasing the affinity of haemoglobin for O_2 (Gladwin and Kim-Shapiro 2008). At the same time superoxide can scavenge this NO˙, and the peroxynitrite generated could contribute to tissue damage.

Concluding remarks and future trends

Oxygen is an extremely dangerous molecule for *E. histolytica* because it is able to oxidize Fe–S centres and Fe^{2+} of its proteins, leading to metabolic failure and death. During the first steps of invasion (before 5 hours) the parasite resists this insult by generation of intracellular hypoxia principally through the couple NADPH:flavin oxidoreductase-peroxiredoxin (Fig. 17.1). The contribution to this process of other uncharacterized antioxidant enzymes present in *E. histolytica* remains to be explored. This efficient constitutive-inducible enzymatic system might be an appropriate target for new prophylactic therapies.

On the other hand, in *in vitro* conditions, $NO^{\cdot}$ affects the endoplasmic reticulum of *E. histolytica* and similar to O_2 it is able to cause metabolic failure and apoptosis. Also, $NO^{\cdot}$ is the main molecule produced by murine macrophages toxic for *E. histolytica*. Since $NO^{\cdot}$ is not in contact, at least in cytotoxic concentrations, with *E. histolytica* during ALAH development, a strategy that specifically stimulate and/or deliver this radical in the parasite microenvironment should be considered when designing therapies for invasive amoebiasis. Currently a strategy that is being pursued to treat some tumours (Eue *et al.*, 1998; Frampton, 2010) could be used to treat invasive amoebiasis: local macrophage stimulation to express iNOS enzyme.

E. histolytica was discovered more than one hundred years ago as the causal agent of human amoebiasis and yet currently the illness remains an important health problem in underdeveloped countries. It is well recognized that the solution to human amoebiasis consists in increased appropriate sanitary conditions and health education in the affected population, but unfortunately they are not feasible in the short term. Instead, a rational alternative is to increase knowledge of both parasite biology and the host–parasite relationship. This review attempts to cover the main molecular mechanisms of human amoebiasis and includes hypotheses that can be tested in the near future.

Acknowledgements

The research done in the author's laboratory is supported by DGAPA Grant IN-204310 and CONACyT-Mexico grants 59175 and 83084 to E.S. We would also like to acknowledge Mr Marco Gudiño for the informatics and art design.

Web resources

The Entamoeba Homepage: http://entamoeba.lshtm.ac.uk

This page provides links to information on *Entamoeba* spp.

NCBI:: http://www.ncbi.nlm.nih.gov/genome

This resource organizes information on genomes including maps, chromosomes, assemblies, and annotations.

EuPathDB: http://amoebadb.org/amoeba/

Allow mining of gene databases for amoeba genomes.

Pathema-Entamoeba: http://pathema.jcvi.org/cgi-bin/Entamoeba/PathemaHomePage.cgi

Allow mining of gene databases for amoeba genomes.

TIGR *Entamoeba histolytica* DB: http://blast.jcvi.org/er-blast/index.cgi?project=eha1

Entamoeba histolytica sequence BLAST search tool.

References

Akbar, M.A., Chatterjee, N.S., Sen, P., Debnath, A., Pal, A., Bera, T., and Das, P. (2004). Genes induced by a high-oxygen environment in *Entamoeba histolytica*. Mol. Biochem. Parasitol. *133*, 187–196.

Al-Abdul-Wahid, M.S., Evanics, F., and Prosser, R.S. (2011). Dioxygen transmembrane distributions and partitioning thermodynamics in lipid bilayers and micelles. Biochemistry *50*, 3975–3983.

Ali, V., Shigeta, Y., Tokumoto, U., Takahashi, Y., and Nozaki, T. (2004). An intestinal parasitic protist, *Entamoeba histolytica*, possesses a non-redundant nitrogen fixation-like system for iron–sulfur cluster assembly under anaerobic conditions. J. Biol. Chem. *279*, 16863–16874.

Almeida C.C., Romão C.V., Lindley P.F., Teixeira M., and Saraiva L.M. (2006). The role of the hybrid cluster protein in oxidative stress defense. J. Biol. Chem. *281*, 32445–32450.

Amyx, H.L., Asher, D.M., Nash, T.E., Gibbs, C.J., Jr., and Gajdusek, D.C. (1978). Hepatic amebiasis in spider monkeys. Am. J. Trop. Med. Hyg. *27*, 888–891.

Anaya-Velázquez, F., and Padilla-Vaca, F. (1992). Effect of intestinal bacteria on the virulence of *Entamoeba histolytica*. Arch. Med. Res. *23*, 183–185.

Andersson, J.O., Hirt, R.P., Foster, P.G., and Roger, A.J. (2006). Evolution of four gene families with patchy phylogenetic distributions: influx of genes into protist genomes. B.M.C. Evol. Biol. *6*, 27.

Arias, D.G., Gutierrez, C.E., Iglesias, A.A., and Guerrero, S.A. (2007). Thioredoxin-linked metabolism in *Entamoeba histolytica*. Free Radic. Biol. Med. *42*, 1496–1505.

Arias, D.G., Carranza, P.G., Lujan, H.D., Iglesias, A.A., and Guerrero, S.A. (2008). Immunolocalization and enzymatic functional characterization of the thioredoxin system in *Entamoeba histolytica*. Free Radic. Biol. Med. *45*, 32–39.

Arnér, E.S., and Holmgren, A. (2000). Physiological functions of thioredoxin and thioredoxin reductase. Eur. J. Biochem. *267*, 6102–6109.

Band, R.N., and Cirrito, H. (1979). Growth response of axenic *Entamoeba histolytica* to hydrogen, carbon dioxide, and oxygen. J Protozool. *26*, 282–286.

Benhar, M., Forrester, M.T., Hess, D.T., and Stamler, J.S. (2008). Regulated protein denitrosylation by cytosolic and mitochondrial thioredoxins. Science *320*(5879), 1050–1054.

Bera, T., Nandi, N., Sudhahar, D., Akbar, M.A., Sen, A., and Das, P. (2006). Preliminary evidence on existence of transplasma membrane electron transport in *Entamoeba histolytica* trophozoites: a key mechanism for maintaining optimal redox balance. J. Bioenerg. Biomembr. *38*, 299–308.

Bernes, S., Siman-Tov, R., and Ankri, S. (2005). Epigenetic and classical activation of *Entamoeba histolytica* heat shock protein 100 (EHsp100) expression. FEBS Lett. *579*, 6395–6402.

Bindoli, A., Fukuto, J.M., and Forman, H.J. (2008). Thiol chemistry in peroxidase catalysis and redox signaling. Antioxid. Redox Signal. *10*, 1549–1564.

Biteau, B., Labarre, J., and Toledano, M.B. (2003). ATP-dependent reduction of cysteine-sulphinic acid by S. cerevisiae sulphiredoxin. Nature *425*(6961), 980–984.

Boschi-Muller, S., Gand, A., and Branlant, G. (2008). The methionine sulfoxide reductases: Catalysis and substrate specificities. Arch. Biochem. Biophys. *474*, 266–273.

Bracha, R., and Mirelman, D. (1984). Virulence of *Entamoeba histolytica* trophozoites. Effects of bacteria, microaerobic conditions, and metronidazole. J. Exp. Med. *160*, 353–368.

Brandt, H., and Pérez-Tamayo, R. (1970). Pathology of human amebiasis. Hum. Pathol. *1*, 351–385.

Bruchhaus, I., and Tannich, E. (1994). Induction of the iron-containing superoxide dismutase in *Entamoeba histolytica* by a superoxide anion-generating system or by iron chelation. Mol. Biochem. Parasitol. *67*, 281–288.

Bruchhaus, I., Richter, S., and Tannich, E. (1997a). Removal of hydrogen peroxide by the 29 kDa protein of *Entamoeba histolytica*. Biochem. J. *326*, 785–789.

Bruchhaus, I., Richter, S., and Tannich, E. (1997b). Characterization of two *E. histolytica* proteins that inactivate reactive oxygen species. Arch. Med. Res. *28* Spec. No. 91–92.

Bruchhaus, I., Richter, S., and Tannich, E. (1998). Recombinant expression and biochemical characterization of an NADPH:flavin oxidoreductase from *Entamoeba histolytica*. Biochem. J. *330*, 1217–1221.

Choi, M.H., Sajed, D., Poole, L., Hirata, K., Herdman, S., Torian, B.E., and Reed, S.L. (2005). An unusual surface peroxiredoxin protects invasive *Entamoeba histolytica* from oxidant attack. Mol. Biochem. Parasitol. *143*, 80–89.

Costa, V., Quintanilha, A., and Moradas-Ferreira, P. (2007). Protein oxidation, repair mechanisms and proteolysis in *Saccharomyces cerevisiae*. IUBMB Life *59*, 293–298.

Crack, J.C., Le Brun, N.E., Thomson, A.J., Green, J., and Jervis, A.J. (2008). Reactions of nitric oxide and oxygen with the regulator of fumarate and nitrate reduction, a global transcriptional regulator, during anaerobic growth of *Escherichia coli*. Methods Enzymol. *437*, 191–209.

Cruz, F., and Ferry, J.G. (2006). Interaction of iron–sulfur flavoprotein with oxygen and hydrogen peroxide. Biochim. Biophys. Acta. *1760*, 858–864.

D'Aquino, M., Bullion, C., Chopra, M., Devi, D., Devi, S., Dunster, C., James, G., Komuro, E., Kundu, S., Niki, *et al.* (1994). Sulfhydryl free radical formation enzymatically by sonolysis, by radiolysis, and thermally: Vitamin A, curcumin, muconic acid, and related conjugated olefins as references. Methods Enzymol. *233*, 34–46.

Davis, P.H., Zhang, X., Guo, J., Townsend, R.R., and Stanley, S.L. (2006). Comparative proteomic analysis of two *Entamoeba histolytica* strains with different virulence phenotypes identifies peroxiredoxin as an important component of amoebic virulence. Mol. Microbiol. *61*, 1523–1532.

Denis, M., and Chadee, K. (1988). In vitro and *in vivo* studies of macrophage functions in amebiasis. Infect. Immun. *56*, 3126–3131.

Ehrenkaufer, G.M., Haque, R., Hackney, J.A., Eichinger, D.J., and Singh, U. (2007). Identification of developmentally regulated genes in *Entamoeba histolytica*: insights into mechanisms of stage conversion in a protozoan parasite. Cell. Microbiol. *9*, 1426–1444.

Eue, I., Kumar, R., Dong, Z., Killion, J.J., and Fidler, I.J. (1998).Induction of nitric oxide production and tumoricidal properties in murine macrophages by a new synthetic lipopeptide JBT3002 encapsulated in liposomes. J. Immunother. *21*, 340–351.

Field, J., Van Dellen, K., Ghosh, S.K., and Samuelson, J. (2000). Responses of *Entamoeba invadens* to heat shock and encystation are related. J. Eukaryot. Microbiol. *47*, 511–514.

Frampton, J.E. (2010). Mifamurtide: a review of its use in the treatment of osteosarcoma. Paediatr. Drugs *12*, 141–153.

García-Calderó, H., Rodríguez-Vilarrupla, A., Gracia-Sancho, J., Diví, M., Laviña, B., Bosch, J., and García-Pagán, J.C. (2011). Tempol administration, a superoxide dismutase mimetic, reduces hepatic vascular resistance and portal pressure in cirrhotic rats. J. Hepatol. *54*, 660–665.

Ghadirian, E., Somerfield, S.D., and Kongshavn, P.A. (1986). Susceptibility of *Entamoeba histolytica* to oxidants. Infect. Immun. *51*, 263–267.

van der Giezen, M., León-Avila, G., and Tovar, J. (2005). Characterization of chaperonin 10 (Cpn10) from the intestinal human pathogen *Entamoeba histolytica*. Microbiology *151*, 3107–3115.

Gilchrist, C.A., Houpt, E., Trapaidze, N., Fei, Z., Crasta, O., Asgharpour, A., Evans, C., Martino-Catt, S., Baba, D.J., Stroup, S., *et al.* (2006). Impact of intestinal colonization and invasion on the *Entamoeba histolytica* transcriptome. Mol. Biochem. Parasitol. *147*, 163–176.

Gillin, F.D., and Diamond, L.S. (1981). *Entamoeba histolytica* and *Giardia lamblia*: effects of cysteine and oxygen tension on trophozoite attachment to glass and survival in culture media. Exp. Parasitol. *52*, 9–17.

Gladwin, M.T., and Kim-Shapiro, D.B. (2008).The functional nitrite reductase activity of the heme-globins. Blood *112*, 2636–2647.

Go, Y.M., and Jones, D.P. (2010). Redox control systems in the nucleus: mechanisms and functions. Antioxid. Redox Signal. *13*, 489–509.

Godoy, J.R., Funke, M., Ackermann, W., Haunhorst, P., Oesteritz, S., Capani, F., Elsässer, H.P., and Lillig, C.H. (2011). Redox atlas of the mouse. Immunohistochemical detection of glutaredoxin-, peroxiredoxin-, and thioredoxin-family proteins in various tissues of the laboratory mouse. Biochim. Biophys. Acta. *1810*, 2–92.

Gonzalez, J., Bai, G., Frevert, U., Corey, E.J., and Eichinger, D. (1999). Proteasome-dependent cyst formation and stage-specific ubiquitin mRNA accumulation in *Entamoeba invadens*. Eur. J. Biochem. *264*, 897–904.

Gu, M., and Imlay, J.A. (2011). The SoxRS response of *Escherichia coli* is directly activated by redox-cycling drugs rather than by superoxide. Mol. Microbiol. *79*, 1136–1150.

Hernández-Campos, M.E., Campos-Rodríguez, R., Tsutsumi, V., Shibayama, M., García-Latorre, E., Castillo-Henkel, C., and Valencia-Hernández, I. (2003). Nitric oxide synthase in *Entamoeba histolytica*: its effect on rat aortic rings. Exp. Parasitol. *104*, 87–95.

Husain, A., Jeelani, G., Sato, D., and Nozaki, T. (2011). Global analysis of gene expression in response to L-Cysteine deprivation in the anaerobic protozoan parasite *Entamoeba histolytica*. BMC Genomics *12*, 275.

Imlay, J.A. (2002). What biological purpose is served by superoxide reductase? J. Biol. Inorg. Chem. *7*, 659–663.

Imlay, J.A. (2003). Pathways of oxidative damage. Annu. Rev. Microbiol. *57*, 395–418.

Imlay, J.A. (2008). Cellular defenses against superoxide and hydrogen peroxide. Annu. Rev. Biochem. *77*, 755–776.

Inai, Y., and Nishikimi, M. (2002). Increased degradation of oxidized proteins in yeast defective in 26 S proteasome assembly. Arch. Biochem. Biophys. *404*, 279–284.

Jacob, C. (2006). A scent of therapy: pharmacological implications of natural products containing redox-active sulfur atoms. Nat. Prod. Rep. *23*, 851–863.

Jang, S., and Imlay, J.A. (2010). Hydrogen peroxide inactivates the *Escherichia coli* Isc iron-sulphur assembly system, and OxyR induces the Suf system to compensate. Mol. Microbiol. *78*, 1448–1467.

Jeelani, G., Husain, A., Sato, D., Ali, V., Suematsu, M., Soga, T., and Nozaki, T. (2010). Two atypical L-cysteine-regulated NADPH-dependent oxidoreductases involved in redox maintenance, L-cystine and iron reduction, and metronidazole activation in the enteric protozoan *Entamoeba histolytica*. J. Biol. Chem. *285*, 26889–26899.

Jensen, D.E., Belka, G.K., and Du Bois, G.C. (1998). S-Nitrosoglutathione is a substrate for rat alcohol dehydrogenase class III isoenzyme. Biochem. J. *331*, 659–668.

Johnson, J.M., Strobel, F.H., Reed, M., Pohl, J., and Jones, D.P. (2008). A rapid LC-FTMS method for the analysis of cysteine, cystine and cysteine/cystine steady-state redox potential in human plasma. Clin. Chim. Acta. *396*, 43–48.

Jones, D.P. (2006). Extracellular redox state: refining the definition of oxidative stress in aging. Rejuv. Res. *9*, 169–181.

Jones, S.A., Chowdhury, F.Z., Fabich, A.J., Anderson, A., Schreiner, D.M., House, A.L., Autieri, S.M., Leatham, M.P., Lins, J.J., Jorgensen, M., *et al.* (2007). Respiration of *Escherichia coli* in the mouse intestine. Infect. Immun. *75*, 4891–4899.

Kästle, M., and Grune, T. (2011). Proteins bearing oxidation-induced carbonyl groups are not preferentially ubiquitinated. Biochimie *93*, 1076–1079.

Kastner, J., Pfeffel, F., Rajek, A., Pezawas, T., Hiesmayr, M., and Eichler, H.G. (1997). Nitric oxide concentration in the gas phase of the gastrointestinal tract in man. Eur. J. Clin. Invest. *27*, 992–996.

Kurtz, D.M., Jr. (2007). Flavo-diiron enzymes: nitric oxide or dioxygen reductases? Dalton Trans. *37*, 4115–4121.

Ladas, S.D., Karamanolism G., and Ben-Soussan, E. (2007). Colonic gas explosion during therapeutic colonoscopy with electrocautery. World J. Gastroenterol. *13*, 5295–5298.

Lauwaet, T., Davids, B.J., Reiner, D.S., and Gillin, F.D. (2007). Encystation of Giardia lamblia: a model for other parasites. Curr. Opin. Microbiol. *10*, 554–559.

Le Moan, N., Clement, G., Le Maout, S., Tacnet, F., and Toledano, M.B. (2006). The *Saccharomyces cerevisiae* proteome of oxidized protein thiols: contrasted functions for the thioredoxin and glutathione pathways. J. Biol. Chem. *281*, 10420–10430.

Lee, B.C., Dikiy, A., Kim, H.Y., and Gladyshev, V.N. (2009). Functions and evolution of selenoprotein methionine sulfoxide reductases. Biochim. Biophys. Acta. *1790*, 1471–1477.

Leitsch D., Kolarich D., Wilson I.B., Altmann F., and Duchêne M. (2007). Nitroimidazole action in *Entamoeba histolytica*: a central role for thioredoxin reductase. PLoS Biol. *5*, e211.

Lin, J.Y., Seguin, R., Keller, K., and Chadee, K. (1994). Tumor necrosis factor alpha augments nitric oxide-dependent macrophage cytotoxicity against *Entamoeba histolytica* by enhanced expression of the nitric oxide synthase gene. Infect. Immun. *62*, 1534–1541.

Little, C., and O'Brien, P.J. (1969). Mechanism of peroxide-inactivation of the sulphydryl enzyme glyceraldehyde-3-phosphate dehydrogenase. Eur. J. Biochem. *10*, 533–538.

Lujan, H.D., and Diamond, L.S. (1997). Cholesterol requirement and metabolism in *Entamoeba histolytica*. Arch. Med. Res. 28S, 96–97.

Lund, E.K., Wharf, S.G., Fairweather-Tait, S.J., and Johnson, I.T. (1999). Oral ferrous sulfate supplements increase the free radical-generating capacity of feces from healthy volunteers. Am. J. Clin. Nutr. *69*, 250–255.

Maisonneuve, E., Ducret, A., Khoueiry, P., Lignon, S., Longhi, S., Talla, E., and Dukan, S. (2009). Rules governing selective protein carbonylation. PLoS One *4*, e7269.

Makioka, A., Kumagai, M., Ohtomo, H., Kobayashi, S., and Takeuchi, T. (2002). Effect of proteasome inhibitors on the growth, encystation, and excystation of *Entamoeba histolytica* and *Entamoeba invadens*. Parasitol. Res. *88*, 454–459.

Maralikova, B., Ali, V., Nakada-Tsukui, K., Nozaki, T., van der Giezen, M., Henze, K., and Tovar, J. (2010). Bacterial-type oxygen detoxification and iron–sulfur cluster assembly in amoebal relict mitochondria. Cell. Microbiol. *12*, 331–342.

Marques, M., Mojzita, D., Amorim, M.A., Almeida, T., Hohmann, S., Moradas-Ferreira, P., and Costa, V. (2006). The Pep4p vacuolar proteinase contributes to the turnover of oxidized proteins but PEP4 overexpression is not sufficient to increase chronological lifespan in *Saccharomyces cerevisiae*. Microbiology *152*, 3595–3605.

Márquez-Monter, H., Fuentes-Orozco, R., Correa-Lemus, I., and Becker, I. (1991). Invasive amebiasis in a spider monkey (*Ateles geoffroyi*). Case report and a short review of the literature of amebiasis in non-human primates. Arch. Invest. Med. *22*, 75–78.

Meerovitch, E., and Ghadirian, E. (1978). Restoration of virulence of axenically cultivated *Entamoeba histolytica* by cholesterol. Arch. Invest. Med. *9 S1*, 253–256.

Mi-Ichi, F., Makiuchi, T., Furukawa, A., Sato, D., and Nozaki, T. (2011). Sulfate activation in mitosomes plays an important role in the proliferation of *Entamoeba histolytica*. PLoS Negl. Trop. Dis. *5*, e1263.

Møller, I.M., Rogowska-Wrzesinska, A., and Rao, R.S. (2011). Protein carbonylation and metal-catalyzed protein oxidation in a cellular perspective. J. Proteomics *74*, 2228–2242.

Muga, A., and Moro, F. (2008). Thermal adaptation of heat shock proteins. Curr. Protein Pept. Sci. *9*, 552–566.

Murdoch, C., Muthana, M., and Lewis, C.E. (2005). Hypoxia regulates macrophage functions in inflammation. J. Immunol. *175*, 6257–6263.

Nandi, N., Sen, A., Banerjee, R., Kumar, S., Kumar, V., Ghosh, A.N., and Das, P. (2010). Hydrogen peroxide induces apoptosis-like death in *Entamoeba histolytica* trophozoites. Microbiology *156*, 1926–1941.

Nauck, M., Wölfle, D., Katz, N., and Jungermann, K. (1981). Modulation of the glucagon-dependent induction of phosphoenolpyruvate carboxykinase and tyrosine aminotransferase by arterial and venous oxygen concentrations in hepatocyte cultures. Eur. J. Biochem. *119*, 657–661.

Nozaki, T., Asai, T., Sanchez, L.B., Kobayashi, S., Nakazawa, M., and Takeuchi,T. (1999). Characterization of the gene encoding serine acetyltransferase, a regulated enzyme of cysteine biosynthesis from the protist parasites *Entamoeba histolytica* and *Entamoeba dispar*. Regulation and possible function of the cysteine biosynthetic pathway in Entamoeba. J. Biol. Chem. *274*, 32445–32452.

O'Leary, V., and Solberg, M. (1976). Effect of sodium nitrite inhibition on intracellular thiol groups and on the activity of certain glycolytic enzymes in Clostridium perfringens. Appl. Environ. Microbiol. *31*, 208–212.

Olivos A., Ramos E., Nequiz M., Barba C., Tello E., Castañón G., González A., Martínez R.D., Montfort I., and Pérez-Tamayo R. (2005). *Entamoeba histolytica*: mechanism of decrease of virulence of axenic cultures maintained for prolonged periods. Exp. Parasitol. *110*, 309–312.

Olivos-García, A., Nequiz-Avendaño, M., Tello, E., Martínez, R.D., González-Canto, A., López-Vancell, R., García de León, M.C., Montfort, I., and Pérez-Tamayo R. (2004). Inflammation, complement, ischemia and amoebic survival in acute experimental amoebic liver abscesses in hamsters. Exp. Mol. Pathol. *77*, 66–71.

Olivos-García, A., Carrero, J.C., Ramos, E., Nequiz, M., Tello, E., Montfort, I., and Pérez-Tamayo R. (2007). Late experimental amebic liver abscess in hamster is inhibited by cyclosporine and N-acetylcysteine. Exp. Mol. Pathol. *82*, 310–315.

Olivos-García, A., Saavedra, E., Ramos-Martínez, E., Nequiz, M., and Pérez-Tamayo, R. (2009). Molecular nature of virulence in *Entamoeba histolytica*. Infect. Genet. Evol. *9*, 1033–1037.

Olson, J.W., Agar, J.N., Johnson, M.K., and Maier, R.J. (2000). Characterization of the NifU and NifS Fe–S cluster formation proteins essential for viability in *Helicobacter pylori*. Biochemistry 39, 16213–16219.

Orenstein, S.J., and Cuervo, A.M. (2010). Chaperone-mediated autophagy: molecular mechanisms and physiological relevance. Semin. Cell. Dev. Biol. *21*, 719–726.

Outten, F.W., and Theil, E.C. (2009). Iron-based redox switches in biology. Antioxid. Redox Signal. *11*, 1029–1046.

Pacheco-Yépez, J., Campos-Rodríguez, R., Shibayama, M., Ventura-Juárez, J., Serrano-Luna, J., and Tsutsumi,

V. (2001). *Entamoeba histolytica*: production of nitric oxide and *in situ* activity of NADPH diaphorase in amebic liver abscess of hamsters. Parasitol. Res. *87*, 49–56.

Pearl, L.H., and Prodromou, C. (2006). Structure and mechanism of the Hsp90 molecular chaperone machinery. Annu. Rev. Biochem. *75*, 271–294.

Pedersen, A., Karlsson, G.B., and Rydström, J. (2008). Proton-translocating transhydrogenase: an update of unsolved and controversial issues. J. Bioenerg. Biomembr. *40*, 463–473.

Pérez-Tamayo, R., Montfort, I., Tello, E., and Olivos, A. (1992). Ischemia in experimental acute amebic liver abscess in hamsters. Int. J. Parasitol. *22*, 125–129.

Pieulle, L., Stocker, P., Vinay, M., Nouailler, M., Vita, N., Brasseur, G., Garcin, E., Sebban-Kreuzer, C., and Dolla, A. (2011). Study of the thiol/disulfide redox systems of the anaerobe Desulfovibrio vulgaris points out pyruvate:ferredoxin oxidoreductase as a new target for thioredoxin 1. J. Biol. Chem. *286*, 7812–7821.

Pineda, E., Encalada, R., Rodríguez-Zavala, J.S., Olivos-García, A., Moreno-Sánchez, R., and Saavedra, E. (2010). Pyruvate:ferredoxin oxidoreductase and bifunctional aldehyde-alcohol dehydrogenase are essential for energy metabolism under oxidative stress in *Entamoeba histolytica*. FEBS J. *277*, 3382–3395.

Poole, L.B., Karplus, P.A., and Claiborne, A. (2004). Protein sulfenic acids in redox signaling. Annu. Rev. Pharmacol. Toxicol. *44*, 325–347.

Raat, N.J., Shiva, S., and Gladwin, M.T. (2009). Effects of nitrite on modulating ROS generation following ischemia and reperfusion. Adv. Drug. Deliv. Rev. *61*, 339–350.

Radcliffe, B.C., Hall, C., and Roediger, W.E. (1989). Nitrite and nitrate levels in ileostomy effluent: effect of dietary change. Br. J. Nutr. *61*, 323–330.

Ramachandran, S., Induruwa, P.A., and Perera, M.V. (1976). pH of amoebic liver pus. Trans. R. Soc. Trop. Med. Hyg. *70*, 159–160.

Ramírez-Emiliano, J., González-Hernández, A., and Arias-Negrete, S. (2005). Expression of inducible nitric oxide synthase mRNA and nitric oxide production during the development of liver abscess in hamster inoculated with *Entamoeba histolytica*. Curr. Microbiol. *50*, 299–308.

Ramos, E., Olivos-García, A., Nequiz, M., Saavedra, E., Tello, E., Saralegui, A., Montfort, I., and Pérez Tamayo, R. (2007). *Entamoeba histolytica*: apoptosis induced *in vitro* by nitric oxide species. Exp. Parasitol. *116*, 257–265.

Ramos, M.A., Mares, R.E., Magaña, P.D., Rivas, I.D., and Meléndez-López, S.G. (2011). *Entamoeba histolytica*: biochemical characterization of a protein disulfide isomerase. Exp. Parasitol. *128*, 76–81.

Ramos-Martínez, E., Olivos-García, A., Saavedra, E., Nequiz, M., Sánchez, E.C., Tello, E., El-Hafidi, M., Saralegui, A., Pineda, E., Delgado, J., *et al.* (2009). *Entamoeba histolytica*: oxygen resistance and virulence. Int. J. Parasitol. *39*, 693–702.

Reeves, R.E. (1984). Metabolism of *Entamoeba histolytica* Schaudinn, 1903. Adv. Parasitol. *23*, 105–142.

Reid, M., Badaloo, A., Forrester, T., and Jahoor, F. (2006). *In vivo* rates of erythrocyte glutathione synthesis in adults with sickle cell disease. Am. J. Physiol. Endocrinol. Metab. *291*, E73–79.

Rigothier, M.C., Khun, H., Tavares, P., Cardona, A., Huerre, M., and Guillén, N. (2002). Fate of *Entamoeba histolytica* during establishment of amoebic liver abscess analyzed by quantitative radioimaging and histology. Infect. Immun. *70*, 3208–3215.

Rocha, E.R., Tzianabos, A.O., and Smith, C.J. (2007). Thioredoxin reductase is essential for thiol/disulfide redox control and oxidative stress survival of the anaerobe *Bacteroides fragilis*. J. Bacteriol. *189*, 8015–8023.

Rydström, J. (2006). Mitochondrial NADPH, transhydrogenase and disease. Biochim. Biophys. Acta *1757*, 721–726.

Saavedra, E., Encalada, R., Pineda, E., Jasso-Chávez, R., and Moreno-Sánchez, R. (2005). Glycolysis in *Entamoeba histolytica*. Biochemical characterization of recombinant glycolytic enzymes and flux control analysis. FEBS J. *272*, 1767–1783.

Sánchez, R., Alagón, A., and Stock, R.P. (2002). *Entamoeba histolytica*: intracellular distribution of the proteasome. Exp. Parasitol. *102*, 187–190.

Santi-Rocca J., Smith S., Weber C., *et al.* (2012). Endoplasmic reticulum stress-sensing mechanism is activated in *Entamoeba histolytica* upon treatment with nitric oxide. PLoS ONE. In press.

Saul, R.L., Kabir, S.H., Cohen, Z., Bruce, W.R., and Archer, M.C. (1981). Reevaluation of nitrate and nitrite levels in the human intestine. Cancer Res. *41*, 2280–2283.

Sawyer, M.K., Bischoff, J.M., Guidry, M.A., and Reeves, R.E. (1967). Lipids from *Entamoeba histolytica*. Exp. Parasitol. *20*, 295–302.

Sen, A., Chatterjee, N.S., Akbar, M.A., Nandi, N., and Das, P. (2007). The 29-kilodalton thiol-dependent peroxidase of *Entamoeba histolytica* is a factor involved in pathogenesis and survival of the parasite during oxidative stress. Eukaryot. Cell *6*, 664–673.

Serrano-Luna, J., Gutiérrez-Meza, M., Mejía-Zepeda, R., Galindo-Gómez, S., Tsutsumi, V., and Shibayama, M. (2010). Effect of phosphatidylcholine-cholesterol liposomes on *Entamoeba histolytica* virulence. Can. J. Microbiol. *56*, 987–995.

Subczynski, W.K., Widomska, J., and Feix, J.B. (2009). Physical properties of lipid bilayers from EPR spin labeling and their influence on chemical reactions in a membrane environment. Free Radic. Biol. Med. *46*, 707–718.

Tannich, E., Bruchhaus, I., Walter, R.D., and Horstmann, R.D. (1991). Pathogenic and nonpathogenic *Entamoeba histolytica*: identification and molecular cloning of an iron-containing superoxide dismutase. Mol. Biochem. Parasitol. *49*, 61–71.

Tazreiter, M., Leitsch, D., Hatzenbichler, E., Mair-Scorpio, G.E., Steinborn, R., Schreiber, M., and Duchêne, M. (2008). *Entamoeba histolytica*: response of the parasite to metronidazole challenge on the levels of mRNA and protein expression. Exp. Parasitol. *120*, 403–410.

Tosatto, S.C., Bosello, V., Fogolari, F., Mauri, P., Roveri, A., Toppo, S., Flohé, L., Ursini, F., and Maiorino, M. (2008). The catalytic site of glutathione peroxidases. Antioxid. Redox Signal. *10*, 1515–1526.

Tovar, J., Cox, S.S., and van der Giezen, M. (2007). A mitosome purification protocol based on percoll density gradients and its use in validating the mitosomal nature of *Entamoeba histolytica* mitochondrial Hsp70. Methods Mol. Biol. *390*, 167–177.

Tovy, A., Hertz, R., Siman-Tov, R., Syan, S., Faust, D., Guillen, N., and Ankri, S. (2011). Glucose Starvation Boosts *Entamoeba histolytica* Virulence. PLoS Negl. Trop. Dis. *5*, e1247.

Townsend, D.M. (2007). S-glutathionylation: indicator of cell stress and regulator of the unfolded protein response. Mol. Interv. *7*, 313–324.

Tsutsumi, V., and Shibayama, M. (2006). Experimental amebiasis: a selected review of some *in vivo* models. Arch. Med. Res. *37*, 210–220.

Uys, J.D., Xiong,Y., and Townsend, D.M. (2011). Nitrosative stress-induced S-glutathionylation of protein disulfide isomerase. Methods Enzymol. *490*, 321–332.

Vabulas, R.M., Raychaudhuri, S., Hayer-Hartl, M., and Hartl, F.U. (2010). Protein folding in the cytoplasm and the heat shock response. Cold Spring. Harb. Perspect. Biol. *2*, a004390.

Vermeiren, J., Van de Wiele, T., Verstraete, W., Boeckx, P., and Boon, N. (2009). Nitric oxide production by the human intestinal microbiota by dissimilatory nitrate reduction to ammonium. J. Biomed. Biotechnol. *2009*, 284718.

Verweij, J.J., Vermeer, J., Brienen, E.A., Blotkamp, C., Laeijendecker, D., van Lieshout, L., and Polderman, A.M. (2003). *Entamoeba histolytica* infections in captive primates. Parasitol. Res. *90*, 100–103.

Vicente, J.B., Ehrenkaufer, G.M., Saraiva, L.M., Teixeira, M., and Singh, U. (2009a). *Entamoeba histolytica* modulates a complex repertoire of novel genes in response to oxidative and nitrosative stresses: implications for amebic pathogenesis. Cell. Microbiol. *11*, 51–69.

Vicente, J.B., Testa, F., Mastronicola, D., Forte, E., Sarti, P., Teixeira, M., and Giuffrè, A. (2009b). Redox properties of the oxygen-detoxifying flavodiiron protein from the human parasite Giardia intestinalis. Arch. Biochem. Biophys. *488*, 9–13.

Villalba-Magdaleno, J.D., Pérez-Ishiwara, G., Serrano-Luna, J., Tsutsumi, V., and Shibayama, M. (2011). *In vivo* programmed cell death of *Entamoeba histolytica* trophozoites in a hamster model of amoebic liver abscess. Microbiology *157*, 1489–1499.

Wardman, P. (1998). Evaluation of the 'radical sink' hypothesis from a chemical-kinetic viewpoint. J. Radioanal. Nucl. Chem. *232*, 23–27.

Weber, C., Guigon, G., Bouchier, C., Frangeul, L., Moreira, S., Sismeiro, O., Gouyette, C., Mirelman, D., Coppee, J.Y., and Guillén, N. (2006). Stress by heat shock induces massive down regulation of genes and allows differential allelic expression of the Gal/GalNAc lectin in *Entamoeba histolytica*. Eukaryot. Cell *5*, 871–875.

Weinbach, E.C., and Diamond, L.S. (1974). *Entamoeba histolytica*. I. Aerobic metabolism. Exp. Parasitol. *35*, 232–243.

Weston, C.J., Venning, J.D., and Jackson, J.B. (2002). The membrane-peripheral subunits of transhydrogenase from *Entamoeba histolytica* are functional only when dimerized. J. Biol. Chem. *277*, 26163–26170.

Wong, C.M., Marcocci, L., Liu, L., and Suzuki, Y.J. (2010). Cell signaling by protein carbonylation and decarbonylation. Antioxid. Redox Signal. *12*, 393–404.

Wong, W.K., Tan, Z.N., Lim, B.H., Mohamed, Z., Olivos-Garcia, A., and Noordin, R. (2011). Comparison of protein-free defined media, and effect of L-cysteine and ascorbic acid supplementation on viability of axenic *Entamoeba histolytica*. Parasitol. Res. *108*, 425–430.

World Health Organization. 1997. Amoebiasis. Wkly. Epidemiol. Rec. *72*, 97–100.

Yousuf, M.A., Mi-ichi, F., Nakada-Tsukui, K., and Nozaki, T. (2010). Localization and targeting of an unusual pyridine nucleotide transhydrogenase in *Entamoeba histolytica*. Eukaryot. Cell *9*, 926–933.

Zhao, R., and Houry, W.A. (2007). Molecular interaction network of the Hsp90 chaperone system. Adv. Exp. Med. Biol. *594*, 27–36.

Index

A

B

Q

R

U

V

W

X

Y

Z